Haematological
Cytochemistry

Haematological Cytochemistry

F. G. J. Hayhoe
MA MD(Cantab) FRCP FRCPath

Leukaemia Research Fund Professor of Haematological Medicine,
Department of Haematological Medicine, University of Cambridge, UK

D. Quaglino
MD PhD(Cantab) FRCPath

Professor of Clinical Medicine, Department of Internal Medicine,
University of L'Aquila, Italy

With contributions to Chapters 3, 8 and 15 by

Dr A. De Pasquale MD

Assistant Director of Research,
University of L'Aquila, Italy

SECOND EDITION

CHURCHILL LIVINGSTONE
EDINBURGH LONDON MELBOURNE AND NEW YORK 1988

CHURCHILL LIVINGSTONE
Medical Division of Longman Group UK Limited

Distributed in the United States of America by
Churchill Livingstone Inc., 1560 Broadway, New York,
N.Y. 10036, and by associated companies, branches
and representatives throughout the world.

First published 1980
Second Edition 1988

ISBN 0-443-03545-8

British Library Cataloguing in Publication Data
Hayhoe, F.G.J.
Haematological cytochemistry.
2nd ed.
1. Blood—Examination 2. Diagnosis,
Cytologic 3. Cytochemistry
I. Title II. Quaglino, D. III. De
Pasquale, A.
616.07′561 RB45

Library of Congress Cataloging in Publication Data
Hayhoe, F. G. J. (Frank George James)
Haematological cytochemistry.
Bibliography: p.
Includes index.
1. Blood cells—Examination. 2. Cytochemistry.
3. Diagnosis, Cytologic. I. Quaglino, Dennis.
II. De Pasquale, A. III. Title. [DNLM: 1. Blood
Cells. 2. Cytodiagnosis. 3. Histocytochemistry.
QY 402 H419h]
RB45.H395 1988 616.07′561 87-7992

Produced by Longman Group (FE) Ltd.
Printed in Hong Kong

Preface to the Second Edition

In preparing the second edition of this book, we have retained the general approach outlined in the preface to the first edition, but the book has been extensively rewritten, and much new material has been added. Reference has been made to over a thousand papers on aspects of haematological cytochemistry published since the first edition appeared in 1980. Advances in haemic cell immunology, immunocytochemistry, and biochemical enzymology have been incorporated in so far as they have directly affected or influenced haematological cytochemical practice. A new Chapter 15 has been added to bring together the major cytochemical and histochemical findings in lymphocyte subpopulations and in normal, reactive and lymphomatous lymph nodes.

We are pleased to welcome Dr Andrea De Pasquale as co-author of Chapters 3, 8 and 15 and acknowledge with appreciation his helpful comments on other chapters and assistance in preparing manuscripts and illustrations and in correcting proofs.

The colleagues whose contributions, direct or indirect, to the production of the book were acknowledged in the preface to the first edition are now joined by others, including many members of the graduate staff who have worked in our respective clinical units and laboratories in the past 8 years. We are indebted to all of them, but would especially mention Dr David Swirsky in relation to work on cytology and cytochemistry of leukaemias and related states and Dr Raafat Khalaf whose help with the cytochemistry of 'other' enzymes, including the kind provision of many photographs used in Chapter 11, is greatly appreciated.

We again express our gratitude to Miss Daniela Quaglino, graduate in biology, for her constant collaboration in searching and filing the innumerable articles necessary to the completion of the reference sections.

We gratefully acknowledge the secretarial assistance of Mrs J. M. Chandler and of Miss Lia Ginaldi.

We again wish to thank the staff of Churchill Livingstone for the immense amount of work they have done to help produce this second edition.

Cambridge and L'Aquila, 1988

F.G.J.H.
D.Q.

Preface to the First Edition

The purpose of this book is to provide a short illustrated survey of the uses and value of cytochemical methods in haematology. Excellent monographs exist on the principles and practice of histochemistry and cytochemistry and several journals are devoted to that subject, but the application of these principles and methods to the study of haemic cells has usually received little attention in general monographs. Moreover, the results of cytochemical studies in haematology over the past 30 years or so have been published in journals of many different disciplines, and we are aware of the difficulties newcomers to the subject experience in determining the range of established findings so far as cytochemistry of haemic cells in health and disease is concerned.

We hope that this monograph may serve as a broad review of at least the majority of cytochemical methods with proven value in haematology. Relatively little space has been devoted to principles underlying the methods, because these do not differ essentially from the principles discussed in general textbooks of histochemistry, but the applications to haemic cells, the results obtained and the significance of these results are discussed at greater length. The main emphasis has been on standard colour reactions suitable for use with light microscopy and having immediate relevance in differential diagnostic cytology, but sufficient reference is made to ultrastructural methods, to autoradiography and to fluorescence techniques to allow the contribution of these approaches to the cytochemical understanding of haemic cell structure and function to be set in perspective. Where appropriate these various techniques together with standard cytochemical ones are considered in relation to cell surface immunology, cytogenetics, and other studies which also contribute to the classification and understanding of normal and pathological haemic cells and their involvement in disease processes. Where it is relevant to do so we have added a technical appendix to each chapter, describing the main cytochemical methods used and giving details of the preparation of reagents and their practical use. The methods described have been chosen for the most part because they are ones which we have used ourselves and found satisfactory.

Colour illustrations have been used extensively because cytochemical reactions are often difficult to portray realistically in black and white, but we have aimed by the combination of illustrations chosen to show the many different patterns of reactivity with the chief cytochemical stains not only in normal haemic cells of all series but also in the full range of cells found in pathological states affecting the blood and bone marrow. Some of the more important of these conditions are brought together in the last two chapters.

PREFACE TO THE FIRST EDITION

We are grateful to our colleagues, past and present, in Cambridge and Modena, who have contributed by discussion and observations to our understanding and interpretation of haematological cytochemistry, and especially to our constant technical collaborators, Mr R. J. Flemans, Signor Marco Mugnaini and Signor Massimo Tonelli. We are much indebted to Miss J. M. Thompson and to Miss Daniela Quaglino for secretarial assistance and for typing the manuscript.

We also wish to acknowledge especially the additional work of Mr Flemans and Dr A. De Pasquale in taking and developing many photomicrographs, and of Mr L. F. H. Beard, of the Medical Photography and Illustration Service, Addenbrooke's Hospital, Cambridge, for providing numerous black and white prints from colour transparencies.

Finally we wish to thank the staff of Churchill Livingstone for their helpful collaboration during the whole process of publication.

1980

F.G.J.H.
D.Q.

Contents

CHAPTER 1

History, general principles and procedures, and potential value of cytochemical methods in haematology

Histochemistry is the application of chemical processes to microscopical preparations, and by its use an attempt is made to reveal the chemical composition of tissues and cells, usually by the development of a colour reaction, without causing the destruction of the tissues or damaging the cells so much that they become unrecognizable. When the chemistry of cells rather than tissues is specifically discussed, as is usually the case in haematology, the word 'cytochemistry' may more correctly be used, but in practice the two words 'histochemistry' and 'cytochemistry' are often employed interchangeably.

Historical survey

Details of the history of histochemistry, with references to the important seminal monographs from around 1830 onwards, are provided in major histochemical treatises such as that of Pearse (1968), where deference is paid to François Vincent Raspail as the father of the science. Raspail had indeed written in 1825 the sentence quoted by Baker (1943) as marking the origin of histochemistry—'Je donnerai ... les résultats que j'ai obtenus par l'application des procédés chimiques aux observations du microscope'. More than a hundred years were to elapse, however, before cytochemical studies came to be pursued in any systematic and productive way, and although a number of sporadic contributions to the subject were made in the intervening years, including the introduction of the Prussian blue method for demonstrating iron (Perls, 1867), the use of Sudan III to stain fat (Daddi, 1896) and the detection of calcium in tissues (von Kossa, 1901), it was still possible for Lison in 1936 to regard the subject of his book *Histochimie Animale* as a new science. In the earlier years of this century biochemical studies of blood and marrow cells were occasionally supplemented by cytochemical ones. In 1905 and 1906 Opie drew attention to the existence of a proteolytic enzyme in neutrophils in inflammatory exudates and showed that this enzyme was active in neutral or slightly alkaline media. This initial observation of a leucocytic proteolytic enzyme was followed in 1907 by Winkler's description of an oxidase in the leucocytes of blood films, detectable by the dimethylparaphenylenediamine reaction. The high oxidative metabolism of leucocytes in leukaemic blood was observed by Grafe (1911) and the intense glycolytic activity of normal leucocytes was described by Levene and Meyer (1912). This interest in the biochemistry of white cells was maintained, and by 1923 Fiessinger was able to produce a monograph entitled 'Les Ferments des Leucocytes en Physiologie, Pathologie et Thérapeutique Générale', including discussions of leucocyte lipase, alkaline lecithinase, amylase and proteolytic ferments (Rebuck, 1947).

The first strictly cytochemical report on leucocytes since the oxidase reaction described in 1907 was provided by Sehrt in 1927. Sehrt used Sudan III and Nile blue sulphate to stain the leucocytes in blood smears, and found that the neutrophil granules stained strongly. He observed the resemblance between this appearance and that of peroxidase preparations and thought that the oxidase reaction of cells containing lipid was proportional to the amount of lipid they contained. A leucocytic nucleotidase was reported by Deutsch and Rosler (1929), who obtained large numbers of leucocytes by intrapleural injection of turpentine in dogs and cows with the production of sterile purulent exudates. The phosphatase content of plasma, white cells and red cells was described by Kay (1930), Umeno (1931) and Roche (1931), and Stern (1932) demonstrated the existence of a catalase in the polymorphs of exudates in the rabbit. Soffer and Wintrobe (1932) investigated the metabolism of leucocytes in normal and leukaemic blood and found that the metabolic activity of granulocytes resembled that of malignant cells, and that their oxygen consumption and glycolytic powers exceeded those of lymphocytes. Much of Menkin's work from 1934 to 1940 was devoted to the biochemistry of leucocyte migration. He found a high percentage of polymorphs in exudates with a slightly alkaline pH, and demonstrated that the mechanism of migration of these cells into an area of inflammation was related to a substance, leukotaxine, produced by the injured tissue. By 1940 Barnes was able to demonstrate in the leucocytes of rabbits a wide variety of enzymes—cathepsin, nuclease, amylase, lipase, lysozyme and adenosinase.

It was against this growing background of biochemical knowledge of leucocyte constitution that the study of the cytochemistry of haemic cells was given a great impetus by the numerous technical contributions of Gomori from 1939 onwards, recapitulated in his monograph *Microscopic Histochemistry* (Gomori, 1952). Gomori (1939) and, independently, Takamatsu (1939) devised elegant methods for the cytochemical demonstration of phosphatases which could be applied with suitable modifications for many other enzymes. Despite drawbacks associated with the lead methods in general these techniques are still very widely used at light microscope level and have been adapted also for ultrastructural use. Their use in haematology is discussed at length in later chapters. Within the next few years azo-dye methods for phosphatases were introduced in which naphthol liberated by the enzyme from α-naphthyl phosphate was captured by a diazonium salt with the production of a brightly coloured dye (Menten *et al.*, 1944). This advance in turn had a major impact on practical haematological enzyme cytochemistry, while a third and equally important step forward came from the use of tetrazolium compounds as hydrogen acceptors in the demonstration of dehydrogenases (Seligman *et al.*, 1949). Some reference is made later in this work to the sequence of developments and modifications of each of these three broad techniques for enzyme demonstration which has led to the current methods recommended for use in haematological material, but detailed consideration of the principles and practice involved in their use would not be appropriate, since the subject has been fully dealt with in general texts on histochemistry. Apart from these methods and others which can be regarded as directly cytochemical, the techniques of autoradiography and immunocytochemistry have contributed enormously to the study of haemic cells. The major applications of the former to the analysis of protein and especially nucleoprotein synthesis and cell kinetics, and of the latter to the study of cytoplasmic and surface antigenic constitution of blood cells, especially cells of the lymphocytic series, are briefly reviewed in later chapters.

Studies of enzyme activity and localization by immunocytochemical methods, making use of the antigenicity of enzymes, and by autoradiography using radiolabelled enzyme inhibitors, have also contributed new data to extend the information derived from conventional cytochemical methods. Details of the variety of autoradiographic techniques available and the theoretical principles underlying them are outside the scope of this work but have been extensively reviewed elsewhere (Rogers, 1967). Methods especially applicable to haematological material are given in Chapter 2. Similarly, techniques for immunofluorescence and immunocytochemical studies used in haematology are given in Chapter 3.

The technical successes achieved by all these approaches at both light and electron microscope level in recent decades have still left many areas of uncertainty, especially over the question of whether cytochemical reactions are providing useful biological information or just irrelevant, if sometimes beautiful, visual patterns. Galjaard (1980) pointed out that histochemical journals contained much reporting of elaborate enzyme cytochemistry reactions of little or no medical or biological importance, but little work that had application to fundamental problems in genetics and cell biology. Borgers and Verheyen (1985), in a recent review of enzyme cytochemistry, devote attention to this problem under the heading 'Useful versus useless cytochemistry'. On the 'useful' side, where cytochemical studies have clearly supplemented classical morphology in the enzyme field, they include important contributions to 'interdisciplinary assessment' in eight areas, which may be briefly summarized as follows:

1. Identification of metabolic routes and sites of activity.
2. Differentiation and dedifferentiation of cells.
3. Identification of cell origins and cell lineages.
4. Identification of subcellular organelles and elucidation of their functions and interrelationships.
5. Processes of cell degeneration and regeneration.
6. Assessment of functional change in pathological states, assisting diagnosis, prognosis and management of treatment.
7. Provision of 'quality control' in demonstrating marker enzymes in, for

example, subcellular fractions used for biochemical quantitative studies.

8. Validation of cell and tissue culture systems, to assess the function of the in-vitro system as compared to intact organs.

We very much endorse this listing. The generalizations drawn from enzyme cytochemistry as a whole apply strongly in the haematological field, and throughout this monograph we have attempted to give particular emphasis to the contribution which the cytochemical study of haemic cells, with regard both to enzymes and to non-enzymic constituents, make in these useful areas.

We also share the disquiet expressed by Borgers and Verheyen that inadequately controlled and validated cytochemical procedures, frequently performed and reported, are not worth while. Lack of specificity of reagents, problems with fixation and with localization or diffusion of reaction products, difficulties in interpreting results and the need for adequate control procedures, all must be carefully assessed if reactions are not to fall into the 'useless cytochemistry' bracket. Nevertheless it must be admitted that some cytochemical reactions which are clearly useful in haematological diagnosis—the Sudan black reaction for example—have an uncertain basis in chemistry and enzymology, while many others—leucocyte alkaline phosphatase cytochemistry, for example—are valuable in diagnosis and probably represent chemically and enzymologically accurate reactions and localizations, but demonstrate enzymes whose exact role is unknown. In these instances the practical value of the reaction in haematology clearly justifies inclusion and discussion.

General principles and procedures in haematological cytochemistry

Haematologists usually deal with smears of blood and bone marrow, or imprint preparations from node biopsies, rather than with sections, and they encounter a number of problems relating to preparation of slides, fixative procedures, and methods of incubation which warrant some general discussion. The cells in smears and imprints are mostly intact, coated with protein and other materials from plasma and relatively impermeable to components of incubation mixtures, whereas the cells of tissue sections are cut across and their cytoplasmic structures are laid open for substrate access. This difference probably explains why many histochemical techniques which give satisfactory results in tissue sections fail to work on blood and bone marrow slides, as has been our own experience and that noted by others (Melnick, 1968). A second serious problem concerns the tendency for haemic cells in unfixed or weakly fixed preparations to be washed off the slides during exposure to incubation media, especially at an acid pH and when the smears are thick. A third arises from the conflicting contribution of fixation and other preparative procedures, which may assist the reaction through rendering cytoplasmic membranes more permeable to substrate and capture agent, but diminish its manifestations by allowing enzymes or reaction product to diffuse more readily out of the cells into the incubation medium.

Preparation of slides. Films of blood or bone marrow for cytochemistry are best made directly from the aspirated sample without the use of anticoagulants. When this cannot easily be done before the specimen clots, or when the expected cellularity of the specimen is low, material may be aspirated into an anticoagulant and slides subsequently made in the laboratory directly from the anticoagulated specimen or from the buffy coat obtained by gentle centrifugation. The common anticoagulants, ethylene diamine tetracetic acid (EDTA), oxalate, citrate and heparin, have relatively little effect on cytochemical positivity, either for enzymes or other cell components, provided that slides are made within one to two hours, or within six to eight hours if the specimen is stored in the refrigerator. Among the anticoagulants, our own preference is for heparin, which appears to cause least morphological disturbance in the leucocytes. Slides should be grease-free and smears should be spread thinly.

Cytochemical reagents are often expensive and scanning poorly cellular films for reaction product is time-consuming: for both reasons we emphasise the value of making smears from buffy coats when the initial level of nucleated cells is much below $50 \times 10^9/1$. The whole of the buffy coat should be pipetted off together with a little plasma and a small volume of red cells; after thorough mixing, films may be made in the usual manner. We have not found preliminary treatment of slides with albumen or serum necessary to aid adhesion, but the practice is recommended by some authors (Melnick, 1968). The tendency for smears to show wrinkling or to be washed off during fixation and incubation may be much reduced if the smears are allowed to dry thoroughly by standing on the bench for 24 to 48 hours before fixation. Unfixed slides may be stored at room temperature, stacked, and wrapped together, and will remain suitable for many cytochemical reactions for long periods. Among the commonest procedures used in haematology laboratories, the periodic-acid Schiff (PAS) reaction works perfectly well on slides many years old, but the enzyme reactions and partially enzyme-dependent stains gradually decrease in positivity with time. Acid and alkaline phosphatases, esterases, peroxidases and the peroxidase-associated Sudan black stain are essentially unaffected at one month, progressively weaker at three, six, 12 and 18 months, but still show detectable positivity in cells which were originally strongly positive even after three years. For any of these reactions, strongly positive control slides may be used satisfactorily for at least a year. There is no need to fix slides before storage, but the appropriate fixative should be used when the cytochemical reaction is eventually to be performed. Preservation of enzyme activity in stored slides may be considerably improved and prolonged by keeping the slides at 4°C or still better at −20°C.

Fixation and stabilising additives. Fixative procedures used in haematological cytochemistry cover a very wide range, most of them derived empirically by trial and error as an effective compromise for the particular reaction concerned between preservation of good morphology, which is best achieved by powerful fixatives, especially alcohols, and optimum preservation *in situ* of the enzyme or other substance being studied, yet with accessibility to the reaction mixture.

While the smears of single-cell thickness mostly used in haematology are relatively susceptible to fixative damage, and the enzyme content of haemic cells is rarely high, there may sometimes be a case for fixative procedures being sufficiently prolonged and vigorous to cause much or even most of an enzyme to be inactivated. This is especially the case where capture agents with low trapping efficiency, such as lead and other heavy metals, are used, because high enzyme activity with release of large amounts of substrate split products may result in their diffusion away from the site of formation, with either artefactual deposition or washing away of the reaction product. When tissue sections are used—lymph node or bone marrow biopsy material, for example—a standard, thorough, fixative procedure may be best, followed by an appropriately adjusted set of conditions, in terms of substrate concentration and time of exposure, for the incubation stage of the reaction.

In general the hydrolases, including certainly most of the phosphatases and esterases, are readily demonstrable in smears fixed with formalin vapour. Alternative fixatives, including short exposure to formalin alcohol mixtures, may in some circumstances give better cellular preservation and reduce the tendency for parts of the film to be washed off the slide, especially when prolonged incubation is required. Dehydrogenases and other oxidative enzymes are more labile than hydrolases and even short exposure to powerful fixatives renders them inactive. Some workers have used unfixed preparations, protected against damage during incubation by the addition of a high molecular weight sucrose polymer (Stuart and Simpson, 1970). Others have used acetone as a gentle fixative in various concentrations or combinations (Quaglino and Hayhoe, 1960; Novikoff *et al.*, 1960; Melnick, 1968). To avoid diffusion of enzyme or reaction products, as tends to occur with prolonged incubation in aqueous buffers, cryoprotective additives, substances which had been found earlier to assist the preservation of cellular integrity during freezing procedures, proved effective (Melnick and Heizer, 1965). In a study of a wide range of such additives Melnick (1968) found polyvinylpyrrolidone (PVP) to yield much the best results. Among several different molecular weights of PVP tested the most effective were those of 10 000 and 40 000, and slides may be exposed for 1 minute to a 5% solution in physiological saline following acetone fixation and before incubation in the cytochemical reaction mixture. Alternatively, PVP may be added to the incubation mixture (Eränkö and Palkama, 1961). The mode of action of PVP and similar compounds, such as polyvinyl alcohol (PVA)—recommended for the same purpose by Altman and Chayen (1965)—appears to be the formation of large numbers of hydrogen bonds, probably to the water component of the lipoprotein complex of ultrastructural cell membranes (Doebbler, 1966), leading to increased stability of these structures.

Incubation procedures. When circumstances permit the simultaneous staining of several slides for any cytochemical reaction the most satisfactory method of incubation is to use Coplin jars or similar containers. Directions for preparing most cytochemical reaction mixtures are given for a volume of incubation medium of 100 ml, enough to allow for the staining of up to 30 slides, placed back to back, 10 to each of three Coplin jars. For smaller numbers of slides an appropriately reduced quantity of incubation medium can be made up, even as little as 10 ml for use in small plastic slide containers, but difficulties arise when only one or two slides need to be stained, yet the ingredients are expensive and the preparation of even 10 ml of medium is unjustified. In these circumstances small volumes of incubation medium, 1 ml or less, may be used, applied directly on the surface of the film and allowed to spread over the slide, or alternatively, restricted to a particular area, by the use of a ring of perspex or wax. A grease-pencilled ring is effective if short incubation periods are required, but for prolonged incubation, 30 minutes or more, the use of a cover slip over the wax or perspex ring prevents loss of medium through evaporation, and this may be especially important when the reaction calls for incubation at 37°C.

Preservation of stained slides. Azo-dye reaction products are soluble in immersion oil, xylol and most mounting media. Slides are probably best kept unmounted, when areas not exposed to immersion oil retain their positivity for two to three years, especially if stored in the dark. In practice cytochemical slides rarely need to be stored for restudy, since they may be examined soon after preparation and the appropriate semi-quantitative score or other assessment made and recorded then. If there is a requirement for preservation, so that several observers may study specimens over a period of days or even weeks, mounting in glycerine jelly or synthetic media such as DPX or Apathy's medium may be recommended.

Histochemistry of bone marrow sections

While histochemical methods have been used in the study of paraffin embedded or frozen sections of solid organs for as long as they have on smears and imprints, sections of bone marrow, of particular interest to the haematologist, have been less easily adaptable, largely because satisfactory methods for fixation and decalcification prior to paraffin embedding have been destructive to most enzymes and to many surface antigens of immunocytochemical importance. The use of modified fixative and decalcification procedures prior to paraffin embedding has certainly enabled some enzymes to be demonstrated (Grouls, 1980; Katayama and Mochino, 1983) but the technical quality of the preparations left much to be desired. The introduction by Block *et al.* (1953), Burkhardt (1956, 1966) and Cathey (1963) of methods using plastic embedding for bone marrow sectioning opened new possibilities, both for detailed cytological study and cytochemistry. Dancey *et al.* (1976) embedded human marrow biopsy material in methyl methacrylate. With prior acrolein fixation excellent Romanowsky staining was achieved, and with buffered formalin fixation neutrophils showed good positivity with the chloroacetate esterase cytochemical reaction. Later cytochemical studies explored different fixatives and modified forms of plastic embedding.

Beckstead *et al.* (1981), for example, used fixation in a mixture of paraformaldehyde, glutaraldehyde and acrolein at 4°C for two hours before washing, dehydrating in graded acetones and embedding in the plastic JB4 with methyl methacrylate monomer for 24 hours. Sections of 1–3 μm thickness proved to give good immunohistochemical results and good cytochemical demonstration of esterases, peroxidase, alkaline and acid phosphatases and lysozyme. The plastic embedding procedures may assist in allowing enzyme histochemistry in other tissues, for example lymph nodes, without the need for freezing. Beckstead (1983) reported good reactions in a large series of lymph nodes for the enzymes 5′-nucleotidase, acid and alkaline phosphatases, adenosine triphosphatase, β-glucuronidase, various esterases and peroxidase.

There seem, therefore, to be considerable advantages in the use of plastic sections, especially for cytochemical but also for basic Romanowsky-type staining of bone marrow sections (Bianco *et al.*, 1984; Burgio *et al.*, 1984), and to some extent also in lymph node cytochemistry, although here simplified tissue embedding procedures have been devised which increasingly permit a range of enzyme cytochemical reactions to be carried out with satisfactory results on paraffin sections (Chilosi *et al.*, 1981).

Quantitation of cytochemical positivity

Throughout this book reference is rarely made to cell positivity in any cytochemical reaction without the addition of some qualifying adjective, weak, strong, intense, scattered, granular, diffuse and so forth. For many practical purposes in diagnostic haematology this degree of crude quantitative assessment is sufficient. Indeed, an accurate measurement of the total reaction product in the average cell of a given population may be less informative than a descriptive account of the pattern of distribution. This is perhaps especially true of the periodic-acid Schiff reaction for glycogen, where the pattern of positivity is of greater diagnostic value in distinguishing cell types among, for example, the acute leukaemias, than measurements of the overall amount of positive material would be.

Semiquantitative scoring methods have been employed in measuring leucocyte alkaline phosphatase (LAP) levels, and similar rough methods have been applied to acid phosphatase, peroxidase, Sudan black, and PAS reactions (Hayhoe *et al.*, 1964; Cassuto *et al.*, 1977), but apart from LAP and perhaps plasma cell acid phosphatase scoring such methods have not found a place in haematological laboratory practice. Nevertheless the search for more reliable and accurate quantitative methods than simple visual assessment has been pursued with vigour since the development of reproducible cytochemical methods made the exercise worth while. An excellent discussion of the problem appeared in Gomori's *Microscopic Histochemistry* in 1952, and subsequent broad approaches to quantitation in histochemistry generally are reviewed in *Practical Histochemistry* by Chayen, Bitensky and Butcher (1973). The latter review discusses the use of elution techniques, allowing the separation and quantitation of such reaction products as the formazans of tetrazolium salts (Altman, 1972), double-beam recording microdensitometry, which gives a measurement of optical density capable of conversion into an accurate quantitation of reaction product localised at different regions traversed (Altman, 1971, 1972), and scanning and integrating microdensitometry, which allows the accurate measurement of reaction product in areas of tissue or in single cells (Bitensky *et al.*, 1972).

These techniques, and others such as ultraviolet spectrophotometry of nucleic acids, have been applied to haemic cells in many studies during the past 40 years. Thus Thorell (1947) used the ultraviolet absorptiometric technique devised by Caspersson (1936) and Caspersson and Santesson (1942) to determine the location and concentration of nucleic acids, proteins and haemoglobin in bone marrow cells. He made use of the respective maximum absorptions shown by nucleic acids at 2600 Å, by tryptophane and tyrosine in proteins at 2800 Å and by haemoglobin at 4050 Å and 4360 Å. Since there is no microspectrophotometric difference between DNA and RNA, Thorell carried out Feulgen reactions for DNA in parallel with the u.v. absorption studies and was thus able to plot the separate distributions of DNA and RNA. The nucleolus and cytoplasm of primitive cells showed a high nucleic acid content due to RNA, since it was found in Feulgen negative areas, while strong absorption over Feulgen positive chromatin and chromosomes allowed a measurement of DNA at these sites. The method allowed the separation of four phases of development among erythroblasts:

1. A phase of rapid growth when the concentration of cytoplasmic RNA was maximal
2. A phase of declining growth associated with a diminishing content of RNA and with the active synthesis of basic cellular proteins such as globin
3. A phase of differentiation with rapid formation of haemoglobin, and
4. A phase of declining differentiation.

These stages, quantitatively assessed, parallel those observed qualitatively with Romanowsky stains; the early cytoplasmic basophilia is due to a high concentration of RNA and in the later stages of development the basophilia is replaced by polychromatic, and later orthochromatic, cytoplasm as the RNA is replaced by haemoglobin: in orthochromatic cells RNA has almost completely disappeared, the last remnants of it being seen in the reticular material of reticulocytes. Despite substantial technical improvements in the equipment available for microspectrophotometry there have been no more than minor changes in the sequence proposed by Thorell: the subject is reviewed by Wickramasinghe (1975).

The stages observed in erythropoiesis correspond roughly to comparable steps in leucopoiesis, although the RNA content of undifferentiated primitive cells of the leucocyte series is slightly less than that of proerythroblasts, and nucleated red cells generally contain more RNA than the granulocyte precursors at a similar stage of maturation.

Gahrton (1964, 1966) used microspectrophotometry to quantitate PAS positivity in neutrophils and established base-line values with which to compare positivity in abnormal states; differences detected in this way could be used in the diagnostic differentiation between CML, with low levels, and other myeloproliferative states, with high levels. Similar studies of nucleic acid and protein levels in leukaemic cells were also carried out by Gahrton and his colleagues (Gahrton and Foley, 1966). This work, taken together with autoradiographic studies, contributed to the analysis of leukaemic cell population dynamics, discussed further in Chapter 14. Studies of DNA synthesis and distribution, using combined techniques of exposure of cell populations to a DNA-specific radioactive precursor, followed sequentially by Feulgen-staining of smears, microphotometry of the amount of stain and subsequent autoradiographic development and grain counting also added a valuable quantitative element to investigation of cell cycle activity in different leucocyte populations (Hale and Cooper, 1963, 1965; Quaglino *et al.*, 1974). The same approach has been helpful in distinguishing proliferating from non-proliferating cell types in lymph node material from patients with lymphomas (Peckham and Cooper, 1969).

With the exception of Gahrton's work on the quantitation of PAS positivity, all these studies have dealt with nucleic acid and protein levels in developing or proliferating cells, and have formed, as it were, a subsection of cytochemistry, dealing essentially with cell population kinetics, closely linked with autoradiographic studies of DNA and RNA synthesis and movement, and in general far more quantitative in practice than any other area of cytochemical study, at least so far as haematology is concerned.

Although the elution technique does not seem to have been applied to blood or marrow smears, integrating microdensitometry has been used as a method of measuring the quantity of formazan deposited and thus assessing dehydrogenase activity in haemic cells from normal and leukaemic patients (Stuart *et al.*, 1970; Stuart and Simpson, 1970). Evidence of increased levels of extramitochondrial dehydrogenase activity but not of intramitochondrial respiratory dehydrogenases or α-glycerophosphate dehydrogenase was found in human leukaemic lymphoblasts. These are findings of considerable interest and importance, but it is relevant here to note that few other haematological cytochemists seem to have made use of this quantitative approach to enzyme cytochemistry. This may be because the technique is laborious and time-consuming and requires large numbers of separate readings on individual cells, while the necessary apparatus is expensive and not available in many laboratories. Nevertheless the method clearly has a place in research work and has occasionally been used in other haematological studies, such as that of Williams (1975) on alkaline phosphatase levels in rat leucocytes.

Technical considerations underlying the successful use of quantitative cytochemistry for assessing alkaline phosphatase activity were summarised by Howat and Chayen (1983). Comparing various methods based on the calcium-trapping of liberated phosphate or diazonium salt coupling with liberated naphthol they found reaction products measurable by microdensitometry, the most even precipitate and therefore the most suitable for microdensitometry being provided by a calcium-trapping method with calcium glucoheptonate.

Schofield *et al.* (1983) used scanning-integrating microdensitometry to study staining reactions in normal and toxic granulation blood neutrophils and in myeloid precursor cells of normal human bone marrow. They found that normal marrow granulocyte maturation was associated with a progressive decrease in azurophilic granule enzymes (myeloperoxidase, β-glucuronidase, acid phosphatase and chloroacetate esterase) and in acid mucosubstances stained by Alcian blue. An increase in the specific granule marker lactoferrin was found. These findings conform very much with the long-recognised sequence of changes appreciated by qualitative morphological and cytochemical studies, but provision of quantitative assessments, with ranges and standard errors for each enzyme at each main maturation stage provides a valuable base-line for subsequent studies. The demonstration that toxic granulation neutrophils showed evidence of cytoplasmic immaturity and a particular increase in acid mucosubstance illustrates the way this approach may be used.

The results of these various quantitative studies of haemic cell enzyme cytochemistry are discussed in the appropriate later chapters, but they have not made as great an impact on haematological research and practice as might perhaps have been expected.

A degree of quantitation is involved in several recent techniques for automatic leucocyte classification using cytochemically stained smears. The approach of Ornstein and Ansley (1974) for example, was to find cytochemical stains with narrow absorption bands, spectrally well separated from each other, and corresponding to cell constituents peculiar to specific cell types. The value of this approach, using a stain incorporating methyl green for DNA, pyronin Y chiefly for RNA, and fluorescein for basic proteins, is discussed by Tycko *et al.* (1976). Other cytophotometric methods based predominantly on fluorochrome dyes appear to be equally successful in allowing the automatic classification of fixed leucocytes as lymphocytes, monocytes, basophils, eosinophils or neutrophils (Shapiro *et al.*, 1976). Azo-dye methods for non-specific esterase have been used for selective staining of monocytes in suspension prior to their automatic identification and enumeration in a flow cytophotometer (Kaplow *et al.*, 1976). In this and some other methods analysis of the oscilloscope absorption display enables a quantitative assessment to be made of the intensity of staining and hence of the amount of intracellular enzyme or other structural component being stained.

Nevertheless the main use of cytophotometry has been in automated differential counting, with the emphasis on counting large numbers of strongly or weakly positive or negative cells with a single cytochemical reaction such as peroxidase, and thus producing percentages of granulocytes, monocytes and lymphocytes. The application of this technique and its variants in laboratory haematology has been reviewed by Pierre (1984). Among earlier differential counters the Hemalog D machine (Ansley and Ornstein, 1970), used an acid peroxidase reaction and also an esterase reaction (for monocytes), and astra blue (for basophils) in three separate continuous flow channels. Analysis of 10 000 leucocytes concurrently in each channel produced an oscilloscope display on which the differential count could be based. This machine has more recently been replaced by the H6000 (both these instruments have been produced by the Technicon Instrument Co., Basingstoke, Hants, UK) where the principle remains similar but the esterase channel is omitted and the peroxidase reaction is carried out at pH 7.3, monocytes being recognised by their weak peroxidase reaction.

The value of these instruments in differential counting and in the recognition of acute leukaemias has been discussed by Patterson *et al.* (1980), Hinchcliffe *et al.* (1981), D'Onofrio and Mango (1984) and Lai *et al.* (1986). They produce reliable differential leucocyte counts so far as neutrophils, eosinophils, basophils, lymphocytes and monocytes are concerned, and are effective in identifying possible acute leukaemias for further study (Drewinko *et al.*, 1987), though their value in determining the nature of blast cells present remains restricted and is in no way a substitute for direct observation of appropriately stained smears. Both the Hemalog D and the H6000 machines, with their ability to produce full differential leucocyte counts on the basis of cytochemical reactions, are clearly superior in diagnostic capacity to machines such as the Coulter S Plus IV, which provides a 3-cell differential count based on cell size. An example of this superiority in the detection of chronic lymphocytic and hairy cell leukaemia is given by Lai et al (1987), but again even the best results need to be supplemented by visual microscopy before a firm diagnosis can be made.

Ultrastructural cytochemistry

The general problems of obtaining cytochemical reaction products which are themselves electron dense, or are strongly osmiophilic, or can otherwise be rendered visible in the electron microscope have been discussed in detail in excellent review articles (Holt and Hicks, 1963; Scarpelli and Kanczak, 1965; Borgers and Verheyen 1985) and in texts of electron microscopy (Sjöstrand, 1968) and histochemistry (Pearse, 1968; Hayat, 1974, 1975). Although a very large number of methods, reviewed recently by Borgers and Verheyen (1985), have been devised for enzyme ultrastructural cytochemistry, relatively few of them have been applied at all extensively in haematology; apart from the method of Barka and Anderson (1962) for acid phosphatase with subsequent modification, and those of Graham and Karnovsky (1966) and Novikoff and Goldfischer (1968) for peroxidase, both in their original forms and with later modifications (Breton-Gorius, 1980) which have been widely used, only restricted studies have been pursued with techniques for arylsulphatases, 5-nucleotidase, alkaline phosphatase and esterases (Wetzel *et al.*, 1967; Bainton and Farquhar, 1968; Monahan *et al.*, 1981; Borgers and Verheyen, 1983). In our own experience (Cawley and Hayhoe, 1973) single cell suspensions of peripheral blood and bone marrow cells are more suitable for ultrastructural cytochemical work than small fragments of solid tissue. Leucocytes may be separated from peripheral blood most satisfactorily by sedimentation of heparinised blood in dextran, the leucocyte rich plasma layer being pipetted off into glutaraldehyde, left for 1 hour and the fixed leucocytes then centrifuged down. After washing the cells may be incubated for cytochemistry in appropriate media. Aspirated bone marrow material may be transferred directly from the syringe into fixative and similarly treated: the resulting single cell suspension usually permits useful cytochemical study. Where ultrastructural cytochemistry has contributed usefully in the study of haemic cells, reference is made and details of the technique are given in later chapters, notably on peroxidase and phosphatases, but for most other cytochemical reactions technical difficulties remain to be overcome before acceptably reproducible results at EM level can be achieved.

A technical advance which seems likely to improve the demonstration of enzyme activity at ultrastructural level is the immunocytochemical approach. The method involves the production of monoclonal antibodies against the enzyme to be detected, coupling of this antibody to an appropriate electron-dense marker such as ferritin or colloidal gold and use of the labelled antibody in a suitable suspension to achieve cell permeation.

Potential value of cytochemical methods in studies of normal and pathological haemopoiesis

Cytochemical demonstration of specifically reacting cell components at an intracellular level provides information on the constitution of cells supplementary to that acquired from conventional stains and electron microscopy, and contributes to knowledge of cellular function. The development and application of cytochemical methods for the various groups of cellular substances, from nucleoproteins, proteins, carbohydrates and lipids to a range of specific enzymes, the advantages and drawbacks of different technical methods, the extent to which they have succeeded in expanding knowledge of cell constitution and function, and some assessment of their theoretical and practical value in haematology, forms the main substance of the next ten chapters.

Since reaction patterns often differ markedly between cell lines recognisably distinct by conventional staining, cytochemistry may provide discriminating cell markers enabling otherwise poorly differentiated members of such cell lines to be recognised. When care is taken not to carry this argument too far, and claim an unjustifiable degree of specificity for any particular reaction, cytochemistry proves of great value in the classification of cell types and in their allocation to particular developmental lines. This is true of normal haemopoietic cells, of haemic cell lines in short and long-term culture, and of cells involved in states of pathological erythropoiesis and leucopoiesis. The contribution of cytochemical studies in these areas, where a range of different cytochemical reactions are considered together and jointly interpreted, is of considerable practical importance in the understanding of cell potentialities for *in vitro* development, and in diagnostic clinical and laboratory haematology. For this reason, the final four chapters are devoted to these broad areas of cytochemical application.

REFERENCES

Altman F P 1971 The use of a recording microdensitometer for the quantitative measurement of enzyme activities inside tissue sections. Histochemie 27: 125

Altman F P 1972 Quantitative dehydrogenase histochemistry with special reference to pentose-shunt dehydrogenases. Progress in Histochemistry and Cytochemistry. Fischer, Stuttgart

Altman F P, Chayen J 1965 Retention of nitrogenous material in unfixed sections during incubation for histochemical demonstration of enzymes. Nature 207: 1205

HAEMATOLOGICAL CYTOCHEMISTRY

Ansley H, Ornstein L 1970 Enzyme histochemistry and differential white cell counts in the Technicon Hemalog D. In: Advances in Automated Analysis. Technicon International Congress, p 437

Bainton D F, Farquhar M G 1968 Differences in enzyme content of azurophil and specific granules of polymorphonuclear leucocytes. I. Histochemical staining of bone marrow smears. II. Cytochemistry and electron microscopy of bone marrow cells. Journal of Cell Biology 39: 286, 299

Baker J R 1943 The discovery of the uses of colouring agents in biological microtechnique. Journal of the Quekett Microscopical Club 4 ser., 1: 256

Barka T, Anderson P T 1962 Histochemical methods for acid phosphatase using hexazonium pararosaniline coupler. Journal of Histochemistry and Cytochemistry 10: 741

Barnes J M 1940 Enzymes of lymphocytes and polymorphonuclear leucocytes. British Journal of Experimental Pathology 21: 264

Beckstead J H 1983 The evaluation of human lymph nodes using plastic sections and enzyme histochemistry. American Journal of Clinical Pathology 80: 131

Beckstead J H, Halverson P S, Ries C A, Bainton D F 1981 Enzyme histochemistry and immunohistochemistry on biopsy specimens of pathologic human bone marrow. Blood 57: 1088

Bianco P, Ponzi A, Bonucci E 1984 Basic and 'special' stains for plastic sections in bone marrow histopathology with special reference to May-Grünwald Giemsa and enzyme histochemistry. Basic and Applied Histochemistry 28: 265.

Bitensky L, Butcher R G, Chayen J 1972 Quantitative cytochemistry in the study of lysosomal function. In: Dingle J T (ed) Lysosomes in Biology and Pathology, Vol. 3. North Holland, Amsterdam

Block M, Smaller V, Brown J 1953 An adaptation of the Maximow technique for preparation of sections of hematopoietic tissues. Journal of Laboratory and Clinical Medicine 42: 145

Borgers M, Verheyen A 1983 Electron microscopic cytochemistry of alkaline phosphatase. Cell Biochemistry and Function 1: 77

Borgers M, Verheyen A 1985 Enzyme Cytochemistry. In: Bourne G H, Danielli J F (eds) International Review of Cytology, vol. 95. Academic Press Inc., New York, p 163

Breton-Gorius J 1980 The value of cytochemical peroxidase reactions at the ultrastructural level in haematology. Histochemical Journal 12: 127

Burgio V L, Magrini V, Ciardini L, Pezzoni G 1984 An enzyme-histochemical approach to the study of the human bone marrow stroma. Acta Haematologica 71: 73

Burkhardt R 1956 Die Myelotomie: eine neue Methode zur kombinierten cytologisch-histologischen Knockhenmarksbiopsie. Blut 2: 267

Burkhardt R 1966 Präparative Voraussetzungen zur klinischen Histologie des menschlichen Knochenmarks. Blut 13: 337

Caspersson T Ø 1936 Über den chemischen Aufbau der Strukturen des Zellkernes. Skandinavisches Archiv für Physiologie 73: Suppl. 8

Caspersson T Ø, Santesson L 1942 Studies on protein metabolism in the cells of epithelial tumours. Acta Radiologica, Stockholm Suppl. 46

Cassuto J P, Hammou J C, Pastorelli E, Dujardin P, Masseyeff R 1977 Plasma cell acid phosphatase, a discriminative test for benign and malignant monoclonal gammopathies. Biomedicine 27: 197

Cathey W J 1963 A plastic embedding medium for thin sectioning in light microscopy. Stain Technology 38:213

Cawley J C, Hayhoe F G J 1973 Ultrastructure of Haemic Cells. Saunders, London, Philadelphia

Chayen J, Bitensky L, Butcher R G 1973 Practical Histochemistry. Wiley, London

Chilosi M, Pizzolo G, Menestrina F, Iannucci A M, Bonetti F, Fiore-Donati L 1981 Enzyme histochemistry of normal and pathological paraffin-embedded lymphoid tissues. American Journal of Clinical Pathology 76: 729

Daddi L 1896 Nouvelle methode pour colorer la graisse dans les tissues. Archivio Italiano di Biologia 26: 143

Dancy J T, Denbelbeiss K A, Harker L A 1976 Section preparation of human marrow for light microscopy. Journal of Clinical Pathology 29: 704

Deutsch W, Rosler K 1929 Zeitschrift fur Physiologische Chemie 185: 146

Doebbler G F 1966 Symposium on cryoprotective additives. Cryobiology 3: 1

D'Onofrio G, Mango G 1984 Automated cytochemistry in acute leukaemias. Acta Haematologica 72: 221

Drewinko B, Bollinger P, Brailas C, et al 1987 Flow cytochemical patterns of white blood cells in human haematopoietic malignancies. 1. Acute leukaemias. British Journal of Haematology 66: 27

Eränkö O, Palkama A 1961 Improved localisation of phosphorylase by the use of polyvinyl pyrrolidone and high substrate concentration. Journal of Histochemistry and Cytochemistry 9: 585

Fiessinger N 1923 Les ferments des leucocytes en physiologie, pathologie et therapeutique generale. Masson et Cie: Paris.
Gahrton G 1964 Microspectrophotometric quantitation of the periodic acid-Schiff (PAS) reaction in human neutrophil leukocytes based on a model system of glycogen microdroplets. Experimental Cell Research 34: 488
Gahrton G 1966 Normal human neutrophil leucocytes as a reference system for the microspectrophotometrically quantitated periodic acid-Schiff reaction. Journal of Histochemistry and Cytochemistry 14: 45
Gahrton G, Foley G E 1966 Cytochemical population analyses of the DNA, RNA and protein content of human leukaemic cells. Acta Medica Scandinavica 180: 485
Galjaard H 1980 General Discussion 1: Validation. In: Trends in Enzyme Histochemistry and Cytochemistry. Ciba Foundation Symposium Publications, Excerpta Medica, Amsterdam, p 53
Gomori G 1939 Histochemical demonstration of phosphatase in tissue sections. Proceedings of the Society for Experimental Biology 42: 23
Gomori G 1952 Microscopic Histochemistry. Univ. of Chicago Press, Chicago, Illinois
Grafe E 1911 Die Steigerung des Stoffwechsels bei chronischer Leukämie und ihre Ursachen. Deutsch Archiv für Klinische Medizin 102: 406
Graham R C, Karnovsky M J 1966 The early stage of absorption of injected horseradish peroxidase in the proximal tubules of mouse kidney: ultrastructural cytochemistry by a new technique. Journal of Histochemistry and Cytochemistry 14: 291
Grouls V 1980 Diagnosis of hairy-cell leukaemia by tartrate-resistant acid phosphatase activity in paraffin-embedded tissue sections. Journal of Clinical Pathology 33: 552
Hale A J, Cooper E H 1963 DNA synthesis in infectious mononucleosis and acute leukaemia. Acta Haematologica 29: 257
Hale A J, Cooper E H 1965 Studies on DNA replication in leukaemias and non-leukaemic leukocytes. In: Hayhoe F G J (ed) Current Research in Leukaemia. Cambridge Univ. Press, Cambridge, p 95
Hayat M A (ed) 1974 Electron Microscopy of Enzymes. Principles and Methods. Vols 1–3 (1974), Vol. 4 (1975) Van Nostrand-Reinhold, Princeton, New Jersey
Hayhoe F G J, Quaglino D, Doll R 1964 The Cytology and Cytochemistry of Acute Leukaemias. Special Report Series of the Medical Research Council No. 304: HMSO: London.
Hinchcliffe R F, Lilleyman J S, Burrows N F, Swan H T 1981 Use fo the Hemalog D automated leucocyte differential counter in the diagnosis and therapy of leukaemia. Acta Haematologica 65: 79.
Holt S J, Hicks R M 1963 Enzyme cytochemistry and electron microscopy. Journal of the Royal Microscopical Society 81: 173
Howat D W, Chayen J 1983 Quantitative cytochemistry of alkaline phosphatase activity. Cell Biochemistry and Function 1: 81
Kaplow L S, Dauber H, Lerner E 1976 Assessment of monocyte esterase activity by flow cytophotometry. Journal of Histochemistry and Cytochemistry 24: 363
Katayama T, Mochino T 1983 Naphthol AS-D chloroacetate esterase reaction. Systemic modifications for optimisation to smears and sections. Acta Pathologica Japonica 33: 381
Kay H D 1930 Plasma phosphatase; method of determination: some properties of enzyme. Journal of Biological Chemistry 89: 235
Kossa J von 1901 Ueber die im Organismus künstlich erzeugbaren Verkalkungen. Beitrage zur Pathologischen Anatomie 29: 163
Lai A P, Martin P J, Richards J D M, Goldstone A H, Cawley J C 1986 Automated leucocyte differential counts in acute leukaemia; a comparison of the Hemalog D, H6000 and Coulter S-plus IV. Clinical and Laboratory Haematology 8: 33
Lai AP, Martin PJ, Cawley JC, Richards JDM, Goldstone AH 1987 Automated leucocyte differential counts in chronic lymphocytic and hairy cell leukaemias: a comparison of the Hemalog D, H6000 and Coulter S Plus IV. Clinical and Laboratory Haematology 9: 169
Levene P A, Meyer G M 1912 On the action of leucocytes on glucose. Journal of Biological Chemistry 13: 265
Lison L 1936 Histochimie Animale. Gauthier-Villars: Paris.
Melnick P J 1968 Histochemical enzymology of leukaemic cells. In: Amromin GD. Pathology of Leukaemia. Harper and Row, New York and London, p 125
Melnick P J, Heizer P 1965 Enzyme histochemistry with cryoprotective additives. Cryobiology 2:27
Menkin V 1934 Studies on inflammation: cytological picture of inflammatory exudate in relation to its hydrogen ion concentration. American Journal of Pathology 10: 193
Menkin V 1938 Studies on inflammation: concerning mechanism of cell migration. Journal of Experimental Medicine 67: 145

Menkin V 1940 Dynamics of Inflammation: An Inquiry into the Mechanism of Infectious Processes. McMillan, New York
Menten M L, Junge J, Green M H 1944 A coupling histochemical azo-dye test for alkaline phosphatase in the kidney. Journal of Biological Chemistry 153: 471
Monahan R A, Dvorak H F, Dvorak A M 1981 Ultrastructural localization of nonspecific esterase activity in guinea pig and human monocytes, macrophages and lymphocytes. Blood 58: 1089.
Novikoff A B, Goldfischer S 1968 Visualization of microbodies for light and electron microscopy. Journal of Histochemistry and Cytochemistry 16: 507
Novikoff A B, Shin W Y, Drucker J 1960 Cold acetone fixation for enzyme localisation in frozen sections. Journal of Histochemistry and Cytochemistry 8: 37
Opie E L 1905 Enzymes and anti-enzymes of inflammatory exudates. Journal of Experimental Medicine 7: 316
Opie E L 1906 The enzymes in phagocytic cells of inflammatory exudates. Journal of Experimental Medicine 8: 410
Ornstein L, Ansley H R 1974 Spectral matching of classical cytochemistry to automated cytology. Journal of Histochemistry and Cytochemistry 22: 453
Patterson K G, Cawley J C, Goldstone A H, Richards J D M, Janossy G 1980 A comparison of automated cytochemical analysis and conventional methods in the classification of acute leukaemia. Clinical and Laboratory Haematology 2: 281
Pearse A G E 1968 Histochemistry, 3rd edn. Churchill, London
Peckham M J, Cooper E H 1969 Proliferation characteristics of the various classes of cells in Hodgkin's disease. Cancer 24: 135
Perls M 1867 Nachweis von Eisenoxyd in gewissen Pigmenten. Virchows Archiv 39: 42
Pierre R V 1984 Differential Counting. In: Koepke JA (ed) Laboratory Hematology. Churchill Livingstone, Edinburgh, p 973
Quaglino D, Hayhoe F G J 1960 Acetone fixation for the cytochemical demonstration of dehydrogenases in blood and bone marrow cells. Nature 187: 85
Quaglino D, De Pasquale A, Zanni G, Zagni G 1974 Autoradiographic and cytophotometric studies on granulocyte precursors in chronic myeloid leukaemia. Haematologica 59: 298
Rebuck J W 1947 Functions of white blood cells. American Journal of Clinical Pathology 17: 614
Roche J 1931 Importance du substrat pour le pH optimum d'action des phosphatases sanguines (globules blancs et rouges, serum). Compte Rendu de la Societe de Biologie 107: 640
Rogers A W 1967 Techniques of Autoradiography. Elsevier, Amsterdam
Scarpelli D G, Kanczak N M 1965 Ultrastructural cytochemistry: principles, limitations and applications. International Reviews of Experimental Pathology 4: 55
Schofield K P, Stone P C W, Beddall A C, Stuart J 1983 Quantitative cytochemistry of the toxic granulation blood neutrophil. British Journal of Haematology 53: 15
Sehrt E 1927 Die histologische Darstellung der Lipoide der weissen Blutzellen (neutro- und eosinophile Leukozyten, Mastzellen, Uebergangszellen und Mononukleäre) und die Beziehung dieser Lipoide zur Oxydasereaktion. Munchener Medizinische Wochenschrift 74: 139
Seligman A M, Gofstein R, Rutenburg A M 1949 Preparation of radioactive iodotetrazolium salt and its distribution in mice. Cancer Research 9: 366
Shapiro H M, Schildkraut E R, Curbalo R, Laird C W, Turner R B, Hirschfeld T 1976 Combined blood cell counting and classification with fluorochrome stains and flow instrumentation. Journal of Histochemistry and Cytochemistry 24: 396
Sjöstrand F S 1968 Electron microscopy of cells and tissues. Academic Press, New York
Soffer L J, Wintrobe M 1932 Metabolism of leucocytes from normal and leukaemic blood. Journal of Clinical Investigation 11: 661
Stern K G 1932 Über die Katalase farbloser Blutzellen. Zeitschrift für Physiologische Chemie 204: 259
Stuart J, Simpson J S 1970 Dehydrogenase enzyme cytochemistry of unfixed leucocytes. Journal of Clinical Pathology 23: 517
Stuart J, Simpson J S, Mann J R 1970 Intracellular hydrogen transport systems in acute leukaemia. British Journal of Haematology 19: 739
Takamatsu H 1939 Histochemical and biochemical studies of phosphatase. Transactions of the Japanese Pathology Society 29: 492
Thorell B 1947 Studies on the formation of cellular substances during blood cell production. Acta Medica Scandinavica Suppl. 200
Tycko D H, Arbalagan S, Liu H C, Ornstein L 1976 Automatic leukocyte classification using cytochemically stained smears. Journal of Histochemistry and Cytochemistry 24: 178
Umeno M 1931 Studien über Phosphatase: über die Glycerophosphatase der Leucocyten im Blute. Biochemistry 231: 339
Wetzel B K, Spicer S S, Horn R G 1967 Fine structural localisation of acid and alkaline

phosphatases in cells of rabbit blood and bone marrow. Journal of Histochemistry and Cytochemistry 15: 311
Wickramasinghe S N 1975 Human Bone Marrow. Blackwell, Oxford, London
Williams D M 1975 Leucocyte alkaline phosphatase as a marker of cell maturity: a quantitative cytochemical and autoradiographic study. British Journal of Haematology 31: 371
Winkler F 1907 Der Nachweis von Oxydase in den Leukozyten mittels der Dimethyl-paraphenylenediamine-Alphanaphthol-Reaktion. Folia Haematologica 4: 323

CHAPTER 2

Nucleoproteins and technical methods of flow cytometry

Amongst the various cellular constituents, nucleoproteins occupy a position of particular prominence, since their presence is of great importance for the structure and biochemistry of the cell as well as for the processes of reproduction, differentiation and maturation.

Nucleoproteins are formed by the combination of a basic protein (protamine or histone) with nucleic acids. The latter consist of long chains of nucleotides (polynucleotides), combined one with another through phosphate diester linkages. Nucleotides are the structural unit of nucleic acids and are composed of a purine or a pyrimidine base attached to a sugar (ribose or 2-deoxyribose) by a glycosidic linkage: the sugar is then combined with phosphoric acid. Two general types of nucleotides are found in nucleic acids: one contains D-ribose, the other contains 2-deoxy-D-ribose. It is therefore customary to refer to these nucleic acids as ribonucleic acid (RNA) or deoxyribonucleic acid (DNA).

Deoxyribonucleic acid (DNA) contains thymidylic acid (thymine + deoxyribose + phosphoric acid) rather than uridylic acid. The DNA molecules of mammalian cells appear to be smaller than those from most organisms, containing only 30 000 nucleotides in a typical molecule (Harper, 1973). X-ray diffraction patterns of fibrous DNA (Watson and Crick, 1953, Wilkins *et al.*, 1953) suggest that its form is that of a double helix. The presence of the double helix can also be shown by studies of ultraviolet light absorption of solutions of DNA, since nucleotides absorb ultraviolet light with wavelengths near 260 nm by virtue of the presence of purine and pyrimidine rings.

The two strands of the double helix of DNA are conceived of as running in opposite directions with respect to the 3,5-deoxyribose linkage. Biochemical studies of DNA obtained from different cell types indicate that the amount of adenine equals the amount of thymine and the amount of guanine equals the amount of cytosine. This 'pairing' of the nitrogenous bases and the X-ray diffraction data constitute the essential foundation of the Watson and Crick model for DNA structure.

There are three characteristic forms of RNA: messenger RNA (mRNA), ribosomal RNA (rRNA) and transfer RNA (tRNA).

mRNA is a single stranded molecule. It is formed in the nucleus of the cell as a complement to one strand of DNA, which may contain information for the synthesis of several proteins. Therefore mRNA serves as a means of transcribing the information stored in DNA, by determining the amino acid sequence and thus the primary structure of proteins being synthesized within the endoplasmic reticulum of the cell. mRNA cannot be formed with a DNA template unless the DNA is exposed by dissociation from the histone. The extent of dissociation of the DNA template from the histone, with which it is associated, thus influences the type of mRNA made in the cell.

rRNA is that associated with the ribosomes of the endoplasmic reticulum in the cytoplasm. Like mRNA, ribosomal RNA is formed so as to be complementary to some region of DNA. rRNA is required to bind mRNA and the specific enzymes utilized for peptide bond synthesis. The subunits of the ribosomes are classified in terms of their sedimentation constants (S): mammalian ribosomes are comprised of 40 S and 60 S components. Each part contains RNA along with several different proteins. Ribosomal RNA is made in the nucleolus as a single chain and subsequently hydrolysed into several pieces, some of which associate with the necessary proteins to form ribosomes.

tRNA (also called soluble RNA—sRNA): this type of RNA accomplishes the transfer of aminoacids to the proper site on the RNA template of mRNA. For each amino acid to be incorporated into the primary structure of a protein molecule there is therefore a specific tRNA molecule to which the amino acid is attached. There are certain similarities in the structure of all the tRNAs so far studied. The amino acid carrying end of all the tRNA molecules contains the same trinucleotide, having the base sequence A-C-C, i.e., an adenylic acid and two cytidylic acid nucleotide residues attached to the terminal adenosine residue. tRNA molecules are presumed to possess a secondary structure with base pairing by means of hydrogen bonds as was described for DNA. Consideration has also been given to the possibility of a tertiary structure wherein the molecule may fold to constitute a more globular form. Purified species of tRNA have been crystallized and subjected to single crystal X-ray diffraction investigations.

There are a number of cytochemical techniques for the study of nucleic acids, and of these methods some are suitable for quantitative assessments.

1. *Ultraviolet spectrophotometry* (Caspersson, 1936; Thorell, 1947) is based on the intense absorption of ultraviolet light of wavelength 260 nm by the purine and pyrimidine components of both DNA and RNA. Such an intense absorption of ultraviolet light is shown in cells only by nucleic acids and a certain number of pigments. Ultraviolet spectrophotometry may be employed for the selective identification of one or other of the two nucleic acids, when cells are subjected to appropriate nuclease digestion.

2. *The gallocyanin chromalum* method was introduced by Einarson in 1932 for the demonstration of Nissl granules. It was subsequently proposed by Einarson (1951) for the histochemical study of nucleic acids. The reaction seems to be based on a stoichiometric combination between basic dyes and the phosphoric group of nucleic acids (Barka and Anderson, 1963). Both DNA and RNA are stained and they cannot be distinguished by this method alone. The technique appears suitable for cytophotometric measurement of the relative amount of total nucleic acids (Sandritter *et al.*, 1954, 1966). A distinctive advantage of the method is the progressive nature

of the staining. When the end part of the reaction has been reached, no further change in staining intensity will occur.

3. *The methyl green pyronine method*: this old empirical method introduced by Pappenheim in 1899 and based on the simultaneous use of methyl green for staining DNA and pyronine for demonstrating RNA, has been extensively employed for the study of nucleic acid distribution, especially after Brachet (1940, 1953) confirmed the specificity of these dyes by abolishing the staining reaction with appropriate nuclease digestion. Experimental evidence suggests that methyl green is specifically bound to highly polymerized DNA, so that nuclei are coloured in green, while pyronine has a special affinity for lower polymer RNA, so that all RNA-containing structures show a pink-red colour. The specificity of pyronine for RNA is less than that of methyl green for DNA. Pyronine staining is also influenced by other factors, such as the pH of the staining solution, so that the specificity of the reaction should be tested by the parallel use of ribonuclease digestion.

Using this cytochemical method, White (1947) found the cytoplasmic content of RNA to decrease as cell differentiation proceeded and concluded that the high levels of RNA in the cytoplasm and nucleoli of young haemopoietic cells were closely related to the proliferative and synthesizing capacities of these cells (Plate 1A). Pyroninophilic RNA is present also in the cytoplasm of mature leucocytes of all kinds, although in smaller amounts than in precursor cells (Plate 1B, C). An intensely positive reaction for RNA may be especially observed in the cytoplasm of plasma cells (Rheingold and Wislocki, 1948) (Plate 1D), though positive reactions for RNA may also be detected in neutrophil granulocytes (Berghe and Hoffman, 1945) in Auer rods of leukaemic myeloblasts (Ackerman, 1950) and in toxic granulations of degenerating leucocytes (Jackson, 1954). Among lymphocytes, cells which have marked basophilia with Romanowsky stains are occasionally found in the peripheral blood, especially in disorders associated with autoimmune disturbance and chronic infection. These cells stain deeply with pyronin and the term 'pyroninophilic cells' is sometimes applied to them (Bessis, 1973). It seems likely that these cells are activated lymphocytes undergoing a transformation preliminary to antibody production or other immunological activity (Plate 1E, F).

4. *Secondary fluorescence* induced in cells by fluorochromes such as acridine orange has been successfully employed for demonstrating nucleic acids (Kosenow, 1956; Marmont, 1960). With this technique DNA gives a green or yellow fluorescence, while RNA stains orange to red. This method has been found useful for the study of reticulocytes (in which case the fluorescent staining replaces the dye brilliant cresyl blue), plasma cells and L.E. cells. In reticulocytes Kosenow (1956) has observed three different fluorescent patterns according to the concentration of acridine orange employed: at a concentration of 1:100 000 a weak grey-green fluorescence was diffusely present in the cell; at a concentration of 1:40 000 it was possible to observe the same diffuse appearance with the addition of one or more yellow-orange granules; finally at a concentration from 1:5000 to 1:2000 a red-orange fluorescent reticulum became apparent on a completely

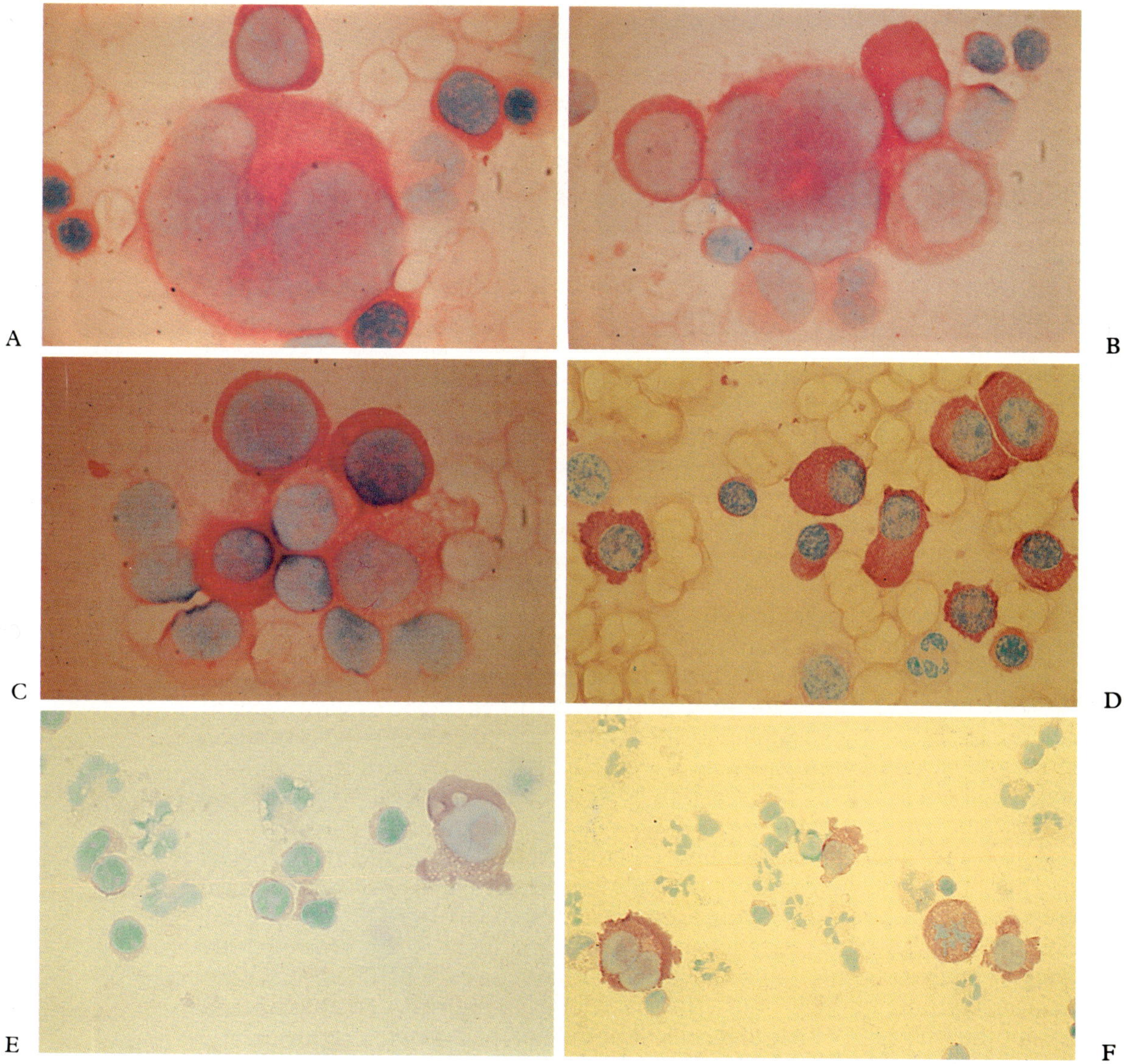

Plate 1

A. Normal human bone marrow. Methyl green pyronin. A megakaryocyte surrounded by erythroblasts with variable pyroninophilia and a neutrophil with weak pyroninophilia.

B. The same preparation. A megakaryocyte with four erythroblasts showing variable pyroninophilia —strong in early stages and weak in later, four granulocyte precursors with weak pyroninophilia and a single strongly reacting plasma cell.

C. The same preparation. Three erythroblasts with strong pyroninophilia, three granulocyte precursors and five lymphocytes with weak positivity.

D. Bone marrow from a patient with multiple myeloma. Methyl green pyronin. Eight strongly reacting plasma cells, an erythroblast of intermediate positivity, a lymphocyte with a rim of pyroninophilia, two almost negative neutrophils and a weakly pyroninophilic monocyte.

E. PHA-transformed lymphocytes. Methyl green pyronin. An intensely positive transformed lymphocyte with prominent nucleolus and various normal lymphocytes and neutrophils.

F. PHA-transformed lymphocytes. Methyl green pyronin. Three intensely reacting transformed lymphocytes, one of which is in mitosis and another binucleated.

negative cellular background. Studies with the electron microscope have confirmed these findings and have provided an explanation for the observed phenomena: at weak concentrations, the fluorochrome attaches itself to the ribosomes; at intermediate concentrations the fluorochrome is concentrated in the Golgi region and gives rise to vacuoles; at high concentrations, ribosomes precipitate (Bessis and Breton-Gorius, 1964). In leucocytes mitochondria as well as cytoplasmic granules show a more or less intense red-orange fluorescent stain after acridine orange (Kosenow, 1956) (Plate 2A, B). Employing atabrine both *in vivo* and *in vitro* leucocyte granules show a yellow gold fluorescence (White, 1954). The early phases of cellular death are indicated by a change from the usual green to a red fluorescent stain (Kosenow, 1956), due to DNA depolymerization (Schümmelfeder *et al.*, 1957). Other fluorochromes, such as ethidium bromide and propidium iodide, have been most commonly employed in flow cytometric studies and will be dealt with later on (see p. 25).

5. *The Feulgen reaction.* The most important reaction for demonstrating DNA is still that introduced by Feulgen and Rossenbeck (1924), which depends on the liberation of free pentose aldehyde groups from deoxyribonucleic acid after hydrolysis with warm hydrochloric acid. The free pentose aldehyde groups combine with Schiff's reagent (leucobasic fuchsin, a rosaniline-pararosaniline mixture decolorized by excess of sulphurous acid) to produce a brilliant magenta coloured substance. To be deoxyribonucleic acid, the substance responsible for a positive Feulgen reaction must give negative results unless previously hydrolysed with hydrochloric acid. The chemistry of the reaction is described by Baker (1942) and by Stacey and others (1946).

The technique has been extensively applied in normal and pathological haemic cells. A positive reaction, indicating the presence of DNA, is found only in the nuclear chromatin and chromosomes or in the immediate products of their disintegration or dissolution, such as the Howell-Jolly body. The cytoplasm of all cell types is negative, as are platelets and non-nuclear erythrocyte inclusions such as siderotic granules, reticulation and basophilic stippling. Nucleated cells of both leucopoietic and erythropoietic series take up the stain with an intensity proportional to the degree of maturity of their nucleus. Immature cells have a relatively low concentration of DNA and stain palely (Fig. 2.1), while mature cells have more concentrated DNA and give a deep nuclear stain (Fig. 2.2). Nucleoli in primitive cells are unstained and show up more clearly than in Romanowsky preparations. A perinucleolar zone of intensely staining chromatin can often be seen.

These observations were essentially qualitative (Pittaluga and Bessis, 1944; Rheingold and Wislocki, 1948; Gardikas and Israels, 1948), but the combination of Feulgen staining with microdensitometry has enabled accurate quantitative assessments of cell DNA content to be carried out (Marinone, 1953, Perugini *et al.*, 1957, Hale and Wilson, 1959, 1961). Mature or resting cells with a normal diploid number of chromosomes have a constant DNA content irrespective of cell type, usually referred to as 'the 2n value', while nuclei engaged in DNA synthesis prior to division have

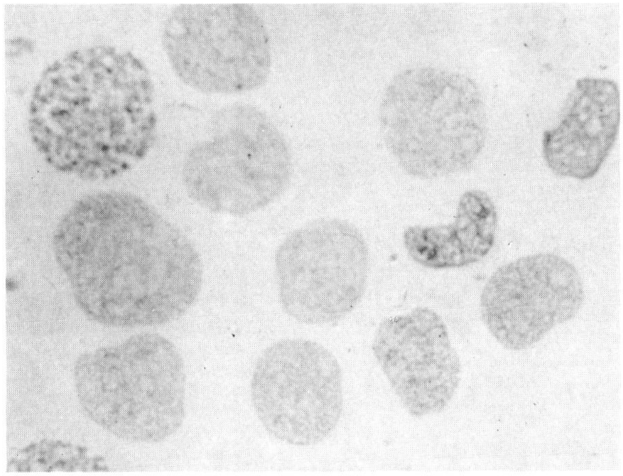

Fig. 2.1 Feulgen reaction. Immature cells from bone marrow in an acute leukaemia stain weakly.

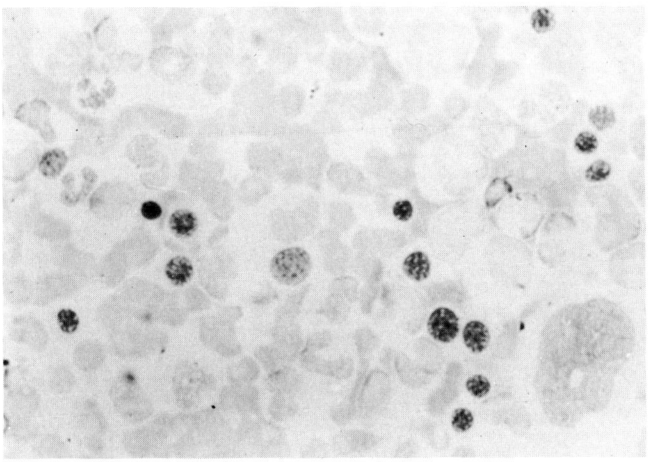

Fig. 2.2 Feulgen reaction. Normal bone marrow cells. Positivity is confined to the nucleus and shows increasing intensity with increasing cell maturity.

increased amounts ranging from 2n to 4n values. Patterns of DNA distribution in normal peripheral blood leucocytes and in mature cells from bone marrow show very little scatter around the predominating 2n value, whereas among early erythroblasts and granulocyte precursors in the marrow a wide scatter from 2n to 4n is characteristically observed. The widespread use in clinical cytokinetic or related studies of many of the chemical and physical methods employed for the determination of the DNA content per cell has been hindered by the requirements of large sample size and by the time-consuming nature of the procedures.

A more rapid determination of DNA per cell on monodisperse cell populations has been obtained by the introduction of laser-based *flow*

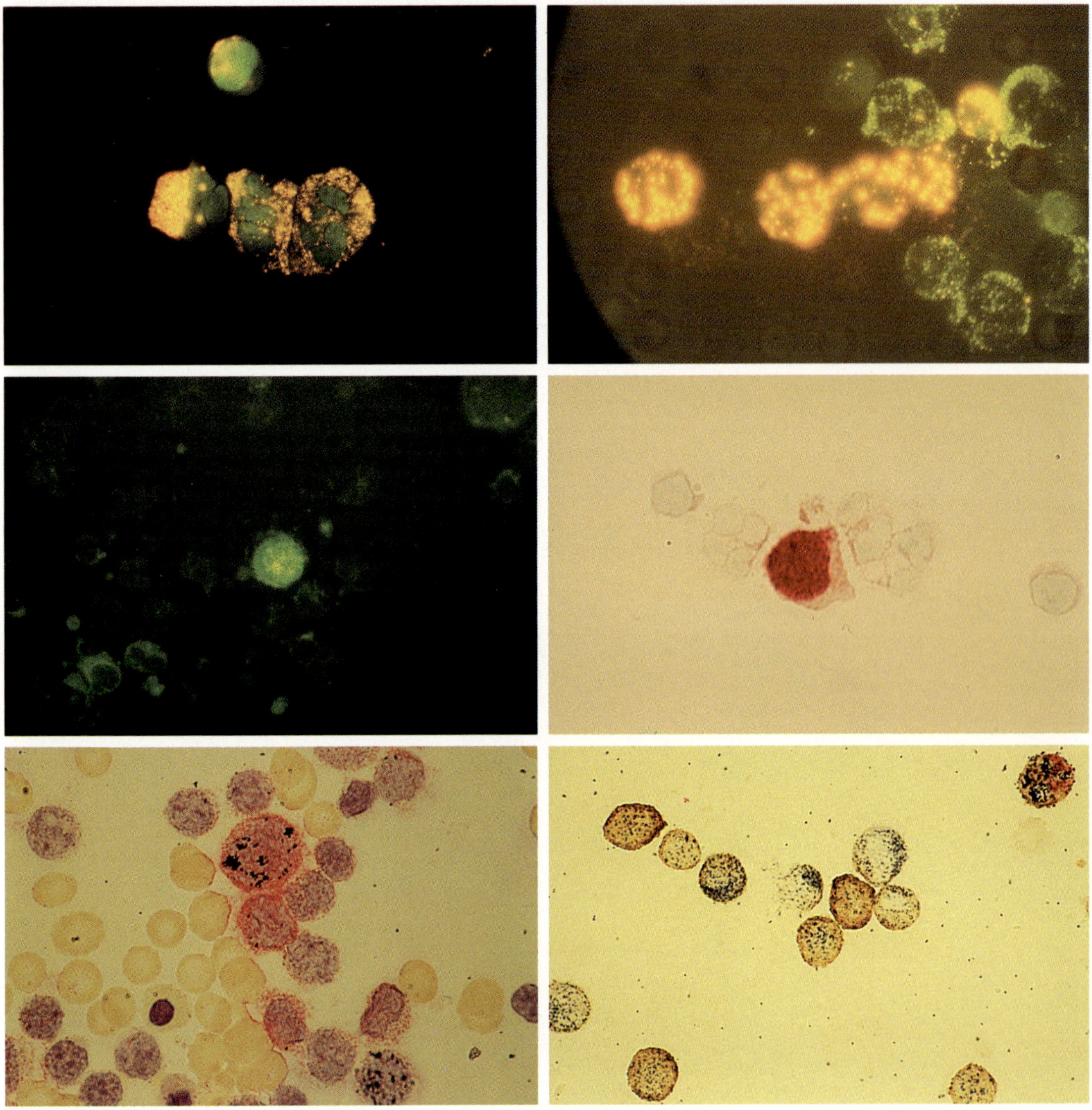

Plate 2

A. Normal blood cells fluoresced with euchrysine (acridine orange). Nuclear DNA stains green, with the lysosomal specific granules of two neutrophils and a basophil fluorescing orange. A single lymphocyte gives weak yellow cytoplasmic fluorescence.

B. Peripheral blood cells from chronic myeloid leukaemia. Group of four basophils, with strong orange fluorescence in active lysosomal specific granules; weak reactions in neutrophil polymorphs.

C. PHA stimulated lymphocytes. BUdR labelling of DNA-synthesizing cells revealed by fluorescent-conjugated monoclonal anti-BUdR antibody; strong nuclear positivity in transformed cells, with negative nuclei in resting lymphocytes.

D. PHA stimulated lymphocytes. BUdR labelling revealed by monoclonal anti-BUdR antibody and immuno-alkaline phosphatase (APAAP) technique.

E. Normal bone marrow, with immunoalkaline phosphatase reactions for anti Leu Ml (in red) plus ^{3}HTdR autoradiography.

F. Acute myelomonocytic leukaemia with dual esterase reaction (chloroacetate esterase with red and acetate esterase with blue granules) plus ^{3}HTdR autoradiography plus immunoperoxidase staining for HLA-DR (brown).

microfluorometry (Van Dilla *et al.*, 1968). Flow system instruments exhibit considerable variety of form but the basic principles of operation are similar in each. Single cells are constrained to flow in fluid suspension so that one cell at a time passes through a high-intensity focused light flux, which is frequently obtained from a laser source. As each cell passes through the exciting beam a pulse of scattered and/or fluorescent light is rendered incident upon a light sensitive detector which responds by giving an electrical pulse that is directly proportional to the quantity of light scattered or emitted from the cell. The electrical pulse is then digitized and stored electronically for computer analysis and display.

A number of different preparative techniques have been employed for cytofluorographic analysis of cell DNA content, and involve the use of either fluorescent dyes, such as acriflavine, auramine 0, etc., after acid hydrolysis by the fluorescent Feulgen method, or the use of DNA intercalating fluorescent compounds, such as ethidium bromide and propidium iodide.

At high concentrations ethidium bromide has been shown (LePecq and Paoletti, 1967) to form a fluorescent complex with helical polynucleotides, which has a strikingly enhanced fluorescent quantum yield. Formation of this complex (complex I) is specific for base-paired regions in DNA, RNA, and DNA:RNA hybrids and the basis for this specificity is intercalation of the dye between base pairs. At low concentrations, a second mode of binding ethidium bromide is observed (complex II): it appears due to electrostatic binding of the cationic dye to negatively charged phosphate groups. Complex II can be readily distinguished from complex I by its fluorescence properties.

Propidium iodide, an analogue of ethidium bromide, intercalates with DNA and RNA and this binding markedly enhances the fluorescence of the dye. In cells incubated with propidium iodide dissolved in an isotonic solution, dye entry into the cell is prevented by the cell membrane, and laser-excited fluorescence is seen only in nuclei of dead cells (Krishan, 1975). However, by incubation of cells in a hypotonic solution of propidium iodide, the structural ability of the plasma membrane for dye exclusion is lost and a brilliant fluorescent staining of the dye-DNA complex is observed within a few minutes of staining. Examination of the nuclei after 5 minutes of staining shows that treatment with hypotonic citrate solution results in isolated nuclei, which do not lose their integrity or staining intensity for up to 60 minutes of incubation at 4°C.

Undoubtedly flow cytometry, based, as we have seen, on the use of fluorescent dyes or intercalating fluorescent compounds, has greatly facilitated interpretation of DNA distribution and has provided useful information on the cell cycle traverse characteristics of normal and malignant cells, by indicating also the fraction of cells with S-phase DNA content. Detection of S-phase cells has in fact considerable relevance in anticancer therapy, since the use of S-phase specific antitumour agents might improve the results of chemotherapy, when the experimentally determined frequency of tumour cells in S-phase is maximal (Barranco *et al.*, 1982).

In addition to the determination of the fraction of cells with S-phase

DNA content obtained by flow cytometry, the evaluation of the amount of radioactive DNA precursor incorporation (namely ^{3}H-thymidine), measured by liquid scintillation spectrometry (Pallavicini *et al.*, 1982) may also have considerable importance in cytokinetic studies. However this method of study, although of considerable interest, fails to provide adequate information on the distribution of radioactivity incorporation among the cells of the population under study.

Similarly, interpretation of DNA distributions, measured by flow cytometry, suffers from the drawback that no information is available on the absolute rate of DNA synthesis. In fact, a cell population proliferating with twice the rate of another might have the same DNA distribution if the two populations spend the same fractional times in the G_1, S and G_2M phases of the cell cycle. Furthermore, discrimination between actively synthesizing and quiescent cells with S-phase DNA content is impossible.

In an attempt to overcome the limitations of both flow cytometry and liquid scintillation spectrometry, Dolbeare *et al.* (1983) have devised a procedure based on the simultaneous flow cytometric measurement of cellular DNA content and amount of incorporated 5-bromo-2′-deoxyuridine (BUdR). The latter represents a thymidine analogue, which in *in vitro* culture is taken up by cells in S-phase and incorporated into their nuclear DNA. For the measurements, propidium iodide is used as the probe for total DNA content and a fluorescein-conjugated monoclonal antibody against BUdR (Gratzner, 1982) is used as the probe for BUdR. The flow cytometric measurements are arranged to form bivariate distributions of DNA content vs BUdR content. These distributions readily show the relative amounts of BUdR incorporated into the DNA of cells whose DNA contents place them in the G_1, S and G_2M phases of the cell cycle. This procedure is sufficiently sensitive to allow quantification of very low levels of BUdR incorporation and furthermore it allows discrimination between DNA synthesizing and non-synthesizing cells with S-phase DNA content.

By a similar method, Danova and his associates (1987) studied the cell kinetics in 18 human haematological and non-haematological tumours. Cell labelling was performed *in vivo* by administering 500 mg of BUdR over a 15–20 min i.v. infusion. The percentage of BRdU-labelled cells and their total DNA content were determined simultaneously with an Ortho 50H cell sorter. The authors claim that with this method labelling index and DNA-synthesis time are similar to those obtained with more complicated and time-consuming autoradiographic techniques.

The critical aspect of the procedure is the use of hydrochloric acid to 'partially' denature cellular DNA, so that the monoclonal antibody can bind to the incorporated BUdR (the antibody does not bind to BUdR in double stranded DNA), while leaving sufficient double-stranded DNA for propidium iodide binding. The HCl treatment also serves to hydrolyze double-stranded RNA, so that digestion with ribonuclease is not required to get highly DNA-specific propidium iodide binding.

More recently, however, an anti-BUdR monoclonal antibody has been produced (Gonchoroff *et al.*, 1985), which does not require DNA denaturation.

The elevated green fluorescence for some cells with G_1 DNA content is a reflection of the sensitivity of the BUdR labelling procedure and is not due to nonspecific antibody binding. This fluorescence results from antibody binding to BUdR incorporated during only a few minutes of DNA synthesis as cells move into S-phase midway during the labelling period. The total DNA content change during this short synthetic period is so small that no change can be detected by propidium iodide binding.

However, non-specific BUdR binding should be kept in mind, especially when dealing with bone marrow populations at different stages of maturation. Dolbeare *et al.* (1983) included in their investigations also mouse bone marrow cells and found a population of G_1 cells which seemed to bind antibody nonspecifically. These cells were studied morphologically and were found to be differentiated cells, which are known to possess Fc receptors and may thus bind immunoglobulins nonspecifically.

BUdR labelling of DNA-synthesizing cells, demonstrated by immunofluorescence and by immunocytochemical methods, is illustrated in Plate 2, C & D.

Long before the introduction of flow cytometric studies based on propidium iodide and BUdR incorporation, *autoradiographic studies*, using radioactive precursors of DNA, RNA and proteins, have provided information of considerable relevance on the metabolic behaviour and proliferative activity of normal and leukaemic cells, and in combination with other cytochemical (qualitative and quantitative) or cytobiological methods of investigation are potentially still capable of furthering our knowledge on particular behaviour patterns of neoplastic cell populations.

It is now widely accepted that the most suitable labelled precursor for the study of DNA synthesis is tritiated thymidine (Fig. 2.3) (Taylor *et al.*, 1957), while the introduction of 5-[^{3}H]uridine by Hayhoe and Quaglino in 1965 provided a specific precursor for the study of RNA metabolism (Fig. 2.4). The study of protein synthesis has been somewhat hampered by the different degree of incorporation of the radioactive aminoacids employed (arginine, lysine, tyrosine, leucine) and this phenomenon is probably related to the influence exerted by the intracellular amino-acid pool.

The first *in vivo* studies on DNA synthesis in haemic cells were those of Cronkite *et al.* (1960), who observed the presence of thymidine incorporation in 35–93% of myeloblasts, in 70% of promyelocytes and 68% of myelocytes. Following *in vitro* incubation, Lajtha (1959) found that 60% of promyelocytes and 30% of myelocytes were labelled with thymidine. Similar percentages (40–50% of labelled myeloblasts and promyelocytes, 10–20% of labelled myelocytes) have been reported by Gavosto *et al.* (1959) and Mauri (1962). The average length of the S phase, during which DNA synthesis takes place, is shorter in immature cells than in the more advanced maturation stages on account of a more intense DNA synthetic activity, as revealed by the average grain count. Bond *et al.* (1959) reported the following average grain counts in primitive erythroid cells: 50–80 in proerythroblasts, 25 in basophilic erythroblasts, 12 in polychromatic erythroblasts. Lajtha (1959) obtained values of 74 in basophilic erythroblasts and 39 in polychromatic erythroblasts, and of 60 in promyelocytes

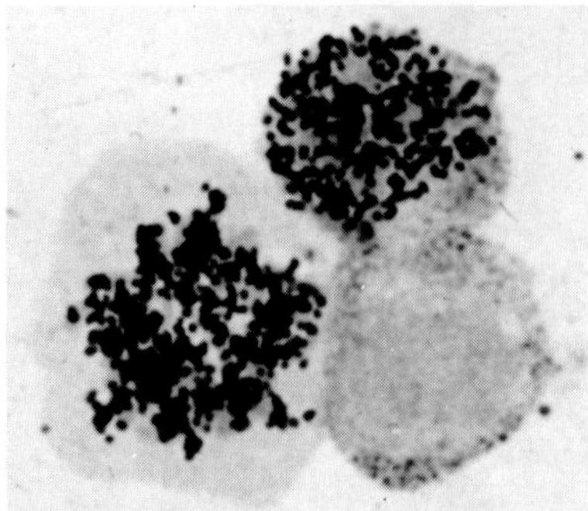

Fig. 2.3 Autoradiograph (ARG) of proliferating cells labelled with tritiated thymidine: the grains are confined to the nucleus, and to the chromosomes in the dividing cell. PAS counterstain.

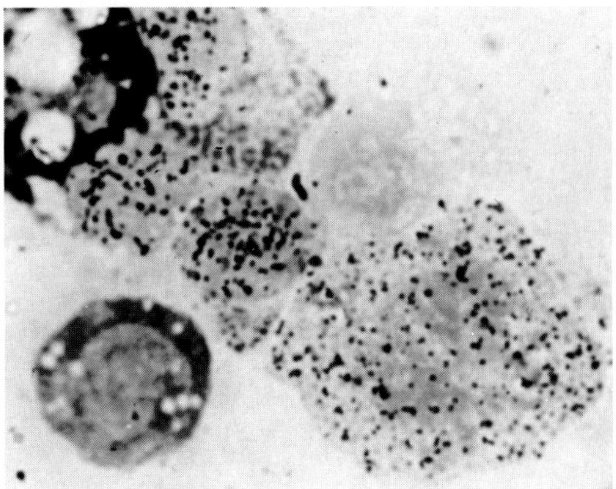

Fig. 2.4 ARG of proliferating cells labelled with 5-[^{3}H] uridine: the grains are localized to RNA and do not label the chromosomes in the dividing cell.

and 37 in myelocytes. Similar findings were reported by Mauri (1961a). Quantitative autoradiographic studies using ^{14}C labelled thymidine (Dörmer, 1973) have again confirmed the progressive lengthening of duration of S phase with increasing maturity among human erythroblasts, with estimated figures of 9.4 $\pm$ 1.9 h in proerythroblasts to 17.0 $\pm$ 3.4 h in early polychromatic normoblasts.

In 1963 and 1965 Hale and Cooper extended the study of DNA distribution and DNA synthetic activity in haemic cells by a combination of microdensitometry and autoradiography. The technique consisted in labelling the cell population under study with a DNA radioactive precursor, fixing the cells, staining them by the Feulgen method, making a photographic map of the stained cells, measuring their DNA content, applying stripping film or emulsion, exposing and developing it and finally counting the number of grains over each individual cell (Figs 2.5, 2.6, 2.7 and 2.8).

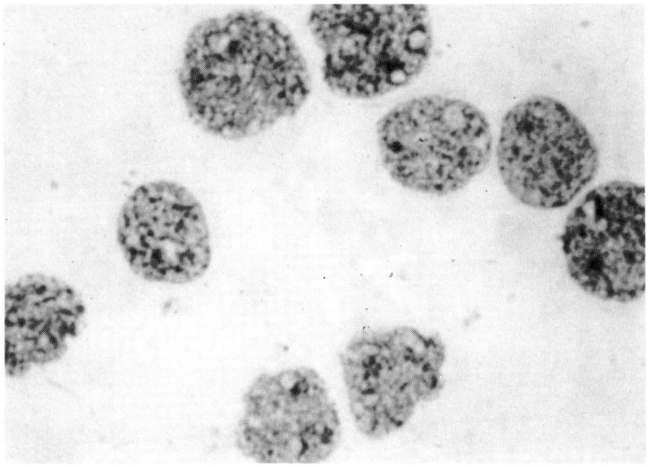

Fig. 2.5 AML bone marrow. Feulgen stain for quantitative evaluation of DNA content of the leukaemic myeloblasts.

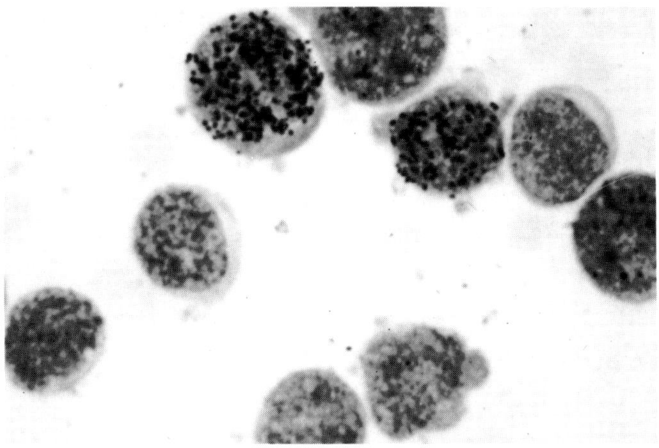

Fig. 2.6 The same field after autoradiography with ^{3}HTdR label. Two blast cells show intense incorporation of the DNA precursor, while two others show weak uptake. Five blast cells are completely negative.

Using this approach DNA replication was studied in infectious mononucleosis cells, in PHA transformed lymphocytes and in leukaemic blast cells. Other studies using the combination of Feulgen stain and autoradiography are discussed below.

RNA and protein synthesis, unlike that of DNA, takes place during the entire intermitotic period of normal myeloid cells (Lajtha *et al.*, 1954). RNA and protein synthesis are less marked in mature cells compared to primitive ones (Lajtha *et al.*, 1954). In both chicken nucleated erythrocytes (Cameron

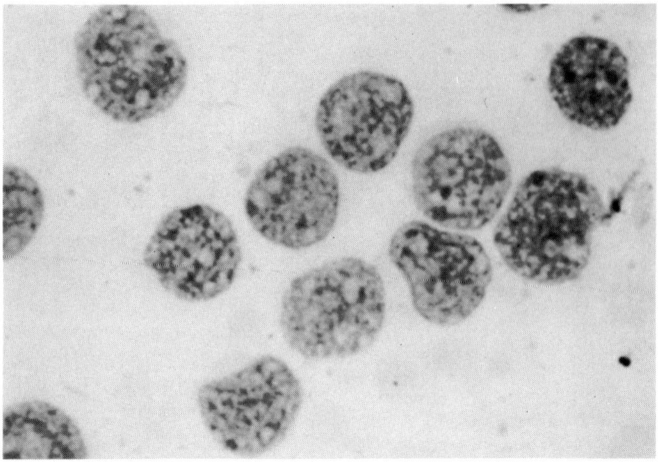

Fig. 2.7 AML bone marrow. Feulgen stain for quantitative evaluation of DNA content of the leukaemic myeloblasts.

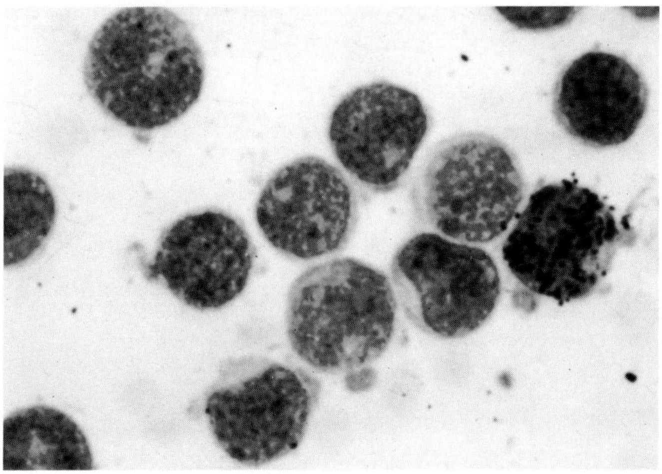

Fig. 2.8. The same field after autoradiography with ^{3}HTdR label. One blast cell shows intense incorporation of the radioactive precursor while the remainder are negative.

and Prescott, 1963) and in rat reticulocytes (Pinheiro *et al.*, 1963) study of RNA and protein precursor incorporation shows that RNA synthesis ceases well before protein synthesis.

Lymphocytes have been the object of several autoradiographic investigations carried out principally with a view to studying the fate as well as the life span of lymphoid cells.

In vivo studies carried out in the guinea pig and rat by Everett *et al.* (1960), by injecting tritiated thymidine, showed, in agreement with the findings of Schooley *et al.* (1959) in mice, that large and medium sized lymphocytes are labelled early in all tissues, while small lymphocytes begin to be labelled only after a certain interval of time. It was also observed that the mean intensity of labelling decreases from large to medium to small lymphocytes.

Little *et al.* (1962) studied the incorporation of DNA in the leucocytes of rats using continuous infusion of tritiated thymidine for 90 days. The authors found that all granulocytes were labelled by the 3rd or 4th day, 20–45% of small lymphocytes by 5 to 10 days and all large lymphocytes by the 60th day, but maximum labelling of small lymphocytes was not achieved until after 100 days of infusion and even at 220 and 271 days 5 to 10% of cells remained unlabelled. The observations of Caffrey *et al.* (1962) are in agreement with these results, which appear to indicate the existence of two populations of small lymphocytes, one with a long circulating life-span (at least 8 weeks) and another with a circulating life-span of less than 14 days. The data of Little *et al.* (1962) and of Caffrey *et al.* (1962) tend to exclude DNA reutilization as an explanation for the prolonged life of lymphocytes. The proportion of long-lived, unlabelled cells progressively falls from thoracic duct lymph to lymph nodes, blood, spleen, the bone marrow and thymus (Ford and Gowans, 1969). This long-lived lymphocyte pool contains not only T cells, as thought at one time, but also B marrow cells (Sprent and Basten, 1973, Ropke *et al.*, 1975), while the short-lived lymphocytes are probably mostly virgin B or T precursors (Strober and Dilley, 1973 and Strober, 1975). Comparable data in man are not available, but a similar pattern of cell distribution seems likely. Estimates of the life span of long-lived T cells in man have given a mean value of over 4 years, with some cells persisting for another 20 years (Nowell, 1965; Buckton *et al.*, 1967).

Both RNA and proteins are synthesized in lymphocytes, but the degree of synthetic activity varies considerably according to the size as well as the morphological and functional features of the different types of lymphocytes (Mauri *et al.*, 1966). Figure 2.9, for example, illustrates a population of variably activated lymphocytes in short-term culture with PHA. Residual resting lymphocytes show little or no uridine uptake, whereas transformed cells are heavily labelled with strong nucleolar labelling in one of them.

In the normal plasma cell series, immature cells show marked RNA synthetic activity, while amino acid incorporation is considerably more intense in mature plasma cells (Quaglino *et al.*, 1967). The intense protein synthesis which takes place in mature plasmacytes in the absence of any significant RNA synthesis might imply that the elaboration of information, i.e. the synthesis of messenger RNA could have occurred earlier during the differentiation process, as suggested also by Mitchell and Nossal (1963) in their study on the RNA metabolism of rat plasma cells.

With regard to DNA synthetic activity in the plasma cell series, Schooley (1961) observed 1 hour after tritiated thymidine injection in immunized rats that 80% of plasmablasts and 45% of proplasmacytes were labelled, whereas no incorporation was present in mature plasma cells. A similar behaviour is present also in the normal human plasma cell series (Quaglino *et al.*, 1967). In myeloma cells thymidine incorporation is significantly decreased and never exceeds 10% even in cells displaying immature features

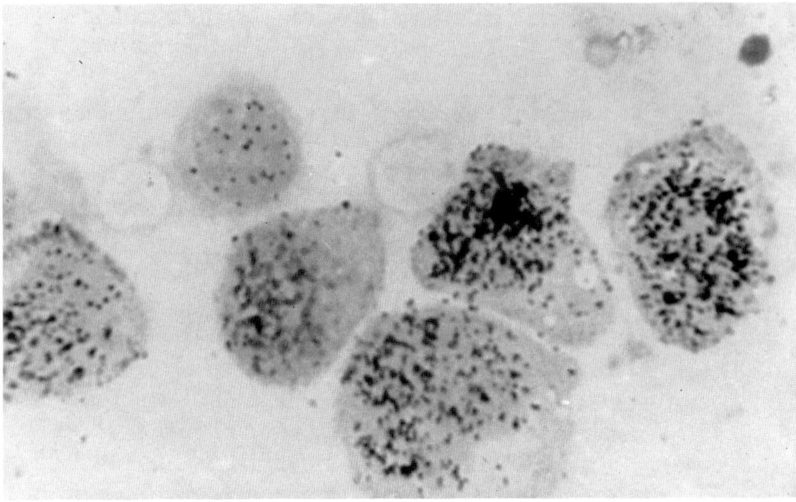

Fig. 2.9 PHA stimulated lymphocytes. ARG with 5-[^{3}H] uridine. Strong uridine incorporation in transformed cells with conspicuous nucleolar uptake in one of them.

(Mauri *et al.*, 1961; Quaglino *et al.*, 1967). The average labelling intensity of myeloma cells does not appear to be comparable to that of cells which are already half-way through the differentiation process, but, on the contrary, in the small percentage of labelled myeloma cells the rate of thymidine incorporation is high and this suggests that these cells go through the S period at a rate similar to that of normal plasmablasts. These findings suggest the existence in multiple myeloma of two cell populations: one, corresponding to a minority of cells, characterized by a normal rate of reduplication and capable of division; the second, comprising the great majority of cells, apparently devoid of any capacity to proliferate. Such an interpretation is also supported by the findings of Killmann *et al.* (1962). More recently still the pattern outlined above has been confirmed by Pileri and Conte (1977), who studied the kinetic behaviour of the proliferating and non-proliferating compartment of the myeloma cell population, also with the aid of immunofluorescence techniques and the use of anti-idiotype sera. The investigations carried out suggest the existence of a B lymphocyte premyelomatous compartment and indicate that the passage from the proliferating to the non-proliferating compartment is associated with reduced RNA synthesis, a less evident decrease in protein synthesis, increased protein and RNA concentration, marked reduction of transcriptional activity and increased messenger RNA half-life. In 1984 Riccardi *et al.* investigated the relationship between proliferative activity, response to therapy and survival in 33 patients with multiple myeloma (25 untreated and 8 at first relapse). Median survival was longer than 50 months for patients with labelling index (LI) less than 3.5%, and 7.5 months for those with LI greater than 3.5%. This difference was independent of clinical stage at diagnosis, M component type, percentage of bone marrow plasma cells and type of first treatment (melphalan or peptichemio).

At first relapse LI was found higher than at diagnosis in 5 out of 9 patients, similar in 3 and lower in one. Median LI of all 17 patients studied at first relapse was 4.4%. In the 9 patients with LI less than 4.4%, relapse occurred slowly after responses which lasted for more than 3 months, while 6 out of 8 patients with LI greater than 4.4 % had rapid relapse after responses which lasted less than 3 months. Median survival from relapse was 16 and 3 months respectively.

The data presented by the authors suggest that proliferative kinetics of bone marrow plasma cells during the course of multiple myeloma is not identical in all patients. The LI appears to have prognostic significance since high values are frequently associated with reduced survival.

Autoradiographic techniques have been employed advantageously especially in acute leukaemias, where early studies of thymidine incorporation indicated the presence of a low DNA synthetic activity (Lajtha *et al.*, 1954; Gavosto *et al.*, 1959 and Mauri, 1961), thus confirming the previously reported observation of a low proliferative activity of leukaemic blast cells studied with the stathmokinetic test (Astaldi and Mauri, 1950).

Subsequent studies have contributed greatly to knowledge of leukaemic cell kinetics: the results and their relevance to the clinical management of leukaemias are briefly reviewed in Chapter 14.

In 1960 Gavosto *et al.* studied the incorporation of RNA precursors in leukaemic blast cells and reported that they were characterized by mean labelling values well below those of normal immature myeloid cells. Mauri *et al.* (1965) confirmed these findings and showed that in leukaemic blast cells there was also a decreased incorporation of labelled protein precursors.

Quaglino and Hayhoe (1965) observed that in one case of acute lymphoblastic leukaemia and one case of plasma cell leukaemia, studied with 5-[^{3}H]uridine, the cells showed a very much lower percentage of uptake and a much slower and less marked nuclear-cytoplasmic shift of label. While this observation was true of most cells, a few in the lymphoblastic case and between 20 to 30% in the plasma cell case showed a degree of uptake and an extent of shift much more closely comparable with that in PHA-stimulated lymphoid blast cells.

Sinks and Hayhoe (1966) reported that, taking cell populations as a whole, the time relationship in the movement of rapidly labelled RNA from nucleus and nucleolus to cytoplasm during a period following exposure to 5-[^{3}H]uridine was essentially similar in leukaemic cells to that seen in PHA-induced lymphoblasts or HeLa cells, although the extent of labelling might vary widely (Fig. 2.10).

In further studies however Quaglino *et al.* (1971, 1972) observed that the degree of nuclear-cytoplasmic transfer of 5-[^{3}H]uridine, taken as an expression at the cytological level of ribosomal formation, was decreased in a variable proportion of the cell population, the difference in nuclear uptake and extent of nuclear-cytoplasmic shift being related to cell size, the larger blast cells showing the greatest degree of transfer (Figs 2.11 and 2.12). The existence of a direct correlation between the rate of ribosomal formation and the proliferative activity of leukaemic cells was thus postulated, since larger

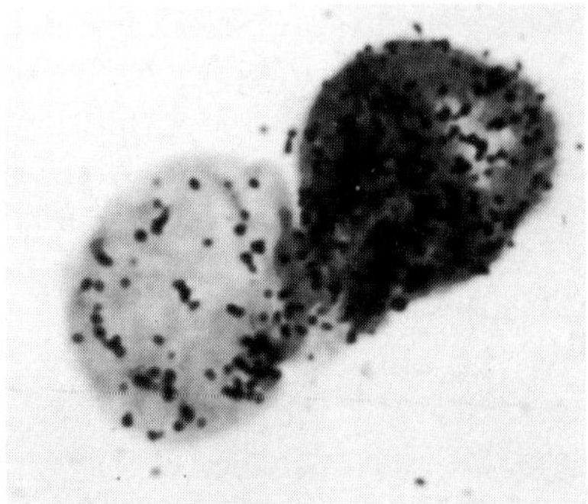

Fig. 2.10 AML bone marrow. ARG with 5-[^{3}H] uridine. Marked difference in RNA labelling in two leukaemic blast cells.

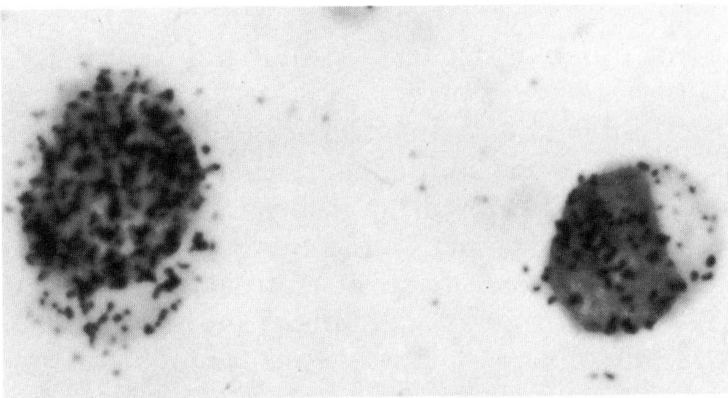

Fig. 2.11 AML bone marrow. ARG with 5-[^{3}H] uridine. RNA transfer from nucleus to cytoplasm is more marked in the larger myeloblast.

cells are known to display greater DNA synthetic activity.

Rather similar findings in childhood leukaemia were reported by Foadi *et al.* (1968), who found, following labelling with tritiated uridine, the highest grain counts over the largest cells. In some cells no RNA synthesis was detected and the authors concluded that cells in cycle had the highest rate of RNA synthesis. In their study Quaglino *et al.* (1972) also suggested that the variable degree of ribosomal formation might be responsible for the different extent of differentiation reached in each particular case by the leukaemic blast cell population. The association in some cases of intense uridine incorporation and of cytochemical features, characteristic of a fairly advanced stage of differentiation, seemed highly suggestive of such an hypothesis.

Also in chronic myeloid leukaemia and in myelofibrosis the autoradiographic study of nucleic acid precursors (Mauri *et al.*, 1972) has indicated that the abnormal myeloid cell population displays, especially in terminal blastic crisis, metabolic features similar to those reported in acute leukaemia, with reduced thymidine incorporation and decreased nuclear-cytoplasmic transfer of tritiated uridine in a proportion of immature cells. A

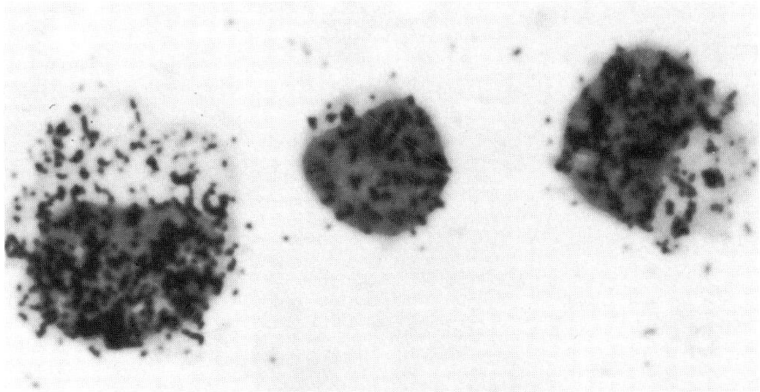

Fig. 2.12 Another field from the same preparation as Fig. 2.10, showing variable transfer of RNA, more marked in larger blast cells.

combined autoradiographic and cytophotometric investigation (Quaglino *et al.*, 1974), carried out in order to correlate the metabolic activities of the granulocyte precursors with their position in relation to cell cycle as determined by estimation of their DNA content, confirmed the preliminary findings of Mauri *et al.* (1972) and indicated that already in the initial stage of chronic myeloid leukaemia there are different cell lines, some of which are metabolically defective, and are present in variable proportions, thus influencing most probably the subsequent clinical course of the disease. In blastic crisis there appears to be an enhancement of the functional alterations present initially in a variable but usually small proportion of cells, so that the salient feature of blastic crisis appears ultimately to be a progressive incapacity for differentiation. Such a possibility seemed to be confirmed in a further study (Quaglino *et al.*, 1977), based on cytochemical, autoradiographic and cytogenetic investigations. The data obtained suggest that in the initial phase of chronic myeloid leukaemia the cell population consists principally of a leukaemic cell clone, which still retains a variable but fairly considerable degree of differentiation capacity and gives rise to mature granulocytes, which display multiple cytochemical abnormalities (Pedersen and Hayhoe, 1971 and Quaglino, 1972). The predominance of this particular cell clone overshadows the presence of a residual normal cell line, and possibly that of a more malignant blast cell population, initially present as a small fraction of the total population. In blastic crisis this latter cell line, incapable of differentiation, predominates over all other cell lines and replaces in particular the differentiating leukaemic cell clone, thus explaining the presence of mature granulocytes with more or less normal cytochemical features. The hypothesis of a progressive derangement of a previously differentiating cell line, rather than the emergence of a new malignant cell line, seems to be supported by the discrepancy, often observed between the development of particular cytochemical features, the autoradiographic behaviour of the granulocyte precursors and the karyotype obtained. This lack of parallelism between the cytochemical, autoradiographic and cytogenetic alterations would be in keeping with the concepts put forward by Stryckmans (1974), according to which blastic crisis is the fully expressed end of a continuous process which from the very beginning of the disease is progressing more or less rapidly.

An *in vitro* evaluation of multiple kinetic parameters by quantitative ^{14}C autoradiography (Ucci *et al.*, 1985) has shown that in chronic myeloid leukaemia DNA synthesis time is normal for all myeloid precursors (11.5 hours for myeloblasts, 11.7 for promyelocytes and 16.1 for myelocytes), while labelling index is reduced at all levels of maturation, leading to a decreased relative production rate. In acute leukaemias the same authors obtained kinetic data (DNA synthesis time, labelling index, growth fraction, cell cycle time, relative production rate, compartment residence time) indicative of a pattern of slow proliferation, when the disease becomes clinically evident.

In recent years attempts have been made to establish possible relationships between cell kinetic patterns and other biological features. For instance, Look *et al.* (1982) have studied cell phenotype in relation to proliferative activity assessment and have indicated the existence of a hierarchy of proliferative activities, with surface immunoglobulin positive acute leukaemias showing the highest proliferative rate, T-cell leukaemias intermediate values and common-antigen (cALLA) positive forms the lowest proliferative activity. In myeloblastic leukaemias, because phenotype grouping has not been properly defined, similar studies are not yet available.

Proliferative activity appears also to be directly related to the extent of the initial white cell count and tumour bulk (Mauer and Look, 1985). A relationship seems to exist also between cell proliferative activity and the likelihood of subsequent treatment failure. Leukaemic cells in relapse tend to have greater proliferative activity than at the time of diagnosis, prior to therapy, probably as a consequence of a higher rate of return of the leukaemic cells to cell cycle following mitosis. Cells remaining in cycle tend more easily to undergo mutation, develop drug resistance and emerge at relapse with a greater growth fraction.

Experiments have been recently carried out with the aim of establishing possible correlations between proliferative activity and chromosomal alterations. Studies by Williams *et al.* (1984) have in fact indicated that the large growth fraction of surface immunoglobulin positive B cell leukaemias is associated with an 8;14 translocation, leading to the speculation that the high proliferative rate connected with this type of leukaemia might be related to the translocation of the *c-myc* gene into the region of the immunoglobulin heavy chain locus. The *c-myc* gene is normally regulated with respect to the cell cycle and is tightly repressed in quiescent cells (Kelly *et al.*, 1983). In cytoplasmic immunoglobulin positive pre-B lymphoblastic leukaemias, the presence of a 1;19 translocation may be responsible for a greater proliferative activity, suggesting that genes in the vicinity or at these sites might have a precise function in regulating proliferative activity. On the other hand T-cell lymphoblastic leukaemia characterized by an 11;14 translocation was found to have a significantly decreased percentage of cells in S-phase, possibly associated with a gene responsible for growth inhibition. Thus the possible existence of a relationship between specific translocations and proliferative activity opens up a whole new field of research, in which there may be an opportunity for studying the role and mode of action of genes in regulating the kinetic behaviour of leukaemic cells.

From this brief survey, it is quite apparent that the cytochemical and autoradiographic study of nucleoproteins still represents a useful means of investigating the metabolic and functional aspects of the various cell types in both normal and pathological conditions and is particularly fruitful when employed in combination with other techniques, such as microdensitometry, electron microscopy, cytogenetics and *in vitro* culture methods. Examples of combined cytochemical or immunocytochemical reactions with autoradiography in the study of haemic cells in normal and leukaemic states are illustrated in Plate 2, E and F.

TECHNICAL METHODS

Feulgen reaction

1. Fix air-dried smears of blood or bone marrow for 1 hour in the following solution: Methanol 85 parts
   Formaldehyde 40% 10 parts
   Glacial acetic acid 5 parts
2. Rinse in running tap water for 5 minutes
3. Transfer to HCl 5 mol/l at 25°C for 20–40 minutes
4. Place slides directly in Schiff's reagent (pH 2) at room temperature in the dark for 90 minutes.
5. Rinse in 3 changes of bisulphite solution for 2 minutes each
6. Wash in running tap water for 15 minutes.

Remarks

A long period of fixation and the use of HCl 5 mol/l at room temperature are recommended because the process of hydrolysis is slowed down, it takes place more gradually and it allows more reproducible results for quantitative analysis.

Study of cells by combined Feulgen microdensitometry and tritiated thymidine autoradiography

1. Cells are labelled with tritiated thymidine while in suspension
2. Smears are made of the labelled cell population
3. Preparations are fixed and stained by the Feulgen method
4. A photographic map is made of the cell population
5. Cells are numbered and the DNA content of the numbered cells is measured by microdensitometry
6. Stripping film or emulsion is applied to the preparations, which are then exposed and developed
7. Grain counting is performed on the previously numbered individual cells.

Methyl Green-Pyronin method

After Kass (1979), modified.

1. Fix air-dried smears of blood or bone marrow in absolute methanol for 10 minutes
2. Wash in running tap water for 5 minutes
3. Immerse slides for 30 minutes at room temperature in the staining solution prepared as follows:

Dissolve separately 0.5 g methyl green and 0.5 g pyronin Y each in 50 ml of distilled water containing 0.1% (w/v) phenol. When both dyes are completely dissolved mix the solutions, then add 2.5 ml ethanol and 20 ml glycerin. Kept in refrigerator, the staining solution is stable for at least one month.

4. Rinse briefly in several changes of distilled water and air dry.

Remarks

Methyl green is usually contaminated with methyl violet, which can be removed by repeated extraction with chloroform.

REFERENCES

Ackerman G A 1950 Microscopic and histochemical studies on the Auer bodies in leukaemic cells. Blood 5: 847

Astaldi G, Mauri C 1950 New criteria for the evaluation of the bone-marrow cells activity. Le Sang 21: 378

Baker J R 1942 In: Bourne G (ed) Cytology and Cell Physiology. Clarendon Press, Oxford, p 21

Barka T, Anderson P J 1963 Histochemistry: theory, practice and bibliography. Harper & Row, New York

Barranco S C, May J T, Boerwinkle W, Nichols S, Hokanson K M, Schumann J, Göhde W, Bryant J, Guseman L F 1982 Enhanced cell killing through the use of cell kinetics-directed treatment schedules for two-drug combinations in vitro. Cancer Research 42: 2894

Berghe L, Hofman J 1945 Présence d'acide ribonucléique dans les granulations neutrophiles des leukocytes de l'homme. Compte Rendu de la Societe de Biologie (Paris), 139: 324

Bessis M 1973 Living Blood Cells and their Ultrastructure. Springer Verlag, Berlin, p 435

Bessis M, Breton-Gorius J 1964 Le réticulocyte. Colorations vitales et microscopie électronique. Nouvelle Revue Francaise d'Hematologie 4: 77

Bond V P, Fliedner T M, Cronkite E P, Rubini J R, Robertson J S 1959 Cell turnover in blood and blood-forming tissues studied with tritiated thymidine. In: Stohlman F Jr (ed) The Kinetics of Cellular Proliferation. Grune & Stratton, New York, p 188

Brachet J 1940 La détection histochimique des acides pentose nucleiques. Compte Rendu de la Societe de Biologie (Paris) 133: 88

Brachet J 1953 The use of basic dyes and ribonuclease for the cytochemical detection of ribonucleic acid. Quarterly Journal of the Microscopical Society 94: 1

Buckton K E, Court-Brown W M, Smith P G 1967 Lymphocyte survival in man treated with x-rays for ankylosing spondylitis. Nature 214: 470

Caffrey R W, Rieke W O, Everett N B 1962 Radioautographic studies of small lymphocytes in the thoracic duct of the rat. Acta Haematologica 28: 145

Cameron I L, Prescott D M 1963 RNA and protein metabolism in the maturation of nucleated chicken erythrocytes. Experimental Cell Research 30: 609

Caspersson T 1936 Über den chemischen Aufbau der Strukturen des Zellkernes. Skandinavisches Archiv für Physiologie 73: Suppl. 8

Cronkite E P, Bond V P, Fliedner T M, Killman S A 1960 The use of tritiated thymidine in the study of haemopoietic cell proliferation. Wolstenholme GEW, O'Connor M (eds) Ciba Foundation Symposium on Haemopoiesis. Churchill, London, p 70

Danova M, Wilson G, Riccardi A, *et al* 1987 In vivo administration of bromodeoxyuridine and flow cytometry for cell kinetic studies in human malignancies. Haematologica 72: 115

Dolbeare F, Gratzner H, Pallavicini M G, Gray J W 1983 Flow cytometric measurement of total DNA content and incorporated bromodeoxyuridine. Proceedings of the National Academy of Science 30: 5573

Dörmer P 1973 Kinetics of erythropoietic cell proliferation in normal and anaemic man. A new approach, using quantitative [^{14}C]-autoradiography. Progress in Histochemistry and Cytochemistry 6: 1

Einarson L 1932 A method for progressive selective staining of Nissl and nuclear substance in nerve cells. American Journal of Pathology 8: 295

Einarson L 1951 On the theory of gallocyaninchromalum staining and its application for quantitative estimation of basophilia. A selective staining of exquisite progressivity. Acta Pathologica et Microbiologica Scandinavica 28: 82

Everett N B, Reinhardt W O, Yoffey J M 1960 The appearance of labeled cells in the thoracic duct lymph of the guinea pig after the administration of tritiated thymidine. Blood 15: 82

Feulgen R, Rossenbeck H 1924 Mikroskopisch-chemischer Nachweis einer Nukleinsäure von Typus der Thymosnukleinsäure und die darauf beruhende elektive Färbung von Zellkernen in mikroskopischen Präparaten. Zeitschrift für Physiologische Chemie 135: 203

Foadi M D, Cooper E H, Hardisty R M 1968 Proliferative activity of leukaemic cells at various stages of acute leukaemia of childhood. British Journal of Haematology 15: 269

Ford W L, Gowans J L 1969 The traffic of lymphocytes. Seminars in Haematology 6: 67

Gardikas C, Israels M C C 1948 The Feulgen reaction applied to clinical haematology. Journal of Clinical Pathology 1: 226

Gavosto F, Pileri A, Maraini G 1959 Incorporazione di timidina marcata con tritio negli elementi del midollo osseo normale e leucemico. Indagini autoradiografiche. Haematologica 44: 977

Gavosto F, Maraini G, Pileri A 1960 Nucleic acids and protein metabolism in acute leukaemia cells. Blood 16: 1555

Gonchoroff NJ, Greipp PR, Kyle RA, Katzmann JA 1985 A monoclonal antibody reactive with 5-bromo-2-deoxyuridine that does not require DNA denaturation. Cytometry 6: 506

Gratzner H G 1982 Monoclonal antibody to 5-Bromo- and 5-Iododeoxyuridine: a new reagent for detection of DNA replication. Science 218: 474

Hale A J, Cooper E H 1963 DNA synthesis in infectious mononucleosis and acute leukaemia. Acta Haematologica 29: 257

Hale A J, Cooper E H 1965 Studies on DNA replication in leukaemic and non-leukaemic leukocytes. In: Hayhoe FGJ (ed) Current Research in Leukaemia, Cambridge University Press, Cambridge, p 95

Hale A J, Wilson S J 1959 The deoxyribonucleic acid content of leucocytes in normal and in leukaemic human blood. Journal of Pathology and Bacteriology, 77: 605

Hale A J, Wilson S J 1961 The deoxyribonucleic acid content of leukocytes in human blood, bone marrow and lymph glands. Journal of Pathology and Bacteriology, 82: 483

Harper H A 1973 Review of Physiological Chemistry. Lange, California

Hayhoe F G J, Quaglino D 1965 Autoradiographic investigations of RNA and DNA metabolism of human leucocytes cultured with phytohaemagglutinin; 5-[^{3}H]-uridine as a specific precursor of RNA. Nature 205: 151

Jackson J F 1954 The cytology of the degenerating leucocyte. Journal of Laboratory and Clinical Medicine 43: 227

Kass L 1979 Bone Marrow Interpretation. J B Lippincott Company, Philadelphia, Toronto.

Kelly K, Cochran B H, Stiles C D, Leder P 1983 Cell-specific regulation of the c-myc gene by lymphocyte mitogens and platelet-derived growth factor. Cell 35: 603

Killmann S A, Cronkite E P, Fliedner T M, Bond V P 1962 Cell proliferation in multiple myeloma studied with tritiated thymidine *in vivo*. Laboratory Investigation 11: 845

Kosenow W 1956 Lebende Blutzellen im Fluoreszenz und Phasenkontrastmikroskop. Karger, Basel

Krishan A 1975 Rapid flow cytofluorometric analysis of mammalian cell cycle by propidium iodide staining. Journal of Cell Biology 66: 188

Lajtha L G 1959 On DNA labelling in the study of the dynamics of bone marrow cell populations. In: Stohlman F Jr (ed) The Kinetics of Cellular Proliferation. Grune & Stratton, New York, p 173

Lajtha L G, Oliver R, Ellis F 1954 Incorporation of ^{32}P and adenine ^{14}C into DNA by human bone marrow cells *in vitro*. British Journal of Cancer 8: 367

LePecq J B, Paoletti C 1967 A fluorescent complex between ethidium bromide and nucleic acids. Journal of Molecular Biology 27: 87

Little J R, Brecher G, Bradley T R, Rose S 1962 Determination of lymphocyte turnover by continuous infusion of ^{3}H-thymidine. Blood 19: 236

Look A T, Melvin S L, Williams D L, Brodeur G M, Dahl G V, Kalwinsky D K, Murphy S B, Mauer A M 1982 Aneuploidy and percentage of S-phase cells determined by flow cytometry correlate with cell phenotype in childhood acute leukemia. Blood 60: 959
Marinone G 1953 Gli acidi nucleici in ematologia. Bibliographia Haematologica, Tip. Viscontea: Pavia
Marmont A 1960 Acridine orange fluorescence microscopy in haematology. Proc. 7th. Congr. Europ. Soc. Haemat., London, 1959, Vol. II, p 361. Karger: Basel
Mauer A M, Look T 1985 Leukemic cell kinetics today. In: Quaglino D, Hayhoe F G J (eds) The cytobiology of leukaemias and lymphomas. Raven Press, New York, p 209
Mauri C 1961 Autoradiographic studies in vitro on the incorporation in bone marrow cells of thymidine labelled with tritium, in health and disease. Proc. 8th Congr. Europ. Soc. Haemat., Wien, paper 65
Mauri C 1961a Die DNA-Synthese in den Knochenmarkerertythroblasten. Autoradiographische Untersuchungen mittels ^{3}H-Thymidin. Folia Haematologica NF 6: 239
Mauri C 1962 DNA synthesis and mitotic rate in leukaemic cells. Cancro 15: 145
Mauri C, Torelli U, Grossi G 1961 Studio citoautoradiografico dell'incorporazione della timidina marcata con tritio nelle cellule mielomatose. Bollettino della Società di Medicina e Chirurgia di Modena 61: 669
Mauri C, Torelli U, Artusi T, Grossi G, Emilia G 1965 Autoradiographische Untersuchungen über den RNA und Proteinstoffwechsel der unreifen Zellen bei akuter Leukämie. Vorläufige Mitteilung. Schweizerische Medizinische Wochenschrift 95: 1485
Mauri C, Torelli U, Quaglino D 1966 Aspetti metabolici della reattività linfocitaria *in vitro*. Minerva Medica 57: 1674
Mauri C, Quaglino D, Emilia G, Zanni G, De Pasquale A 1972 Ricerche citoautoradiografiche sulle cellule immature della leucemia mieloide cronica e della mielofibrosi. Haematologica 57: 865
Mitchell J, Nossal G J V 1963 Ribonucleic acid metabolism in the plasma cell sequence. Nature 197: 1121
Nowell P C 1965 Unstable chromosome changes in tuberculin stimulated leucocyte cultures from irradiated patients: evidence for immunologically committed, long-lived lymphocytes in human blood. Blood 26: 789
Pallavicini M G, Gray J W, Folstad L J 1982 Quantitative analysis of the cytokinetic response of KHT tumors in vivo to 1 β-D-Arabinofuranosylcytosine. Cancer Research 42: 3125
Pappenheim A 1899 Vergleichende Untersuchungen über die elementare Zusammansetzung des roten Knochenmarkes einiger Säugetiere. Virchows Archiv 157: 19
Pedersen B, Hayhoe F G J 1971 Cellular changes in chronic myeloid leukaemia. British Journal of Haematology 21: 251
Perugini S, Torelli U, Soldati M 1957 Differences in the desoxyribonucleoprotein complex of normal and leukaemic lymphocytes. Experientia 13: 441
Pileri A, Conte P F 1977 The biological and clinical features of human myeloma. Haematologica 62: 202
Pinheiro P, Leblond C P, Droz B 1963 Synthetic capacity of reticulocytes as shown by radioautography after incubation with labelled precursors of protein and RNA. Experimental Cell Research 31: 517
Pittaluga G, Bessis M 1944 Sur la structure des nucléoles dans les cellules normales et pathologiques. Bulletin d'Histologie Appliqué 2: 23
Quaglino D 1972 Chronic myeloid leukaemia: advances in cytochemistry. Haematologica 57: 650
Quaglino D, Hayhoe F G J 1965 Metabolic changes in short-term in vitro cultures of normal and leukaemic cells: studies with cytochemistry and autoradiography. In: Hayhoe FGJ (ed) Current Research in Leukaemia, Cambridge University Press, Cambridge, p 124
Quaglino D, Torelli U, Sauli S, Mauri C 1967 Cytochemical and autoradiographic investigations on normal and myelomatous plasma cells. Acta Haematologica 38: 79
Quaglino D, Torelli U, Emilia G, De Pasquale A, Mauri C 1971 Autoradiographic studies on the kinetics of nuclear-cytoplasmic transfer of RNA in blast cells of acute leukaemia. Acta Haematologica 45: 192
Quaglino D, Emilia G, De Pasquale A, Ferreira E, Mauri C 1972 Relationship between RNA metabolism and cell cycle in acute leukaemia. Acta Haematologica 47: 326
Quaglino D, De Pasquale A, Zanni G, Zagni G 1974 Autoradiographic and cytophotometric studies on granulocyte precursors in chronic myeloid leukaemia. Haematologica 59: 298
Quaglino D, Emilia G, Piccinini L, De Pasquale A, Zanni G, Guerzoni O, Giuliani M 1977 Relationship between autoradiographic, cytochemical and cytogenetic findings in chronic myeloid leukaemia. Haematologica 62: 360
Rheingold J J, Wislocki G B 1948 Histochemical methods applied to haematology. Blood 3: 641
Riccardi A, Montecucco C M, Merlini C, Ucci C, Cremonini N, Gobbi P G, Ascari E 1984

Proliferative activity, response to therapy and survival in multiple myeloma. Haematologica 69: 148
Ropke C, Hanger H P, Everett N B 1975 Long-lived T and B lymphocytes in the bone marrow and thoracic duct lymph of the mouse. Cell Immunology 15: 82
Sandritter W, Diefenbach H, Krantz F 1954 Über die quantitative Bindung von Ribonukleinsäure mit Gallozyaninchromalum. Experientia 10: 210
Sandritter W, Kiefer G, Rick W 1966 Gallocyanin chrome alum. In: Wied GL (ed) Introduction to Quantitative Cytochemistry, Academic Press, New York
Schooley J C 1961 Autoradiographic observations on plasma cell formation. Journal of Immunology 86: 331
Schooley J C, Bryant B J, Kelly L S 1959 Preliminary autoradiographic observations of cellular proliferation in lymphoid tissues using tritiated thymidine. In Stohlman F Jr (ed) The Kinetics of Cellular Proliferation. Grune and Stratton, New York, p 208
Schümmelfeder N, Ebschner K, Krogh E 1957 Die Grundlage der differenten Fluorochromierung von Ribo-und Desoxyribonukleinsäure mit Akridinorange. Naturweiss 44: 467
Sinks L F, Hayhoe F G J 1966 Kinetics of rapidly labelled RNA in human leukaemic cells. British Journal of Haematology 12: 759
Sprent J, Baston A 1973 Circulating T and B lymphocytes of the mouse. II. Lifespan. Cellular Immunology 7: 40
Stacey M, Deriaz R E, Teece E G, Wiggins L F 1946 Chemistry of the Feulgen and Dische nucleal reaction. Nature 157: 740
Strober S 1975 Maturation of B lymphocytes in the rat. II. Subpopulations of virgin B lymphocytes in the spleen and thoracic duct lymph. Journal of Immunology 114: 877
Strober S, Dilley J 1973 Biological characteristics of T and B memory lymphocytes in the rat. Journal of Experimental Medicine 137: 1275
Stryckmans P A 1974 Current concepts in chronic myelogenous leukaemia. Seminars in Haematology 11: 101
Taylor J H, Woods P S, Hughes W L 1957 The organization and duplication of chromosomes as revealed by autoradiographic studies using tritium labeled thymidine. Proceedings of the National Academy of Sciences 48: 471
Thorell B 1947 The relation of nucleic acids to the formation and differentiation of cellular proteins. Cold Spring Harbor Symposia on Quantitative Biology 12: 247
Ucci G, Riccardi A, Danova M, Montecucco C M, Ascari E 1985 In vitro evaluation of multiple kinetic parameters in human leukaemia by quantitative 14-C autoradiography. Haematologica 70: 101
VanDilla M A, Trujillo T T, Mullaney P F, Coulter J R 1968 Cell microfluorometry: a method for rapid fluorescence measurement. Science 163: 1213
Watson I D, Crick F H C 1953 Molecular structure of nucleic acids. Structure for deoxyribose nucleic acid. Nature 171: 1953
White J C 1947 The cytoplasmic basophilia of the bone marrow cell. Journal of Pathology and Bacteriology 59: 223
White L P 1954 The intravascular lifespan of transfused leukocytes tagged with Atabrine. Blood 9: 73
Wilkins M H F, Seeds W E, Stokes AR, Wilson HR 1953 Helical structure of crystalline deoxypentose nucleic acid. Nature 172: 759
Williams D L, Look A T, Melvin S L, Roberson P K, Dahl G, Flake T, Stass S 1984 New chromosomal translocations correlate with specific immunophenotypes of childhood acute lymphoblastic leukemia. Cell 36: 101

CHAPTER 3

Other proteins, amino acids and immunocytochemical methods for antigenic proteins

Haemic cells and tissues such as bone marrow, spleen and lymph nodes contain a wide range of simple and conjugated proteins. The simple proteins, made up chiefly of amino acids, include fibrous molecules such as reticulin, elastin, fibrin, collagen and keratin and globular molecules such as albumens, globulins, haemoglobin and histones. The conjugated proteins, apart from nucleoproteins, include lipoproteins, glycoproteins, phosphoproteins and mucoproteins. These liberate on hydrolysis considerable quantities of non-protein substances as well as amino acids.

Cytochemical colour reactions for proteins

Since free amino acids are not usually found in cells and tissues, cytochemical methods developed from biochemical colour reactions specific for amino acids may be used to demonstrate the presence of protein containing these amino acids. Tyrosine, for example, is present in most proteins apart from collagen, and may be demonstrated cytochemically by various modifications of the classical Millon's reaction depending on the formation of a red compound when the hydroxyphenyl group of tyrosine reacts with mercuric and nitrite ions. The most widely used and reliable of the cytochemical modifications is that described by Baker (1956). Similarly, cytochemical applications of the Sakaguchi (1925) reaction for arginine have been devised. This reaction depends on the linkage of α-naphthol to the terminal guanidine grouping of arginine, with the subsequent development of a colour reaction on the addition of hypochlorite. The histochemical method proposed by McLeish *et al.* (1957), using dichloronaphthol, gives generally satisfactory results, although different fixatives may be needed with different tissues (La Cour *et al.*, 1958; McLeish, 1959).

Fast green is an acidic dye used for staining proteins, the reaction depending on the presence of ionised basic groups, especially the guanidine group of arginine, and the ε-amino group of lysine (Klotz, 1950; Alfert and Geschwind, 1953). Nucleic acids, which tend to bind these basic amino acid groupings, are removed with hot trichloroacetic acid before the fast green reaction is performed.

A less specific method, giving colour reactions with tyrosine, tryptophane and histidine, is the tetrazotized dianisidine method (Danielli, 1953). The weak colour produced when diazonium hydroxides react with the respective phenol, indole and imidazole groups of the three amino acids is intensified in this method by adding a synthetic amine or phenol to the free diazo group of the bis-diazonium salt used. A further method applicable to most amino acids is the ninhydrin-Schiff reaction (Yasuma and Ichikawa, 1953), which depends on the capacity of ninhydrin to react with free NH_2 groups of amino acids and liberate an aldehyde which in turn reacts with Schiff's reagent to give a brightly-coloured magenta product. The proportion of free NH_2 groups in protein chains is not great, however, and Glenner (1963) concluded that only the ε-amino groups of lysine and hydroxylysine and the α-amino groups of terminal γ-glutamyl and β-aspartyl peptides were likely to be oxidised by ninhydrin.

Reactions for protein bound sulphydryl and disulphide groups have also been applied cytochemically to demonstrate the presence of the sulphur-containing amino acids cystine and cysteine. Among these the ferric ferricyanide method of Chèvremont and Fréderic (1943) depends on the capacity of sulphydryl groups to reduce ferricyanide to ferrocyanide, which then reacts with ferric ions to give an insoluble precipitate of Prussian blue. Disulphide groups need to be reduced to SH by thioglycolic acid or ascorbic acid before the reaction is applied. Other strongly reducing substances will give positive reactions, but the specificity can be checked by control preparations in which the sulphydryl groups have been blocked by exposure for 48–72 hours to a saturated solution of phenyl mercuric chloride in butyl alcohol (Adams, 1956).

A more specific method developed by Barrnett and Seligman (1952, 1954) uses dihydroxy-dinaphthyl-disulphide (DDD) to combine with protein-SH groups with the production of a protein-naphthyl disulphide. Free naphthyl mercaptan liberated during the reaction is removed with alcohol and ether and the residual naphthyl compound reacts with a diazonium salt such as Fast blue B to form an azo-dye.

Several of these cytochemical methods were used by Hertl (1959) in a study of bone marrow cells. Diffuse staining was found in both nucleus and cytoplasm of all cell types with most reactions, although protein-bound sulphydryl groups were chiefly confined to the granules of granulocytic cells and to the cytoplasm of erythroblasts. Not unexpectedly, there were few differences in staining intensity between cells of different series or of different stages of maturity, and other approaches such as autoradiography and immunofluorescence have proved more helpful in analysing patterns of protein metabolism in haemic cells and in identifying particular protein molecules of antigenic importance. Nevertheless, simple protein stains may be useful in preliminary exploration of the constitution of unusual inclusions. For example, rod-shaped cytoplasmic inclusions in the lymphocytes of a patient with chronic lymphocytic leukaemia failed to stain with Romanowsky dyes, Sudan black, PAS, or with methods for peroxidase or phosphatases, but were positive for protein with the ninhydrin-Schiff reaction. Their immunoglobulin nature and specificity was later determined by immunofluorescence methods (Cawley *et al.*, 1973).

Immunocytochemical methods for antigenic proteins

Most proteins are actively antigenic, and when injected into a suitable animal of species different from that of their origin will stimulate the formation of specific antibodies. Such antibodies, present in the immunoglobulin fraction of the injected animal's serum, will subsequently combine specifically with the antigens against which they were formed, giving precipitation, agglutination or complement fixation reactions in appropriate circumstances.

The first successful attempts at developing cytochemical applications of immunological reactivity date back to the early 1940s (Coons *et al.*, 1941, 1942). Since then, immunocytochemical methods have been extensively employed in the haematological field, and have provided an invaluable help in the understanding of many functional aspects of haemic cells and haemopoietic tissues under physiological and pathological conditions. The wealth of information obtained by these methods has shown an impressive growth especially during the last decade, following the development of hybridoma technology (Köhler and Milstein, 1975) and the consequent widespread use of monoclonal antibodies. According to the most commonly employed technique, splenic cells from an immunized mouse are fused by means of appropriate concentrations of polyethylene glycol with cells of an established non-secreting myeloma cell line, such as NS/1-Ag4, P3-X63-Ag8, SP-2/0-Ag14, etc. The murine myeloma cell lines employed carry a metabolic defect, which prevents their growth in HAT media (i.e. culture media containing Hypoxanthine, Aminopterin and Thymidine). Hybrid cells, resulting from fusion, inherit from myeloma cells the capacity for unlimited *in vitro* growth, and from spleen cells the ability for secreting specific immunoglobulins and for growing in HAT media. Passage in the latter will therefore allow only the survival and growth of hybrid cell clones ('hybridomas'). Since a single fusion usually produces many hybrid cell clones, each capable of synthesizing a monoclonal immunoglobulin of defined specificity, further culture of highly diluted hybrid cell suspensions in microplate wells is required in order to separate and select individual hybridomas. Methods closely related to that described above have been devised for obtaining monoclonal antibodies of animal species different from mouse.

Monoclonal antibodies, being homogeneous and highly specific for individual antigenic determinants, have proved capable of overcoming most of the technical problems inherent in the use of heteroclonal antisera (such as the need for multiple absorptions, the variable degree of specificity and affinity for antigens, etc.), thus greatly enhancing the reliability and reproducibility of immunological methods. Moreover, the use of monoclonal antibodies has enabled in many instances the isolation and structural analysis of the specific antigens against which they are directed. Therefore, it is not surprising that in immunocytochemistry, during the past few years, monoclonal antibodies have progressively replaced most 'conventional' antisera, and are currently regarded as more reliable first-layer reagents.

Although the term *immuno*cytochemistry most properly refers to the cytochemical detection of binding of antibodies to specific antigens, a large number of closely similar methods can be employed for demonstrating the binding of lectins to cell glycoproteins.

Lectins are proteins of non-immune origin (Kocourch and Horejsi, 1981), mostly extracted from plants or invertebrate animals, which specifically interact with mono- or oligosaccharide residues of glycoproteins (see Table 3.1).

Since a single lectin molecule frequently has more than one sugar-binding site, binding of lectins to plasma membrane glycoproteins of haemic cells may cause cell agglutination (hence the name 'agglutinins'). Lectins can be employed for identifying specific cell types, and also as markers of cell differentiation and maturation (Ponder, 1983). Moreover, since several glycoproteins play an important role as membrane receptors, especially in lymphocytes, binding of lectins to the surface of living cells may result in immunological activation and cell proliferation.

As the result of a very large number of investigations carried out in recent years on haemic cells both with heteroclonal and especially with monoclonal antibodies, as well as with lectins, a comprehensive and up-to-date survey of all the relevant findings obtainable by immunocytochemical techniques seems to be hardly conceivable, and probably falls beyond the boundaries of cytochemistry itself. Nonetheless, immunocytochemistry retains close links with 'conventional' cytochemistry, for example as a tool for the detection of enzyme proteins (Borgers and Verheyen, 1985).

The present Chapter will mainly deal with the technical aspects of immunocytochemical methods, while some important applications of haematological cytoimmunology are given, when required, in other Chapters (especially 14 and 15).

Amongst other means for visualizing at the microscopical level the binding of antibodies to cell surface antigens, it is worth mentioning some rapid and sensitive rosetting techniques applicable to the analysis of cell suspensions. According to the most frequently employed methods, specific antibodies can be either coupled with chromium chloride to ox indicator erythrocytes (Molinaro and Dray, 1974; Ling *et al.*, 1977), or else covalently bound via a peptide bond to polyacrylamide beads (Ammann *et al.*, 1977).

More closely related to cytochemistry, and more widely employed, are immunostaining methods, in which the site of antigen–antibody binding is revealed by a coloured reaction product at the light microscopical level, or by an electron-dense deposit at the ultrastructural level.

Immunofluorescence methods

Fluorescent dyes were introduced by Coons and co-workers (1941) as tracers for detecting at the microscopical level sites of antigen–antibody binding. Immunofluorescence methods usually involve the chemical binding of the dye to immunoglobulin molecules: to this purpose, fluorochromes have been mostly employed in the form of isocyanate compounds, since the latter are capable of combining with proteins, probably via the ε-amino groups of lysine, without substantially altering their chemical and biological

Table 3.1 Sugar specificity of some lectins

Lectin	Source (and common name)	Specificity
Abrin	*Abrus precatorius* (Jequirity bean)	β-D-Gal
BS-I	*Bandeiraea simplicifolia*	α-D-Gal α-D-Gal NAc
BS-II	*Bandeiraea simplicifolia*	α-D-Glc NAc
Concanavalin-A	*Canavalia ensiformis* (Jack bean)	α-D-Man α-D-Glc
Dolichos biflorus agglutinin	*Dolichos biflorus* (Horse gram)	α-D-Gal NAc
Helix pomatia agglutinin	*Helix pomatia* (Edible snail)	α-D-Gal NAc
Lens culinaris agglutinin	*Lens culinaris* (Lentil)	α-D-Man α-D-Glc
Limulin III	*Limulus polyphemus* (Horseshoe crab)	NeuAc NeuAc-α-2, 3(6)Gal NAc
Lotus agglutinin	*Tetragonolobus purpureus* (Asparagus pea)	α-L-Fuc
Peanut agglutinin	*Arachis hypogaea* (Peanut)	β-D-Gal β-D-Gal(1→3)-D-Gal NAc
Phytohaemagglutinin-E	*Phaseolus vulgaris* (Red kidney bean)	α-D-Gal NAc (NeuAc)Gal-Glc NAc-Man
Phytohaemagglutinin-L	*Phaseolus vulgaris* (Red kidney bean)	(NeuAc)Gal-Glc NAc-Man
Pokeweed mitogen	*Phytolacca americana* (Pokeweed)	Glc NAc-β-1, 4Glc NAc
RCA_I	*Ricinus communis* (Castor bean)	β-D-Gal α-D-Gal
Soybean agglutinin	*Glycine max* (Soybean)	α-D-Gal NAc β-D-Gal NAc
Ulex europaeus agglutinin	*Ulex europaeus* (Gorse)	α-L-Fuc
Wheat germ agglutinin	*Triticum vulgaris* (Wheat germ)	β-D-Glc NAc NeuAc

characteristics. The theoretical and practical aspects of fluorescent labelling of immunoglobulins have been reviewed by Pearse (1980). The efficiency of the current methods for conjugation is usually such as to ensure the coupling of 2–3 molecules of fluorochrome to each immunoglobulin molecule.

Fluorescein isothiocyanate (FITC) is the most widely employed fluorescent label: when excited at an appropriate wavelength (450–500 nm with a maximum at 488 nm), it shows a bright yellow-green fluorescence (with a wavelength range of 500–550 nm). Immunoglobulins labelled with other fluorescent dyes have been employed in the haematological field mainly for dual staining methods, as for example for the simultaneous detection of κ and λ light chains, or of different lymphocyte subsets. Tetramethylrhodamine isothiocyanate (TRITC) has been long used to this purpose

(Pryzwansky, 1982): its fluorescence spectrum shows maximal excitation at 555 nm, with the emission maximum at 580 nm (red fluorescence). FITC-conjugated and TRITC-conjugated immunoglobulins of different specificity can be simultaneously applied with a single incubation to the same cell or tissue specimen, but since the useful excitation range of the two fluorochromes is completely different, they cannot be visualized simultaneously. For this reason, in recent years TRITC has been increasingly replaced by phycoerythrin (PE) for the labelling of antibodies (Oi *et al.*, 1982). The useful absorption spectrum of PE ranges from about 480 to 570 nm, with an excitation maximum at 545 nm, while its maximum fluorescence emission is at 575 nm (red fluorescence). Since the absorption spectrum of PE partially overlaps that of FITC, excitation at an appropriate wavelength (488 nm, UV light) is capable of visualizing simultaneously sites of FITC-labelled and PE-labelled antibody binding, respectively as a green and as an orange fluorescent staining.

In so-called 'direct' methods, fluorochrome-conjugated antibodies specific for cell or tissue antigens are used for visualizing directly the site of reactivity in cell suspensions, fixed smears or tissue sections. As an alternative, 'indirect' fluorescence methods can be used in which the antigen is allowed to combine first with unlabelled antibody and the site of the localized antibody molecules defined by the application of a fluorescent-labelled antibody prepared against immunoglobulins of the original antibody-forming species. This indirect or 'sandwich' technique enables fluorescence to be enhanced by the availability of more combining sites for the final fluorescent layer of the 'sandwich'. Moreover, a single fluorochrome-conjugated second-layer antibody can be used for visualizing multiple antigens, when the latter are detected by unlabelled first-layer antibodies from the same animal species. Further modifications of the basic system have been devised for example for use in the detection of antibody in cells by similar direct and indirect methods. Complement may be added to antigen–antibody complexes in cell preparations and labelled antibody against the complement used to provide the fluorescence.

Successful use of immunofluorescence methods depends on meticulous preparation and use of reagents. Optimal dilution of individual antisera and the regular employment of control systems are necessary to ensure that non-specific reactions are excluded. The subject has been well discussed in critical detail in the monograph 'Fluorescent protein tracing' by Nairn (1976). Since the use of multiple immunoglobulin layers somewhat increases the possibility of non-specific staining, mainly due to unwanted cross-reactivity, alternative indirect methods have been devised, such as that based on the high chemical affinity existing between avidin (a glycoprotein found in egg white) and the small molecule vitamin, biotin (Bayer and Wilchek, 1978). Thus, biotinylated first-layer antibodies can be revealed by fluorochrome-conjugated avidin, or else avidin can be used as a bridge between biotinylated immunoglobulins and fluorochrome-conjugated biotin, in order to enhance fluorescence intensity. These methods, in combination with a direct method employing a different fluorochrome, are also useful for visualizing simultaneously pairs of antigens detected by (monoclonal) antibodies from the same animal species (Pizzolo and Chilosi, 1984).

The alternative choice of direct or indirect fluorescence methods for detecting antigenic proteins depends on several considerations. In general, direct methods are more easy and quick to perform, while indirect methods are more sensitive: therefore the former appear to be adequate for detecting highly-expressed cell antigens (e.g. some membrane antigens on cell suspensions or cytoplasmic immunoglobulins on fixed smears or cryostat sections), while indirect methods are usually required for visualizing weakly-expressed membrane antigens, and for use with fixed smears and tissue sections. On the other hand, when the distribution of membrane antigens is evaluated on cell suspensions the antigen density on individual cells is better shown by direct methods, since in indirect methods the proportionality between the amount of antigen and the fluorescence intensity per cell is altered. For this reason, direct methods are preferably employed for quantitative studies by automated flow cytofluorimetry.

Indirect methods (excluding those based on avidin-biotin binding) also involve a proportionally higher risk of non-specific membrane staining of cells bearing receptors for the Fc portion of aggregated immunoglobulins, because first- and second-layer antibodies form an immune complex. This problem can however be minimized by using as immunological reagents $F(ab)_2$ fractions of immunoglobulins, or else monoclonal antibodies whose Fc fragment cannot be bound by the cell receptors.

The sensitivity and specificity of individual immunofluorescence methods may be greatly influenced by the quality and the preparation of cell or tissue samples. Thus, membrane fluorescence is best performed on suspensions of viable cells (Plate 3A), while cytoplasmic fluorescence may be demonstrated on smears fixed in dry acetone at −20°C. A rapid and simple direct micro-method, originally developed in our laboratories for demonstrating intracytoplasmic immunoglobulins (Worman, 1978) is given at the end of this Chapter. As far as tissue sections are concerned, the successful demonstration and localization of membrane or cytoplasmic antigens depends on a long series of preparative procedures, including fixation, storage of the samples, unmasking of antigens, etc. The subject has been discussed in some detail by Mason and Biberfeld (1980) and by Brandtzaeg (1982). According to the former authors, the use of unfixed or acetone-fixed cryostat sections, pioneered by Jaffe *et al.* (1974), seems to be most suitable for immunofluorescent analysis of lymph nodes, since it allows maximal preservation of antigenic reactivity, while preservation of cytological detail is not a major requirement of immunofluorescence methods. Cryostat sections have also been employed for investigating bone-marrow samples (Chilosi *et al.*, 1983).

Immunoenzyme methods

Immunofluorescence methods are of invaluable help in the study of haemic cells, and in some cases can hardly be replaced by other methods, as for

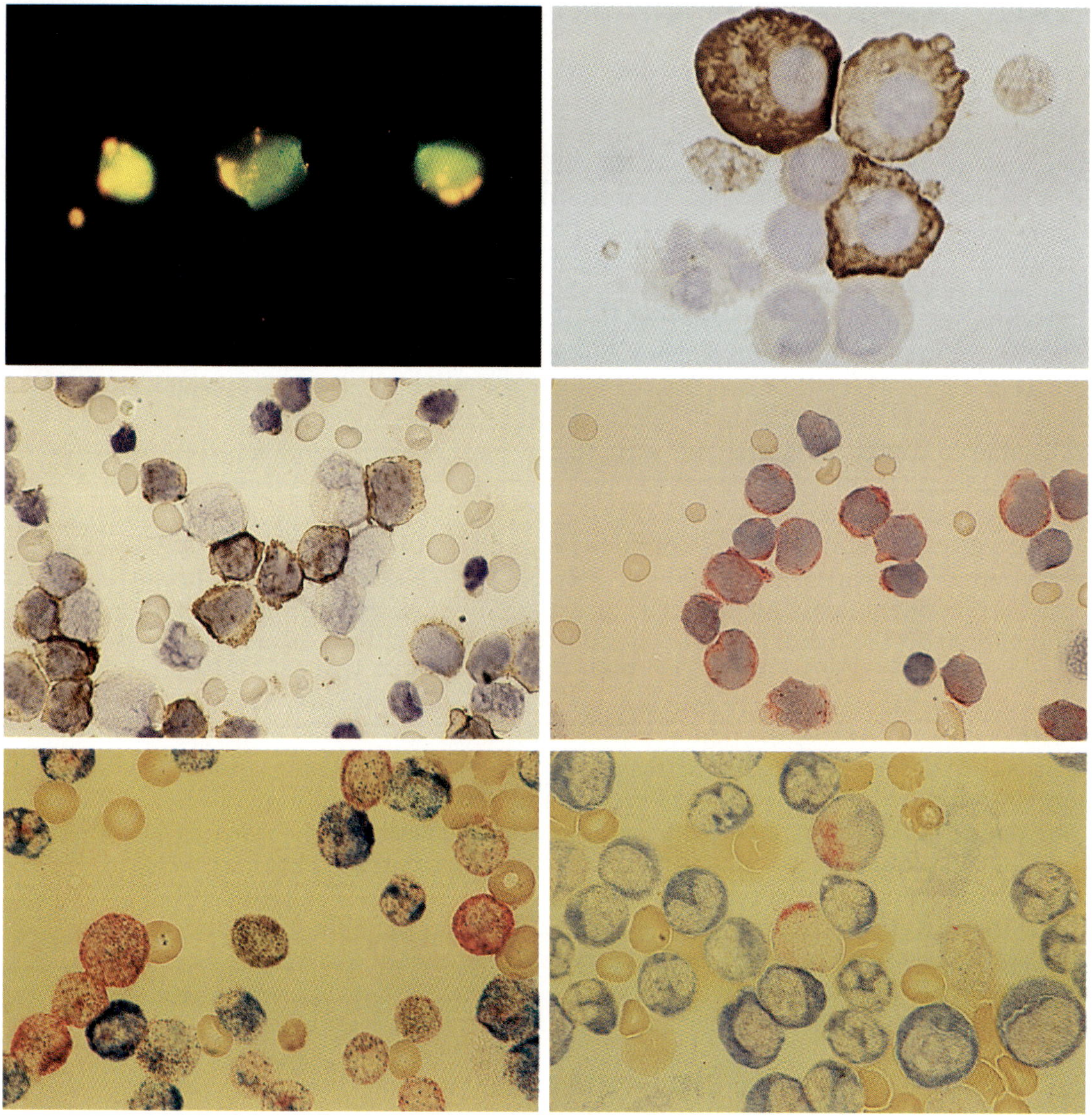

Plate 3
A. CLL: immunofluorescence of lymphocyte surface light chains reacting with fluorescent-labelled antibody.
B. Multiple myeloma: plasma cells reacting strongly with a specific McAb; immunoperoxidase method.
C. AML: reaction with anti-HLA-DR McAb, demonstrated by immunoperoxidase (PAP). About two thirds of the blast cells are positive.
D. The same case and reaction as in C, but using the immuno-alkaline phosphatase (APAAP) method; here virtually all the blast cells show reactivity.
E. AMML: dual esterase (chloroacetate, blue; butyrate, brown) together with anti-Leu M1 demonstrated by APAAP (red reaction).
F. CML: chloroacetate esterase (blue) together with anti-HLA-DR demonstrated by APAAP (red reaction).

example when they are employed for quantitative studies of membrane antigens or for fluorescence-activated cell sorting. Nevertheless, fluorescent methods suffer from two major drawbacks, which are especially relevant when smears or tissue sections are investigated. First, morphological details of cells and tissues cannot be observed simultaneously with the fluorescent staining and are in many instances poorly preserved, and second, long-term recording of the immunochemical reaction is impossible, due to rapid fluorescence fading.

Mainly for these reasons, other methods have been developed, exactly comparable to immunofluorescence staining in their capacity to localize the site of antigen–antibody reactions, in which the place of the fluorescent label is taken by some other molecule attached to antibody globulin. Small molecules endowed with enzyme activity are particularly useful for this purpose, since their site of deposit can be visualized by suitable cytochemical staining procedures.

The most typical and widely employed enzyme label is represented by horse-radish peroxidase (Nakane and Pierce, 1966; Nakane, 1968). The basic immunoperoxidase method, like direct immunofluorescence, involves conjugation of the peroxidase to specific immunoglobulins, usually by means of glutaraldehyde (Avrameas, 1969; Avrameas and Ternynck, 1971). The sites of antigen–antibody reaction can then be made apparent by using the enzymatic activity of the peroxidase on a chromogenic substrate such as diaminobenzidine (DAB) (Plate 3B). The method has limitations partly similar to those of immunofluorescence, and requires equally rigorous controls if specificity is to be achieved. Enhanced sensitivity is shown by indirect techniques, based either on immunoglobulin binding (Sternberger, 1979) or on avidin–biotin complex (Guesdon *et al.*, 1979). While some of these indirect methods are closely related to corresponding immunofluorescent methods, others have been originally proposed and specifically employed as immunoenzyme methods. This is the case with the peroxidase–anti-peroxidase (PAP) method introduced by Sternberger *et al.* (1970). According to the original report, an excess of pig anti-rabbit immunoglobulin was used to link the primary rabbit antibody with a PAP immune complex made up of rabbit antiperoxidase combined with peroxidase antigen. The method has been subsequently adapted for use with antisera from other animal species and with monoclonal antibodies.

The PAP method, and a variety of other indirect methods involving multiple antigen–antibody layers, with the addition of free horse-radish peroxidase in the final stages (Mason *et al.*, 1969; Ford and Stoward, 1974; Klockars and Reitano, 1975) or the use of hybrid antibodies having bispecific activity against both the antigen under study and horse-radish peroxidase (Sternberger, 1974), have been described in detail and reviewed by several authors (Taylor, 1978; Sternberger, 1974, 1979; Pearse, 1980).

An immunoperoxidase method based on the use of an avidin–biotin–peroxidase complex (ABC) was proposed by Hsu *et al.* (1981) as an alternative to methods based on the use of unlabelled antibodies. The ABC method involves the sequential application of unlabelled first-layer antibodies, biotin-labelled secondary antibodies, and a complex created by the incubation of a relative excess of avidin with biotin-labelled peroxidase. In a study carried out on acetone-fixed, frozen sections of tonsils (Hsu *et al.*, 1983), the ABC method was compared with both unlabelled antibody and conjugated antibody methods. ABC and PAP methods produced a similar staining intensity, superior to that obtainable by conjugated antibody methods; however less non-specific background staining was observed when the former method was employed.

Other peroxidase substrates, such as 4-chloro-1-naphthol (Nakane, 1968), p-phenylenediamine and pyrocatechol (Hanker *et al.*, 1977), 3-amino-9-ethylcarbazole (Boenisch, 1980) etc., may replace DAB for immunoperoxidase staining, but the respective reaction products are comparatively less stable and less clearly localized.

Variations in the method of exposure and the use of living or fixed cells allow for detection of either surface membrane or cytoplasmic antigens with the immunoperoxidase techniques. Living cells in suspension will not admit immunoglobulins into the cytoplasm and the application of antisera to viable cells accordingly detects only surface antigens, such as SIg of B cells (Bernard *et al.*, 1975; Stanislawski *et al.*, 1976). Different fixative procedures cause varying degrees of damage to cytoplasmic membranes, as discussed in Chapter 1, and accordingly allow penetration of macromolecules to a greater or less extent (Feltkamp-Vroom, 1975; Nakane, 1975). Gentle fixation of cells in suspension still permits the detection of surface antigens, while fixed cell smears and imprints may demonstrate both plasma membrane and cytoplasmic antigens (Gourdin *et al.*, 1976; Mason *et al.*, 1975, 1977).

In tissue sections after standard formalin fixation and paraffin embedding, cytoplasmic and nuclear antigens can be demonstrated by immunoperoxidase reactions, while counterstaining with standard methods such as haematoxylin and eosin allows for morphological distinction of different cells and tissues (Taylor and Burns, 1974; Taylor, 1978). Search for cell surface antigens has usually been carried out on gently fixed cryostat sections, both of lymphoid tissues (Hoffman-Fezer *et al.*, 1979; Janossy *et al.*, 1980; Stein *et al.*, 1980; Borowitz *et al.*, 1982) and of bone marrow (Pizzolo *et al.*, 1982; Falini *et al.*, 1984). Recently, however, Beckstead *et al.* (1986) have been able, employing the ABC method, to demonstrate membrane and cytoplasmic antigens of megakaryocytes even on suitably prepared plastic-embedded sections of bone marrow.

The use of immunoperoxidase techniques has proved extremely profitable especially for the study of normal and neoplastic lymphoid tissues (see Chapter 15), in which endogenous peroxidase activity is usually absent or confined to a low number of easily identifiable cells. In comparison, more problems may be encountered in the study of myeloid cells and tissues, since the endogenous peroxidase activity of granulocytes and red cells can be revealed as a positive staining by the final step of immunoperoxidase procedures. Therefore, inhibition of endogenous peroxidase activity, e.g. by methanol/H_2O_2 treatment (Streefkerk, 1972), is needed prior to immunoperoxidase staining. This procedure is currently employed for the demonstration of 'stable' antigens, such as cytoplasmic immunoglobulins, in paraffin embedded tissues (Burns *et al.*, 1974), and has been successfully used also for detecting with monoclonal antibodies surface and cytoplasmic antigens of haemic cells on smears (Swirsky *et al.*, 1983) and on plastic-embedded bone marrow sections (Beckstead *et al.*, 1986). Nevertheless, methods for inhibiting endogenous peroxidase activity are generally inimical to antigenic reactivity, and are better avoided when low amounts of antigen or easily denaturable antigenic determinants are to be detected (Plate 3C). Consequently, when cells or tissues rich in endogenous peroxidase activity are to be investigated, immunoperoxidase methods can be advantageously replaced at the light microscopical level by immuno-alkaline phosphatase methods (Avrameas, 1969), in which the enzyme label is alkaline phosphatase, most commonly from calf intestine. Visualization of enzyme activity is then achieved by cytochemical methods which utilize naphthol AS phosphate derivatives as the substrates and diazonium salts as the coupling agents (Mason and Woolston, 1982). 1 mM levamisole, added to the staining solution, inhibits endogenous alkaline phosphatase activity (Ponder and Wilkinson, 1981), while leaving unaffected the enzyme activity of calf intestinal alkaline phosphatase (Plate 3D). Direct and indirect alkaline phosphatase-conjugated antibody methods exist, as well as unconjugated antibody methods. Of these, the APAAP (alkaline phosphatase–anti-alkaline phosphatase) method, closely similar in its basic principles to the PAP method, displays a very high sensitivity, and has been proposed as the method of choice for investigating with monoclonal antibodies tissues rich in endogenous peroxidase (Cordell *et al.*, 1984). Moreover, combined use of alkaline phosphatase and peroxidase allows for double immunoenzymatic labelling of pairs of antigens (Mason and Sammons, 1978; Mason and Woolston, 1982).

Immuno-alkaline phosphatase techniques appear to be particularly suited for the study of blood cells on smears (Yam *et al.*, 1983; Moir *et al.*, 1983; Li *et al.*, 1984; Erber *et al.*, 1984) and on frozen bone marrow sections (Falini *et al.*, 1984), and can to some extent be combined with cytochemical staining methods (Li *et al.*, 1984; Chilosi *et al.*, 1985) (Plate 3, E & F). The potentialities of the APAAP method in the immunophenotyping of haematological malignancies are well illustrated by the paper by Erber *et al.*, (1986) who employed a large panel of monoclonal antibodies, as well as polyclonal anti-TdT and anti-myeloperoxidase antisera, for the study of malignant cells in 259 patients. Cell samples, either from bone marrow or peripheral blood, were mostly prepared as for conventional cytological analysis. 149 cases of acute leukaemia, including all the best-known varieties of both AML and ALL as well as one case of mixed leukaemia, 18 cases of chronic myeloid leukaemia in blast crisis and 92 cases of chronic lymphoproliferative disorders (CLL, PLL, HCL, Sezary syndrome and multiple myeloma) could all be reliably characterized by means of the APAAP staining, and according to the authors the results corresponded to those obtained by immunofluorescent techniques. Details of the APAAP method are given at the end of this Chapter.

Immunocytochemical methods for electron microscopy

The immunoperoxidase methods can be applied to tissues at ultrastructural level either by staining thick cryostat sections for antigen first and then processing the tissue and cutting for EM, or by allowing the reaction to operate at the surface of ultrathin sections, when little penetration beyond the surface can be expected. Improved penetration into the tissues can be achieved by the use of reagents of low molecular weight, such as a haem-bearing octapeptide from cytochrome c, which stains positively with the DAB reaction, conjugated to Fab fragments of immunoglobulins (Kraehenbul *et al.*, 1974). However, unlabelled antibody methods have also been employed with satisfactory results in post-embedding immunostaining techniques (Ordronneau, 1982).

In dealing with material from blood and bone marrow, or from lymph nodes when a cell suspension can be obtained, the whole reaction from exposure to the first antibody layer to the cytochemical demonstration of the final layer of peroxidase by the use of DAB is best performed on fixed and washed cells in suspension, the material then being again washed, post-fixed, dehydrated, infiltrated, embedded and sectioned in the usual way (see Technical Methods, below). The electron density of the staining product can be enhanced, when required, by osmium tetroxide post-treatment (Pearse, 1980).

An alternative to the immunoperoxidase method for particular use with the electron microscope involves conjugating the antibody protein with the electron dense marker, ferritin (Takamiya *et al.*, 1975; Abbas *et al.*, 1976). Technical details of preparation and handling of ferritin-conjugates have been extensively reviewed by Williams (1977).

In 1971, Faulk and Taylor first reported the use of colloidal gold as a label for immunocytochemical investigations. Colloidal gold is usually prepared by treating chloroauric acid with suitable reducing agents: depending on the preparation technique, the diameter of colloidal gold particles ranges from 2–3 to 150 nm (Roth, 1982). Since the hydrophobic gold particles carry a net negative surface charge, electrically charged hydrophilic substances, such as immunoglobulins, are adsorbed onto their surface by electrostatic forces, resulting in the formation of a stable complex. Immunogold methods have encountered in recent years increasing favour, especially as a tool for investigating surface antigens of haemic cells with monoclonal antibodies. The latter can be either directly coupled to colloidal gold, or detected by a gold-conjugated second-layer antibody (De Mey, 1983). The very high electron density of colloidal gold and the possibility of selecting particles of appropriate size, make these methods especially suitable for ultrastructural studies, both by transmission (Matutes and Catovsky, 1982; Matutes *et al.*, 1983) and by scanning (Lambertenghi-Deliliers *et al.*, 1985) electron microscopy. Moreover, immunogold staining of cell membrane antigens may be easily combined with enzyme cytochemistry, at the light microscopical (Crockard and Catovsky, 1983) as well as at the ultrastructural level (Breton-Gorius *et al.*, 1984; Polli *et al.*, 1984; Robinson *et al.*, 1984), thus providing not only a refined research tool, but also an important diagnostic help in selected leukaemia cases (Lambertenghi-Deliliers *et al.*, 1985).

Autoradiography

Surprisingly few cytochemical studies have been carried out on the synthesis and catabolism of proteins in haemic cells, although autoradiographic methods comparable to those used so widely in studies of DNA and RNA metabolism are available. Gavosto *et al.* (1960) and Pileri *et al.* (1960) studied in parallel the pattern of uptake of cytidine^{3}H and DL-leucine ^{3}H as respective markers for RNA and general protein synthesis in bone marrow cells. The most active uptake occurred in the more primitive cells with both labelled precursors, with mean grain counts per cell decreasing successively in the order: proerythroblasts, myeloblasts, promyelocytes, basophilic erythroblasts, myelocytes, lymphocytes, polychromatic erythroblasts. A close inter-relationship between RNA and protein synthetic activity was found, assessed by the close parallelism in uptake of the two precursors in all the different cellular types from normal marrow at all stages of maturation. The findings were similar when DL-p-phenylalanine ^{3}H was used as the protein precursor. This evidence accords with the biochemical and cytological data from other mammalian tissues suggesting that RNA is closely involved in protein synthesis, both as a template in the specific synthetic mechanism and as an intermediary in the presynthetic transfer of activated amino acids to the microsomes.

Leucocytes in acute and chronic leukaemias have been found by similar techniques to incorporate radioactive amino acids more rapidly than normal mature leucocytes do (Baker *et al.*, 1957); with ^{14}C-leucine and ^{14}C-alanine as precursors leukaemic blasts showed some sixfold increase over the uptake of normal polymorphs, while the peripheral blood cells of chronic myeloid and lymphocytic leukaemias showed an intermediate value. These differences probably do no more than reflect the relative immaturity of leukaemic cells.

The synthesis of proteins in haemic cells, including the non-haem proteins of erythroblasts, appears to occur at all interphase stages of the cell cycle, with probable depression during mitosis (Wickramasinghe and Pratt, 1970; Yataganas *et al.*, 1970). This is in keeping with the general finding in most growing mammalian cells (Mitchison, 1971).

Autoradiographic studies of protein synthesis, using leucine^{3}H as precursor, have shown that, whereas erythroblasts from normal bone marrow almost all acquire label after one hour incubation, the megaloblasts of pernicious anaemia appear to be much less active, with some 10% of basophilic megaloblasts, 30% intermediate megaloblasts and over 50% of late megaloblasts remaining unlabelled (Wickramasinghe *et al.*, 1969). Parallel studies with ^{3}H-TdR labelling and microdensitometric measurements of DNA content indicate that many megaloblasts fail to incorporate either label yet show a continuous distribution of DNA content between the 2n and 4n levels, thus indicating an apparent arrest of cell activity at any stage in the cell cycle. Failure of protein synthesis may be the primary disturbance leading to this arrest (Wickramasinghe and Chalmers, 1968) since nucleohistone synthesis normally accompanies DNA replication (Bloch and Godman, 1955) and inhibition of the former leads to arrest of the latter. A marked increase in the incorporation of ^{3}H-arginine into nuclear histones follows the parenteral injection of cyanocobalamin (Kass, 1973).

Studies with tritiated leucine show that myeloma cells are unable to reach and maintain a high rate of protein synthetic activity, such as usually shown by normal plasmacytes, and this reduced but persistent activity might be related to a defective elaboration of messenger RNA fractions which, in turn, may be responsible for the observed presence of nuclear-cytoplasmic asynchronism, of developmental arrests at varying stages of maturation, and of nuclear, nucleolar and cytoplasmic deformities demonstrated by the ultrastructural studies of Bessis *et al.* (1963).

Haemoglobin synthesis

Thorell (1947) used the quantitative cytochemical technique of microspectrophotometry to measure the concentrations of haemoglobin and nucleic acids in avian and rat erythroblasts (see p. 0). Haemoglobin synthesis began in basophilic erythroblasts in which the RNA concentration had already fallen below the level in proerythroblasts. The level of haemoglobin rose sharply with increasing maturity of erythroblasts while RNA values fell almost to zero, until in late erythroblasts there was no detectable RNA and haemoglobin concentration had risen to 20–25%. These findings showed an obvious parallel with the increasing eosinophilia and decreasing basophilia manifest in the erythroblast maturation sequence stained by Romanowsky methods.

The picture is generally similar in human erythroblasts, although there is evidence that haemoglobin synthesis may start at an earlier stage than in the rat erythroblasts. Carvalho (1955), for example, found considerably more haemoglobin in early basophilic erythroblasts from human marrow than Thorell had found in the rat. More recently, Yataganas *et al.* (1970) have confirmed the main findings of Thorell but detected initial haemoglobin formation even in proerythroblasts, while some residual cytoplasmic RNA was found in late normoblasts. Yataganas *et al.* confirmed that synthesis of protein and RNA occurred continuously throughout interphase, but showed also that the haemoglobin content of basophilic and early polychromatic normoblasts increased only during the G1 phase and the earlier part of the S phase of the cell cycle. During the latter part of the cycle haemoglobin synthesis did not occur, and indeed the concentration of haemoglobin fell at this time as other cytoplasmic constituents continued to be formed.

Several other less quantitative cytochemical methods are available for the demonstration and study of haemoglobin and its components. The Lepehne peroxidase reaction (Undritz, 1973), using benzidine and perhydrol alcohol, gives a positive yellow to dark green-brown stain with increasing concentration of haemoglobin. The reaction is negative in leucocytes, apart from occasional weak positivity in eosinophils, and is helpful in differentiating erythrocytes and erythroblasts from other haemic cells especially in certain pathological states. Macario *et al.* (1984) have recently studied the role of the erythroblastic nest (EN) in the haemoglobinization of maturing red cells, employing Lepehne's cytochemical method and an experimental model consisting of highly erythropoietic spleens from C57B1/6J mice, from which cell suspensions were prepared, significantly enriched in ENs and suitable for direct morphological and cytochemical analysis. The distribution of erythroblasts was examined in ENs and compared with that elsewhere in the spleen. Erythroblasts were counted and classified daily on the basis of their load in haemoglobin, for a week after erythropoiesis induction. Haemoglobin-rich consistently predominated over haemoglobin-poor erythroblasts around the EN's central cell from the beginning of the observation period. This was in sharp contrast with the picture outside ENs, where haemoglobin-poor erythroblasts were more abundant than haemoglobin-rich cells up to day 6. Only on day 7, a prevalence of the latter cells was observed outside ENs. This shifting from low to high haemoglobin contents was accelerated by iron administration, but never attained the same magnitude in as short time as that found to occur in ENs. The data obtained by the authors support the notion that ENs are centres for erythroblasts to become loaded with haemoglobin, or perhaps to obtain the raw materials they need for its synthesis.

Haemoglobin may also be detected cytochemically, at the EM level, using the technique, reported by Breton-Gorius (1970), based on the use of diaminobenzidine at a pH of 9. The main feature of this method is the presence of high concentrations of H_2O_2, in excess of those necessary for detecting granulocyte peroxidase. It is the pseudoperoxidase activity of haemoglobin which is responsible for the staining reactivity of these methods.

Supravital staining with methyl violet is useful in demonstrating the intracytoplasmic inclusions due to precipitated α chains in homozygous β-thalassaemia (Yataganas and Fessas, 1969), and other inclusions resulting from instability of haemoglobin due either to primary defects in haemoglobin formation or deficiencies of enzymes such as glucose-6-phosphate dehydrogenase.

The Kleihauer (1957) method for assessing the proportion of erythrocytes containing substantial amounts of foetal haemoglobin (Hb F) depends on the elution of adult haemoglobin from the erythrocytes by treatment with a citric acid-phosphate buffer at low pH, followed by counterstaining with eosin. Cells with residual HbF stain with the eosin, whereas those without appear as colourless ghosts.

Autoradiographic studies of iron uptake, or ultrastructural search for the presence of ferritin may provide indirect evidence on haemoglobin formation and its disorders.

The contribution of these cytochemical methods to the interpretation of disorders of erythropoiesis is further discussed in Chapter 13.

TECHNICAL METHODS

Ninhydrin-Schiff method for amino-acid NH_2 groups

Yasuma and Ichikawa, 1953

Reagents

1. Ninhydrin. Dissolve 0.5 g ninhydrin in 100 ml absolute ethanol.
2. Schiff's reagent. Dissolve 5 g basic fuchsin in 500 ml hot distilled water. Filter when cool. Saturate with SO_2 gas by bubbling for 1 hour. Shake with 2 g activated charcoal in a conical flask for a few seconds and filter immediately through a Whatman No. 1 filter into a dark bottle. The charcoal extraction should be done in the fume cupboard. (Keeps for 6 months).
3. Carazzi's haematoxylin. Dissolve 75 g potassium aluminium sulphate in 1200 ml warm distilled water. Add 1.5 g haematoxylin powder (Gurr) dissolved in 300 ml glycerol by grinding with pestle and mortar. Dissolve 0.3 g sodium iodate in a little water and add gradually.

Technique

1. Fix air-dried smears for 10 minutes in a solution of 10 ml 40% formalin and 90 ml ethanol
2. Wash briefly in tap water
3. Immerse slides in ninhydrin solution at 37°C overnight
4. Wash gently in tap water and blot dry
5. Immerse in Schiff's basic fuchsin in a Coplin jar for 30 minutes. (N.B. the fuchsin solution should be returned to the stock bottle immediately after use).
6. Wash in tap water 5–10 minutes
7. Counterstain in Carazzi's haematoxylin 10–15 minutes.

Ferric ferricyanide method for sulphydryl groups

After Chèvremont and Frèderick, 1943

Reagents

1. 1% ferric sulphate $Fe_2(SO_4)_3$ in distilled water.
2. 0.1% potassium ferricyanide $K_3Fe(CN)_6$ in distilled water.

The reaction medium is prepared by adding 3 parts of (1) to one part of (2) immediately before use and adjusting the pH to 2.4.

Technique

1. Fix air-dried smears in formalin vapour for 5–10 minutes
2. Wash briefly in tap water
3. Immerse in three changes of the ferricyanide reagent for a total of 10 minutes

4. Wash briefly in distilled water
5. Counterstain nuclei with nuclear fast red for 10 minutes and blot dry.

Protein-bound sulphydryl groups

After Barrnett and Seligman, 1952

Reagents

1. DDD reagent. Dissolve 25 mg of dihydroxy-dinaphthyl-disulphide (DDD) in 15 ml of absolute ethanol and add 35 ml of 0.1 mol/l veronal acetate buffer (pH 8.5).

Technique

1. Fix air-dried smears in formalin vapour for 5–10 minutes
2. Wash briefly in tap water
3. Incubate for 30 minutes at 50°C in the DDD solution
4. Cool and wash briefly in distilled water
5. Wash for 10 minutes in two changes of distilled water acidified with acetic acid to pH 4–4.5
6. Extract free naphthols by passing briefly through 50, 70 and 90% ethanol and wash twice in absolute ether for 5 minutes
7. Wash briefly in distilled water
8. Immerse for 2 minutes at room temperature in a freshly prepared solution of Fast blue B salt, 50 mg in 50 ml of 0.1 mol/l phosphate buffer, pH 7.4
9. Wash in tapwater, and blot dry.

Immunofluorescence for intracytoplasmic Ig

After Worman, 1978

Reagents

FITC-labelled antisera, monospecific for γ, α, μ and δ heavy chains and for κ and λ light chains are available from commercial sources (e.g. Dakopatts reagents from Mercia Brocades Ltd., Watford, U.K.; Behring reagents from Hoechst Pharmaceuticals, Hounslow, U.K.).

Technique

1. Incubate washed mononuclear cell suspensions in serum free medium for 60 minutes at 37°C to remove extrinsic Ig
2. Prepare cytocentrifuged cell deposits and fix in acetone at −20°C overnight
3. Place a 3 µl drop of monospecific FITC-labelled anti-serum on a circular coverslip 13 mm in diameter and invert over the cytocentrifuged cells. Incubate for 30–40 minutes
4. Wash in phosphate-buffered saline (PBS) (the coverslip is floated off)

5. Counterstain for 5 minutes, at room temperature with 0.2% w/v Evan's Blue (British Drug Houses, Poole, England) in PBS
6. Rinse off counterstain and mount in glycerol-PBS (1:1)
7. Read under combined phase contrast and u.v. epifluorescence. The cytoplasm of negative cells fluoresces a bright red colour, thus greatly facilitating percentage cell counts; the counterstained nucleus also allows identification of any polymorphs present. Positive cells show a green fluorescence, with some red fluorescence confined to the Golgi region.

Immunoperoxidase—the peroxidase–antiperoxidase (PAP) method

After Sternberger, 1974

Reagents

1. Rabbit anti-human Ig antisera, specific for γ, α, μ and δ heavy chains and for κ and λ light chains, swine anti-rabbit immunoglobulin and peroxidase–rabbit antiperoxidase immune complex (PAP), are available from commercial sources, e.g. Dakopatts reagents from Mercia Brocades Ltd., Watford, U.K. and Behring reagents from Hoechst Pharmaceuticals, Hounslow, U.K.

The PAP solution should be centrifuged for 5 minutes at 500***g*** before use to remove insoluble aggregates.

2. DAB peroxidase reagent. A solution of 5 mg 3,3′diaminobenzidine (DAB) in 10 ml tris-saline solution plus 1 drop of 100 vol hydrogen peroxide is freshly prepared and filtered before use.

Technique

(a) *for smears and imprints*

1. Fix air-dried smears in methanol for 10 minutes to block endogenous peroxidase
2. Immerse in normal swine serum, diluted 1 in 2 in tris-saline buffer, for 10 minutes then tip off excess
3. Expose to rabbit anti-human Ig antiserum 1 in 500* for 15 minutes
4. Wash in tris-saline solution for 5 minutes
5. Expose to swine anti-rabbit Ig 1 in 100 for 15 minutes
6. Wash as in (4) above
7. Expose to PAP complex 1 in 40 for 15 minutes
8. Wash as in (4) above
9. Expose to freshly prepared DAB peroxidase reagent for 5 minutes
10. Counterstain with Leishman, diluted immediately with buffer for 5 minutes.

(b) *for cell suspensions—for use in electron microscopy*

1. Fix cell rich concentrates (from bone marrow, lymph node cell suspensions, or buffy coat, followed dextran sedimentation of red cells from

*A titration effect may be achieved by using concentrations ranging from 1 in 20 to 1 in 2000.

peripheral blood) by pipetting into an excess volume of freshly prepared 1% glutaraldehyde in 0.1 mol/l cacodylate-HCl buffer (pH 7.4) containing 1% sucrose, leaving for 30 minutes at room temperature and then centrifuging at 200***g*** for 5 minutes.

2. Wash the deposit of fixed cells 3 times with repeated suspension in sucrose-buffer followed by centrifugation
3. Expose to procedures 2 to 9 as in (a) above, resuspending and centrifuging for each step
4. Post fix for 1 hour at room temperature in 2% osmium tetroxide in cacodylate-HCl (pH 7.4)
5. Rapidly dehydrate by centrifugation and resuspension in single changes of 70 and 90% ethanol, 4 changes of absolute ethanol, and 2 changes of propylene oxide
6. Infiltrate for 6–36 hours in 3 changes of low-viscosity Taab or Spurr resins, with appropriate accelerator added
7. Embed in capsules in freshly prepared resin and polymerize at 60°C for 24 hours
8. Cut thin sections on an ultratome and mount on clear uncoated copper 200 mesh grids
9. Examine without further staining, or after staining for 30 minutes with lead citrate only.

Immunoalkaline phosphatase—The alkaline phosphatase-antialkaline phosphatase (APAAP) method for use with murine monoclonal antibodies

After Cordell *et al.*, 1984

Reagents

1. Any murine monoclonal antibody can be employed as first-layer reagent. Rabbit anti mouse immunoglobulins antiserum and APAAP complex, consisting of alkaline phosphatase and mouse monoclonal antibody specific for alkaline phosphatase, are commercially available from Dakopatts.
2. Incubation medium for alkaline phosphatase. 2 mg of naphthol AS-MX (or AS-BI) phosphate are dissolved in 0.2 ml of N,N-dimethylformamide, to which 9.8 ml of 0.1 M Tris-HCl buffer, pH 8.2, are subsequently added. Immediately prior to use, add 10 mg of Fast Red TR salt (or of another suitable diazonium salt), and filter directly onto the slides. When necessary, endogenous alkaline phosphatase activity can be inhibited by adding to the Tris-HCl buffer 0.24 mg/ml (1 mmol/l) of levamisole.

Technique for smears and imprints

1. Fix air-dried smears in buffered formalin-acetone solution (see p. 210), for 30 sec at room temperature, then wash briefly in distilled water and allow to dry

2. Immerse in Tris-saline buffer (TBS) for 10 minutes, then tip off excess
3. Expose to mouse monoclonal antibodies. Optimal dilution in TBS has to be selected for individual antibodies. An incubation time of 30 minutes is adequate for most monoclonal antibodies, but longer incubation may be required in some cases
4. Wash slides in TBS (3 changes, 3 minutes each)
5. Expose to rabbit anti-mouse Ig, 1 in 25 in TBS, for 15–30 minutes. Cross-reactivity with human proteins present on the smears can be prevented by adding 5% normal human serum
6. Wash as in (4) above
7. Expose to APAAP complex, 1 in 50 in TBS, for 15–30 minutes
8. Wash as in (4) above
9. Expose to freshly prepared incubation medium for 15–30 minutes.
10. Wash briefly in running tap water, then rinse in distilled water and allow to dry
11. Counterstain with haematoxylin for 2 minutes.

Lepehne peroxidase reaction for erythroid cells

After Undritz, 1973

Reagents

Lepehne reagent: Mix 2 ml of a 0.6% solution of benzidine in 96% ethanol with 5 ml of perhydrol-alcohol (0.5 ml of 30% perhydrol to 4.5 ml of 70% ethanol).

Technique

1. Fix air-dried smears in methanol for 10 minutes
2. Immerse for 5 minutes in Lepehne reagent
3. Rinse with tap water and blot dry
4. Counterstain with Leishman or Giemsa solution diluted immediately with buffer.

Kleihauer reaction for foetal haemoglobin

Reagents

Eluting solution Add 37.7 ml of 0.1 mol/l citric acid to 12.3 ml of 0.2 mol/l Na_2HPO_4. Adjust pH to 3.3.

Technique

1. Fix air-dried blood films within 2 hours of preparation in 80% ethanol for 5–10 minutes
2. Rinse with tap water and dry

3. Immerse in the eluting solution in a Coplin jar at 37°C for 5 minutes
4. Rinse with tap water and blot dry
5. Stain with haematoxylin 1 minute
6. Rinse with tap water
7. Stain with eosin 1 minute
8. Rinse with tap water and dry.

REFERENCES

Abbas A K, Dorf M E, Karnovsky M J, Unanue E R 1976 The distribution of Ia antigens on the surface of lymphocytes. Journal of Immunology 116: 371

Adams C W M 1956 A stricter interpretation of the ferric ferricyanide reaction with particular reference to the demonstration of protein-bound sulphydryl and disulphide groups. Journal of Histochemistry and Cytochemistry 4: 23

Alfert M, Geschwind I I 1953 A selective staining method for the basic proteins of cell nuclei. Proceedings of the National Academy of Science 39: 991

Ammann A J, Borg D, Kondo L, Wara D W 1977 Quantitation of B cells in peripheral blood by polyacrylamide beads coated with anti-human chain antibody. Journal of Immunological Methods 17: 365

Avrameas S 1969 Indirect immunoenzyme techniques for the intracellular detection of antigens. Immunochemistry 6: 825

Avrameas S, Ternynck T 1971 Peroxidase labelled antibody and Fab conjugates with enhanced intracellular penetration. Immunochemistry 8: 1175

Baker J R 1956 The histochemical recognition of certain phenols, especially tyrosine. Quarterly Journal of the Microscopical Sciences 97: 161

Baker W H, Zamecnik P C, Stephenson M L 1957 In vitro incorporation of ^{14}C-DL-leucine into normal and leukaemic white cells. Blood 12: 822

Barrnett R J, Seligman A M 1952 Demonstration of protein-bound sulfhydryl and disulfide groups by two new histochemical methods. Journal of the National Cancer Institute 13: 215

Barrnett R J, Seligman A M 1954 Histochemical demonstration of sulfhydryl and disulfide groups of protein. Journal of the National Cancer Institute 14: 769

Bayer E, Wilchek M 1978 The avidin-biotin complex as a tool in molecular biology. Trends in Biochemical Sciences 3: 257

Beckstead J H, Stenberg P E, McEver R P, Shuman M A, Bainton D F 1986 Immunohistochemical localization of α-granule proteins in human megakaryocytes: application to plastic-embedded bone-marrow biopsy specimens. Blood 67: 285

Bernard J, Jaennesson P, Zagury D, Fridman W H, Ternynck T, Avrameas S 1975 Biology of isolated immunocytes. II. Simultaneous detection of cell surface Ig and theta-antigen by immunoperoxidase staining at the ultrastructural level. Annals of Immunology 126: 565

Bessis M, Breton-Gorius J, Binet J L 1963 Étude comparée du plasmacytome et du syndrome de Waldenström: examen au microscope électronique. Nouvelle Revue Francaise d'Hematologie 3: 159

Bloch D P, Godman G C 1955 A microphotometric study of the syntheses of desoxyribonucleic acid and nuclear histone. Journal of Biophysical and Biochemical Cytology 1: 17

Boenisch T, ed. 1980 PAP/Immunoperoxidase. Reference Guide Series 1. DAKO Corporation, Santa Barbara

Borgers M, Verheyen A 1985 Enzyme cytochemistry. International Review of Cytology 95: 163

Borowitz M J, Croker B P, Burchette J 1982 Immunocytochemical detection of lymphocyte surface antigens in fixed tissue sections. Journal of Histochemistry and Cytochemistry 30: 171

Brandtzaeg P 1982 Tissue preparation methods for immunohistochemistry. In: Bullock G R, Petrusz P (eds) Techniques in immunocytochemistry, Volume 1. Academic Press, London, p 1

Breton-Gorius J 1970 Utilisation de la diaminobenzidine pour la mise en évidence au microscope électronique de l'hémoglobine intracellulaire. La réactivite des différents organelles des érythroblasts. Nouvelle Revue Francaise d'Hematologie 10: 249

Breton-Gorius J, Vanhaeke D, Pryzwansky K B, Guichard J, Tabilio A, Vainchenker W, Carmel R 1984 Simultaneous detection of membrane markers with monoclonal antibodies and peroxidatic activities in leukaemia: ultrastructural analysis using a new method of fixation preserving the platelet peroxidase. British Journal of Haematology 58: 447

Burns J, Hambridge M, Taylor C R 1974 Intracellular immunoglobulins. A comparative study on

three standard tissue processing methods using horse-radish peroxidase and fluorochrome conjugates. Journal of Clinical Pathology 27: 548
Carvalho S de 1955 Absorption microspectroscopy of bone marrow cells. Blood 10: 453
Cawley J C, Barker C R, Britchford R, Smith J 1973 Intracellular IgA immunoglobulin crystals in chronic lymphocytic leukaemia. Clinical and Experimental Immunology 13: 407
Chèvremont M, Fréderick J 1943 Une nouvelle méthode histochimique de mise en evidence des substances à fraction sulfhydrile. Archives of Biology 54: 589
Chilosi M, Pizzolo G, Fiore-Donati L, Bofill M, Janossy G 1983 Routine immunofluorescent and histochemical analysis of bone marrow involvement of lymphoma/leukaemia: the use of cryostat sections. British Journal of Cancer 48: 763
Chilosi M, Pizzolo G, Ambrosetti A, Morittu L, Pasini F, Martinelli L, Montagna M, Lestani M, Fiore-Donati L 1985 Bone marrow immunohistology: detection of normal and neoplastic lymphoid populations in lymphoma. In: Quaglino D, Hayhoe F G J (eds) The cytobiology of leukaemias and lymphomas. Raven Press, New York p 47
Coons A H, Creech H J, Jones R N 1941 Immunological properties of antibody containing fluorescent group. Proceedings of the Society for Experimental Biology and Medicine 47: 200
Coons A H, Creech H J, Jones R N, Berliner E 1942 The determination of pneumococcal antigen in tissues by the use of fluorescent antibody. Journal of Immunology 45: 159
Cordell J L, Falini B, Erber W N, Ghosh A K, Abdulaziz Z, MacDonald S, Pulford K A F, Stein H, Mason D Y 1984 Immunoenzymatic labeling of monoclonal antibodies using immune complexes of alkaline phosphatase and monoclonal anti-alkaline phosphatase (APAAP complexes). Journal of Histochemistry and Cytochemistry 32: 219
Crockard A D, Catovsky D 1983 Cytochemistry of normal lymphocyte subsets defined by monoclonal antibodies and immunocolloidal gold. Scandinavian Journal of Haematology 30: 433
Danielli J F 1953 Cytochemistry. A critical approach. Biological Research Series. Wiley, New York
De Mey J 1983 Colloidal gold probes in immunocytochemistry. In: Polak J M, Van Noorden S (eds) Immunocytochemistry—Practical applications in pathology and biology. Wright, Bristol, p 82
Erber W N, Pinching A J, Mason D Y 1984 The immunocytochemical detection of T and B cell populations in routine blood smears. Lancet i, 1042
Erber W N, Mynheer L C, Mason D Y 1986 APAAP labelling of blood and bone-marrow samples for phenotyping leukaemia. Lancet i: 761
Falini B, Martelli M F, Tarallo F, Moir D J, Cordell J L, Gatter K C, Loreti G, Stein H, Mason D Y 1984 Immunohistological analysis of human trephine biopsies using monoclonal antibodies. British Journal of Haematology 56: 365
Faulk W P, Taylor G M 1971 An immunocolloidal method for the electron microscope. Immunocytochemistry 8: 1081
Feltkamp-Vroom T M 1975 Preparation of tissues and cells for immunohistochemical processing. Annals of the New York Academy of Sciences 254: 21
Ford P M, Stoward P J 1974 The detection of autoantibodies with an enzyme bridge method. Journal of Clinical Pathology 27: 118
Gavosto F, Maraini G, Pileri A 1960 Nucleic acids and protein metabolism in acute leukaemia cells. Blood 16: 1555
Glenner G G 1963 A re-evaluation of the ninhydrin-Schiff reaction. Journal of Histochemistry and Cytochemistry 11: 285
Gourdin M F, Reyes F, Lejoue J I, Mannoni P, Dreyfus B 1976 Surface features of cells in human lymphoproliferative disorders. An immunoelectron microscopic study. Hamatologie und Bluttransfusion 19: 207
Guesdon J L, Ternynck T, Avrameas S 1979 The use of avidin-biotin interaction in immunoenzymatic techniques. Journal of Histochemistry and Cytochemistry 27: 1131
Hanker J S, Yates P E, Metz C B, Rustioni A 1977 A new specific, sensitive and non-carcinogenic reagent for the demonstration of horseradish peroxidase. Histochemical Journal 9: 789
Hertl M 1959 Cytochemische Untersuchungen an Knochenmarkzellen. Folia Haematologica 3: 269
Hoffmann-Fezer G, Thierfelder S, Pielsticker K, Rodt H 1979 Immunohistochemical demonstration of cell surface antigens on tissue sections of lymphomas. Leukemia Research 3: 297
Hsu S M, Raine L, Fanger H 1981 Use of avidin-biotin-peroxidase complex (ABC) in immunoperoxidase techniques: a comparison between ABC and unlabeled antibody (PAP) procedures. Journal of Histochemistry and Cytochemistry 29: 577
Hsu S M, Cossman J, Jaffe E S 1983 A comparison of ABC, unlabeled antibody and conjugated immunohistochemical methods with monoclonal and polyclonal antibodies—An examination of germinal center of tonsils. American Journal of Clinical Pathology 80: 429
Jaffe E S, Shevach E M, Frank M M, Berard C W, Green I 1974 Nodular lymphoma. Evidence for origin from follicular B lymphocytes. New England Journal of Medicine 290: 813
Janossy G, Thomas J A, Pizzolo G, Granger S, McLaughlin J, Habeshaw J A, Stansfeld A G,

Sloane J 1980 Immunohistological diagnosis of lymphoproliferative diseases by selected combinations of antisera and monoclonal antibodies. British Journal of Cancer 42: 224
Kass L 1973 Histone biosynthesis in pernicious anaemia proerythroblasts and megaloblasts. Blood 41: 549
Kleihauer E, Braun H, Betke K 1957 Demonstration von fetalem Hämoglobin in den Erythrozyten eines Blutausstrichs. Klinische Wochenschrift 35: 637
Klockars M, Reitano S 1975 Tissue distribution of lysozyme in man. Journal of Histochemistry and Cytochemistry 23: 932
Klotz I M 1950 The nature of some ion-protein complexes. Cold Spring Harbour Symposia on Quantitative Biology 14: 97
Kocourch J, Horejsi V 1981 Defining a lectine. Nature 290: 188
Köhler G, Milstein C 1975 Continuous cultures of fused cells secreting antibody of predefined specificity. Nature 256: 495
Kraehenbul J P, Galardy R E, Jamieson J D 1974 Preparation and characterization of an immunoelectron microscope tracer consisting of a haem-octapeptide coupled to Fab. Journal of Experimental Medicine 139: 208
La Cour L F, Chayen J, Gahan P B 1958 Evidence for lipid material in chromosomes. Experimental Cell Research 14: 469
Lambertenghi-Deliliers G, Soligo D, Polli N, Nava M, Pozzoli E, Polli E 1985 Ultrastructural immunocytochemistry of leukaemic cells. In: Quaglino D, Hayhoe F G J (eds) The cytobiology of leukaemias and lymphomas. Raven Press, New York, p 263
Li C Y, Ziesmer S C, Yam L T, English M C, Janckila A J 1984 Practical immunocytochemical identification of human blood cells. American Journal of Clinical Pathology 81: 204
Ling N R, Bishop S, Jeffris R 1977 Use of antibody-coated red cells for the sensitive detection of antigen and in rosette tests for cells bearing surface immunoglobulin. Journal of Immunological Methods 15: 279
Macario A J L, Dugan C B, Conway de Macario E 1984 The erythroblastic nest (EN): cytochemical evidence for its role in the haemoglobinization of maturing erythroid cells. Haematologica 69: 521
Mason D Y, Biberfeld P 1980 Technical aspects of lymphoma immunohistology. Journal of Histochemistry and Cytochemistry 28: 731
Mason D Y, Sammons R 1978 Alkaline phosphatase and peroxidase for double immunoenzymatic labelling of cellular constituents. Journal of Clinical Pathology 31: 454
Mason D Y, Woolston R E 1982 Double immunoenzymatic labelling. In: Bullock G R, Petrusz P (eds) Techniques in immunocytochemistry Volume 1. Academic Press, London, p 135
Mason D Y, Farrell C, Taylor C R 1975 The detection of intracellular antigens in human leucocytes by immunoperoxidase staining. British Journal of Haematology 31: 361
Mason D Y, Labaume S, Preud'homme J L 1977 The detection of membrane and cytoplasmic immunoglobulin in human leucocytes by immunoperoxidase staining. Clinical and Experimental Immunology 20: 1
Mason T E, Phifer R F, Spicer S S, Swallow R A, Dreskin R B 1969 An immunoglobulin-enzyme bridge method for localising tissue antigens. Journal of Histochemistry and Cytochemistry 17: 563
Matutes E, Catovsky D 1982 The fine structure of normal lymphocyte subpopulations. A study with monoclonal antibodies and the immunogold technique. Clinical and Experimental Immunology 50: 416
Matutes E, Robinson D, O'Brien M, Haynes B F, Zola H, Catovsky D 1983 Candidate counterparts of Sézary cells and adult T-cell lymphoma-leukemia cells in normal peripheral blood: an ultrastructural study with the immunogold method and monoclonal antibodies. Leukemia Research 7: 787
McLeish J 1959 Comparative microphotometric studies of DNA and arginine in plant nuclei. Chromosoma 10: 686
McLeish J L, Bell L G E, La Cour L F, Chayen J 1957 The quantitative cytochemical estimation of arginine. Experimental Cell Research 12: 120
Mitchison J M 1971 The Biology of the Cell Cycle. Cambridge Univ. Press, Cambridge
Moir D J, Ghosh A K, Abdulaziz Z, Knight P M, Mason D Y 1983 Immunoenzymatic staining of haematological samples with monoclonal antibodies. British Journal of Haematology 55: 395
Molinaro G A, Dray S 1974 Antibody coated erythrocytes as a manifold probe for antigens. Nature 248: 515
Nairn R C 1976 Fluorescent Protein Tracing, 2nd Edn. Churchill, Edinburgh
Nakane P K 1968 Simultaneous localisation of multiple tissue antigens using the peroxidase-labelled antibody method: a study on pituitary glands of the rat. Journal of Histochemistry and Cytochemistry 16: 557
Nakane P K 1975 Recent progress in the peroxidase-labeled antibody method. Annals of the New York Academy of Sciences 254: 203

Nakane P K, Pierce G B Jr 1966 Enzyme-labelled antibodies: preparation and application for the localization of antigens. Journal of Histochemistry and Cytochemistry 14: 929

Oi V T, Glazer A N, Stryer L 1982 Fluorescent phycobiliprotein conjugates for analyses of cells and molecules. Journal of Cell Biology 93: 981

Ordronneau P 1982 Post-embedding unlabeled antibody techniques for electron microscopy. In: Bullock G R, Petrusz P (eds) Techniques in immunocytochemistry. Volume 1. Academic Press, London, p 269

Pearse A G E 1980 Histochemistry theoretical and applied, 4th edn. Volume 1—Preparative and optical technology. Churchill Livingstone, Edinburgh

Pileri A, Maraini G, Gavosto F 1960 Inter-relationship between RNA and protein metabolism in human normal bone marrow cells. Experimental Cell Research 20: 645

Pizzolo G, Chilosi M 1984 Double immunostaining of lymph node sections by monoclonal antibodies using phycoerythrin labeling and haptenated reagents. American Journal of Clinical Pathology 82: 44

Pizzolo G, Chilosi M, Cetto G L, Fiore-Donati L, Janossy G 1982 Immunohistological analysis of bone marrow involvement in lymphoproliferative disorders. British Journal of Haematology 50: 95

Polli N, Zola H, Catovsky D 1984 Characterization by ultrastructural cytochemistry of normal and leukaemic myeloid cells reacting with monoclonal antibodies. American Journal of Clinical Pathology 82: 389

Ponder B A J 1983 Lectin histochemistry. In: Polak J M, Van Noorden S (eds) Immunocytochemistry—practical applications in pathology and biology. Wright, Bristol, p 129

Ponder B A, Wilkinson M M 1981 Inhibition of endogenous tissue alkaline phosphatase with the use of conjugates in immunohistochemistry. Journal of Histochemistry and Cytochemistry 29: 981

Pryzwansky K B 1982 Applications of double-label immunofluorescence. In: Bullock G R, Petrusz P (eds) Techniques in immunocytochemistry. Volume 1. Academic Press, London, p 77

Robinson D, Tavares de Castro J, Polli N, O'Brien M, Catovsky D 1984 Simultaneous demonstration of membrane antigens and cytochemistry at ultrastructural level: a study with the immunogold method, acid phosphatase and myeloperoxidase. British Journal of Haematology 56: 617

Roth J 1982 The protein A-gold (pAg) technique—a qualitative and quantitative approach for antigen localization on thin sections. In: Bullock G R, Petrusz P (eds) Techniques in immunocytochemistry. Volume 1. Academic Press, London, p 107

Sakaguchi S 1925 Über eine neue Farbreaktion von Protein and Arginin. Journal of Biochemistry (Tokyo) 5: 25

Stanislawski M, Mitard M, Stanislawski L 1976 Immunoenzymatic methods for the study of surface bound immunoglobulins in mice. Journal of Immunological Methods 11: 231

Stein H, Bonk A, Tolksdorf G, Lennert K, Rodt H, Gerdes J 1980 Immunohistologic analysis of the organization of normal lymphoid tissue and non-Hodgkin's lymphomas. Journal of Histochemistry and Cytochemistry 28: 746

Sternberger L A 1974 Immunocytochemistry. Prentice-Hall, New Jersey

Sternberger L A 1979 Immunocytochemistry, 2nd edn. John Wiley and Sons, New York.

Sternberger L A, Hardy P H Jr, Cuculis J J, Meyer H G 1970 The unlabeled antibody enzyme method of immunohistochemistry. Preparation and properties of soluble antigen-antibody complex (horseradish peroxidase-antihorseradish peroxidase) and its use in identification of spirochetes. Journal of Histochemistry and Cytochemistry 18: 315

Streefkerk J G 1972 Inhibition of erythrocyte pseudoperoxidase activity by treatment with hydrogen peroxide following methanol. Journal of Histochemistry and Cytochemistry 20: 829

Swirsky D M, Watt S M, Gilmore D J, Hayhoe F G J, Waldmann H 1983 The characterization of monoclonal antibodies against haemopoietic cells: comparison of an immunoperoxidase method with fluorescence activated cell sorting. Journal of Immunological Methods 61: 171

Takamiya H, Shimizu F, Vogt A 1975 A two-stage method for cross-linking antibody globulin to ferritin by glutaraldehyde. III. Size and antibody activity of the conjugates. Journal of Immunological Methods 8: 301

Taylor C R 1978 Immunoperoxidase techniques: practical and theoretical aspects. Archives of Pathology and Laboratory Medicine 102: 113

Taylor C R, Burns J 1974 The demonstration of plasma cells and other immunoglobulin-containing cells in formalin-fixed paraffin-embedded tissues using peroxidase-labelled antibody. Journal of Clinical Pathology 27: 14

Thorell B 1947 Studies on the formation of cellular substances during blood cell production. Acta Medica Scandinavica, Suppl. 200

Undritz E 1973 Sandoz Atlas of Haematology. Sandoz, Basle, p 35

Wickramasinghe S N, Chalmers D G 1968 Protein synthesis in megaloblastic erythropoiesis caused by vitamin B12 deficiency. Nature 218: 463

Wickramasinghe S N, Pratt J R 1970 Myelocyte proliferation in pernicious anaemia. Acta Haematologica 44: 37

Wickramasinghe S N, Chalmers D G, Cooper E H 1969 Arrest of cell proliferation and protein synthesis in megaloblasts of pernicious anaemia. Acta Haematologica 41: 65

Williams M A 1977 Autoradiography and immunocytochemistry. In: Glauert A M (ed) Practical methods in electron microscopy, volume 6, part 1. North-Holland, Amsterdam

Worman C P 1978 Cytoplasmic immunofluorescent staining: a micromethod with counterstain. Journal of Immunological Methods 23: 193

Yam L T, English M C, Janckila A J, Ziesmer S, Li C Y 1983 Immunocytochemical characterization of human blood cells. American Journal of Clinical Pathology 80: 314

Yasuma A, Ichikawa T 1953 Ninhydrin-Schiff and alloxan-Schiff staining: a new histochemical staining method for protein. Journal of Laboratory and Clinical Medicine 41: 296

Yataganas X, Fessas Ph 1969 The pattern of hemoglobin precipitation in thalassemia and its significance. Annals of the New York Academy of Sciences 165: 270

Yataganas X, Gahrton G, Thorell B 1970 DNA, RNA and haemoglobin during erythroblast maturation. A cytophotometric study. Experimental Cell Research 62: 254

CHAPTER 4

Carbohydrates

The carbohydrate or more especially the glycogen content of leucocytes has been studied by microchemical and histochemical methods. With the former, the average glycogen content has been found to be 4.23 micrograms per 10^6 granulocytic leucocytes (Wagner, 1946). In cases of neutrophilic leucocytosis associated with infection, average unit granulocyte glycogen is significantly above the observed normal mean, although there is some overlap with the normal range. The leucocytes from subjects with polycythaemia vera with leucocytosis or leukaemoid features possess the highest glycogen content of any disease studied. In sharp contrast, low mean values are found in chronic granulocytic leukaemia (Valentine *et al.* 1953).

Wagner and his associates (1950, 1952), studying the breakdown of glycogen in separated horse leucocytes, identified glucose-1-phosphate, glucose-6-phosphate, fructose-6-phosphate, phosphoglyceric acid and fructose 1,6-diphosphate among the intermediate products. The breakdown of glycogen to lactic acid follows a pattern similar to that in other tissues of the body, and is the result of phosphorylating glycogenolysis.

The first attempt to demonstrate glycogen histochemically in human neutrophil leucocytes was that of Neukirch (1910), who applied Best's carmine method of staining to the polymorphonuclear cells present in blood and inflammatory exudates. The effect of iodine staining was also studied. By both methods, the presence of granules in the cytoplasm of neutrophil leucocytes was revealed. As these granules were found to be soluble in salivary amylase, Neukirch concluded that they represented glycogen or a closely related carbohydrate. Iodophilic and carmine stainable structures were also found in human blood platelets, but they were not soluble in saliva. However, the iodine staining technique is based on a rather imperfectly understood chemical reaction and succeeds in detecting glycogen when present in considerable quantities, as in the later granulocytes and in megakaryocytes and platelets, but fails to detect the small quantities of glycogen in the more primitive cells.

In 1946 Wislocki and Dempsey, using the Bauer-Feulgen method and an ammoniacal silver nitrate staining technique on blood cells of Rhesus monkey, obtained positive results with both methods. Glycogen was observed in the cytoplasm of neutrophil leucocytes and metamyelocytes of the bone marrow, but none in other cell types. Megakaryocytes did not stain positively.

In 1949 Gibb and Stowell applied three different histochemical methods to the study of glycogen in normal and abnormal peripheral blood and bone marrow films. The techniques employed were the chromic acid-silver-methenamine method of Gomori (1946), the periodic acid-fuchsin-sulphurous acid technique of Hotchkiss (1948), modified from the method of McManus (1946), and the Bauer-Feulgen stain (1933). Control smears were treated with saliva. Gibb and Stowell (1949) found that the most sensitive and reliable results were obtained with the first two procedures, which also showed essentially similar staining patterns. The Bauer–Feulgen technique proved less satisfactory, since a good deal of non-specific colouration of the films occurred and this made it difficult to assess the presence or absence of small quantities of glycogen. Gibb and Stowell could not detect any glycogen in the red cells or their precursors. All cells of the myeloid series contained significant amounts of glycogen, the most intense staining being found in the segmented polymorphonuclear leucocytes. The large specific granules of eosinophilic cells were negative, but the cytoplasm around these granules contained large amounts of glycogen. Lymphocytes and monocytes contained small quantities of glycogen in the cytoplasm. Both megakaryocytes and platelets were found to possess glycogen.

Other and later studies (Wislocki *et al.*, 1949; Storti and Zubiani, 1949; Storti *et al.*, 1953a) confirmed that the most reliable and reproducible results were obtained with the periodic acid–Schiff (PAS) reaction and showed that the staining pattern observed was substantially similar to that described by Gibb and Stowell.

Chemical basis and specificity of PAS reaction

As all subsequent observations on glycogen distribution in normal and leukaemic cells have been obtained chiefly with this reaction, it seems appropriate at this stage to discuss briefly the chemical basis of the reaction and its specificity.

Periodic acid (HIO_4) acts specifically to oxidize the 1–2 glycol grouping (CHOH—CHOH) or its amino- or alkylamino-derivatives to produce a dialdehyde. The oxidation does not proceed further, and the resulting aldehydes give with Schiff's reagent a substituted dye of magenta colour. According to Hotchkiss (1948) a positive reaction may be given by several classes of naturally occurring carbohydrates, including monosaccharides, polysaccharides, glycoprotein and mucoprotein conjugates, phosphorylated sugars, inositol derivatives and cerebrosides. Substances of low molecular weight are likely to diffuse away during the course of fixation or after oxidation, and glycogen can be distinguished from other positively reacting compounds of high molecular weight by the use of control material subjected to salivary amylase digestion. Compounds such as ribose or deoxyribose nucleic acids and the hydroxyamino acid residues of proteins are chemically substituted so that the free glycol group is not present and they do not give the reaction.

Distribution of positivity in normal haemic cells
(Plate 4A, B, C, D, E & F)

Positive reactions are localized to the cell cytoplasm and extracellular staining is only present in very thick areas of the film.

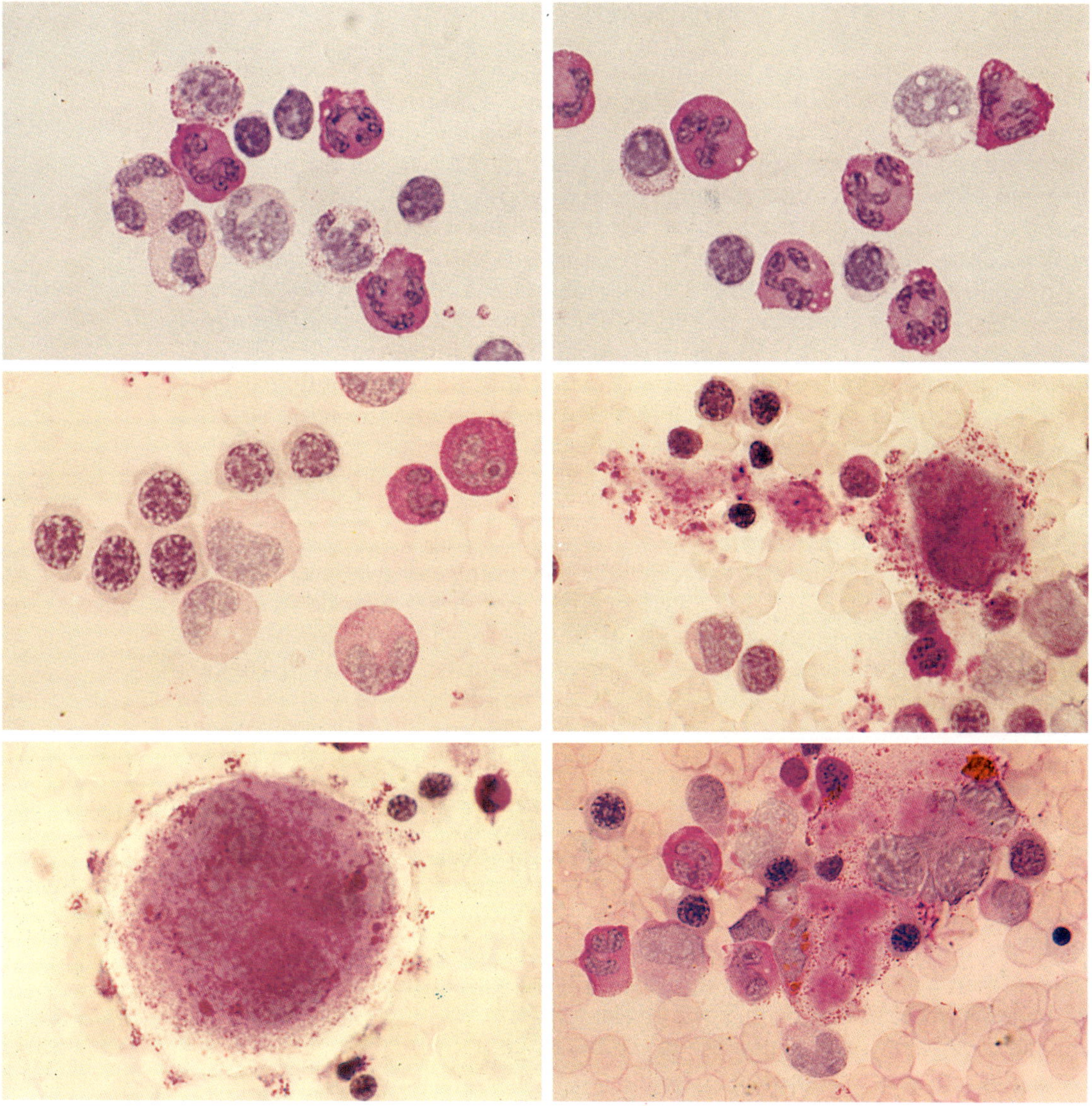

A B C D E F

Plate 4

A. Normal buffy coat. PAS stain. Three strongly positive neutrophils, four lymphocytes with negative reaction, a basophil with coarsely granular positivity, two monocytes one showing scattered granules and the other negative, and two eosinophils, with negative granules against a positive background.

B. Normal buffy coat. PAS stain. Six positive neutrophils, one monocyte with weak diffuse and finely granular reaction and three lymphocytes, one showing + + positivity.

C. Normal bone marrow. PAS stain. Granulocyte stages from myelocyte to neutrophil polymorph, showing increasing positivity with increasing cell maturity: six negative intermediate normoblasts.

D. Normal bone marrow. PAS stain. Active platelet formation, with platelet masses showing dense central positivity and diffuse peripheral reactions. Various normoblasts and granulocytes show normal reactions.

E. Normal bone marrow, PAS stain. A megakaryocyte showing diffuse and finely granular reaction over nucleus and cytoplasm, coarse glycogen inclusion bodies, a clear 'separation' zone towards the periphery and a ring of newly formed platelets.

F. Normal bone marrow. PAS stain. A macrophage with ingested megakaryocyte nuclear material, PAS positivity and free iron deposits (staining brown). There are also erythroblasts (PAS negative) and normally reacting granulocyte series.

The erythropoietic series does not normally show detectable amounts of glycogen at any stage of cellular development. The myeloid series gives positive results in all identifiable cells. Neutrophil polymorphonuclear leucocytes contain very large amounts of PAS positive material, usually in the form of moderately fine granules packed so closely that the cytoplasmic background cannot be distinguished. Granules can be seen over the nuclear surface, but they rarely obscure its outline and detail. Neutrophil metamyelocytes and myelocytes contain many glycogen droplets, though not so many as in the mature polymorphonuclear cells. The granules show some variation in size and arrangement, being sometimes relatively few and coarse, sometimes numerous and fine. Eosinophil polymorphonuclear cells show a very distinctive appearance in PAS preparations: the large specific eosinophilic granules remain unstained and sharply outlined against a background of diffusely positive intergranular cytoplasm. This appearance is also found in the recognizable eosinophil precursors, eosinophil metamyelocytes and myelocytes.

With regard to basophils, there have been many contradictory results concerning the localization of the PAS positive and diastase labile structures in these cells. However the debate about the nature of the PAS reaction in basophilic cells has come to a close following the investigations of Inagaki (1968) who found PAS positive deposits in the intergranular space to be regularly lysed by diastase treatment, and of Kaung (1969) who clearly showed that the specific granules of blood basophils always gave a PAS negative reaction. These results were obtained both with the light and the electron microscope. The inconsistent reports regarding glycogen in blood basophils are mostly due to the fact that the glycogen content depends on the momentary metabolic condition of the cell and is therefore subject to great fluctuations (Parwaresch, 1976). Metachromatic basophil granules definitely do not contain glycogen. The PAS positive reaction occasionally observed in specific granules is essentially due to the presence of acid mucopolysaccharides and phospholipids.

Early granulocyte precursors, myeloblasts and promyelocytes show a diffuse cytoplasmic colouring, usually faint, but sometimes more intense with a tendency to granule formation like that seen in myelocytes. Lymphocytes have a much smaller glycogen content than granulocytes, but PAS positive granules can often be demonstrated in the cytoplasm. They may be fine or coarse, but are usually discrete, varying greatly in number. From 10 to 40% of lymphocytes in peripheral blood normally show positivity with one or two rings of perinuclear granules, but more intense staining with 3 or more concentric rings of granules or with heavy clumps of positive material occurs in only 1 to 2% of lymphocytes (Quaglino and Hayhoe, 1959). PAS positivity does not appear to differ significantly in either frequency or extent among the T- or B-cell components of the peripheral blood lymphocyte population (Higgy *et al.*, 1977). Monocytes appear normally to contain small amounts of finely scattered granules of glycogen. PAS positive material in platelets appears in two different forms: the one finely scattered, lightly staining and peripherally distributed, and the other densely clumped, heavily positive and centrally situated. The reaction in megakaryocytes is discussed in detail below.

Changes in PAS positivity in pathological states

The patterns of PAS positivity in the erythroid and leucocytic series in haemopoietic disorders are often of considerable value in differential diagnosis and are discussed in some detail in relation to other cytochemical findings in Chapters 13, 14 and 15.

It may be appropriate in the present chapter, however, to recapitulate the essential variations in a number of pathological states, where the PAS reaction is markedly subject to alterations. Among these conditions are included the main varieties of acute leukaemia, disorders of the megakaryocyte—platelet system, thalassaemia and some lymphoproliferative states. We also discuss the underlying mechanisms which may influence changes in PAS positivity and review studies with, for example, the skin window technique, which have helped to elucidate these mechanisms.

Patterns of PAS positivity in leukaemic blast cells

The PAS reaction is certainly of considerable importance in the differential diagnosis of the various forms of acute leukaemia and will be discussed in some detail, in relation to other cytochemical techniques, in Chapter 14.

The major characteristic aspects of PAS positivity in the different types of blast cells may, however, be summarized briefly as follows: myeloblasts and promyelocytes from acute leukaemia tend to be substantially PAS negative, showing at most a diffuse cytoplasmic tinge (Plate 5A), while lymphoblasts of acute lymphoblastic leukaemia show, in a percentage of cells highly variable from one case to another, very strong PAS positivity in the form of several concentric rings of coarse granules or as heavy blocks of glycogen against a negative background (Plate 5B). Acute leukaemic monoblasts, stained with the PAS reaction, show a mixed pattern, some being completely negative, others displaying a rather strong positivity in the form of reddish diffuse background staining with superimposed fine to moderately coarse or coarse granules (Plate 5C, D). Finally erythroid precursors from cases of erythraemic myelosis and erythroleukaemia often show in a variable proportion of cells concentric annular rings of moderately coarse to coarse PAS positive cytoplasmic granules (Plate 5E).

In some particular forms of acute myeloid leukaemia associated with high percentages of atypical bone marrow eosinophils, such as acute myelomonocytic leukaemia with pericentric inversion of chromosome 16 (Bitter *et al.*, 1984; Schmitz *et al.*, 1984) or acute granulocytic leukaemia with 8;21 translocation (Swirsky *et al.*, 1984), the eosinophil granules frequently display a variable degree of abnormal PAS positivity. This finding, together with other morphological and cytochemical abnormalities, suggests that in these forms the eosinophil lineage is part of the leukaemic cell population.

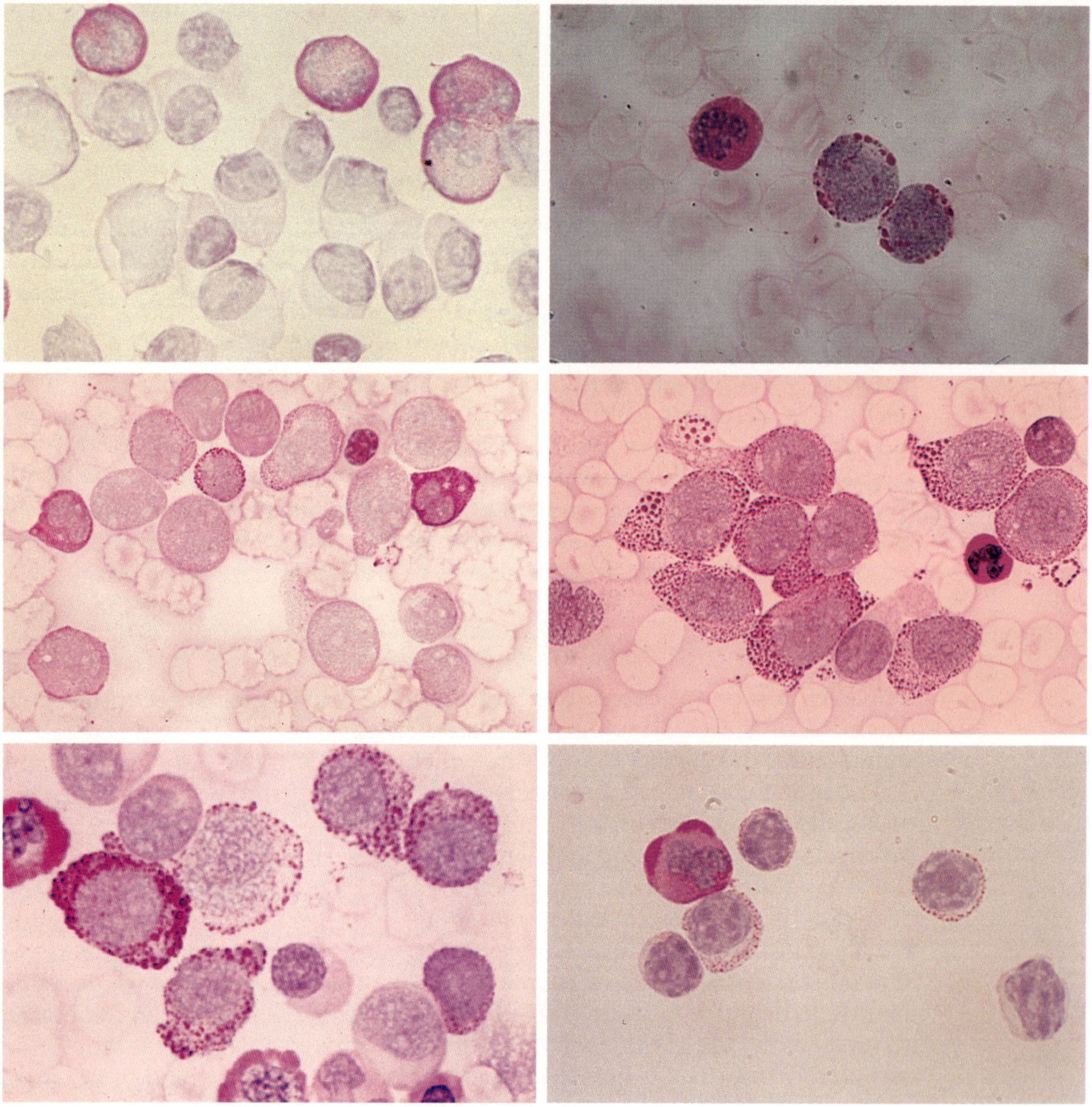

Plate 5

A. AML marrow. PAS stain. Myeloblasts show negative reactions, while positivity is seen in four mature neutrophils.
B. ALL marrow. PAS stain. A normally positive neutrophil and two lymphoblasts with coarse granular positivity.
C. AMML marrow. PAS stain. Mixed granulocyte and monocyte precursors with reaction ranging from weak tinge to moderately coarse granular positivity against diffusely staining background. Two later granulocytes show strong reaction.
D. AMonL marrow. PAS stain. Fine to coarse granular positivity on a diffusely positive background.
E. Erythroleukaemic marrow. PAS stain. Intense granular and diffuse positivity in erythroblasts at all stages of maturity. Three myeloblasts and a plasma cell appear almost negative; a metamyelocyte shows weak positivity.
F. CLL peripheral blood. PAS stain. A normally positive neutrophil and five lymphocytes, two negative and three showing crowns of positive granules, respectively of 1, 2 and 3 plus rating.

The megakaryocyte–platelet system

The early papers on the subject were almost unanimous in reporting the presence of glycogen in megakaryocytes, a finding which was obtained using different cytochemical techniques. Normally the PAS reaction shows diffuse and granular positivity scattered over the nucleus and cytoplasm of the megakaryocyte, with variable numbers of densely positive 'glycogen inclusion bodies', a clearer and sometimes negative peripheral zone with a fringe of positive granularity indicating platelet separation (Plate 4D & E). Platelets are strongly positive.

The cytochemical investigations of Storti *et al.* (1953b) showed that a close similarity in chemical composition existed between the thrombocytic granules and the granulations present at the periphery of megakaryocytes. Storti *et al.* (1953a) observed in two cases of idiopathic thrombocytopenic purpura (ITP) a decrease in the amount of megakaryocytic glycogen associated with platelet formation and an increase in the number of intracytoplasmic glycogen bodies present in a proportion of the megakaryocytes. In the only case of ITP examined after splenectomy, the opposite picture was observed. Examples of intracytoplasmic staining with minimal peripheral granularity, and the reverse pattern, are shown in Figs. 4.1 to 4.3.

A general decrease in megakaryocytic glycogen was reported by Heckner (1956) in cases of thrombocytopenia of different nature: in patients with ITP, in the initial phases of acute leukaemia, in pernicious anaemia and hypersplenism. The author believed that the study of the glycogen content of megakaryocytes was a most useful means for evaluating their activity and the consequent formation of platelets.

Daniell in 1959, using the PAS reaction, applied a semi-quantitative method for assessing the glycogen content of megakaryocytes. This method was based on the considerable variation in single megakaryocytes of the 'glycogen inclusion bodies' and the 'mean glycogen score' was obtained by evaluating the amount of inclusion bodies in 100 consecutive megakaryocytes. In his studies on bone marrows from normal individuals, Daniell observed that persons who have undergone repeated phlebotomies tend to have higher mean glycogen scores than subjects without repeated phlebotomies. The implications drawn from these observations were that as a consequence of blood loss there is a stimulus to increased red cell formation and that the need for glycogen, which represents a potential source of energy to be utilized in platelet production, is increased in the presence of enhanced erythropoiesis because of lowered oxygen tension (Fig. 4.4).

A further study on megakaryocytic glycogen was carried out by Jamra and Lorenzi (1961) in 61 cases of ITP. In the chronic forms of the disease, in agreement with the observations of Storti *et al.* (1953a), there was usually an increase in coarse positivity, while in acute ITP there was no such increase, in accordance with the findings of Heckner (1956). These results would explain the discrepant results obtained in acute and chronic forms of ITP and could signify that in acute ITP the developmental failure occurs before

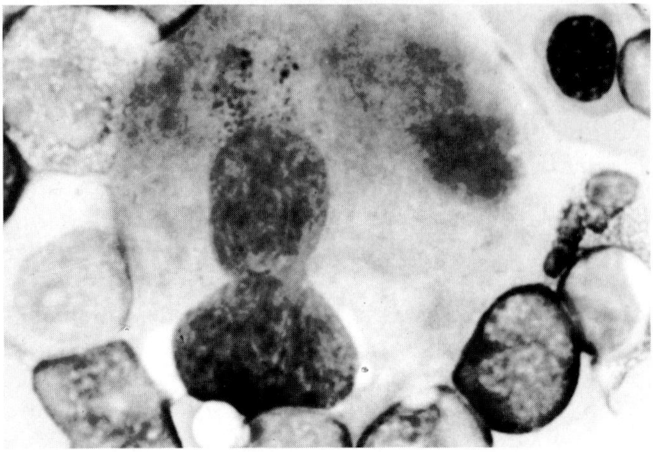

Fig. 4.1 Megakaryocyte stained with the PAS reaction, showing clumped diffuse and granular intracytoplasmic pattern but negative peripheral granularity and no platelet formation. Surrounding cells include an eosinophil with negative granules, a negative normoblast and several positively reacting neutrophil granulocytes.

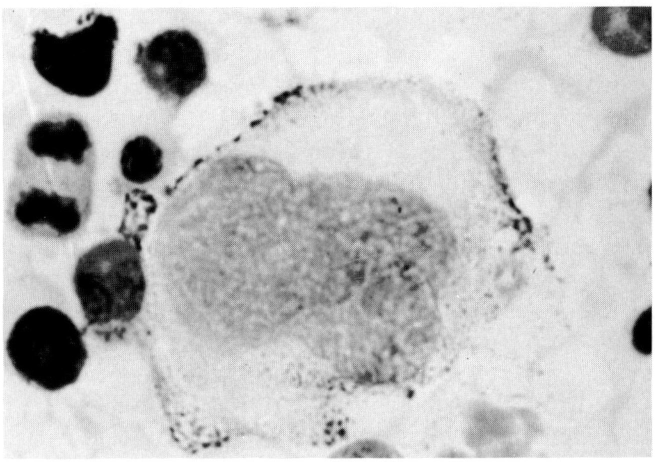

Fig. 4.2 Megakaryocyte stained with the PAS reaction, showing little intracytoplasmic granular positivity and no diffuse reaction or clumped glycogen bodies, but with strong peripheral granularity and early platelet formation.

the appearance of glycogen (Fig. 4.3); whereas in chronic ITP, in which megakaryocytes reach a more advanced stage of differentiation, the possible failure of platelet formation leads to an excessive accumulation of intracytoplasmic glycogen in the form of coarse 'inclusion bodies'. This hypothesis does not exclude the possibility that in chronic ITP the increase in glycogen may represent a condition of enhanced metabolic activity, which occurs as a compensatory mechanism because of excessive platelet destruction.

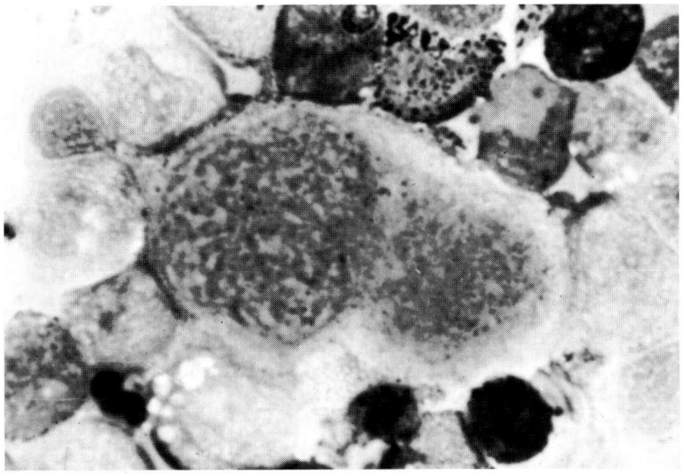

Fig. 4.3 Megakaryocyte in acute ITP: PAS stain shows weak diffuse cytoplasmic reaction only. Surrounding granulocytes show normal PAS reaction, increasing with increasing cell maturity. There is a coarsely positive basophil.

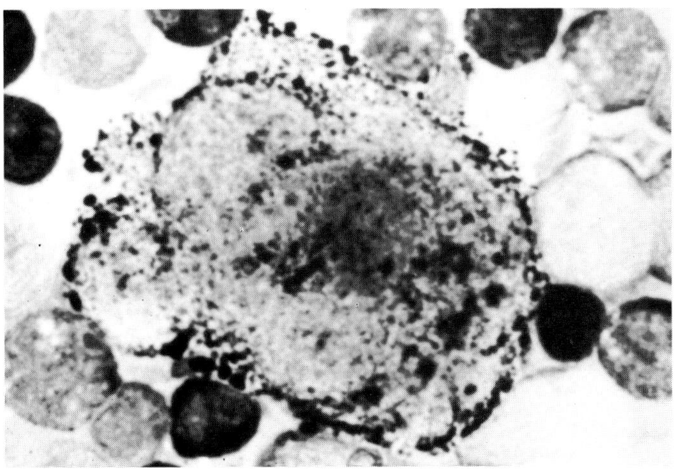

Fig. 4.4 Megakaryocyte from patient with haemorrhagic anaemia: PAS stain shows strong and coarse granularity with glycogen inclusion bodies and marked peripheral platelet formation.

In 1966 Headington reported high average phosphorylase scores in patients with ITP before splenectomy in association with increased numbers of coarse PAS positive granules.

Another study concerning megakaryocytic glycogen is that of Shammas and Engeset (1986), who, employing a quantitative microspectrophotometric technique, determined the intensity of the PAS reaction in these cells and reported that glycogen was present also in megakaryocytes with a diffuse staining pattern, but generally the mean glycogen content and glycogen per unit area was higher in megakaryocytes with inclusion bodies.

They concluded that differences in the total amount of glycogen between different subjects become significant only when the proportion of megakaryocytes with inclusion bodies exceeds 25%.

Erythroid series

Another field where the techniques for demonstrating glycogen have been most useful has been the study of the carbohydrate content of red cells in normal and pathological conditions.

A substantial agreement exists amongst the early workers on the absence of glycogen in the normal erythropoietic series. In 1954 Astaldi *et al.* reported that in cases of thalassaemia major a variable proportion of erythrocyte precursors and erythrocytes showed a positive reaction with the techniques of Hotchkiss, Bauer, Gomori, and others. The staining patterns observed led the authors to believe that the PAS positive substance present in the erythroblasts of thalassaemic patients was a gluco- or muco-protein or a neutral polysaccharide. Other conditions were also examined, amongst which were cases of acute and chronic leukaemia, erythroleukaemia, haemolytic anaemia, and blood from premature infants. Only a small proportion of PAS positive erythroblasts were found in one case of chronic myeloid leukaemia; the reactivity of this PAS positive substance towards the various blocking and enzymatic degradation tests was such as to reveal a different nature from that present in thalassaemic erythoblasts and erythrocytes.

However, the results of other investigators (Stegagno and Pollitzer, 1953; Sansone and Bertolotti, 1958; Matioli and Del Piano, 1961), the detailed study of Fessas and Papayannopoulou (1965) and the observations of Yataganas *et al.* (1973) support the contention that the PAS positive material in thalassaemic erythroblasts is glycogen.

A more comprehensive account of the cytochemical changes present in thalassaemia and other pathological conditions of the erythroblastic tissue will be given in Chapters 13 and 14.

Lymphoproliferative diseases

All the early cytochemical studies, mentioned at the beginning of this chapter, agreed that a variable proportion of normal lymphocytes showed the presence of cytoplasmic granules of glycogen. According to Gibb and Stowell (1949) no substantial difference was noticeable between normal and leukaemic lymphocytes, while later investigations (Storti *et al.*, 1953a; Astaldi and Verga, 1957) indicated in chronic lymphocytic leukaemia an increase in the proportion of lymphocytes displaying PAS positive granules in the cytoplasm. However, such changes in the glycogen content of lymphocytes may not be peculiar to leukaemia, according to the studies of Mitus and his associates (1958). The application of a semiquantitative scoring method (see below: technical methods) for the evaluation of PAS positivity in the lymphocytes of a variety of lymphoproliferative diseases revealed high levels of glycogen not only in chronic lymphocytic leukaemia,

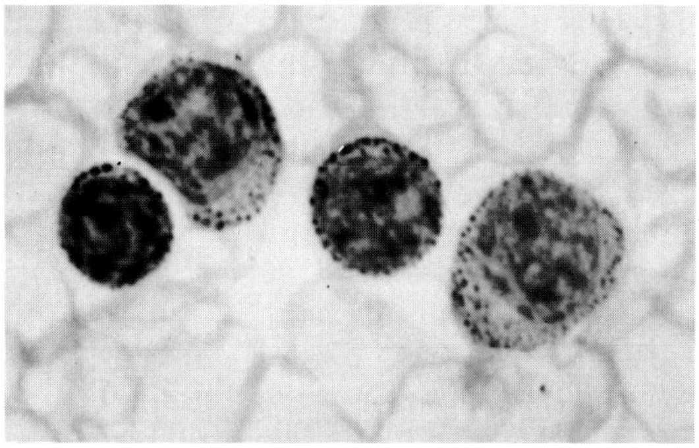

Fig. 4.5 Group of four lymphocytes, two probably lymphomatous, showing strong PAS positivity.

but also in lymphomas and in other non-malignant proliferations. The authors suggested that the increase in glycogen was a non-specific consequence of increased metabolic and proliferative activity in the lymphocytic tissue.

However, both interpretations might be valid. Quaglino and Hayhoe (1959) confirmed the existence of increased levels of glycogen in chronic lymphocytic leukaemia (Plate 5F) and lymphoma (Fig. 4.5) and observed a close relationship between the level of the PAS scores and the effect of treatment. However, the presence of elevated PAS scores in Hodgkin's disease, uninfluenced by treatment, as well as in other conditions associated with lymphocytosis suggested that the increased amount of glycogen could take place through different mechanisms. The strong PAS positivity in chronic lymphocytic leukaemia could be due to an enzyme defect or could be related to the low proliferative activity of these cells, while in other conditions of lymphoid hyperplasia the same cytochemical pattern could reflect a condition of increased metabolic activity. The combination of autoradiography, employing both DNA and RNA radioactive precursors, with the PAS reaction may throw additional light on the proliferative and metabolic behaviour of normal and leukaemic cells and on the possible mechanisms which are responsible for apparently similar PAS reactivity (Quaglino *et al.*, 1964).

Quantitation of PAS positivity in granulocytes

All the cytochemical studies so far mentioned have been based on simple visual semiquantitative scoring methods, which, besides providing a fairly accurate picture of the glycogen distribution in the different types of haemic cells, have contributed in no small way to the understanding of certain metabolic aspects in normal and pathological states. However, the diffuse cytoplasmic positivity present in certain cell types, such as neutrophil granulocytes, prevents a similar quantitative evaluation of the glycogen changes occurring in physiological conditions and in disease states. Such an assessment may be obtained only by the use of quantitative micro-spectrophotometric techniques. The application of such a method has enabled Gahrton (1966a and b) to study the mean amount of PAS reactive material (mainly glycogen) in neutrophil leucocytes and the frequency distribution of the same per cell in 20 normal subjects and in patients with chronic myeloid leukaemia. The mean amount of PAS reactive material in neutrophils from untreated and relapse cases of chronic myeloid leukaemia was 43% less than in normal neutrophils. In clinical remission after myleran treatment the neutrophils contained normal amounts of PAS reactive material. The reduced glycogen content of chronic myeloid leukaemia leukocytes is consistent with the amount of chemically estimated glycogen in separated myeloid cells, as we have mentioned earlier in this chapter (Valentine *et al.*, 1953). The reduction cannot be ascribed to the increased number of band forms in chronic myeloid leukaemia, because there is no significant difference between the PAS reaction in these cells and in cells with a different number of nuclear lobes. It is also unlikely that the decreased glycogen content of leukaemic leucocytes is due to a defect in some of the enzymes involved in glycogen synthesis, since a 3 to 10 fold increase from normal in the incorporation of glucose into glycogen in myeloid cells from chronic myeloid leukaemia has been observed (Vanderwende *et al.*, 1964). According to Pedersen and Hayhoe (1971a), the reduced glycogen content of polymorphs in chronic myeloid leukaemia is compatible with the assumption of relative cytoplasmic immaturity, as indicated also by the results of ultrastructural studies (Bessis, 1968; Kakefuda, 1968). The change from low to normal amounts of glycogen after treatment is probably related to the emergence of a predominantly normal neutrophil population, an hypothesis which is supported by the behaviour of the alkaline phosphatase reaction, which tends to reach normal values in remission (Hayhoe and Quaglino, 1958; Xefteris *et al.*, 1961; Linke and Löffler, 1963) and by the parallel increase in neutrophil phagocytic activity (Pedersen and Hayhoe, 1971b).

In 1980 Efrati and Yaari, using a scanning microdensitometer, confirmed the presence of decreased amounts of glycogen in the neutrophils of cases of chronic myeloid leukaemia, but in contrast with the findings of Gahrton, a return to normal amounts could not be demonstrated in busulphan-treated cases which went into remission. Efrati and Yaari also reported the presence of increased amounts of glycogen in neutrophils from patients with different chronic myeloproliferative disorders (myelofibrosis and polycythaemia vera). In the latter condition glycogen content of neutrophil granulocytes was reported as invariably increased, irrespective of the number of granulocytes, a finding which is at variance with the observations of Valentine *et al.* (1953), who, employing biochemical methods, found increased glycogen content of white blood cells only in cases associated with leucocytosis. In cases of secondary polycythaemia, usually due to chronic lung disease, the values were also high, but did not reach the mean values found in polycythaemia vera. In B-lymphocyte disorders (non-Hodgkin's lymphomas, chronic lymphocytic leukaemia and multiple myeloma) only the glycogen content of neutrophil granulocytes was found increased, whereas in another group of diseases (Hodgkin's disease, infectious mononucleosis) the relation of glycogen density to cell area was also high.

The PAS reaction and cell metabolism

It is now well established both from biochemical and histochemical studies that glycogen plays an important part in the primary function of different types of haemic cells. In physiological terms (Scott, 1968a) the glycogen particle represents an organelle, a 'machine', whose function is to store glucose and make it available on demand. Macromolecules such as proteins and nucleic acids have a fixed structure, whereas the glycogen molecule is in a state of constant turnover of glucose units and its size varies depending upon the glucose needs of the cell at a given time. Glycogen does not exist as a naked molecule but is apparently associated with enzymes concerned with its synthesis and degradation. Whereas the glycogen portion of this 'machine' is in constant turnover, the enzyme proteins are stable, i.e. glycogen synthesis and degradation do not require continued synthesis of the enzymes associated with it (Scott, 1968b; Scott and Still, 1968).

The presence of large stores of glycogen in such cells as neutrophils, transforming T lymphocytes in PHA cultures and platelets is related to their highly specialised function. In fact, although the mature neutrophil is relatively inactive metabolically as it circulates, when it performs its specific role of phagocytosis and destruction of engulfed material, it does so with a relatively brief burst of activity that requires considerable energy. In addition, when the cell has migrated out of the circulation, it functions in an environment relatively deficient in oxygen and glucose. Therefore it requires a large store of glucose which can be rapidly released, even under anaerobic conditions.

Using the 'skin window' technique Wulff (1962) observed changes in the glycogen content of inflammatory neutrophils and macrophages. In the skin lesions some of the first neutrophils to appear contained less glycogen than the neutrophils in the blood and the finding seemed consistent with an increased energy requirement during the migratory process. In later preparations leucocytes showed an increased PAS staining reaction, similar to that observed in leucocytes from acute infections (Wachstein, 1955). An increasing glycogen content was also observed in macrophages. Phosphorylase activity was demonstrated by Wulff (1962) in both neutrophils and macrophages.

A functional situation similar to that of leucocytes in inflammatory exudates occurs when normal lymphocytes are incubated in culture with PHA and are transformed into 'blast-like' cells. The initial rise in glycogen and its subsequent decrease may be interpreted as the storing and utilisation of energy for the mitotic process and as an aid for the function of lysosomes which appear simultaneously with the onset of mitoses (see also Chapter 12).

Finally the platelet provides another example of a cell type which circulates without great energy needs until the brief period when it performs a specific function in haemostasis. Once in a haemostatic plug or fibrin clot, the platelet must continue to support such energy-requiring functions as clot retraction even though blood glucose may be less readily available.

Thus, it is necessary to maintain large glucose reserves in the form of glycogen.

In this chapter we have given an account of the various cytochemical techniques used for the demonstration of glycogen in haemic cells, its distribution in the principal cell types and some of the changes which take place in certain physiological and experimental conditions.

Only a brief mention has been made of the alterations in PAS staining reactivity present in some pathological conditions, since a more extensive and detailed review of the glycogen changes present in pathological erythropoiesis and leucopoiesis will be given in Chapters 13, 14 and 15, together with other major cytochemical findings.

Cytochemical demonstration of glycoconjugates

Carbohydrates other than glycogen have not been very widely studied cytochemically in haemic cells, but the presence of acid mucosubstances in certain leucocyte granules is of considerable interest. These anionic glycoconjugates or complex carbohydrates contain carboxyl or sulphate groups linked to glycoprotein. They have been found in neutrophil granules by light microscope histochemistry (Horn and Spicer, 1964; Dunn and Spicer, 1969) using dialysed iron staining, and identified at EM level as sulphated compounds by the high iron diamine method of Spicer (1965). Ultrastructural cytochemistry (Parmley *et al.*, 1976) has further localised the positivity to primary granules and suggested that the function of the sulphated complex carbohydrate may relate to storage of acid hydrolases. Parmley *et al.* (1978) overcame the fixative difficulties which had previously prevented the demonstration of acid mucosubstance in monocyte granules at EM level by using brief fixation in a weak glutaraldehyde-formalin mixture. Exposure of specimens fixed in this way to dialyzed iron solution or to a high iron diamine solution showed that some 10% of monocyte granules from normal blood reacted positively with the former, but the granules of leukaemic monocytes were more often positive, with over 90% of the granules in two of seven cases studied reacting intensely. The monocyte granule reaction differed from that in neutrophils by failing to stain with the high iron diamine technique, presumably reflecting a lack of sulphate esters, or possibly their masking or extraction during processing. The same group of authors (Parmley *et al.*, 1979; Parmley *et al.*, 1980) subsequently developed a periodic acid-thiocarbohydrazide-silver proteinate (PA-TCH-SP) method for the demonstration at the ultrastructural level of glycol-containing hexoses in glycoconjugates. The authors claim that the PA-TCH-SP procedure has the same cytochemical significance in visualizing complex carbohydrates with vicinal glycol groups as the PAS method.

Eguchi *et al.*, (1986) have recently applied the PA-TCH-SP method to the ultrastructural study of periodate-reactive glycoconjugates in human leukaemic cells. The PA-TCH-SP technique stained cytoplasmic granules, Auer bodies, Golgi cisternae and glycogen.

Blast cells in acute lymphoblastic leukaemia (ALL) without T or B cell markers showed a variable distribution of positive cytoplasmic granules, and could be classified into four groups according to PA-TCH-SP staining pattern. In the first group cells exhibited clustering of positive granules in the notch of the nucleus or near the Golgi zone. In the second group, cells revealed single granules scattered throughout the cytoplasm and not clustered. In the third group, cells showed abnormal granules, possibly related to the presence of lysosomes incorporating glycogen, whereas, in the last group, cells displayed large granules, more than 0.5 μm in diameter, scattered in the cytoplasm, separate from the other granules. It is not known whether differences exist in the origin and biologic nature of the four types of cytoplasmic granules, identified according to the pattern of PA-TCH-SP positivity. In addition to morphological differences between the cell types, the carbohydrate content of clustered cytoplasmic granules appeared to differ from that of scattered cytoplasmic granules, since the dialyzed iron reaction for localizing acid mucosubstances was positive in scattered cytoplasmic granules in contrast to the lack of dialyzed iron reactivity in clustered cytoplasmic granules. This difference, in the opinion of the authors, could reflect a difference in origin and functional activity. Glycogen in the cytoplasm of ALL cells showed two different patterns of distribution. In one type, the small glycogen particles measuring 10 nm were scattered evenly in the cytoplasm. In the other type, glycogen particles formed large clusters. All cases lacking T and B markers could thus be classified into 3 groups, one group containing clustered glycogen, the second group containing scattered glycogen and the third group lacking glycogen. The first distribution pattern, consisting of clusters of glycogen, appeared to correspond to the well-known clustered PAS positivity in ALL. However, these latter aggregates with strong PAS staining may not be the same as the clustered cytoplasmic granules with PA-TCH-SP positivity, because cells in some cases of ALL lacked glycogen and contained clustered cytoplasmic granules with only faint clustered PAS reactivity. Thus the PAS reactivity of ALL cells, when compared to PA-TCH-SP staining for electron microscopy, represents different reactive intracellular sites.

To determine whether PAS reactivity was correlated to the prognosis of ALL in children, the authors followed up patients with non-B, non-T ALL for two or more years and correlated response to treatment with PAS reactivity. All patients having abnormal granules with glycogenoid substances in the granule had a poor prognosis, while those lacking glycogen and those containing clustered glycogen tended to be responsive to treatment. Leukaemic T cells differed from normal T cells in that they contained fewer clustered cytoplasmic granules with PA-TCH-SP reactivity.

Leukaemic B cells in Burkitt type ALL contained moderate amounts of glycogen scattered in the cytoplasm as normal B cells. The PA-TCH-SP positive clustered granules were more frequent in leukaemic than in normal cells.

In acute myeloid leukaemia and acute promyelocytic leukaemia cytoplasmic granules were PA-TCH-SP positive and thus resembled immature primary granules in early granulocyte precursors, except that the staining persisted in granules of peripheral blood leukaemic cells, indicating that leukaemic cells and their primary granules do not undergo normal maturation (Parmley *et al.*, 1980), which involves loss of granular PA-TCH-SP reactivity. Auer bodies exhibited striated PA-TCH-SP positivity in the matrix of the body. In acute promyelocytic leukaemia PA-TCH-SP reactivity was observed in the cisternae of the rough endoplasmic reticulum in three out of four cases, a finding not reported in other cases of leukaemia. This positively stained material may be related to increased synthesis of glycoprotein, which is the main source of PA-TCH-SP reactive substances, except for glycogen (Parmley *et al.*, 1979). Cells in acute monocytic leukaemia showed PA-TCH-SP reactivity in small granules. Golgi cisternae in these cells and those of acute myeloblastic leukaemia also reacted positively. In erythroleukaemia PA-TCH-SP positive glycogen particles appeared to be increased in size and number.

Granulocytes of chronic myeloid leukaemia resembled normal granulocytes in PA-TCH-SP reactivity, while distinctive staining was observed in blastic crisis. Small leukaemic megakaryocytes in a megakaryocytic crisis of chronic myeloid leukaemia exhibited PA-TCH-SP reactivity similar to that of normal megakaryocytes, characterized by positivity of the cell surface and demarcation membranes, as well as by an eccentric focus of reactivity within the alpha granules.

TECHNICAL METHODS

PAS Reaction

Reagents

1. Schiff's reagent
2. Carazzi's haematoxylin

} prepared as described on p. 57

Technique

1. Fix air-dried smears of blood or bone marrow by exposure to formalin vapour for 5 minutes
2. Wash in running tap water for 15 minutes
3. Treat with 1% periodic acid solution for 10 minutes
4. Immerse in Schiff's leucobasic fuchsin solution for 30 minutes
5. Rinse in several changes of SO_2 water for 3 minutes
6. Wash in distilled water for 10 minutes
7. Counterstain with aqueous haematoxylin for 10 minutes, and blue in tap water for 5 minutes.

Remarks

Control smears are exposed to salivary amylase digestion for 30 minutes before treatment with periodic acid.

Scoring methods

The intensity of the PAS reaction in lymphocytes and primitive cells may be evaluated semiquantitatively according to the following individual cell ratings:

0 cells with cytoplasm devoid of PAS-positive granules.
\+ cells with scattered granules or one annular ring of PAS-positive granules.
\+ + cells with two concentric annular rings of PAS-positive granules in the cytoplasm.
\+ + + cells with three or more concentric annular rings of granules or with heavy clumps of PAS-positive material.

The sum of 100 consecutive individual cell ratings gives a total score, with a possible range of 0–300.

In erythroblasts PAS positivity may be assessed semiquantitatively according to the following criteria:

0 erythroblasts with no positive reaction.
\+ erythroblasts with a few scattered granules or a pale reddish tinge in the cytoplasm.
\+ + erythroblasts with one or two concentric rings of granules or a moderately deep diffuse red tinge in the cytoplasm.
\+ + + erythroblasts with a massive concentration of granules or a very deep red tinge, comparable in intensity to the staining observed in polymorphs.

The sum of 100 consecutive cell ratings provides the score for each specimen, with a possible variation from 0 to 300.

REFERENCES

Astaldi G, Verga L 1957 The glycogen content of the cells of lymphatic leukaemia. Acta Haematologica 17: 129

Astaldi G, Rondanelli E G, Bernardelli E, Strosselli E 1954 An abnormal substance present in the erythroblasts of Thalassaemia major. Cytochemical investigations. Acta Haematologica 12: 145

Bauer H 1933 Mikroskopisch-chemischer Nachweis von Glykogen und einigen anderen Polysacchariden. Zeitschrift für Mikroskopich-Anatomische Forschung 33: 143

Bessis M 1968 Ultrastructure of normal and leukemic granulocytes. In: Zarafonetis CJD (ed) Proc. Int. Conf. on Leukaemia-Lymphoma. Lea & Febiger, Philadelphia , p.281

Bitter M A, Le Beau M M, Larson R A, Rosner M C, Golomb H M, Rowley J D, Vardiman J W 1984 A morphologic and cytochemical study of acute myelomonocytic leukemia with abnormal marrow eosinophils associated with inv(16)(p13q22). American Journal of Clinical Pathology 81: 733

Daniell H W 1959 Studies of megakaryocyte glycogen. I. A semiquantitative method of measurement; effect of phlebotomies in young adults. Blood 14: 60

Dunn WB, Spicer SS 1969 Histochemical demonstration of sulfated mucosubstances and cationic proteins in human granulocytes and platelets. Journal of Histochemistry and Cytochemistry 17: 668

Efrati P, Yaari A 1980 Evaluation of the periodic-acid-Schiff (PAS) reaction in mature neutrophil granulocytes from peripheral blood by a scanning microdensitometer (M-85, Vickers). Haematologica 65: 168

Eguchi M, Furukawa T, Sakakibara H, Ishikawa K, Sugita K, Sakamaki H, Spicer SS 1986

Ultrastructural evaluation of periodate-reactive glycoconjugates in human leukaemia cells. Scandinavian Journal of Haematology 36: 385
Fessas P, Papayannopoulou T 1965 Cytochemical observations on beta-Thalassaemia. I. The PAS positive substance of erythroblasts. Acta Haematologica 34: 1
Gahrton G 1966a Normal human neutrophil leucocytes as a reference system for the microspectrophotometrically quantitated periodic acid-Schiff reaction. Journal of Histochemistry and Cytochemistry 14: 45
Gahrton G 1966b The periodic acid-Schiff reaction in neutrophil leukocytes in untreated and myleran-treated chronic myelocytic leukaemia. Blood 28: 544
Gibb R P, Stowell R E 1949 Glycogen in human blood cells. Blood 4: 569
Gomori G 1946 A new histochemical test for glycogen and mucin. American Journal of Clinical Pathology 10: 177
Hayhoe F G J, Quaglino D 1958 Cytochemical demonstration and measurement of leukocyte alkaline phosphatase activity in normal and pathological states by a modified azo-dye coupling technique. British Journal of Haematology 4: 375
Headington J T 1966 Phosphorylase activity in megakaryocytes. Nature 209: 928
Heckner F 1956 Zytochemische Darstellung der Polysaccharide in den Zellen des Blutes und der blutbildenden Gewebe. Acta Haematologica 16: 1
Higgy K E, Burns G F, Hayhoe F G J 1977 Discrimination of B, T and Null lymphocytes by esterase cytochemistry. Scandinavian Journal of Haematology 18: 437
Horn R G, Spicer S S 1964 Sulfated mucopolysaccharide and basic protein in certain granules of rabbit leukocytes. Journal of Laboratory Investigation 13: 1
Hotchkiss RD 1948 A microchemical reaction resulting in the staining of polysaccharide structures in fixed tissue preparations. Archives of Biochemistry 16: 131
Inagaki S 1968 Cytochemical reactions of basophil leucocytes. VII. Periodic acid-Schiff (PAS) reaction. Acta Haematologica Japonica 31: 169
Jamra M, Lorenzi TF 1961 Megakaryocytic formula. Morphological and cytochemical study of the megakaryocytes in thrombopenic purpura. Revisto do Hospital das Clinicas 16: 274
Kakefuda T 1968 Electron microscopy of normal and leukemic cells. In: Amromin G D Pathology of Leukaemia. Harper & Row, New York, p.182
Kaung D T 1969 Periodic acid-Schiff reaction in human basophilic leucocytes. Acta Haematologica 42: 269
Linke A, Löffler H 1963 Über die alkalische Leukozyten-phosphatase während der Therapie der reifzelligen myeloischen Leukämie mit Trenimon. In: Merker H (ed) Zyto-und Histochemie in der Hämatologie, Neuntes Freiburger Symposium. Springer, Berlin, p.306
McManus J F A 1946 Histological demonstration of mucin after periodic acid. Nature 158: 202
Matioli G, Del Piano E 1961 Su due casi di talassemia minima a monoespressività genica. Haematologica 46: 83
Mitus W J, Bergna L J, Mednicoff I B, Dameshek W 1958 Cytochemical studies of glycogen content of lymphocytes in lymphocytic proliferations. Blood 13: 748
Neukirch P 1910 Über die jodophile Substanz der Leukocyten und ihr Verhalten zur bestchen Färbung. Zeitschrift für Klinische Medizin 70: 251
Parmley R T, Ogawa M, Spicer S S, Wright N J 1976 Ultrastructure and cytochemistry of bone marrow granulocytes in culture. Experimental Haematology 4: 75
Parmley R T, Spicer S S, O'Dell RF 1978 Ultrastructural identification of acid complex carbohydrate in cytoplasmic granules of normal and leukaemic human monocytes. British Journal of Haematology 39: 33
Parmley R T, Eguchi M, Spicer S S 1979 Ultrastructural cytochemistry of complex carbohydrates in leukocyte granules. Journal of Histochemistry and Cytochemistry 27: 1167
Parmley R T, Eguchi M, Spicer S S, Alavarez C J, Austin R L 1980 Ultrastructural cytochemistry and radioautography of complex carbohydrates in heterophil granulocytes from rabbit bone marrow. Journal of Histochemistry and Cytochemistry 28: 1067
Parwaresch M R 1976 The Human Blood Basophil. Morphology, origin, kinetics, function and pathology. Springer-Verlag, Berlin
Pedersen B, Hayhoe F G J 1971a Cellular changes in chronic myeloid leukaemia. British Journal of Haematology 21: 251
Pedersen B, Hayhoe F G J 1971b Relation between phagocytic activity and alkaline phosphatase in chronic myeloid leukaemia. British Journal of Haematology 21: 257
Quaglino D, Hayhoe F G J 1959 Observations on the periodic acid-Schiff reaction in lymphoproliferative diseases. Journal of Pathology and Bacteriology 78: 521
Quaglino D, Cowling D C, Hayhoe F G J 1964 Cytochemical and autoradiographic studies on normal and leukaemic cells in short-term tissue cultures. British Journal of Haematology 10: 417
Sansone G, Bertolotti E 1958 Ricerche citochimiche sugli eritroblasti di prematuri anemici. Atti 7

Congr. Soc. Int. Emat., Vol. 2. II Pensiero Scientifico, Roma, p 1586
Schmitz N, Gödde-Salz E, Gassmann W, Löffler H 1984 Acute myelomonocytic leukemia with involvement of eosinophils and inversion of chromosome 16. Blut 48: 263
Scott R B 1968a The role of glycogen in blood cells. New England Journal of Medicine 278: 1436
Scott R B 1968b Glycogen in human peripheral blood leukocytes. I. Characteristics of the synthesis and turnover of glycogen in vitro. Journal of Clinical Investigation 47: 344.
Scott R B, Still W J 1968 Glycogen in human peripheral blood leukocytes. II. The macromolecular state of leukocyte glycogen. Journal of Clinical Investigation 47: 353
Shammas F V, Engeset A 1986 Glycogen content and PAS staining pattern of human megakaryocytes. Scandinavian Journal of Hematology 37: 237
Spicer S S 1965 Diamine methods for differentiating mucosubstances histochemically. Journal of Histochemistry and Cytochemistry 13: 211
Stegagno G A, Pollitzer C 1953 Ricerche citochimiche sui polisaccaridi e lipidi delle cellule ematiche e midollari di bambini affetti da Talassemia major morbo di Cooley. Archivio Italiano di Pediatria e Puericoltura 16: 247
Storti E, Zubiani M 1949 Ricerche di istochimica ematologica. Il glicogeno nelle cellule ematiche normali. Bollettino della Societa di Medicina e Chirurgia di Pavia 63: 3
Storti E, Perugini S, Soldati M 1953a Le modificazioni del contenuto polisaccaridico delle cellule ematiche in alcune emopatie e malattie infettive. (Ricerche citochimiche). Publ. Chim. Biol. Med., I, 475, C. Erba: Milano.
Storti E, Perugini S, Soldati M 1953b Cytochemical investigations of normal megakaryocytes and platelets. Acta Haematologica 10: 144
Swirsky D M, Li Y S, Matthews J G, Flemans R J, Rees J K H, Hayhoe F G J 1984 8: 21 translocation in acute granulocytic leukaemia: cytological, cytochemical and clinical features. British Journal of Haematology 56: 199
Valentine W N, Follette J H, Lawrence J S 1953 The glycogen content of human leukocytes in health and in various disease states. Journal of Clinical Investigation 32: 251
Vanderwende C, Miller W L, Gaiss S A 1964 Glycogen metabolism of normal and leukaemic leukocytes. Life Sciences 3: 223
Wachstein M 1955 Histochemistry of leukocytes. Annals of the New York Academy of Sciences 59: 1052
Wagner R 1946 The estimation of glycogen in whole blood and white blood cells. Archives of Biochemistry 11: 249
Wagner R, Reinstein J E 1950 Enzyme studies on white blood cells. II. Phosphorylating glycogenolysis, etc. Archives of Biochemistry 29: 260
Wagner R, Yourke A 1952 Enzyme studies on white blood cells. III. Phosphorylating glycogenolysis and phosphorylated intermediates. Archives of Biochemistry 39: 174
Wislocki G B, Dempsey E W 1946 Observations on the chemical cytology of normal blood and hemopoietic tissues. Anatomical Record 96: 249
Wislocki G B, Rheingold J J, Dempsey E W 1949 The occurrence of the periodic acid-Schiff reaction in various normal cells of blood and connective tissue. Blood 4: 582
Wulff H R 1962 Histochemical studies of leukocytes from an inflammatory exudate. Glycogen and phosphorylase. Acta Haematologica 28: 86
Xefteris E, Mitus W J, Mednicoff J B, Dameshek W 1961 Leukocytic alkaline phosphatase in busulphan-induced remissions of chronic granulocytic leukemia. Blood 18: 202
Yataganas X, Gahrton G, Fessas P H, Kesse-Elias M, Thorell B 1973 Proliferative activity and glycogen accumulation of erythroblasts in β-thalassaemia. British Journal of Haematology 24: 651

CHAPTER 5

Lipids and sudanophilia

The chemistry of lipids is complex. A wide range of substances share the lipid characteristics of relative insolubility in water but solubility in chloroform, ether and benzene, and the incorporation of fatty acids in their structure. A widely used classification, proposed by Bloor (1943) separates three broad groups.

Simple lipids: esters of fatty acids with various alcohols. This group includes the 'neutral fats', composed of glycerol esters with saturated or unsaturated fatty acids, and the esters of the steroid alcohol cholesterol.

Compound lipids: esters of fatty acids with alcohols (usually glycerol) containing other additional groups

1. Phospholipids—the additional groups include a phosphoric acid residue and usually nitrogenous bases such as choline. Important members of the phospholipids include phosphatidyl choline, phosphatidyl ethanolamine, phosphatidyl serine and the sphingomyelins.
2. Glycolipids (Cerebrosides)—the extra groups include carbohydrate, but no phosphoric acid.
3. Sulpholipids and aminolipids.

Derived lipids: substances derived from simple or compound lipids by hydrolysis.

Lipids are commonly associated with protein in cells, and the term 'lipoprotein' is used for substances which generally possess characteristics of protein but from which lipids can be extracted. These lipoproteins occur in cell membranes, in mitochondria and other organelles, and play an important part in lipid transport in the plasma.

General principles of lipid cytochemistry

Cytologically, lipids may appear as microscopically obvious droplets, as in fat cells, anoxic hepatic cells and lipid-laden macrophages, but this appearance is uncommon in most cells of the haemopoietic series. Here the lipids are finely dispersed in droplets too small to be recognised under the light microscope without selective staining, or are physically or chemically masked in membranes or by combination with protein as lipoproteins. Cytochemical methods used to demonstrate the presence of lipids are generally thought to depend on the capacity of lipophilic dyes to leave their solvent and concentrate preferentially in the intracellular lipid. Although there is some evidence (see below) that chemical binding may contribute to sudanophilia, especially of polymorphs, partition is still regarded as the main *modus operandi* of staining with the range of Sudan dyes, the Oil Reds,

Scharlach R and the fluorescent dyes 3,4 benzpyrene and Rhodamine B. Aqueous solutions of Nile blue contain an oxidation product, Nile red, which similarly partitions into lipids and colours them red, but there is also a more directly chemical component of the Nile blue reaction, in that both the blue cation and an imino-base present in the dye link with acidic groups to produce a blue colour, which indicates the presence of fatty acids or phospholipids when found in sites giving a general lipid stain. A further chemical method, claimed to be specific for phospholipids, is Baker's acid haematin test (Baker, 1946; Chayen, 1968).

Lipids are present in many blood cells and their precursors, but their cytochemical demonstration requires the use of lipophilic dyes capable of penetrating lipoprotein complexes or other forms of masking, or the preliminary use of procedures which produce some degree of unmasking of lipid molecules. As Chayen (1968) has pointed out, the degree to which lipids are bound to masking substances may vary considerably in different physiological and pathological circumstances, and may be influenced by drugs, hormones and other agents, so that apparently erratic results obtained with a given staining method may reflect different degrees of lipid binding, rather than an inherent variability in the method.

Staining of haemic cells by Sudan black

The early cytochemical studies of leucocytes with the lipophilic dyes Sudan III and Scharlach R and with Nile blue sulphate showed positive reactions in neutrophil polymorphs (Savini, 1921; Sehrt, 1927; Bacsich, 1936), but the results were variable and inconsistent, and it was not until some years after Sudan black B had been introduced by Lison (1934, 1936) that reliably repeatable staining patterns were described. Sheehan (1939) first used Sudan black B to stain leucocytes in air-dried blood films fixed in methyl alcohol for 30 seconds and then exposed to a saturated solution of the dye in 70% alcohol for 30 minutes. He used an eosin-methylene blue counterstain to aid in cell identification and briefly reported that neutrophil polymorphs were filled with deeply-stained granules, eosinophil granules appeared to have only a surface layer of sudanophilia, transitional cells showed variable granular content, while large and small lymphocytes were always quite free from stainable material. Myelocytes showed many granules, myeloblasts usually had a small number of obvious granules but were sometimes clear; lymphoblasts showed no granules at all. Similar findings were reported by McManus (1945) and by Wislocki and Dempsey (1946), who studied the blood cells of monkeys and noted that monocytes contained a few black granules and that a very finely granular positivity could be detected in megakaryocytes and platelets. Further studies on leucocyte granules after staining wth Sudan black B and May-Grünwald-Giemsa were reported by Bailiff and Kimbrough (1947), who were able to confirm the results of McManus: they stated that monocytes, lymphocytes and basophil leucocytes showed no sudanophilia.

Sheehan and Storey (1947) described an improved method of staining leucocytes with Sudan black B, in which the usual methyl alcohol fixation was replaced by formalin vapour fixation, and the Sudan black solution was maintained in a neutral or slightly alkaline pH with a small proportion of phenol added to act as a mordant.

Rheingold and Wislocki (1948) made detailed observations on the sudanophilia of human peripheral blood and bone marrow cells. They confirmed previous findings in neutrophil and eosinophil granulocytes but found also that basophil granules were positive, though ranging in intensity from pale grey to deep black. Since previous reports had differed over the sudanophilia of monocytes, Rheingold and Wislocki studied the blood of patients with high monocyte counts in which these cells were readily identifiable. Variable cytoplasmic sudanophilia was found, ranging from a few faint dots up to evident particulate dusting of the cytoplasm. The use of a counterstain tended to obscure this positivity and monocytes were studied after staining with Sudan black alone. Lymphocytes contained no sudanophilic material, nor was there any granular positivity in megakaryocytes and platelets in human bone marrow and blood.

Bloom and Wislocki (1950) observed the reactions of blood and bone marrow cells by the acid haematin method of Baker, and compared the results with those seen after staining with Sudan black B. They concluded that Baker's method gave more satisfactory staining for the finer particles, probably phospholipid mitochondrial components, present in myeloblasts, lymphoblasts, megakaryocytes and monocytes, but noted that these particles could also be seen, although less distinctly, with Sudan black B. This increased sensitivity over earlier results with Sudan black was explained as being due to the use of formalin calcium fixation instead of methyl alcohol fixative, and to the avoidance of any obscuring counterstain.

Storti and Perugini (1951) first drew attention to the value of Sudan black B staining in the differentiation of primitive haemic cells in normal and pathological states. They used a Giemsa counterstain to assist in the identification of cells. In normal material they found all cells of the erythroid and lymphoid series, plasma cells, megakaryocytes, platelets, most basophil leucocytes and most marrow macrophages or reticulo-endothelial cells to be completely sudanophobic. The granular positivity of neutrophil and eosinophil cells from the promyelocyte stage onwards was confirmed, and the finer and more scattered type of positivity characteristic of monocytes was emphasised. Primitive cells without evidence of maturation in Romanowsky-stained preparations, both in normal marrows and in acute leukaemias, could often be differentiated by Sudan black B staining, since the typical pattern of positivity characteristic of granulocytes or monocytes could sometimes already be discerned.

Nature of sudanophilia and its relation to peroxidase activity

Although Sheehan and Storey (1947) had observed that the sudanophilic granules of leucocytes were resistant to decolorization with 100% alcohol or xylene, the Sudan black stain was generally regarded as specific for free or combined lipids, but this belief was challenged by Lillie and Burtner (1953), who found that Sudan-stained granules of polymorphs resisted dye extraction not only with alcohol and xylene, but also with acetone, chloroform, carbon tetrachloride, mineral oil and other dye solvents. If sudanophilia depended on a simple physical solution of the dye in the lipid, then the dye should be extractable by an excess of solvent, and this did not occur. Lillie and Burtner also reported that esterification of the naphthol moiety of Sudan dyes blocked staining of leucocytes although ordinary lipids such as myelin, ceroid, cardiac lipofuscin, retinal lipids and adrenal fats were still readily stained. They therefore argued that the sudanophilia of leucocytes resulted from a chemical combination of the dyes with non-lipid cytoplasmic constituents. Nevertheless, Perugini (1954) showed that sudanophilic granules of polymorphs also gave positive reactions with methods dependent on the solution of fluorescent material in lipids, including the benzpyrene-caffeine method of Berg (1951), and it seemed most likely that the primary localisation of all these dyes was in phospholipid or lipoprotein complexes, although the relative roles of physical partition and chemical combination were uncertain. Lillie and Burtner coined the term 'stable sudanophilia' to differentiate the reaction of leucocyte granules from the 'unstable' or readily reversible positivity of acceptable lipids, and investigated the relation of stable sudanophilia to benzidine peroxidase and indophenol oxidase staining because of the obvious similarities in distribution of positivity to all three reactions in leucocytes. Their behaviour in response to heat and various enzyme inhibitors suggested that these two enzymes were probably identical, but differed from the sudanophil substance of leucocytes. Terner *et al.* (1963), however, claimed from experiments with blocking reactions such as methylation and deamination and the use of cyanide, fluoride and other enzyme inhibitors that Sudan black B staining of polymorph granules was an enzymatically dependent process, related to peroxidase activity, and that other components of Sudan black B positivity in leucocytes might result from the affinity of the two amino groups of the dye for acid groupings such as those of the acid-bearing phospholipids. In their view, neither reaction was specific for lipids. Schaefer and Fischer (1970) elaborated the peroxidase link, showing that after addition of horse-radish peroxidase to Sudan black B solutions the reaction in granulocytopoietic cells in smears was completely inhibited, whereas stainability of the lipids in fat cells and macrophages in the bone marrow remained unchanged. They suggested that in combination with phenols, as used in the revised method of Sheehan and Storey, and traces of peroxides originating in the staining solution, Sudan black B might serve as a substrate for endogenous myeloperoxidases. Inhibition of granule staining by the addition of horse-radish peroxidase was attributed to substrate consumption.

These arguments have not been entirely resolved by attempted correlations between patterns of sudanophilia and either the biochemically demonstrable distribution of lipids (Gottfried, 1972) or the ultrastructural localization of peroxidases (Cawley and Hayhoe, 1973) in leucocytes. Normal lymphocytes and polymorphs contain some 5% of their wet weight as lipid, about a third of this being phospholipid, a third neutral lipid (chiefly triglyceride and free cholesterol) and about a sixth glycolipid. The cholesterol to phosphatidyl choline ratio is about 1:4 for mature polymorphs and lymphocytes. Leukaemic cells and cultured leucocyte lines have about 40% less lipid, the loss being mostly of neutral lipid, with cholesterol to phosphatidyl choline ratio of 1.0 or less and as low as 0.5 to 0.7 in acute leukaemic blast cells. Although there are characteristic differences between lymphocytes and polymorphs in the detailed patterns of lipid components, the overall lipid content is sufficiently similar to suggest that, in both cell types, the lipid molecules play roughly similar roles, probably chiefly in the constitution of structural membranes. While some degree of masking undoubtedly exists, there seems no obvious reason why this should differ so markedly between lipids of lymphocytes and polymorphs as to render the first inaccessible to Sudan black staining and the second readily accessible, if the dye were simply specific for lipids. Staining differences would be explicable, however, if Sudan black operated through one of the postulated biochemical mechanisms, whether enzymatic or otherwise, if the target substance were absent from cells of the lymphocyte series. Identification of the target as peroxidase seems less certain. Ultrastructural demonstration of peroxidase by the method of Graham and Karnovsky (1966) suggests that myeloperoxidase is contained in primary or azurophil granules and not in specific neutrophil granules. The difficulties in reconciling this distribution with the progressive increase in peroxidase positivity with increasing cell maturity seen in light microscopic cytochemistry, and the relevance of ultrastructural cytochemistry by the Novikoff and Goldfischer technique (1968, 1969), which we have found to show peroxidase positivity not only in primary but also in secondary, though not tertiary, specific granules (Cawley and Hayhoe, 1973), are discussed in Chapter 10. It is appropriate to point out here that considerable doubts exist over the specificity of benzidine peroxidase staining methods. To link stable sudanophilia with peroxidase positivity is therefore coupling two reactions of uncertain specificity. Some obvious differences in the patterns of sudanophilia and benzidine peroxidase positivity exist: sudanophilia is usually more strongly manifest in leukaemic myeloblasts and shows a different distribution from peroxidase positivity in monocytes and monoblasts. Nevertheless, close parallels in positivity of Auer rods, azurophil granules and the granules of later neutrophil and eosinophil cells exists.

Comparison of Sudan black and Oil Red O positivity in haemic cells

Many of the experimental results contravening the supposition that Sudan staining resulted from a simple physical solution of the dye in lipids apply to Sudan IV and Oil Red O, and in the case of Oil Red O the positive granularity in polymorphs is extremely fine in character after brief exposure and becomes coarser on prolonged staining, rather than showing a constant granule size with progressively deepening staining as might be expected if absorption of dye into a preformed lipid structure was involved (Lillie and Burtner, 1953). But Oil Red O stains cells of the lymphocytic series more avidly than granulocytes. Bennett and Dutcher (1969) even found it helpful in differentiating leukaemic lymphoblasts from myeloid precursors, claiming that blast cells in acute myeloid leukaemias were devoid of staining while those of acute lymphoblastic leukaemias mostly showed clear globular positivity with Oil Red O. Although Shaw and Klemp (1974) found the reaction less valuable in separating the forms of acute leukaemias, they confirmed that in lymphoblastic cases positivity tended to be more globular than in myeloblastic cases, where a more finely granular positivity occurred. If this reaction, like that with Sudan black B, is chemically based and not a consequence of solubility partition, then the target substance giving localization cannot be peroxidase in both cell types. Perhaps, with this dye, both mechanisms operate, and the globular positivity of lymphoid cells results from physical partition while the granular positivity of myeloid cells is a consequence of chemical reaction.

In the case of Sudan black B, the partition mechanism seems to be inoperative in the main haemic cell lines, where the dual chemical mechanisms operate as suggested by Terner *et al.* (1963), the one possibly, but by no means certainly, linked with enzymes of the peroxidase group, and the other resulting from the acidophil nature of the amino groups of the dye.

Sudanophilia of normal haemic cells
(Plate 6A, B, C, D and Figs 5.1 to 5.5)

Whatever the exact nature of the reaction, the distribution of sudanophilia in cells of the blood and bone marrow in normal and pathological states is now well established (Hayhoe, 1969). Normally, cells of the granulocytic series show increasing positivity with progressive maturity. Myeloblasts may be negative or possess a few small granules localised near the nucleus, probably in the region of the Golgi apparatus. This form of positivity becomes coarser and more extensive in promyelocytes, while myelocytes and metamyelocytes of the neutrophil series have increasingly numerous sudanophilic granules, apparently corresponding to the specific granules of Romanowsky-stained preparations although generally somewhat larger in diameter. Neutrophil polymorphs show very strong positivity, the granules filling the cytoplasm and often obscuring part of the nucleus. Eosinophil cells of all stages of maturity show a strongly positive reaction at the periphery of the specific granules with a negative or lightly-staining central core. The reaction in basophils is variable, some being negative and others having positivity related to the specific granules. Positivity in the basophil series tends to decrease with increasing cell maturity, most early basophil precursors and myelocytes being strongly positive. The granular reaction may be of a reddish metachromatic colour rather than black, especially in pathological states with basophilia (see Chapter 14) (Plate 8A).

There seems little doubt that the sudanophilic granules in these myeloid cells are identical with the specific granules. Their occurrence, number and distribution closely parallels that of primary and secondary specific granules in Romanowsky-stained smears, and the specific granules of a Leishman-stained preparation become obscured if, after short exposure to the Leishman stain, the slide is stained with Sudan black B. Again, Sudan black

A B C D E F

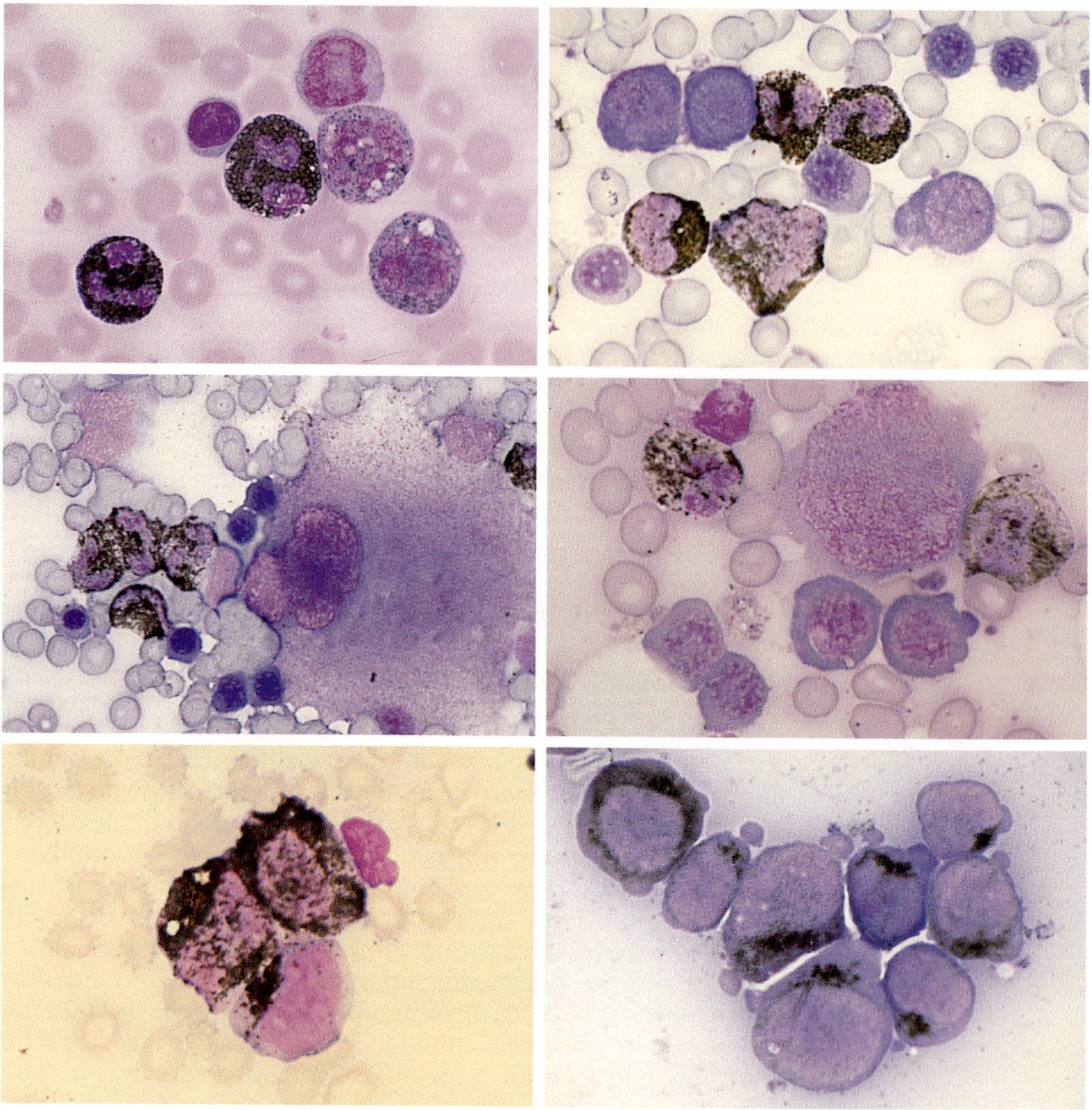

Plate 6

A. Normal buffy coat. SB stain. Two negative lymphocytes, 2 monocytes with scattered granules, a strongly positive neutrophil and an eosinophil.

B. Normal marrow. SB stain. Sudanophilia increasing from myelocyte to neutrophil polymorph; a proerythroblast and various later normoblast stages all negative.

C. Same preparation. Negative reaction in megakaryocyte; negative normoblasts; positive neutrophils.

D. Same preparation. Negative early megakaryocyte; negative proerythroblasts and early normoblasts; a negative lymphocyte; a monocyte with discretely scattered positive granules and a basophil with irregular coarse sudanophilia, not of the metachromatic pattern sometimes seen.

E. AML. SB stain. Blast cells showing strong localized and dense cytoplasmic positivity of granulocyte type.

F. AML, t(8;21) SB stain. Strong localized reaction with hollow staining Auer rod.

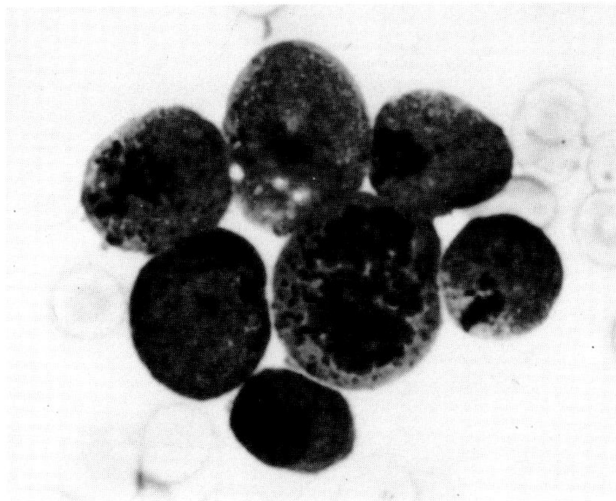

Fig. 5.1 Group of early granulocyte precursors from normal bone marrow. SB stain. Localized positivity in promyelocytes.

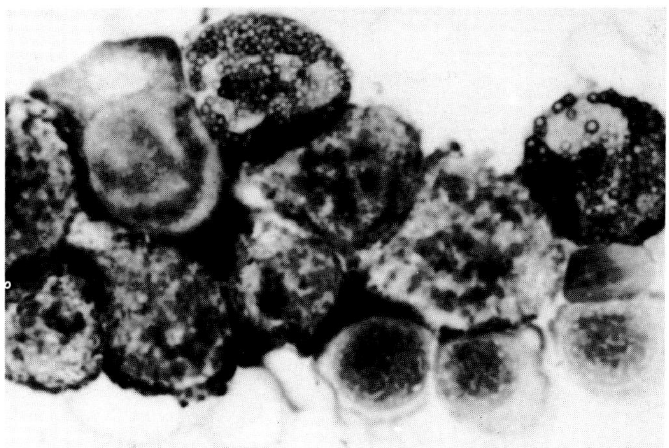

Fig. 5.2 Various myelocytes from normal bone marrow. SB stain. Neutrophil myelocytes, including one in mitosis, show moderate positivity: two eosinophils have the typical positive granules with negative centre: a proerythroblast and three normoblasts are negative.

preparations reveal no specific granules distinct from the sudanophilic granules when counterstained with Leishman's stain. Nevertheless there is a considerable discrepancy in size between sudanophilic and specific granules, the former being much larger and almost filling the mature neutrophil polymorph. This difference was discussed by Discombe (1946), who concluded that the apparent small size of specific granules is an artefact due to the absorption of one of the Leishman azures (probably Bernthsen's violet) on the surface of the granule, with the production of a discoid image of much smaller diameter than the granule itself when light from the microscope condenser is focussed by the granule.

Lymphocytes and their precursors are almost invariably negative, except for occasional very fine paranuclear rods or dots visible in lymphoblasts only

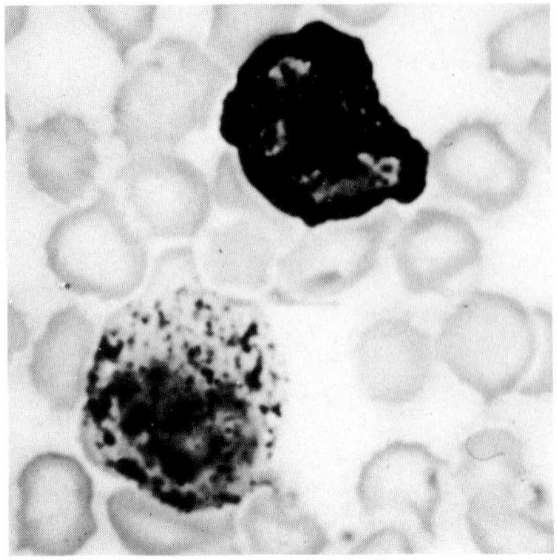

Fig. 5.3 Normal peripheral blood: SB stain. Very strong positivity, obscuring the nucleus and appearing almost confluent in a neutrophil polymorph: a contrasting monocyte shows discrete scattered sudanophilic granules.

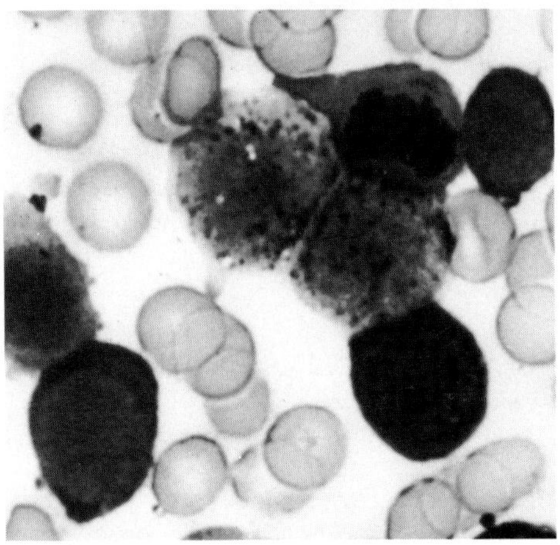

Fig. 5.4 Normal bone marrow: SB stain. Two monocytes with discrete scattered granules: the surrounding early normoblasts are negative, but show cytoplasmic basophilia.

in uncounterstained preparations, and probably representing mitochondrial phospholipids. Similar fine positivity may be seen in most primitive cells of all series but is obscured by counterstaining and can be neglected in the differentiation of primitive cells where coarser sudanophilia only is assessed. In normally counterstained preparations lymphoblasts usually appear quite negative.

Monocytes and their precursors may sometimes be quite negative, but frequently show variable numbers of fine or moderately coarse granules scattered discretely over the cell, with little tendency to concentration in a paranuclear zone or in the cytoplasmic rim as occurs in early myelocytes. This pattern of discretely scattered granules is very characteristic of cells of

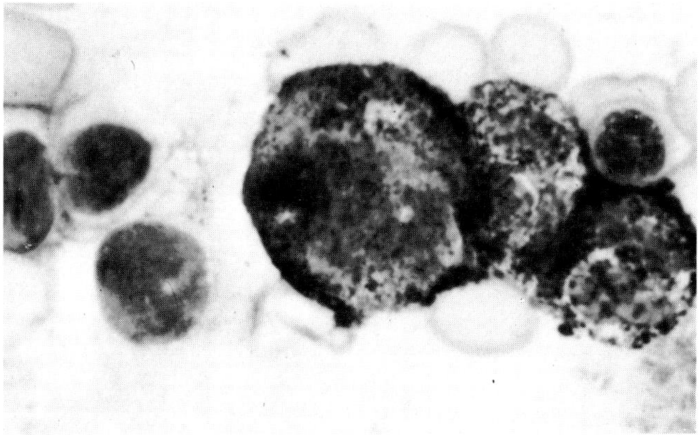

Fig. 5.5 Normal bone marrow. SB stain. Granular positive reaction in a neutrophil myelocyte and two polymorphs: a late normoblast and three lymphocytes are negative.

the monocyte series and has no counterpart in the distribution of peroxidase positivity in the cell series.

Individual platelets are usually negative as are megakaryocytes, although the latter may rarely show diffuse background staining with very fine sudanophilic granules scattered throughout the cytoplasm and over the nucleus. Proerythroblasts and later cells of the erythroid series, including reticulocytes and mature erythrocytes, never show granular positivity, apart from the very fine dots seen in primitive cells of the series, as in others, in uncounterstained preparations.

It is necessary to emphasise that the appearances described above and those to be outlined later in haemic cells involved in pathological states are those found after standard methods of preparation and staining with Sudan black B as described in the appendix on technical methods. More elaborate preparative techniques may reveal sudanophilia in other sites. Gaham and his associates (1974), for example, made chromosome preparations from HeLa cells and human lymphocyte cultures, treated them with 5% trichlor-acetic acid at 95°C for 15 minutes to unmask lipoproteins and remove nucleic acids, fixed with calcium formalin and carried out reactions with Baker's acid haematin method and with Sudan black B. Positive reactions with both stains gave cytochemical evidence interpreted as indicating the presence of phospholipids on human chromosomes. Without such special unmasking and extractive procedures neither nuclei nor chromosomes show sudanophilia.

Sudan black B reaction in haemopoietic disorders

In pathological states the Sudan black B reaction has particular value in differentiating immature cells in acute leukaemias, since, as Storti and Perugini (1951) first pointed out, cells which appear quite undifferentiated in smears stained by Romanowsky methods and have negative peroxidase reactions sometimes show a pattern of sudanophilia resembling that seen in normal promyelocytes and myelocytes or that seen in normal monocytes. Sudan black staining was used by Hayhoe *et al.* (1964) as one component of an extensive objective study designed to clarify the classification of acute leukaemias. Sudanophilia proved generally more helpful than peroxidase positivity in distinguishing varieties of the disease. Of 131 cases without a prominent erythraemic component, 25 showed some degree of localized positivity in a proportion of the primitive cells resembling that of the normal granulocyte series (Plate 6E), 54 showed a mixed pattern of sudanophilia, some cells having a granulocytic type of positivity and others the discrete scattered finely granular pattern typical of some normal monocytes (Plate 7A), 9 showed the monocytic pattern only (Plate 7B) and 43 were completely negative (Plate 7C, D). The proportion of primitive cells showing positivity in the first three groups varied widely, from almost all cells to very few of them, but these findings taken together with other observations in the study strongly suggested that the first group of 25 cases were acute myeloblastic or promyelocytic leukaemias, the second group of 54 were mixed mono-myeloblastic cases, the third group of 9 were more purely monoblastic, and the negative group of 43 were lymphoblastic. In the first two groups Auer rods could often be seen very conspicuously in Sudan black preparations (Plate 7A).

The presence of sudanophilic granules in leukaemic cells which showed no sign of azurophilic or specific granulation in Romanowsky preparations was clearly helpful in differentiating varieties of acute leukaemia. The authors suggested that leukaemic primitive cells tended to show a nuclear-cytoplasmic asynchronism in maturation, and that cytoplasmic sudanophilia of a kind specific to the cell line involved but normally found at a later stage of maturity was acquired while the nucleus remained still leptochromatic and nucleolated.

Among polymorphs in pathological states variations in sudanophilia appear to be confined chiefly to the myeloid leukaemias. A reduction in sudanophilia, with occasional negative polymorphs or polymorphs with negative areas in the cytoplasm (Fig. 5.6), has been noted in chronic myeloid leukaemia (Storti and Perugini, 1951; Hayhoe, 1953) and similarly negative polymorphs (Fig. 5.7), sometimes also showing absence of specific granules, negative peroxidase and negative alkaline phosphatase reaction were observed in a minority of cases of acute myeloid leukaemias (Hayhoe *et al.*, 1964), though not in acute lymphoblastic cases. The possibility certainly exists that these abnormal polymorphs may be derived from leukaemic precursors, even if the majority of mature polymorphs in acute leukaemias are more likely to have been derived from surviving normal precursors.

This hypothesis has recently been confirmed by Fearon and his associates (1986), who employed recombinant-DNA techniques and demonstrated that DNA probes are capable of detecting clonal populations of granulocytes in acute myeloid leukaemias and of providing evidence that in some patients leukaemic blast cells differentiate to form mature cells.

Anomalous findings of significant granular sudanophilia in cells of the lymphoid or erythroid series are uncommon, but in a cytochemical survey of

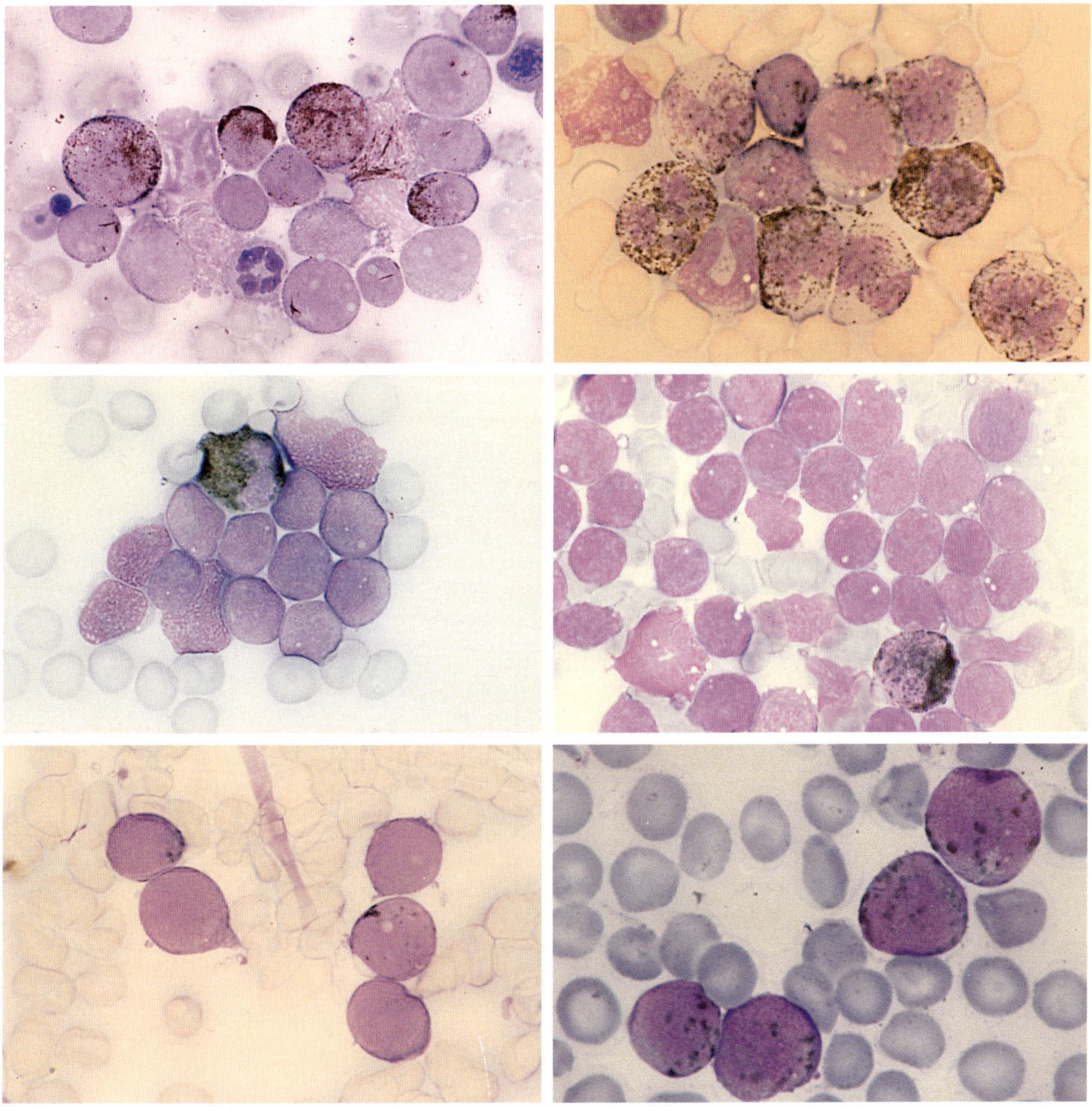

A B C D E F

Plate 7

A. AMML. SB stain. An example of mixed reactions, with both granulocytic and monocytic type of positivity present. Auer rods are conspicuously positive.

B. AMonL. SB stain. The monocytic pattern of discrete scattered granules is present in nearly all the leukaemic cells.

C. ALL. SB stain. A positive eosinophil and negative lymphoblasts of T-cell lineage.

D. ALL. SB stain. A single positive neutrophil contrasts with uniformly negative blasts of pre-B lineage.

E. ALL. SB stain. Rare example of coarse globular SB reaction in some 20% of lymphoblasts (cALL antigen positive).

F. ALL. SB stain. A second very rare example of similar sudanophilia in most cALL positive lymphoblasts.

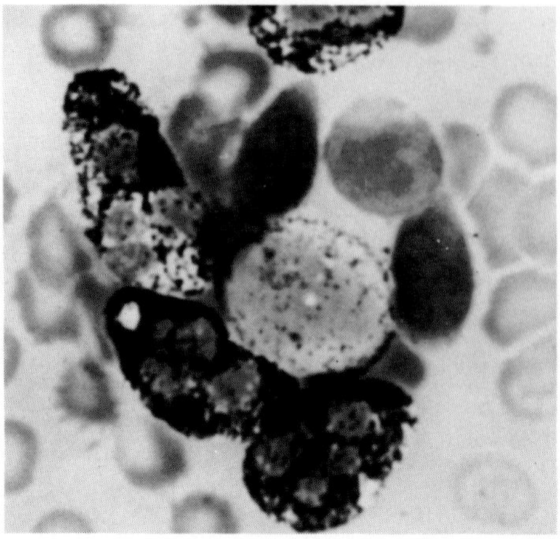

Fig. 5.6 CML bone marrow. SB stain. Three negative lymphocytes, a monocyte with discrete scattered granules and three neutrophil polymorphs all showing strong granular positivity, but with occasional negative areas.

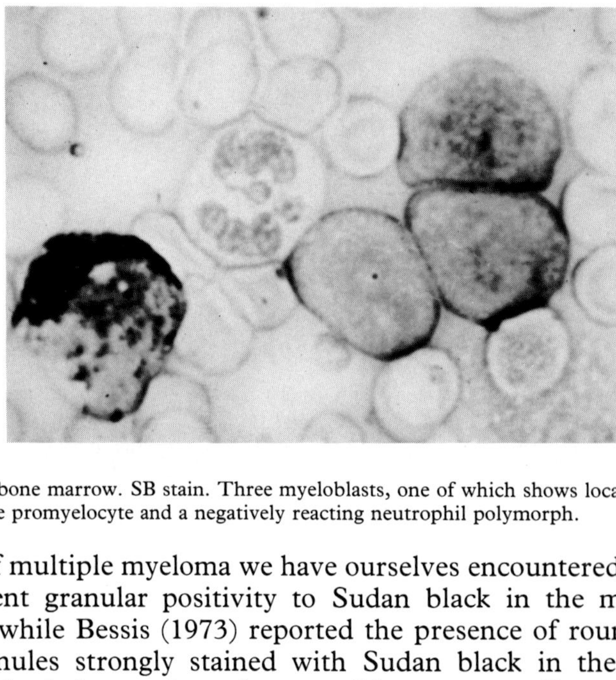

Fig. 5.7 AML bone marrow. SB stain. Three myeloblasts, one of which shows localized positivity, a strongly positive promyelocyte and a negatively reacting neutrophil polymorph.

200 cases of multiple myeloma we have ourselves encountered a single case with frequent granular positivity to Sudan black in the myeloma cells (Plate 8B), while Bessis (1973) reported the presence of round or comma shaped granules strongly stained with Sudan black in the red cells of patients with thalassaemia major, possibly corresponding to the myelin forms seen in these cells under the electron microscope.

More important, because it detracts from the specificity of sudanophilia to the myeloid lines among acute leukaemic blast cells, is the observation first reported by Tricot and his colleagues (1982) of Sudan black B positivity in the cells of an undoubted case of acute lymphoblastic leukaemia. Surface antigen studies confirmed the cells to be of a typical non-T, non-B ALL phenotype (cALL$^+$, Tdt$^+$, Ia$^+$), yet some 30% showed strong sudanophilic globules. The cells were negative for myeloperoxidase, naphthol AS-D chloroacetate and naphthyl butyrate esterase by standard cytochemical methods. PAS staining showed block positivity in some 6% of cells, typical of ALL.

Ultrastructural studies showed multiple lysosomal inclusions containing small vesicles and fragments of mitochondria and other organelles in more than 25% of the blast cells sectioned. The authors thought it likely that these inclusions, probably autophagosomes containing large amounts of phospholipid, explained the sudanophilia. We have seen these preparations and confirm the appearances as described and illustrated. In our view the globular concentration of sudanophilia entirely accords with the interpretation given by Tricot *et al.*, but we note that this globular appearance is somewhat unlike the more finely granular sudanophilia seen in granulocytes and their precursors, and more like the positivity seen in lymphoblasts stained with Oil Red O.

Occasional additional sporadic reports of sudanophilia in leukaemic blast cell populations clearly marked as lymphoblastic by immunological reactivity were subsequently reported (Ho *et al.*, 1983; Savage and Hoffman, 1983) and Stass *et al.* (1984) reported a substantial retrospective study on the matter. Among the bone marrow smears from 350 children newly diagnosed as ALL a total of six (1.6%) were found to show Sudan black B positivity in 5% or more of the blast cells. In each case, the diagnosis of ALL was supported by other morphological, cytochemical and immunological features and by therapeutic response. Although the authors say that the pattern of positivity was generally indistinguishable from that seen in acute myeloid leukaemias, the photomicrographs in their report again suggest a more globular type of reaction. This is also our own experience (Plate 7E, F).

These examples of sudanophilia in the leukaemic blasts in a small proportion of ALL cases are not to be confused with the rare cases of mixed lymphoid and myeloid phenotypes in acute leukaemia where markers of both lineages are expressed either in different blast cell components of the leukaemic population (Janossy *et al.*, 1978; Chessells *et al.*, 1979; Mertelsmann *et al.*, 1978; Hull and Grief, 1980) or within the same, doubly marking, cells (Bettelheim *et al.*, 1982; Pui *et al.*, 1984). The absence of myeloid markers other than sudanophilia, including myeloperoxidase, and chloroacetate esterase, make dual expression unlikely.

We think it probable that the type of SB positivity seen in this small percentage of ALL cases does not represent the staining of primary or secondary granules as in the stable sudanophilia of granulocytes and their precursors, but the staining of neutral fat and phospholipids within membrane-bound phagosomes. Such lipids are commonly demonstrable in lymphoblasts by Oil Red O staining, but are usually dissolved out by the solvent ethanol during the Sudan black B staining procedure. Lipids occasionally persist to give sudanophilia in lymphoid cells in conditions other than acute leukaemia; they have long been observed in the cells of Burkitt's lymphoma (Wright, 1970) and have sometimes been reported in large-cell lymphomas (Savage and Hoffman, 1983).

If the possibility that Sudan black positivity, especially of a scattered,

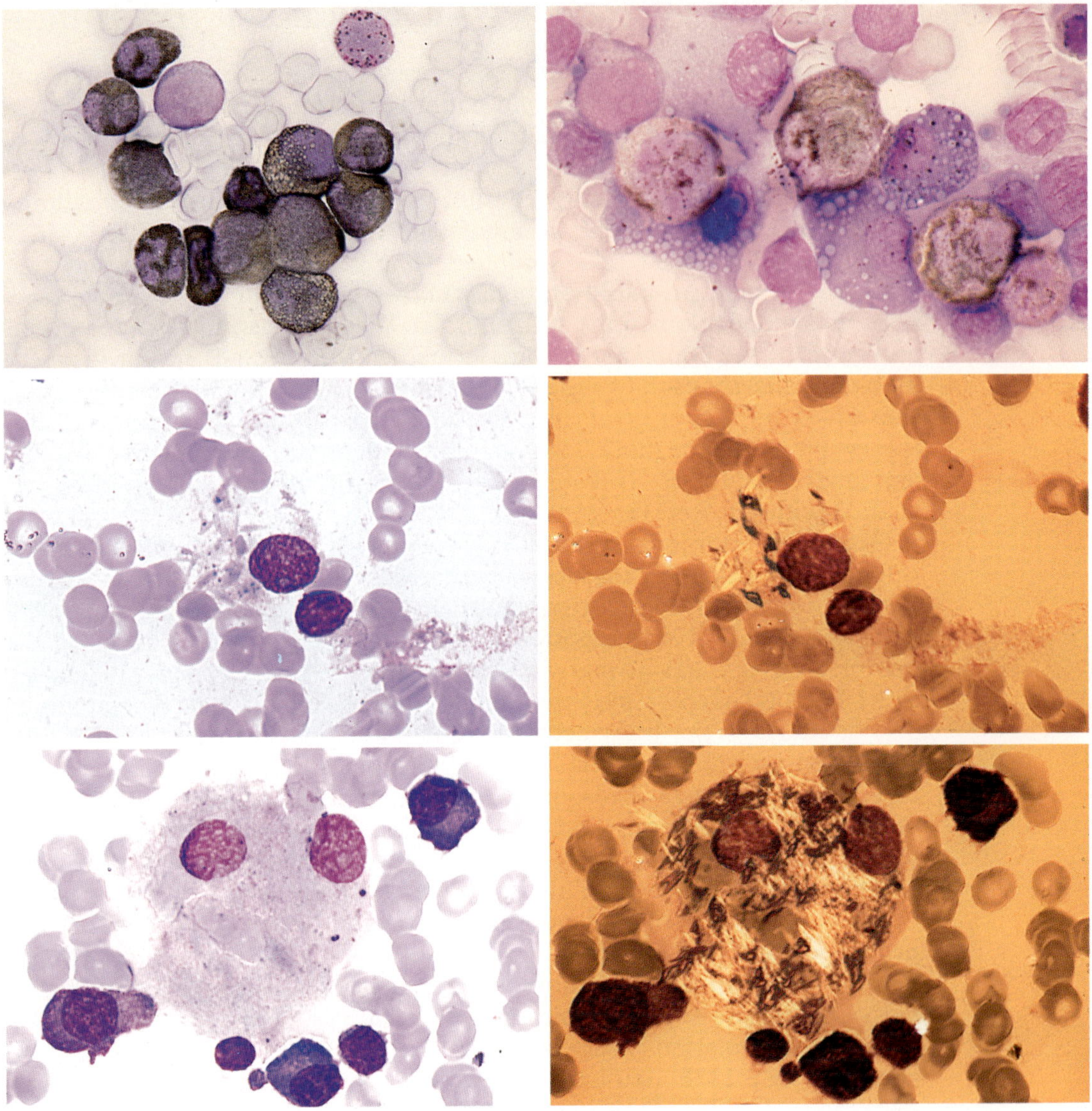

A B C D E F

Plate 8

A. CML. SB stain. A negative myeloblast, various positive neutrophil precursors, 2 eosinophils and a metachromatically staining basophil.

B. Multiple myeloma. SB stain. Scattered positive granules in abnormal plasma cells: a very rare finding.

C. Blue crystals in a fragmented macrophage from a patient with CML. Bone marrow. Leishman stain.

D. The same field as 8C under polarised light showing birefringence of crystals.

E. A pseudo-Gaucher cell containing phagocytosed cell debris from a patient with aplastic anaemia. Bone marrow. Leishman stain.

F. The same field as 8E under polarised light showing numerous coarse birefringent crystals.

globular, nature, may rarely be seen in lymphoblasts is borne in mind, this staining method retains its general usefulness as a more sensitive marker of early granulocyte precursors than conventional peroxidase stains or chloroacetate esterase reactions.

Lipids in macrophages

The lipid laden macrophages found in the bone marrow, spleen and other organs in Gaucher's disease, Niemann-Pick disease and the clinical syndromes of eosinophilic granuloma and Hand-Schüller-Christian disease generally show some degree of positivity to lipid stains. Although Rozenszajn and Efrati (1961) found Gaucher cells usually negative to Sudan black, Fisher and Reidbord (1962) and Volk *et al.* (1972) reported them to be positive and also reactive to Baker's acid haematin method. They are strongly birefringent in polarised light (Figs 5.8 and 5.9). Ultrastructurally these cells contain numerous rod-shaped inclusions, 0.5 to 3.0 μm in length, with an internal structure of smooth-walled tubes, presumably representing an organised pattern of deposit of the accumulated glucocerebroside. Stronger sudanophilia is found in the sphingomyelin of Niemann-Pick foam cells (McCusker and Parsons, 1962) and Lynn and Terry (1964) found this material also positive to Sudan IV, Oil Red O, Nile blue sulphate, Baker's acid haematin and the mercuric nitrate method for phospholipids. Electron microscopy here shows the lipid inclusions to be organized partly as lamellated multiple membranes with periodicity of 4.5 to 5 nm, and partly as closely packed granules. These structures appear likely to be abnormal cytosomes resulting from the spatial orientation of complex lipid molecules in relation to water rather than a degradation of mitochondria or lysosomes (Lynn and Terry, 1964).

There have been many reports of compound lipid inclusions occurring as a secondary phenomenon in marrow macrophages, especially in myeloid leukaemias (Dosik *et al.*, 1972) and dyserythropoietic states (Lewis and Verwilghen, 1977), and less commonly in idiopathic thrombocytopenic purpura. The subject has been reviewed recently by Hayhoe *et al.* (1979), who grouped the inclusion material into three categories, distinguishable by Romanowsky staining, ultraviolet fluorescence and the use of polarised light. Birefringent blue crystals and Gaucher-like cells formed one variety (Plate 8C, D, E, F and Fig. 5.10), where the composition of the inclusion material seems likely to be the same as in true Gaucher's disease, namely glucosyl ceramide, and the ultrastructure shows a similar pattern of parallel bundles of microtubules (Fig. 5.11) which probably account for the striking birefringence seen in both true and pseudo Gaucher cells.

The second variety of inclusion consists of granular sea-blue material (Plate 9A, B), essentially the same as that found in the macrophages of the 'sea-blue histiocyte syndrome', a relatively benign genetic disorder characterised by hepatosplenomegaly, thrombocytopenia, frequent lung and eye changes and accumulation of sea-blue histiocytes in spleen, liver and bone marrow (Sawitsky *et al.*, 1954; Silverstein *et al.*, 1964, 1970). In this case the inclusion material contains phospholipid and glycosphingolipid, and there

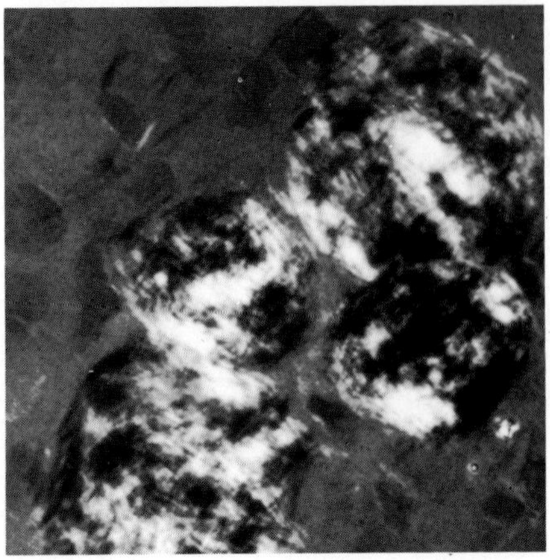

Fig. 5.8 Gaucher cells in polarized light, showing birefringence.

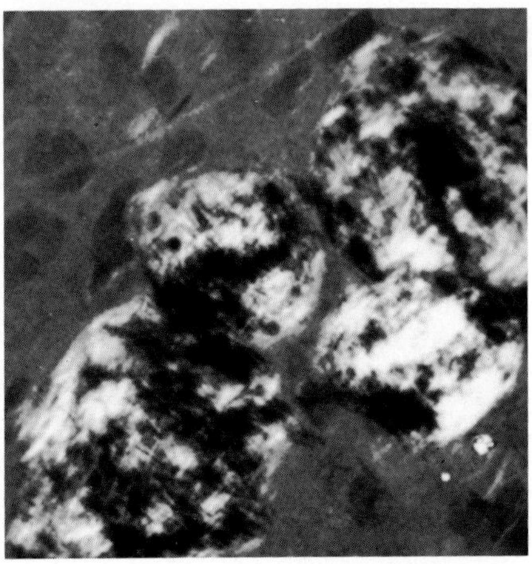

Fig. 5.9 The same cells with polarized light rotated at right angles to show reversal of birefringence.

are some parallels with the sphingomyelin accumulation of Niemann-Pick disease and the ceroid material commonly found in liver cells (Fig. 5.12), in that all these forms of inclusion show weak autofluorescence in u.v. light and may have an ultrastructure of lamellar 'fingerprint' appearance, while they do not show birefringence.

The third type of lipid inclusion, manifest as grey-green crystals in Romanowsky preparations (Plate 9C, E, F, Fig. 5.13), shows no autofluorescence or birefringence, and is of unknown chemical composition, though probably lipoprotein and has an amorphous ultrastructure with no suggestion of periodicity.

Cytochemical studies of the secondary crystalline and granular inclusions with a variety of techniques showed no more than surface positivity with the lysosomal enzymes acid phosphatase and α-naphthyl butyrate esterase

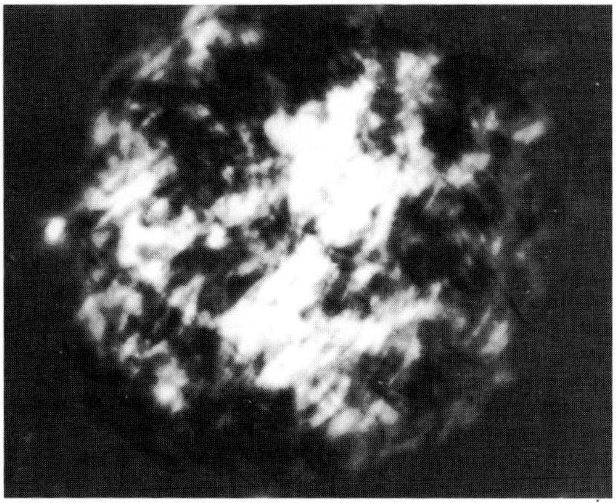

Fig. 5.10 A Gaucher-like cell from a patient with CML, showing birefringence and resemblance to a true Gaucher cell.

(Plate 9D), and weaker or negative reactions to peroxidase, PAS (Fig. 5.14), alkaline phosphatase and Sudan black B, except for occasional sudanophilia of sea-blue granules (Hayhoc *et al.*, 1979). In the sea-blue histiocyte syndrome itself cytochemical studies of the histiocytic inclusion material, reviewed by Sawitsky *et al.* (1972), have generally given very similar results, although sudanophilia has usually been observed and PAS positivity, resistent to diastase digestion, has also been reported.

TECHNICAL METHODS

Sudan black B

After Sheehan and Storey, 1947

Reagents

1. Sudan black B (Gurr) 0.3 g in 100 ml absolute ethanol.
2. Buffer: dissolve 16 g crystalline phenol in 30 ml absolute ethanol. Add to 100 ml distilled water in which 0.3 g hydrated disodium hydrogen phosphate ($Na_2HPO_4 \cdot 12H_2O$) has been dissolved.
3. Working stain: add 40 ml buffer to 60 ml Sudan black B solution and filter by suction. Keeps 2–3 months if stored in the refrigerator. May be used cold.

Technique

1. Fix air-dried smears in formalin vapour 5–10 minutes
2. Wash briefly in tap water
3. Immerse in working stain 1 hour
4. Wash off with 70% ethanol
5. Counterstain with Leishman or MGG stains.

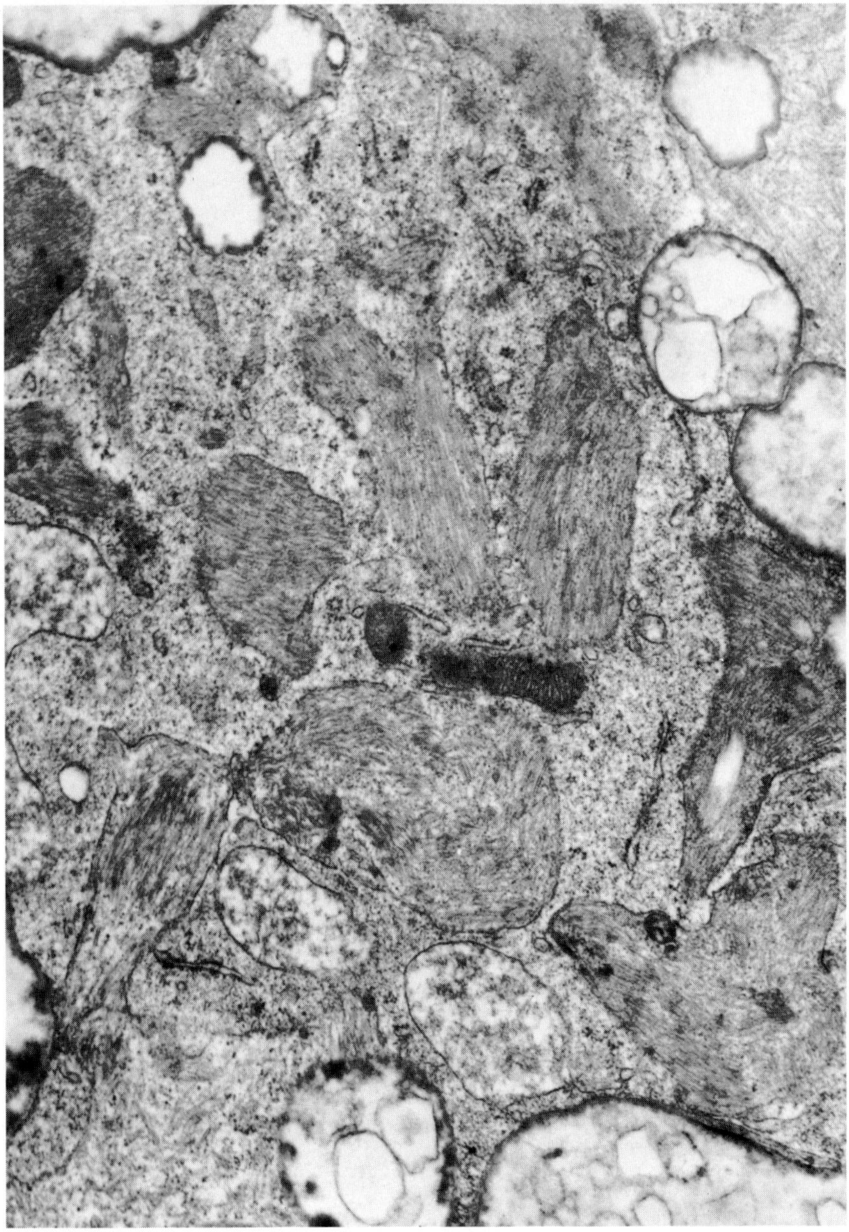

Fig. 5.11 Ultrastructure of birefringent crystalline material from a patient with thrombocytopenic purpura, showing parallel bundles of microtubules.

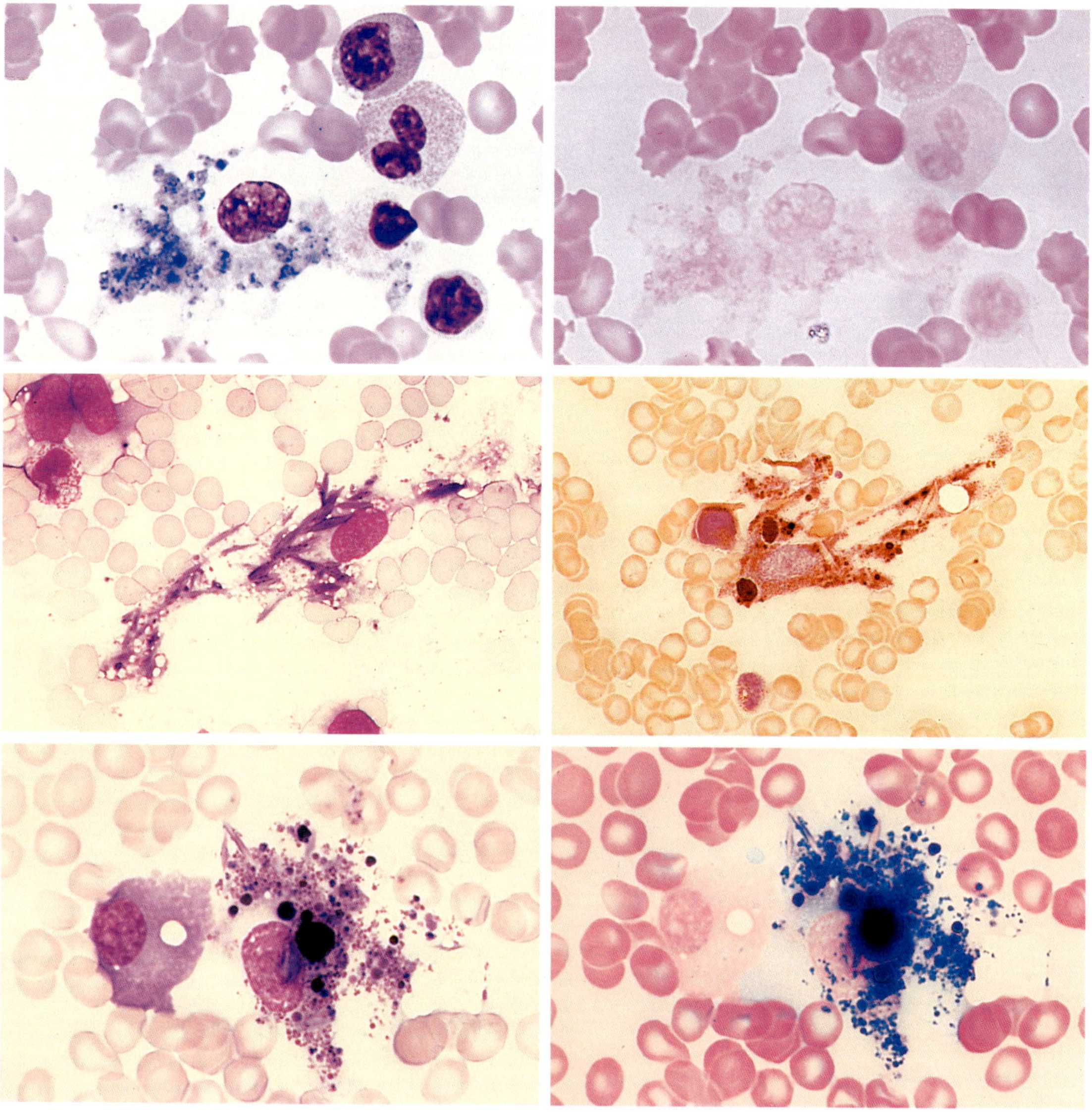

A B C D E F

Plate 9

A. A sea-blue histiocyte in the marrow of a patient with myeloid leukaemia. Leishman stain.

B. The same field as 9A after restaining with Perls stain. The histiocyte does not contain free iron.

C. Grey-green crystals in a disrupted macrophage from a patient with AML. Bone marrow. Leishman stain.

D. Similar crystals, showing surface reaction only to butyrate esterase.

E. A macrophage containing coarse purple granules and grey-green crystals, from the same patient. Bone marrow. Leishman stain.

F. Same field as in 9E, restained with Perls stain. The coarse granules are seen to contain free iron: the crystals remain negative.

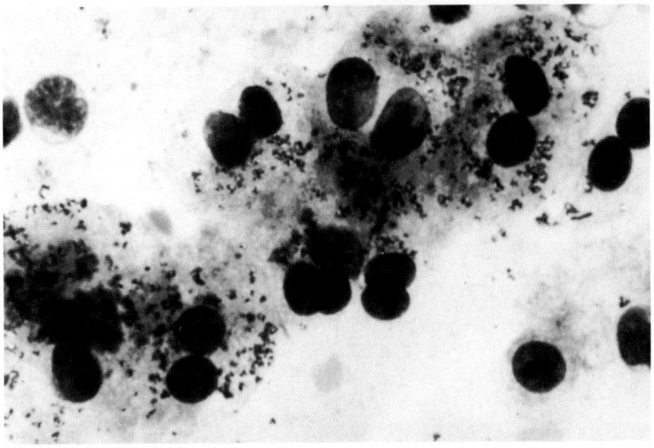

Fig. 5.12 'Ceroid' inclusion material in liver cells from an impression smear. Leishman stain.

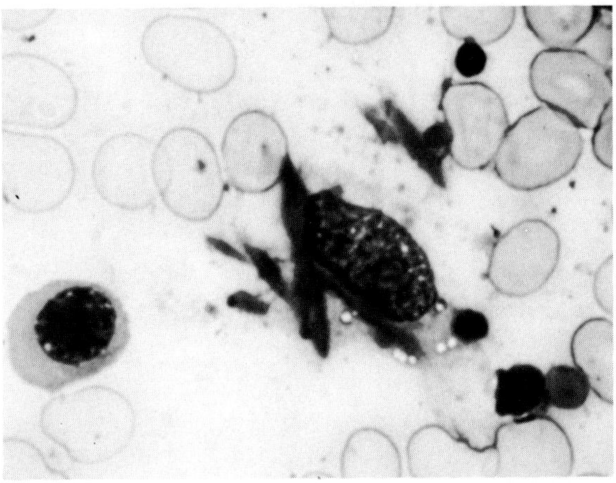

Fig. 5.13 Grey-green crystals in a macrophage from the bone marrow of a patient with AML. Leishman stain.

Oil Red O

After Lillie, 1944

Reagents

1. Oil Red O stock solution is prepared as a saturated solution of the dye, Oil Red O or Oil Red 4B, about 0.5 g in 100 ml absolute isopropyl alcohol. The solution keeps indefinitely.
2. Working stain: add 60 ml stock solution to 40 ml distilled water; leave to stand for 10 minutes and filter immediately before use.

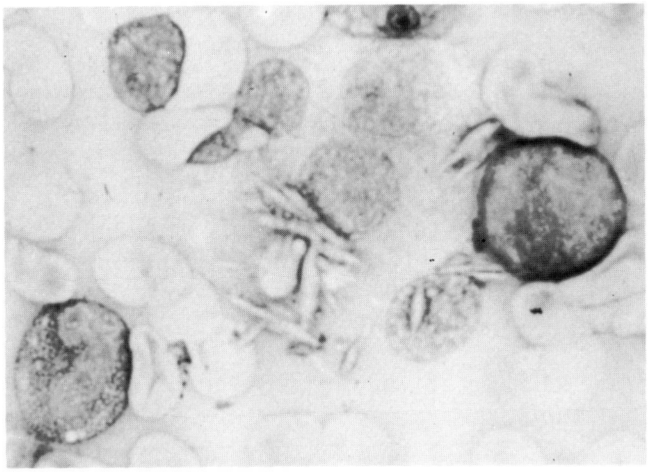

Fig. 5.14 Similar crystals to those in Fig. 5.13, stained with the PAS reaction: the crystals are negative, although some surface absorption of PAS positive material is seen.

Technique

1. Fix air-dried smears in formalin vapour 5–10 minutes
2. Wash briefly in 60% isopropyl alcohol
3. Immerse in working stain for 10 minutes
4. Rinse briefly in 60% isopropyl alcohol
5. Counterstain with Leishman or MGG stains.

Acid haematin method

Baker, 1946

Reagents

1. Dichromate-calcium solution:

Potassium dichromate	5 g
$CaCl_2$ (anhydrous)	1 g
Distilled water	100 ml.

2. Acid haematin

To 0.05 g haematoxylin add 48 ml distilled water and 1 ml of 1 per cent sodium iodate ($NaIO_4$). Heat to boiling, cool and add 1 ml glacial acetic acid. Prepare immediately before use.

3. Borax-ferricyanide solution

Potassium ferricyanide	0.25 g
Sodium tetraborate $10H_2O$	0.25 g
Distilled water	100 ml

The solution should be kept in the dark.

HAEMATOLOGICAL CYTOCHEMISTRY

Technique

1. Fix air-dried smears in formalin vapour 5–10 minutes
2. Transfer to dichromate-calcium solution at room temperature for 18 hours
3. Transfer to dichromate-calcium solution at 60°C for 24 hours
4. Wash in tap water for 6 hours
5. Transfer to dichromate-calcium solution at 60°C for 1 hour
6. Wash in several changes of distilled water for 5 minutes
7. Place in acid haematin at 37°C for 5 hours
8. Rinse in distilled water and place in borax-ferricyanide at 37°C for 18 hours
9. Wash in several changes of distilled water for 10 minutes and examine without counterstaining.

REFERENCES

Bacsich P 1936 On the staining of lipoid granules in leucocytes. Journal of Anatomy 70: 267

Baillif R N, Kimbrough C 1947 Studies on leucocyte granules after staining with Sudan black B and May-Grünwald-Giemsa. Journal of Laboratory and Clinical Medicine 32: 155

Baker J R 1946 The histochemical recognition of lipine. Quarterly Journal of the Microscopical Sciences 87: 441

Bennett J M, Dutcher T F 1969 The cytochemistry of acute leukaemias: observations on glycogen and neutral fat in bone marrow aspirates. Blood 33: 341

Berg N O 1951 A histological study of masked lipids. Acta Pathologica et Microbiologica Scandinavica Suppl. 40

Bessis M 1973 Living Blood Cells and their Ultrastructure. Springer-Verlag, Berlin, p 229

Bettelheim P, Paietta E, Majdic O, Gadner H, Schwarzmeier J, Knapp W 1982 Expression of a myeloid marker on Tdt-positive acute lymphocytic leukaemia cells: evidence by double fluorescence staining. Blood 60: 1392

Bloom M L, Wislocki G B 1950 The localization of lipids in human blood and bone marrow cells. Blood 5: 79

Bloor W R 1943 Biochemistry of the Fatty Acids. John Wiley and Sons, New York

Cawley J C, Hayhoe F G J 1973 Ultrastructure of Haemic Cells. Saunders, London

Chayen J 1968 Histochemistry of phospholipids and its significance in the interpretation of the structure of cells. In: Ross K F, McGee-Russell S (eds) The Interpretation of Cell Structure. Arnold, London

Chessells J M, Janossy G, Lawler S D, Secker Walker L M 1979 The Ph^1 chromosome in childhood leukaemia. British Journal of Haematology 41: 25

Discombe G 1946 The nature of neutrophilic granulation. Journal of Pathology and Bacteriology 58: 572

Dosik H, Rosner F, Sawitsky A 1972 Acquired lipidoses: Gaucher-like cells and 'blue cells' in chronic myeloid leukaemia. Seminars in Haematology 9: 309

Fearon E R, Burke P J, Schiffer C A, Zehnbauer B A, Vogelstein B 1986 Differentiation of leukaemia cells to polymorphonuclear leukocytes in patients with acute non-lymphocytic leukemia. New England Journal of Medicine 315:15

Fisher E R, Riedbord H 1962 Gaucher's disease: pathogenetic considerations based on electron microscopic and histochemical observations. American Journal of Pathology 41: 679

Gaham P B, Bartlett R, Cleland L, Olsen K 1974 Cytochemical evidence for the presence of phospholipids on human chromosomes. Histochemical Journal 6: 219

Gottfried E L 1972 Lipid patterns of leucocytes in health and disease. Seminars in Haematology 9: 241

Graham R C, Karnovsky M J 1966 The early stages of absorption of injected horseradish peroxidase in the proximal tubules of mouse kidney: ultrastructural cytochemistry by a new technique. Journal of Histochemistry and Cytochemistry 14: 291

Hayhoe F G J 1953 The cytochemical demonstration of lipids in blood and bone marrow cells. Journal of Pathology and Bacteriology 65: 413

Hayhoe F G J 1969 The cytochemistry of haemopoiesis. In: Britton C J C (ed) Disorders of the Blood. Churchill, London, p 123
Hayhoe F G J, Quaglino D, Doll R 1964 The Cytology and Cytochemistry of Acute Leukaemias. A study of 140 cases. MRC Special Report Series no. 304. HMSO: London.
Hayhoe F G J, Flemans R J, Cowling D C 1979 Acquired lipidosis of marrow macrophages: birefringent blue crystals and Gaucher-like cells: sea-blue histiocytes: grey-green crystals. Journal of Clinical Pathology 32: 420
Ho F C S, Chan G T C, Todd D 1983 Non-specificity of Sudan black B in the diagnosis of acute myeloid leukaemia. British Journal of Haematology 53: 171
Hull M I, Grief J A 1980 Mixed leukemia, lymphatic and myelomonocytic. American Journal of Clinical Pathology 74: 473
Janossy G, Woodruff R K, Paxton A, Greaves M F, Capellaro D, Kirk B, Innes E M, Eden O B, Lewis C, Catovsky D, Hoffbrand AV 1978 Membrane marker and cell separation studies in Ph^1-positive leukemia. Blood 51: 861
Lewis S M, Verwilghen R L 1977 Dyserythropoiesis. Academic Press, London, p 17
Lillie R D 1944 Various oil soluble dyes as fat stains in supersaturated isopropanol technic. Stain Technology 19: 55
Lillie R D, Burtner H J 1953 Stable sudanophilia of human neutrophil leucocytes in relation to peroxidase and oxidase. Journal of Histochemistry and Cytochemistry 1: 8
Lison L 1934 Sur des nouveaux colorants histologiques specifiques des lipides. Comptes Rendus des Séances de la Societe de Biologie 115: 202
Lison L 1936 Histochimie Animale. Gauthier-Villars, Paris
Lynn R, Terry R D 1964 Lipid histochemistry and electron microscopy in adult Niemann-Pick disease. American Journal of Medicine 37: 987
McCusker J J, Parsons D B 1962 Niemann-Pick disease. Report of two cases in siblings including necropsy and histochemical findings in one. Archives of Pathology 74: 127
McManus J F A 1945 Granules of the human polymorphonuclear leucocyte. Nature 156: 173
Mertelsmann R, Koziner B, Ralph P, Filippa D, McKenzie S, Arlin Z A, Gee TS, Moore M A S, Clarkson B D 1978 Evidence for distinct lymphocytic and monocytic populations in a patient with terminal transferase-positive acute leukemia. Blood 51: 1051
Novikoff A B, Goldfischer S 1968 Visualisation of microbodies for light and electron microscopy. Journal of Histochemistry and Cytochemistry 16: 507
Novikoff A B, Goldfischer S 1969 Visualization of peroxisomes (microbodies) and mitochondria with diaminobenzidine. Journal of Histochemistry and Cytochemistry 17: 675
Perugini S 1954 Sul significato dell'affinità dei granuli neutrofili dei leucociti per i coloranti tipo Sudan. Experientia 10: 381
Pui C-H, Dahl G V, Melvin S, Williams D L, Peiper S, Mirro J, Murphy S B, Stass S 1984 Acute leukaemia with mixed lymphoid and myeloid phenotype. British Journal of Haematology 56: 121
Rheingold J J, Wislocki G B 1948 Histochemical methods applied to haematology. Blood 3: 641
Rozenszajn L, Efrati P 1961 Cytochemical and phase contrast observations on Gaucher cells. Acta Haematologica 25: 43
Savage R A, Hoffman G C 1983 Sudan black positivity in non-myeloid leukaemia. British Journal of Haematology 54: 494
Savini E 1921 Die Lipoide der Leukocytengranula. Wiener Medizinische Wochenschrift 46: 1964.
Sawitsky A, Hyman GA, Hyman J B 1954 An unidentified reticuloendothelial cell in bone marrow and spleen: Report of two cases with histochemical studies. Blood 9: 877
Sawitsky A, Rosner F, Chodsky S 1972 The sea-blue histiocyte syndrome. A review: genetic and biochemical studies. Seminars in Haematology 9: 285
Schaefer H E, Fischer R 1970 Peroxydaseaktivitat als Ursache der stabilen Sudanophilie in Granulocyten. Klinische Wochenschrift 48: 1424
Sehrt E 1927 Die histologische Darstellung der Lipoide der weissen Blutzellen (neutro-und eosinophile Leukozyten, Mastzellen, Uebergangszellen und Mononukleäre) und die Beziehung dieser Lipoide zur Oxydasereaktion. Münchener Medizinische Wochenschrift 74: 139
Shaw M T, Klemp V A 1974 Cytochemical staining with Oil Red and for β-glucuronidase in the diagnosis of acute leukaemia. American Journal of Clinical Pathology 61: 169
Sheehan H L 1939 The staining of leucocyte granules by Sudan black B. Journal of Pathology and Bacteriology 49: 580
Sheehan H L, Storey G W 1947 An improved method of staining leucocyte granules with Sudan black B. Journal of Pathology and Bacteriology 59: 336
Silverstein M N, Young D G, Remine W H, Pease G L 1964 Splenomegaly with rare morphologically distinct histiocyte: a syndrome. Archives of Internal Medicine 114: 251
Silverstein M N, Ellefson R D, Ahern E J 1970 The syndrome of the sea-blue histiocytes. New England Journal of Medicine 282: 1

Stass S A, Pui C-H, Melvin S, Rovigatti O, Williams D, Motroni T, Kalwinsky D, Dahl G V 1984 Sudan black B positive acute lymphoblastic leukaemia. British Journal of Haematology 57: 413
Storti E, Perugini S 1951 Cytochemical researches on the lipids of the haematic cells with particular attention to those of acute leucosis. Acta Haematologica 5: 321
Terner J Y, Schmur J, Gurland J 1963 Stable Sudanophilia. Contributions to the histochemistry of the Sudan dyes. Laboratory Investigation 12: 405
Tricot G, Broeckart-van Orshoven A, van Hoof A, Verwilghen R L 1982 Sudan black B positivity in acute lymphoblastic leukaemia. British Journal of Haematology 51: 615
Volk B W, Adachi M, Schneck L 1972 The pathology of sphingolipidoses. Seminars in Haematology 9: 317
Wislocki G B, Dempsey E W 1946 Observations on the chemical cytology of normal blood and haemopoietic tissues. Anatomical Record 96: 249
Wright D H 1970 Microscopic features, histochemistry, histogenesis and diagnosis. In: Burkitt D P, Wright D H (eds) Burkitt's Lymphoma. Livingstone, Edinburgh, p 82

CHAPTER 6

Iron and other inorganic elements

The histochemical demonstration of inorganic elements in animal tissues has been hampered by a series of technical problems, such as their extremely low concentrations, which may account for the difficulty in obtaining accurate histochemical localizations, the fact that some of the inorganic constituents occur in masked non-reactive forms and need to be solubilized either by unmasking techniques or in the course of the histochemical processing, and finally the fact that some of the staining methods employed, such as those for demonstrating sodium, potassium and chloride have in the past been largely unsatisfactory, both from the point of view of sensitivity and of preciseness of localization. However, reasonably satisfactory methods exist for the demonstration in human tissues of various metals, such as zinc, copper, magnesium, iron, lead and manganese.

The present chapter will only be concerned with those histochemical procedures which may be applied to study the localization and distribution of inorganic elements in normal and pathological blood and bone marrow cells. In haemic cells trace elements have been demonstrated in leucocytes and more precisely in myeloid cells.

Zinc, copper, magnesium, manganese and cobalt

Vallee and coworkers (1954) isolated *zinc* from leucocytes in the form of a zinc protein complex. It has been suggested that this zinc protein compound is a metaloenzyme. Zinc ions increase the activity of many enzymes and there seems to be little doubt that this metal has important biological functions relating to cellular oxidation and perhaps other aspects of metabolism.

In 1960 Dennes *et al.*, employing a method of neutron activation analysis, carried out zinc determinations on blood samples from normal subjects and patients with chronic lymphatic and chronic myeloid leukaemia. They found that leukaemic leucocytes contained much less zinc than the normal leucocytes. The relatively low zinc content of leukaemic leucocytes could be due, in the opinion of the authors, either to the predominant immaturity of the leukaemic population or to the fact that the synthetic processes concerned with the formation of a zinc protein or some other zinc compound associated with leucocytes may be deficient in the leukaemic subject.

The histochemical demonstration of zinc relies principally on the use of the dithizone method. Dithizone forms an insoluble coloured inner complex salt with a number of heavy metals (Zn, Pb, Ag, Cu, Hg, Au, Cd). The zinc salt is reddish purple and can be formed in neutral, alkaline and acid solutions. This salt is soluble in carbon tetrachloride without change in colour. The test is very sensitive for zinc in neutral or acetic acid solution. Although the sensitivity is less, specificity is greater at alkaline pH. The specificity of the dithizone reaction may be increased by employing complex-forming buffers. Although dithizone is not specific for zinc, its reaction with other heavy metals seldom presents difficulties since these elements are not usually present in sufficient quantities to give positive staining. The reaction products with other ions have different colours and different pH optima.

In a histochemical study on human leucocytes McNary (1960) showed the presence of zinc and small amounts of copper in neutrophils, while larger amounts of copper and smaller quantities of zinc were reported in eosinophils. In neutrophils, employing dithizone-acetone-water and dithizone-complex-forming buffer staining methods, McNary reported the presence of zinc in both the cytoplasm and granules. The zinc associated with the granules was soluble and could be removed, while that associated with the cytoplasm was firmly fixed. Therefore, those staining methods which require a long reaction period result in dissolution of the zinc-dithizone complex in the myeloid granules, while the cytoplasmic stain remains. On the contrary, those staining methods which act rapidly such as zincon, dithizone-acetone-water and magnesium dithizonate, all demonstrate both cytoplasmic and granular zinc. The different chemical reactivity of granular and cytoplasmic zinc may be confirmed by soaking blood smears in acetone-water or dilute NH_4OH for one hour following staining with dithizone-acetone-water. Following this procedure, myeloid granules fail to show the dithizone zinc colour, while the cytoplasm retains its stain. This loss of the coloured complex from the myeloid granules has also been reported by Amann and Wolf (1956). Hoch and Vallee (1952) observed that zinc occurred in the human leucocyte in firm combination with protein as two types of zinc-protein complexes: one a compound soluble in phosphate buffer at pH 7.2, the other insoluble in this fluid. From the foregoing it seems possible to conclude that granular zinc is weakly bound in the leucocyte, while cytoplasmic zinc is strongly bound. This would allow the zinc-dithizone coloured complex to be removed on soaking but prevent cytoplasmic zinc from being removed from the cell.

The zinc content of granulocytes, employing McNary's method, was estimated semiquantitatively in normal subjects and leukaemic patients by Saito (1971). The zinc score in normal peripheral neutrophil granulocytes averaged 200, while lower than normal values were found in acute myeloblastic leukaemia and especially in chronic myeloid leukaemia. In acute lymphoblastic and chronic lymphatic leukaemia zinc scores were substantially normal. The low zinc values in the acute and chronic forms of myeloid leukaemia tended to revert to normal following cytostatic treatment. Experimental evidence in mice suggests that the decrease in zinc might be somehow related to the onset and development of leukaemia.

Copper will react with dithizone, zincon and rubeanic acid. The most commonly used cytochemical method is based on the latter compound. An alcoholic solution of rubeanic acid gives a dark green, almost black precipitate of copper rubeanate with copper salts. This precipitate, similar to those formed with other metals (Ni, Co) has the constitution of an inner complex salt of the diimido form of rubeanic acid. Copper may be demonstrated in haemic cells employing wet preparations prepared by adding one drop of a solution of one part concentrated NH_4OH and one part saturated alcoholic solution of rubeanic acid. As previously mentioned copper is mainly detected in eosinophil leucocytes.

The staining reactions employed for the histochemical demonstration of manganese, magnesium and cobalt in leucocytes are the least dependable, since they are influenced by the presence of other metal ions, such as zinc and copper, which are present in leucocytes. Attempts made to complex and remove interfering ions usually result in loss of sensitivity of the methods employed. To some extent the addition of potassium cyanide may prevent most ions from interfering with the above histochemical reactions. Magnesium and manganese are present in both the neutrophil and eosinophil, while cobalt is found only in the eosinophil.

Magnesium may be demonstrated histochemically in leucocytes by three different techniques: (a) the zincon method, (b) the titan yellow method, and (c) p-nitrobenzeneazoresorcinol method.

In the first reaction magnesium will react with zincon to produce a green complex.

The titan yellow method depends on a change in colour from yellow to red when magnesium hydroxide is precipitated in the presence of the dye. Blood smears were prepared by McNary (1960) by placing a drop of 0.1% titan yellow in 1 mol/l NaOH on the air dried smear, covering and sealing with paraffin wax. This procedure, however, led to a confused microscopic picture since the magnesium in the plasma and erythrocytes also reacted and laking of the erythrocytes occurred. If the smear was quickly rinsed in distilled water and dried before the application of the titan yellow, the leucocyte reaction could be clearly observed. Magnesium is demonstrated by the flame-red colour of the magnesium-titan yellow compound.

In the third reaction magnesium will stain with p-nitrobenzeneazoresorcinol in basic solutions containing NaOH. It is believed that such a staining results from the physical absorption of the sodium salt of p-nitrobenzeneazoresorcinol on the hydroxide of the magnesium, resulting in a blue magnesium compound. An excess of potassium cyanide decreases the interference of copper, zinc and other metals which form complex cyanide compounds. Wet preparations may be prepared by placing a drop of a 1% solution of p-nitrobenzeneazoresorcinol in 0.5% NaOH to which an excess of potassium cyanide is added on the slide.

Magnesium, together with adenylic acid, has been shown to be necessary as a coenzyme associated with phosphorylation and acts as a strong activator in the histochemical reaction for the demonstration of alkaline phosphatase.

Manganese is an essential trace element in many common animals and may be shown histochemically in leucocytes either by the dithizone method or the silver amine salt method, described by Feigl (1954). This technique is based on the chemical reaction:

$$Mn^{++} + 2Ag(NH_3)_2^+ + 4OH \rightarrow MnO_2 + 2Ag + 4NH_3 + 2H_2O$$

The formation of a dark precipitate of MnO_2 and finely divided silver is a positive test. In blood smears McNary (1960) carried out the silver amine reaction by placing a drop of silver amine salt solution on an air-dried blood smear and sealing a cover slip over this drop with melted paraffin. The precipitate formed in about 10 minutes. Because of the high alkalinity of the silver amine solution red blood cells were almost all destroyed and a granular mass appeared on the slide. To prevent this the slide was rinsed in distilled water and dried before staining. This step laked the erythrocytes and prevented their interference when observing the leucocyte reaction.

Arginase activity is closely associated with manganese ions in both the rat and rabbit (Underwood, 1956). Manganese ions in leucocytes may be associated with protease activity (Lillie, 1954).

For the demonstration of *cobalt* there are four different methods, based respectively on the use of dithizone, zincon, rubeanic acid and EDTA + H_2O_2. Cobalt will react with dithizone to form a violet complex, while the cobalt-zincon complex is blue green in colour and unstable in acid. Cobalt will form a yellow-brown complex with rubeanic acid. When a wet preparation is prepared by placing a drop of 10% of ethylene diamine tetra acetic acid (EDTA) in 1% aqueous Na_2CO_3 on the slide and a drop of 1% H_2O_2 is added, cobalt and iron will produce a blue to violet compound. The most sensitive of all the methods is the rubeanic acid technique.

Although, as we have already mentioned, considerable difficulty has been experienced in the histochemical demonstration of trace metals, their role in biological fluids and cellular structures has become increasingly important. In fact trace metals have been shown to be essential metabolic factors, since they act as activators to enzyme systems and are present as a component part of many enzymes, such as carbonic anhydrase, alcohol dehydrogenase, glutamic dehydrogenase, lactic dehydrogenase, carboxypeptidase and polyphenol oxidase (Underwood, 1956).

Presence and demonstration of iron in haemic cells

Iron in either the ferric or ferrous state is not usually observed in leucocytes, whereas its demonstration in the red cell series provides a useful guideline for the study and differentiation of a number of pathological conditions affecting the erythropoietic system

The iron resulting from the breakdown of haemoglobin, if not immediately used for synthesis, is stored chiefly in bone marrow macrophages in the form of *haemosiderin* (ferric hydroxide polymer) or *ferritin* (a ferrous iron protein complex) in which iron is strongly bound and cannot be released by acid treatment (masked iron).

Ferritin contains approximately 20% iron by dry weight. It is localized in the form of micells within the interior of a homogeneous protein called apoferritin. Ferritin present in erythroblasts may be derived from macrophages by a process of rhopheocytosis (Bessis and Breton-Gorius, 1961). This phenomenon, present also in normal conditions, is particularly conspicuous in certain pathological states, such as acute infections, haemolytic anaemias and hypersiderotic conditions. However, according to other authors, ferritin might be synthesized by erythroblastic precursors (Mazur and Carleton, 1963) or by reticulocytes (Eylar and Matioli, 1965). Indeed studies by Tanaka *et al.* (1966) have confirmed the presence of rhopheocytosis on the membrane of those erythroblasts situated in the proximity of reticuloendothelial cells, rich in ferritin; the authors, however, were also able to observe the formation of ferritin on the membrane of two adjoining erythroblasts. These latter findings suggest that ferritin may be synthesized, following contact with the erythroblast membrane.

The presence of ferritin can be detected only with the electron microscope in the form of characteristic dense granules which may be observed at the corners of an octahedron (Bessis and Breton-Gorius, 1960).

There are three classical methods for the demonstration of inorganic iron: Perls' Prussian blue, Turnbull blue and the iron sulphide staining techniques. Two additional 'rapid' cytochemical methods have been proposed by Kass (1975) and by Kass and Eickholt (1978), based respectively on the use of Alizarin red S and Bromochlorophenol blue. According to the authors, both dyes are capable of reacting with ferric iron, producing insoluble precipitates. With Alizarin red S, the presence of iron is revealed as a purple perinuclear halo or as discrete blocks of purple-staining material, whereas following the use of Bromochlorophenol blue, iron appears as discrete dark blue particles (Kass, 1979).

In haematology the most widely applied reaction is the Prussian blue staining method, which depends on the production of ferric ferrocyanide, when ferric ions (Fe^{+++}) react with ferrocyanide in acid solution (Perls, 1867). Treatment with dilute acid is an essential preliminary in the performance of the Prussian blue reaction since by this means the ferric ion is liberated from unreactive loose combinations with protein, such as haemosiderin. In carrying out the reaction the two reagents are invariably employed simultaneously since, if treatment with ferrocyanide follows treatment with acid, diffusion of iron is considerable.

The reaction sites are marked by the presence of a blue or blue-green precipitate of ferric ferrocyanide.

Examination of the bone marrow for iron, employing the Prussian blue reaction, is valuable in assessing the adequacy of body iron stores. Large granules or aggregates of blue-staining iron may be observed in normal bone marrow within the reticulo-endothelial cells (sometimes surrounded by normoblasts, Plate 10A), or they may appear to be lying free between cells in aspirated material (Plate 10B). Smaller granules can be seen within developing red cells in properly prepared smears of normal marrow. The iron present in reticulo-endothelial cells and in the form of loose aggregates is obviously not identical with the granules visible within maturing red cells (Plate 10C, D). It may be regarded as a storage form of iron (haemosiderin)

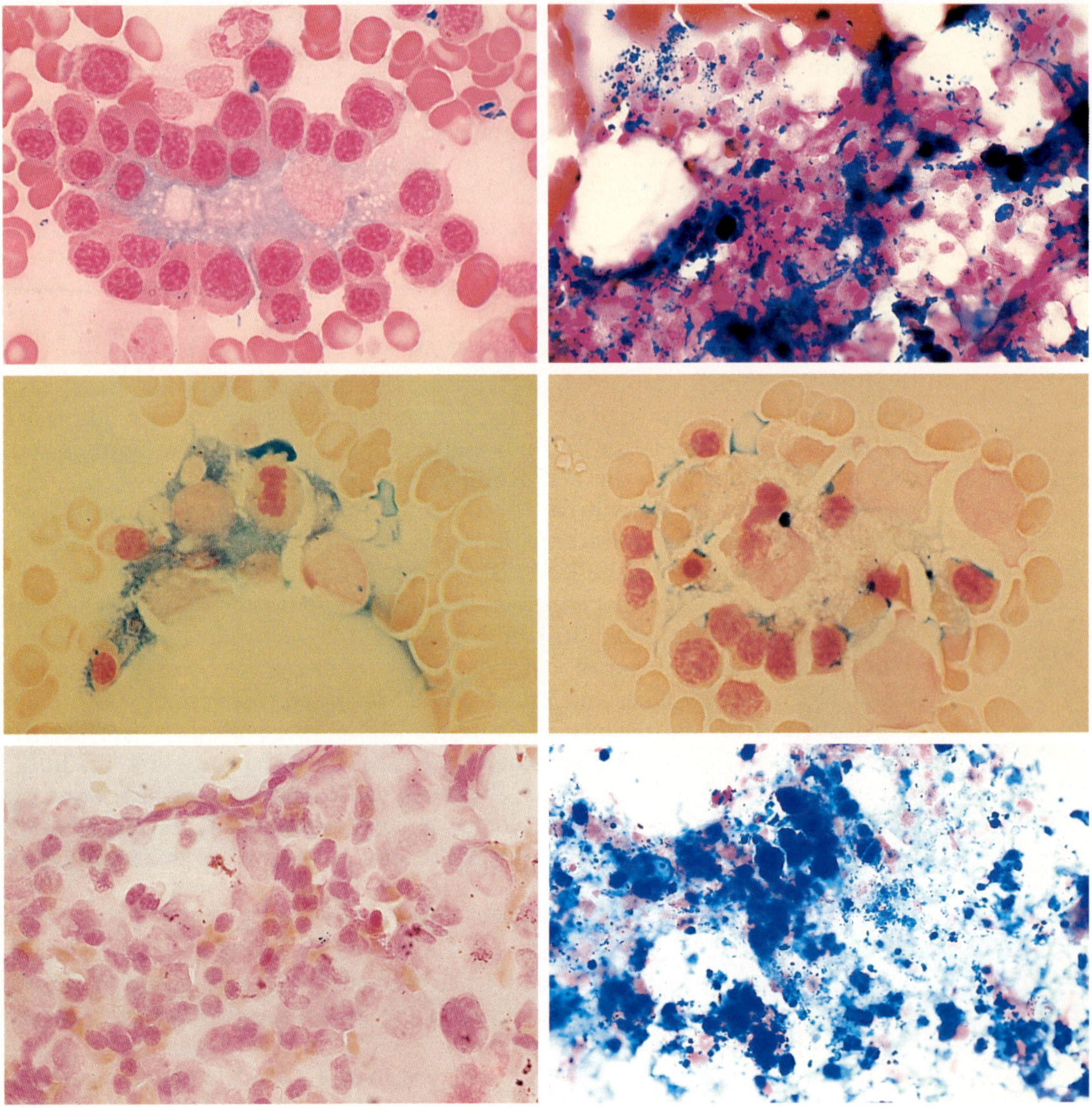

A B C D E F

Plate 10

A. Marrow showing erythroblastic hyperplasia. Prussian blue stain for free iron. A central macrophage laden with free iron is surrounded by a nest of normoblasts.

B. Normal bone marrow. Prussian blue stain showing coarse particles containing free iron scattered in and between the cells of a marrow fleck.

C. Normal bone marrow. Prussian blue stain showing a macrophage containing free iron and occasional iron granules in surrounding normoblasts and erythrocytes.

D. An area of the same preparation as in C, showing a similar appearance, with close apposition of iron-containing. macrophage cytoplasm to neighbouring normoblasts.

E. Bone marrow. Iron deficiency anaemia. Prussian blue stain, showing total absence of free iron from a marrow fleck.

F. Bone marrow. Sideroblastic anaemia. Prussian blue stain. Excess storage of free iron in a marrow fleck.

and has been used as an index of iron available for haemoglobin synthesis. A depletion of these iron deposits occurs in iron deficiency anaemia (Plate 10E), while an excess storage takes place in idiopathic and acquired haemochromatosis, chronic haemolytic anaemias, thalassaemia, and sideroblastic anaemia (Plate 10F).

Incidence and significance of siderocytes and sideroblasts

The term *siderocyte* was introduced in 1941 by Gruneberg to indicate the presence of erythrocytes which contain iron granules stainable with the Prussian blue reaction (Plate 11A). Doniach *et al.* (1943) and Pappenheimer *et al.* (1945) reported the occurrence of siderocytes in the blood of adults, chiefly in splenectomized patients (Plate 11B). Dacie and Doniach (1947) drew attention to the presence of similar iron granules in the normoblasts of human bone marrow. In the same year MacFadzean and Davis reported the presence of abundant iron granules in the bone marrow normoblasts of patients with lead poisoning and certain haemolytic anaemias, whereas in normal individuals and in patients with various haematological disorders iron granules were present in fewer than 10% normoblasts. On the contrary Kaplan *et al.* (1954) observed sideroblasts in appreciable numbers in normal individuals and in a wide variety of haematological conditions, with the single exception of iron deficiency anaemia, in which they were consistently and markedly reduced (Plate 11C). The prompt rise in the number of sideroblasts following the administration of iron to subjects with iron deficiency suggested that the iron granules represented iron taken up from the available stores rather than a degradation product of haemoglobin. The experimental results obtained with marrow cultures from iron deficient individuals when iron was added to the medium paralleled the experiments *in vivo*. According to Kaplan and his coworkers iron granules represent either an intracellular depot for heme synthesis or they indicate an excess of iron which has entered the cells but cannot be utilized. The sequence of events in treated iron deficiency anaemia favours the first of these hypotheses, while the presence of abundant iron granules in thalassaemia and in lead poisoning shows that the accumulation of intracellular iron in marrows with active erythropoiesis does not in itself assure proper utilization, perhaps on account of an intracellular block in heme synthesis which interferes with the utilization of iron.

The presence of large numbers of sideroblasts and siderocytes, mainly normoblastic, but with occasional megaloblastic features, has been reported by different authors (Björkman, 1956; Heilmeyer *et al.*, 1958; Dacie *et al.*, 1959; Heilmeyer *et al.*, 1960) in a condition called 'refractory sideroblastic anaemia'. The clinical and haematological features of this condition appear to be reasonably well established, were it not for the occurrence in a considerable proportion of cases of a terminal transformation to a frankly leukaemic, erythroleukaemic or erythraemic state. Hayhoe and Quaglino (1960) studied a number of cases of refractory sideroblastic anaemia as well as cases of erythroleukaemia, employing the Prussian blue reaction, the PAS reaction and the alkaline phosphatase technique. While the latter reaction

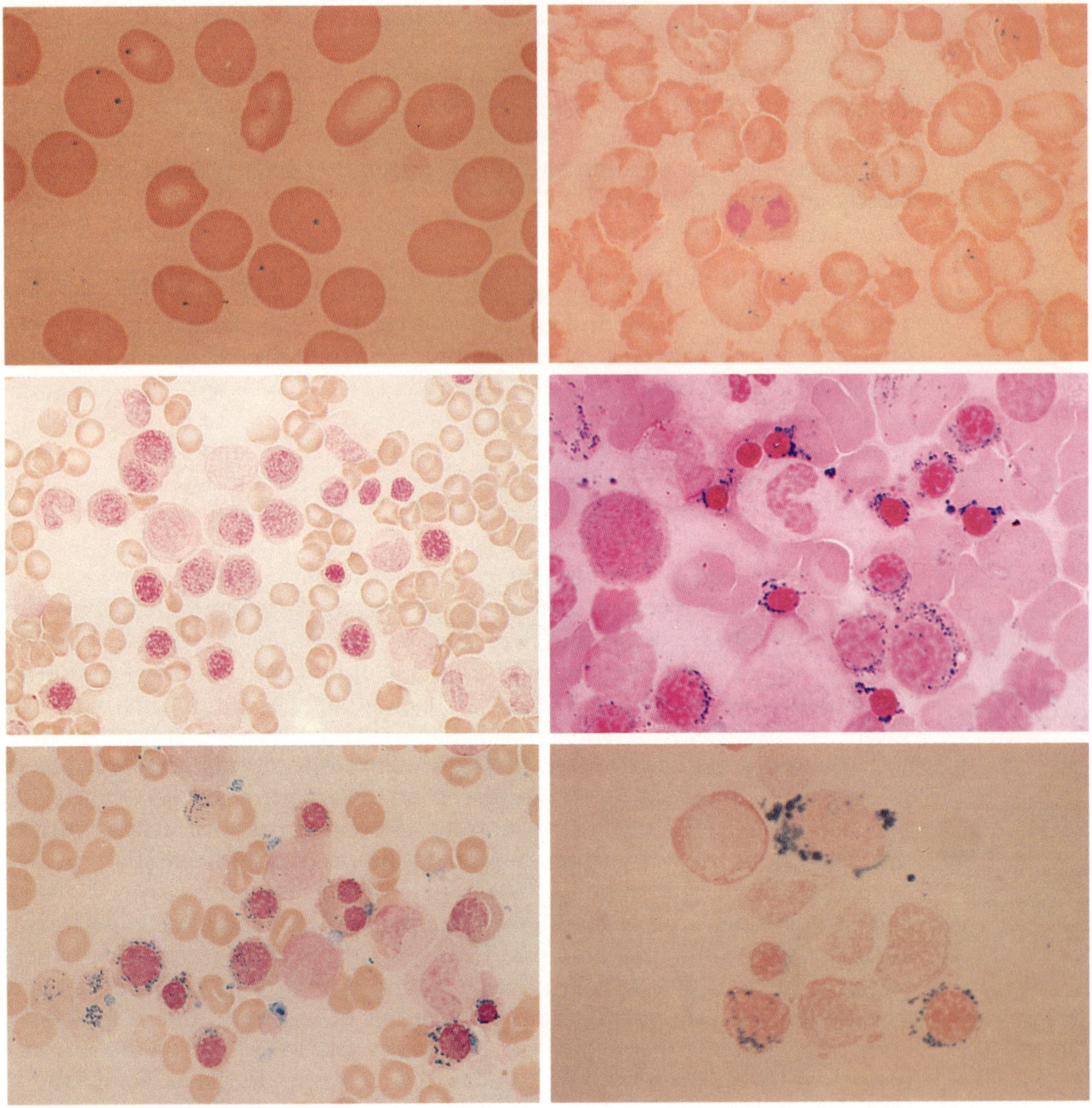

A B C D E F

Plate 11

A. Peripheral blood. Prussian blue stain. Free iron granules in red cells—siderocytes.

B. Peripheral blood, post-splenectomy. Gross poikilocytosis. Prussian blue stain. Free iron granules in a late dividing normoblast—sideroblast—and in several mature red cells—siderocytes.

C. Bone marrow. Iron deficiency anaemia. Prussian blue stain. Absence of sideroblasts among the red cell precursors.

D. Bone marrow. Refractory sideroblastic anaemia. Prussian blue stain. Numerous ringed sideroblasts.

E. Bone marrow. Erythroleukaemia. Prussian blue stain. Free iron granules numerous, especially in late erythroblasts, with some ringed sideroblasts.

F. Bone marrow. Erythroleukaemia. Prussian blue stain. A macrophage with heavy iron granules, a negative proerythroblast, two strongly positive intermediate normoblasts (sideroblasts), four negative myeloblasts and a negative lymphocyte.

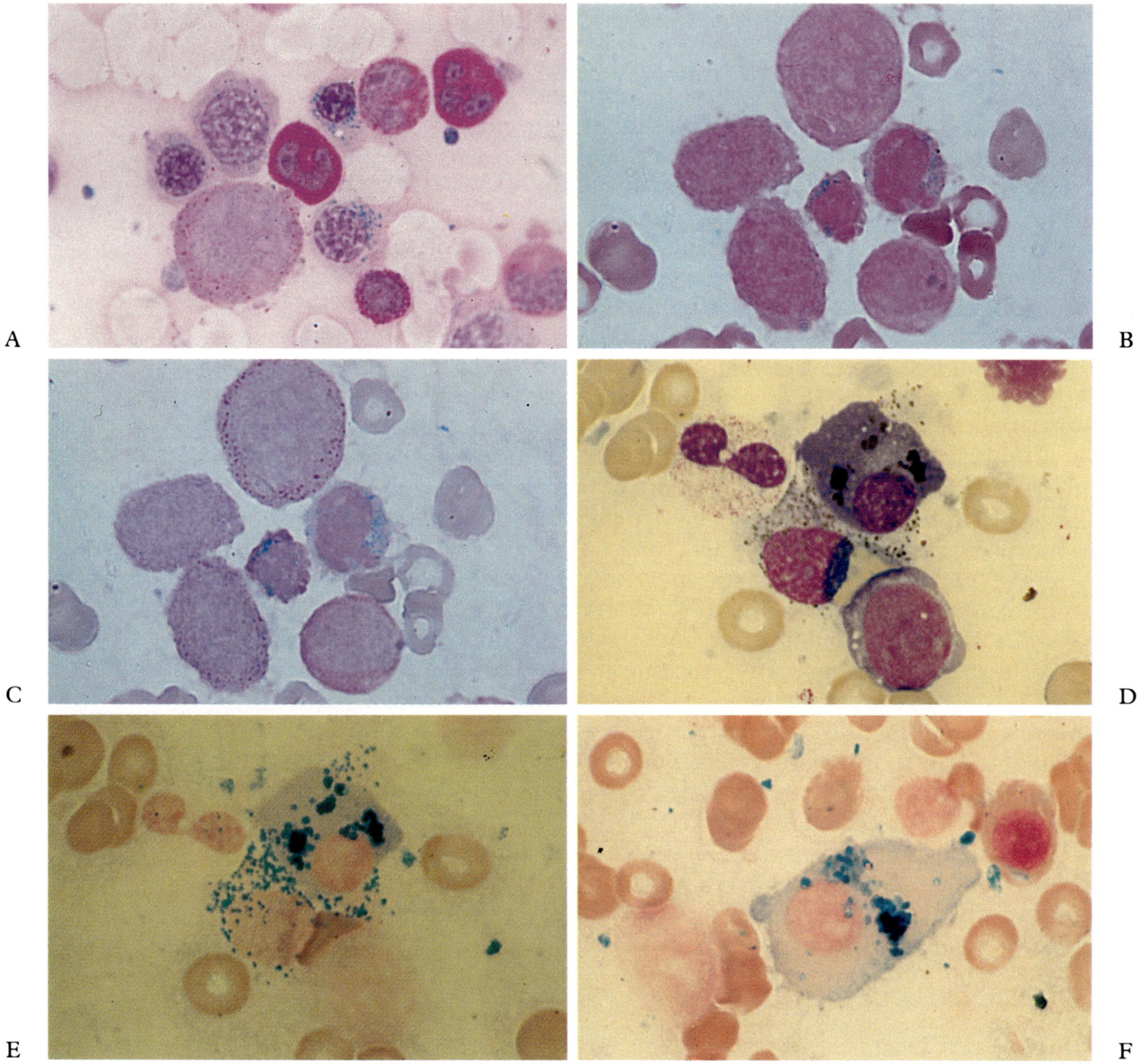

Plate 12

A. Bone marrow. Refractory sideroblastic anaemia. Combined Prussian blue stain and PAS reaction. PAS positivity conspicuous in neutrophils and in a proerythroblast, with free iron in later erythroblasts including a ringed sideroblast.

B. Bone marrow. Erythroleukaemia. Prussian blue stan. Positivity marked in later erythroblasts.

C. Same field as B. Combined Prussian blue stain and PAS reaction. PAS positivity is seen in proerythroblasts and in the central sideroblast.

D. A plasma cell with a neighbouring RE cell. Romanowsky stain.

E. Duplicate free iron stain on same field as in D; both plasma cell and RE cell show positivity.

F. Free iron positivity in another plasma cell from the same preparation.

did not provide a consistent pattern, the levels of the score being extremely variable from one case to another, the Prussian blue reaction showed, as previously reported in refractory sideroblastic anaemia, an intense degree of positivity in the form of one or more concentric rings of iron granules in the cytoplasm of intermediate and late erythroblasts (Plate 11D). The intensity of the Prussian blue reaction was assessed by a semi-quantitative scoring method (see below). In normal subjects the scoring method gave values ranging from 6 to 76, but it is noteworthy that, although the percentage of erythroblasts showing a few fine granules was sometimes high, cells with positivity stronger than + were uncommon. The iron scores obtained in refractory sideroblastic anaemia were invariably high with individual cell ratings of + to + + +. In erythroleukaemia the scores were moderately high. Free iron granules were numerous and sometimes coarse in late erythroblasts, where a typical perinuclear annular disposition was often observed (Plate 11E), and also in many of the earlier erythroblasts, though not in proerythroblasts (Plate 11F). Iron granules were often conspicuous in morphologically abnormal cells.

Studies were also carried out in refractory sideroblastic anaemia and erythroleukaemia to assess whether the erythroblasts containing glycogen were the richer in iron and these investigations were made possible by the use of a three-step consecutive technique (Hayhoe and Quaglino, 1960). With this method no direct relationship emerged in refractory sideroblastic anaemia between the pattern of PAS and Prussian blue positivity (Plate 12A), but in erythroleukaemia the later erythroblasts displaying strong PAS positivity often contained more numerous and coarser granules of free iron (Plate 12B, C).

The implications which may be drawn from these cytochemical studies on the nature of refractory sideroblastic anaemia and the relationship between this condition on the one hand and erythraemic myelosis and erythroleukaemia on the other are more fully discussed in Chapter 14.

Plasma cell iron

In addition to the presence of iron in maturing erythroid cells and macrophages, iron has also been found in the plasma cell series (Jaffe, 1938; Koszewski, 1952; Izarn and Emberger, 1958; Goodman and Hall, 1966; Karcioglu and Hardison, 1978; Cook and Madden, 1982). The mechanism of iron uptake by plasma cells is unknown. The demonstration of intra-cytoplasmic bacteria, haemopoietic elements and treponemes in plasma cells (Abramson *et al.*, 1970; Azar *et al.*, 1970; Ludwig and Pavelka, 1980; Raman *et al.*, 1983) suggests the possibility that the plasma cell may possess phagocytic properties. Studies by Schoenberg *et al.* (1963), describing cytoplasmic connections between macrophages and plasma cells, raise the possibility of direct transfer of cytoplasmic contents from macrophages to plasma cells, and the cytological appearances are often compatible with this view (Plate 12D, E). Ultrastructural studies by Shanmugathasa *et al.* (1979) confirmed the presence of cytoplasmic bridges between plasma cells and macrophages and also demonstrated plasma cell uptake of ferritin by micropinocytosis.

In 1984 McCurley *et al.* reviewed the medical records and marrow aspirates of 53 patients with plasma cell iron hospitalized over a seven-year period. Plasma cell iron was associated most commonly with alcoholism and occurred in marrows with increased, normal or decreased iron stores. In patients with decreased marrow iron, plasma cells were the major site of stainable iron. Plasma cell iron was found in patients without other morphologic changes of alcoholism such as megaloblastosis, erythroid vacuolization and ringed sideroblasts (Plate 12F). Plasma cell iron could be demonstrated also in biopsy and autopsy material from extra-marrow sites.

The effects of alcoholism on iron metabolism are complex. How plasma cell iron relates to the sideroblastic anaemia of alcoholism is unclear. Some authors have suggested that plasma cell iron is in fact located in mitochondria and accumulates as a result of the same defects in heme biosynthesis thought responsible for formation of ringed sideroblasts. The ultrastructural studies of McCurley *et al.* (1984) however showed iron always located in membrane bound lysosomal vesicles of plasma cells, without ultrastructural features of mitochondria.

TECHNICAL METHODS

Perls' reaction

1. Fix air-dried bone marrow or peripheral blood smears with 10% formalin in ethanol for 10 minutes.
2. Place slides in a mixture of equal volumes of 2% freshly prepared solution of potassium ferrocyanide in distilled water and 2% (0.25 mol/l) HCl for 1 hour at room temperature.
3. Rinse thoroughly in distilled water for 5 minutes.
4. Counterstain with 0.1% nuclear fast red for 10 minutes.

With a view to defining more precisely the relative localization of free iron and PAS-positive material within cells at different stages of maturation, it is possible to carry out the following three-step consecutive staining procedure:

1. Slides are stained by the Leishman or May-Grünwald-Giemsa methods.
2. Representative fields from such preparations are photographed and the location of each field is noted using a Vernier scale.
3. The slides are then decolorized for 24 hours in absolute methanol and subsequently dried and immersed in the Prussian-blue solution for 1 hour. Slides are then rinsed with distilled water and counterstained with nuclear fast red.
4. Photographs are again taken of the same fields.
5. The smears are then subjected to the PAS reaction. The periodic acid removes the counterstain completely without affecting the staining results of the Prussian blue reaction. After treatment with periodic acid the slides are immersed in leucobasic fuchsin for 30 minutes, rinsed in tap water for

15 minutes and counterstained with haematoxylin for 10 minutes.

6. The previous fields are again examined and photographed.

Scoring method

The intensity of the Prussian-blue reaction may be assessed in the following way:

0	no positivity
+	a few scattered granules in the cytoplasm or one crown of fine granules
+ +	one crown of moderately coarse granules or two crowns of fine granules
+ + +	two or more concentric rings of moderately coarse granules

The sum of 100 consecutive cell ratings provides the score for each specimen, with a possible variation from 0 to 300.

REFERENCES

Abramson N, Kapff C, Ginsburg A D 1970 The phagocytic plasma cells. New England Journal of Medicine 283: 248

Amann R, Wolf H P 1956 Schwermetalle in Leukocyten. Zeitschrift fur die Gesamte Experimentelle Medizin 127: 281

Azar H A, Pham T D, Kurban A 1970 An electron microscopic study of a syphilitic chancre. Archives of Pathology 90: 143

Bessis M, Breton-Gorius J 1960 Aspects de la molécule de ferritine et d'apoferritine au microscope électronique. Comptes Rendus des Seances de L'Academie des Sciences 250:1360

Bessis M, Breton-Gorius J 1961 L'ilot érythroblastique et la rhophéocytose de la ferritine dans l'inflammation. Nouvelle Revue Francaise d'Hematologie 1: 569

Björkman S E 1956 Chronic Refractory Anemia with sideroblastic bone marrow: a study of 4 cases. Blood 11: 250

Cook M K, Madden M 1982 Iron granules in plasma cells. Journal of Clinical Pathology 35: 172

Dacie J V, Doniach I 1947 Basophilic property of iron-containing granules in siderocytes. Journal of Pathology and Bacteriology 59: 684

Dacie J V, Smith M D, White J C, Mollin D L 1959 Refractory normoblastic anaemia: a clinical and haematological study of seven cases. British Journal of Haematology 4: 56

Dennes E, Tupper R, Wormall A 1960 Zinc content of erythrocytes and leucocytes of blood of normal and leukaemic subjects. Nature 187: 302

Doniach I, Gruneberg H, Pearson J E G 1943 The occurrence of siderocytes in adult human blood. Journal of Pathology and Bacteriology 55: 23

Eylar E H, Matioli G 1965 Apoferritin synthesis in human erythroid cells in thalassemia. Nature 208: 661

Feigl F 1954 Spot Tests. Vol. I. Inorganic Applications. Elsevier, New York

Goodman J R, Hall S G 1966 Plasma cells containing iron: an electron micrographic study. Blood 28: 83

Gruneberg H 1941 Siderocytes: a new kind of erythrocyte. Nature 148: 114

Hayhoe F G J, Quaglino D 1960 Refractory sideroblastic anaemia and erythraemic myelosis: possible relationship and cytochemical observations. British Journal of Haematology 6: 381

Heilmeyer L, Keiderling W, Bilger R, Bernauer H 1958 Über chronische refraktäre Anämien mit sideroblastichem Knochenmark (Anaemia refractoria sideroblastica). Folia Haematologica Frankfurt, 2: 49

Heilmeyer L, Keiderling W, Merker H, Clotten R, Schubothe H 1960 Die Anaemia refractoria sideroblastica und ihre Beziehungen zur Lebersiderose und Hämochromatose. Acta Haematologica 23: 1

Hoch F L, Vallee B L 1952 Extraction of zinc-containing protein from human leukocytes. Journal of Biological Chemistry 195: 531

Izarn P, Emberger J 1958 Plasmocytes avec inclusions d'hemosiderine. Sang 29: 701

Jaffe J H 1938 The reticulo-endothelial system. In: Downey H (ed) Handbook of Hematology. Harper and Brothers, New York, p 1170
Kaplan E, Zuelzer W W, Mouriquand C 1954 Sideroblasts. A study of stainable nonhemoglobin iron in marrow normoblasts. Blood 9: 203
Karcioglu G L, Hardison J E 1978 Iron-containing plasma cells. Archives of Internal Medicine 138: 97
Kass L 1975 Rapid detection of ringed sideroblasts in erythremic myelosis. Archives of Pathology 99: 225
Kass L 1979 Bone marrow interpretation. J B Lippincott Company, Philadelphia
Kass L, Eickholt M 1978 Rapid demonstration of ringed sideroblasts with bromochlorophenol blue. American Journal of Clinical Pathology 70: 738
Koszewski B J 1952 The occurrence of megaloblastic erythropoiesis in patients with hemochromatosis. Blood 7: 1182
Lillie R D 1954 Histopathologic Technic and Practical Histochemistry. Blakiston, New York
Ludwig H, Pavelka M 1980 Phagocytic plasma cells in a patient with multiple myeloma. Blood 56: 173
McCurley T L, Cousar J B, Graber S E, Glick A D, Collins R D 1984 Plasma cell iron. Clinical and morphological features. American Journal of Clinical Pathology 81: 312
MacFadzean A J S, Davis L J 1947 Iron staining erythrocyte inclusions with special reference to acquired hemolytic anaemia. Glasgow Medical Journal 28: 237
McNary W F Jr 1960 The histochemical demonstration of trace metals in leukocytes. Journal of Histochemistry and Cytochemistry 8: 124
Mazur A, Carleton A 1963 Relation of ferritin ion to heme synthesis in marrow and reticulocytes. Journal of Biological Chemistry 238: 1817
Pappenheimer A M, Thompson W P, Parker D D, Smith K E 1945 Anemia associated with unidentified erythrocytic inclusions, after splenectomy. Quarterly Journal of Medicine 14: 75
Perls M 1867 Nachweis von Eisenoxyd in gewissen Pigmenten. Virchows Archiv 39: 42
Raman S, Frame B, Saeed S M, Tolia K, Usha R, Kottamasu S 1983 Diffuse nonsecretory osteosclerotic myeloma with extensive erythrophagocytosis. American Journal of Clinical Pathology 80:84
Saito K 1971 Studies on zinc metabolism in leukemia. I. Estimation of zinc score in granulocytes of leukemic patients by McNary's method. Okayama Igaikkai Zasshi 83: 185
Schoenberg M D, Mumaw V R, Moore R D, Weisberger A S 1963 Cytoplasmic interaction between macrophages and lymphocytic cells in antibody synthesis. Science 143: 964
Shanmugathasa M, Sobel H J, Marquet E, Trubowitz S 1979 Plasma cells with inclusions containing iron. Archives of Pathology and Laboratory Medicine 103: 577
Tanaka Y, Brecher G, Bull B 1966 Ferritin localization on the erythroblast cell membrane and ropheocytosis in hypersiderotic human bone marrows. Blood 28: 758
Underwood E J 1956 Trace Metals in Human and Animal Nutrition. Academic Press, New York
Vallee B L, Hoch F L, Hughes W L 1954 Studies on metalo-proteins. Soluble zinc-containing protein extracted from human leukocytes. Archives of Biochemistry and Biophysics 48: 347

CHAPTER 7

Phosphatases

The acid and alkaline phosphatases (EC 3.1.3.2; EC 3.1.3.1) include a range of isoenzymes with the general ability to liberate phosphate from many alcoholic or phenolic phosphomonoesters at pH levels respectively below and above neutrality. The enzymes are widely distributed in cells and tissues and have proved especially interesting and valuable in haematology. The cytochemical demonstration of phosphatases under light microscopy depends on the formation of an insoluble coloured precipitate at sites of substrate hydrolysis, and the methods used fall into two broad categories.

The Gomori–Takamatsu methods

In the first, cellular preparations are allowed to act on a suitable phosphomonoester in the presence of ions capable of forming an insoluble compound with the released phosphate at the selected pH. Gomori (1939) and Takamatsu (1939) independently devised a method for achieving this with the alkaline enzyme and their technique with minor modifications (Wachstein, 1946; Danielli, 1946; Gomori, 1952) has been used extensively. The principle involves the use of a glycerophosphate or other suitable phosphomonoester as substrate at alkaline pH in the presence of calcium and activating magnesium ions. The phosphate liberated is trapped as insoluble calcium phosphate, which is converted to cobalt phosphate and then to black cobalt sulphide. A typical reaction sequence is as follows

1. sodium β-glycerophosphate $\xrightarrow{\text{Alk. phosphatase } Ca^{++}\ Mg^{++}}$ glycerine + phosphate
2. phosphate ions + Ca^{++} $\xrightarrow{\text{at pH 9.2}}$ calcium phosphate
3. calcium phosphate + cobalt nitrate → cobalt phosphate
4. cobalt phosphate + ammonium sulphide → cobalt sulphide

A similar technique may be used for the acid enzyme (Gomori, 1952), but since calcium phosphate is soluble at acid pH, lead nitrate is used as the phosphate trapping agent, insoluble lead phosphate being deposited at sites of enzymic activity. The lead phosphate is then rendered visible by conversion to lead sulphide, using either ammonium sulphide or hydrogen sulphide.

With both enzymes these techniques initially gave somewhat erratic results. In blood and bone marrow cells, for example, the use of the Gomori-Takamatsu procedure for alkaline phosphatase by Wachstein (1946), Rheingold and Wislocki (1948), Rabinovitch and Andreucci (1949),

Kerppola (1951), Brodell and Swisher (1954) and Wiltshaw and Moloney (1955) led to considerable disagreement as to the cell types showing positive staining and the location of the reaction sites in individual cells. Heavy nuclear staining was commonly found as well as variable cytoplasmic granularity, and Martin and Jacoby (1949) and Pearse (1960) found evidence that at least the nuclear positivity might be an artefact due to diffusion of the enzyme or of the liberated phosphate ions. This problem may be partly overcome by the use of higher concentrations of calcium ions, since Butcher and Chayen (1966) were able to show that the enzyme was immobilised by a sufficient concentration of these ions. Despite these difficulties, early observations of cytoplasmic positivity in granulocytes already revealed striking variations in disease states, notably a sharp depression in chronic myeloid leukaemias (Wachstein, 1946) and marked elevation in myeloid metaplasia and pyogenic infections (Wiltshaw and Moloney, 1955). Artefactual complications were even more prominent in the acid phosphatase method using lead salts. The studies of Rheingold and Wislocki (1948), Rabinovitch and Andreucci (1949) and Storti (1952) all showed predominantly nuclear staining in haemic cells, although with some cytoplasmic granular positivity in different cell types. Diffusion of enzyme or of liberated phosphate may again have contributed, but two further features were of importance in this reaction. First, as shown by Holt (1959), the concentration of lead used in the incubation mixture was critical, the normal concentration leading to instability and non-specific deposition of lead, or its absorption on to sites, especially nuclear, with a 'metallophilia' or affinity for lead. Second, the acid enzyme, unlike the alkaline, is localised in lysosomes, and may therefore be inaccessible to the substrate unless the lysosomal membrane has been rendered permeable by fixation or low pH.

Although the Gomori methods for phosphatases have largely been superseded in diagnostic haematology by the azo-dye methods to be discussed below, they possess certain advantages, notably relative permanence of the preparations and potential suitability for ultrastructural work, which have led to continued efforts to overcome the problems of artefactual staining.

With the alkaline technique, for example, a direct lead capture method might avoid any false localization caused by diffusion of calcium phosphate or incomplete conversion into cobalt phosphate. The difficulty here lies in the tendency for lead ions to precipitate, especially at high pH levels. The problem is less severe at acid pH, and Barka and Anderson (1962) were able to prevent the development of artefactual large granular or crystalline deposits and nuclear staining in the acid phosphatase reaction by using a tris-maleate buffer instead of veronal acetate, and reducing the concentration of lead nitrate to 2.4 mmol/l. They concluded that maleic ions appeared to play an important role in determining the concentration of lead ions held in solution, and emphasised that the maleic acid used in making the buffer should be of high purity. Hugon and Borgers (1966) applied a very similar technique for demonstrating alkaline phosphatase, with sodium β-glycerophosphate as substrate, lead nitrate as capture agent and magnesium chloride as activator, and with tris-maleate buffer. The reaction was carried out at pH 8.2 and at pH 9, with stronger positivity at the lower pH, presumably because they found it necessary to halve the relative concentrations of lead to buffer at pH 9 to prevent deposition of lead hydroxide. Satisfactory results with freedom from artefact were found when freshly prepared medium was used for short incubation periods of 5 to 15 minutes. This method has been widely used in the ultrastructural demonstration of alkaline phosphatase. In an attempt to improve the stability of the lead medium and allow a sufficient concentration of lead ions to be retained in solution at pH 9 for the demonstration of the alkaline enzyme activity at its optimum pH, El-Aaser and his colleagues (1975) adapted a method previously used by Mölbert *et al.* (1960) in which a tris buffer containing tartrate as chelating agent for the lead is used. Mölbert and her colleagues added potassium sodium tartrate in a concentration of about 4% to tris buffer and used disodium phenyl phosphate as substrate and lead nitrate as capture agent. The medium proved unstable at pH values above 7.6 and the positivity found may not have been wholly attributable to alkaline phosphatase. El-Aaser *et al.* used a much lower concentration of tartrate, 0.025 per cent of L-tartaric acid, in tris buffer and found this a considerable improvement, providing a stable medium which retained the lead in solution even at pH 11.0 when sodium β-glycerophosphate was used as substrate. Other substrates gave a less stable medium and the authors found evidence to support the suggestion that the hydroxyl group of β-glycerophosphate played a part in retaining the lead in solution by chelation.

We have used this method ourselves on smears of blood and bone marrow cells and obtained results which compare very well in localization of positivity with those found using standard azo-dye methods. Positivity tends to be more sharply defined in a discretely granular pattern than with most of the azo-dye techniques, and the stained preparations store well for long periods without fading (Plate 13A).

Attempts have also been made to stabilise the incubation medium for alkaline phosphatase by adding high molecular weight inert polymers. Thus Szmigielski (1971) explored the effect of adding varying percentages of dextrans with molecular weights of 110 000 and 250 000, and Ficoll with molecular weight 400 000. Precipitation of lead from the medium was prevented and satisfactory staining at both light and electron microscope level was achieved with dextran of MW 250 000 at concentrations between 3 and 15%.

Improvements in the lead methods for localization of acid phosphatases at both light and electron microscope levels have resulted from the use of substrates other than β-glycerophosphate. Among these, p-nitrophenyl phosphate (pNPP) was shown by Barka and Anderson (1962) to be hydrolysed more rapidly than β-glycerophosphate by rat liver acid phosphatases, thus increasing the sensitivity of the cytochemical reaction. Ryder and Bowen (1975) used pNPP as substrate in the study of acid phosphatase activity during cellular autolysis and obtained satisfactory ultrastructural localization of the enzyme when the standard Gomori media had proved unsuccessful. They found that the pH optimum for reactions with β-glycerophosphate in biochemical tests was 3.8, whereas several optima over a fairly wide pH range were found with pNPP, and they concluded that the cytochemical differences in reactivity might reflect different substrate specificities of various acid phosphatase isoenzymes. This question is discussed in relation to haemic cell reaction patterns on pp. 153–8 below.

The ultrastructural demonstration of alkaline phosphatase in haemic cell preparations, especially in cell suspensions, has presented the usual difficulties associated with poor intracellular penetration of the reaction medium. Satisfactory results have been claimed using lead-based reactions similar to those for light microscopy, with lead nitrate plus tartaric acid or with a saturated solution of lead citrate as the capture agents (Mayahara *et al.*, 1967). Wilson *et al.* (1981) found this approach satisfactory, using either β-glycerophosphate or an alternative, and perhaps more physiological, substrate, pyridoxal-5′-phosphate (Smith and Peters, 1981; Wilson *et al.*, 1983).

More recently a method originally proposed by Veenhuis *et al.* (1980) in which the metal cerium is used as capture agent instead of lead has aroused considerable interest (Robinson and Karnovsky, 1983; Hulstaert *et al.*, 1983; Angermuller and Fahimi, 1984). Robinson (1985) used cerium capture together with the addition of the detergents, Triton X-100 and saponin, to the cytochemical media to increase the permeability of the cells. This combination permitted good intracellular localization of acid phosphatase and also of alkaline phosphatase within neutrophils without the loss of cell surface reaction. In the absence of the detergents reactivity to alkaline phosphatase was confined to the cell surface. The cerium-based procedures are claimed to offer improvement over lead-based methods in stability of substrates and ease of handling, but have the drawback of being limited to electron microscopy since the cerium reaction product is scarcely visible in the light microscope (Borgers and Verheyen, 1983).

Azo-dye methods

The second group of techniques for demonstrating phosphatases cytochemically involve the use of substrates containing the phosphate of naphthol or a naphthol derivative. As enzymatic hydrolysis occurs in the presence of activating magnesium ions the liberated naphthol, rather than the phosphate, is captured by a suitable diazotised amine, the coupling resulting in the formation of an insoluble brightly coloured precipitate of azo-dye.

The steps in the reaction may be summarized as follows:

(a)

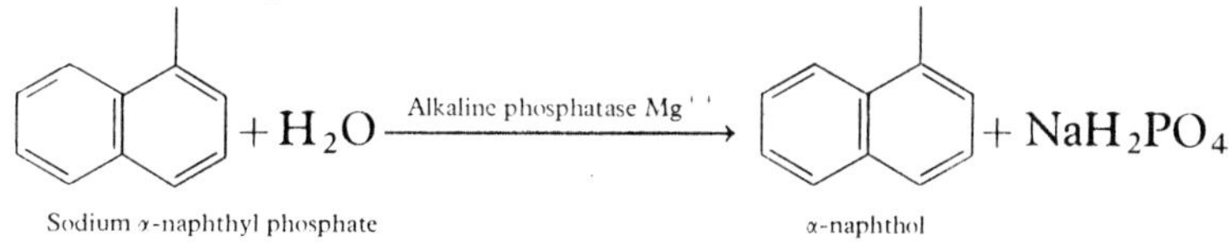

(b)

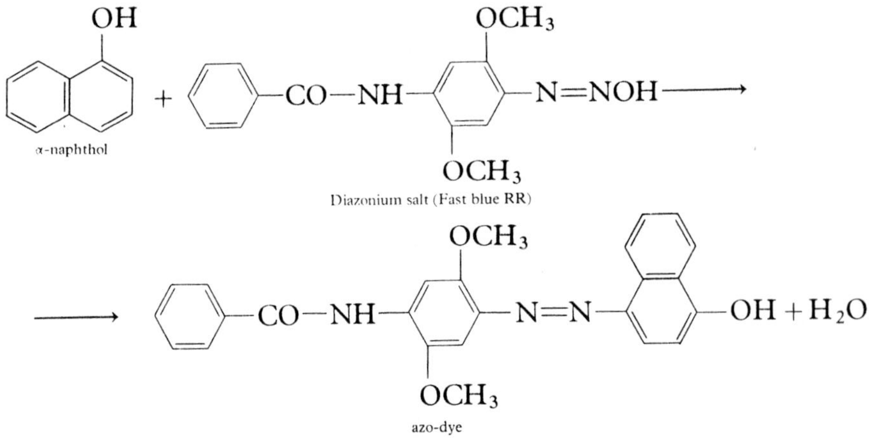

The method was first introduced for the demonstration of alkaline phosphatase by Menten *et al.* (1944) and a modification suitable for use with smear preparations of blood and bone marrow was described by Kaplow (1955) and further developed by Hayhoe and Quaglino (1958), Kaplow (1963) and Rutenburg *et al.* (1965). The various methods used in haematology differ in detail with regard to fixative procedures, the choice of naphthol derivative—whether sodium α-naphthyl phosphate, or more complex naphthols such as naphthol AS (3-hydroxy-2-naphthanilide), naphthol AS-D (3-hydroxy-2-naphtho-*o*-toluidide), naphthol AS-BI (7-bromo-3-hydroxy-2-naphtho-*o*-anisidide) or naphthol AS-LC (4′-chloro-3-hydroxy-2′5′-dimethoxy-2-naphthanilide)—and the particular diazonium salt used—Fast blue RR, Fast Garnet GBC or Fast blue BBN—but all give essentially similar results, with only minor variations in sensitivity and discreteness of granular reaction in positive cells. An important point to be made about the practical use of these methods is that commercially available naphthol derivatives vary in their purity and the extent of their contamination with trace amounts of sodium phosphate. In our own experience pure α-naphthyl phosphate synthesised in the laboratory gives totally negative results unless a small amount of sodium phosphate is added. This additive appears to act as a necessary primer to institute enzymic activity. Of all the azo-dye methods we have used perhaps the clearest and strongest positivity is found with pure α-naphthyl phosphoric acid, as substrate, but with the addition of a small quantity (0.1 mg per 100 ml of medium) of Na_2HPO_4. Diazonium salts also show some variability from batch to batch in their stability and solubility in the incubation medium, their avidity of coupling with released α-naphthol and the intensity of colour of the azo-dye produced. Most batches of the diazonium salts commonly used for alkaline phosphatase cytochemistry, Fast blue RR, Fast Garnet GBC and Fast blue BB, give satisfactory results, but adjustments in the concentrations of diazonium salt added to the incubation medium may be necessary when a fresh batch of salt is brought into use if optimum staining is to be achieved. The resulting azo-dyes differ in their solubility in alcohol, xylol, and common mounting media and immersion oils. If stained slides are scored for positivity (see below) and then discarded these differences are unimportant and in our experience Fast Garnet gives the most easily scored staining patterns (Plate 13B), but if slides are to be preserved and reviewed repeatedly preparations made with Fast Garnet need to be mounted in glycerine jelly or in PVP as recommended by Burstone (1962). Alternatively, Fast blue BB may be used, since, as Rutenburg *et al.* (1965) pointed out, the azo-dye in this case is resistant to fading when preparations are mounted in Permount.

A similar technique for cytochemical demonstration of acid phosphatases was described initially by Seligman and Manheimer (1949), who used calcium α-naphthyl phosphate as substrate at pH 5.0 and anthraquinone-1-diazonium chloride as coupling agent. Long incubation periods were required and most workers found the method unsatisfactory, with diffuse background staining and non-specific precipitation of dye. Grogg and Pearse (1952) studied the capacity of a wide selection of diazonium salts to couple with α-naphthol at pH 5.0, the extent to which they inhibited enzyme activity, and their stability in the incubation medium. Of 23 diazonium salts tested six gave reasonably satisfactory results, the best being Fast Garnet GBC, which was stable in solution, gave minimal enzyme inhibition, and coupled effectively to give an insoluble highly coloured azo-dye at the acid pH. We have for many years used sodium α-naphthyl phosphate and more recently naphthol AS-BI phosphoric acid as substrate and Fast Garnet as capture agent in demonstrating acid phosphatase in haemic cells and have found the method generally satisfactory, giving clear cut granular positivity with reproducible distribution. Variations in substrate include the use of different naphthol AS or AS-D phosphates, with which several diazonium salts such as Fast Garnet, Fast blue BB and freshly hexazotised pararosaniline (HPR) have proved to be effective coupling agents. The combination of naphthol AS-BI phosphate and HPR as advocated by Barka and Anderson (1962) and applied to the study of human blood cells by Goldberg and Barka (1962) has been widely used in haematology and we have found it to give well defined and reproducible patterns of positivity. Nevertheless, the patterns are not identical with those shown when Fast Garnet is used as capture agent, in particular diffuse rather than particulate positivity being shown in granulocytes. Moreover the incubation period of two hours recommended by Goldberg and Barka may be reduced to 30 to 60 minutes with advantage if more generalised diffuse reactions in most leucocytes are to be avoided and cell preservation improved. Although several of these methods give satisfactory results in our experience, we currently prefer naphthol AS-BI phosphoric acid as substrate and Fast Garnet as coupling agent. Further details are given in the technical methods section of this Chapter. The reaction product may be protected from fading by mounting in PVP or glycerine jelly.

The differing reaction patterns obtained by the use of different substrates and coupling agents seem unlikely to be due simply to varying sensitivity of substrates to the same enzyme, but more probably result from heterogeneity of phosphatase enzymes, with different isoenzymes predominating in certain cell types and perhaps in certain subcellular structures (Li *et al.*,

1970a and b). Differences in the substrate specificities of these isoenzymes may account for the variation of reaction pattern seen with different technical methods. The subject is further discussed below in relation to the distribution of acid phosphatase positivity in blood and bone marrow cells in health and in disease states.

The range of satisfactory simultaneous coupling azo-dye techniques probably provides an adequate means for exploring acid phosphatase positivity, and post-incubation coupling methods are now rarely used. Perhaps the best of these was the method developed by Rutenburg and Seligman (1955) which employed sodium 6-benzoyl-2-naphthyl phosphate as substrate at pH 5 with post-incubation coupling with Fast blue B salt at pH 7 to 8. The results with this method proved somewhat variable and post-incubation coupling methods have generally now been superseded by simultaneous coupling techniques.

CYTOCHEMICAL REACTION PATTERNS IN HAEMIC CELLS AND THEIR INTERPRETATION

Alkaline phosphatase (EC 3.1.3.1)

The early studies of Wachstein (1946), using the Gomori technique, had suggested that alkaline phosphatase activity of blood cells was confined to the mature cells of the granulocyte series, and that the cell content of enzyme was much reduced in chronic myeloid leukaemia but elevated in inflammatory leucocytosis, leukaemoid reactions and myelofibrosis. These results were in agreement with those obtained by quantitative biochemical analyses of plasma-free leucocyte suspensions (Haight and Rossiter, 1950; Valentine and Beck, 1951; Valentine *et al.*, 1952; Wiltshaw and Moloney, 1955; Valentine *et al.*, 1957a) but the problems of artefactual staining with the Gomori methods discussed earlier led to considerable confusion in interpreting cytochemical results and most later studies of alkaline phosphatase cytochemistry in haemic cells have been carried out with azo-dye methods. The positivity shown by these methods appears confined to the cytoplasm and is virtually restricted in normal blood to a proportion of the neutrophils (Plate 13A, B, C). Eosinophils rarely show some background cytoplasmic staining with negative specific granules but basophils are normally quite negative and indeed remain so in pathological states, although Parwaresch (1976) has reported a positive reaction in basophils together with enhanced activity in neutrophils in a single case of basophilia associated with osteomyelofibrosis without myeloid metaplasia. Other blood cells, including lymphocytes, monocytes, erythrocytes and platelets are almost invariably negative, although faint staining is sometimes observed in occasional platelets, and a very few normal lymphocytes, less than 1%, may show weak or moderate cytoplasmic reaction (Kaplow, 1968; Dikow *et al.*, 1985). In bone marrow smears the only haemic cells normally showing any positive staining are late granulocytes, starting at the metamyelocyte stage, where a very occasional cell shows faint positivity, up to segmented neutrophils which show variability in their intensity of staining. Reticulo-

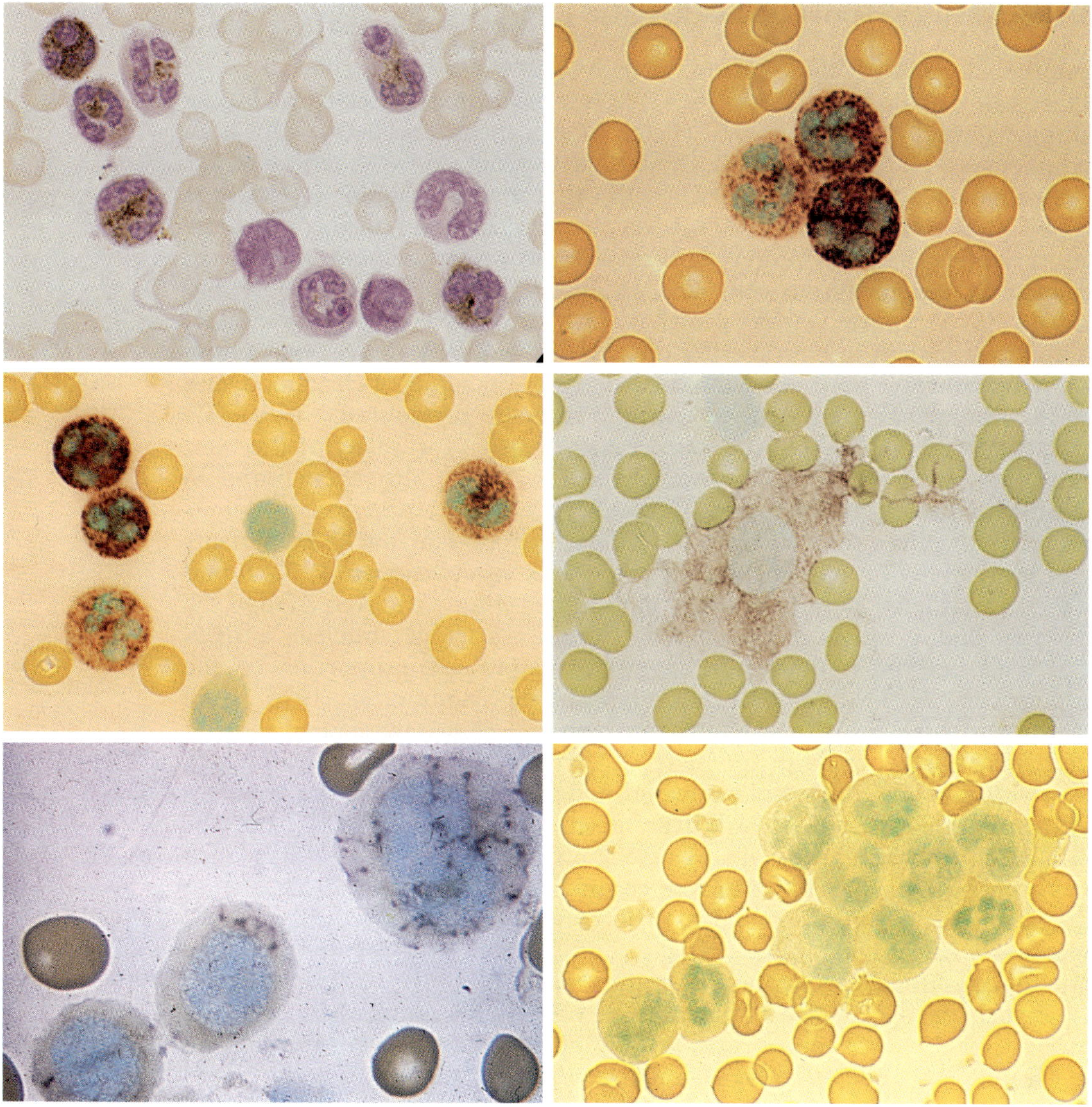

Plate 13

A. Alkaline phosphatase positivity in neutrophils stained by the lead method. Preparation photographed after storage for 12 months. Three neutrophils show 1 plus, two 2 plus, and one 3 plus positivity.

B. Alkaline phosphatase positivity in neutrophils stained by the azo-dye method with Fast Garnet as diazonium salt. The three cells show respectively 2, 3, and 4 plus positivity.

C. Alkaline phosphatase positivity of 2, 3 and 4 plus positivity in neutrophils, with two negative lymphocytes, in a preparation stained by the azo-dye method.

D. Alkaline phosphatase positivity in a marrow macrophage. Azo-dye method.

E. Alkaline phosphatase positivity in the form of rods and dots in three megaloblasts. Azo-dye method.

F. Alkaline phosphatase reaction in neutrophils in chronic myeloid leukaemia, showing negative reaction. Azo-dye method.

endothelial cells of the marrow usually show strong enzyme activity (Plate 13D), as do osteoblasts (Fig. 7.1) but other marrow cells, including younger granulocyte precursors, erythroblasts, megakaryocytes, almost all lymphocytes, and monocytes, normally fail to react.

Alkaline phosphatase positivity in cells other than neutrophil polymorphs has been relatively little studied, at least by cytochemical means, but there are variations of some interest and importance in pathological states. In leukaemias the primitive cells are usually negative, but Schubert and Martin (1965) reported strong positivity in 19% of myeloblasts in the blood of a unique patient in blastic crisis of CML, Tomonaga *et al.* (1984) found 24% of monocytes positive in an unusual acute leukaemia of childhood, and we have ourselves seen a single example of acute myeloblastic leukaemia with leucocyte alkaline phosphatase (LAP) positivity in many of the blast cells. Positive cytochemical reactions for LAP have also been found in cultured clones of promyelocytes (Greenberger *et al.*, 1975). In continuous tissue culture lines derived from the blood or bone marrow of 28 patients with haematological malignancies, Karpas and his colleagues (1977) found cytochemical LAP activity in many of the lines (Fig. 7.2), even though the cytology and positive reactions for Epstein-Barr (EB) nuclear antigen suggested that most lines were 'lymphoblastoid'. A proportion of the tumour cells in one case of Burkitt's lymphoma (Mann *et al.*, 1976), have been found LAP positive, while in our own experience occasional cases of CLL may show positivity in many of the lymphocytes (Fig. 7.3). LAP positivity in many 'prolymphocytes' in node imprints made from a patient with prolymphocytic leukaemia was observed by Leder (1978), but the circulating neoplastic cells in this case showed no positivity. The enzyme may thus be more prominent or more readily demonstrable in cells from lymph nodes than in blood and marrow lymphocytes. Nanba *et al.* (1975) observed LAP activity on the surface membranes of lymphocytes in certain cases of non-Hodgkin's lymphoma and a systematic study was later made of the cytochemically or histochemically demonstrable enzyme in neoplastic lymphocytes in sections or smears of biopsied material from a total of 60 lymphomas or leukaemias (Nanba *et al.*, 1977). Positivity was found only in certain B-cell malignancies, including three of six lymphocytic lymphomas of 'intermediate differentiation', four of 13 nodular lymphomas and one of seven Burkitt's lymphomas. Well differentiated lymphocytic malignancies, including CLL and lymphocytic lymphomas were negative (14 cases), as were eight other diffuse B-cell tumours, eight lymphoblastic lymphomas and the T cells of four cases of Sezary's syndrome. In normal control lymph nodes LAP positive lymphocytes were confined to primary follicles and the mantle zones of secondary follicles; and the authors suggest that positively reacting lymphoid neoplasms are derived from these sites.

A further study of 87 cases of non-Hodgkin's lymphoma was reported by Poppema *et al.* (1981). Again, positivity for alkaline phosphatase was found in only a minority of cases, 10 in total, and all these showed surface Ig indicating their B-cell origin. Most were follicle-centre cell tumours, but they did not show common characteristics, either pathological or clinical,

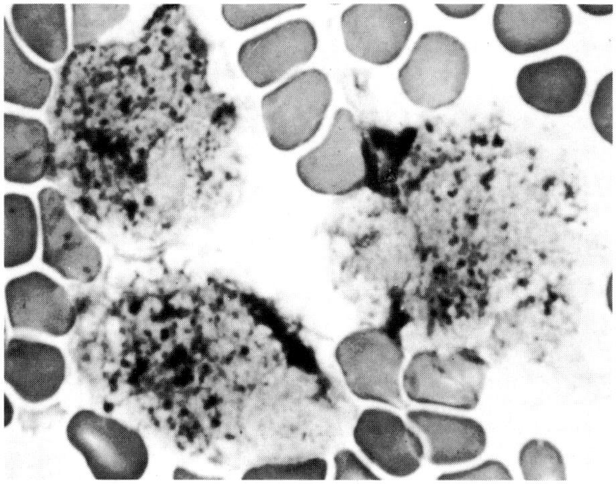

Fig. 7.1 Group of three osteoblasts in normal bone marrow. Alkaline phosphatase, azo-dye method. Strong granular positivity.

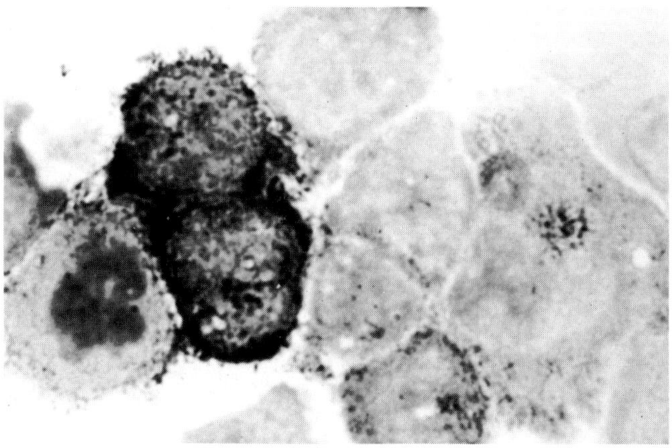

Fig. 7.2 Cultured haemic cells, EBNA positive, stained for alkaline phosphatase by the lead method. Strong positivity present in two cells and weaker reactions in several others, including one in metaphase.

sufficient to imply that they represented a separate entity.

Goffinet *et al.* (1983) described cytochemical studies of alkaline phosphatase at both light and electron microscope level in the thymus of normal C57BL/ka mice. Enzyme was found in lymphoblasts of fetal thymus, suggesting that expression was limited to an early stage of the T-cell differentiation pathway. Following inoculation of post-radiation leukaemia virus an increase in enzyme level occurred as malignant transformation emerged. These observations refer, of course, to T cells in the mouse, but the authors note that high alkaline phosphatase activity has been recorded in some pre-B and B-cell lymphomas in mice, so that a parallel with the

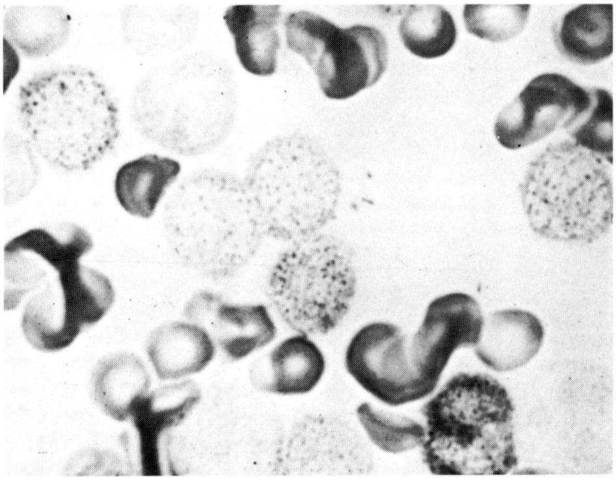

Fig. 7.3 Buffy coat; CLL in good clinical and haematological control. Alkaline phosphatase, azo-dye method. Most lymphocytes in this very unusual case show scattered positive granules; a single neutrophil shows + + positivity.

findings in man exists, although hitherto neither normal T cells nor the cells of T-cell lymphomas have been reported to show enzyme positivity in man.

These cytochemical reactions in lymphoid cells may be demonstrating an alkaline phosphatase, the so-called N-APase, described by Neumann and her colleagues (1976) as peculiar to the lymphoid cell series, which differs biochemically from the usual granulocytic LAP in its substrate specificity, reacting with monoesters of orthophosphoric acid but not with S-substituted monoesters of thiophosphoric acid, whereas the granulocytic LAP, like most other alkaline phosphatase isoenzymes, reacts with both classes of substrate. This lymphoid enzyme has been found by biochemical assays in extracts of CLL lymphocytes and in human lymphoblastoid cell lines and its recognition would add a dimension to the study of alkaline phosphatase in haemic cells, if a simple cytochemical technique for its demonstration could be developed.

For example, in one of our patients with a leukaemic phase of non-Hodgkin's lymphoma, follicular centrocytic-centroblastic in type, 20% of the circulating lymphocytes (centrocytes) showed positivity for alkaline phosphatase. The cells also manifested scattered weak butyrate esterase and acid phosphatase, showed surface Ig with kappa chains and failed to show mouse erythrocyte rosetting. Biochemically the alkaline phosphatase was increased 10-fold over normal lymphocyte levels and 65 to 70% of the activity was attributable to the N-APase isoenzyme (Flemans *et al.*, 1980).

Unfortunately neither Gomori methods nor azo-dye techniques are easily adaptable to the detection of reaction products arising from the splitting of thiophosphate monoesters and a suitable method for identifying N-APase activity cytochemically has not yet been devised.

In erythroblasts positivity has been observed, usually in the form of dots and rods of reaction product in the cytoplasm and often overlying the nucleus, especially in conditions of erythroblastic hyperplasia (Plate 13E). Quaglino (1964) noted such positivity in from 12 to 50% of bone marrow erythroblasts in two cases of haemolytic anaemia and five cases of pernicious anaemia. Again, Löffler and Pralle (1970) found alkaline phosphatase positivity both cytochemically and biochemically in erythroblasts and erythrocytes in members of a family with erythrocytosis. Strongly positive R-E cells and surrounding erythroblast islands are usually prominent in the marrow when erythroblastic positivity is observed, and there is a possibility that the enzyme may be transferred from the R-E cells to the erythroblasts by rhopheocytosis. Marked increase in numbers of alkaline phosphatase positive R-E cells has been noted by Beckstead *et al.* (1981) in sections of plastic embedded bone marrow from cases of early myelofibrosis.

The intracellular localization of alkaline phosphatase in human haemic cells has been relatively little studied at the ultrastructural level. Nakatsui (1969) worked with unfixed neutrophils from peripheral blood and thought that the enzyme reactivity was localised in the secondary granules, but the cell preservation was very poor and the reaction product appeared in only a few vacuolar structures of very variable size. Bainton and her associates (1971) also concluded that LAP activity was ultrastructurally localised in cisternae of the Golgi apparatus and in immature secondary granules. In the micrographs illustrated, however, the positive reactions appear chiefly in small granules of irregular or elongated shape which resemble the tertiary granules as described by Scott and Horn (1970) and Cawley and Hayhoe (1973). This localization would conform with the light microscope finding of positivity occurring normally in a proportion of granulocytes only at a late stage of development, beyond the metamyelocyte, when tertiary granules are being formed in increasing numbers. Confirmation that the enzyme is chiefly present in small granules comes from biochemical studies with subcellular fractions. Young and Prager (1962), for example, found alkaline phosphatase activity in a more slowly sedimenting fraction of human neutrophils than acid phosphatase, as would be expected if the former were in tertiary granules and the latter in primary granules, while Clarke (1965) found maximum alkaline phosphatase activity in a 27 000 g fraction composed mainly of granules.

West *et al.* (1974), Bretz and Baggiolini (1974) and Spitznagel *et al.* (1975), using cell fractionation techniques found the enzyme-rich component to be distinct from that containing the specific (secondary) granules of polymorphs. Borgers *et al.* (1978) provided some cytochemical confirmation of the suggestion made by Bretz and Baggiolini (1974) that the alkaline phosphatase was contained chiefly in plasmalemma fragments, by finding reaction product in the membranes of endocytic vacuoles, in plasma membranes and in some vesicular cytoplasmic structures. Rustin *et al.* (1979) used both analytical subcellular fractionation by density gradient centrifugation and ultrastructural cytochemistry to identify the intracellular localization of alkaline phosphatase in neutrophil polymorphs. They claimed the separability of particles bearing this enzyme from primary and secondary granules, plasma membranes and mitochondria, and showed EM localisation of enzyme positivity in structures which again closely resemble tertiary granules.

In their detailed studies of the ultrastructural distribution of alkaline phosphatase Rustin *et al.* (1979) and Wilson *et al.* (1981) using essentially the Mayahara technique, found some reaction on nuclear, intracytoplasmic and plasma membranes, but strongest activity in irregularly shaped vesicular structures which they believed to be unique and previously undescribed, and for which they coined the term 'phosphasomes'. Whether these structures are indeed unique, or the same as the tertiary granules which develop in later granulocytes, is perhaps of no more than semantic importance, but it is of considerable interest to note that the increase in neutrophil alkaline phosphatase activity in pregnancy (see below) was found by Wilson *et al.* to be associated with more granules, whereas the neutrophils of patients with chronic myeloid leukaemia had fewer granular 'phosphasomes' per cell but normal intensity of positivity within them.

A semiquantitative assessment of overall positivity in the circulating neutrophils is commonly made by rating the intensity of staining and number of positive granules (Plate 13A, B, C) in each of a hundred consecutive neutrophils in the manner described by Kaplow (1955) and Hayhoe and Quaglino (1958) (see Technical Methods).

The sum of 100 individual cell ratings provides the score for any specimen, with a possible range from 0 to 400. Such a score is referred to as the leucocyte or neutrophil alkaline phosphatase (LAP or NAP) score.

Peripheral blood LAP scores show some variability in normal range according to the method used: Hayhoe and Quaglino (1958) obtained a range in 50 normal subjects from 14 to 100 with a mean of 46, whereas Kaplow (1963) found a range from 13 to 160 with a mean of 61, and Rutenburg *et al.* (1965) a range from 37 to 98 with a mean of 68. The differences are of little importance, since any laboratory is likely to use a single method and to establish their own normal range as a background against which to assess abnormally low or high scores in pathological states. As Kaplow (1976) has very properly pointed out, increased sensitivity or stronger staining from prolonged incubation is not necessarily a desirable feature, because the best staining conditions are those which will demonstrate the widest spectrum of scores as found in physiological and pathological states.

The pattern of scores found in a wide range of conditions has become sufficiently established to provide valuable help in differential diagnosis. The original observations of Wachstein (1946) on the increased levels of activity in polycythaemia vera, myelofibrosis, leukaemoid reactions and inflammatory leucocytosis, contrasting with the sharp decrease in activity in chronic myeloid leukaemia, were confirmed by azo-dye scoring methods by Hayhoe and Quaglino (1958), Koler *et al.* (1958), Mitus *et al.* (1958) and by many subsequent authors. High scores were also found by Hayhoe and Quaglino (1958) in Hodgkin's disease in relapse and in carcinomatosis, the former observation being later confirmed by Bennett *et al.* (1968), Sforza *et al.* (1971) and Simmons *et al.* (1973), and the latter by Lille-Szyszkowicz *et al.* (1966). Less markedly elevated scores, with a range overlapping the upper levels of normal, are found in chronic lymphatic leukaemia (Hayhoe and Quaglino, 1958) and acute lymphoblastic leukaemia (Hayhoe *et al.*, 1964). Other haematological states often associated with high LAP scores include aplastic anaemia (Lewis and Dacie, 1965), multiple myeloma (Hayhoe and Neuman, 1976; Brook and Dreisbach, 1977) and hairy cell leukaemia (Fig. 7.4) (Hayhoe *et al.*, 1977).

Scores generally within the normal range have been found in non-Hodgkin's lymphoma of all varieties (Hayhoe and Quaglino, 1958) in chronic myelomonocytic leukaemia and in most anaemias due to deficiencies of haematinic factors or to haemolysis not secondary to leukaemias or Hodgkin's disease.

Low scores are almost invariably found in typical Ph^1 positive CML (Plate 13F) and also in the Ph^1 negative juvenile form, but in atypical Ph^1 negative CML in adults scores may sometimes be normal or high (Tanzer *et al.*, 1964; Hayhoe and Hammouda, 1965) and in every reported case of chronic neutrophilic leukaemia in which an LAP score has been carried out the level has been notably high, usually from 300 to 400 (You and Weisbrot, 1979). Even Ph^1 positive cases may sometimes show raised scores in the chronic phase, but they are usually atypical in some other respect; for example, there are now six reported examples of CML developing after prolonged ingestion of lithium, and although all were Ph^1 positive they all showed normal LAP scores (Moreb and Herschko, 1985).

High scores are frequently found during terminal metamorphosis of CML especially when blastic crisis emerges (Quaglino, 1961; Hammouda *et al.*, 1964). In acute myeloid leukaemias LAP scores are low when myeloblasts or promyelocytes predominate but usually normal or high when monocytic precursors predominate. Similarly in erythroleukaemias of erythromyeloblastic pattern low scores are usual whereas higher scores are found when the pattern is erythromonoblastic (Hayhoe *et al.*, 1964). Two further haematological disorders with conspicuously low LAP scores are paroxysmal nocturnal haemoglobinuria and infectious mononucleosis (Tanaka *et al.*, 1960; Sramkova *et al.*, 1965; Lewis and Dacie, 1965).

The scoring procedure as commonly used is quick and easy, but other methods of assessing neutrophil alkaline phosphatase activity, emphasising the distribution of reaction intensity among single cells, have been claimed to be equally or more rapid and superior in discrimination. Bendix-Hansen and Helleberg-Rasmussen (1985) for example applied such a method to the neutrophils of 50 healthy subjects and 89 leukaemics. Their findings were essentially in accord with those of the more customary scoring method but their technique may have something new to contribute if it is widely applied in a range of haematological disorders (Bendix-Hansen, 1987).

LAP scores in all the conditions listed above provide information helpful in diagnosis when taken in conjunction with other findings. The contribution they may offer in practical diagnosis and in monitoring the progress of disease and response to therapy is discussed in Chapter 14, but it is appropriate here to consider in general terms what is known or conjectured as to the mechanism by which changes in LAP activity are brought about.

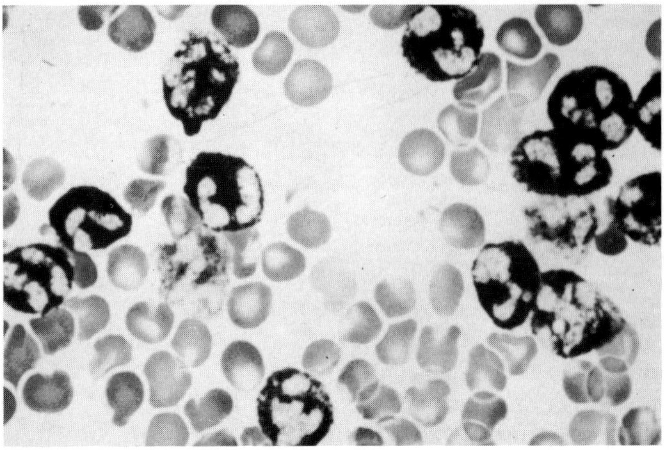

Fig. 7.4 Buffy coat, HCL. Alkaline phosphatase, azo-dye method. Very strong positivity, 3 to 4 plus, in most neutrophils.

Most tissues having a sufficient content of alkaline phosphatase to be analysed appear to possess tissue-specific isoenzymes, separable from each other in biochemical, immunological and electrophoretic properties and presumably differing in their amino acid sequences. Each of these isoenzymes may in turn comprise a family of isozymes (Markert and Moller, 1959) having essentially the same structure but differing in their electrophoretic mobility. Thus, isoenzymes of alkaline phosphatase peculiar to liver, bone, intestine, placenta, HeLa cells and certain tumours have been recognised. The subject was reviewed by Fishman (1974), but his otherwise comprehensive and informative article contained little reference to the alkaline phosphatase of leucocytes. Nevertheless, sufficient work has been done on the characterization of leucocyte alkaline phosphatases to suggest that they represent a separable isoenzyme, less severely inactivated by heat than the liver or bone isoenzymes but not so resistant as the placental isoenzyme, and less inhibited by phenylalanine than the placental or intestinal isoenzymes but more so than the liver or bone isoenzymes (Bottomley *et al.*, 1969; Pralle *et al.*, 1973).

Miller *et al.* (1983), using enzyme inhibitors and heat inactivation, confirmed these findings and showed leucocyte alkaline phosphatase to be distinct from other organ specific alkaline phosphatases, especially the placental enzyme. They showed that the enzyme in leucocytes with normal levels was similar or identical to that in leucocytes with high levels. The results of a study of the biochemical and immunological properties of human neutrophil alkaline phosphatase in comparison to other established alkaline phosphatase isoenzymes, carried out by Smith *et al.* (1984) using affinity column chromatography and inhibition with L-phenylalanine, L-homoarginine and levamisole gave confirmation that the leucocyte enzyme was quite different from both placental and intestinal alkaline phosphatase isoenzymes. A resemblance was found, however, with the liver, kidney and bone isoenzymes in biochemical parameters, and a monoclonal antibody raised to human liver alkaline phosphatase crossreacted with isoenzymes from kidney, bone and neutrophils. The authors concluded that these four isoenzymes were probably essentially identical, possibly the product of the same structural gene coding for all four. This proposal might imply the existence of three main alkaline phosphatase genes in man, respectively for placental, intestinal and the liver, kidney, bone and leucocyte isoenzymes. In this context it is of interest to note that Stinson (1984), studying the size and stability to sodium dodecyl sulphate of the three main alkaline phosphatases from human placenta, liver and small intestinal mucosa found all the isoenzymes to be natural dimers, with subunit molecular weights respectively 68 300, 80 800 and 79 400.

Ueda (1983), using polyacrylamide gel electrophoresis, noted that leucocyte alkaline phosphatase moved more anodal to the origin than did other phosphatases, but again noted general similarity to liver and bone enzymes. How extensive a family of alkaline phosphatase isozymes exists is not entirely clear. On starch gel electrophoresis two or three bands have been derived from butanol extracts of normal neutrophil leucocytes. Thus Lyons *et al.* (1968) found three major isozymes, F (fast), I (intermediate) and S (slow) in extracts of normal leucocytes. The patterns were unchanged in mongols, where a high LAP score is usually found, and in polycythaemia, but the F band was absent in CML and in post-partum mothers and reduced in patients treated with prednisone. The raised LAP levels in pregnancy appeared to be due to an increase in the specific I leucocyte isozyme and not related to the elevated serum enzyme levels derived from placental isoenzyme.

More widely variable patterns with extra bands have been reported by some workers in various conditions such as CML, pregnancy, and paroxysmal nocturnal haemoglobinuria (Klein *et al.*, 1966; Robinson *et al.*, 1966) whereas Bottomley *et al.* (1969) found only two bands with no significant differences in electrophoretic patterns between enzyme derived from normal cells, the strongly positive cells of reactive granulocytosis or the nearly negative cells of CML. The variability in electrophoretic results may perhaps be due to differing amounts of sialic acid residues. In their very detailed series of experiments Bottomley and his colleagues (1969) compared the alkaline phosphatase of leucocytes in these three conditions using column chromatography, sucrose density ultracentrifugation, starch gel electrophoresis, determination of Michaelis constant for p-nitrophenyl phosphate, heat inactivation, inhibition by L-phenylalanine, Ouchterlony double immunodiffusion precipitation and immunoelectrophoresis using rabbit anti-LAP serum. The antiserum was also used to quantitate the amount of alkaline phosphatase protein in purified enzyme preparations. Purified enzyme preparations from each of the three leucocyte sources, CML, normal blood and reactive leucocytosis, were similar in most biochemical and biophysical characteristics, with molecular weights between 5 and 600 000, comparable Km values for p-nitrophenyl phosphate and similar susceptibility to inhibition by heat and phenylalanine. Immunologically they appeared identical. The enzyme specific activities were markedly different, however, in material from the three sources, with low specific activity in CML leucocytes, intermediate activity in normal leucocytes and very high activity in leucocytes of reactive granulocytosis. At least in the circumstances of these experiments, it would appear that variations in alkaline phosphatase activity demonstrable in leucocytes did not reflect alterations in the amount of enzyme synthesis but rather changes in the specific activity of the enzyme molecules.

Conflicting results were later reported by some other workers using similar approaches to the same problem. Rosenblum and Petzold (1975) found reduced amounts of a normally active enzyme protein in CML leucocytes, and Rustin and Peters (1979), using more sensitive assay techniques, confirmed that the alkaline phosphatase isoenzymes in CML were not qualitatively different from normal but were quantitatively deficient.

Nevertheless, the finding of two leucocyte isoenzymes has more recently been confirmed by Miller *et al.* (1984), using polyacrylamide gel electrophoresis with Triton X-100. The two isoenzymes were designated LAPs (slow) and LAPf (fast) respectively. Thirteen of 15 normal leucocyte extracts contained the LAPs isoenzyme only, the remaining two having predominantly LAPs together with 10% or less of LAPf. In eight cases of CML the neutrophils contained only small amounts of LAPs and no LAPf, and in six reactive leucocytoses with increased LAP levels the isoenzyme was again entirely LAPs. In three of five cases of polycythaemia vera there was again LAPs only but the two other cases showed both LAPs and LAPf. The most striking finding in these studies occurred in blastic crisis of CML, where 10 of 11 cases showed both LAPs and LAPf, with the latter sometimes even predominant. LAPf may perhaps emerge as a possible marker for blast crisis, especially if it predominates over the usual common LAPs isoenzyme.

Further studies of these two isoenzymes by Miller *et al.* (1985), were based on the extraction and purification of alkaline phosphatase from normal granulocytes and from the granulocytes of a single case of CML in blastic crisis and with a high LAP score. In the 'normal' material a single protein with LAP activity was found, with molecular weight 61 000 determined by SDS gel electrophoresis, while the LAP activity in the extract from CML in blastic phase was contained in two proteins with respective molecular weights 61 000 (LAPs) and 45 000 (LAPf). These two proteins showed differing sensitivity to heat and distinct isoelectric points. In general these findings provide a compromise between the views that changes in LAP score in pathological states reflect on the one hand altered enzyme activity or, on the other, altered enzyme quantity. In some cases, for example the lowered scores of CML or the raised ones of reactive leucocytosis, the change is quantitative with altered amounts of the slow LAPs, whereas in others, especially the blastic crisis of CML, a qualitative change occurs, with presumed derepression of the allele, normally largely inactive in health, which regulates the synthesis of the fast LAPf.

Whatever the outcome of this controversy, the possibility that enzyme activity may vary independently of enzyme synthesis agrees with the studies of Griffin and Cox (1966), who found that prednisolone caused an increase in alkaline phosphatase activity in HeLa cells in culture, but that although protein and RNA synthesis were involved there was no increase in the rate of enzyme synthesis or decrease in enzyme catabolism. Cox *et al.* (1971) suggested that the enhanced enzyme activity resulted from the synthesis of a modifier substance which reacted with the enzyme to affect the binding of zinc. Zinc is certainly one of the most important metallic activators of alkaline phosphatase activity in leucocytes as in various other tissues, although magnesium is also necessary and is usually added to cytochemical incubation media for demonstrating enzyme activity. Calcium and cobalt may also activate alkaline phosphatase (Nguyen-van Thoai *et al.*, 1947) as may some amino acids such as glycine (Bodansky, 1936). The possible relevance of zinc concentrations to LAP activity is further demonstrated by the presence of considerable amounts of protein-bound zinc in normal leucocytes (Vallee, 1959), with substantial reduction in CML and higher levels in infections (Gibson *et al.*, 1950). Moreover Trubowitz *et al.* (1959) and Valentine *et al.* (1960) showed that LAP activity could usually be inhibited by EDTA and reactivated by zinc but not by magnesium, although, as Perillie and Finch (1967) demonstrated, the LAP level in some patients with non-leukaemic granulocytosis is not affected by EDTA and may depend on other factors for primary activation.

Bottomley *et al.* (1969) discussed the possibility that the changes in specific activity of LAP found in their study might be due to differences in the zinc to enzyme ratio, but was unable to produce supporting evidence. In particular, the addition of zinc failed to increase the specific enzyme activity in leucocytes from CML. Whether or not the concentration of zinc is important, there has long been strong evidence from biochemical studies that a relationship exists between adrenal cortical activity and LAP levels. Valentine *et al.* (1954) found marked increases in phosphatase levels in conditions associated with stress, such as myocardial infarction, diabetes, infections, haemorrhage and cerebral vascular accidents, the levels falling with recovery or response to treatment. The administration of ACTH or cortisone to normal subjects led to comparable increases in LAP. Conditions of clinical stress not showing the feature included rheumatoid arthritis, acute rheumatic fever, and, of course, CML. ACTH injections usually failed to increase the low LAP levels in CML, although normal responses occurred in occasional CML patients in good remission (Valentine *et al.*, 1957b).

Another example of stress increase in LAP activity, perhaps mediated by steroids, is that following surgery. Valentine *et al.* (1954) using biochemical enzyme assays, and Wyllie (1962) and Kaplow and Beck (1964) using cytochemical methods, demonstrated a sharp rise in LAP levels with a maximum at 2 to 3 days postoperatively. The levels returned gradually to normal over the next 4 to 6 days. The changes in LAP levels were delayed as compared with the postoperative leucocytosis, which reached a maximum within 24 hours and returned to normal levels within 3 or 4 days. A very similar pattern of response, with neutrophil leucocytosis and left shift preceding by 24 to 36 hours the development of maximum increase in LAP score, occurs after injection of bacterial endotoxin in the 'pyrogen test'

(Merker, 1961, 1965). The lack of coincidence between the neutrophil peak and the LAP activity peak makes it unlikely that a release of sequestrated positive neutrophils can be responsible for the changes following operation or endotoxin injection, and a more likely explanation is that young granulocytes are released from the marrow and subjected to a phosphatase-inducing influence. Their initial appearance in the peripheral blood while they still retain the negative or weak LAP activity of stab cells and young polymorphs produces the neutrophil peak with low LAP score, while the subsequent induction of enzyme as the cells age and approach the end of their circulating phase gives the delayed LAP activity peak, with the majority of neutrophils showing intense positivity. Some support for this proposed mechanism comes from the demonstration by Boggs and his colleagues (1968) of a 'neutrophil-releasing' factor in human plasma following endotoxin injection. The phosphatase-inducing factors remain unidentified, but corticosteroid hormone seems likely to be among them. Comparable findings were reported by Mishler and Williams (1980), using aetiocholanolone and prednisolone as the leucocytosis-inducing agents, the rise in neutrophil numbers again preceding the rise in LAP score. The authors argued that the findings could be explained by the delayed emergence of LAP-rich juveniles from the bone marrow, but direct examination of LAP reactions in bone marrow cells shows that juvenile and stab cells are only weakly positive, if at all, and the findings described by Mishler and Williams seem more likely to be due to phosphatase induction than to a single cell-redistribution mechanism.

That other endocrine factors influence LAP activity is strongly suggested by the pattern of variation of the enzyme activity in relation to age and sex and during pregnancy. Studies of human embryo neutrophils were carried out by Keleman and his colleagues (1978) who found mature neutrophils in liver and bone marrow to be LAP negative during the first trimester, with occasional strongly positive cells appearing in the blood and haemopoietic organs in the second trimester and quite strong positivity in blood neutrophils in the third trimester, when LAP scores averaged 169, well above the normal adult range of 14 to 100. Newborn infants have long been known to have high LAP scores (Scorza *et al.*, 1961; O'Kell and Axon, 1965; Halbrecht and Shabtay, 1972; Corberand *et al.*, 1973) and these high levels and the rising fetal LAP in the second and third trimesters have been thought to reflect the progressive increase in the amount of progesterone produced in pregnancy and its free passage through the placental barrier. The levels continue high during early childhood, without significant differences between the sexes (Rosner and Lee, 1965). The levels fall off gradually to reach the adult range around puberty, although Bailer (1972), in a study of 179 healthy girls aged from 7 to 15 years, found a substantial fall in LAP score between the 7th and 10th years—from a mean figure of 67.6 to 16.2—with increase to 58.2 by 15 years. Rosner and Lee (1965) showed that in adult life average LAP levels in men are about two thirds those in premenopausal women, although after the menopause the values in women fell to the male level.

An extensive study by Stavridis *et al.* (1981) involving 440 subjects, including 20 males and 20 females in each age decade, 40 pregnant women and 40 samplings of fetal circulating cells generally confirmed these findings. LAP scores were higher in children than in adults but showed no differences between male and female children. After puberty and before 50 to 60 females had higher scores than males, but subsequently the scores levelled. These authors carried out parallel studies of LAP levels and nitroblue tetrazolium reduction activity in the peripheral blood neutrophils in these subjects and found a remarkably close concordance between them. They postulate that the sex differences in both are due to progesterone effects.

Treatment with androgens also leads to decreased LAP values, and there is some *in vitro* evidence that androgens may inhibit LAP activity when added to leucocyte suspensions. No sex differences have been found in leucocyte zinc or magnesium values (Rosner *et al.*, 1968). During pregnancy significant elevation of LAP scores occurs within the first month, and mean scores around twice the normal average persist throughout most of pregnancy, with a further rise during the last week or two and a fall back to normal levels within a few weeks of delivery (Pritchard, 1957; Quigley *et al.*, 1960), although the levels tend to remain elevated longer after delivery in women who breast-feed (Haering, 1964).

Climie *et al.* (1962), confirming in a study of 162 women the close association between elevated LAP score and pregnancy of at least 7 weeks gestation, suggested that progesterone or a related substance might be exercising a stimulatory effect, and noted that increased LAP scores were found in women taking synthetic progesterones as oral contraceptives. *In vitro* studies of a wide range of oestrogens showed no specific effects on LAP levels, but it seems clear that, *in vivo*, female sex hormones influence leucocyte enzyme activity. Part of the increase in LAP scores in pregnancy may be due to absorption of placental isoenzyme on the cell surface. Wilson *et al.* (1981) in an ultrastructural study of neutrophil alkaline phosphatase showed that the neutrophils in the third trimester of pregnancy not only showed more intracytoplasmic granules (phosphasomes) with enzyme activity, but also had large deposits of reaction product on the external surface of the plasma membrane, with properties of the placental isoenzyme found in the serum during pregnancy.

The high LAP levels in childhood suggest a possible activating role for growth hormone (Rosner and Lee, 1965). Thus in man we have evidence, mostly of an indirect nature, to suggest that LAP levels may be influenced by several hormones, being raised by ACTH and corticosteroids, and perhaps by growth hormone and by oestrogens and progesterone, and reduced by androgens. The mechanism by which this influence is exerted remains unclear, and in particular whether increases in LAP levels may sometimes be due to increased synthesis of enzyme, or whether, as the studies of Griffin and Cox (1966) and Bottomley *et al.* (1969) suggest, the quantity of enzyme remains constant but its specific activity changes. If the latter hypothesis is true, are the differences in activity mediated by a newly synthesised modifier substance which unmasks active sites on the alkaline phosphatase molecule or alters zinc binding as suggested by Cox *et al.* (1971)? These questions remain unanswered as yet.

An alternative approach to the understanding of LAP variations has involved the analysis of enzyme levels in the cells of CML, *in vivo* and *in vitro*, in the chronic phase and in blastic crisis, and the possible link between LAP activity and alterations in the small acrocentric chromosomes. The existence in the chronic phase of CML of very low LAP levels and a chromosomal abnormality, the Philadelphia chromosome (Ph^1), thought originally to represent a partial deletion of the long arm of chromosome 21, and the observation that LAP activity was high in Down's syndrome, where the karyotype contains an extra chromosome 21, led Alter and his associates (1963) to postulate that a specific gene regulating LAP activity was present in that chromosome. The return of LAP levels towards normal during periods of successful control of CML under treatment (Hayhoe and Quaglino, 1958), despite the continued Ph^1 positivity of virtually all marrow cells (Tough *et al.*, 1963) made this hypothesis unlikely, and the co-existence of high LAP scores in some cases of blastic crisis of CML with persisting Ph^1 positivity (Hammouda *et al.*, 1964) appeared to rule out any simple causal relationship between the Ph^1 deletion and depression of LAP activity. The argument that an LAP gene-dose effect might be operating in one direction in Down's syndrome and in the other in CML was based on the presumed identity of the affected chromosome in the two conditions. Once the Ph^1 chromosome was shown to involve a small acrocentric chromosome of a different pair from that involved in the trisomy of Down's syndrome (O'Riordan *et al.*, 1971) the argument could no longer be maintained.

Nevertheless, the Ph^1 abnormality, now recognised to represent a reciprocal translocation between the long arm of a 22 chromosome and usually the long arm of a 9 chromosome, might still be linked with decreased LAP activity through some mechanism other than the existence of a LAP-regulating gene on the affected chromosome. Hook and Engel (1964) put forward an hypothesis that the Ph^1 chromosomal defect might be responsible for prolongation of the intravascular life span of mature granulocytes, and that the lowered LAP activity might be due to cellular ageing. There is good evidence that the intravascular life span of granulocytes is indeed prolonged in CML (Raeb *et al.*, 1962; Mauer and Jarrold, 1963; Athens *et al.*, 1965, 1968; Galbraith, 1966; Perry, 1968), especially when there are immature neutrophils present, but the same is true in polycythaemia vera and myelofibrosis, and in these latter conditions the LAP levels are usually much elevated above normal. The hypothesis that increased circulation time *per se* leads to low LAP scores as a result of cellular ageing is therefore untenable. Moreover, the weight of evidence suggests that, in man, young granulocytes have little LAP activity and that the enzyme develops or is activated as the granulocyte ages. Thus the neutrophils in the bone marrow normally show LAP scores less than half those in peripheral blood; circulating stab cells, normally almost devoid of LAP activity, show lower levels of activity than segmented neutrophils even in the leucocytosis of infection with its associated left shift and high LAP score (Hayhoe and Quaglino, 1958). The suggestion of Keleman (1973) that the elevated LAP scores found in drug-induced neutropenia and some cases of idiopathic aplastic anaemia are attributable to release of newly formed neutrophils with high LAP activity seems less likely than the explanation for these old observations offered originally by Trubowitz *et al.* (1958), namely, that the marrow depression in these states reduces the influx of cells into the polymorph population and thus leads to an increase in the mean age of the polymorphs, with consequent rise in the LAP score. In any event, attempts to explain average LAP levels in different disease states as a reflection of the age of the predominating cell population are undoubtedly over-simplistic. Normally, stab cells are usually negative and only rarely have one plus positivity, while segmented polymorphs in the bone marrow are also mostly quite negative, with only a few showing one plus positivity, despite the fact that the mean value for transit time through the marrow stab cell stage is 66 hours and through the segmented stage to release is 95 hours, as measured by tritiated thymidine studies (Cronkite and Vincent, 1970). Circulating neutrophils appear to be destroyed or leave the bloodstream in a random fashion, with a half-disappearance time of about 7 hours, yet the level of LAP during this stage is about double that seen in the marrow phase, although few neutrophils show more than two plus positivity and the majority remain negative or only weakly positive. There is no significant extravascular sequestration of labelled neutrophils in normal subjects (McMillan and Scott, 1968) and once neutrophils leave the blood and cross the endothelial barrier into the tissues they do not return to the circulation in significant numbers (Mauer *et al.*, 1959). The oldest neutrophils are therefore to be found in the tissues, where Kaplow (1976) noted stronger LAP positivity than in the blood, but under normal circumstances cells with three or four plus positivity are very rare.

In acute infectious leucocytoses, however, despite a shortening of the marrow transit time, the peripheral blood contains numerous neutrophils with two or three plus positivity, and this degree of positivity is found even in stab cells. If account is also taken of the striking differences in LAP scores in conditions which are not associated with major changes in granulocyte numbers or kinetics such as Hodgkin's disease or diabetes, with high scores, paroxysmal nocturnal haemoglobinuria with low levels or hereditary hypophosphatasia with absent LAP, there can be no question that influences other than cell ageing affect the synthesis or activation of the enzyme. The external stimulus that leads to inflammatory leucocytosis may well also activate LAP, while the internal cellular defect in the leukaemic granulocyte clone of CML which leads to imbalance in production and release of granulocytes may also render the production or activation of LAP defective.

Some evidence for the latter hypothesis has been put forward by Pedersen and Hayhoe (1971a,b). They compared LAP activity and phagocytic capacity in neutrophils from patients with CML and found that the cells with demonstrable enzyme were more active phagocytes than the LAP negative ones. Successful treatment in CML leads to increase in the percentage of LAP positive cells (Hayhoe and Quaglino, 1958; Xefteris *et al.*, 1961; Linke and Löffler, 1963; Elves *et al.*, 1963) and also to improvement in phagocytic activity (Brandt, 1965; Marmont and Damasio, 1965; Rosner *et al.*, 1970), while in relapse the rising leucocyte numbers in the peripheral blood contain a predominance of poorly phagocytic, LAP negative, cells (Hayhoe and Quaglino, 1958; Koler *et al.*, 1958). Pedersen and Hayhoe linked the two phenomena of low LAP activity and defective phagocytic capacity as manifestations of cytoplasmic immaturity in the segmented neutrophils of CML. Whereas acid phosphatase is prominent in primary neutrophil granules, the azurophil granules of Romanowsky preparations, alkaline phosphatase, as we have seen earlier, is not found in these granules on ultrastructural examination of human leucocytes but only in more mature cells (Bainton *et al.*, 1971), being confined chiefly to small specific granules which resemble morphologically tertiary granules (Cawley and Hayhoe, 1973). Ultrastructural studies of CML polymorphs reveal that, while some do not appear to differ in nuclear and cytoplasmic structure from normal polymorphs (Anderson, 1966; Tanaka and Goodman, 1972), others show evidence of unbalanced development, with nuclear lobulation preceding chromatin condensation, and with numerous primary but few secondary or tertiary granules, and even an abundance of endoplasmic reticulum, usually lost by the polymorph stage (Bessis, 1968; Kakefuda, 1968).

Moreover, in CML the glycogen content of segmented neutrophils is low, resembling that of the normal myelocyte stage (Valentine *et al.*, 1952; Bauer-Sic and Lambers, 1963; Gahrton, 1966), and the acid phosphatase level is high, reflecting the persistence of primary, azurophilic granules (Lambers and Bauer-Sic, 1963; Rosales *et al.*, 1965).

These findings are not entirely explicable on the basis of a shortened marrow transit time in CML, with premature release of polymorphs, although this pattern of cell kinetics may well apply (Stryckmans *et al.*, 1977). The cytological disorder is not a simple immaturity, but an imbalance or asynchronism between several morphological aspects of maturation, such as to imply an intrinsic cellular defect.

Nevertheless, Chikkappa *et al.* (1973) found that CML leucocytes grown in diffusion chamber cultures showed a capacity to mature to segmented neutrophils and to express LAP activity. Whereas LAP scores in the implanted cells were initially nil, neutrophils were first noted to show positivity after two days and reached a peak activity of 200 between days 6 and 14, subsequently falling off. Most positive cells showed two plus reactions, but some three or four plus reactions were found. This work has been confirmed in part by Moloney (1977) who noted recovery of LAP activity in 4 of 11 cases. Chikkappa and his associates thought their results to be compatible with the suggestion of Rosen and Teplitz (1968) that LAP activity might be controlled by a derepressor/repressor system, and that the inactivity of the derepressor occurring in CML was correctable by a strong enough stimulus *in vivo*, as by the coexistence of an infective or stressful state, or by the particular milieu of the diffusion chamber cultures.

Chiyoda and Muira (1977) and Dallegri *et al.* (1979) again confirmed the re-emergence of normal alkaline phosphatase expression in the neutrophils of CML in culture. The latter authors cultured bone marrow cells from eight Ph^1 positive CML patients *in vitro*. The mean viable neutrophil counts peaked at day 8 of culture, while the Ph^1 chromosome was still present until day 10. The LAP score reached a maximum at day 8. The authors attempted to reconcile these findings with the notion that young neutrophils have higher scores and that the low scores in CML are due to post-maturity of the neutrophils by suggesting that the positive neutrophils in their cultures were newly formed cells, but the findings could be explained equally well, or better, as due simply to ageing of the pre-existing and emerging neutrophils *in vitro*.

That other internal biochemical factors may play a part in controlling neutrophil alkaline phosphatase synthesis in CML is suggested by Das and Sandhya (1980) who found both LAP and cyclic adenosine monophosphate (AMP) to be low in the neutrophils in CML. Addition of 3′5′ cyclic AMP to cell suspensions led to an enhancement of LAP activity, suggesting that the low LAP scores in CML might have been due to low cyclic AMP content in the cells.

Extrinsic factors have also been demonstrated to affect LAP levels in the neutrophils of CML. Rustin *et al.* (1980) collected CML neutrophils by leucapheresis and transfused them into three neutropenic recipients. After 14 to 17 hours the LAP scores were much elevated, and the authors concluded that an extrinsic factor, reduced in CML patients but present in the neutropenic recipients, was responsible for this increase. More direct evidence of extrinsic factor activity in increasing alkaline phosphatase levels in the post-mitotic granulocytes from normal individuals and from patients with CML was provided by Sato *et al.* (1982). An extract from the cystic fluid of a squamous cell carcinoma, known to be rich in granulocyte/macrophage colony stimulating factor, was found to increase the LAP levels in neutrophils from normal bone marrow and from CML. These findings especially suggest that increases in LAP activity reflect maturation of granulocytes, that the low LAP scores of CML are due to immaturity of the circulating neutrophils, and that external factors may stimulate maturation and with it a rise in LAP levels.

Reviewing all these features we may conclude that the low LAP levels in the chronic phase of CML are due in part to an intrinsic insensitivity of the CML neutrophil to enzyme-stimulating factors, and in part to defective maturation of circulating cells, possibly consequent on rapid transit through the marrow maturation compartment. The former defect can be overcome by sufficiently strong stimuli, especially if the latter is also corrected as a consequence of effective therapy restoring the cell kinetics closer to normal.

There has been argument over the role of the spleen in the chronic phase of the disease, Spiers *et al.* (1975) suggesting that newly formed neutrophils emerge from the bone marrow with strongly positive LAP reactions, are trapped in the spleen and retained until their LAP activity has disappeared, and are then released into the peripheral blood. The removal of the spleen was noted in eight patients to be followed by a sharp neutrophil leucocytosis and a parallel rise in LAP scores—up to a mean of 221 (N.R. 37–98), with an equally sharp drop over the ensuing 5 to 10 days. Persistence of some increase in LAP level over the preoperative figure for a period of six months or longer in half the patients studied was claimed by the authors in a subsequent letter (Spiers and Williams, 1976), but the data given in their original paper suggest that any such increase can hardly be significant. In a reasoned critique of this work Kaplow (1976) pointed out that the effects of splenectomy on the leucocyte count and LAP score in CML closely paralleled the well known effects of surgical operation in non-leukaemic subjects and that newly formed neutrophils were more likely to be weak than strong in LAP activity. The splenic role postulated by Spiers *et al.* seems an unlikely one, and it is further negated by the authors' own observations (Spiers and Williams, 1976) that the LAP scores in imprints of spleen from CML are low, whereas on their hypothesis of splenic capture and retention of LAP positive cells the scores should be relatively high.

A striking increase in LAP scores sometimes occurs in the phase of terminal metamorphosis, especially in blastic crisis, of CML. Pedersen and Hayhoe (1971a) suggested that this rise might be due to an influx of fully matured neutrophils whose site of origin is the spleen rather than the bone marrow. The supposition was based on the observation of Clarkson *et al.* (1963) that in blastic crises most of the circulating granulocytes are produced in the spleen, and the laboratory work of Lahini and Van Putten (1969) who showed that colonies formed by cells derived from the spleen after injection in irradiated mice were more committed to differentiation than were those derived from bone marrow cells. As Brandt and Schnell (1969) demonstrated, a reduced tendency to granulocyte differentiation in the bone marrow in CML, with more primitive cells, appears to be compensated by an increased maturation of the splenic granulocyte population. At the extreme where the marrow is entirely blastic, effective granulopoiesis with the production of polymorphs may still occur in the spleen. Indeed, there may be a degree of independence of clonal development in spleen and bone marrow, as instanced by the report of a patient with myeloid metaplasia in whom chromosomal changes associated with blastic metamorphosis were found in marrow cells but not in splenic myeloid cells (Brandt *et al.*, 1975). Nevertheless, in patients who have been splenectomised during the chronic phase of their CML a rise in LAP score may still occur in blastic crisis (Goldstone *et al.*, 1976), so that the spleen is clearly not the only potential source of strongly positive neutrophils at that stage of the disease, and if the marrow is not itself producing them, other extramedullary sites of granulopoiesis may have a contribution to make. The altered isoenzyme pattern in CML in blastic crisis, described by Miller *et al.* (1984) and discussed earlier, may of course contribute to this sharp change in LAP level. Since the emergence of blastic crisis is commonly associated with the development of fresh chromosomal abnormalities, it may well prove that the apparent derepression of the gene which determines synthesis of the fast LAP isoenzyme, that leads to the increase in LAP score at this stage of the disease, may be a consequence of cytogenetic change.

The studies of LAP levels in CML have thus provided data capable of different interpretations, and the mechanisms by which the LAP score is changed in CML as in other diseases are still uncertain, but external factors, hormonal or otherwise, and the ability of the cell to activate its LAP under their influence, seem likely to be more important than changes in cell kinetics and associated variations in average age of circulating cells.

Unfortunately, animal experiments are of little help in solving the problems of LAP activity and its control in man. This is because there is no animal, at least among the eighteen species previously studied by various authors and listed by Kaplow (1968) in his informative review of LAP cytochemistry, which has LAP levels comparable to those in man. There are extreme differences, with very high levels (LAP scores 300 to 400) in hamster, guinea pig and most species of monkeys, high levels (150 to 200) in rat, rabbit, steer, burro, ground hog, horse, cow, sheep and goat; and very low levels, with virtual absence of enzyme, in mouse, dog, cat, chicken, pigeon, and at least one species of monkey (Macaca mulatta). As Kaplow points out, 'the extreme differences in animal LAP activity make it doubtful whether the interpretation of the results of studies on animals may be applied to humans.' The study of Williams (1975), for example, suggested that in the rat the LAP levels in neutrophils in the blood fell significantly with increasing age, though still remaining very high, and Spiers and Williams (1976) used this finding as an argument to support their view that in man LAP 'is present in the young mature neutrophil and absent from the ageing post-mature neutrophil'. But the pattern of LAP positivity in the rat is quite different from that in man. As Moloney (1958) pointed out, in the normal rat myeloblasts, promyelocytes, myelocytes and metamyelocytes show strong LAP positivity, while nearly all the circulating neutrophils have two or three plus positivity. There is no parallel with the distribution and levels of enzyme activity in man and it is unwise to attempt to draw inferences from so misleading an animal model.

Acid phosphatase (EC 3.1.3.2)

As we have seen above, the earlier studies of the distribution of acid phosphatase in haemic cells carried out by Rheingold and Wislocki (1948), Rabinovitch and Andreucci (1949) and Storti (1952) showed predominance of nuclear staining, probably artefactual, and when azo-dye techniques were introduced most subsequent cytochemical studies at the light microscope level used the new methods. Rozenszajn *et al.* (1963), for example, found positive granules in the cytoplasm of most nucleated haemic cells with stronger positivity in granulocytes than in lymphocytes (Plate 14A). Eosinophils, monocytes and platelets were usually quite strongly positive. Among bone marrow cells the most intense reactions were found in certain reticulum cells (Plate 14C), plasma cells (Plate 15A, B), megakaryocytes (Plate 14D), platelets (Plate 14E) and tissue mast cells, while granulocyte precursors were more strongly positive than were mature polymorphs (Plate 14B). The latter observation is in accord with the ultrastructural demonstration using modified Gomori techniques such as the method of Barka and Anderson (1962), that acid phosphatase is principally located in primary neutrophil granules and is absent from secondary and tertiary

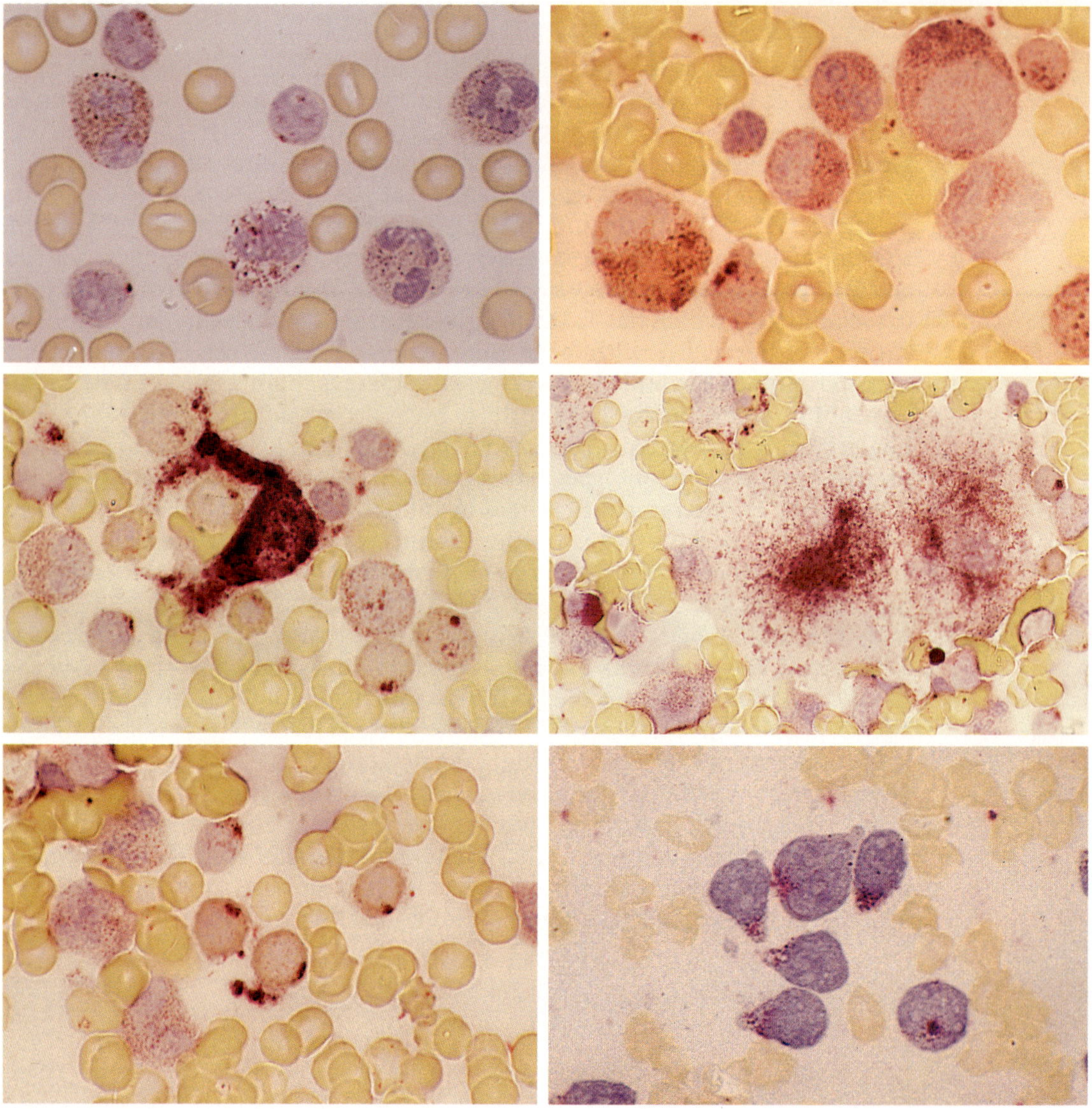

A B

C D

E F

Plate 14

A. Acid phosphatase, azo-dye method, normal buffy coat. Three neutrophils, a basophil and three lymphocytes show typical reactions.

B. Normal bone marrow. Acid phosphatase, azo-dye method. Granulocyte precursors, with stronger reactions in earlier cells, strong paranuclear reaction in a normoblast and localised cytoplasmic reactions in two lymphocytes.

C. Normal bone marrow. Acid phosphatase, azo-dye method. Heavily positive macrophage, with six neighbouring normoblasts of varying stages of maturity, all showing strong paranuclear positivity. Three lymphocytes show localised or scattered reactions and there is a positive eosinophil.

D. Normal bone marrow. Acid phosphatase, azo-dye method. Two megakaryocytes, showing strong reactions.

E. Normal bone marrow. Acid phosphatase, azo-dye method. Strong localized paranuclear reaction in normoblasts. Heavy positivity in platelets. Weaker granular reaction in mature neutrophils.

F. T-cell ALL. Acid phosphatase, azo-dye method. Localized reaction in region of Golgi in all cells.

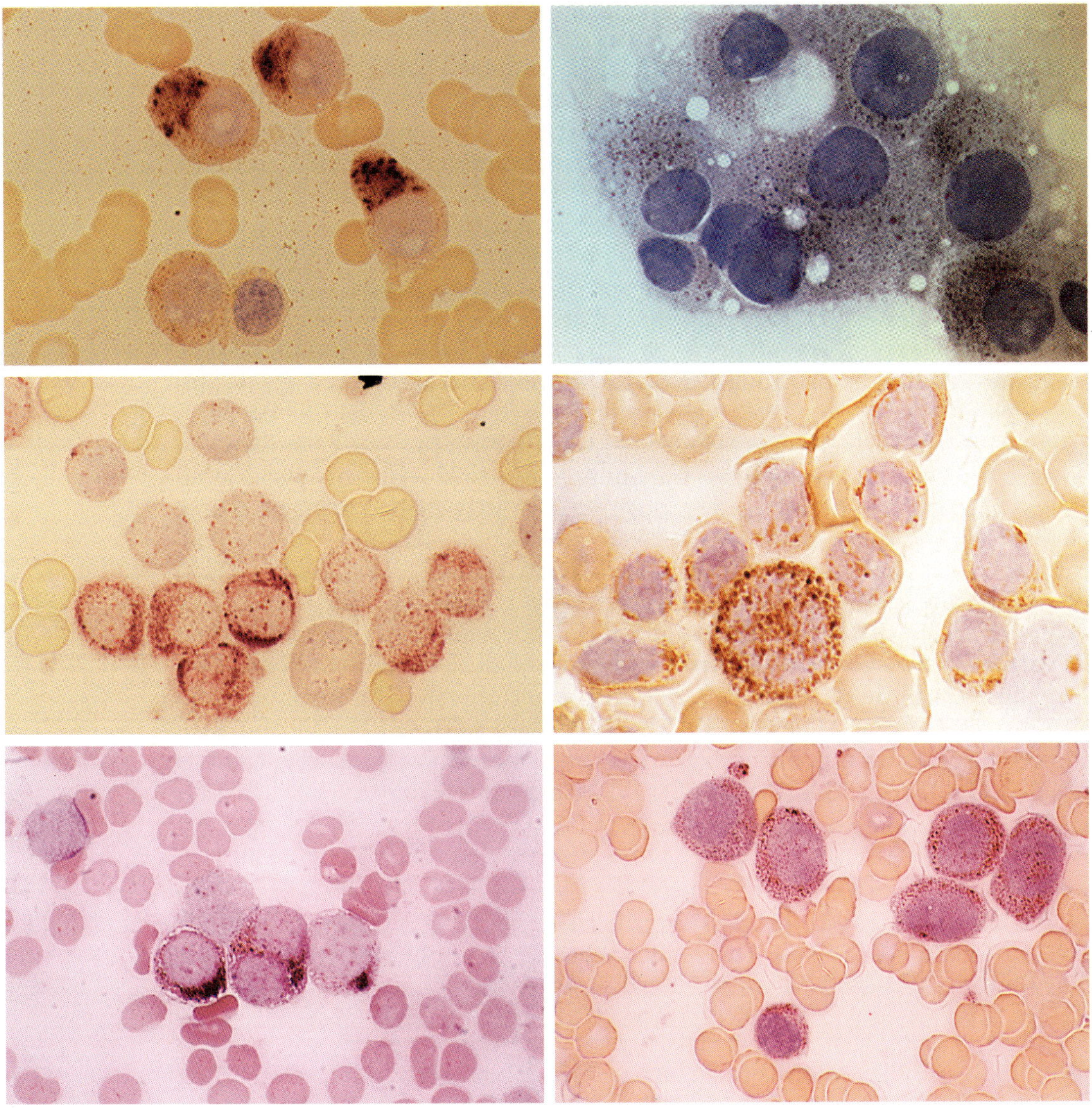

A B C D E F

Plate 15
A. Multiple myeloma bone marrow. Acid phosphatase, azo-dye method. Three strongly positive plasma cells, a weakly positive erythroblast and a normally reacting neutrophil promyelocyte.
B. Multiple myeloma bone marrow. Acid phosphatase, azo-dye method. A multinucleated plasma cell, with moderately strong positivity, and a single strongly reacting plasma cell.
C. HCL. Buffy coat. Acid phosphatase, azo-dye method. Seven HCs show strong positivity, tartrate resistant; other mononuclear cells present are only feebly reactive.
D. CLL bone marrow. Acid phosphatase, azo-dye method. A strongly positive myelocyte and numerous lymphocytes, mostly showing moderately strong cytoplasmic reactions.
E. Erythroleukaemia. Acid phosphatase, azo-dye method. Strong paranuclear reaction in erythraemic erythroblasts.
F. AMonL bone marrow. Acid phosphatase, azo-dye method. Strong positivity is shown in the leukaemic blast cells, and moderate reaction in a lymphocyte.

granules (Wetzel *et al.*, 1967; Bainton and Farquhar, 1968; Farquhar *et al.*, 1972; Cawley and Hayhoe, 1973). Thus, promyelocytes contain only acid phosphatase positive granules, while later granulocytes contain fewer, since the primary azurophilic granules are distributed in successive mitoses between daughter cells which develop increasing numbers of secondary and tertiary acid phosphatase negative granules. Acid phosphatase activity in most cells appears to be largely confined to lysosomes (De Duve, 1959) and there can be little doubt that leucocytic primary granules represent a form of primary lysosome since they contain not only acid phosphatase, but also other lysosomal enzymes (Bainton and Farquhar, 1968; Baggiolini, 1972). Acid phosphatase is also found in both primary and secondary eosinophil granules (Wetzel *et al.*, 1967; Cawley and Hayhoe, 1973) and in basophil granules (Plate 14A) (Parwaresch, 1976), and, at least in some specimens, in cisternae of the Golgi apparatus (Farquhar *et al.*, 1972). Erythroblasts at all stages of maturity show frequent positivity, mostly in the form of coarse paranuclear granules (Plate 14C, E).

Li *et al.* (1970a, b) made an important contribution to the study of acid phosphatase distribution in haemic cells by correlating the patterns of cytochemical positivity and the results of polyacrylamide gel electrophoresis of cell concentrates from normal and pathological blood samples. Two cytochemical methods were used, one with naphthol AS-BI phosphoric acid as substrate and Fast Garnet GBC as diazonium salt, and the second with the same substrate but with hexazotised pararosanilin as coupling agent as used by Goldberg and Barka (1962). The same fixative method was used for each reaction, namely 30 second exposure to buffered methanol acetone. A total of 400 smears of blood, bone marrow, serous effusions, phytohaemagglutinin (PHA)-stimulated lymphocytes and imprints of lymph nodes and spleen from 230 patients with various haematological disorders were studied, and compared with the findings in blood and bone marrow smears from 20 haematologically normal subjects.

The results of the cytochemical study were generally similar with the two methods, except that both granulocytes and monocytes showed chiefly granular positivity with Fast Garnet GBC but predominantly diffuse activity with the hexazotized pararosaniline method. Earlier findings of positivity in most nucleated blood cells with variable intensity of reaction as reported by Goldberg and Barka (1962) and Rozenszajn *et al.* (1963) were confirmed, but in addition the authors noted a frequent decrease in acid phosphatase activity of lymphocytes in CLL and non-Hodgkin's lymphoma—i.e. in B-cell tumours, but increased activity in the lymphocytes of infectious mononucleosis, and, as previously reported by Quaglino and Hayhoe (1965), in lymphocytes transforming under the influence of PHA. These two conditions represent proliferations chiefly of T cells, and it is of interest that the blast cells of T-cell acute lymphoblastic leukaemia (T-ALL) usually show strong localised cytoplasmic positivity, especially in the region of the Golgi body (Plate 14F), whereas the blast cells of non-T, non-B (null)-ALL show similar positivity in only a few cases (Catovsky *et al.*, 1974; Brouet *et al.*, 1976).

This observation of strong polar reactivity for acid phosphatase in T-cell ALL provided the first clear example of a correlation between a cytochemical property and a well-defined immunological phenotype; the correlation has been repeatedly confirmed at both light microscope and ultrastructural levels (Davey *et al.*, 1980; Savage *et al.*, 1981; Telek *et al.*, 1983; Veerman *et al.*, 1983; Boesen, 1984). This enzymic localization in malignant T lymphoblasts parallels that in normal thymocytes. Basso *et al.* (1980) for example reported a cytochemical study of fetal and post-natal thymocytes and T lymphocytes in which all three cell types commonly showed positivity to acid phosphatase although reactions for esterases and glucosaminidase become positive only in more mature cells and were negative in fetal thymocytes. Similarly, Hirokawa *et al.* (1983) found polar localization of acid phosphatase in normal thymocytes.

Attempts to establish the polar acid phosphatase reaction as a criterion for recognising mature T lymphocytes have given less clearcut results. Higgy *et al.* (1977), using light microscopy, found little difference between the acid phosphatase positivity of T and B-cell concentrates obtained by separating E-rosetting from non-rosetting cells. Ultrastructural studies of similarly separated T and B lymphocytes from normal human blood (Poore *et al.*, 1981) showed acid phosphatase positivity, mostly in secondary lysosomes, in 80% of T cells but also in 45% of B cells. Davey *et al.* (1980) and Yang *et al.* (1982), on the other hand, found a focal paranuclear reaction to characterize most normal or pathological T cells, whereas B lymphocytes were either negative or showed a diffuse granular positivity. In chronic T-lymphocytic and prolymphocytic leukaemias localised acid phosphatase positivity seems to occur with greater intensity in T-CLL than in T-PLL (Crockard *et al.*, 1982; Boesen *et al.*, 1984).

The discrepancies in reported findings may be due partly to differences in methodology. We have already noted that the use of hexazotized pararosaniline as capture agent gives more diffuse localisation of reaction product at light microscope level than does the use of Fast Garnet. Boesen (1984) compared two substrates, napthol AS-BI phosphoric acid, and sodium β-glycerophosphate, in ultrastructural acid phosphatase cytochemistry, in each case using lead capture agents. With the first substrate, reaction product in cases of T-ALL, T-CLL and hairy cell leukaemia (HCL) was variably found in granules, vesicles, the Golgi apparatus, rough endoplasmic reticulum and nuclear envelope, whereas with the second substrate reaction product was seldom seen in organelles or membranes other than the granular and vesicular lysosomal sites.

Another source of confusion lies in the possibility that subpopulations of the T lymphocytes, defined either functionally as helper, suppressor and other subsets, or by monoclonal antibodies (T4, T8, etc.) may show different degrees of reactivity for the enzyme. Binkley *et al.* (1983) carried out cytochemical and ultrastructural studies on human T-cell subsets with particular reference to helper, OKT4 positive cells, versus suppressor, OKT8 positive cells. In the former group 62% showed acid phosphatase positivity (56% granular and 6% diffuse patterns) whereas the latter showed 70% positive (38% granular and 32% diffuse patterns). These authors concluded that T4 and T8 subsets could not be distinguished by acid phosphatase cytochemistry.

Armitage *et al.* (1982) distinguished a dot-like pattern of localised paranuclear staining for acid phosphatase (as for non-specific acetate esterase) and regarded this as characteristic of 'true' T cells as defined by the UCHT1 monoclonal antibody. They noted that more positivity for both enzymes occurred in the helper subset than in the suppressor subset. Bernard and Dufer (1983), using different technical methods and different monoclonals, reached a similar conclusion so far as dot-like esterase positivity was concerned, but could not distinguish between T cells and non-T cells, or between T-cell subsets in acid phosphatase reactivity. They were unable, however, to identify the dot-like pattern with this enzyme and in an exchange of published correspondence Armitage *et al.* (1983) agreed that this identification was sometimes difficult. Hirt *et al.* (1984) looked particularly for 'focal' positivity in helper and suppressor subsets defined by Leu3a and Leu2a monoclonal antibodies respectively and were able to observe a significantly greater positivity in the helper subset (63.3% $\pm$ 10.9) than in the suppressor (52.3% $\pm$ 10), but the reactions showed far too much overlap for focal acid phosphatase positivity to be in any way definitive of one subtype rather than another.

As far as acid phosphatase activity in B lymphocytes is concerned, while the polar localization shown in some T cells is less often seen, strong positivity emerges as immunological stimulation and transformation to B immunoblasts and to plasma cells occurs. Dufer and Bernard (1985) used short term culture of B-CLL lymphocytes in the presence of the differentiation-inducing agent TPA to demonstrate the emergence of acid hydrolase activity. Among other enzymes showing increased activity, acid phosphatase positivity rose sharply, with the percentage of positive cells rising from 0 to 93 by the fourth day of culture. Grossi *et al.* (1982) produced further evidence that acid phosphatase might be a marker of maturation in the B-cell series from a study of B-CLL. Of 30 cases, 5 were negative and 25 positive for acid phosphatase, and the positive cells showed more mature features on ultrastructural examination.

Plasma cells have long been noted (Li *et al.*, 1970a, b) to show very strong acid phosphatase activity, sometimes especially concentrated in the perinuclear area and Golgi zone. Similar positivity occurs in the plasma cells of myeloma (Plate 15A, B) while the abnormal lymphocytic cells of macroglobulinaemia also show strong positivity with some tendency to perinuclear concentration. Hayhoe and Neuman (1976) studied the acid phosphatase activity of plasma cells in 80 cases of myeloma in relation to the variety of immunoglobulin produced. The levels were generally very high and did not differ according to the M protein patterns. There is evidence that the acid phosphatase levels in the plasma cells of myeloma are considerably higher than in normal plasma cells and perhaps also than in the plasma cells of reactive polyclonal plasmacytosis and benign monoclonal paraproteinaemia (Cassuto *et al.*, 1977; Tortarolo *et al.*, 1981; Hoffman *et al.*, 1985). Further details of these findings and their value in differential diagnosis are described and discussed in Chapter 14.

The hairy cells of 10 cases of leukaemic reticulo-endotheliosis (LRE) or hairy cell leukaemia (HCL) were studied by Li *et al.* (1970a), and found invariably to show strong enzyme positivity. Later studies (Yam *et al.*, 1971) suggested that the acid phosphatase isoenzyme in these cells was tartrate resistant and characteristic of HCL, but subsequent reports have indicated that the percentage of positive cells varies considerably from case to case and that the enzyme when present is not always tartrate resistant (Burke *et al.*, 1974; Catovsky, 1977; Higgy *et al.*, 1978).

Moreover tartrate resistant acid phosphatase (TRAP) has been reported to occur in the neoplastic cells from some differentiated lymphoid leukaemias including cases of B-CLL, T-CLL and adult T cell leukaemia-lymphoma, and even occasionally in T-ALL cells (Usui *et al.*, 1982). It has also been reported in the cells of cutaneous histiocytic lymphomas (Krishnan *et al.*, 1983), and in some Sezary cells (Naeim *et al.*, 1979) although this last finding seems contrary to most experience (Flandrin and Brouet, 1974; Krishnan *et al.*, 1983). Nevertheless the presence of tartrate resistant isoenzyme remains a diagnostically helpful, though not pathognomonic, feature of HCL (Plate 15C). Indeed, if only intense TRAP positivity in cireulating leucocytes is counted then the finding is virtually diagnostic. Yam *et al* (1987) reviewed experience with over 1000 patients and found only 2 of 200 HCL cases without such positivity whereas only 3 of over 800 non-HCL cases showed comparable positivity. The authors recommend Fast Garnet as receptor for highest sensitivity.

Further cytochemical results from the study of Li *et al.* (1970a) included the demonstration of a slight increase in acid phosphatase activity in neutrophils in some cases of chronic myeloid leukaemia, polycythaemia vera or myelofibrosis, and in infections. Platelets and megakaryocytes showed strong positivity, not significantly altered in disease states. Erythroblasts and reticulocytes usually showed positivity, but erythrocytes were normally negative although occasional granules of positivity were seen in erythrocytes of splenectomised patients. Our own experience of acid phosphatase distribution in haemic cells, illustrated in Plates 14 and 15, is in general agreement with that described by others, although like Grossi *et al.* (1982) we have found CLL lymphocytes more often positive than negative (Plate 15D). Osteoblasts show moderate positivity (Fig. 7.5) while macrophages show much variation (Fig. 7.6).

These patterns of cytochemically demonstrable acid phosphatase activity were correlated with the results of acrylamide gel electrophoresis (Li *et al.*, 1970a, b). Initially five and later seven isoenzymes of acid phosphatase were separated by electrophoresis and by ion exchange chromatography of leucocyte extracts, and numbered 0, 1, 2, 3, 3b, 4 and 5. Studies of preparations derived from lymphocytes of normal subjects and of patients with CLL, myeloblasts from patients with AML, platelets from various subjects and neutrophils from patients with polycythaemia vera were carried out. Spleen cell suspensions from a patient with Gaucher's disease were also studied. The distribution of the various electrophoretically separable isoenzymes in these preparations suggested that some isoenzymes

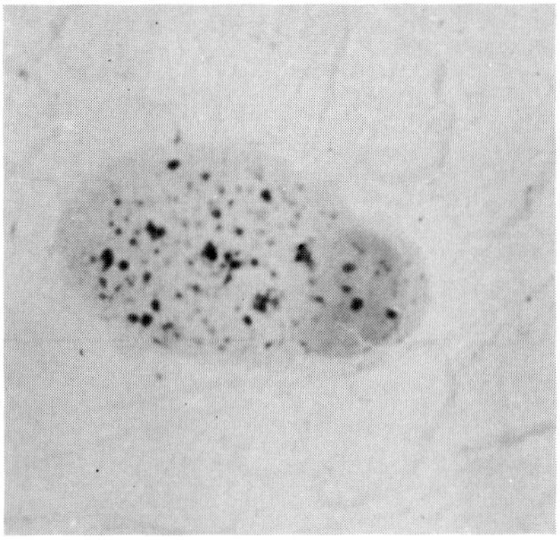

Fig. 7.5 An osteoblast from normal bone marrow. Acid phosphatase, azo-dye method. Moderate granular positivity, much less than shown by plasma cells.

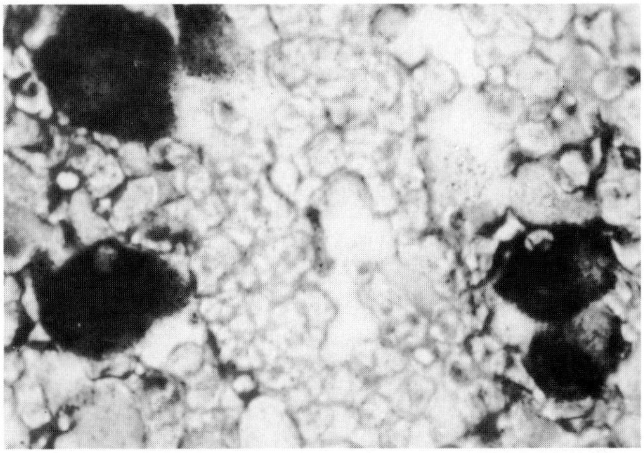

Fig. 7.6 Macrophages from a splenic aspirate in malignant histiocytosis. Acid phosphatase, azo-dye method. Exceptionally strong positivity is seen.

were specific to certain cell types and that in general the isoenzymes found in myeloid cells differed from those in lymphoid cells. Thus the Gaucher cell extracts contained chiefly isoenzyme 0 which seemed not to be present in any leucocyte preparations except in the cells of acute monocytic leukaemia, and then only in small amounts, while isoenzyme 5 was initially confined to the hairy cells of HCL. The tartrate resistance of this isoenzyme provided a further distinction from other isoenzymes. In neutrophils isoenzymes 1, 2 and 4 predominated, in monocytes 1 and 4, in lymphocytes and platelets 3, in myeloblasts 3b, and in monocytes and monoblasts of acute monocytic leukaemia 3b together with 1 and 4.

Each of the isoenzyme fractions demonstrated by acrylamide gel electrophoresis could also be separated by column chromatography, and studies on these separated fractions showed the molecular sizes of the enzyme molecules to be similar, thus excluding heterogeneity due to aggregation of enzyme subunits. Moreover, the fractions had different activities to different substrates, naphthol AS-BI phosphoric acid being the preferred substrate for isoenzymes 3, 3b and 5, sodium α-naphthyl acid phosphate for 1, 2, and 4, and p-nitrophenylphosphate for 0, and also to an extent almost equal to sodium α-naphthyl phosphate for 1, 2 and 4. Differences in Km values and in sensitivity to inhibitors further suggested that the fractions may indeed represent functionally different isoenzymes.

Kass and Munster (1979) used analytical isoelectric focusing in polyacrylamide gels (IEF) to study acid phosphatase isoenzymes in normal human blood cells. They found distinctive isoenzyme patterns for each of the main cell types, lymphocytes, neutrophil granulocytes, and monocytes and also in platelets. More bands were detected by this technique than had been resolved by conventional electrophoretic methods, with 6 in monocytes, 4 in lymphocytes, 10 in granulocytes and 7 in platelets, but the authors did not label the bands or compare them with the nomenclature proposed by Li *et al.* (1970a, b). A similar but more extensive study by Radzun and Parwaresch (1980) included also thymocytes and separated T and B subsets of lymphocytes. Characteristic patterns were again found for each cell type, although they differed considerably from those previously described by Kass and Munster (1979), perhaps because different methods of extraction—by the detergent Triton X-100 instead of ultrasonication—were used, and the bands were differently stained, Radzun and Parwaresch using α-naphthol AS-BI phosphate and Kass and Munster the simpler α-naphthyl phosphate. The band patterns found by Radzun and Parwaresch included 3 common bands with isoelectric points between pH 6.3 and 6 in monocytes, granulocytes and platelets, with monocytes showing a further 8 bands between pH 9.95 and 4, granulocytes another 9 between pH 5.9 and 4.65, and platelets a single additional band at pH 5.95. T lymphocytes and thymocytes showed one band at pH 6.3 and B lymphocytes no fewer than 11 bands between pH 5.3 and 3.85, plus a single cathodal band. Again, these authors did not propose a new classification or relate their findings clearly to those of Li *et al.* (1970a, b).

Gaedicke *et al.* (1979) used a similar technique of IEF to study the acid phosphatase isoenzymes in the cells of acute lymphoid and myeloid leukaemias (ALL and AML) and non-Hodgkin's lymphoma (NHL). Five groups of at least 8 isoenzymes were found. All five groups were found in normal thymocytes, peripheral blood lymphocytes and T-ALL cells. Patterns in AML and in mediastinal (?T cell) lymphoma were identical with those in normal lymphocytes, but in c-ALL variable patterns were found, one group of cytochemically negative cases having the main isoenzyme Group I band, and another with weakly positive cytochemistry showing bands of Group I-III. Gaedicke *et al.* noted that Li *et al.* (1970a, b) had found isoenzyme 3b to be characterised by high substrate specificity for naphthol AS-BI phosphate, to have a molecular weight of 110 000, and to be found chiefly in lymphocytes and leukaemic lymphoblasts. They thought it to be equivalent to their Group I and II isoenzymes, although these were of lower molecular weight and might be subunits. These authors thought that further comparative studies would be necessary before a proper classification of acid phosphatase isoenzymes could be achieved.

In later studies Radzun *et al.* (1981) again confirmed that each blood cell type showed a specific constellation of isoenzymes, and that leukaemic monocytes showed the same pattern as normal monocytes. Drexler and Gaedicke (1983) further agreed that analysis of leukaemic cell acid phosphatase showed no additional or leukaemia-specific isoenzymes beyond the 5 groups previously described in normal blood cells. They did, however, report a marked heterogeneity among the patterns in the various immunologically defined subclasses of ALL, with increasingly complex and complete isoenzyme patterns, as compared with normal lymphocytes, emerging with increasing maturation stage of the involved leukaemic cell. These leukaemic cell patterns of isoenzyme activity appeared to correlate with the particular parallel stage of normal cell development (Drexler *et al.*, 1984). In quantitative terms only acute leukaemias with monocyte line involvement (AMML and AMonL) showed acid phosphatase levels much higher than normal; other cases of AML and all types of ALL showed no significant differences either quantitatively or qualitatively from their normal counterparts (Hultberg and Sjoergen, 1980; Broadhead *et al.*, 1981; Tanaka *et al.*, 1981; Besley *et al.*, 1983).

Whatever the nomenclature the existence of different isoenzyme patterns in the main haemic cell types raises the possibility that certain isoenzymes of acid phosphatase are confined to specific subcellular organelles. Thus, to follow the original suggestion of Li *et al.* (1970a, b), and to use their original isoenzyme numbering, the ultrastructural localization of acid phosphatase in granulocytes in the primary granules may involve specifically isoenzymes 1, 2 and 4, which are not present in lymphocytes. Isoenzymes 0, 3, 3b and 5 appear to be concentrated in lysosomes very different in structure from the primary granules of either granulocytes or monocytes, in the region of the Golgi apparatus, in the rough endoplasmic reticulum (RER) and along the perinuclear cistern, in Gaucher's cells, lymphocytes and their activated derivatives including plasma cells, as well as in leukaemic blast cells and hairy cells (Streifler, 1970; Katayama *et al.*, 1972; Cawley and Hayhoe, 1973).

The acid phosphatase positivity of erythroblasts and reticulocytes appears ultrastructurally to be chiefly in cytoplasmic granules, presumed to be primary lysosomes (Cawley and Hayhoe, 1973). This enzyme, like leucocyte isoenzymes 0 and 5, is resistant to tartrate, and appears to contain three erythroid isoenzymes, which combine to form five genetically controlled electrophoretic patterns, designated A, BA, B, CA and CB. The variations in erythroid acid phosphatase in disease, especially its increase in megaloblastic anaemia and decrease in Caucasians with glucose-6-phosphate dehydrogenase deficiency have been analysed by biochemical techniques but not cytochemically although we have noted strong positivity in the erythroblasts of erythroleukaemia (Plate 15E). The subject was discussed by Yam (1974) in the course of an informative general review of serum and tissue acid phosphatases in man.

The changes involved in the maturation of monocytes to macrophages include increases in the size and complexity of the Golgi apparatus and in the numbers of lysosomes, with a parallel increase in the activity of acid phosphatase which is often conspicuous in the blast cells of acute monocytic leukaemia (Plate 15F). The enzyme is presumably chiefly composed of isoenzymes 1 and 4 as in the original monocytes, but there may well be a further family of isozymes in macrophages. In lysosomal fractions of rabbit alveolar macrophages, for example, 5 electrophoretically distinct acid phosphatase isozymes have been described (Axline, 1968).

Again, the acid phosphatase activity of bone tissue seems chiefly to reside in osteoclasts, and the increased serum enzyme activity found in diseases associated with resorption of bone is thought to be of osteoclastic origin. The enzyme has not been clearly characterised in terms of leucocyte isoenzymes, but appears to be tartrate resistant (Burstone, 1959; Yam, 1974).

The contribution of acid phosphatase cytochemistry to haematological differential diagnosis is further discussed in Chapter 14, where the findings in leukaemic blast cells are especially considered, but some generalisations may be made here on the role of the enzyme and the factors which appear to influence its formation or activation. Together with other lysosomal enzymes, acid phosphatase seems to be formed in the endoplasmic reticulum in metabolically active cells and then packaged into membrane bound primary lysosomal particles in the Golgi apparatus.

Novikoff (1964) used the acronym GERL for a specialized region of Golgi associated endoplasmic reticulum active in lysosome formation in dorsal root ganglion neurons. Many subsequent studies of the structure and function of the GERL system have shown wide variations among different cell types and in different animal species. The subject is reviewed by Borgers and Verheyen (1985). A distinction needs to be made between the GERL system on the inner, concave, 'trans' face of the Golgi apparatus and collections of saccules and vesicles on the outer, convex, 'cis', face (Hand, 1980). As Bainton and Farquhar (1966) had shown in a study of granulogenesis in rabbit neutrophils, at the myelocyte stage the Golgi ceases production of azurophil granules (acid phosphatase positive) from the trans face and begins producing specific granules (acid phosphatase negative) from the cis face. Nevertheless, in haemic cells the distinction is not always clearcut. Bentfeld and Bainton (1975), using ultrastructural cytochemistry of aryl sulphatase and acid phosphatase, showed reaction product in megakaryocytes and platelets of normal and thrombocytopenic rats to be confined to vesicles, initially formed in the earliest megakaryocytes apparently at the cis face of the Golgi, rather than at the GERL complex.

Once formed, these hydrolytic lysosomal enzymes may have multiple functions, but are concerned especially with intracellular digestive processes such as follow phagocytosis, the enzymes being discharged into phagocytic vacuoles (North, 1966). Fusion between primary lysosomes and phagocytic vacuoles or smaller pinocytotic vesicles may occur to produce secondary lysosomes, and there is evidence that the formation of lysosomes and their enzymes is influenced, at least in cells with phagocytic capacity, by the activity of the phagocytic process and the nature of the phagocytosed material. Axline and Cohn (1969), for example, showed that phagocytosis stimulated synthesis of lysosomal enzymes in macrophages to different degrees according to the digestibility of the material engulfed; digestible erythrocytes were more active stimulants than indigestible polystyrene.

In eosinophils it appears that acid phosphatase is held in specific granules in a latent, inactive form prior to degranulation. Bass *et al.* (1981) found that these cells responded to membrane stimulation with opsonized staphylococci or the surface-active agent phorbol myristate acetate or the calcium ionophore A23187 by sharp activation of acid phosphatase, with an increase, for example, from 7.8 ± 1.2% of positive granules to 78.9 ± 5%.

Other stimuli to acid phosphatase production include immunological and mitogenic activation; thus circulating lymphocytes, whether B or T, have only moderate amounts of cytochemically demonstrable acid phosphatase, but, as we have seen, activation of B cells along the immunological pathway towards plasma cells is accompanied by increase in acid phosphatase content, while the stimulation of T cells by PHA leads to a similar increase. The Reed-Sternberg cells of Hodgkin's disease, still of uncertain origin, usually show scattered granular positivity for acid phosphatase (Fig. 7.7). The genesis of these cells is further discussed in Chapter 15.

TECHNICAL METHODS

Alkaline phosphatase: lead methods

1. After Gomori (1952)

Reagents

Incubation medium

Calcium chloride	1 g
Magnesium chloride	100 mg
Sodium β-glycerophosphate	300 mg
Sodium barbital	500 mg
Distilled water	100 ml

Adjust pH to 9.2

Technique

1. Fix air-dried smears in formalin vapour 5–10 minutes
2. Wash briefly in tap water
3. Incubate in freshly prepared medium at 37°C for 1 hour
4. Rinse in tap water
5. Immerse in 2% cobalt chloride for 5 minutes and rinse
6. Immerse in 5% (v/v) ammonium sulphide solution for 2 minutes
7. Wash in distilled water and examine.

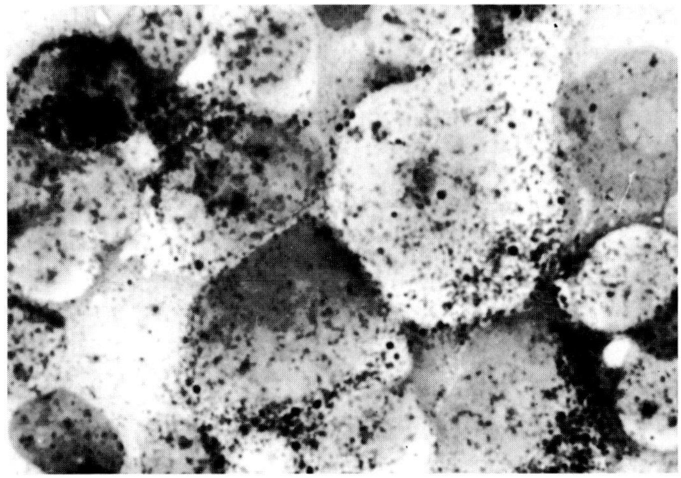

Fig. 7.7 Lymph node impression smear from Hodgkin's disease. Acid phosphatase, azo-dye method. There is strong positivity in a probable basophil, scattered granules in several lymphocytes, and moderately strong granular positivity in three Reed-Sternberg cells.

2. After El-Aaser *et al.* (1975)

Reagents

1. *Stock buffer*: pH 7.8–9.0
   0.2 mol/l tris buffer 150 ml
   0.05 mol/l sodium β-glycerophosphate 50 ml
   1% L-tartaric acid 6 ml
   Distilled water 25 ml
   Store at 4°C.
2. *Incubation medium*: To 45 ml of stock buffer add 4.5 ml 1% lead nitrate followed by 2 ml 0.1 mol/l magnesium nitrate.

Technique

1. Fix air-dried smears in formalin vapour 4 minutes
2. Immerse in freshly prepared incubation medium for 45 to 60 minutes at room temperature
3. Rinse in 5% (v/v) ammonium sulphide solution for 2 minutes
4. Rinse well and counterstain with 0.1% aqueous neutral red for 3 minutes, or with Carazzi's aqueous haematoxylin for 5 minutes

Alkaline phosphatase: azo-dye methods

1. After Hayhoe and Quaglino (1958), modified from Kaplow (1955)

Reagents

1. *Stock propanediol buffer solution* (0.2 mol/l): Dissolve 10.5 g of 2-amino-2-methyl-propane-(1,3)-diol in 500 ml distilled water. Store at 4°C. Discard after 3 months.

2. *Working buffer*: Mix 25 ml of stock buffer with 5 ml of 0.1 mol/l HCl and dilute with distilled water to 100 ml.

3. *Substrate solution*

Sodium α-naphthyl phosphate	35 mg
Brentamine Fast Garnet	35 mg
Working 0.05 mol/l propanediol buffer	35 ml

Mix immediately before use.

4. *Methyl green*: Prepare a 2% solution in distilled water and free from methyl violet by extraction with one-half its volume of chloroform for 48 hours. Store at 20°C in continuous contact with the chloroform.

Technique

1. Fix air-dried smears in 10% formalin in absolute methanol for 30 seconds (use stop watch) at 0 ± 5°C
2. Filter freshly prepared substrate directly on to slides and incubate at room temperature for 5–10 minutes. The substrate must be used within 5 minutes of preparation
3. Rinse in tap water briefly
4. Counterstain with methyl green for 15 minutes.

2. After Kaplow (1963)

Reagents

1. *Fixative*: Stir together 300 ml of absolute acetone and a mixture of 32 ml 0.03 mol/l sodium citrate and 168 ml 0.03 mol/l citric acid.
2. *Stock and working propanediol buffer*: As above in Method 1.
3. *Substrate solution*: Dissolve 5 mg naphthol AS-BI phosphate in 0.25 ml of dimethyl formamide. Add 60 ml working buffer and 40 mg fast red violet salt LB. Prepare and filter immediately before use.

Technique

1. Fix air-dried smears for 30 seconds in the fixative solution at room temperature
2. Wash briefly in tap water
3. Immerse in substrate solution for 10 minutes at room temperature
4. Wash briefly in tap water
5. Counterstain in aqueous haematoxylin 5 minutes
6. Wash briefly and examine.

3. After Rutenburg *et al.* (1965)

Reagents

1. *Stock substrate solution*: Dissolve 30 mg naphthol AS phosphate in 0.5 ml N,N-dimethyl formamide; add 100 ml 0.2 mol/l tris buffer, pH 9.1. Store in refrigerator at 2 to 4°C (stable for several months).

2. *Working substrate*: Add 10 mg Fast blue BBN to 10 ml of stock substrate solution, stir to dissolve diazonium salt and filter rapidly before use.

Technique

1. Fix air-dried smears in 10% formalin in absolute methanol for 30 seconds at 0 to 4°C
2. Rinse with tap water and dry
3. Incubate in freshly prepared substrate solution for 15 minutes at room temperature
4. Wash in tap water
5. Counterstain with a 0.1% aqueous solution of neutral red for 3 minutes.

Assessment and scoring of alkaline phosphatase positivity

The intensity of staining and number of positive granules in each of a hundred consecutive neutrophils may be scored as follows:

0 negative or colourless
1 faint diffuse positivity or occasional granules
2 diffuse positivity with moderate numbers of granules
3 strong positivity with numerous granules
4 very strong positivity with very numerous coarse granules, often obscuring the nucleus.

The sum of 100 cell ratings gives a score with a possible range from 0 to 400. The normal ranges with the azo-dye methods described are as follows:

Hayhoe and Quaglino—14 to 100 (mean 46)
Kaplow —13 to 160 (mean 61)
Rutenburg *et al.* —37 to 98 (mean 68).

Acid phosphatase: lead method

1. Modified from the original lead method of Gomori (1952)

Reagents

Incubation medium: To 100 ml tris-maleate buffer, 0.02 mol/l, pH 5.0 add 80 mg lead nitrate and 120 mg sodium β-glycerophosphate. Leave for 24 hours at 37°C, cool and filter. Use immediately after filtration.

Technique

1. Fix air-dried smears in formalin vapour 4 minutes
2. Immerse in incubation medium for 45 to 60 minutes at room temperature
3. Wash thoroughly in distilled water
4. Immerse in 5% (v/v) ammonium sulphide solution
5. Rinse well and counterstain with 0.1% aqueous neutral red for 3 minutes or with Carazzi's aqueous haematoxylin for 5 minutes

Acid phosphatase: azo-dye method

2. After Burstone (1958) and Li *et al.* (1970)

Reagents

1. *Substrate solution:* Dissolve 10 mg naphthol AS-BI phosphoric acid and 10 mg Fast Garnet GBC Salt in 60 ml acetate buffer, 0.1 mol/l, pH 5.0. Mix well. Filtration is not required.
2. Acetate buffer: Stock solution A: 60 ml glacial acetic acid in 1 l distilled water. Stock solution B: 82.04 g sodium acetate in 1 l distilled water.

Working buffer should be made up freshly before use by diluting the stock buffer solution 1 in 10 and for pH 5.0, mixing 59 parts of A to 141 parts of B. For example, for one Coplin jar (60 ml) take 3 ml A and dilute to 30 ml with distilled water, take 5 ml B and dilute to 50 ml with distilled water, then add 18 ml diluted A to 43 ml diluted B.

3. Carazzi's aqueous haematoxylin.

Technique

1. Fix air-dried smears in formalin vapour for 4 minutes
2. Wash briefly in tap water
3. Incubate in substrate solution for 60–90 minutes at 37°C
4. Wash briefly in tap water
5. Counterstain with Carazzi's aqueous haematoxylin for 10 minutes.

For assessment of tartrate resistance of acid phosphatase, add 60 mg L-tartaric acid to 30 ml substrate working solution and incubate slides in this medium in parallel with tartrate-free medium at Step 3.

Ultrastructural cytochemistry of acid phosphatase

Modified from the original lead method of Gomori (1952). (After Holt, 1959; Barka and Anderson, 1962; Cawley and Hayhoe, 1973).

Leucocyte rich buffy coat from dextran sedimented peripheral blood, or directly aspirated bone marrow material, is transferred to an excess volume of freshly prepared 1.5% glutaraldehyde (Taab) in 0.1 mol/l cacodylate-HCl buffer containing 1% sucrose (pH 7.4, 390 mosmol) (Bainton and Farquhar, 1968).

The resulting single cell suspensions are fixed for 1 hour at room temperature and then centrifuged at 200 **g** for 5 minutes to deposit the nucleated cells. The deposit is washed 3 times by repeated centrifugation and suspension in 0.1 mol/l cacodylate-HCl containing 7% sucrose and then incubated at 37°C for 1 hour in the medium described above for light microscope cytochemistry of acid phosphatase by the lead method, containing sodium β-glycerophosphate as substrate and lead nitrate as capture agent at pH 5.0.

After incubation the cells are again washed in cacodylate-HCl-sucrose buffer, post fixed for 1 hour at room temperature in 2% osmium tetroxide in cacodylate-HCl (pH 7.4), rapidly dehydrated by centrifugation and resuspension in single changes of 70% and 90% ethanol, four changes of absolute ethanol, and two 15 minute changes of propylene oxide.

The deposited cells are next infiltrated for 6–36 hours in three changes of low viscosity Taab or Spurr resins, embedded in capsules in freshly prepared resin and polymerised at 60°C for 24 hours.

Thin sections are then cut and mounted on clear uncoated copper mesh grids. Since uranyl acetate stains certain leucocyte granules intensely and may obscure the results of cytochemistry, preparations are best studied either without counterstaining, or after staining with Reynolds' lead citrate only for 30 minutes.

Further modification using cerium and cell permeabilization (after Robinson and Karnovsky, 1983; Robinson 1985)

Incubation in the medium containing lead nitrate as capture agent described above is replaced by incubation for 1 hour at 37°C in a medium containing 2 mM $CeCl_3$, 2 mM β-glycerophosphate and 0.1 M sodium acetate buffer (pH 5.0) with 5% sucrose and 0.0001% Triton X-100.

Following incubation the cells are washed, post-fixed, and further prepared for microscopy as decribed previously.

REFERENCES

Alter A A, Lee S L, Pourfar M, Dobkin G 1963 Studies of leucocyte alkaline phosphatase in mongolism: a possible chromosome marker. Blood 22: 165

Anderson D R 1966 Ultrastructure of normal and leukaemic leukocytes in human peripheral blood. Journal of Ultrastructural Research (Suppl) 9: 5

Angermuller S, Fahimi H 1984 A new cerium-based method for cytochemical localization of thiamine pyrophosphate in the Golgi complex of rat hepatocytes. Comparison with the lead technique. Histochemistry 80: 107

Armitage R J, Linch D C, Worman C P, Cawley J C 1982 The morphology and cytochemistry of human T-cell subpopulations defined by monoclonal antibodies and Fc receptors. British Journal of Haematology 51: 605

Armitage R J, Linch D C, Worman C P, Cawley J C 1983 Cytochemistry of human lymphocyte subpopulations delineated by monoclonal antibodies. British Journal of Haematology 53: 352

Athens J W, Raeb S O, Haeb O P, Boggs D R, Ashenbrucker H, Cartwright G E, Wintrobe M M 1965 Leukokinetic studies. X. Blood granulocyte kinetics in chronic myelocytic leukemia. Journal of Clinical Investigation 44: 765

Athens J W, Bishop C R, Cartwright G E 1968 The kinetics of neutrophilic granulocytes in chronic myelocytic leukaemia—a review. In: Zarafonetis C J D (ed) Proc. Internat. Conf. Leukaemia-Lymphoma. Lea & Febiger, Philadelphia, p 219

Axline S G 1968 Isozymes of acid phosphatase in normal and Calmette-Guerin bacillus-induced rabbit alveolar macrophages. Journal of Experimental Medicine 128: 1031

Axline S G, Cohn Z A 1969 The in vitro induction of lysosomal enzymes by phagocytosis. Journal of Clinical Investigation 48: 4a

Baggiolini M 1972 The enzymes of the granules of polymorphonuclear leucocytes and their functions. Enzyme 13: 132

Bailer P 1972 Das Verhalten der alkalischen Leucocyten-phosphatase bei Mädchen im Kindesalter und in der Pubertät. Monatsschrift für Kinderheilkunde 120: 441

Bainton D F, Farquhar M G 1966 Origin of granules in polymorphonuclear leucocytes. Two types derived from opposite faces of the Golgi complex in developing granulocytes. Journal of Cell Biology 28: 277

Bainton D F, Farquhar M G 1968 Differences in enzyme content of azurophil and specific granules of polymorphonuclear leukocytes. II. Cytochemistry and electron microscopy of bone marrow cells. Journal of Cell Biology 39: 299

Bainton D F, Ullyot J L, Farquhar M G 1971 The development of neutrophilic polymorphonuclear leucocytes in human bone marrow. Journal of Experimental Medicine 134: 907

Barka T, Anderson P J 1962 Histochemical methods for acid phosphatase using hexazonium pararosanilin as coupler. Journal of Histochemistry and Cytochemistry 10: 741

Bass D A, Lewis J C, Szejda P, Cowley L, McCall C E 1981 Activation of lysosomal acid phosphatase of eosinophil leukocytes. Laboratory Investigation 44: 403

Basso G, Cocito M G, Semenzato G, Pezzutto A, Zanesco L 1980 Cytochemical study of thymocytes and T lymphocytes. British Journal of Haematology 44: 577

Bauer-Sic P, Lambers K 1963 Zur Glycogen-verteilung in Blutzellen. In: Merker H (ed) Zyto-und Histochemie in der Hämatologie. Neuntes Freiburger Symposium. Springer, Berlin, p 426

Beckstead J H, Halverson P S, Ries C A, Bainton D F 1981 Enzyme histochemistry and immunohistochemistry on biopsy specimens of pathological human bone marrow. Blood 57: 1088

Bendix-Hansen K 1987 Enzyme cytochemistry of neutrophil granulocytes. British Journal of Haematology 65: 127.

Bendix-Hansen K, Helleberg-Rasmussen I 1985 Evaluation of neutrophil alkaline phosphatase (NAP) activity: untreated myeloid leukaemia, lymphoid leukaemia and normal humans. Scandinavian Journal of Haematology 34: 264

Bennett J M, Nathanson L, Rutenburg A M 1968 Significance of leukocyte alkaline phosphatase in Hodgkin's disease. Archives of Internal Medicine 121: 338

Bentfeld M E, Bainton D F 1975 Cytochemical localization of lysosomal enzymes in rat megakaryocytes and platelets. Journal of Clinical Investigation 56: 1635

Bernard J, Dufer J 1983 Cytochemistry of human lymphocyte subpopulations delineated by monoclonal antibodies. British Journal of Haematology 53: 351

Besley G T, Moss S E, Bain A D, Dewar A E 1983 Correlation of lysosomal enzyme abnormalities in various forms of adult leukaemia. Journal of Clinical Pathology 36: 1000

Bessis M 1968 Ultrastructure of normal and leukaemic granulocytes. In: Zarafonetis C J S (ed) Proc. Internat. Conf. Leukemia-Lymphoma. Lea & Febiger, Philadelphia, p 281

Binkley B, Valenzuela R, Sheibani K, Savage R, McMahon J, Turinic R, Antuinez A R 1983 Cytochemical and ultrastructural studies on human T cell subsets separated with the fluorescence activated cell sorter. American Journal of Clinical Pathology 80: 121

Bodansky O 1936 The accelerant effect of α-amino acids on the activity of bone phosphatase. Journal of Biological Chemistry 114: 273

Boesen A M 1984 Ultrastructural localization of acid phosphatase in immunologically defined neoplastic lymphocytic cells and hairy cells. Scandinavian Journal of Haematology 32: 245

Boesen A M, Hokland P, Jensen O M 1984 Acid phosphatase and acid esterase activity in neoplastic and non-neoplastic lymphoid cells. A semiquantitative evaluation related to immunological markers in 112 cases. Scandinavian Journal of Haematology 32: 313

Boggs D R, Marsh J C, Chervenich P A, Cartwright G E, Wintrobe M M 1968 Neutrophil releasing activity in plasma of normal human subjects injected with endotoxin. Proceedings of the Society for Experimental Biology and Medicine 127: 689

Borgers M, Verheyen A 1983 Electron microscopic cytochemistry of alkaline phosphatase. Cell Biochemistry and Function 1: 77

Borgers M, Verheyen A 1985 Enzyme Cytochemistry. In: Bourne GH, Danielli JF (eds) International Review of Cytology, Vol. 95. Academic Press Inc, London, p 163

Borgers M, Thone F, De Cree J, De Cock W 1978 Alkaline phosphatase activity in human polymorphonuclear leukocytes. Histochemical Journal 10: 31

Bottomley R H, Loring C A, Holt R, Griffin M J 1969 Comparison of alkaline phosphatase from human normal and leukaemic leukocytes. Cancer Research 29: 1866

Brandt L 1965 Adhesiveness to glass and phagocytic activity of neutrophilic leukocytes in myeloproliferative diseases. Scandinavian Journal of Haematology 2: 126

Brandt L, Schnell C-R 1969 Granulopoiesis in bone marrow and spleen in chronic myeloid leukaemia. Scandinavian Journal of Haematology 6: 65

Brandt L, Mitelman F, Panani A 1975 Cytogenetic differences between bone marrow and spleen in a case of agnogenic myeloid metaplasia developing blast crisis. Scandinavian Journal of Haematology 15: 187

Bretz U, Baggiolini M 1974 Biochemical and morphological characterization of azurophil and specific granules of human neutrophilic polymorphonuclear leukocytes. Journal of Cell Biology 63: 251

Broadhead D M, Besley G T, Moss S E, Bain A D, Eden O E, Sainsbury C P 1981 Recognition of abnormal lysosomal enzyme patterns in childhood leukaemia by isoelectric focusing with special

reference to some properties of abnormally expressed components. Leukemia Research 5: 29
Brodell H, Swisher S N 1954 Studies of leucocyte alkaline phosphatase determined by a clinically applicable histochemical method. Clinical Research Proceedings 2: 58
Brook J, Dreisbach P B 1977 Leucocyte alkaline phosphatase levels in multiple myeloma. Journal of Laboratory and Clinical Medicine 90: 114
Brouet J C, Valensi F, Daniel M-T, Flandrin G, Preud'homme, J-L, Seligmann M 1976 Immunological classification of acute lymphoblastic leukaemias: evaluation of its clinical significance in a hundred patients. British Journal of Haematology 33: 319
Burke J S, Byrne G E Jr, Rappaport H 1974 Hairy cell leukaemia (leukaemic reticuloendotheliosis). I. A clinical pathological study of 21 patients. Cancer 33: 1399
Burstone M S 1959 Chemical demonstration of acid phosphatase activity in osteoclasts. Journal of Histochemistry and Cytochemistry 7: 1
Burstone M S 1962 Enzyme histochemistry and its application in the study of neoplasms. Academic Press, New York
Butcher R G, Chayen J 1966 Quantitative studies on the alkaline phosphatase reaction. Journal of the Royal Microscopical Society 85: 111
Cassuto J P, Hammou J C, Pastorelli E, Dujardin P, Massayeff R 1977 Plasma cell acid phosphatase, discriminative test for benign and malignant monoclonal gammopathies. Biomedicine 27: 197
Catovsky D 1977 Hairy cell leukaemia and prolymphocytic leukaemia. Clinics in Haematology 6: 245
Catovsky D, Galetto J, Okos A, Miliani E, Galton D A G 1974 Cytochemical profile of B and T leukaemic lymphocytes with special reference to acute lymphoblastic leukaemia. Journal of Clinical Pathology 27: 767
Cawley J C, Hayhoe F G J 1973 Ultrastructure of Haemic Cells. Saunders; London, p 54, 55, 77
Chikkappa G, Boecker W R, Borner G, Carsten A L, Conkling K, Cook L, Cronkite E P, Dunwoody S 1973 Return of alkaline phosphatase in chronic myelocytic leukaemia cells in diffusion chamber cultures. Proceedings of the Society for Experimental Biology and Medicine 143: 212
Chiyoda S, Muira Y 1977 Alkaline phosphatase activity in chronic myelogenous leukaemic cells in culture. Acta Haematologica Japonica 40: 172
Clarke A E 1965 Hydrolytic enzymes of human polymorphonuclear leucocytes and rat monocytes. Australian Journal of Experimental Biology and Medical Science 43: 201
Clarkson B, Ota K, O'Connor A, Karnofsky D A 1963 Production of granulocytes by the spleen in chronic granulocytic leukaemia. Journal of Clinical Investigation 42: 924
Climie A R W, Henrichs W L, Foster I J 1962 Neutrophil alkaline phosphatase test. A review with emphasis on findings in pregnancy. American Journal of Clinical Pathology 38: 95
Corberand J, Pris J, Regnier C 1973 Cytochemical leukocyte reactions in normal newborn infants. Biology of the Neonate 22: 280
Cox R P, Elson N A, Tu S, Griffin M J 1971 Hormonal induction of alkaline phosphatase activity by an increased catalytic efficiency of the enzyme. Journal of Molecular Biology 58: 197
Crockard A, Chalmers D, Matutes E, Catovsky D 1982 Cytochemistry of acid hydrolases in chronic B- and T-cell leukemias. American Journal of Clinical Pathology 78: 437
Cronkite E P, Vincent P C 1970 Granulocytopoiesis. In: Stohlman F Jr (ed) Hemopoietic Cellular Proliferation. Grune & Stratton, New York, p 211
Dallegri F, Sassarego M, Ghio R, Patrone F 1979 Alkaline phosphatase activity in neutrophils of chronic myelocytic leukemia grown in liquid culture. Acta Haematologica 62: 12
Danielli J F 1946 A critical study of techniques for determining the cytological position of alkaline phosphatase. Journal of Experimental Biology 22: 110
Das U N, Sandhya P 1980 Effect of cyclic AMP on leukocyte alkaline phosphatase activity in chronic myeloid leukaemia. Indian Journal of Experimental Biology 18: 1325
Davey F R, Dock N L, McCallum J 1980 Cytochemical reactions in resting and activated T lymphocytes. American Journal of Clinical Pathology 74: 174
Dikow A, Meschkov T, Radeva M, Pfleiderer G 1985 Die alkalische Phosphatase der Lymphozyten des Menschen. 1. Zytochemischer Nachweis der alkalischen phosphatase in den Lymphozyten aus dem peripheren Blut gesunder Menschen. Folia Haematologica 112: 38
Drexler H G, Gaedicke G 1983 Analysis of isoenzyme patterns of acid phosphatase in acute leukaemias. Blut 47: 105
Drexler H G, Gaedicke G, Minowada J 1984 Enzyme markers in acute leukaemias, advances during the last decade. Journal of the National Cancer Institute 72: 1283
Duve C de 1959 Lysosomes a new group of cytoplasmic particles. In: Hayashi T (ed) Subcellular Particles. Ronald, New York
Dufer J, Bernard J 1985 The cytochemistry of B-chronic lymphocytic leukaemic cells: correlation of

acid hydrolases expression with the maturation of B-CLL cells stimulated by TPA. In: Quaglino D, Hayhoe F G J (eds) The Cytobiology of Leukaemias and Lymphomas. Raven Press, New York, p 245
El-Aaser A A, Hassanain S, Hassanain M 1975 A new direct lead technique for histochemical demonstration of alkaline phosphatase activity. Acta Biologica Hungarica 26: 105
Elves M W, Roath S, Isräels M C G 1963 A modified azo-dye method for the localisation and assessment of leucocyte alkaline phosphatase. Acta Haematologica 29: 141
Farquhar M G, Bainton D F, Baggiolini M, de Duve C 1972 Cytochemical localisation of acid phosphatase activity in granule fractions from rabbit polymorphonuclear leukocytes. Journal of Cell Biology 54: 141
Fishman W H 1974 Perspectives on alkaline phosphatase isoenzymes. American Journal of Medicine 56: 617
Flandrin G, Brouet J C 1974 The Sezary cell; cytologic, cytochemical and immunologic studies. Mayo Clinic Proceedings 49: 573
Flemans R J, Neumann H, Reynolds D, Worman C, Hayhoe F G J 1980 N-alkaline phosphatase activity in the peripheral blood lymphocytes and serum of patients with malignant lymphoma and with cytochemically demonstrable lymphocyte alkaline phosphatase. Leukemia Research 4: 351
Gaedicke G, Arndt R, Kabisch H, Landbeck G 1979 Isoenzyme studies in human leukaemia. 1. Acid phosphatase. Leukemia Research 3: 417
Gahrton G 1966 The periodic acid-Schiff reaction in neutrophil leukocytes in chronic myeloproliferative diseases. A microspectrophotometric study. Scandinavian Journal of Haematology 3: 106
Galbraith P R 1966 Studies on the longevity, sequestration and release of the leukocytes in chronic myelogenous leukaemia. Canadian Medical Association Journal 95: 511
Gibson J G II, Vallee B L, Fluharty R G, Nelson J E 1948–50 Studies of the zinc content of the leukocytes in myelogenous leukaemia. Acta Universitatis Internationalis Contra Carcinoma 6: 1102
Goffinet G, Houben-Defresne M P, Boniver J 1983 Correlation of alkaline phosphatase activity to normal T-cell differentiation and to radiation leukemia virus-induced preleukemic cells in the C57BL mouse thymus. Cancer Research 43: 5416
Goldberg A F, Barka T 1962 Acid phosphatase activity in human blood cells. Nature 195: 297
Goldstone A H, Cawley J C, Hayhoe F G J 1976 Leukaemias, Lymphomas and Allied Disorders. Saunders, London, p 108
Gomori G 1939 Histochemical demonstration of phosphatase in tissue sections. Proceedings of the Society for Experimental Biology (New York) 42: 23
Gomori G 1952 Microscopic histochemistry. Principles and Practice. Univ. of Chicago Press, Chicago, Illinois
Greenberger J S, Aaronson S A, Rosenthal D S, Moloney W C 1975 Continuous production of peroxidase, esterase, alkaline phosphatase and lysozyme by clones of promyelocytes. Nature 257: 143
Griffin M J, Cox R P 1966 Studies on the mechanism of hormonal induction of alkaline phosphatase in human cell cultures. Journal of Cell Biology 29: 1
Grogg E, Pearse A G E 1952 A critical study of the histochemical techniques for acid phosphatase. Journal of Pathology and Bacteriology 64: 627
Grossi C E, Zicca A, Leprini A, Cadoni A, Pistoia V, Ferrarini M 1982 Acid hydrolases as markers of maturation in B-cell chronic lymphocytic leukemia. Blood 60: 220
Haering M 1964 Das Verhalten der alkalischen Leukozytenphosphatase (ALP) in der Schwangerschaft, unter der Geburt und im Wochenbett. Zeitschrift fur Geburtshilfe und Gynäkologie 162: 132
Haight W F, Rossiter R J 1950 Acid and alkaline phosphatase in white cells. Data for the lymphocyte and the polymorphonuclear leukocyte of man and the rabbit. Blood 5: 267
Halbrecht I, Shabtay F 1972 Leucocyte alkaline phosphatase activity in pre-term, term and post-term infants. Israel Journal of Medical Science 8: 1956
Hammouda F, Quaglino D, Hayhoe F G J 1964 Blastic crises of chronic granulocytic leukaemia. Cytochemical, cytogenetic and autoradiographic studies in four cases. British Medical Journal 1: 1275
Hand A R 1980 Cytochemical differentiation of the Golgi apparatus from GERL. Journal of Histochemistry and Cytochemistry 28: 82
Hayhoe F G J, Hammouda F 1965 Cytogenetic and metabolic observations in leukaemias and allied states. In: Hayhoe F G J (ed) Current Research in Leukaemia. Camb. Univ. Press, Cambridge, p 55
Hayhoe F G J, Neuman Z 1976 Cytology of myeloma cells. Journal of Clinical Pathology 29: 916
Hayhoe F G J, Quaglino D 1958 Cytochemical demonstration and measurement of leucocyte alkaline phosphatase activity in normal and pathological states by a modified azo-dye coupling technique. British Journal of Haematology 4: 375

Hayhoe F G J, Quaglino D, Doll R 1964 The Cytology and Cytochemistry of Acute Leukaemias. Spec. Rep. Ser. Med. Res. Counc. No. 304: HMSO, London
Hayhoe F G J, Flemans R J, Burns G F, Cawley J C 1977 Leucocyte alkaline phosphatase scores in hairy cell leukaemia. British Journal of Haematology 37: 158
Higgy K E, Burns G F, Hayhoe F G J 1977 Discrimination of B, T and null lymphocytes by esterase cytochemistry. Scandinavian Journal of Haematology 18: 437
Higgy K E, Burns G F, Hayhoe F G J 1978 Identification of the hairy cells of leukaemic reticulo-endotheliosis by an esterase method. British Journal of Haematology 38: 99
Hirokawa K, Saitoh K, Hotakayama S 1983 Enzyme histochemical study on human thymus and its age change. Acta Pathologica Japonica 33: 275
Hirt A, Auner H L, Wagner H P 1984 Cytochemistry of human T lymphocyte subpopulations in peripheral blood. Scandinavian Journal of Haematology 32: 190
Hoffman J J M, Breed W P M, Swaak-Lammers N 1985 Plasma cell acid phosphatase score in multiple myeloma and related disorders. British Journal of Haematology 59: 67
Holt S J 1959 Factors governing the validity of staining methods for enzymes and their bearing upon the Gomori acid phosphatase technique. Experimental Cell Research (Suppl) 1: 1
Hook E B, Engel R R 1964 Leucocyte life span, leucocyte alkaline phosphatase and the 21st chromosome. Lancet i: 112
Hugon J, Borgers M 1966 A direct lead method for the electron microscopic visualisation of alkaline phosphatase activity. Journal of Histochemistry and Cytochemistry 14: 429
Hulstaert C E, Kalicharan D, Hardonk M J 1983 Cytochemical demonstration of phosphatases in the rat liver by a cerium-based method in combination with osmium tetroxide and potassium ferrocyanide post-fixation. Histochemistry 78: 71
Hultberg B, Sjoergen U 1980 Diagnostic significance of lysosomal enzymes in different types of leukemias. Acta Medica Scandinavica 207: 105
Kakefuda T 1968 Electron microscopy of normal and leukaemic cells. In: Amromin GD (ed) Pathology of Leukemia. Harper & Row, New York, p 82
Kaplow L S 1955 A histochemical procedure for localising and evaluating leucocyte alkaline phosphatase activity in smears of blood and marrow. Blood 10: 1023
Kaplow L S 1963 Cytochemistry of leucocyte alkaline phosphatase. Use of complex naphthol AS phosphates in azo-dye coupling techniques. American Journal of Clinical Pathology 39: 439
Kaplow L S 1968 Leucocyte alkaline phosphatase cytochemistry: application and methods. Annals of the New York Academy of Sciences 155: 911
Kaplow L S 1976 Changes in neutrophil alkaline phosphatase following splenectomy. Journal of Clinical Pathology 29: 172
Kaplow L S, Beck R E 1964 Leukocyte alkaline phosphatase response to surgical stress. Surgery, Gynecology and Obstetrics 119: 97
Karpas A, Hayhoe F G J, Greenberger J S, Barker C R, Cawley J C, Lowenthal R M, Moloney W C 1977 The establishment and cytological, cytochemical and immunological characterisation of human haemic cell lines: evidence for heterogeneity. Leukemia Research 1: 35
Kass L, Munster D 1979 Acid phosphatase in normal human blood cells: characterisation by isoelectric focusing in polyacrylamide gel. Experimental Haematology 7: 272
Katayama I, Li C Y, Yam L T 1972 Histochemical study of acid phosphatase isoenzyme in leukaemic reticulo-endotheliosis. Cancer 29: 157
Keleman E 1973 Granulocyte alkaline phosphatase activity: a measure of the emergence time of mature marrow neutrophils? Acta Haematologica 50: 19
Keleman E, Gulya E, Vass K 1978 Ontogeny of human neutrophil granulocyte alkaline phosphatase. Journal of Cell Physiology 95: 353
Kerppola W 1951 Observations on the phosphatase content of blood and bone marrow cells in normal and pathological haemopoiesis. Blood 6: 454
Klein U E, Löeffler H, Leuckfeld E 1966 Elektrophoretische Trennung von Milchsäuredehydrogenase-Isoenzymen, unspezifischen Esterasen und álkalischen Phosphatasen aus menschlichen Blutzellen. Klinische Wochenschrift 44: 637
Koler R D, Seaman A J, Osgood E E, van Bellingham P 1958 Myeloproliferative diseases: diagnostic value of the leukocyte alkaline phosphatase test. American Journal of Clinical Pathology 30: 295
Krishnan J, Li C Y, Su W P D 1983 Cutaneous lymphomas: correlation of histochemical and immunohistochemical characteristics and clinicopathological features. American Journal of Clinical Pathology 79: 157
Lahini S K, van Putten L M 1969 Distribution and multiplication of colony forming units from bone marrow and spleen after injection in irradiated mice. Cell and Tissue Kinetics 2: 21
Lambers K 1962 Übersichten. Zur Zytochemie der Blutzellen. Deutsche Medizinische Wochenschrift 87: 1913
Lambers K, Bauer-Sic P 1963 Hydrolytische Enzyme in der klinischen Diagnostik. In: Merker H

(ed) Zyto-und Histochemie in der Hämatologie. Neuntes Freiburger Symposium. Springer, Berlin, p 254
Leder L-D 1978 Alkaline phosphatase and tartrate resistant acid phosphatase activity in cells of prolymphocytic leukaemia. Klinische Wochenschrift 56: 313
Lewis S M, Dacie J V 1965 Neutrophil (leucocyte) alkaline phosphatase in paroxysmal nocturnal haemoglobinuria. British Journal of Haematology 11: 549
Li C Y, Yam L T, Lam K W 1970a Acid phosphatase isoenzyme in human leukocytes in normal and pathological conditions. Journal of Histochemistry and Cytochemistry 18: 473
Li C Y, Yam L T, Lam K W 1970b Studies of acid phosphatase isoenzymes in human leukocytes: demonstration of isoenzyme cell specificity. Journal of Histochemistry and Cytochemistry 18: 901
Lille-Szyszkowicz I, Gabay P, Saracino R T, Bourdin J S 1966 La phosphatase alcaline leucocytaire chez les malades atteints de tumeurs ou d'hémopathies malignes. Nouvelle Revue Francaise d'Hematologie 6: 187
Linke A, Löffler H 1963 Über die alkalische Leukozyten-phosphatase wahrend der Therapie der reifzelligen myeloischen Leukämie mit Trenimon. In: Merker H (ed) Zyto-und Histochemie in der Hämatologie. Neuntes Freiburger Symposium. Springer, Berlin, p 306
Löffler H, Pralle H 1970 Cytochemical and biochemical demonstration of an alkaline phosphatase in human red cells. Klinische Wochenschrift 48: 763
Lyons R B, Weaver D D, Beck J H 1968 Isoenzymes of human leukocyte alkaline phosphatase. Annals of the New York Academy of Sciences 155: 948
Mann R B, Jaffe E S, Braylan R C, Nanba K, Frank M M, Ziegler J L, Berard C W 1976 Non-endemic Burkitt's lymphoma. New England Journal of Medicine 295: 685
Markert C L, Moller F 1959 Multiple forms of enzymes: tissue, ontogenic and species specific patterns. Proceedings of the National Academy of Sciences 45: 753
Marmont A, Damasio E 1965 The phagocytic response to the L E factor of leucocytes from chronic granulocytic and acute leukaemias. In: Hayhoe F G J (ed) Current Research in Leukaemia. Camb. Univ. Press, Cambridge, p 16
Martin B F, Jacoby F 1949 Diffusion phenomenon complicating the histochemical reaction for alkaline phosphatase. Journal of Anatomy 83: 351
Mauer A M, Jarrold T 1963 Granulocyte kinetic studies in patients with proliferative disorders of the bone marrow. Blood 22: 125
Mauer A M, Athens J W, Ashenbrucker H, Cartwright G E, Wintrobe M M 1959 An analysis of leukocyte radioactivity curves obtained with radioactive diisopropylfluorophosphate (DFP^{32}). In Stohlman F Jr (ed) The Kinetics of Cellular Proliferation. Grune & Stratton, New York, p 231
Mayahara H, Hirano H, Saito T, Ogawa K 1967 The new lead citrate method for the ultracytochemical demonstration of activity of nonspecific alkaline phosphatase (ortho-phosphoric monoester phosphohydrolase). Histochemie 11: 88
McMillan R, Scott J L 1968 Leukocyte labeling with ^{51}chromium. 1. Technic and results in normal subjects. Blood 32:738
Menten M L, Junge J, Green M H 1944 A coupling histochemical azo-dye test for alkaline phosphatase in the kidney. Journal of Biological Chemistry 153: 471
Merker H V 1961 Der Aktivitätswechsel der alkalischen Neutrophilenphosphatase im Pyrexalreizversuch und die Bedeutung dieses Testes für die Klinik einiger Blutkrankheiten. Folia Haematologica 6: 153
Merker H V 1965 The significance of cytochemical and cytogenetic findings in chronic granulocytic leukaemia and related diseases. In: Hayhoe F G J (ed) Current Research in Leukaemia. Camb. Univ. Press, Cambridge, p 1
Miller D M, Yang A, Liepman M 1983 Leukocyte alkaline phosphatase: another organ-specific alkaline phosphatase. American Journal of Hematology 15: 171
Miller D M, Yang A, Liepman M 1984 Altered isozyme patterns of leucocyte alkaline phosphatase in disease states. British Journal of Haematology 57: 145
Miller D M, Yang A, Liepman M 1985 Evidence for two isoenzymes of leucocyte alkaline phosphatase in leukemic leukocytes. American Journal of Hematology 18: 159
Mishler J M, Williams D M 1980 Alkaline phosphatase as a marker of maturity in human neutrophils. Journal of Clinical Pathology 33:555
Mitus W J, Bergna L J, Mednicoff I B, Dameshek W 1958 Alkaline phosphatase of mature neutrophils in chronic forms of the myeloproliferative syndrome. American Journal of Clinical Pathology 30: 285
Mölbert E R G, Duspiva F, Deimling O H von 1960 The demonstration of alkaline phosphatase in the electron microscope. Journal of Biophysical and Biochemical Cytology 7: 387
Moloney W C 1958 Leucocyte alkaline phosphatase activity in the rat. Annals of the New York Academy of Sciences 75: 31
Moloney W C 1977 Natural history of chronic granulocytic leukaemia. Clinics in Haematology, 6: 41

Moreb J, Herschko C 1985 Increased leukocyte alkaline phosphatase and transcobalamin III in chronic myeloid leukemia associated with lithium therapy. Scandinavian Journal of Haematology 34: 238
Naeim F, Capostagno V J, Johnson C E Jr, Lehrer R, Gatti R A 1979 Sezary syndrome: tartrate-resistant acid phosphatase in the neoplastic cells. American Journal of Clinical Pathology 71: 528
Nakatsui T 1969 Ultrastructural localisation of alkaline phosphatase in human neutrophils. Acta Haematologica 42: 275
Nanba K, Itagaki T, Iijima S 1975 Enzyme histochemical investigations of human malignant lymphomas. Beitrage Zur Pathologie 154: 233
Nanba K, Jaffe E S, Braylan R C, Soban E-J, Bernard C W 1977 Alkaline phosphatase-positive malignant lymphoma. American Journal of Clinical Pathology 68: 535
Neumann H, Klein E, Hanck-Granoth R, Yachmin S, Ben-Hassat H 1976 Comparative study of alkaline phosphatase activity in lymphocytes, mitogen-induced blasts, lymphoblastoid cell lines, acute myeloid leukaemia and chronic lymphatic leukaemia cells. Proceedings of the National Academy of Sciences 73: 1432
Nguyen-van Thoai, Roche J, Roger M 1947 Inactivation et réactivation complètes de la phosphomonoesterase alcaline et interchangeabilité des métaux actifs. Biochimica et Biophysica Acta 1: 61
North R J 1966 The localization by electron microscopy of acid phosphatase activity in guinea pig macrophages. Journal of Ultrastructural Research 16: 96
Novikoff A B 1964 GERL, its form and function in neurons of rat spinal ganglia. Biological Bulletin 127: 358A
O'Kell R T, Axon L L 1965 Leucocyte alkaline phosphatase in the newborn infant. American Journal of Obstetrics and Gynaecology 93: 1181
O'Riordan M L, Robinson J A, Buckton K E, Evans H J 1971 Distinguishing between the chromosomes involved in Down's syndrome (trisomy 21) and chronic myeloid leukaemia (Ph^1) by fluorescence. Nature 230: 167
Parwaresch M R 1976 The Human Blood Basophil. Springer-Verlag: Berlin, p 49, 217
Pearse A G E 1960 Histochemistry: Theoretical and Applied. 2nd Edn. Churchill, London
Pedersen B, Hayhoe F G J 1971a Cellular changes in chronic myeloid leukaemia. British Journal of Haematology 21: 251
Pedersen B, Hayhoe F G J 1971b Relation between phagocytic activity and alkaline phosphatase content of neutrophils in chronic myeloid leukaemia. British Journal of Haematology 21: 257
Perillie P E, Finch S C 1967 Histochemical studies on leucocyte alkaline phosphatase in non-leukaemic granulocytosis: effects of anticoagulants and role of zinc and magnesium. British Journal of Haematology 13: 289
Perry S 1968 Leukocyte kinetics in leukaemia. In: Zarafonetis C J D (ed) Proc. Internat. Conf. Leukaemia-Lymphoma. Lea & Febiger, Philadelphia, p 299
Poore T E, Barrett S G, Kadin M E, Bainton D F 1981 Ultrastructural localization of acid phosphatase in rosetted T and B lymphocytes of normal human blood. American Journal of Pathology 102: 72
Poppema S, Elema J D, Halie M R 1981 Alkaline phosphatase positive lymphomas: A morphologic, immunologic and enzyme histochemical study. Cancer 47: 1303
Pralle H, Schnellbacher E, Löffler H 1973 Anreicherung und Eigenschaften von alkalischer Phosphatase aus Erythrocyten bei familiäres Erythrocytose. Clinica Chimica Acta 46: 363
Pritchard J A 1957 Leukocyte phosphatase activity in pregnancy. Journal of Laboratory and Clinical Medicine 50: 432
Quaglino D 1961 Cytochemical Methods in Haematology. PhD Thesis, Cambridge Univ.
Quaglino D 1964 Alkaline phosphatase positivity in erythroblasts. Haematologica Latina, 7: 263
Quaglino D, Hayhoe F G J 1965 Metabolic changes in short term *in vitro* cultures of normal and leukaemic cells: studies with cytochemistry and autoradiography. In: Hayhoe F G J (ed) Current Research in Leukaemia. Camb. Univ. Press; Cambridge, p 124
Quigley H J, Dawson E A, Hyun B H, Custer R P 1960 The activity of alkaline phosphatase in granular leukocytes during pregnancy and the puerperium. A preliminary report. American Journal of Clinical Pathology 33: 109
Radzun H J, Parwaresch M R 1980 Isoelectric focusing pattern of acid phosphatase and acid esterase in human blood cells, including thymocytes, T lymphocytes and B lymphocytes. Experimental Hematology 8: 737
Radzun H J, Parwaresch M R, Feller A C 1981 Cell-specific polymorphism of acid phosphatase in human blood cells: their functional and leukemic variants. Proceedings of the Society for Experimental Biology and Medicine 167: 394
Rabinovitch M, Andreucci D 1949 A histochemical study of acid and alkaline phosphatase distribution in normal human bone marrow smears. Blood 4: 580

Raeb S O, Athens I W, Haeb O P, Boggs D R, Cartwright G E, Wintrobe M M 1962 Leukokinetic studies in chronic myelocytic leukaemia (CML) using radioactive diisopropylfluorophosphate (DFP^{32}). Blood 20: 803
Rheingold J J, Wislocki J B 1948 Histochemical methods applied to haematology. Blood 3: 641
Robinson J C, Pierce J E, Goldstein D P, Rosse W R 1966 Leucocyte-alkaline-phosphatase isozymes. Lancet ii: 805
Robinson J M 1985 Improved localization of intracellular sites of phosphatases using cerium and cell permeabilization. Journal of Histochemistry and Cytochemistry 8: 749
Robinson J M, Karnovsky M J 1983 Ultrastructural localization of several phosphatases with cerium. Journal of Histochemistry and Cytochemistry 31: 1197
Rosales C L, Bennett J M, Rutenburg A M 1965 An azo-dye method for the histochemical demonstration of leukocyte acid phosphatase. Journal of Histochemistry and Cytochemistry 13: 714
Rosen R B, Teplitz R L 1968 Chronic granulocytic leukaemia complicated by ulcerative colitis; elevated leucocyte alkaline phosphatase and possible modifier gene deletion. Blood 26: 148
Rosenblum D, Petzold S J 1975 Neutrophil alkaline phosphatase: comparison of enzymes from normal subjects and patients with polycythaemia vera and chronic myelogenous leukaemia. Blood 45: 335
Rosner F, Lee S L 1965 Endocrine relationships of leukocyte alkaline phosphatase. Blood 25: 356
Rosner F, Lee S L, Schultz F S, Gorfian P C 1968 The regulation of leukocyte alkaline phosphatase. Annals of the New York Academy of Science 155: 902
Rosner F, Valmont I, Kozinn P J, Caroline L 1970 Leukocyte function in patients with leukaemia. Cancer 25: 835
Rozenszajn L, Marshak G, Efrati P 1963 Acid phosphatase activity in normal human blood and bone marrow cells as demonstrated by the azo-dye method. Acta Haematologica 30: 310
Rustin G J S, Peters T J 1979 Studies on the subcellular organelles of neutrophils in chronic granulocytic leukaemia with special reference to alkaline phosphatase. British Journal of Haematology 41: 533
Rustin G J S, Wilson P D, Peters T J 1979 Studies on the subcellular localization of human neutrophil alkaline phosphatase. Journal of Cell Science 36: 401
Rustin G J S, Goldman J M, McCarthy D, Mees S, Peters T J 1980 An extrinsic factor controls neutrophil alkaline phosphatase synthesis in chronic granulocytic leukaemia. British Journal of Haematology 45: 381
Rutenburg A M, Seligman A M 1955 Acid phosphatase by post-incubation coupling. Journal of Histochemistry and Cytochemistry 3: 456
Rutenburg A M, Rosales C L, Bennett J M 1965 An improved histochemical method for the demonstration of leukocyte alkaline phosphatase activity in clinical application. Journal of Laboratory and Clinical Medicine 65: 698
Ryder T A, Bowen I D 1975 A method for the fine structural localization of acid phosphatase activity using p-nitrophenyl phosphate as substrate. Journal of Histochemistry and Cytochemistry 23: 235
Sato N, Mori M, Oshimura M, Ueyama Y, Miwa T, Ohsawa N, Kosaka K, Asano S 1982 Factor(s) responsible for the increase in alkaline phosphatase activity of postmitotic granulocytes from normal individuals and patients with chronic myeloid leukaemia. Blood 59: 141
Savage R A, Valenzuela R, Hoffman G C 1981 Acid phosphatase staining pattern as an indication of T-cell acute leukemia. American Journal of Clinical Pathology 76: 760
Schubert J C F, Martin H 1965 Zytochemisches Nachweis alkalischer Phosphataseaktivität in Myeloblasten. Medizinische Welt 13: 531
Scorza P, Paolucci G, Cacciari E 1961 Alcuni enzimi leucocitari nel parto a termine e nello immaturo nei primi due mesi di vita. Bollettino Società Italiana Biologia Sperimentale 37: 323
Scott R E, Horn R G 1970 Ultrastructural aspects of neutrophil granulocyte development in humans. Laboratory Investigation 23: 292
Seligman A M, Manheimer L H 1949 A new method for the histochemical demonstration of acid phosphatase. Journal of the National Cancer Institute 9: 427
Sforza M B, De Somoza N, Morgenfeld C M 1971 Significado de la fosfatase alcalina leucocitaria en la enfermedad de Hodgkin. Sangre. 16: 411
Simmons A V, Spiers A S D, Fayers P M 1973 Haematological and clinical parameters in assessing activity in Hodgkin's disease and other malignant lymphomas. Quarterly Journal of Medicine 42: 111
Smith G P, Peters T J 1981 Subcellular localization and properties of pyridoxal phosphate phosphatases of human polymorphonuclear leukocytes and their relationship to acid and alkaline phosphatase. Biochimica et Biophysica Acta 661: 287
Smith G P, Harris H, Peters T J 1984 Studies of the biochemical and immunological properties of

human neutrophil alkaline phosphatase with comparison to the established alkaline phosphatase isoenzymes. Clinica Chimica Acta 142: 221
Spitznagel J K, Dalldorf F G, Leffel M S, Folds J D, Welsh R H, Cooney M H, Martin L E 1975 Character of azurophil and specific granules purified from human polymorphonuclear leukocytes. Laboratory Investigation 30: 774
Spiers A S D, Williams D M 1976 Changes in neutrophil alkaline phosphatase following splenectomy. Journal of Clinical Pathology 29: 173
Spiers A S D, Liew A, Baikie A G 1975 Neutrophil alkaline phosphatase score in chronic granulocytic leukaemia: effects of splenectomy and anti-leukaemic drugs. Journal of Clinical Pathology 28: 517
Srámková L, Kouba K, Bendová N 1965 Alkaline phosphatase in neutrophil leucocytes of patients with infectious mononucleosis and the effect of corticosteroid therapy. Blood 26: 479
Stavridis J, Creatsas G, Lolis D, Traga G, Antonopoulos M, Kaskarelis D 1981 Relationship between leucocyte alkaline phosphatase and nitroblue tetrazolium reduction activities in the peripheral blood polymorphonuclear leucocytes in normal individuals. British Journal of Haematology 47: 157
Stinson R A 1984 Size and stability to sodium dodecyl sulfate of alkaline phosphatases from their three established human genes. Biochimica et Biophysica Acta 790: 268
Storti E 1952 L'istochimica applicata all' ematologia. C R 3rd Congr. Soc. Europ. Hemat. Rome. p 79. EMES
Streifler C 1970 Study of acid phosphatase in sera of Gaucher's disease patients, subcellular tissue fractions and platelet extracts. Israel Journal of Medical Science 6: 479
Stryckmans P A, Debusscher L, Collard E 1977 Cell kinetics in chronic granulocytic leukaemia. Clinics in Haematology 6: 21
Szmigielski S 1971 The use of dextran in phosphatase techniques employing lead salts. Journal of Histochemistry and Cytochemistry 19: 504
Takamatsu H 1939 Histochemical and biochemical studies of phosphatase. Transactions of the Japanese Pathology Society 29: 492
Tanaka Y, Goodman JR 1972 Electron Microscopy of Human Blood Cells. Harper & Row, New York
Tanaka K R, Valentine W N, Fredericks R E 1960 Diseases or clinical conditions associated with a low leukocyte alkaline phosphatase. New England Journal of Medicine 262: 912
Tanaka T, Kobayashi M, Saito O, Kamada N, Kuramoto A, Usui T 1981 Biochemical activities of lysosomal acid hydrolases in leukemic cells. Clinica Chimica Acta 117: 121
Tanzer J, Harel P, Boiron M, Bernard J 1964 Cytochemical and cytogenetic findings in a case of chronic neutrophilic leukaemia of mature cell type. Lancet i: 387
Telek B, Kiss A, Pecze K, Rak K 1983 Differential diagnostic value of acid phosphatase and beta-glucuronidase in acute leukaemia. Blut 46: 67
Tomonaga M, Ichimaru M, Tsuji Y 1984 Alkaline phosphatase-positive leukemic monocytes in a child with acute monocytic leukaemia. Acta Haematologica 71: 66
Tortarolo M, Cantore N, Grande M, Buonanno G, Dammacco F 1981 Plasma cell acid phosphatase as an adjunct in the differential diagnosis of monoclonal immunoglobulinaemias. Acta Haematologica 65: 103
Tough I M, Jacobs P A, Court Brown W M, Baikie A G, Williamson E R D 1963 Cytogenetic studies on bone marrow in chronic myeloid leukaemia. Lancet i: 844
Trubowitz S, Feldman D, Benante C, Kirman D 1957 Metal requirements of alkaline phosphatases of human and rabbit leukocytes. Proceedings of the Society for Experimental Biology 93: 35
Trubowitz S, Feldman D, Benante C, Hunt V M 1958 The alkaline phosphatase content of human leukocytes following marrow suppression by nitrogen mustard. Journal of Laboratory and Clinical Medicine 52: 259
Trubowitz S, Feldman D, Benante C, Hunt V M 1959 The alkaline phosphatase content of the human polymorphonuclear leucocyte in blood and marrow. American Journal of Clinical Pathology 31: 483
Ueda N 1983 Physicochemical studies on leucocyte alkaline phosphatase. American Journal of Clinical Pathology 80: 342
Usui T, Konishi H, Sawada H, Uchino H 1982 Existence of tartrate-resistant acid phosphatase activity in differentiated lymphoid leukemic cells. American Journal of Haematology 12: 47
Valentine W N, Beck W S 1951 Biochemical studies on leukocytes. I. Phosphatase activity in health, leukocytosis and myelocytic leukemia. Journal of Laboratory and Clinical Medicine 38: 39
Valentine W N, Beck W S, Follette J H, Mills H, Lawrence J S 1952 Biochemical studies in chronic myelocytic leukaemia, polycythaemia vera and other idiopathic myeloproliferative disorders. Blood 7: 959
Valentine W N, Follette J H, Hardin E B, Beck W S, Lawrence J S 1954 Studies on leukocyte

alkaline phosphatase activity: relation to stress and pituitary-adrenal activity. Journal of Laboratory and Clinical Medicine 44: 219
Valentine W N, Follette J H, Solomon D H, Reynolds J 1957a Biochemical and enzymatic characteristics of normal and leukaemic leucocytes with particular reference to leucocyte alkaline phosphatase. In: Henry Ford Hospital International Symposium. The leukaemias: aetiology, pathophysiology and treatment. Academic Press, New York, p 457
Valentine W N, Follette J H, Solomon D H, Reynolds J 1957b The relationship of leukocyte alkaline phosphatase to 'stress', to ACTH and to adrenal 17-OH-corticosteroids. Journal of Laboratory and Clinical Medicine 49: 723
Valentine W N, Tanaka K R, Fredericks R E 1960 Studies on leukocyte alkaline phosphatase: role of zinc and magnesium. Journal of Laboratory and Clinical Medicine 55: 303
Vallee B L 1959 Biochemistry, physiology and pathology of zinc. Physiological Reviews 39: 443
Veenhuis M, van Dijken J P, Harder W 1980 A new method for the cytochemical demonstration of phosphatase activities in yeasts based on the use of cerous ions. Microbiology Letters 9: 285
Veerman A J P, Huismans D R, Van Zantwijk C H 1983 Immunological phenotype related to acid alpha-naphthyl acetate esterase and acid phosphatase in childhood acute lymphoblastic leukaemia. Acta Haematologica 69: 32
Wachstein M 1946 Alkaline phosphatase activity in normal and abnormal human blood and bone marrow cells. Journal of Laboratory and Clinical Medicine 31: 1
West B C, Rosenthal A S, Gelb N A, Kimball H R 1974 Separation and characterisation of human neutrophil granules. American Journal of Pathology 77: 41
Wetzel B K, Spicer S S, Horn R G 1967 Fine structural localization of acid and alkaline phosphatase in cells of rabbit blood and bone marrow. Journal of Histochemistry and Cytochemistry 15: 311
Williams D M 1975 Leucocyte alkaline phosphatase as a marker of cell maturity: a quantitative cytochemical and autoradiographic study. British Journal of Haematology 31: 371
Wilson P D, Rustin G J S, Peters T J 1981 The ultrastructural localization of human neutrophil alkaline phosphatase in normal individuals during pregnancy and in patients with chronic granulocytic leukaemia. Histochemical Journal 13: 31
Wilson P D, Smith C P, Peters T J 1983 Pyridoxal-5′-phosphate: a possible physiological substrate for alkaline phosphatase in human neutrophils. Histochemical Journal 15: 257
Wiltshaw E, Moloney W C 1955 Histochemical and biochemical studies on leucocyte alkaline phosphatase activity. Blood 10: 1120
Wyllie R G 1962 Neutrophil alkaline phosphatase. Response to acute inflammation. Australian Annals of Medicine 11: 118
Xefteris E, Mitus W J, Mednicoff I B, Dameshek W 1961 Leukocytic alkaline phosphatase in busulfan induced remissions of chronic granulocytic leukaemia. Blood 18: 202
Yam L T 1974 Clinical significance of the human acid phosphatases—a review. American Journal of Medicine 56: 604
Yam L T, Li C Y, Lam K W 1971 Tartrate resistant acid phosphatase isoenzyme in the reticulum cells of leukaemic reticuloendotheliosis. New England Journal of Medicine 284: 357
Yam L T, Janckila A J, Li C Y, Lam K W 1987 Cytochemistry of tartrate-resistant acid phosphatase: 15 years experience. Leukemia 1: 285
Yang K, Bearman R M, Pangalis G A, Zelman R J, Rappaport H 1982 Acid phosphatase and alpha-naphthyl acetate esterase in neoplastic and non-neoplastic lymphocytes. A statistical analysis. American Journal of Clinical Pathology 78: 141
You W, Weisbrot I M 1979 Chronic neutrophilic leukemia. Report of two cases and review of the literature. American Journal of Clinical Pathology 72: 233
Young J E, Prager M D 1962 Some hydrolytic enzymes in normal and leukemic human leukocytes. Journal of Laboratory and Clinical Medicine 60: 385

CHAPTER 8

Esterases

The biochemistry of esterases is very complex and the nomenclature confused, but in cytochemistry the term esterase, or 'non-specific' esterase is generally restricted to enzymes capable of hydrolysing the simpler esters of N-free alcohols and organic acids. Carboxyl esterases (EC 3.1.1.1), aryl esterases (EC 3.1.1.2) and acetyl esterases (EC 3.1.1.6) all belong to this group (Lojda *et al.*, 1979).

Enzymes which split esters of choline are given the specific names acetylcholinesterase and cholinesterase (EC 3.1.1.7 and 3.1.1.8); they are inhibited by eserine and by organophosphorus compounds. They are of little interest in haematological cytochemistry. Zajicek *et al.* (1954) used a substrate of acetylthiocholine in the presence of copper glycinate in an attempt to demonstrate acetyl cholinesterase by observing the distribution of crystalline copper thiocholine at presumed sites of enzymic activity with a phase contrast microscope. In human haemic cells positivity was restricted to megakaryocytes and red cells; all leucocytes were negative.

Esters of long chain fatty acids are hydrolysed by so-called lipases, not inhibited by eserine and not regularly affected by organophosphorus compounds. Esters formed between inorganic acids and alcohols, e.g. between phosphoric acid and ethyl alcohol, are hydrolysed by specific types of esterase, e.g. phosphomonoesterase or phosphatase.

These specific types of esterase are dealt with elsewhere in so far as they are of interest in haematology, and the present chapter is concerned with the main leucocytic esterases demonstrable by their activity on simple acyl or chloracyl esters of α-naphthol or naphthol AS. Thiolacyl esters have also proved suitable substrates for the ultrastructural demonstration of non-specific esterase activity of blood cells (Payne *et al.*, 1980), while methods based on the use of esters of indoxyl compounds (Delellis and Fishman, 1965) have been little exploited in haematological cytochemistry. The manner in which these enzymes react with their substrates appears to be similar to that described in some detail for aliphatic esters by Dixon and Webb (1964). The enzymes possess sites of affinity for the respective acyl and alkyl groups of the substrate, with an intermediate site for the ester bond. When esters with varying lengths of acyl and alkyl groups were studied it emerged that affinity and reactivity was increased for both groups when extended to 4–6 carbon atoms, but further extension led to reduced reaction, presumably because larger groups could no longer be accommodated on the enzyme affinity sites. Following alignment on the enzyme, the acyl group becomes linked to the esterase, while the alkyl group is liberated as the free alcohol or naphthol. Hydrolysis of the acylated enzyme then yields free acid and enzymes.

Development of cytochemical methods for esterases

Naphthyl esters have been used predominantly in esterase cytochemistry because the liberated naphthol will couple with an appropriate diazotized amine to give a brightly coloured azo-dye, as described in the chapter on Phosphatases, enzymes which may be demonstrated cytochemically by a similar technique. This azo-dye method was first introduced for esterases by Nachlas and Seligman (1949) using β-naphthyl acetate as substrate, and various modifications with different substrates and adjustments of incubation period, pH and temperature were subsequently devised to eliminate diffusion artefacts and give sharply granular pictures. In haemic cells a number of different reaction patterns came to be recognised as characteristic of granulocytes and monocytes when particular substrates or inhibitors were used.

Gomori (1953) first described the distribution of esterase positivity when the chloroacetates of α-naphthol and naphthol AS were used as substrates. With either of these substrates mast cells stained intensely and granulocytes at all stages of maturation from early myelocytes onwards showed strong positivity. Myeloblasts, monocytes and lymphocytes were reported to show no reaction. Moloney and his colleagues (1959, 1960) confirmed and extended this work, using naphthol AS-D chloroacetate, a stable, highly chromogenic substrate, which gave sharply localised granular staining in the cytoplasm of nearly all neutrophilic granulocytes. Although myeloblasts and other stem cells were negative, promyelocytes showed strong positivity. Eosinophils, basophils, monocytes, lymphocytes, plasma cells, nucleated red cells and megakaryocytes showed no activity (Figs 8.1 and 8.2).

In contrast to this pattern of positivity, the distribution of enzyme reaction product when α-naphthyl acetate was used as substrate (Braunstein, 1959; Leder, 1966, 1967) was largely confined to monocytes (Fig. 8.3), while the use of naphthol AS or AS-D acetate (Wachstein and Wolf, 1958; Báguena and Forteza, 1961; Shnitka and Seligman, 1961; Hayhoe *et al.*, 1964) gave some degree of positivity in most haemic cells, but with especially strong reactions in monocytes.

Rozenszajn and his colleagues (1968) compared the results of esterase cytochemistry with each of these three substrates in normal haemopoietic cells and in megaloblastic and leukaemic cells. They confirmed the results previously described in normal haemic cells and noted strong α-naphthyl acetate esterase positivity in megaloblasts whether from cases of vitamin B_{12} or folic acid deficiency or in Di Guglielmo's disease or megaloblastic hypoplastic states. These findings were confirmed by Kass (1977a) and Kass and Peters (1977), who also found this non-specific esterase activity in erythroblasts to be moderately fluoride-resistant (see below). In erythroleukaemia some abnormal erythroid precursors were also noted to show chloroacetate esterase positivity (Kass, 1977a). In the differentiation of acute leukaemias the reactions proved helpful (Rozenszajn *et al.*, 1968); α-

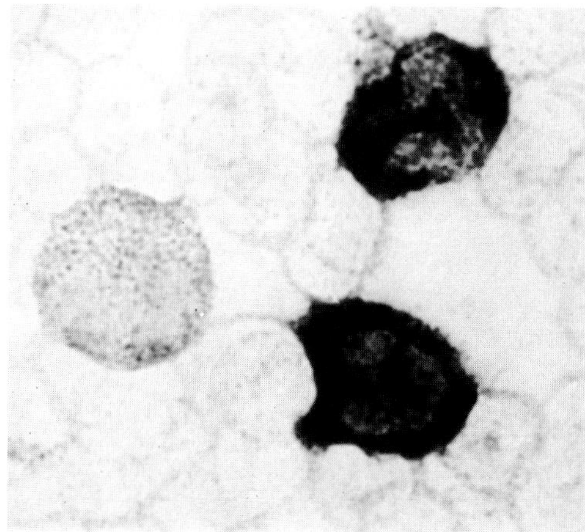

Fig. 8.1 Normal buffy coat. Chloroacetate esterase. Strong positivity in two neutrophil polymorphs, with negative reaction in an eosinophil.

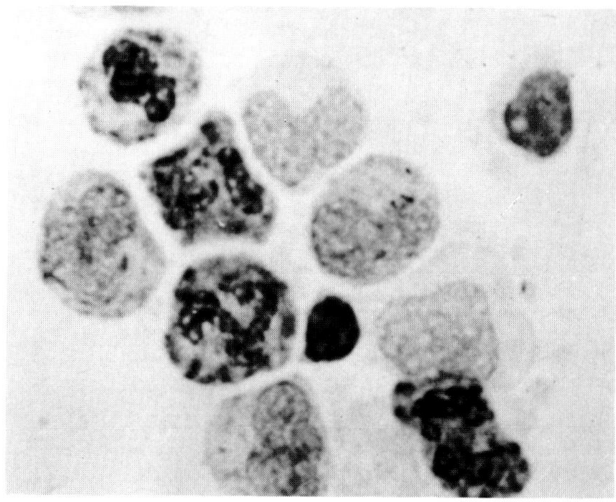

Fig. 8.2 Normal buffy coat. Chloroacetate esterase. Irregular granular positivity in four neutrophil polymorphs; negative reaction in five monocytes and two lymphocytes.

naphthyl acetate esterase showed little or no reactivity in acute myelocytic and lymphoblastic leukaemias, whereas monocytes and monoblasts from acute monocytic leukaemias showed strong activity; naphthol AS-D chloroacetate esterase was active in leukaemic myelocytes and some myeloblasts (Fig. 8.4), while monoblasts and lymphoblasts were negative.

Chloroacetate esterase positivity of Auer rods, already noted by Rozenszajn *et al.* (1968), seems actually to be confined to some peculiar forms of acute myeloid leukaemia, such as the acute granulocytic leukaemia with 8;21 translocation (Swirsky *et al.*, 1984) and acute promyelocytic leukaemia. In the latter, we have observed that the conspicuous Auer rods often show a positively stained periphery with an unstained centre (Plate 19C). On the contrary, in a case of acute myeloid leukaemia with pseudo-Chediak-

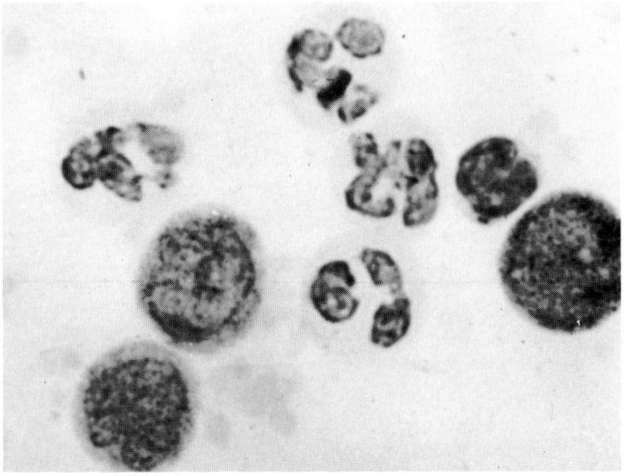

Fig. 8.3 Normal buffy coat. Naphthyl acetate esterase. Negative reaction in four neutrophils and one lymphocyte; three monocytes, one positive, one weakly positive and one negative.

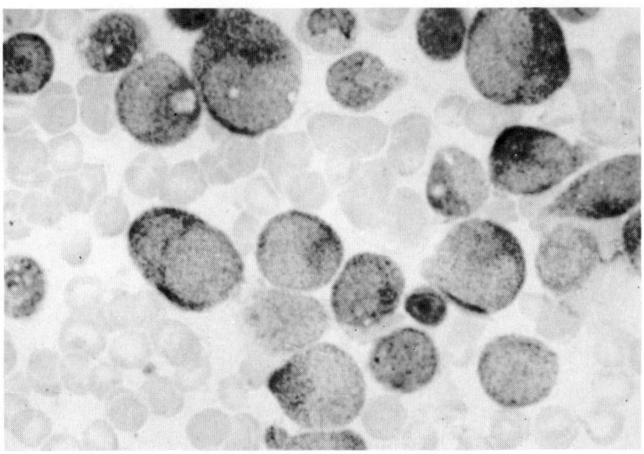

Fig. 8.4 AGL bone marrow. Chloroacetate esterase. Most myeloblasts and promyelocytes show a positive reaction.

Higashi anomaly, similar to that described by Tulliez *et al.* (1979), we found the giant granules to be frequently characterized by a chloroacetate positive core surrounded by a negative peripheral area (Plate 19D).

Although in the study of Rozenszajn *et al.* (1968), naphthol AS-D acetate showed positivity of some degree in most haemic cells, so that the reaction did not permit the clear differentiation of various cells of the blood or bone marrow, Fischer and Schmalzl (1964) had shown that non-specific esterase activity in the monocytes is inhibited by fluoride, whereas that in granulocytes is not. Schmalzl and Braunsteiner (1971) and Daniel *et al.* (1971) found strong naphthol AS-D esterase reactions with fluoride inhibition to

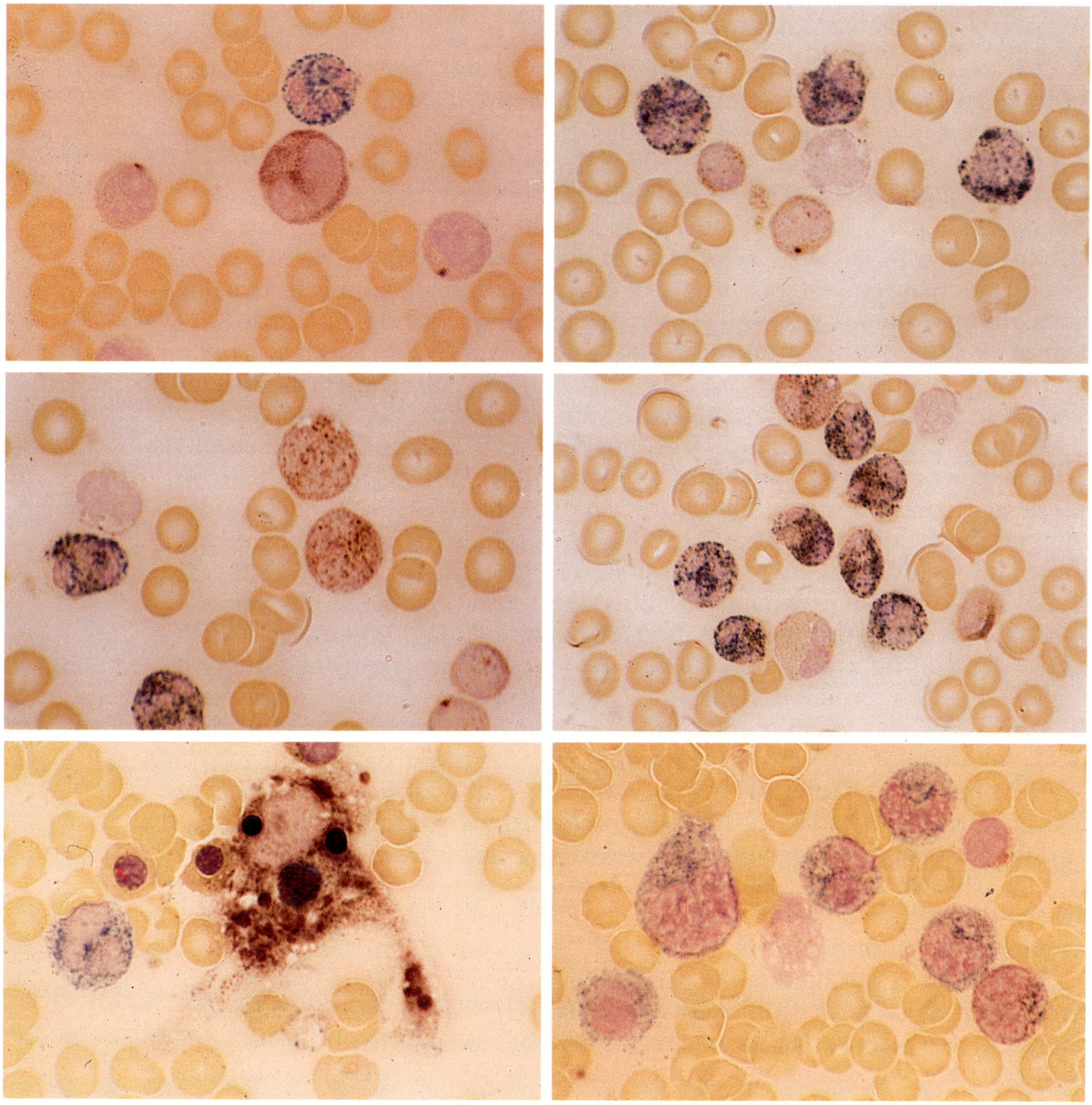

A B C D E F

Plate 16

A. Normal buffy coat. Dual esterase (chloro-acetate esterase (CE) with Fast blue BB, and butyrate esterase (BE) with Fast Garnet). One CE positive neutrophil, one BE positive monocyte and two lymphocytes with localised positive BE reaction of the T cell type.

B. Normal buffy coat. Dual esterase. Three CE positive neutrophils and three lymphocytes, one negative (B cell type), one with a coarse block of localised BE reaction (T cell type) and one with scattered BE granules (null cell type).

C. Normal buffy coat. Dual esterase. Two CE positive neutrophils, two monocytes with predominant BE positivity but a few granules of CE reaction, and three lymphocytes with single dots of BE positivity of T cell type.

D. Normal buffy coat. Dual esterase. Seven CE positive neutrophils, one monocyte with mixed but chiefly BE reaction, a negative eosinophil and two lymphocytes, one with negative, B cell, reaction and one with localised, partly obscured, T cell type of BE reaction.

E. Normal bone marrow. Dual esterase. A macrophage with strong BE positivity and an ingested CE positive neutrophil; a CE positive metamyelocyte and two negative normoblasts.

F. Normal bone marrow. Dual esterase. Six granulocyte precursors, chiefly myelocytes and metamyelocytes, with predominant CE positivity, although one myelocyte also shows BE reaction. A single lymphocyte shows the T cell pattern of localised BE positivity.

be characteristic of the cells of acute monocytic leukaemias (Figs 8.5 and 8.6), whereas leukaemic lymphoblasts, myeloblasts and promyelocytes showed weaker reactions, resistant to fluoride.

The non-specific esterase activity of normal and neoplastic megakaryocytes (Polli *et al.*, 1985) and of abnormal red cell precursors (Kass and Peters, 1977) has also been reported as at least partially fluoride sensitive. Moreover, significant activity of both fluoride-sensitive non-specific esterase and chloroacetate esterase has been described in malignant cells of some cases of myelomonocytic leukaemia (Tavassoli *et al.*, 1979; Huhn and Twardzik, 1983) and promyelocytic leukaemia (Golomb *et al.*, 1980; Tomonaga *et al.*, 1985), as well as in granulocyte precursors of myelodysplastic syndromes and megaloblastic anaemias (Scott *et al.*, 1983).

Owing to the wide differences existing between these studies with regard to non-specific esterase substrates and staining methods, an investigation has been carried out in our laboratories in order to ascertain whether under comparable staining conditions the use of different non-specific esterase substrates could influence the results of fluoride inhibition tests. α-Naphthyl acetate and α-naphthyl butyrate were considered comparable substrates, since both represent short-chain simple naphthyl esters, which can be split quite rapidly by the cell enzymes. Normal and pathological specimens, after buffered formalin-acetone fixation, were incubated for 30 min at 25°C in a staining solution containing 0.5 mmol/l substrate (from stock solutions of 0.05 mol/l in acetone) and 3.3 mol/l diazonium salt in 0.067 M phosphate buffer pH 7.2. Fast blue BB salt was used throughout as the coupler, after it had been shown by preliminary experiments that the use of different coupling reagents, such as hexazonium pararosaniline or Fast Garnet GBC salt, did not affect the proportional affinity of the cell enzymes for the two substrates. A concentration of 40 mmol/l sodium fluoride was employed for inhibition tests.

Under these controlled staining conditions, the esterase activity of normal circulating monocytes produced a clearly stronger positivity when α-naphthyl butyrate was used as substrate rather than α-naphthyl acetate: both reactions were inhibited by the addition of fluoride. Also the dot-like staining of lymphocytes (see below) was better demonstrated by α-naphthyl butyrate, but the reaction was unaffected by fluoride. In both neutrophil granulocytes and platelets, esterase activity was practically detectable only by α-naphthyl acetate: fluoride did not alter the weak stainability of granulocytes, but it apparently reduced the percentage of positive platelets.

In non neoplastic bone-marrow samples, α-naphthyl acetate esterase was demonstrable in a large percentage of myeloid cells, including, in addition to monocytes and macrophages, the majority of promyelocytes and myelocytes, a high proportion of megakaryocytes and a considerable number of the less mature red cell precursors, usually in the form of a paranuclear cluster of granules. The esterase activity of granulocyte precursors and megakaryocytes appeared to be consistently weaker when α-naphthyl butyrate was employed as substrate, and the detection of positive erythroblasts was very rare indeed. Inhibition tests showed in most cell types a

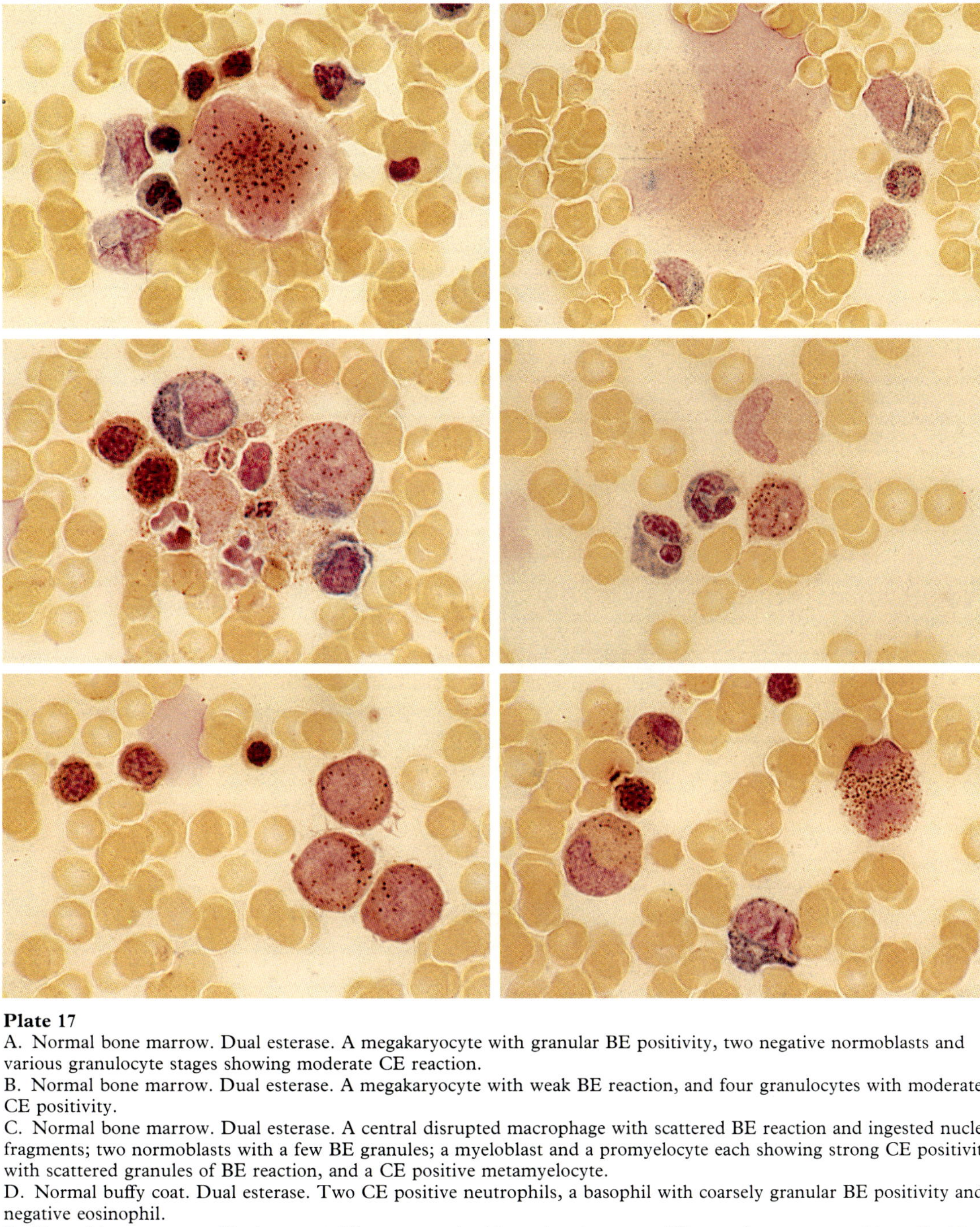

A B C D E F

Plate 17

A. Normal bone marrow. Dual esterase. A megakaryocyte with granular BE positivity, two negative normoblasts and various granulocyte stages showing moderate CE reaction.

B. Normal bone marrow. Dual esterase. A megakaryocyte with weak BE reaction, and four granulocytes with moderate CE positivity.

C. Normal bone marrow. Dual esterase. A central disrupted macrophage with scattered BE reaction and ingested nuclear fragments; two normoblasts with a few BE granules; a myeloblast and a promyelocyte each showing strong CE positivity with scattered granules of BE reaction, and a CE positive metamyelocyte.

D. Normal buffy coat. Dual esterase. Two CE positive neutrophils, a basophil with coarsely granular BE positivity and a negative eosinophil.

E. Normal bone marrow. Dual esterase. Three proerythroblasts showing coarse BE granules, two normoblasts B with a few granules of CE reaction, and three lymphocytes with single dots of BE positivity of T cell type.

F. Normal bone marrow. Dual esterase. A CE positive myelocyte, two eosinophil metamyelocytes with a few BE positive granules, a late normoblast with BE positivity at the nuclear margin and a probable monocyte precursor at anaphase, with strong BE reaction.

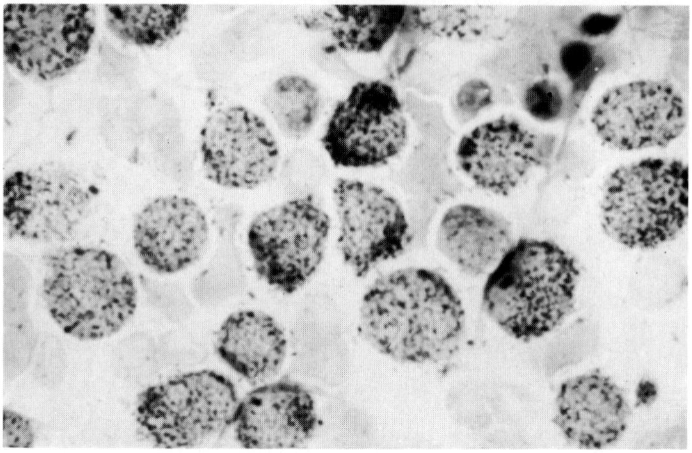

Fig. 8.5 AMML bone marrow. Naphthol AS-D acetate esterase. Strong positivity in leukaemic monoblasts.

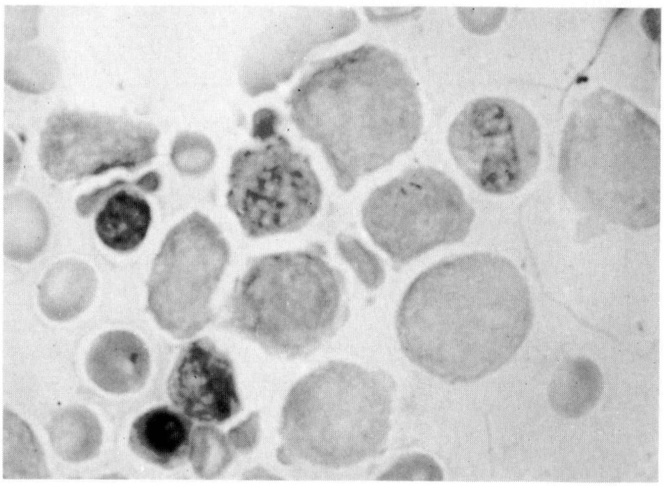

Fig. 8.6 The same specimen and reaction but with added fluoride. The reaction is inhibited in the blast cells.

variable degree of fluoride sensitivity: this was minimal in granulocyte and red cell precursors, intermediate in megakaryocytes and maximal in recognizable monocytes.

The cytochemical behaviour of leukaemic cells was found to be in most instances consistent with that of their normal counterparts, both in terms of different degree of stainability with the two substrates and of fluoride sensitivity. Thus, in acute granulocytic leukaemias esterase activity, when detectable, was better shown by α-naphthyl acetate. The same was true both for leukaemic erythroblasts and malignant megakaryocytes: the latter cell type displayed an evident degree of fluoride sensitivity. With regard to neoplastic monocytes, only in about two thirds of cases was the reactivity pattern similar to that of normal monocytes: a variable proportion of cells from the remaining cases showed an α-naphthyl acetate esterase staining similar to, or even stronger than, butyrate esterase. In some of these cases fluoride inhibition was clearly less effective in reducing acetate esterase activity in comparison with butyrate esterase. This pattern of reactivity was frequently, but by no means constantly associated with a rather immature cell morphology, and with a proportionally stronger myeloperoxidase activity, thus suggesting a less differentiated monocytic phenotype.

In some cases of acute promyelocytic leukaemia and of myelodysplastic syndrome the staining intensity of the abnormal cells was found to be fairly high, and similar with either acetate or butyrate, but in our experience the fluoride sensitivity of this esterase activity, even if present, was by no means comparable with that of normal monocytes.

In conclusion, according to this study a fluoride sensitive butyrate esterase activity, unmatched by an acetate esterase activity of equal intensity, could be confidently assumed as a marker of monocytic differentiation, while on the contrary a fluoride sensitive acetate esterase activity unmatched by butyrate esterase seemed more in keeping with a megakaryocytic origin of leukaemic cells, than with an immature monocytic phenotype as suggested by Cerezo and Head (1982). An esterase activity, which is equally well shown by both substrates, even if more or less sensitive to fluoride inhibition, does not allow any safe conclusion about the real nature of leukaemic primitive cells.

Esterase isoenzymes in haemic cells

The differing reaction patterns with these various substrates and the observation that the enzyme in monocytes was fluoride sensitive whereas in most other cells it was not, strongly suggested the existence of multiple esterases or esterase isoenzymes in human haemic cells. Ansley and his colleagues (Ansley and Ornstein, 1971; Ansley *et al.*, 1971) exploited the use of various esterase staining procedures as potential markers for different leucocyte classes in an automated differential leucocyte counter, the Haemalog D, and confirmed the practical usefulness of the two broad classes of esterase activity, monocytic and granulocytic, in achieving discrimination between these cell types. Nevertheless the various substrates used do not provide an absolute specificity for one or other cell type, but, especially with high concentration, long incubation periods and at certain pH values, tend to give overlapping positivity. Li *et al.* (1973), following the suggestion of Hunter and Burstone (1960) that zymogram studies might help to characterize enzyme substrate specificity, sought to determine whether different types of esterase were involved and their degree of substrate and cell specificity, using a combination of cytochemistry and polyacrylamide gel electrophoresis. Leucocyte enzymes from normal and a variety of pathological states were extracted with 0.1% cetyltrimethylammonium bromide and separated by electrophoresis using the method of Reisfeld *et al.* (1962) with the tray buffer at pH 4.0. This technique proved suitable for separating isoenzymes of both chloroacetate esterase and α-naphthyl acetate esterase.

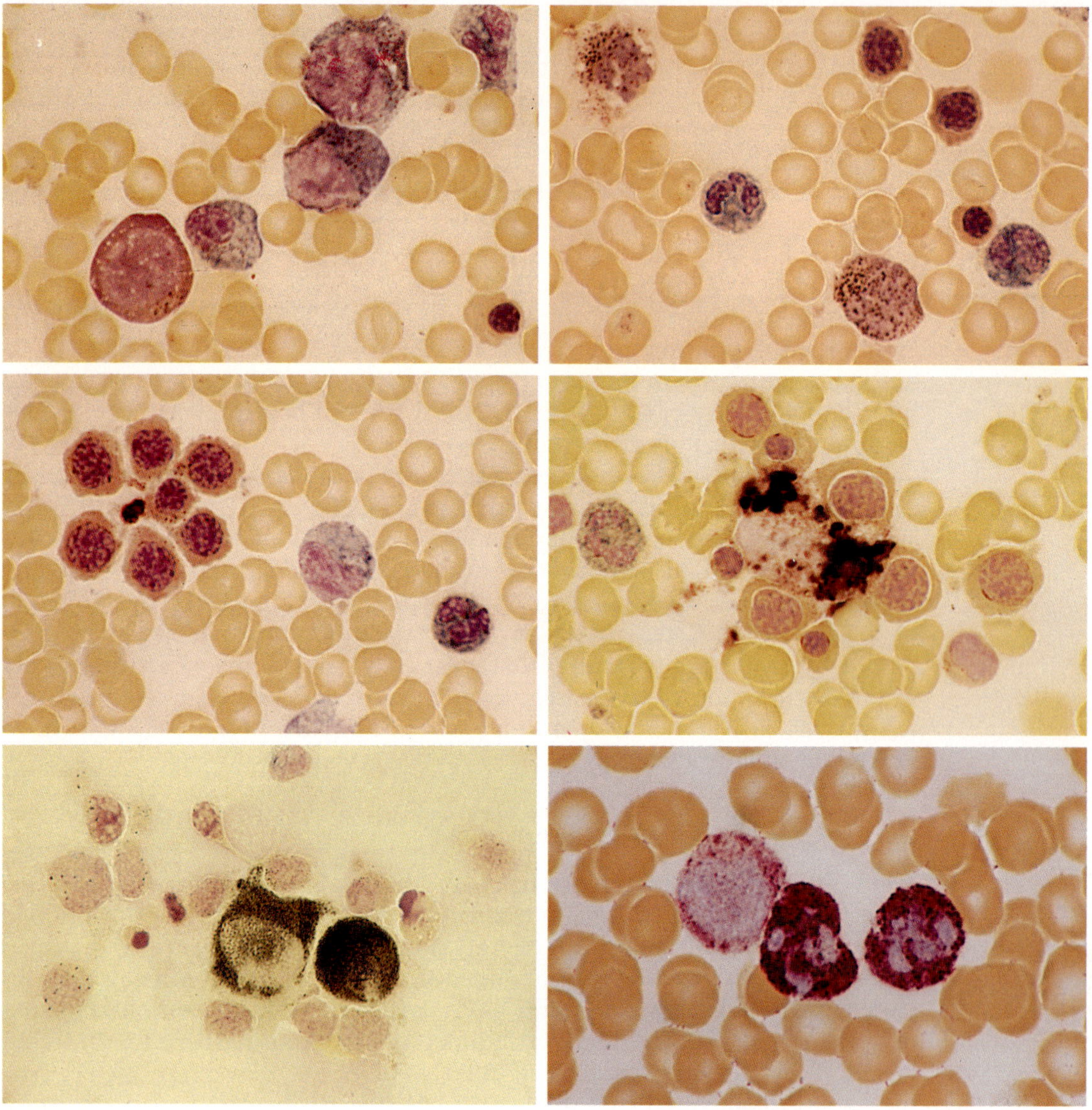

A B

C D

E F

Plate 18

A. Normal bone marrow. Dual esterase. Two myelocytes and two segmented neutrophils with CE positivity, a proerythroblast with a few BE positive granules, and a negative late normoblast.

B. Normal bone marrow. Dual esterase. Three negative normoblasts, two CE positive segmented neutrophils and two monocytes with strong BE positivity.

C. Normal bone marrow. Dual esterase. A nest of seven normoblasts, with occasional BE positive granules, a myelocyte and a segmented neutrophil both showing CE positivity.

D. Normal bone marrow. Dual esterase. Intensely BE positive macrophage with seven surrounding normoblasts, several showing crescents of BE positivity at an arc of the nuclear rim. Two lymphocytes, one negative and one with weak T-cell type localised BE positivity. A CE positive neutrophil.

E. Splenic imprint. Butyrate esterase reaction. Positivity in two RE cells.

F. Normal blood. Chloroacetate esterase reaction with new fuchsin as receptor. Prolonged incubation (2 hours) shows positivity in a monocyte as well as neutrophils.

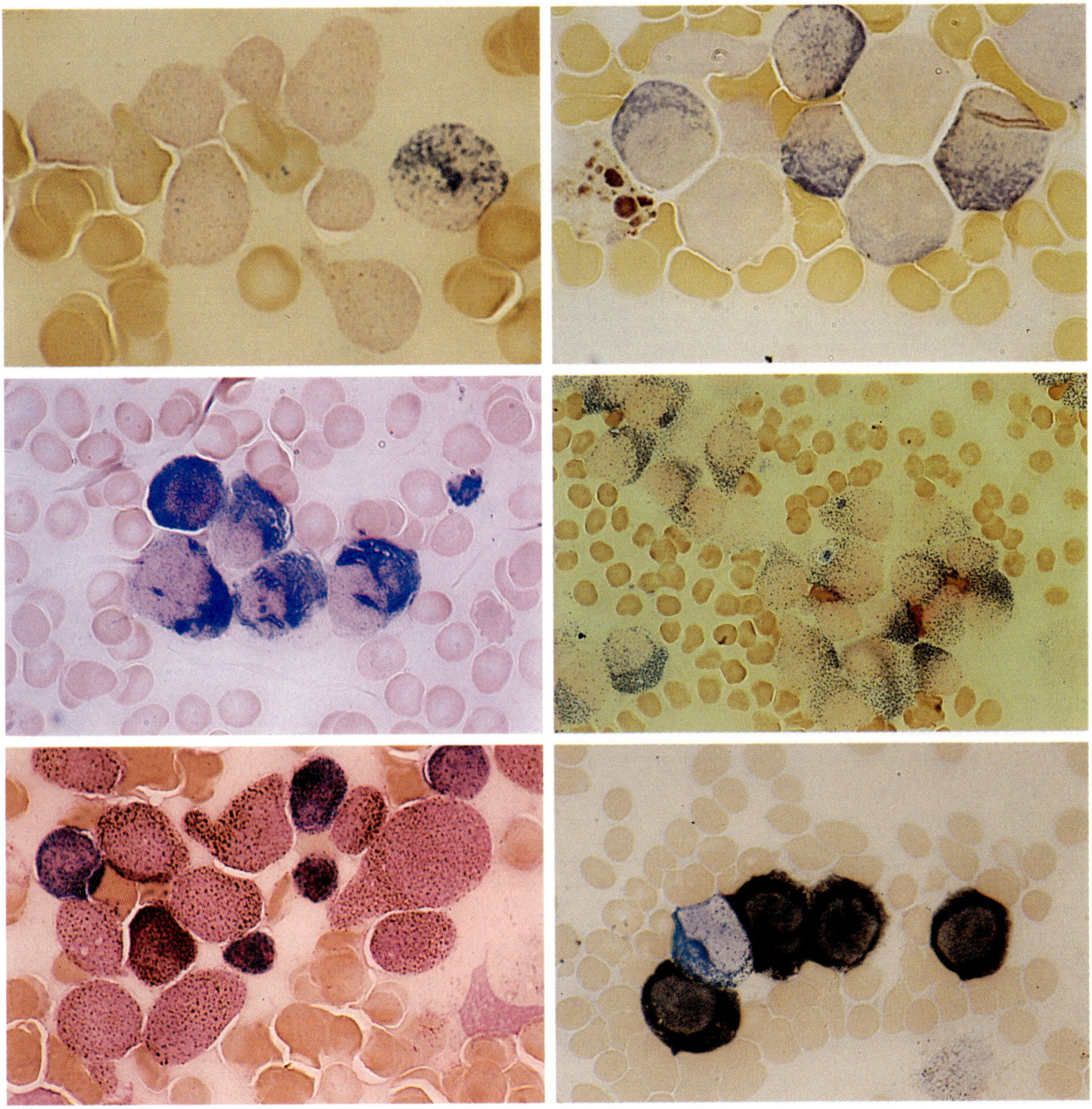

Plate 19

A. ALL (non-T, non-B) bone marrow. Dual esterase. A CE positive neutrophil and seven lymphoblasts showing weak scattered reaction, chiefly BE, as shown by null lymphocytes.

B. AML bone marrow. Dual esterase. A group of myeloblasts or promyelocytes, five showing positive CE reaction. A neighbouring macrophage contains coarsely granular BE activity.

C. APL bone marrow. Dual esterase. Five promyelocytes with strong CE positivity. Frequent Auer rods show negative core with positive outline.

D. AML bone marrow with pseudo-Chediak-Higashi inclusions. CE reaction showing central core CE positivity with negative periphery in the inclusions.

E. AMonL bone marrow. Dual esterase. Five strongly CE positive neutrophils and eleven BE positive monocyte precursors. These blast cells appear to contain only BE and no granules of CE reactivity can be detected.

F. AMonL bone marrow. Dual esterase. Four BE positive monoblasts and a CE positive promyelocyte. Single Fast blue B receptor method.

Separated bands were demonstrated using a variety of α- and β-naphthyl esters, naphthol AS esters and naphthol AS chloroacyl esters with acyl groups of different chain lengths, Fast Garnet GBC being used routinely as coupler, and the reactions allowed to take place at a range of pH values from 6.0 to 7.6. Parallel cytochemical staining procedures were carried out on smears of blood or bone marrow fixed in buffered formalin acetone at pH 6.6 for 30 secs at 4° to 10°C. Fast Garnet GBC or Fast blue RR gave good staining patterns but in view of the solubility of their reaction products in xylene and immersion oil the more permanent stain resulting from the use of hexazonium pararosaniline in acid pH and hexazonium fuchsin in alkaline pH were preferred.

Nine bands of isoenzymes were separated by electrophoresis, bands 1, 2, 7, 8 and 9 being best demonstrated by naphthol AS-D chloroacetate and presumably representing the activity of chloroacetate esterase as found cytochemically in granulocytes. The remaining bands 3, 4, 5 and 6 reacted best with α-naphthyl esters, acetate, propionate or butyrate, and were thought to represent the non-specific esterases demonstrable cytochemically especially in monocytes, platelets and plasma cells. Prolonged staining led to some reaction at the position of bands 2, 7, and 9. Naphthol AS acetate and its derivatives were hydrolysed more slowly than α-naphthyl acetate, gave chief staining reactions in band 5, but also hydrolysed bands 2, 7 and 9.

The zymogram and enzyme cytochemical studies ran closely parallel. Zymogram patterns in extracts from different leucocyte concentrates showed predominance of the expected bands in relation to cytochemical positivity, namely bands 2, 7 and 9 in extracts of cells which contained cytochemical chloroacetate esterase, and bands 3, 4, 5 and 6 in extracts of cells containing α-naphthyl acetate esterase. Extracts from leukaemic myeloblasts and promyelocytes showed strong bands at positions 2 and 7 with chloroacetate and some diffuse staining at band 6 with α-naphthyl acetate, while extracts from leukaemic monoblasts and monocytes gave bands 4, 5 and 6, band 4 being best demonstrated with α-naphthyl butyrate and band 5 with acetate. The activity of the monocytic enzymes was inhibited by sodium fluoride. Megakaryocytes and platelets showed very strong activity in band 5 and some activity in bands 3 and 4. Cytochemically megakaryocytes were strongly positive with α-naphthyl acetate but negative or only weakly positive with butyrate as substrate, while platelets showed a similar differential reactivity, with moderate positivity to acetate but negative reactions with butyrate. Preparations from the bone marrow in multiple myeloma were also studied. Esterase activity at band 6 was found with α-naphthyl acetate as substrate; the reaction was inhibited by fluoride and was not demonstrable with α-naphthyl butyrate.

In lymphocyte preparations from various subjects including a patient with CLL, only weak activity was found, situated at bands 2 and 6. Cytochemically, a single granule of reaction product could sometimes be demonstrated in lymphocytes using either α-naphthyl acetate or butyrate as substrate. The authors noted that the enzyme activity of band 2 appeared to be less specifically located in a particular cell type than other bands, having been found in various types of cell preparation. Moreover it hydrolyzed both acyl and chloroacyl esters, though with a substrate preference for α-naphthyl propionate or butyrate, and was resistant to fluoride inhibition. Apart from the two main leucocytic esterases, a third enzyme was demonstrable in human mast cells; this was an aminocaproate esterase, with trypsin-like properties, not present in other leucocytes.

The electrophoretic analysis of chloroacetate esterases has been further elaborated in both normal granulocytes and in chronic myeloid leukaemia by Kass *et al.* (1978). Using polyacrylamide disc gel electrophoresis of cetyltrimethylammonium bromide extracts, 8 bands of specific granulocytic esterases were found in material derived from normal polymorphonuclear cells, while 11 bands were found in extracts from CML cells. The extra bands might reflect abnormal metabolism of malignant granulocytes, or perhaps represent enzyme components present in immature cells.

By the same method, Kass and Peters (1978) could not reveal any specific or non-specific esterase activity in two cases of acute lymphoblastic leukaemia. When cases of acute myelomonocytic or histiomonocytic leukaemia were examined, a prominent fluoride-sensitive esterase activity of band 5 was revealed with either α-naphthyl acetate or α-naphthyl butyrate as the substrate, while the same band was reported as undetectable in cases of acute myeloblastic leukaemia.

Yourno and Mastropaolo (1981) used polyacrylamide slab gel electrophoresis at pH 9.5 to obtain zymograms of purified blood cells, including granulocytes, monocytes, T and B lymphocytes, erythrocytes and platelets. Two bands of low electrophoretic mobility, identified as AB1 and AB2, were equally well demonstrated employing either α-naphthyl acetate or α-naphthyl butyrate as substrate, while nine bands, named A1–A9, were better shown by the use of α-naphthyl acetate. One additional band, B1, was more clearly revealed by α-naphthyl butyrate. AB1 and AB2, both highly fluoride-sensitive, were detected in zymograms of purified granulocytes and monocytes, but in the former cell type the per cell isoenzyme activity appeared to be only 2 to 3% that of monocytes. Band B1 was present in all cell preparations, being especially intense in red cell zymograms. A2 was found in all cell types, and A1 in all except erythrocytes. A3 and A4 were visualized in erythrocyte zymograms, while bands A5 through A9, which all showed partial fluoride-resistance, were present in platelet preparations.

The high resolution technique of isoelectric focusing in polyacrylamide gel, which had been already proposed by Young and Bittar (1973) for the study of tissue esterases in Hodgkin's disease and other tumours, was used by Kass and Munster (1979a) in order to investigate the isoenzyme pattern of non-specific esterases from purified normal peripheral blood cells. After ultrasonic disruption of the cells, samples of enzyme solutions were incorporated in polyacrylamide gel containing an appropriate concentration of ampholytes, and subjected to isoelectric focusing. The isoenzyme bands were subsequently revealed by the use of either α-naphthyl acetate or α-naphthyl butyrate as the substrate, at a neutral pH, with Fast Garnet GBC salt as coupler. By employing α-naphthyl acetate, 12 bands were detected in monocytes, 12 in lymphocytes, 8 in granulocytes and 14 in platelets. The presence of fluoride-sensitive isoenzymes was reported only in monocytes and lymphocytes. With α-naphthyl butyrate, 6 isoenzymes were demonstrated in monocytes, 8 in lymphocytes, 3 in granulocytes and 8 in platelets, all of them allegedly being fluoride sensitive.

Radzun *et al.* (1980) used a similar technique in an attempt to study the lysosomal acid esterase activity of isolated normal blood cells. After cell disruption, produced by a cavitation bomb, lysosomes were separated by ultracentrifugation, and their enzyme content was extracted by a detergent treatment. Following isoelectric focusing, α-naphthyl acetate was employed in the presence of hexazonium pararosaniline at pH 5.8 in order to visualize the isoenzyme bands. Each individual cell type displayed a characteristic isoenzyme pattern. In monocytes 4 bands were present, with isoelectric points ranging from 6.2 to 5.86, the most anodic of which accounted for 85% of the total enzyme activity of these cells. Granulocytes showed 11 bands distributed through a pH range of 7.9–4.8. T lymphocytes and thymocytes revealed similar enzyme patterns, characterized by respectively 13 and 11 bands grouped in three distinct complexes, while only 7 bands were detected in B lymphocytes. Platelets showed 10 bands, and erythrocytes 5 ill-defined bands.

Oertel and Kastner (1984) attempted a two-dimensional separation of α-naphthyl acetate esterase isoenzymes, by employing sequentially isoelectric focusing and polyacrylamide gradient gel electrophoresis at pH 9.0. By this method, extractable enzymes of normal isolated blood cells were separated according to both isoelectric point and molecular weight (MW). Visualization of isoenzyme bands was achieved using Fast blue BB salt as coupler at pH 8. Granulocyte extracts contained an esterase activity characterized by a MW of 60 000 and isoelectric point in the range 4.5–5.0, while in lymphocytes an esterase with isoelectric point 4.3–4.8 and MW 50 000 was detected. Monocytes, in addition to the two bands already mentioned, contained 5 isoenzyme bands, all with isoelectric points in the range 5.7–6.2, whose MW varied from a minimum of 145 000 to a maximum of 340 000. Platelet esterase activity was represented by a group of isoenzymes with isoelectric points varying from 6.5 to 8, and MW ranging in direct proportion from 150 000 to 400 000. Since most of the α-naphthyl acetate esterase isoenzymes detected in leucocytes and platelets could be inhibited by 0.1 mM diisopropyl fluorophosphate, but not by 0.1 mM eserine or 1 mM p-hydroxymercuribenzoate, they were identified as probable carboxyl esterases (EC 3.1.1.1). The five isoenzymes typical of monocytes were additionally found to be inhibited by 0.1 mM bis-(4-nitrophenyl)-phosphate and by 40 mM sodium fluoride.

The technique of isoelectric focusing has been exploited, in recent years, also for the study of esterase isoenzymes of leukaemic cells. Kass and Munster (1979b) investigated by this method patients affected by acute lymphoblastic, myeloblastic or monocytic leukaemia. Isoenzymatic patterns appeared to be typical for each cytologic type. When α-naphthyl acetate was employed as substrate, up to 13 bands were detected in lymphoblasts, up to 20 bands in myeloblasts and 16 bands in neoplastic monocytes. With α-naphthyl butyrate the same cell types showed respectively 3 to 5, 5 to 7 and

6 to 8 isoenzyme components. Only esterases extracted from leukaemic monocytes were found sensitive to fluoride inhibition.

The utility of isoelectric focusing of esterase isoenzymes in discriminating between myeloid and lymphoid varieties of acute leukaemia was confirmed by Yourno *et al.* (1982); however, according to these authors, while the zymogram demonstrated in a case of myelomonocytic leukaemia was similar to those shown by normal granulocytes and monocytes, the isoenzyme pattern of leukaemic lymphoblasts bore little resemblance to that of normal B or T lymphocytes.

In an extensive study carried out by Drexler and Gaedicke (1983), isoelectric focusing and polyacrylamide gel electrophoresis were combined in order to obtain a two-dimensional separation of carboxyl esterase isoenzymes. As the result of a series of experiments with different substrates and couplers, best results were obtained by using α-naphthyl acetate and Fast blue RR and pH 7.2 for the demonstration of isoenzyme bands. According to isoelectric points and electrophoretic mobility, four main groups of isoenzymes, named L, M, T and U were distinguished in leukaemic cells. Group M could be further subdivided into M_1 and M_2, the latter comprising a fluoride-sensitive isoenzyme band specific for monocytes. L and U isoenzyme complexes were found in malignant cells from most cases of differentiated leukaemias, including all cases of acute myeloid leukaemias and the majority of cases of lymphoblastic leukaemias, both of the common type and with immunological evidence of differentiation along the T cell axis. In addition, leukaemic myeloblasts were found to display strong M_1 and M_2 isoenzymes; a strong fluoride-sensitive monocytic enzyme was however only detectable in typical cases of myelomonocytic leukaemia. T and pre-T lymphoblastic leukaemias usually showed, along with L and U isoenzymes, a strong activity of esterases of the T group, while cells from eight cases of acute undifferentiated leukaemia were apparently devoid of any esterase activity. The isoenzyme pattern of acute lymphoblastic leukaemias of the common type showed the highest degree of variability, ranging from a complete lack of esterase activity to the presence of variable proportions of L, U and T isoenzymes.

The same group of authors (Drexler *et al.*, 1985a) applied the isoelectric focusing technique to the study of the esterase isoenzyme patterns in a large number of human lymphoma and leukaemia cell lines, whose immunological phenotype had been extensively investigated by the use of polyclonal and monoclonal antibodies. Although, as a general rule, the number and the staining intensity of the isoenzyme bands paralleled, in both myeloid and lymphoid cell lines, the degree of maturity as defined by the immunological phenotype, with the single exception of the above mentioned monocytic enzyme no cell type specific or leukaemia specific isoenzyme or isoenzyme pattern could be detected. According to the same study, the isoenzyme pattern was stably expressed in any given cell line and in its subcultures, and did not show any cell cycle dependency. However, the expression of new isoenzyme bands has been described (Drexler *et al.*, 1985b) in the erythroleukaemia cell line K-562 after induction of differentiation by 12-0-tetradecanoylphorbol-13-acetate (TPA) or by other differentiation inducing factors. In a further investigation (Drexler et al., 1987) the expression of the monocyte-specific, fluoride-sensitive acetate esterase isoenzyme was studied by isoelectric focusing in a panel of cultured leukaemia-lymphoma cell lines, in freshly obtained leukaemia-lymphoma cells and in fresh and cultured leukaemic cells treated with TPA. This esterase isoenzyme was found in 92% of cases of acute myelomonocytic or monocytic leukaemias and in only 3% of non-monocytic acute myeloid leukaemias. Treatment with TPA of AML and CML cells, which commonly differentiate to monocyte/macrophage-like cells, induced *de novo* expression of the monocyte-specific isoenzyme.

Isoelectric focusing of α-naphthyl acetate esterases was employed also by Scott *et al.* (1984) for investigating the malignant cells from cases of acute myeloid leukaemia. Four main isoenzyme patterns emerged from this study. The first of these, described as G-type, was characterized by the presence of a variable number of isoenzyme bands with isoelectric points in the range 6.2–8.1, and was similar to that shown by normal granulocytes. The other three types, designated Mon1, Mon2 and Mon3, showed a progressively increasing number (up to six) and staining intensity of bands with isoelectric points between 5.5 and 6.1, in addition to those detectable in the G-type pattern. Normal control monocytes displayed a Mon2 pattern. Out of 16 cases of myeloblastic leukaemia, 12 showed a G pattern and 4 a Mon1 pattern. All four types of isoenzyme pattern were present in 18 cases of acute myelomonocytic leukaemia, while 14 cases of pure monocytic leukaemia were characterized by a Mon1, Mon2 or Mon3 pattern. When the isoenzyme patterns were compared with the intensity of cytochemical α-naphthyl acetate esterase staining and with the expression of monocyte-specific membrane antigens, as detected by monoclonal antibodies, a good degree of correlation was found only in monocytic leukaemias, while contrasting results were reported in cases of myeloblastic, and especially of myelomonocytic leukaemia. Moreover, the fluoride sensitivity of the cytochemical reaction was found of little value in assessing a monocytic involvement in acute myeloid leukaemias.

Another study by the same group of authors (Scott *et al.*, 1985) carried out on 39 cases of acute myelomonocytic leukaemia further stressed the extreme heterogeneity of this type of leukaemia, suggesting that the blasts in this form represent an early, transitional and perhaps reversible stage of monocytic differentiation.

It is quite evident, from the conflicting observations reported, that at present the results of different studies on esterase isoenzymes, partly as a consequence of technical variations, are hardly comparable to each other, and that the links with conventional enzyme cytochemistry are still fairly hazy.

Nevertheless, these investigations seem to be worthy of adequate attention, since they are potentially capable of investigating in some detail cytochemical diversity, and of producing in turn modifications of cytochemical techniques more specific for certain cell types.

Moreover, the use of antisera capable of recognizing single esterase isoenzymes may be helpful in identifying some cell types with a high degree of specificity. Such an approach has been recently employed by Lam *et al.* (1985) for the immunohistochemical detection of normal and malignant monocytes and histiocytes.

Applications of esterase cytochemistry in haematological practice

The cytochemical demonstration of esterases in leucocytes has chiefly been of practical use in the past in distinguishing granulocytes and their precursors from monocytes and monoblasts. The reactions have been of value in differentiating forms of acute myeloid leukaemia and in distinguishing granulocyte and monocyte precursors in otherwise poorly differentiated leukaemic cell populations. Smears can be exposed consecutively to chloroacetate with one diazonium salt, Fast blue BB, and then to α-naphthyl butyrate with a different diazonium salt, Fast Garnet GBC, to give dual esterase staining in the same preparation, sites of chloroacetate esterase activity staining blue and those of butyrate esterase staining red.

Since several coupling reagents are capable of producing differently coloured staining products when reacting with the split products of different substrates, a single incubation double esterase reaction has been proposed by Swirsky (1984). According to this method, dual esterase staining is obtained by incubation in a medium containing both naphthol AS-D chloroacetate and α-naphthyl butyrate in the presence of Fast blue BB salt. Chloroacetate esterase positivity is thus revealed as a bright blue staining, while butyrate esterase activity produces a brown colour (Plate 20E). As an alternative, the use of hexazonium new fuchsin as the coupler produces a bright red and a brown colour (Plate 20F).

Although the chloroacetate esterase reaction predominates in early leukaemic granulocyte precursors and butyrate esterase in monocyte precursors, intermediate mixed patterns are not infrequent in myelomonocytic leukaemias and, indeed, some admixture of the two esterases is commonly found in normal monocytes and in granulocyte precursors, especially at the earlier stages. Reaction patterns in most normal blood and bone marrow cells are illustrated and described in some detail in Plates 16 to 18, and the patterns in typical leukaemic cells in Plates 19 and 20. The value of esterase reactions in differentiating leukaemias is more fully discussed in Chapter 14.

Esterase staining has also been used as an aid to automatic differential leukocyte counting (Ansley and Ornstein, 1971). Kaplow *et al.* (1976) confirmed the value of the α-naphthyl acetate esterase reaction as a specific stain for monocytes, which allowed these cells to be identified and rapidly counted in a flow cytophotometer, with results much more reproducible than those derived from manual differential cell counting of stained blood smears. The intensity of stain could be used to obtain data on the activity of the monocytic esterase in any particular specimen by appropriate analysis of the oscilloscope display, thus providing a method, potentially applicable to any suitably cell-selective cytochemical technique, by which intracellular enzyme activity of a single class of cells in a suspension of mixed cell constitution could be measured.

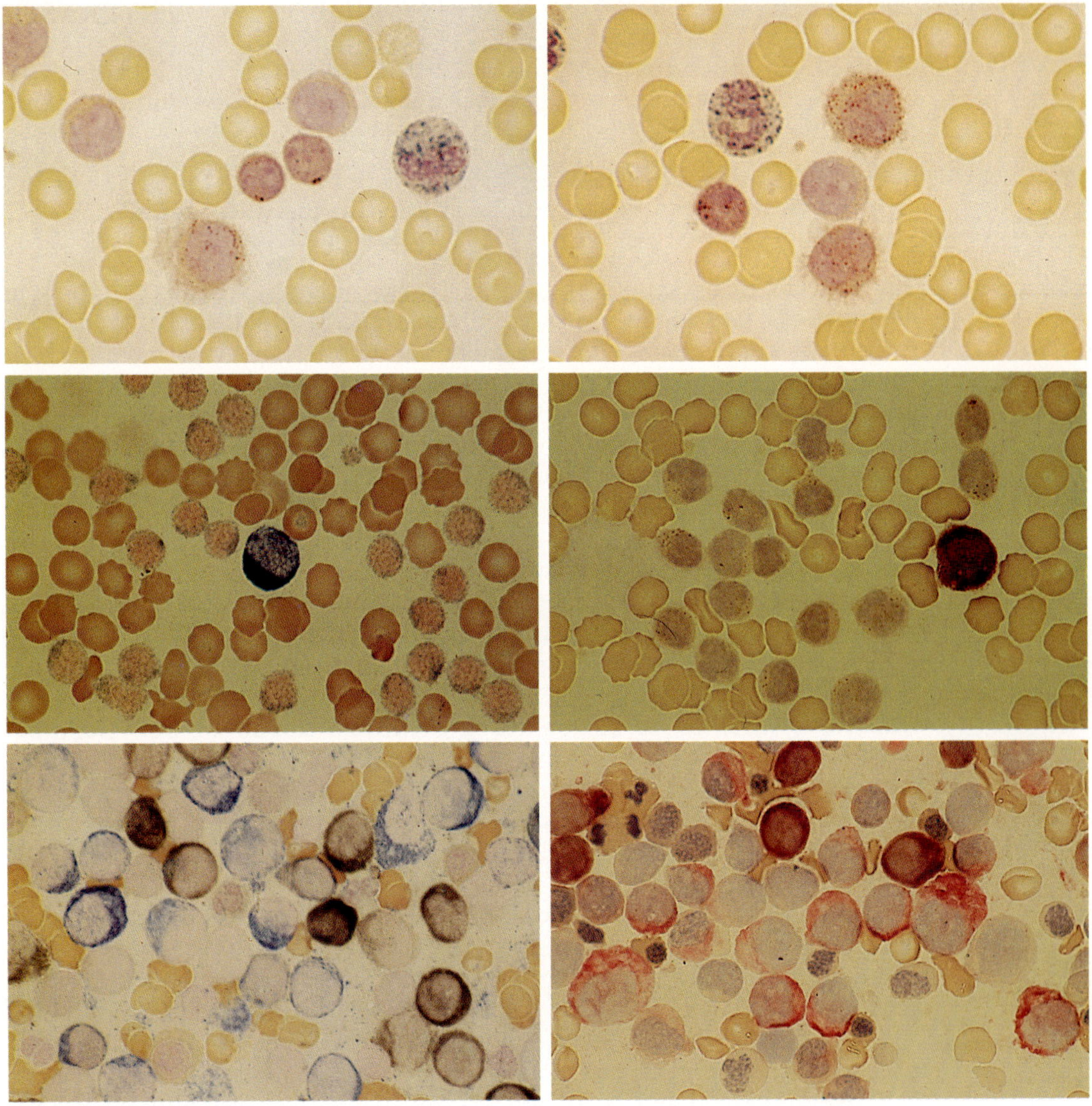

A B C D E F

Plate 20

A. HCL buffy coat. Dual esterase. One CE positive neutrophil, five lymphocytes, two with T-type BE positivity, and a hairy cell showing the crescentic disposition of BE reaction, characteristic of this cell type.

B. HCL buffy coat. Dual esterase. One CE positive neutrophil, two lymphocytes, one negative (B) and one with localized BE reaction (T) and two hairy cells with scattered BE reaction, though without conspicuous crescent localization.

C. T-8 CLL. CE reaction showing a positive myelocyte and numerous less strongly positive lymphocytes.

D. The same case. BE reaction showing a positive monocyte and weak scattered reaction in most lymphocytes. Only one or two lymphocytes show localized dot-like reaction.

E. AMML. Dual esterase with single incubation, Fast blue B as receptor. CE positivity in granulocyte precursors shows blue, with BE positivity in monocyte precursors as brown.

F. AMML, same case. Similar reaction with new fuchsin as receptor. CE positivity shows red and BE positivity brown.

Automated differential leucocyte counting, together with other methods of study, comprising esterase staining of cytospins of peripheral blood monocytes and isoenzyme analysis of mononuclear cell extracts, has been employed by Markey and his associates (1986) to detect inheritable esterase deficiency. In a population survey the authors reported an incidence of 1.7% of monocyte esterase deficiency as compared to 7.1% in a survey of non-Hodgkin's lymphoma and B-CLL patients. The authors postulate, on the basis of these findings, that inheritable monocyte esterase deficiency might be a factor predisposing to the ultimate development of lymphoproliferative diseases.

Positivity in a proportion of lymphocytes had been noted in many reports of esterase reactions with substrates other than chloroacetate (Wachstein and Wolf, 1958; Braunstein, 1959; Quaglino *et al.*, 1962; Rozenszajn *et al.*, 1968, Li *et al.*, 1973), but the possibility that this enzymic reaction might serve to differentiate immunologically separable varieties of lymphocyte has been explored only recently. Mueller *et al.* (1975), using α-naphthyl acetate as substrate, at pH 5.8 with a prolonged incubation period of 21 hours, described a dot-like localized positivity in the cytoplasm of lymphocytes in the paracortical areas of lymph nodes in mice, presumed to be T cells. Fixation in Baker's formol calcium was used. Ranki *et al.* (1976) applied a similar acid esterase reaction to human lymphocytes and found intense localized positivity confined to T cells, this observation being later confirmed by Kulenkampff *et al.* (1977), who also noted faint localized activity in a minor subpopulation of thymocytes and in blasts from 4 of 5 cases of T-ALL, but negative reactions in B lymphocytes, cells from CLL, mitogen-activated T and B lymphoblasts from peripheral blood, and the blast cells from the common form of childhood ALL (non-T, non-B).

Acid esterase positivity of a proportion of normal thymocytes has been confirmed by subsequent studies (Ranki, 1978; Basso *et al.*, 1980) while in normal peripheral blood the association between percentage of T lymphocytes and acid esterase positivity has been found to be highly significant (Giorno and Beverly, 1982; Yang *et al.*, 1982). The same reaction was found useful in the diagnosis of chronic T-cell disorders, and proportionally more intense in T-prolymphocytic leukaemia than in T-chronic lymphocytic leukaemia (Crockard *et al.*, 1982), but a considerable activity of acid esterase has been reported also in cases of B cell neoplasms (Jaffe *et al.*, 1980).

As far as acute leukaemias are concerned, Veerman *et al.* (1983) and Boesen *et al.* (1984) could not confirm the acid esterase positivity of T lymphoblasts. On the contrary a positive reaction was found in the vast majority of cases by Huhn (1983), who employed fixation in formalin vapour. Indeed, the methods used to demonstrate the so-called acid esterase have generally involved fixative procedures which are inimical to enzyme preservation, and long incubation periods have been necessary to obtain positive reactions.

While the first studies of acid esterase reactions were being pursued, work in our own laboratories had compared a range of fixative procedures, a series of substrates and substrate concentrations, and different periods of incubation at a range of temperatures and pH values, to provide a technique capable of optimum demonstration of non-specific esterase activity in human blood lymphocytes. This study (Higgy *et al.*, 1977) showed that exposure to formalin vapour for four minutes allowed good enzyme activity with a clearly defined granular disposition, while providing adequate cytological fixation. Shorter fixation was cytologically unsatisfactory and longer fixation reduced esterase reactivity. Glutaraldehyde and paraformaldehyde gave less satisfactory preservation of cell structure with a reduction in enzyme reaction product, especially in lymphocytes. High concentrations of substrate (20 mg per 100 ml buffer), prolonged incubation (60 min or more) and pH levels below 7.0 all tended to give mixed diffuse and granular reactions involving to variable degrees most cell lines in buffy coat preparations (Plate 18F). The most selective reaction patterns in terms of cytological differentiation among lymphocytes and monocytes, with absence of reaction in granulocytes, were found with α-naphthyl butyrate as substrate, at a concentration of 2.33×10^{-4} mol/l (5 mg per 100 ml buffer) at pH 8, using either Fast Garnet GBC or Fast blue BB as coupling agents. The reaction gave optimal results after 15–30 min incubation, when most monocytes showed a strong granular positivity scattered all over the cytoplasm. This reaction could be inhibited by sodium fluoride. Granulocytes were negative for butyrate esterase and a number of distinct reaction patterns, resistant to sodium fluoride, were found in lymphocytes. Comparative studies of reactions in T-cell enriched, B-cell enriched and null-cell enriched lymphocyte preparations from normal peripheral blood (Plate 16) showed that most T cells contained dense, localized, 'dot-like' esterase positivity (Figs 8.7 and 8.8), B cells were esterase negative (Fig. 8.9), and null cells commonly showed a scattered finely granular positivity.

To some extent these discriminatory patterns persisted in pathological proliferations of the three cell types. Thus in two cases of the T-cell disorder, Sézary's syndrome, the T-cell esterase pattern was found in the Sézary cells, while a high proportion of the blast cells in six cases of non-T, non-B ALL showed the null cell pattern of scattered granular esterase positivity (Plate 19A) (Higgy *et al.*, 1978). In 1978, Grossi *et al.* suggested that the dot-like pattern of esterase reaction found in many T cells was in fact confined to the normally predominant subpopulation of Tμ cells, which possess a receptor for the Fc portion of IgM and predominantly act as 'helper' cells in B-cell immune responses. The functionally separate Tγ subpopulation, with receptors for Fcγ, was initially reported as devoid of enzyme activity, but further investigations by the same group (Ferrarini *et al.*, 1980) detected in this T-cell subset a paranuclear distribution of the α-naphthyl acetate esterase reaction product. Although this finding has been broadly confirmed by several authors, it might have been influenced by the rosetting methods employed for identifying T lymphocytes, since exposure to IgG complexes causes enzyme redistribution (Bevan *et al.*, 1980; Quesada and Murphy, 1982; Crockard, 1984). Only a dot-like and a scattered granular non-specific esterase staining of lymphocytes were reported by Armitage *et al.* (1982): according to these authors the first staining pattern predominates in Tμ lymphocytes, and the second in Tγ cells, but a minority of cells in both subsets show a different cytochemical

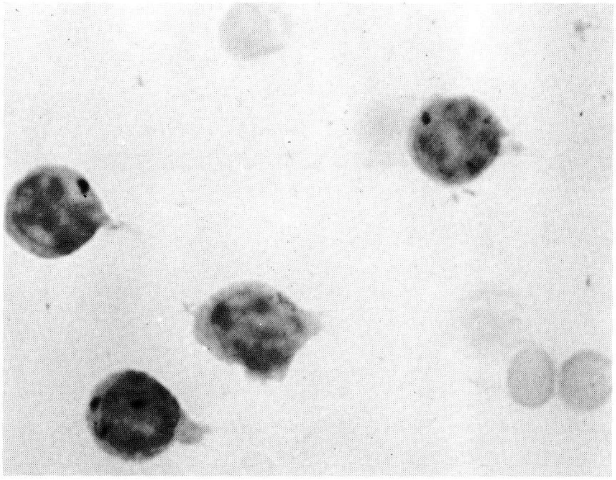

Fig. 8.7 T-cell enriched preparation. α-Naphthyl butyrate esterase. Localized dot-like BE positivity in three of four lymphocytes.

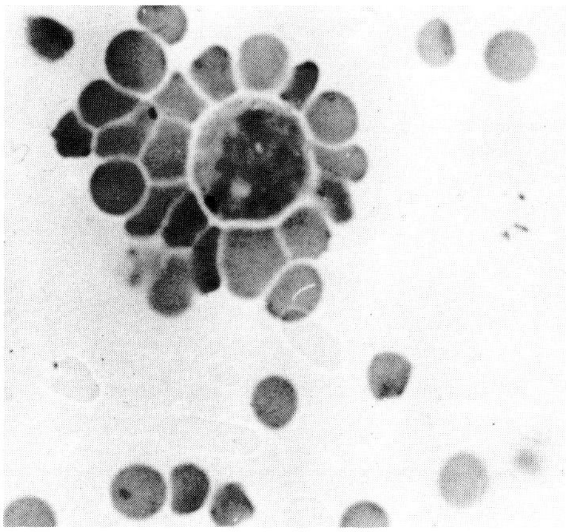

Fig. 8.8 Sheep erythrocytes rosetting around a T-cell from normal peripheral blood. α-Naphthyl butyrate esterase. The T cell has a block of BE positivity.

behaviour, i.e. respectively a granular and a dot-like staining pattern. It was therefore concluded that no direct relationship exists between dot-like positivity and Fcμ receptor expression.

In recent years, a large series of studies has been carried out, based on the use of monoclonal antibodies. T4, anti-Leu 3a or similar monoclonal antibodies have been more often employed in order to identify the helper/inducer T cell subset, while the T8 or Leu 2a positive subset has been commonly regarded as responsible for suppressor/cytotoxic functional activity. However, the expression of membrane antigens recognizable by these monoclonal antibodies only partially overlaps the presence of receptors for the Fc portion of IgM or IgG, and the relationship between the immunological phenotype and the functional activity of T-cell subsets is far

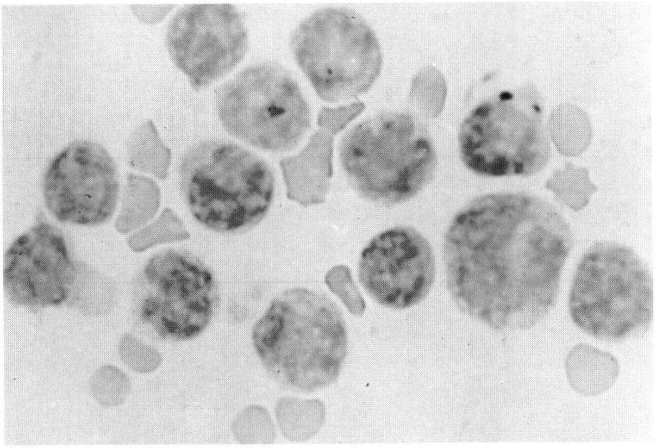

Fig. 8.9 B-cell enriched preparation. α-Naphthyl butyrate esterase. All lymphocytes except one, presumably a contaminating T cell, are negative.

from being fully understood. Therefore, it seems at present more appropriate to refer to the esterase activity of T4/Leu 3a (CD4) and of T8/Leu 2a (CD8) positive lymphocytes, rather than of helper/inducer or suppressor/cytotoxic T-cell subsets. A high percentage (usually more than 80%) of dot-like positive lymphocytes has been reported by most authors in the CD4 positive subpopulation (Armitage *et al.*, 1982; Bernard and Dufer, 1983; Crockard and Catovsky, 1983; De Waele *et al.*, 1983; Crockard, 1984). A localized non-specific esterase activity is also detectable in a proportion of CD8 positive cells, ranging, according to different authors, from 25% (Bernard and Dufer, 1983) to 65% (Crockard and Catovsky, 1983). In any case, the wide margin of overlapping precludes the use of non-specific esterase reactivity as a means of discriminating between the respective T-cell subsets (Crockard, 1984).

Bernard and Dufer (1983) described also naphthol AS-D chloroacetate esterase positivity in a minority (13%) of T8 positive lymphocytes from normal subjects. In our experience, when low substrate concentrations and short incubation period are employed, all lymphocytes are consistently negative for the chloroacetate esterase reaction, but some degree of enzyme activity is demonstrable in the same cells, usually in the form of a weak diffuse cytoplasmic staining, with longer staining times and higher substrate concentrations, similar to those commonly employed for histochemical purposes. Although a higher staining intensity was found in some cases of T8 positive chronic lymphocytic leukaemia (Plate 20C), we have been unable to confirm a well defined relationship between any particular lymphocyte subset and chloroacetate esterase activity.

Recently, Bernard (1984) compared the pattern of acid α-naphthyl acetate esterase staining with the expression of another membrane antigen (Leu 8), in combination with Leu 3a. Leu $3a^{+}8^{-}$ lymphocytes (so-called 'true helper' cells) were found 100% positive for esterase activity, with 96% of cells displaying a dot-like staining pattern. In contrast, dot-like esterase positivity was only detected in 42% of Leu 8^{+} lymphocytes and in 23% of

Leu 3a$^-$8$^-$ cells. These results could not be confirmed by Armitage and Cawley (1985), who used a different rosetting technique to identify the various lymphocyte subsets. According to these authors, no significant difference exists, both in terms of percentage and positivity pattern, between the acetate esterase reactivity of Leu 3a$^+$8$^-$ and Leu 3a$^+$8$^+$ T-cell subsets.

Although esterase activity of lymphocytes, following conventional cytochemical studies, appears to be consistently resistant to fluoride inhibition, a fluoride-sensitive acid α-naphthyl acetate esterase was reported by Grossi *et* al. (1982) in so-called large granular lymphocytes. These cells were characterized by cytoplasmic azurophilic granules, avid membrane receptors for the Fc portion of IgG and cytotoxic (natural killer or antibody dependent) functional activity: according to the authors these criteria could be applied to the majority of both Tγ and non-T non-B ('third population') lymphocytes. For demonstration of acetate esterase activity, purified large granular lymphocytes from peripheral blood of healthy donors were fixed in suspension with 1.25% glutaraldehyde in cacodylate buffer. Esterase staining was carried out either on cell suspensions or on cytocentrifuged samples, employing 1.3 mM substrate and 6% hexazotized pararosaniline in 0.067 M phosphate buffer at pH 5.8. The results appeared to be different according to the two techniques. When cells had been stained in suspension, the reaction product was mainly localized on the plasma membrane, while only a weak granular positivity was detectable in the cytoplasm. Staining of cytospin preparations resulted in positivity localized to a paranuclear area and to the surrounding cytoplasmic granules. In both instances, the cytoplasmic positivity was found fluoride sensitive, while the membrane staining shown by the first method was described as only partially sensitive to inhibition. Heumann *et al.* (1983) reported in some large granular lymphocytes also an evident degree of naphthol AS-D chloroacetate esterase activity.

We tried to reproduce these results on conventionally smeared lymphocyte enrichments, both from healthy subjects and from patients with abnormal expansions of large granular lymphocytes. The matter of chloroacetate staining of lymphocytes has been discussed already (see p. 196). As far as non-specific esterase activity is concerned, a wide range of suitable fixatives, substrates, couplers and staining conditions were tested, which variously affected the pattern of positivity and the staining intensity of morphologically recognizable large granular lymphocytes, but in all instances the results were substantially unaltered by the addition of 40 mM sodium fluoride. Therefore, in our experience, the chloroacetate esterase and fluoride-sensitive non-specific esterase activity of this lymphocyte subset, even if existing, cannot be easily exploited for its recognition by routine cytochemical methods (Plate 20C, D).

The sensitivity of lymphocyte esterase activity to both sodium fluoride and to the serine protease inhibitor phenyl methyl sulfonyl fluoride (PMSF) was investigated by Dufer *et al.* (1984). In all instances, lymphocyte esterase was found resistant to conventional concentrations (50 mmol/l) of both inhibitors, whereas partial inhibition of the dot-like lymphocytic acetate esterase was observed with higher concentrations (500 mmol/l) of sodium fluoride. In contrast, the acetate esterase activity of plasma cells, similar to that of platelets, was sensitive to 50 mM sodium fluoride, but resistant to PMSF.

A characteristic pattern of butyrate esterase reaction was observed in our laboratories in the course of a study of leukaemic reticuloendotheliosis or hairy cell leukaemia (Higgy *et al.*, 1978). A high proportion of morphologically recognisable 'hairy cells' showed this pattern, which consisted of a mixture of scattered fine granules together with coarser granules, often localised in a conspicuous crescentic configuration to one side of the nucleus (Plate 20A, B). The positivity was not inhibited by sodium fluoride. Esterase positivity in hairy cells had been noted previously by Yam *et al.* (1972), Schnitzer and Kass (1974), and Kass (1977b), but, contrary to our findings, the latter authors found the reaction to be of the monocytic fluoride sensitive type, and the precise pattern of crescentic positivity had not been recognized. The usefulness of butyrate esterase in the diagnosis of hairy cell leukaemia has been subsequently confirmed by Katayama and Aiba (1980) who found the reaction more sensitive than the acid phosphatase reaction in identifying hairy cells, whereas according to Variakojis *et al.* (1980) non specific esterase activity of hairy cells was better demonstrated by α-naphthyl acetate than by α-naphthyl butyrate, and was inhibited by sodium fluoride. Partial sensitivity of acid α-naphthyl acetate esterase of hairy cells to fluoride inhibition was also reported by Boesen (1984), both at the light microscopical and at the ultrastructural level (see below). Although there has been considerable argument over the origin of hairy cells, whether from monocytic or lymphocytic cell lines, the weight of evidence at present strongly favours a B-cell origin (Burns *et al.*, 1978; Korsmeyer *et al.*, 1983), and it is therefore of interest that among 40 cases of CLL, all immunologically confirmed as B-cell tumours, we have found four cases, each with otherwise typical clinical and haematological features of CLL, showing the crescent pattern of esterase positivity of the hairy cell type. Since hairy cells do not possess a receptor for fixed C3 (Burns *et al.*, 1977a) it may not be entirely coincidental that these four cases were the only ones among the 40 in which the leukaemic cells were devoid of a C3 receptor. Six other cases showed a scattered granular esterase reaction of the null-cell type, and were also notable for showing substantial organomegaly, especially lymphadenopathy, and being difficult to control clinically, with poor responses to cytotoxic therapy. The remaining 30 cases of CLL showed the negative esterase reaction characteristic of normal B cells.

While patterns of esterase reactivity in pathological states would hardly be expected always to conform to the patterns shown in normal cells of the same lineage, they certainly have something to contribute to the problem of normal and neoplastic cell inter-relationships, not only in the lymphoid but also in the myeloid context. The strong α-naphthyl acetate or butyrate esterase positivity of Gaucher's cells (Rozenszajn and Efrati, 1961) and its partial inhibition by sodium fluoride, taken with surface marker features (Burns *et al.*, 1977b) confirm the assumed origin of Gaucher's cells from the monocyte-histiocyte series. The reaction in Reed-Sternberg (R-S) cells is variable, and casts a less clear light on the origin of these abnormal cells.

Our own experience has varied. In a patient with Hodgkin's disease showing the unusual development of terminal R-S cell leukaemia the circulating R-S cells were negative for both α-naphthyl butyrate and naphthol AS-D chloroacetate esterases, and these reactions, taken with other cytochemical, ultrastructural and surface immunological studies, suggested that the R-S cells were of B-cell lineage (Hayhoe *et al.*, 1978). However, we have encountered a range of reactions to α-naphthyl butyrate esterase in R-S cells in lymph node imprints, with some cells negative and others showing positivity, generally of a weak scattered granular type (see Chapter 15), somewhat resembling the pattern observed in null-lymphocytes, and much weaker and less susceptible to fluoride inhibition than the positivity in macrophages.

Ultrastructural demonstration of esterases in haemic cells

In 1980, Payne *et al.* published the results of a study, in which non-specific esterase activity of lymph node and peripheral blood cells had been investigated at the ultrastructural level. For this purpose, buffy coat preparations from peripheral blood and samples of reactive lymph nodes had been fixed, respectively for 1 and 4 hours at 4°C, in a buffered solution containing 2.5% glutaraldehyde and 2% paraformaldehyde. For the cytochemical demonstration of enzyme activity, 2-naphthylthiol acetate was used as the substrate, and Fast blue BB as the coupling agent, for an average incubation time of 6 hours, followed by postincubation with osmium tetroxide, according to the method of Hanker *et al.* (1966). The authors reported a good degree of correlation between the ultrastructural pattern of positivity shown by this method, and the non-specific esterase staining, as revealed at the light microscopical level by α-naphthyl acetate or naphthol AS acetate. The reaction product was evident in the cytoplasm of lymph node histiocytes in the form of electron-dense deposits of varying sizes and shapes. Although in some cases positivity was observed in close association with membrane-bound granules, in most instances esterase staining was unrelated to cytoplasmic organelles (lysosomes, mitochondria, etc.). In comparison, positive granules in circulating monocytes were found much smaller both in number and in size range. Granulocytes were consistently unreactive, whereas two patterns of reactivity were observed in lymphoid cells: those of the germinal centres were always negative, while a proportion of lymph node cells outside the germinal centres and some circulating lymphocytes contained 1–4 small, dot-like granules. Ultrastructural reactivity patterns comparable to those described in normal cells could be detected by the same group of authors (Kim *et al.*, 1982) in malignant histiocytes, both from fresh lymph node tissue and from a cultured cell line, as well as in leukaemic monocytes.

Quite different findings were reported by Bozdek and Bainton (1981), who studied the ultrastructural localization of α-naphthyl butyrate esterase by employing hexazonium pararosaniline as the coupling agent. Normal peripheral blood cells were fixed in suspension, for 10 min at 4°C, with a solution containing 0.25% glutaraldehyde and 1% sucrose in cacodylate buffer. Optimal staining of esterase activity was reported with an incubation time of 20–90 min, followed by postfixation with OsO_4. With this method, the reaction product in monocytes was found to be localized over the entire external surface of the plasma membrane, and no positivity at all was observed within these cells. A small percentage of monocytes (5%), identifiable by their morphology and peroxidase content, were apparently devoid of esterase activity. The same method showed in some lymphocytes esterase-positive intracellular structures, corresponding to the content of round or oval organelles, 0.3 to 1.2 μm in size, limited by a single unit membrane. These organelles were found, singly or occasionally in groups of two or three, near the Golgi region, often adjacent to multivesicular bodies, and were thought to correspond to Gall bodies. As the result of inhibition tests, membrane staining of monocytes was completely abolished by sodium fluoride, whereas the reactivity of lymphocytes was apparently unaffected.

Zicca *et al.* (1981) investigated at the ultrastructural level the α-naphthyl acetate esterase of human Tμ lymphocytes from healthy subjects. Prior to staining, cells were fixed for 30 min in 1.25% buffered glutaraldehyde solution. The cytochemical reaction was carried out by incubating the cell suspension for 45 min in a medium containing 1.3 mM α-naphthyl acetate and 6% hexazotized pararosaniline in phosphate buffer at pH 5.8. Postfixation with osmium tetroxide was omitted. The product in Tμ lymphocytes appeared to be localized in restricted cytoplasmic areas, around single or groups of vesicle-like structures. The electron-density of the reaction product was rather weak, and decreased from the centre towards the periphery of the reaction site, and no staining was ever observed within the vesicles. Since ultrastructural investigations carried out in parallel indicated that acid phosphatase activity was localized in the same cell structures as acid α-naphthyl acetate esterase, the authors suggested that the esterase-reactive organelles were primary lysosomes. The possibility was considered that the observed staining pattern was the result of a reaction occurring at or near the membrane of the granules, with subsequent diffusion of the reaction product.

Another ultrastructural study of the localization of non-specific esterase activity was carried out by Monahan *et al.* (1981) on guinea pig and human haemic cells, by employing α-naphthyl acetate as the substrate with hexazotized pararosaniline as the coupling agent. In guinea pig, esterase activity was detected on the cell surface of lymphocytes, monocytes, macrophages, neutrophils, eosinophils, megakaryocytes and platelets. In addition, in monocytes, macrophages and lymphocytes a cytoplasmic acetate esterase activity was associated with the external (cytoplasmic) surface of clustered vesicles, localized to the adjacent cytoplasm. Demarcation channels of megakaryocytes and perinuclear cisternae of normoblasts were also found positive. By employing α-naphthyl butyrate as the substrate, an identical distribution of the reaction product was observed. In man, non-specific esterase activity was found both on the cell surface and in the cytoplasm of circulating monocytes, with a distribution pattern similar to that observed in guinea pig monocytes. Granulocytes and platelets displayed surface staining for acetate esterase. In addition, neutrophils contained several small positive cytoplasmic granules, and eosinophils showed single or multiple large esterase-reactive lipid structures.

In a proportion of human lymphocytes, esterase activity was detectable both on the cell surface and within the cytoplasm. The bulk of cytoplasmic staining was associated with clusters of vesicles, extending to the adjacent cytoplasm. Moreover, positivity was present in Gall bodies, entirely localized to their inner whorled membranes.

Matutes *et al.* (1983) investigated the ultrastructural cytochemistry of four acid hydrolases, including α-naphthyl acetate esterase, in chronic T-cell leukaemias. In four cases of T-prolymphocytic leukaemia, acetate esterase activity was detectable as an homogeneous reaction deposit in single, large- to medium-sized granules, possibly representing Gall bodies. A minority of cells also showed positivity of one or several clustered small granules. In a single patient, the majority of leukaemic prolymphocytes showed esterase positivity on the plasma membrane, as well as in lysosomal structures; perinuclear staining was also detectable in a small percentage of cells.

In two out of three patients with T-CLL no acetate esterase staining of lymphocytes could be detected, whereas in the third a positive reaction was observed in approximately 40% of cells, localized in 1–4 small or medium-sized cytoplasmic granules. In all cases, lysosomal granules and parallel tubular arrays were unreactive for acetate esterase, but strongly positive for the other hydrolases (acid phosphatase, β-glucuronidase, β-glucosaminidase).

In one out of two patients with adult T-cell lymphoma/leukaemia, 40% of cells showed acetate esterase activity, localized in a single or few medium-sized granules. On the basis of the results of this investigation, the authors suggested that the differences in the type and enzyme content of cytoplasmic structures might be attributable to normal and neoplastic lymphocytes with different immunological phenotypes.

Ultrastructural localization of acid α-naphthyl acetate esterase was detected also by Boesen (1984), both in normal and in neoplastic lymphocytes and monocytes, as well as in hairy cells. Normal and neoplastic monocytes displayed a similar reactivity pattern, with reaction products localized to the external surface of the plasma membrane and to small vesicles close to the membrane, or more rarely to larger cytoplasmic vesicles containing endocytosed material. Cytoplasmic granules of monocytes were found unreactive.

A proportion of normal lymphocytes and cells from one case of T-CLL showed a positivity pattern similar to that described by Monahan *et al.* (1981): the reaction product was observed either associated with paranuclear clusters of vesicles, extending from the membranes into the adjacent cytoplasm, or localized within Gall bodies. Plasma membrane staining was also occasionally observed in lymphocytes, but never as prominent as in monocytes. Lymphocytes from three cases of B-CLL and blast cells from four cases of acute lymphoblastic leukaemia (two of the common type and two T-ALL) were completely devoid of esterase reactivity, or showed a faint plasma membrane staining. In the hairy cells from two patients, an intense esterase staining was observed on the plasma membrane: a considerable proportion of cells also showed positivity on the membranes of small vesicles in the proximity of the cell surface, or of intracytoplasmic vesicles containing phagocytosed material. The addition of sodium fluoride to the incubation medium was found to inhibit the membrane staining of all types of cells, whereas the cytoplasmic staining of lymphocytes was unaffected. The ultrastructural patterns of non-specific esterase activity in normal and leukaemic monocytes and lymphocytes have been investigated also by Fenoglio *et al.* (1985), who employed α-naphthyl butyrate as the substrate. A membrane reactivity was detected in most normal monocytes and lymphocytes. The activity in monocytes was very strong and fluoride-sensitive, while that in lymphocytes was less intense and fluoride-resistant. Only in a small percentage of lymphocytes the reaction product was localized in the cytoplasm. Leukaemic lymphocytes and monocytes showed a pattern of reactivity similar to that found in their normal counterparts, although mostly of lower intensity. In a few cases of hairy cell leukaemia, neoplastic cells showed a discrete distribution of the fluoride-resistant reaction product on the plasma membrane.

In close similarity with the results of isoenzyme studies, it is quite evident that in many instances, partly due to technical variability, the findings so far obtained by ultrastructural investigations of esterase activity are hardly comparable to each other and with the results of conventional cytochemistry. Nevertheless, a better knowledge of the subcellular localization of enzyme activity seems to be of crucial importance in order to understand the functional significance of esterases in individual cell types.

Functional significance of esterases in haemic cells

Like acid phosphatase, the non-specific esterases have been regarded for a long time as lysosomal enzymes. The chloroacetate esterases, for example, show strong concentration in the lysosomal primary granules and in Auer rods. Among lymphocytes, a parallel was drawn (Kulenkampff *et al.*, 1977) between the stronger esterase positivity of T-cells and the reportedly stronger acid phosphatase, β-glucuronidase and N-acetyl-β-glucosaminidase activity of T cells or T cell regions of lymph nodes as compared with B cells (Tamaoki and Essner, 1969). Although the parallel has not always been supported by published reports of enzyme activity—Barr and Perry (1976), for example, found the activity of acid phosphatase in T cells to be negative or weak whereas that in B and null cells was moderate or strong, while Higgy *et al.* (1977), found no significant difference in the extent of acid phosphatase positivity in the lymphocyte subgroups from normal peripheral blood—there does seem to be a greater prominence of several hydrolases in normal T-cells and in activated or malignant cells of T-cell origin than in their B cell and null cell counterparts. Nevertheless, these enzymes clearly do not show increased or decreased activity in parallel. A strong increase in acid phosphatase occurs in T-cells stimulated to transform by polyclonal activators such as PHA (Quaglino *et al.*, 1962) whereas the α-naphthyl butyrate and acetate esterase reactions are negative in PHA-transformed lymphocytes (Higgy *et al.*, 1977; Kulenkampff *et al.*, 1977; Manconi *et al.*, 1979). Moreover, the results of ultrastructural studies (Matutes *et al.*, 1983) suggest that in many instances the esterase activity of lymphocytes is localized outside the lysosomes and is unrelated to the activity of other hydrolases. The lack of any relationship between esterases and lysosomal structures is even more apparent in monocytes, in which, according to the results of most ultrastructural studies (see p. 000), the bulk of enzyme activity is localized on the external surface of plasma membrane.

The esterases may well have different functions in different cell types. In granulocytes the chloroacetate esterases were shown by Li *et al.* (1973) to have features in common with chymotrypsin, and those workers suggested that the enzymes might perhaps be identical, so that the esterase demonstrable with chloroacetate as substrate might in fact be responsible for the proteolytic and debris-digesting functions of neutrophils. This suggestion was further supported by the studies of Dufer *et al.* (1984), who identified chloroacetate esterase as a serine protease on the basis of its sensitivity to PMSF inhibition. As for the T-lymphocytic esterases, there may be relevance in the demonstration by Ferluga *et al.* (1972), that cytotoxic effects mediated by T-cells against target cells were inhibited by organophosphorus compounds which are strong esterase inhibitors, thus implying that esterases might have a contributory role in the killing function of T-cells.

The fluoride sensitive α-naphthyl butyrate esterase of monocytes shows some variation in activity in the normal monocyte population from peripheral blood. In our experience some 80% of monocytes have strong or moderately strong positivity, while about 20% show weak granular reactions, often accompanied, when dual esterase reactions are performed, by weak positivity to chloroacetate. This heterogeneity of monocytic esterase activity has been described by Kaplow (1975), who observed that esterase reactions varied more widely among individual monocytes from normal peripheral blood than did other enzymes such as acid phosphatase, generally moderately strong and with little variability, and peroxidase, which showed variations of intermediate degree. Kaplow suggested that changes in esterase activity might especially reflect physiological or immunological stimulation of monocytes, and this seems a very reasonable speculation, since monocytes are clearly capable of 'activation' by environmental factors, including infective, parasitic and perhaps antigenic stimuli, along a maturational pathway to give rise to macrophages. Among other changes occurring as this maturation progresses is a general increase in most hydrolytic enzymes (Cline and Golde, 1973).

The possible significance of α-naphthyl acetate esterases towards various monocyte functions has been recently investigated by Oertel *et al.* (1985) by employing the specific inhibitor bis (4-nitrophenyl)-phosphate (BNPP). At a non toxic concentration of 2 mmol/l, BNPP irreversibly inhibited acetate esterase activity without affecting cell viability, and thus enabled subsequent functional studies. BNPP inhibition did not affect adhesion and spreading of monocytes on glass surfaces, or phagocytosis of IgG-coated

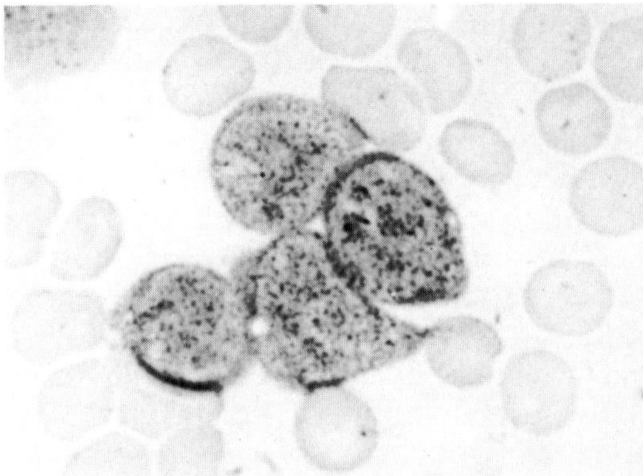

Fig. 8.10 ALL (common non-T, non-B type) bone marrow. α-Naphthyl butyrate esterase. Scattered fine granular positivity in blast cells.

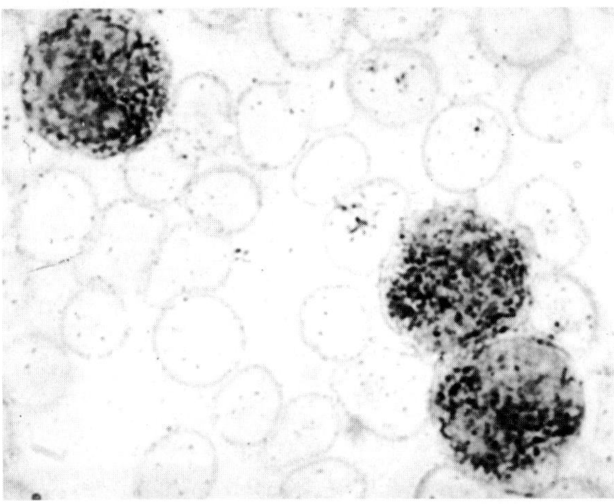

Fig. 8.11 AMML bone marrow. α-Naphthyl butyrate esterase. Moderately strong reaction in monocyte precursors.

latex particles. In contrast, BNPP was found to inhibit the spontaneous cytotoxicity, but not the antibody dependent cell-mediated cytotoxicity, towards the erythroleukaemic cell line K562. On the basis of these results, the authors suggested that the acetate esterases of monocytes might be involved in spontaneous cytotoxicity against tumour cells.

We have seen earlier that strong positivity for α-naphthyl butyrate esterase of the monocytic type has often proved a valuable criterion in the recognition of acute monocytic and myelomonocytic leukaemias with poorly differentiated cells. This is certainly the common pattern in such cases, but the variability in extent of positivity described above for normal monocytes is also manifest in leukaemic monocyte precursors, which may occasionally show no more than a weak, though characteristically fluoride sensitive, esterase reaction. Perhaps in the case of acute leukaemias the degree of

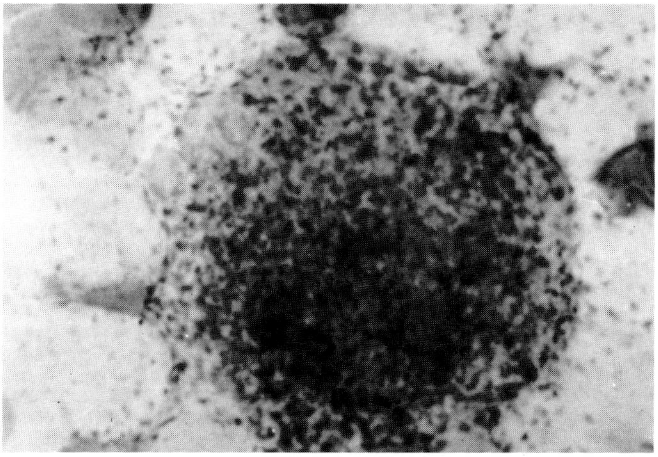

Fig. 8.12 Normal bone marrow. α-Naphthyl acetate esterase. A megakaryocyte showing strong positivity. The surrounding granulocyte precursors are negative.

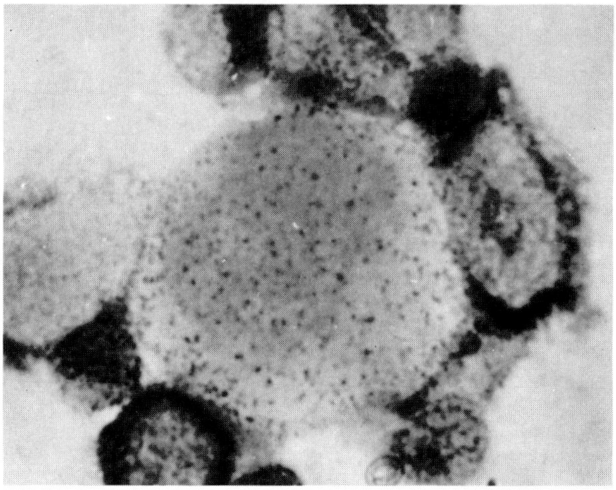

Fig. 8.13 Normal bone marrow. Chloroacetate esterase. A megakaryocyte showing only a faint scattering of granules, while the surrounding granulocyte precursors are strongly positive.

enzyme activity is related to the capacity of the stem cell involved for functional differentiation along the monocytic line, the less the capacity for differentiation the less the expression of esterase activity.

The significance of esterase positivity of the α-naphthyl acetate type observed in erythroblasts in untreated pernicious anaemia, erythraemic myelosis and in iron deficiency (Kass and Peters, 1977) is uncertain, as is the presence of chloroacetate esterase positivity in erythroblasts of erythraemic myelosis (Kass, 1977a). Electrophoretic patterns and fluoride sensitivity of the former esterase were apparently similar to those of enzyme derived from monocytes. Possibly, under conditions of special metabolic stress, erythroblasts may express enzyme potentialities shared with monocytes and granulocytes, since all these cells are derived from a common functional myeloid stem cell.

The patterns of esterase positivity with different substrates in different cell lines in both normal and pathological, especially leukaemic, states, may be summarized in simplified form as shown in Table 8.1.

TECHNICAL METHODS

Esterase reactions

Naphthol AS (or AS-D) acetate esterase

After Wachstein and Wolf, 1958

Reagents

Substrate solution

40 ml of 1% propylene glycol in 0.2 mol/l phosphate buffer (pH 6.9)
0.4 ml of 1% naphthol AS (or AS-D) acetate in acetone
80 mg Fast blue BB salt
(for testing fluoride inhibition, add NaF 1.5 mg/ml to buffer).

Technique

1. Fix air-dried smears in formalin vapour—5 minutes
2. Wash in distilled water
3. Immerse in freshly prepared substrate at room temperature for 30–60 minutes
4. Rinse in distilled water
5. Counterstain for 10 minutes with 0.1% nuclear fast red in 5% aluminium sulphate solution.

α-Naphthyl butyrate esterase

After Higgy, Burns and Hayhoe, 1977

Reagents

Substrate solution

0.1 mol/l phosphate buffer (pH 8.0)	10 ml
α-naphthyl butyrate	0.5 mg* (2.33×10^{-4} mol/l)

[*The α-naphthyl butyrate is liquid at room temperature. To obtain 0.5 mg of the butyrate, 0.01 ml (= 10 mg) of the anhydrous liquid is taken into 0.5 ml of acetone and then 0.025 ml of this solution contains 0.5 mg butyrate.]

Fast blue BB salt *or*	15 mg (5×10^{-3} mol/l)
Fast Garnet GBC salt	3 mg (9×10^{-10} mol/l)

(for testing fluoride inhibition, add NAF 1.5 mg/ml to buffer).

Technique

1. Fix air-dried smears in formalin vapour for 4 minutes
2. Wash briefly in distilled water and blot dry
3. Incubate in freshly prepared substrate at room temperature, away from direct light, for 15–30 minutes
4. Wash briefly in distilled water
5. Counterstain in aqueous haematoxylin for 5 minutes or in nuclear Fast Red for 10 minutes
6. Wash in distilled water, blot dry and examine.

Naphthol AS-D chloroacetate esterase

After Higgy, Burns and Hayhoe, 1977

Reagents

Substrate solution

0.1 mol/l phosphate buffer (pH 8.0)	10 ml
naphthol AS-D chloroacetate	0.25 mg in 0.1 ml acetone (0.7×10^{-4} mol/l)
Fast blue BB salt	15 mg (5×10^{-3} mol/l)

Technique

1. Fix air-dried smears in formalin vapour for 4 minutes
2. Wash briefly in distilled water and blot dry
3. Incubate in freshly prepared substrate at room temperature, away from direct light, for 5 to 15 minutes
4. Wash briefly in distilled water
5. Counterstain in aqueous haematoxylin for 5 minutes or in nuclear Fast Red for 10 minutes
6. Wash in distilled water, blot dry and examine.

Dual technique for α-naphthyl butyrate and naphthol AS-D chloroacetate esterases

Reagents

Substrates as for separate reactions, described above, but with Fast Garnet GBC as the chosen option in the butyrate substrate mixture.

Table 8.1 Esterase cytochemistry of normal and leukaemic cells

Substrate	α-naphthyl acetate or butyrate	Naphthol AS-D acetate	Naphthol AS-D chloroacetate
Lymphocytic cells	Normal B cells negative, 'null' cells show some fine positive granules and T cells may show strong localized dots of positivity. Lymphoblasts in ALL may reflect their lineage—B cells negative, T cells with localised reaction and the common null cell type showing scattered fine granules (Fig. 8.10). Lymphocytes of CLL usually negative, but may show null cell or hairy cell patterns. Characteristic crescentic positivity in hairy cells.	The patterns in lymphocytes are generally similar to those revealed by butyrate or acetate as substrate, but additional scattered granules tend to occur and the reaction is less discriminatory.	Negative.
Granulocytic cells	Cells of the granulocytic series are usually negative, but occasionally may show a few scattered positive granules.	Positive. Promyelocytes and myelocytes show the strongest reaction. Blast cells may show some positivity. Positivity not inhibited by sodium fluoride. Reaction of Auer rods not reported.	Activity chiefly found in the granulocyte series. Myeloblasts may show some positivity. Auer rods positive.
Monocytic cells	Most normal monocytes and leukaemic monoblasts show strong positivity, but some 20% have weak reactions (Fig. 8.11). Monocyte reactions are fluoride sensitive.	Monocytes and monoblasts probably show the strongest reaction of all haemic cells, but the difference is not sufficiently striking to be diagnostic. However, reaction is inhibited by sodium fluoride.	Mostly negative, but some normal monocytes and some leukaemic monocytes have positivity to both butyrate and chloroacetate esterases.
Erythroid cells	Normal erythroblasts negative or weakly positive. Red cell precursors in Di Guglielmo's disease strongly positive. Megaloblasts usually positive.	Proerythroblasts and basophilic erythroblasts show some positivity which decreases with increasing maturity. Cells of Di Guglielmo's disease similar.	Normally negative, may show positivity in Di Guglielmo's disease.

Megakaryocyte platelet series	Positivity in both megakaryocytes and platelets stronger with acetate than butyrate (Fig 8.12). Fluoride sensitive.	Both megakaryocytes and platelets show granular positivity of variable intensity. Fluoride sensitive.	Negative. Occasional weak granular reaction (Fig. 8.13).
Comments	Fluoride sensitive positivity in monocytes of value in differentiating myelomonocytic leukaemia from acute granulocytic leukaemia. Discriminating reaction for normal T, B and null cells may be helpful in subclassifying ALL and CLL. Marker pattern in hairy cells.	Although monocytoid cells show an especially strong reaction, all haemic cells show some activity. This reaction is therefore of less value in differentiating between leukaemic types, although inhibition by sodium fluoride is characteristic of monocytic and not of granulocytic or lymphocytic positivity.	Partially but not completely specific marker for granulocytes. Of value in differentiating granulocytic from monocytic components of acute myeloid leukaemias.

Technique

1. Fix air-dried smears in formalin vapour for 4 minutes
2. Wash briefly in distilled water and blot dry
3. Incubate in freshly prepared chloroacetate substrate, containing Fast blue BB salt, for 5–15 minutes, at room temperature and away from light
4. Wash briefly in distilled water and blot dry
5. Incubate in freshly prepared butyrate substrate, containing Fast Garnet GBC salt, for 15 to 30 minutes, at room temperature and away from light
6. Wash briefly in distilled water
7. Counterstain in aqueous haematoxylin for 5 minutes
8. Wash in distilled water, blot dry and examine.

Single incubation double esterase reaction

After Swirsky, 1984

Reagents

Buffered formalin-acetone

Sodium phosphate (Na_2HPO_4)	20 mg
Potassium phosphate (KH_2PO_4)	100 mg
Distilled water	30 ml
Acetone	45 ml
Concentrated formalin	25 ml

Substrate solution

0.1 mol/l phosphate buffer (pH 8)	50 ml
naphthol AS-D chloroacetate	2.5 mg in 1 ml acetone
α-naphthyl butyrate	4 mg in 1 ml acetone
Fast blue BB salt	80 mg

Technique

1. Fix air-dried smears in buffered formalin-acetone for 30 sec
2. Wash briefly in distilled water and air dry
3. Incubate for 15–20 minutes in the freshly prepared substrate solution, at room temperature in the dark
4. Wash in distilled water
5. Counterstain in aqueous haematoxylin for 1 minute
6. Wash in distilled water and blot dry.

Notes

The olive green to dark brown product with α-naphthyl butyrate is soluble in organic solvents. As an alternative to Fast blue BB salt, 0.4 ml of freshly prepared hexazotized new fuchsin may be used as the coupling agent.

REFERENCES

Ansley H R, Ornstein L 1971 Enzyme histochemistry and differential white cell counts on the Technicon Hemalog D. Advances in Automated Analysis 1: 437

Ansley H R, Sweetman F R, Ornstein L 1971 Leukocyte esterase cytochemistry. Journal of Histochemistry and Cytochemistry 19: 711

Armitage R J, Cawley J C 1985 The esterase reactivity of subpopulations of helper (Leu 3a$^+$) and suppressor (Leu 2a$^+$) T cells. British Journal of Haematology 59: 729

Armitage R J, Lynch D C, Worman C P, Cawley J C 1982 The morphology and cytochemistry of human T-cell subpopulations defined by monoclonal antibodies and Fc receptors. British Journal of Haematology 51: 605

Bàguena R, Forteza B G 1961 El cloroacetato de naftol AS-D para la demostracion de la actividad de las esterasas inespecificas en los leucocitos. Revista Clinica Espanola 82: 324

Barr R D, Perry S 1976 Lysosomal acid hydrolases in human lymphocyte subpopulations. British Journal of Haematology 32: 565

Basso G, Cocito M G, Semenzato G, Pezzutto A, Zanesco L 1980 Cytochemical study of thymocytes and T lymphocytes. British Journal of Haematology 44: 577.

Bernard J 1984 All human 'true' helper cells (Leu 3^+8^-) express acid alpha naphthyl acetate esterase. British Journal of Haematology 58: 381

Bernard J, Dufer J 1983 Cytochemical analysis of human peripheral blood lymphocyte subsets defined by monoclonal antibodies. Scandinavian Journal of Immunology 17: 89

Bevan A, Burns G F, Gray L, Cawley J C 1980 Cytochemistry of human T-cell subpopulations. Scandinavian Journal of Immunology 11: 223

Boesen A M 1984 Ultrastructural localization of acid alpha-naphthylacetate esterase in human normal and neoplastic lymphocytic and monocytic cells and in hairy cells. Scandinavian Journal of Haematology 32: 367

Boesen A M, Hokland P, Jensen O M 1984 Acid phosphatase and acid esterase activity in neoplastic and non-neoplastic lymphoid cells. Scandinavian Journal of Haematology 32: 313

Bozdek M J, Bainton D F 1981 Identification of α-naphthyl butyrate esterase as a plasma membrane ectoenzyme of monocytes and as a discrete intracellular membrane-bounded organelle in lymphocytes. Journal of Experimental Medicine 153: 182

Braunstein H 1959 Esterase in leucocytes (letter). Journal of Histochemistry and Cytochemistry 7: 202

Burns G F, Cawley J C, Barker C R, Hayhoe F G J 1977a Absence of a receptor for C3 on the hairy cells of leukaemic reticuloendotheliosis. Clinical and Experimental Immunology 29: 442

Burns G F, Cawley J C, Flemans R J, Higgy K E, Worman C P, Barker C R, Roberts B, Hayhoe F G J 1977b Surface marker and other characteristics of Gaucher's cells. Journal of Clinical Pathology 30: 981

Burns G F, Cawley J C, Higgy K E, Barker C R, Edwards M, Rees J K H, Hayhoe F G J 1978 Hairy cell leukaemia: a B cell neoplasm with a severe deficiency of normal B lymphocytes. Leukaemia Research 2: 33

Cline M J, Golde D W 1973 A review and reevaluation of the histiocytic disorders. American Journal of Medicine 55: 49

Cerezo L, Head D R 1982 Nonspecific esterase cytochemistry in acute myeloblastic leukemia: alpha-naphthyl acetate versus alpha-naphthyl butyrate. American Journal of Clinical Pathology 77: 248

Crockard A D 1984 Cytochemistry of lymphoid cells: a review of findings in the normal and leukaemic state. Histochemical Journal 16: 1027

Crockard A D, Catovsky D 1983 Cytochemistry of normal lymphocyte subsets defined by monoclonal antibodies and immunocolloidal gold. Scandinavian Journal of Haematology 30: 433

Crockard A D, Chalmers D, Matutes E, Catovsky D 1982 Cytochemistry of acid hydrolases in chronic B and T cell leukaemias. American Journal of Clinical Pathology 78: 437

Daniel M Th, Flandrin G, Lejeune F, Liso P, Lortholary P 1971 Les estérases spécifiques monocytaires. Utilisation dans la classification des leucémies aigues. Nouvelle Revue Francaise d'Hematologie 11: 233

Delellis R, Fishman W H 1965 The variable of pH in the bromoindoxyl acetate method for the demonstration of esterase. Journal of Histochemistry and Cytochemistry 13: 297

De Waele M, Demey J, Moeremans M, Smet L, Broodiaerts L, Van Camp B 1983 Cytochemical profile of immunoregulatory T lymphocyte subsets defined by monoclonal antibodies. Journal of Histochemistry and Cytochemistry 31: 471

Dixon M, Webb E C 1964 Enzymes. 2nd edn. Longman, London

Drexler H G, Gaedicke G 1983 Isoenzyme studies in human leukemia. II. Carboxylic esterase (E. C. 3.1.1.1). Leukemia Research 7: 599

Drexler H G, Gaedicke G, Minowada J 1985a Isoenzyme studies in human leukemia-lymphoma cell lines. 1. Carboxylic esterase. Leukemia Research 9: 209
Drexler H G, Otsuka K, Martin P J, Gaedicke G, Minowada J 1985b Changes in isoenzyme patterns expressed by the erythroleukemia cell lines K-562 and HEL after induction of differentiation. In: Neth R, Gallo R C, Greaves M F, Janka C (eds) Modern trends in human leukemia VI. Springer-Verlag, Berlin, p 430
Drexler H G, Gaedicke G, Hoffbrand A V, Minowada J 1987 Occurrence of particular isoenzymes in fresh and cultured leukemia-lymphoma cells. Cancer 59:77
Dufer J, Trentesaux C, Desplaces A 1984 Differential effect of the serine protease inhibitor phenyl methyl sulfonyl fluoride on cytochemically detectable esterases in human leucocytes and platelets. Scandinavian Journal of Haematology 32: 25
Fenoglio C, Invernizzi R, Gerzeli G 1985 Ultrastructural patterns of the α-naphthyl butyrate esterase activity in normal and leukemic peripheral blood leucocytes. Histochemical Journal 17:595
Ferluga J, Asherson G L, Becker E L 1972 The effect of organophosphorus inhibitors, p-nitrophenol and cytochalasin-B on cytotoxic killing of tumour cells by immune spleen cells and the effect of shaking. Immunology 23: 577
Ferrarini M, Cadoni A, Franzi A T, Ghigliotti C, Leprini A, Zicca A, Grossi C E 1980 Ultrastructure and cytochemistry of human peripheral blood lymphocytes. Similarities between the cells of the third population and T_G lymphocytes. European Journal of Immunology 10: 562
Fischer R, Schmalzl F 1964 Über die Hemmbarkeit der Esteraseaktivität in Blutmonocyten durch Natriumfluorid. Klinische Wochenschrift 42: 751
Giorno R, Beverly S 1982 A rapid method for determining human T lymphocyte levels using acid alpha-naphthyl acetate esterase. American Journal of Clinical Pathology 77: 240
Golomb H M, Rowley J D, Vardiman J W, Testa J R, Butler A 1980 'Microgranular' acute promyelocytic leukemia: a distinct clinical, ultrastructural, and cytogenetic entity. Blood 55: 253
Gomori G 1953 Chloroacyl esters as histochemical substrates. Journal of Histochemistry and Cytochemistry 1: 469
Grossi C E, Webb S R, Zicca A, Lydyard P M, Moretta L, Mingari M C, Cooper M D 1978 Morphological and histochemical analyses of two human T-cell subpopulations bearing receptors for IgM or IgG. Journal of Experimental Medicine 147: 1405
Grossi C E, Cadoni A, Zicca A, Leprini A, Ferrarini M 1982 Large granular lymphocytes in human peripheral blood: ultrastructural and cytochemical characterization of the granules. Blood 59: 277
Hanker J S, Katzoff L, Rosen H R, Seligmann M L, Ueno H, Seligman A M 1966 Design and synthesis of thiolesters for the histochemical demonstration of esterase and lipase via the formation of osmiophilic diazo thioethers. Journal of Medicinal Chemistry 9: 288
Hayhoe F G J, Quaglino D, Doll R 1964 The cytology and cytochemistry of acute leukaemias. A study of 140 cases. MRC Special Report Series No. 304. HMSO, London
Hayhoe F G J, Burns G F, Cawley J C, Stewart J W 1978 Cytochemical, ultrastructural and immunological studies of circulating Reed-Sternberg cells. British Journal of Haematology 38: 485
Heumann D, Colombatti M, Mach J P 1983 Human large granular lymphocytes contain an esterase activity usually considered as specific for the myeloid series. European Journal of Immunology 13: 254
Higgy K E, Burns G F, Hayhoe F G J 1977 Discrimination of B, T and null lymphocytes by esterase cytochemistry. Scandinavian Journal of Haematology 18: 437
Higgy K E, Burns G F, Hayhoe F G J 1978 Identification of the hairy cells of leukaemic reticulo-endotheliosis by an esterase method. British Journal of Haematology 38: 99
Huhn D 1983 Acid esterase (ANAE) in acute lymphoblastic leukemia (ALL). Acta Haematologica 70: 209
Huhn D, Twardzik L 1983 Acute myelomonocytic leukemia and the French-American-British classification. Acta Haematologica 69: 36
Hunter R L, Burstone M S 1960 The zymogram as a tool for the characterisation of enzyme substrate specificity. Journal of Histochemistry and Cytochemistry 8: 58
Jaffe E S, Cossman J, Callihan T R 1980 Presence of acid alpha-naphthyl acetate esterase (ANAE) in human B cell lymphomas. Laboratory Investigation 42: 125
Kaplow L S 1975 Cytochemical heterogeneity of human circulating monocytes. Acta cytologica 19: 358
Kaplow L S, Dauber H, Lerner E 1976 Assessment of monocyte esterase activity by flow cytophotometry. Journal of Histochemistry and Cytochemistry 24: 363
Kass L 1977a Esterase activity in erythroleukaemia. American Journal of Clinical Pathology 67: 368
Kass L 1977b Non-specific esterase activity in 'hairy cells'. Acta Haematologica 58: 103
Kass L, Munster D 1979a Nonspecific esterases in normal human blood cells: characterization of isoenzymes by isoelectric focusing in polyacrylamide gel. Experimental Hematology 7: 535
Kass L, Munster D 1979b Nonspecific esterases in leukemic blasts. American Journal of Clinical Pathology 72: 611

Kass L, Peters C L 1977 Nonspecific esterase activity in pernicious anaemia and chronic erythremic myelosis. American Journal of Clinical Pathology 68: 273
Kass L, Peters C L 1978 Esterases in acute leukemias. A cytochemical and electrophoretic study. American Journal of Clinical Pathology 69: 57
Kass L, Peters C L, Vugrin D 1978 Electrophoretic properties of esterases in chronic granulocytic leukaemia. American Journal of Clinical Pathology 69: 329
Katayama I, Aiba M 1980 Tartrate-resistant acid phosphatase reaction. American Journal of Clinical Pathology 73: 143
Kim H, Pangalis G A, Payne B C, Kadin M E, Rappaport H 1982 Ultrastructural identification of neoplastic histiocytes-monocytes. An application of a newly developed cytochemical technique. American Journal of Pathology 106: 204
Korsmeyer S J, Green W C, Cossman J, Hsu S M, Jansen J P, Neckers L M, Marshall S L, Bakhshi A, Depper J M, Leonard W J, Jaffe E S, Waldman T A 1983 Rearrangement and expression of immunoglobulin genes and expression of TAC antigen in hairy cell leukemia. Proceedings of the National Academy of Sciences 80: 4522
Kulenkampff J, Janossy G, Greaves M F 1977 Acid esterase in human lymphoid cells and leukaemic blasts: a marker for T lymphocytes. British Journal of Haematology 36: 231
Lam K W, Li C Y, Siemens M, Yam L T 1985 Immunohistochemical detection of monocytes by the antiserum specific to monocytic esterase. Journal of Histochemistry and Cytochemistry 33: 379
Leder L D 1966 Fermentcytochemische Untersuchungen zur Herkunst der Blutmonocyten. Klinische Wochenschrift 44: 25
Leder L D 1967 The origin of blood monocytes and macrophages. Blut 16: 86
Li C Y, Lam K W, Yam L T 1973 Esterases in human leukocytes. Journal of Histochemistry and Cytochemistry 21: 1
Lojda Z, Gossrau R, Schiebler T H 1979 Enzyme histochemistry. A laboratory manual. Springer-Verlag, Berlin
Manconi P E, Marrosu M G, Paghi L, Correale G, Zaccheo D 1979 Alpha-naphthyl acetate esterase activity in human lymphocyte subpopulations and in mitogen activated cells. Scandinavian Journal of Immunology 9: 99
Markey G M, Morris T C M, Alexander H D, Kyle A, Thompson A, Robertson J H, Turner A, Burnside P, Drexler H G, Gaedicke G 1986 Monocyte esterase deficiency in the population and in lymphoproliferative disease. British Journal of Haematology 64: 838
Matutes E, Crockard A D, O'Brien M, Catovsky D 1983 Ultrastructural cytochemistry of chronic T-cell leukaemias. A study with four acid hydrolases. Histochemical Journal 15: 895
Moloney W C, Fliegelman L, McPherson K 1959 The use of naphthol AS-D chloroacetate as a substrate for the study of leucocyte esterase activity. Journal of Histochemistry and Cytochemistry 7: 306
Moloney W C, McPherson K, Fliegelman L 1960 Esterase activity in leucocytes demonstrated by the use of naphthol AS-D chloroacetate substrate. Journal of Histochemistry and Cytochemistry 8: 200
Monahan R A, Dvorak H F, Dvorak A M 1981 Ultrastructural localization of nonspecific esterase activity in guinea pig and human monocytes, macrophages, and lymphocytes. Blood 58: 1089
Mueller J, Brun Del Re G, Buerki H, Keller H U, Hess M W, Cottier H 1975 Non-specific acid esterase activity: a criterion for differentiation of T and B lymphocytes in mouse lymph nodes. European Journal of Immunology 5: 270
Nachlas M M, Seligman A M 1949 The histochemical demonstration of esterase. Journal of the National Cancer Institute 9: 415
Oertel J, Kastner M 1984 Two-dimensional separation of α-naphthyl acetate esterases in human leucocytes and platelets. British Journal of Haematology 57: 431
Oertel J, Hagner G, Kastner M, Huhn D 1985 The relevance of α-naphthyl acetate esterases to various monocyte functions. British Journal of Haematology 61: 717
Payne B C, Kim H, Pangalis G A, Rothman A, Rappaport H 1980. A method for the ultrastructural demonstration of non-specific esterase in human blood and lymphoid tissue. Histochemical Journal 12: 71
Polli N, O'Brien M, Tavares de Castro J, Matutes E, San Miguel J F, Catovsky D 1985 Characterization of blast cells in chronic granulocytic leukaemia in transformation, acute myelofibrosis and undifferentiated leukaemia. I. Ultrastructural morphology and cytochemistry. British Journal of Haematology 59: 277
Quaglino D, Hayhoe F G J, Flemans R J 1962 Cytochemical observations on the effect of phytohaemagglutinin in short-term tissue cultures. Nature 196: 338
Quesada J R, Murphy S G 1982 Histochemical patterns of human T lymphocyte subpopulations with nonspecific esterase staining. International Archives of Allergy and Applied Immunology 68: 138
Radzun H J, Parwaresch M R, Kulenkampff C, Staudinger M, Stein H 1980 Lysosomal acid

esterase: activity and isoenzymes in separated normal human blood cells. Blood 55: 891
Ranki A 1978 Nonspecific esterase activity in human lymphocytes. Clinical Immunology and Immunopathology 10: 47
Ranki A, Tötterman T H, Häyry P 1976 Identification of resting T and B lymphocytes by acid α-naphthyl acetate esterase staining combined with rosette formation with staphylococcus aureus strain Cowan 1. Scandinavian Journal of Immunology 5: 1129
Reisfeld R A, Lewis U J, Williams D E 1962 Disc electrophoresis of basic proteins and peptides on polyacrylamide gels. Nature 195: 281
Rozenszajn L, Efrati P 1961 Cytochemical and phase-contrast observations on Gaucher cells. Acta Haematologica 25: 43
Rozenszajn L, Leibovich M, Shoham D, Epstein J 1968 The esterase activity in megaloblasts, leukaemic and normal haemopoietic cells. British Journal of Haematology 14: 605
Schmalzl F, Braunsteiner H 1971 The application of cytochemical methods to the study of acute leukaemias. A review. Acta Haematologica 45: 209
Schnitzer B, Kass L 1974 Hairy cell leukemia: a clinicopathological and ultrastructural study. American Journal of Clinical Pathology 61: 176
Scott C S, Cahill A, Bynoe A G, Ainley M J, Hough D, Roberts B E 1983 Esterase cytochemistry in primary myelodysplastic syndromes and megaloblastic anaemias: demonstration of abnormal staining patterns associated with dysmyelopoiesis. British Journal of Haematology 55: 411
Scott C S, Linch D C, Bynoe A G, Allen C, Hogg N, Ainley M J, Hough D, Roberts B E 1984 Electrophoretic and cytochemical characterization of alpha-naphthyl acetate esterases in acute myeloid leukemia: relationship with membrane receptor and monocyte-specific antigen expression. Blood 63: 579
Scott C S, Morgan M, Limbert H J, MacKarill I D, Roberts B E 1985 Cytochemical, immunological and ANAE-isoenzyme studies in acute myelomonocytic leukaemia: a study of 39 cases. Scandinavian Journal of Haematology 35: 284
Shnitka T K, Seligman A M 1961 Role of esterase inhibition on localization of esterase and the simultaneous cytochemical demonstration of inhibitor sensitive and resistant enzyme species. Journal of Histochemistry and Cytochemistry 9: 504
Swirsky D M 1984 Single incubation double esterase cytochemical reaction using a single coupling reagent. Journal of Clinical Pathology 37: 1187
Swirsky D M, Li Y S, Matthews J G, Flemans R J, Rees J K H, Hayhoe F G J 1984 8; 21 translocation in acute granulocytic leukaemia: cytological, cytochemical and clinical features. British Journal of Haematology 56: 199
Tamaoki N, Essner E 1969 Distribution of acid phosphatase, β-glucuronidase and N-acetyl-β-glucosaminidase activities in lymphocytes of lymphatic tissues of man and rodents. Journal of Histochemistry and Cytochemistry 17: 238
Tavassoli M, Shaklai M, Crosby W H 1979 Cytochemical diagnosis of acute myelomonocytic leukemia. American Journal of Clinical Pathology 72: 59
Tomonaga M, Yoshiharu Y, Tagawa M, Jinnai I, Kuriyama K, Amenomori T, Yoshioka A, Matsuo T, Nonaka H, Ichimaru M 1985 Cytochemistry of acute promyelocytic leukemia (M3): leukemic promyelocytes exhibit heterogeneous patterns in cellular differentiation. Blood 66: 350
Tulliez M, Vernant J P, Breton-Gorius J, Imbert M and Sultan C 1979 Pseudo-Chediak-Higashi anomaly in a case of acute myeloid leukemia: electron microscopic studies. Blood 54: 863
Variakojis D, Vardiman J W, Golomb H M 1980 Cytochemistry of hairy cells. Cancer 45: 72
Veerman A J P, Huismans D R, Van Zantwijk C H 1983 Immunological phenotype related to acid alpha-naphthyl acetate esterase and acid phosphatase in childhood acute lymphoblastic leukaemia. Acta Haematologica 69: 32.
Wachstein M, Wolf G 1958 The histochemical demonstration of esterase activity in human blood and bone marrow smears. Journal of Histochemistry and Cytochemistry 6: 457
Yam L T, Li C Y, Finkel H E 1972 Leukemic reticuloendotheliosis. The role of tartrate resistant acid phosphatase in diagnosis and splenectomy in treatment. Archives of Internal Medicine 130: 248
Yang K, Bearman R M, Pangalis G A, Zelman R J, Rappaport H 1982 Acid phosphatase and alpha-naphthyl acetate esterase in neoplastic and non-neoplastic lymphocytes. A statistical analysis. American Journal of Clinical Pathology 78: 141
Young C W, Bittar E S 1973 Analysis of tissue esterases from patients with Hodgkin's disease and other types of advanced cancer by isoelectric focusing in acrylamide gel. Cancer Research 33: 2692
Yourno J, Mastropaolo W 1981 Nonspecific esterases of the formed elements: zymograms produced by pH 9.5 polyacrylamide gel electrophoresis. Blood 58: 939
Yourno J, Burkart P, Lizzi F, Tartaglia A 1982 Enzymologic classification of acute leukemias: nonspecific esterase markers distinguish myeloid and lymphoid varieties. Blood 60: 304
Zajicek J, Sylven B, Datta N 1954 Attempts to demonstrate acetyl-cholinesterase activity in blood

and bone marrow cells by a modified thiocholine technique. Journal of Histochemistry and Cytochemistry 2: 115
Zicca A, Leprini A, Cadoni A, Franzi A T, Ferrarini M, Grossi C E 1981 Ultrastructural localization of alpha-naphthyl acid esterase in human T_M lymphocytes. American Journal of Pathology 105: 40

CHAPTER 9

DEHYDROGENASES

Dehydrogenases belong to the large and important class of enzymes which catalyze biological oxidations. These enzymes, the oxidoreductases, may be broadly divided into two groups according to the cytochemical methods used for their demonstration (Borgers and Verheyen, 1985). One group, including peroxidase, catalase and the cytochrome c-oxidase and peroxidase systems, is demonstrable by the diaminobenzidine (DAB) procedure, introduced by Graham and Karnovsky in 1966. These enzymes are dealt with in Chapter 10. The second group, demonstrable by other methods, includes especially enzymes which accomplish the removal of hydrogen from the substrate undergoing oxidation, 'the hydrogen donor', to another substrate, 'the hydrogen acceptor'. The dehydrogenases, which catalyze reductions of this kind are not only specific for the hydrogen donor but also for the hydrogen acceptor. Consequently, in the absence of an appropriate hydrogen acceptor, any dehydrogenase, capable though it may be of activating the hydrogen donor, cannot give rise to a chemical change.

Dehydrogenases are present in cells and tissues of many different kinds and their presence may be revealed by histochemical methods. The demonstration and localization of a number of these enzymes within cells has been achieved cytochemically by exposing the cells to appropriate substrates upon which the dehydrogenases may act in the presence of tetrazolium compounds capable of accepting hydrogen with the production of coloured insoluble compounds.

Development of cytochemical methods for dehydrogenases

The first histochemical studies on dehydrogenases of animal tissues were carried out in 1949 by Seligman *et al.*, using a monotetrazolium salt, triphenyl tetrazolium chloride (TTC). Other compounds, such as neotetrazolium or NT (Antopol *et al.*, 1948), the methoxy derivative of NT, called blue tetrazolium or BT (Rutenburg *et al.*, 1950) and finally an iodinated neotetrazolium derivative, INT (Atkinson *et al.*, 1950) were found to give better histochemical results and the majority of the early work on dehydrogenase histochemistry was carried out with either NT or BT. Since 1956 two other tetrazolium salts have been used in dehydrogenase histochemistry. These are 2,2′-di-p-nitro-phenyl-5,5′-diphenyl-3,3′-(3,3′-dimethoxy-4,4′-biphenylene) ditetrazolium chloride or Nitro-BT (Tsou *et al.*, 1956), and 3-(4,5-dimethylthiazolyl-2)-2,5-diphenyl tetrazolium bromide or MTT (Pearse, 1957).

No endeavour will be made in this chapter to review the principles of dehydrogenase cytochemistry and to consider the physico-chemical characteristics of the tetrazolium compounds, as these fundamental aspects are extensively and fully discussed in standard works on histochemistry (e.g. Pearse, 1960; Lojda *et al.*, 1979). The chapter will be mainly concerned in reviewing the attempts made by different investigators to demonstrate cytochemically the presence of dehydrogenases in blood and bone marrow cells and to study the variations of enzyme activity in normal and pathological states.

In haematology, the demonstration of dehydrogenases has been hampered by problems of fixation, since fixatives giving adequate cell preservation generally inactivate the enzymes. Attempts were made to overcome the difficulties arising from the use of fixatives by incubating fresh suspensions of heparinized blood and bone marrow cells with the enzyme substrate and a tetrazolium compound. Wachstein (1950) made use of such a technique, but omitting the substrate, in an attempt to demonstrate the presence of a reducing activity in haemic cells. Despite considerable difficulties in the identification of the different cell types, a large proportion of normal and pathological haemic cells were found to show some evidence of reducing activity. In normal blood and bone marrow suspensions, Wachstein found that polymorphonuclear cells, lymphocytes and monocytes contained one to several purplish granules. In conditions of leucocytosis of different nature, Wachstein claimed that polymorphonuclear cells showed coarser and more numerous granules than commonly seen in normal leucocytes. In acute and chronic myeloid leukaemia primitive cells were negative, while positive granules were visible in myelocytes and in later granulocytes. In chronic lymphocytic leukaemia positivity was extremely variable.

In 1956, in a series of papers (Farber *et al.*, 1956; Sternberg *et al.*, 1956; Farber and Louviere, 1956) it was claimed that all attempts to localize specific dehydrogenases by tetrazolium methods resulted only in the demonstration of one or other of the two pyridine-nucleotide-linked diaphorases. These flavoprotein enzymes transfer hydrogen from the substrate to the acceptor, which in the particular case of enzyme cytochemistry, is one of the tetrazolium compounds. The succinic dehydrogenase system forms an exception since it contains its own 'built-in' flavoprotein and does not require the intervention of a diaphorase in transferring electrons from succinate to any of the tetrazolium salts.

Such a view, i.e. that tetrazolium methods can demonstrate only the two diaphorases and not the specific dehydrogenases was not shared by Hess *et al.* (1958), who believed that it was possible to demonstrate cytochemically the presence of nine of the pyridine-nucleotide-linked dehydrogenases, together with succinic dehydrogenase.

However, a pitfall in dehydrogenase histochemistry which cannot be overlooked is the fact that it is not always possible to choose the optimum pH for a given reaction, especially if this is in the alkaline range, as coenzyme-linked dehydrogenases show above pH 7.0 what Zimmermann and Pearse (1959) called the 'nothing dehydrogenase activity'. These authors observed, using either MTT or Nitro-BT, the production of formazan in a system devoid of any specific substrate and containing only coenzymes, buffer and tissue section. This effect progressively increased from pH 7.0 to pH 9.0. Various considerations put forward by Zimmermann and Pearse suggest that protein-bound SH or glutathione are responsible for this phenomenon.

In 1959 De Souza and Kothare described a method for demonstrating succinic dehydrogenase in human leucocytes, involving the use of Nitro-BT. According to these investigators this salt, in contrast to other tetrazolium compounds which produce large crystal deposits, yields a finely granular formazan deposit and has the added advantage of being practically insoluble in lipids. In the technique published by De Souza and Kothare, fresh suspensions of heparinized blood or bone marrow cells were allowed to react with the enzyme substrate (sodium succinate) and nitro-BT in a phosphate buffer at pH 7.4 in test-tubes for one hour at 37°C. Smears were then made of the centrifuged deposit and afterwards fixed and counterstained with safranin for 10–15 seconds. Formazan deposition was present in heparinized and non-heparinized blood and was scanty in control smears obtained by incubating blood and bone marrow cells in buffer with tetrazolium but without addition of substrate. However, identification of the different cell types remained difficult and only the study of specimens showing a predominance of one cell type (leukaemias and infections) facilitated the recognition of cells and allowed a certain degree of reliability.

The need to possess a method suitable for accurate and detailed dehydrogenase cytochemistry prompted further research in an attempt to find a fixative capable of preserving cell morphology without inhibiting enzymatic activity. The application of fixatives such as methanol, ethanol, formalin solutions and formalin vapour to air-dried smears of blood and bone marrow, while giving good cell preservation, invariably inactivated the enzymes.

In 1959 Willighagen and Planteydt reported the use of acetone fixation of cryostat sections prior to incubation for aminopeptidase activity.

Cold acetone fixation was also used by Novikoff *et al.* (1960) for the study of oxidative enzymes in cryostat sections. The investigators used Nitro-BT and reported that following cold acetone fixation the activities of dehydrogenases remained sufficiently high for ready detection of the formazan that resulted from tetrazolium reduction. Novikoff and his colleagues further emphasised that acetone not only reduced enzyme diffusion and cell disintegration but also had the additional advantage of removing most lipids upon which formazan might otherwise have crystallized.

Distribution of dehydrogenases in haemic cells

Consistent and reproducible results were obtained in haemic cells (Quaglino and Hayhoe, 1960; Quaglino, 1961) employing 60% acetone fixation at room temperature. The distribution pattern of succinic and lactic dehydrogenases was studied in both normal and pathological conditions. In the blood and bone marrow cells of normal individuals enzyme activity was particularly prominent in the more primitive cells of all three myeloid series and tended to decrease in the later maturation stages (Plate 21A and Figs

9.1, 9.2 and 9.3). Succinic dehydrogenase was weak or absent in mature polymorphonuclear cells, although these cells showed a fair degree of lactic dehydrogenase activity in the form of irregularly scattered cytoplasmic granules of formazan deposition, especially noticeable in infectious leucocytosis. Sometimes strong positivity of exceptional intensity may be seen in megakaryocytes (Figs 9.4 and 9.5, Plate 21B). In chronic lymphocytic leukaemia positivity in individual cells, with both sodium succinate and lactate, appeared to be more intense with increasing leucocyte counts and in those cells displaying immature features (Figs 9.6 and 9.7). The primitive cells of the different cytological varieties of acute leukaemia (Figs 9.8, 9.9 and 9.10) and the plasma cells of myeloma all showed variable but sometimes intense succinic and lactic dehydrogenase (EC 1.1.1.27) activity (Plate 21C, D, E, F). Strong enzymatic granular reactions were observed also in phytohaemagglutinin-transformed lymphocytes after 48 hours of culture (Quaglino *et al.*, 1962), a finding later confirmed by Stuart *et al.* (1970).

Ackerman (1960), using vital preparations, demonstrated dehydrogenase activity in white cells from peripheral blood with buffered Nitro-BT-succinate incubating medium.

In 1963 Wulff reported increased activity of succinic dehydrogenase, mitochondrial α-glycerophosphate dehydrogenase (EC 1.1.1.8), NAD and NADP diaphorases in inflammatory leucocytes obtained by the 'skin window' technique.

The increased succinic dehydrogenase activity observed in neutrophils after migration indicated, in the opinion of the author, that the Krebs tricarboxylic acid cycle was much more active in such conditions, whereas the increase in mitochondrial α-glycerophosphate dehydrogenase, which oxidizes α-glycerophosphate to dihydroxyacetone phosphate, suggested that this enzyme provided a substrate for glycogen synthesis, since there was a parallel increase also in glycogen content and phosphorylase activity. According to Wulff the greater diaphorase activity in inflammatory neutrophils, in lymphocytes and monocytes suggests that these cells are less dependent on glycolytic metabolism than are blood neutrophils. The phagocytes (neutrophils and monocytes) present in skin window preparations were also found to show increased choline dehydrogenase (EC 1.1.99.1) activity compared to non-phagocytic blood and bone marrow cells (Merker *et al.*, 1965).

In 1967 Rozenszajn and Shoham made use of a supravital method to demonstrate dehydrogenase and diaphorase activity in living unfixed cells, but concluded that such an approach was inadequate for demonstrating specific enzymatic activity. Rozenszajn and Shoham also employed Balogh and Cohen's method (1961) in unfixed smears and were able to show the presence of specific activity due to NADP linked glucose-6-phosphate dehydrogenase (EC 1.1.1.49), and also NADH diaphorase and succinic dehydrogenase, which are not dependent on the presence of NAD or NADP. The investigators could not demonstrate glutamic (EC 1.4.1.2), betahydroxybutyric (EC 1.1.1.30) and alcohol dehydrogenases, (EC 1.1.1.1), all NAD linked. In fixed smears the activities of lactic and succinic dehydrogenases were shown, whereas those of NADH and NADPH

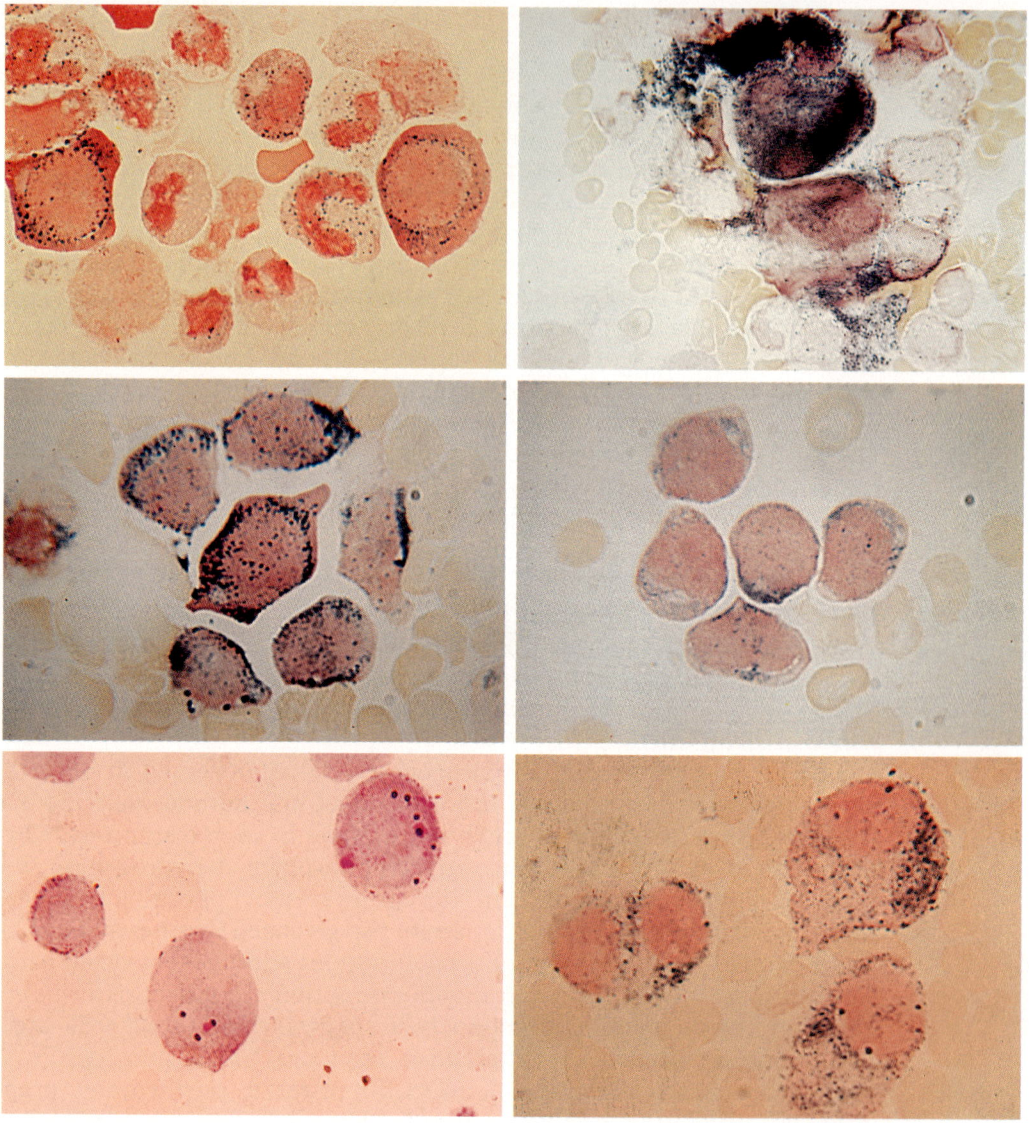

A B C D E F

Plate 21

A. Normal bone marrow. Lactic dehydrogenase. Enzyme activity is prominent in two proerythroblasts but less conspicuous in the later granulocytes present.
B. Normal bone marrow. Lactic dehydrogenase. Strong positivity in megakaryocytes.
C. ALL bone marrow. Lactic dehydrogenase. Coarse granular positivity in blast cells.
D. AMLbone marrow. Lactic dehydrogenase. Moderately strong reaction in a group of five leukaemic myeloblasts.
E. ALL bone marrow. Lactic dehydrogenase and consecutive PAS reaction. Lymphoblasts show a scattering of coarse granules of dehydrogenase reaction product, which sometimes appear to coincide with PAS positive blocks.
F. Multiple myeloma bone marrow. Lactic dehydrogenase. Strong granular positivity in three myeloma cells.

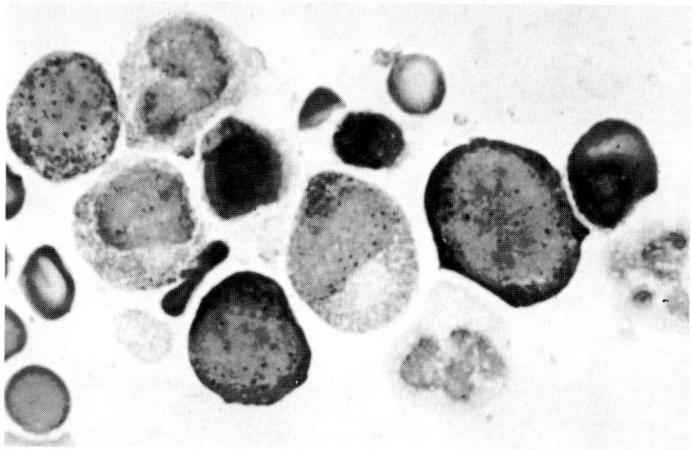

Fig. 9.1 Normal bone marrow. Lactic dehydrogenase. Granular positivity more marked in a proerythroblast, an early normoblast and a promyelocyte than in late normoblasts and granulocytes.

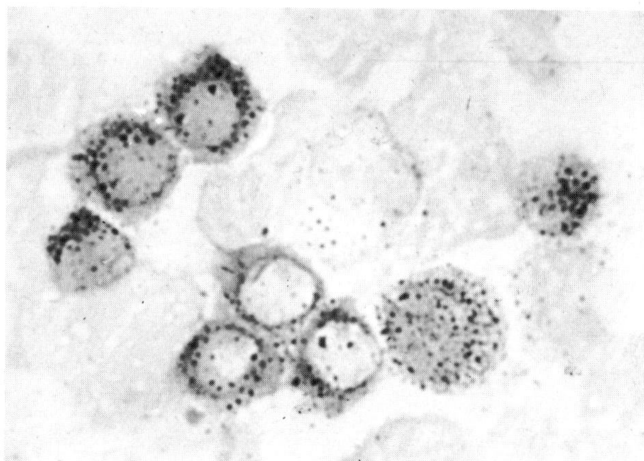

Fig. 9.2 Normal bone marrow. Succinic dehydrogenase. A central weakly reacting myelocyte and a series of red cell precursors, showing strong reaction in early normoblasts and weaker reaction in later stages.

diaphorases were demonstrated only in the presence of NADH and NADPH. According to these authors dehydrogenase and diaphorase activity was higher in younger cells and decreased progressively with maturation.

In 1968 Castoldi *et al.* described a histochemical method for the demonstration in blood and bone marrow cells of dihydroorotic dehydrogenase (EC 1.3.3.1), a flavoprotein which catalyzes the oxidation of dihydro-orotic acid and the reduction of orotic acid. The enzyme is therefore directly involved in pyrimidine synthesis and takes part in nucleic acid metabolism. Biochemical assays (Smith and Baker, 1959) have shown a significant increase in the activity of dihydroorotic dehydrogenase in the cells of chronic lymphocytic leukaemia and acute leukaemias.

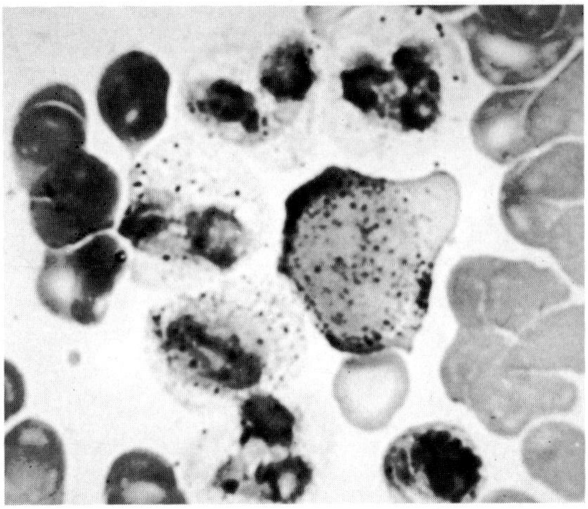

Fig. 9.3 Normal bone marrow. Lactic dehydrogenase. A strongly positive proerythroblast, a metamyelocyte and a stab cell with moderate numbers of granules, three neutrophil polymorphs and a late normoblast with a few granules.

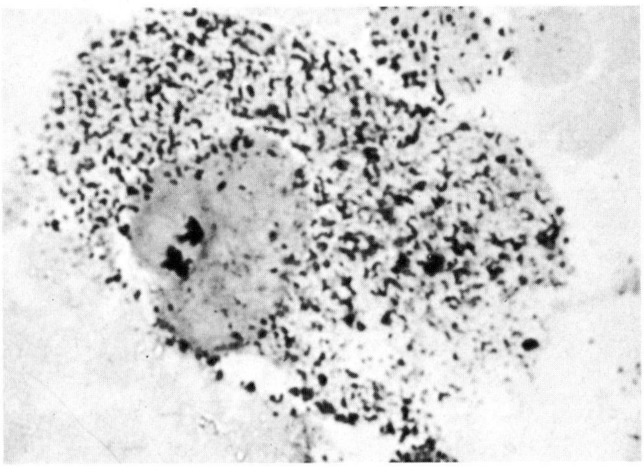

Fig. 9.4 Normal bone marrow. Lactic dehydrogenase. Strong granular reaction in a megakaryocyte.

In normal conditions the enzyme may be demonstrated cytochemically in all blood and bone marrow cells with the exception of mature lymphocytes. A significant increase of dihydroorotic dehydrogenase activity, evaluated by a semiquantitative scoring method, was observed in the neutrophils of chronic myeloid leukaemia and in the lymphocytes of chronic lymphocytic leukaemia, non-Hodgkin's lymphomas, Hodgkin's disease and aplastic anaemia. The significance of the increased levels of this enzyme is not yet fully understood but it has been postulated that a block in pyrimidine synthesis, as in megaloblastic anaemia where levels of lactate and other dehydrogenase enzymes are generally increased (Stuart and Skowron

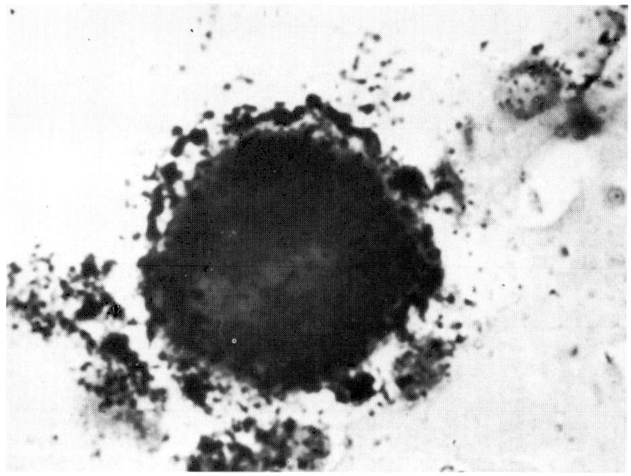

Fig. 9.5 The same preparation. Exceptionally intense positivity in another megakaryocyte.

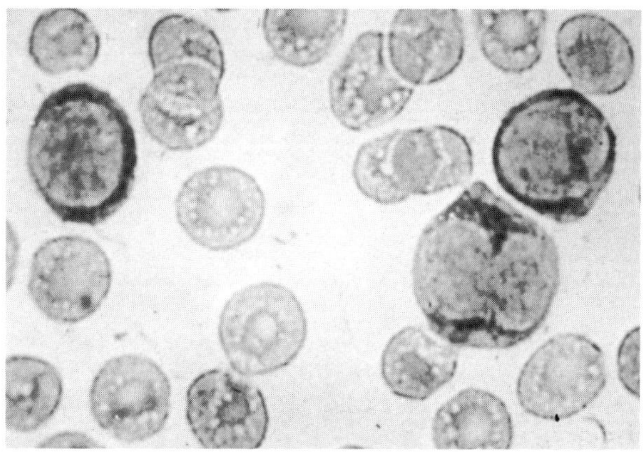

Fig. 9.6 CLL peripheral blood. Succinic dehydrogenase. Strong positivity in lymphocytes.

1968, Fig. 9.11), may lead to an accumulation of active enzyme prior to the site of the block.

In 1970 Stuart and Simpson developed a new technique for the demonstration of dehydrogenases, based on the use of an inert, synthetic sucrose polymer (Ficoll), which was found to have a protective effect on leucocytes during incubation for cytochemical reactions at 37°C. The addition to the incubation solution of this high molecular weight sucrose polymer allows unfixed smeared cells to be retained on glass slides and enables dehydrogenase cytochemistry to be performed on blood and bone marrow cells and also on leucocytes cultured *in vitro*, thus avoiding the use of a fixative, which renders cell membranes highly permeable and inhibits a variable amount of enzyme activity. The membrane stabilizing effect of Ficoll does not block the entry of substrate into mitochondria and therefore does not prevent the cytochemical demonstration of intramitochondrial

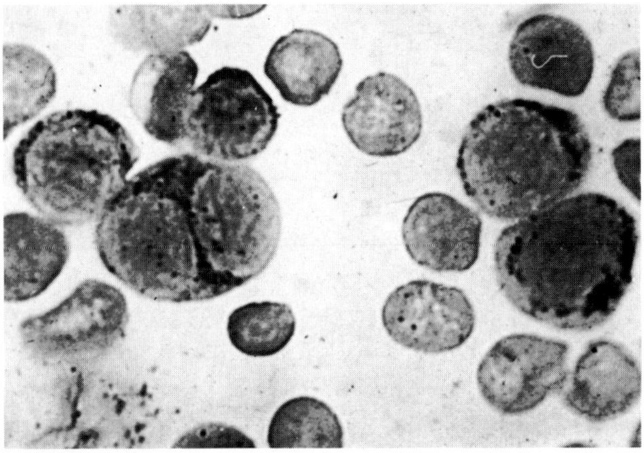

Fig. 9.7 CLL peripheral blood. Lactic dehydrogenase. Similar strong reaction in lymphocytes, especially in cells with prolymphocytic features.

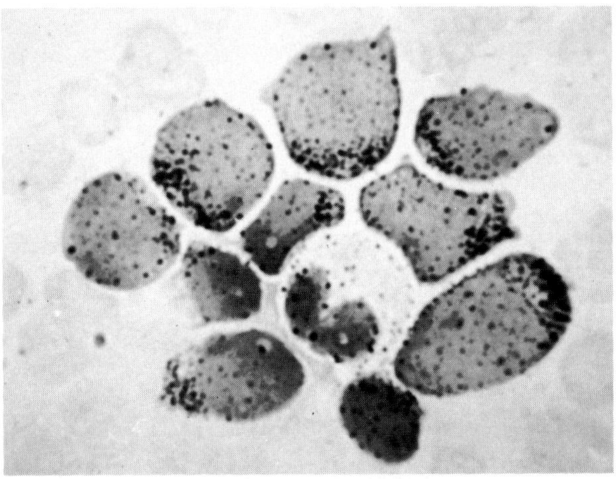

Fig. 9.8 AML bone marrow. Lactic dehydrogenase. Group of blast cells with moderate to strong cytoplasmic positivity surrounding a weakly reacting neutrophil polymorph.

enzymes (succinate dehydrogenase), but it does prevent the cytochemical demonstration of lysosomal enzymes; reactions for β-glucuronidase and aryl sulphatase in unfixed bone marrow films were negative when a 40% concentration of Ficoll was added to the incubation medium.

Quantitation of dehydrogenase cytochemistry in haemic cells and effects of oxamate inhibition

In a further investigation Stuart *et al.* (1972) studied cytochemically a number of intra- and extramitochondrial dehydrogenases as well as a mitochondrial membrane shuttle system in normal myeloid cells, chronic

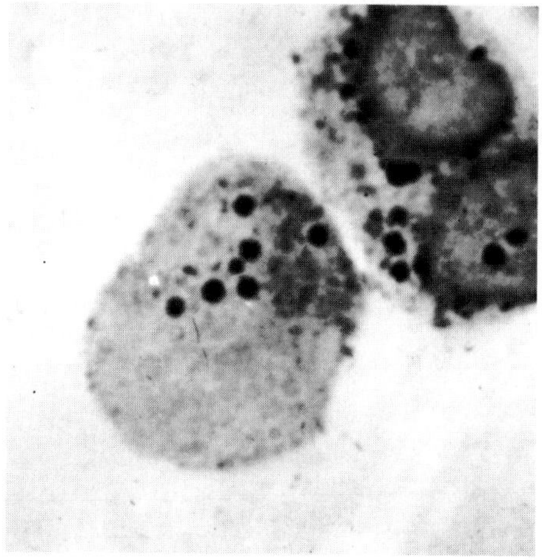

Fig. 9.9 ALL bone marrow. Lactic dehydrogenase. Two blast cells showing positivity in the form of very coarse granules or blocks.

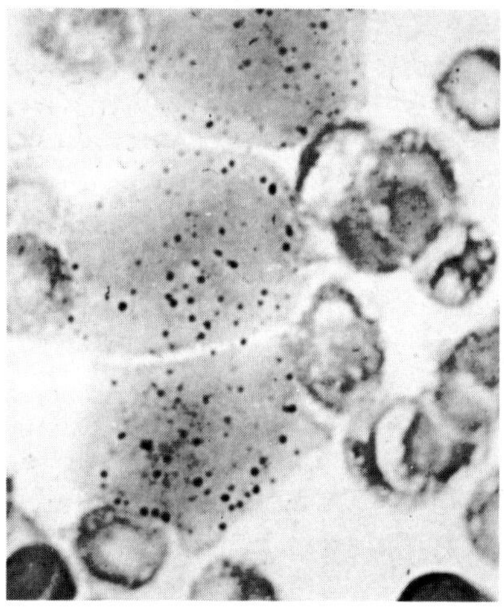

Fig. 9.10 AMML bone marrow. Succinic dehydrogenase. A pattern of discretely scattered positive granules is shown in this case.

myeloid leukaemia, acute myeloblastic and myelomonocytic leukaemia, acute lymphoblastic and chronic lymphocytic leukaemia and in the atypical mononuclear cells of infectious mononucleosis. The variations in enzyme activity in all the conditions examined were evaluated employing a quantitative cytophotometric method. The results obtained for both normal and chronic myeloid leukaemia leucocytes showed a progressive decrease in enzyme activity with maturation from the promyelocyte-myelocyte stage onwards. The progressive decrease in activity of mitochondrial and shuttle system enzymes with maturation was believed to be partly related to

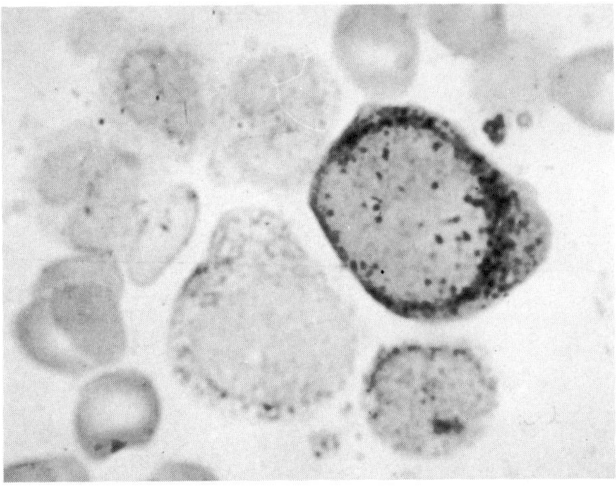

Fig. 9.11 Megaloblastic anaemia. Lactic dehydrogenase. Strong positivity in an early megaloblast, with weak to negative reaction in neighbouring granulocytes.

mitochondrial number since developing cells of the myeloid series show a decrease in the number of mitochondria with differentiation (Pease, 1956; Bessis, 1965; Kakefuda, 1968).

The authors observed in the blast cells of their cases of monocytic and myelomonocytic leukaemia an increased activity of the intramitochondrial enzymes α-glycerophosphate dehydrogenase and succinic dehydrogenase and of the α-glycerophosphate shuttle enzyme α-GPD-NAD, compared to the activity shown in myeloblasts, probably in relation to the greater number of mitochondria present in monoblasts (Freeman and Journey, 1971).

On the contrary, lymphoblasts showed significantly lower activity of mitochondrial and α-glycerophosphate shuttle enzymes than myeloblasts, perhaps on account of the fewer number of mitochondria (Freeman and Journey, 1971). Burkitt lymphoma cell lines also showed very low intramitochondrial enzyme activity and virtually no α-glycerophosphate shuttle enzyme activity.

The characteristic pattern of relatively high extramitochondrial glycolytic enzyme activity, in particular lactate dehydrogenase, with low intramitochondrial and α-glycerophosphate shuttle enzyme activity thus appears to be confined to the lymphoblast and Burkitt lymphoma cells, although the increase in activity of lactic dehydrogenase and malic dehydrogenase in the chronic lymphocytic leukaemia lymphocyte, relative to the normal small lymphocyte, suggests a similar tendency. This pattern of enzyme activity seems unrelated to cell proliferative activity, in view of the high mitotic activity of the Burkitt lymphoma cell (Epstein and Barr, 1965; Cooper *et al.*, 1966) and of the low activity of the lymphoblast (Mauer and Fisher, 1962; Gavosto *et al.*, 1964; Foadi *et al.*, 1968). It seems more likely to be a consequence of the small number of mitochondria, thus implying a metabolic dependence on extramitochondrial glycolysis for energy production.

This finding, together with reports of systemic lactic acidosis in acute leukaemia (Field *et al.*, 1966; Roth and Porte, 1970) and in Burkitt's lymphoma (Block *et al.*, 1966) prompted Mann *et al.* (1972) to investigate the possible existence of a glycolytic enzyme inhibitor, which would display a more or less selective cytotoxic effect on immature lymphoid cells.

A specific lactate dehydrogenase inhibitor, sodium oxamate, which prevents glycolytic resynthesis of ATP, was found to possess a significant *in vitro* cytotoxic effect for EB_2 cells, a cell line derived from Burkitt's lymphoma, and for the acute leukaemia lymphoblast. This selectivity for immature lymphoid cells, which contain small numbers of mitochondria, is in accordance with the results of the enzyme studies of Stuart *et al.* (1972), which showed disproportionately high glycolytic enzyme activity in these cells. Since sodium oxamate blocks cell glycolysis, it is assumed that the cell becomes an obligatory aerobe in respect of ATP synthesis; since mitochondria are scanty, the cell may be unable to utilize mitochondrial respiration as an alternative energy source. Sodium oxamate was not found to exert a significant cytotoxic effect on normal marrow myeloid cells or on blood lymphocytes and neutrophils, although a significant reduction in glycolysis-dependent neutrophil phagocytosis was observed. These cells, according to the authors, appeared to be less dependent on glycolysis for ATP synthesis and viability, showing a greater potential for oxidative phosphorylation due, in the case of myeloid cells, to a greater number of mitochondria per cell.

However from the experiments of Mann and her associates it was not possible to determine whether cell damage by sodium oxamate was specific for leukaemic, as opposed to normal precursor, lymphoblasts: it is possible that susceptibility to lactate dehydrogenase inhibition is related more to mitochondrial number and glycolytic enzyme activity than to leukaemic changes *per se*.

A decrease in lactic and to a lesser extent in malic dehydrogenase activity has been reported in childhood leukaemia following conventional chemotherapy (Vyalushkina, 1973), suggesting that extramitochondrial dehydrogenases might be a preferential or at least more accessible target for cytotoxic drugs.

Dehydrogenase isoenzyme patterns in haemic cells

In an attempt to explore in greater detail possible differences in energy metabolism between normal and leukaemic cells, studies have been carried out on the dehydrogenase isoenzyme patterns in normal and various pathological conditions.

De Contardi and Navarro (1973) studied the distribution in normal and pathological cells of the five lactic dehydrogenase isoenzymes, distinguishable by their electrophoretic mobility.

In normal neutrophil granulocytes there is a predominance of fraction V, a finding which appears to be characteristic of fully mature and differentiated cells. Lymphocytes, instead, irrespective of whether they derive from peripheral blood, spleen or lymph nodes, are characterized by the presence of fraction III, a feature which seems also to be typical for immature cells.

Leucocytes obtained from patients with chronic myeloid leukaemia show an isoenzyme pattern similar to that of normal neutrophils, while lymphocytes from chronic lymphocytic leukaemia and blast cells from acute leukaemia are similar to normal lymphocytes, exhibiting a predominance of fraction III. In the course of blastic crisis in patients with chronic myeloid leukaemia, the isoenzyme pattern, previously typical of the neutrophil granulocyte, changes into that of the lymphocyte, characterized by the predominance of fraction III.

Dioguardi and his associates (1963, 1965, 1966) reported a predominance of fraction II in normal lymphocytes, while in their leukaemic counterparts and in myeloblasts the main isoenzyme component corresponded to fraction III.

In neoplastic tissues Goldman and Kaplan (1963) found an increase in isoenzyme V, a finding which differs from that reported in acute leukaemia cells, in which fraction III predominates. Dawson and Kaplan (1964) suggested that the increase in fraction V in neoplastic tissues was related to their anaerobic metabolism. However this latter fraction is absent from erythrocytes, cells which are characteristically anaerobic.

Investigations by Rabinowitz and Dietz (1968) and Bloom *et al.* (1967) have shown that lymphocytes stimulated by phytohaemagglutinin exhibit an increase in fraction V during the course of their process of transformation; this phenomenon is probably due to increased energy requirements of the cells prior to the process of mitotic division. The changes in isoenzyme pattern may therefore be related to the variable metabolic requirements of the cells rather than to the particular kind of metabolic behaviour of each different cell type.

Normal human erythrocytes also contain five isoenzymes of lactic dehydrogenase. Their variations in a number of haematological disorders have been studied by Starkweather *et al.* (1965), who also paid particular attention to the relationship of these changes to the dynamic state of erythropoiesis. LDH-5 was found exclusively in younger cells, while LDH-1 was the predominant fraction of aged erythrocytes. The apparent increase of LDH-1 in aged cells represents loss of activity of isoenzymes containing labile M subunits and not synthesis of H subunits.

In patients with myeloid metaplasia, Hodgkin's disease, chronic lymphocytic leukaemia and multiple myeloma, two phases of erythropoiesis could be distinguished. In one instance there was a suggestion of active erythropoiesis as reflected by an increase in LDH-5. In the other situation LDH-1 was increased and LDH-5 was absent, indicating lack of erythropoiesis. The continued elevation of LDH-5 in erythrocytes from patients with chronic erythraemic myelosis, throughout the disease, indicated persistent active erythropoiesis. However a gradual decrease of this fraction was noticed in patients with erythroleukaemia during the transitional phase of the disease from active erythropoiesis to the anaemia of the myeloblastic phase.

The increase in LDH-5 in haemolysates from patients with chronic myeloid leukaemia, in the absence of reticulocytosis, may be explained by the presence of nucleated erythrocytes in the peripheral blood.

The alteration of LDH-5 during chemotherapy directed at Hodgkin's disease or multiple myeloma, indicated that erythropoiesis reappeared in certain patients. The increase in LDH-5 was usually, but not always, associated with an increase of reticulocytes.

Data from patients with various anaemias further supported the hypothesis that LDH-5 and LDH-1 could be parameters reflecting the dynamic state of erythropoiesis, so that the analysis of erythrocyte LDH isoenzymes may have clinical value in the rapid assessment of erythropoietic activity.

Another dehydrogenase enzyme which has been the subject of a considerable number of investigations on the part of haematologists is glucose-6-phosphate dehydrogenase. Rozenszajn and Joseph (1974) studied the isozyme patterns of this enzyme in blood and bone marrow cells and in other human tissues. The study was carried out employing polyacrylamide gel disc electrophoresis.

Nucleated cells from bone marrow and spleen show the presence of five enzymatic bands (I to V), liver homogenates show four bands (II, III, IV and V), lymph nodes three bands (I, II and V) and brain homogenates exhibit two bands (I and II).

In all haemic cells bands I and II were found to be specific for glucose-6-phosphate dehydrogenase, while bands III, IV and V appeared to be due to hexose-6-phosphate dehydrogenase activity. No difference in glucose-6-phosphate dehydrogenase isozyme pattern was detected in peripheral blood leucocytes, platelets or bone marrow cells from healthy subjects and patients with glucose-6-phosphate dehydrogenase deficiency. Rozenszajn and Joseph believe that the different isozyme distribution of each tissue is probably an expression of the adaptation of the enzyme structure to the metabolic requirements of the different cells.

Recently, Morellini *et al.* (1985), employing biochemical methods, compared the glucose-6-phosphate dehydrogenase activity of red cells and different types of leucocytes in normal and enzyme deficient individuals. In normal subjects, erythrocytes and B lymphocytes showed similar levels of enzyme activity, whereas the activity of T lymphocytes appeared to be 4 times higher, that of monocytes 10 times higher, and that of granulocytes 30 times higher, compared to that of the former cell types. In enzyme deficient patients, erythrocytes were almost completely devoid of enzyme activity. In the same subjects, glucose-6-phosphate dehydrogenase activity of leucocytes was also reduced, but in a variable degree, compared to the values observed in the corresponding normal cell types, being 51% in B lymphocytes, 27% in T lymphocytes, 15% in monocytes and 22% in granulocytes.

Some differences have been reported by the same authors in normal, but not in enzyme deficient subjects, between the biochemical characteristics of the glucose-6-phosphate dehydrogenase of erythrocytes and that of leucocytes. The most striking difference concerned the affinity for glucose-6-phosphate, which was found twice as high in red cell enzyme compared to leucocyte enzyme. The authors suggest that this diversity might result from post-translational modifications of glucose-6-phosphate dehydrogenase, which do not take place at the same rate in enzyme-deficient patients.

Alternative approaches to dehydrogenase cytochemistry

Although from the foregoing pages it appears quite clear that dehydrogenase cytochemistry has reached a high level of sophistication, various problems still exist concerning especially the preservation of the quantities of enzymes in a native state inside the tissues and the localization of this enzyme activity maintained *in situ*. In relation to these technical aspects, Andersen and Høyer (1974) carried out a series of controlled experiments in order to study the effects of fixation, cryoprotection, osmolar protection and substantivity of formazans. Furthermore, in order to overcome the problem of the 'nothing dehydrogenase' reaction, the authors balanced against each other the concentrations of Nitro-BT, NADP and phenazine methosulphate (PMS) and reduced the incubation time to 10 minutes. The problem of controlling the diffusion of the different components operative in the histochemical dehydrogenase procedure was tackled by the authors by introducing a double section incubation method. The investigators stressed the importance of recording the appearance of the two formazans (rose, violet and blue) with the purpose of following the enzymatic reaction rate and the effects of different procedures as well as the rates of 'nothing dehydrogenase' reaction, diffusion of reduced coenzyme and/or reduced PMS.

Although various methods are available for the histochemical demonstration of dehydrogenases at the ultrastructural level, no attempt has yet been made to study their localization in haemic cells. The electron microscopic demonstration of dehydrogenases has been achieved by conventional tetrazolium methods (Baba and Sharma, 1971) as well as with the use of ferricyanide as an electron acceptor (Benkoël *et al.*, 1976). The various technical problems which have plagued tetrazolium methods have prompted the search for alternative methods in the EM demonstration of dehydrogenases.

For instance lactic dehydrogenase has been demonstrated (Baba *et al.*, 1976) by an immunoelectron microscope method, based on the production of anti-subunit B antibody obtained by immunizing rabbits with purified pig LDH-I(B_4) isoenzyme. Immunoglobulin fragment Fab was obtained from anti-LDH serum immunoglobulin G and conjugated with horseradish peroxidase. The LDH distribution observed in pig heart muscle with the immunoelectron microscope method was essentially similar to that obtained using thiocarbamylnitro-blue tetrazolium.

With the advent in 1957 of new tetrazolium salts, dehydrogenase histochemistry has thus advanced considerably. Instead of coarse ill-defined reactions we may now obtain precise intracellular localization of enzyme activity and in place of a single dehydrogenase and two diaphorases it is possible to demonstrate a spectrum of different specific dehydrogenases, such as those concerned with energy metabolism (Zurwehme *et al.*, 1971).

As we have just seen, the 1970's were characterized by the appearance of a great number of studies concerned with dehydrogenase cytochemistry; following this outburst of interest in the cytochemical demonstration of various dehydrogenase systems, which has provided information of great biological interest but has not exerted the expected positive repercussions on routine haematological practice, in recent years efforts have been mainly directed to improving the sensitivity and reliability of existing methods.

For instance, Van Noorden and Tas (1982) studied the role of exogenous electron carriers in NAD(P) dependent dehydrogenase cytochemistry. The authors investigated quantitatively the applicability of phenazine methosulphate, 1-methoxyphenazine methosulphate, menadione and meldola blue as exogenous electron carriers for the cytochemical staining of NAD(P)-dependent dehydrogenases, using tetranitro-BT in vitro and a model system of polyacrylamide films incorporating glucose-6-phosphate or intact rat liver parenchymal cells. Menadione did not appear to be useful, because electrons did not seem to be transferred directly from NAD(P)H to this compound. Phenazine methosulphate at higher concentrations and meldola blue at concentrations optimal for carrying electrons to tetrazolium salts yielded a high level of 'nothing dehydrogenase activity'. According to the authors, 1-methoxyphenazine methosulphate appears to be the exogenous electron carrier of choice. In 1983 Raap and Van Duijn studied the function of phenazine methosulphate (PMS) and established that PMS acts, not on its own, but as an efficient acceptor of NAD(P)H-oxidizing flavoproteins and thus increases the local NAD(P)H tetrazolium oxidoreductase activity. As escape of reducing equivalents from the site might also occur at the level of reduced PMS (PMSH), the solubility properties of PMSH were studied. It was found that PMSH has a low solubility in aqueous media and is hydrophobic.

In 1984 Van Noorden and Butcher compared the properties of the four most commonly employed tetrazolium salts, i.e. neotetrazolium, nitro-BT, tetranitro-BT and 2-(2′-benzothiazolyl)-3-(4′-phthalhydrazidyl)-5-styryl tetrazolium (BPST) on the localization of NADP-dependent dehydrogenases under optimal incubation conditions in cryostat sections of rat liver. Glucose-6-phosphate dehydrogenase was selected as an example of these dehydrogenases. It was found that the use of nitro-BT and tetranitro-BT, unlike neotetrazolium and BPST, in combination with an exogenous electron carrier and azide, resulted in localization patterns in agreement with the sites of activity, as determined by microchemical techniques. In the absence of a carrier, the localization was very similar to that of NADPH cytochrome c (P450) reductase.

More recently Van Noorden and Vogels (1985) have described a sensitive cytochemical staining method for glucose-6-phosphate dehydrogenase activity in individual erythrocytes. Many attempts have been made in the past (Fairbanks and Lampe, 1968; Rozenszajn *et al.*, 1972; Gordon and Stuart, 1974; Stuart *et al.*, 1975) to develop a reliable cytochemical method for glucose-6-phosphate dehydrogenase activity in erythrocytes in order to detect the most common congenital enzyme deficiency in man. The most widely employed technique has been that based on the use of a tetrazolium salt, but the relatively low amount of enzyme activity per red cell and the lack of accuracy in the localization of the coloured end-product within the cells has so far prevented the routine application of a suitably reliable cytochemical method.

The method reported by Van Noorden and Vogels (1985) is a further improvement of the staining procedure described by Van Noorden *et al.* in 1982. This technique in its optimal form consists in treating erythrocytes with sodium nitrite, and in fixation of the preparation with 0.025% glutaraldehyde (under $NADP^+$ protection of the active site of the enzyme), followed by incubation of the cells in suspension in the presence of tetranitro-BT, 1-methoxyphenazine methosulphate and polyvinyl alcohol. A sharp localization was obtained of glucose-6-phosphate dehydrogenase activity which enabled discrimination between red cells with different levels of enzyme activity. The improvements in the technique, introduced by Van Noorden and Vogels (1985), consisted in adopting a higher incubation temperature (46°C), higher concentrations of polyvinyl alcohol (35–45%), substrate (10 mM glucose-6-phosphate), coenzyme (0.8 mM $NADP^+$) and electron carrier (0.32 mM 1-methoxyphenazine methosulphate). With this technique, a sample from a normal person shows a distribution pattern of activity per red cell which is related to the age of the cells. Young erythrocytes usually display high activity, whereas during ageing these values slowly decrease to undetectable levels. In a blood sample of a heterozygote for glucose-6-phosphate dehydrogenase deficiency, two different populations of red cells can be observed, one with virtually no activity and one with a range of activity comparable with that of cells from healthy persons. The vast majority (78–96%) of cells from homo- or hemizygous patients show no activity, although there are always a few cells present with some degree of activity. These, presumably young, cells may be responsible for the minimal activity measured by spectrophotometric assays in haemolysates of such patients.

Since the method is based on the microscopic analysis of stained individual red blood cells, the various forms of the disease may be well distinguished. Moreover, the overlap of glucose-6-phosphate activity observed with spectrophotometric assays in haemolysates from control subjects and heterozygote patients does not occur in the cytochemical test, which is therefore the first useful tool obtained from dehydrogenase cytochemistry for routine diagnostic purposes.

TECHNICAL METHODS

Cytochemical demonstration of dehydrogenases

Stuart and Simpson, 1970

1. Air-dry fairly dilute bone marrow or buffy coat peripheral blood smears at room temperature for approximately 60 minutes
2. Attach a perspex ring (1.5 × 0.5 cm) by a thin film of paraffin wax to the area proximal to the tail of the film

3. Fill the ring with the following freshly prepared incubation medium:

Buffer glycylglycine	0.05 mol/l pH 7.6
Nitro-BT salt	1 mg/ml buffer
Calcium salt $CaCl_2$ $2H_2O$	13 mg/ml buffer
Inert polymer, Ficoll	400 mg/ml buffer
Coenzyme NAD or NADP	1 mg/ml buffer
Substrate	0.05–0.1 mol/l

(0.005 mol/l for enzymes of the pentose phosphate shunt)

The final pH is 7.3

4. Carry out incubation for 90 minutes in an atmosphere of nitrogen in a 37°C humidity incubation box
5. After incubation, remove ring and tip off solution
6. Terminate reaction by fixing films in 40% formalin for 10 minutes
7. Counterstain with 2% aqueous methyl green for 10 minutes.

Remarks

Air-dried bone marrow or peripheral blood films may be fixed in 60% acetone at room temperature for 60 seconds.

Incubation solution is the same as above, but with the omission of the inert sucrose polymer Ficoll.

Cell preservation with acetone fixation is better but enzyme inhibition is greater, especially for those enzymes with weak activity, e.g. glucose-6-phosphate dehydrogenase.

Cytochemical demonstration of glucose-6-phosphate dehydrogenase in red cells

Van Noorden and Vogels, 1985

1. Following collection in tubes containing lithium heparin, add 1 ml of whole blood to 9 ml of a freshly prepared solution of sodium nitrite (180 mmol/l in normal saline) and incubate for 8 min at room temperature
2. After centrifugation (15 min at 1000 *g* at room temperature) mix 0.075 ml of packed red cells with 0.125 ml of $NADP^+$ solution ($NADP^+$ 20 mmol/l in 0.1 M phosphate buffer pH 7), then incubate for 10 min at room temperature
3. Following centrifugation (15 min at 1000 *g*), discard the supernatant and add 2 ml of 0.025% glutaraldehyde in phosphate buffer to the packed cells. During fixation for 30 min at room temperature, continuous rotation of the tubes is recommended
4. Wash cells three times for 1 min with cold phosphate buffer. Each washing is followed by rapid centrifugation at 2000 *g*. Cells are subsequently resuspended in 0.5 ml of buffer
5. Add 0.1 ml of red cell suspension to 1 ml of the following incubation medium

Glucose-6-phosphate	10 mmol/l
$NADP^+$	0.8 mmol/l

$MgCl_2$	4 mmol/l
1-methoxyphenazine methosulphate	0.32 mmol/l
Sodium azide	5 mmol/l
Tetranitro-BT	1 mmol/l

in 0.1 M phosphate buffer pH 7.

Prior to incubation add 35–45% (w/v) polyvinyl alcohol to the above medium. Incubate for 90 min at 46°C in the dark, under continuous rotation

6. Stop the reaction by adding 2.5 ml of ice-cold buffer, then wash the cells three times with cold buffer as in step 4. Finally, resuspend washed red cells in 1 ml of phosphate buffer

7. Transfer 3 microlitres of cell suspension onto a clean glass slide, and allow to air-dry for 30 min at 37°C prior to mounting in water-free glycerol.

Remarks

Pre-treatment with sodium nitrite produces the conversion of all oxyhaemoglobin into methaemoglobin, thus avoiding non-specific formazan production, while pre-incubation with $NADP^+$ is required in order to prevent as much as possible the inactivation of glucose-6-phosphate dehydrogenase activity by fixation with glutaraldehyde.

REFERENCES

Ackerman G A 1960 Histochemical demonstration of dehydrogenase activity in the cells of normal human blood and bone marrow. Journal of Biophysical and Biochemical Cytology 8: 61

Andersen H, Høyer P E 1974 Simplified control experiments in the histochemical study of coenzyme-linked dehydrogenase. Histochemistry 38: 71

Antopol W, Glaubach S, Goldman L 1948 Effects of a new tetrazolium derivative on tissue, bacteria and onion root tips. Public Health Report, Washington 63: 1231

Atkinson E, Melvin S, Fox S W 1950 Some properties of 2, 3, 5 triphenyltetrazolium chloride and several iodo derivatives. Science, 111: 385

Baba N, Sharma H M 1971 Histochemistry of lactic dehydrogenase of heart and pectoralis muscle of rat. Journal of Cell Biology 51: 621

Baba N, Hurtubise P, Myser T 1976 Immunofluorescence and immunoperoxidase observations of anti-lactic dehydrogenase-1 antibody. Journal of Histochemistry and Cytochemistry 24: 572

Balogh K, Cohen R B 1961 Histochemical demonstration of diaphorases and dehydrogenases in normal human leucocytes and platelets. Blood 17: 491

Benkoël L, Chamlian A, Barrat E, Laffargue P 1976 The use of ferricyanide for the electron microscopic demonstration of dehydrogenases in human steroidogenic cells. Journal of Histochemistry and Cytochemistry 24: 1194

Bessis M 1965 Cytology of the blood and blood-forming organs. Grune & Stratton, New York

Block J B, Bronson W R, Bell W R 1966 Metabolic abnormalities of lactic acid in Burkitt-type lymphoma with malignant effusions. Annals of Internal Medicine 65: 101

Bloom A D, Tsuchioka M, Wajima T 1967 Lactic dehydrogenase and metabolism of human leukocytes 'in vitro'. Science 156: 979

Borgers M, Verheyen A 1985 Enzyme cytochemistry. International Review of Cytology 95: 163

Castoldi G L, Yam L T, Mitus W J 1968 Cytochemical demonstration of dihydroorotic dehydrogenase in blood and bone marrow cells. Acta Haematologica 39: 203

Cooper E H, Hughes D T, Topping N E 1966 Kinetics and chromosome analyses of tissue culture lines derived from Burkitt lymphomata. British Journal of Cancer 20: 102

Dawson D M, Kaplan N D 1964 Lactic dehydrogenase, functions of the two types. Science 14: 929

De Contardi N W, Navarro E 1973 Estudio bioquimico del comportamiento de las cinco isoenzimas

de la lactato dehidrogenasa en leucocitos normales y patològicos. Sangre 18: 135
De Souza E J, Kothare S N 1959 A method for the cytochemical demonstration of succinic dehydrogenase in human leucocytes. Journal of Histochemistry and Cytochemistry 7: 77
Dioguardi N, Agostoni A 1965 Multiple Molekularformen der Lactatdehydrogenase normaler und leukämischer menschlichen Lymphozyten. Enzymologia Biologia et Clinica 5: 3
Dioguardi N, Agostoni A 1966 Lactic dehydrogenase, malic dehydrogenase, and G. U. transaminase in lymphocytes of thymus, fetal spleen, adult spleen and peripheral blood. Enzymologia Biologia et Clinica 6: 324
Dioguardi N, Agostoni A, Fiorelli G, Lamento B 1963 Characterization of lactic dehydrogenase of normal human granulocytes. Journal of Laboratory Medicine 61: 713
Epstein M A, Barr Y M 1965 Characteristics and mode of growth of a tissue culture strain (EB) of human lymphoblasts from Burkitt's lymphoma. Journal of the National Cancer Institute 34: 231
Fairbanks V F, Lampe L T 1968 A tetrazolium-linked cytochemical method for estimation of glucose-6-phosphate dehydrogenase activity in individual erythrocytes: applications in the study of heterozygotes for glucose-6-phosphate deficiency. Blood 31: 589.
Farber E, Louviere C D 1956 Histochemical localization of specific oxidative enzymes. IV. Soluble oxidation-reduction dyes as aids in the histochemical localization of oxidative enzymes with tetrazolium salts. Journal of Histochemistry and Cytochemistry 4: 347
Farber E, Sternberg W H, Dunlap C E 1956 Histochemical localization of specific oxidative enzymes. III. Evaluation of tetrazolium staining methods for DPN diaphorase, TPN diaphorase and the succinic dehydrogenase system. Journal of Histochemistry and Cytochemistry 4: 254
Field M, Block J B, Levin R, Rall D P 1966 Significance of blood lactate elevations among patients with acute leukaemia and other neoplastic proliferative disorders. American Journal of Medicine 40: 528
Foadi M D, Cooper E H, Hardisty R M 1968 Proliferative activity of leukaemic cells at various stages of acute leukaemia of childhood. British Journal of Haematology 15: 269
Freeman A I, Journey L J 1971 Ultrastructural studies on monocytic leukaemia. British Journal of Haematology 20: 225
Gavosto F, Pileri A, Bachi C, Pegoraro L 1964 Proliferation and maturation defect in acute leukaemia cells. Nature 203: 92
Goldman R D, Kaplan N D 1963 Alterations of tissue lactate dehydrogenase in human neoplasma. Biochimica et Biophysica Acta 77: 518
Gordon P A, Stuart J 1974 Red cell cytochemistry in G-6-PD deficiency. British Journal of Haematology 27: 358
Graham R C Jr, Karnovsky M J 1966 The early stages of absorption of injected horseradish peroxidase in the proximal tubules of mouse kidney: ultrastructural cytochemistry by a new technique. Journal of Histochemistry and Cytochemistry 24: 291
Hess R, Scarpelli D G, Pearse A G E 1958 Cytochemical localization of pyridine nucleotide-linked dehydrogenases. Nature 181: 1531
Kakefuda T 1968 Electron microscopy of normal and leukemic cells. In: Amromin G D. Pathology of leukemia. Hoeber, New York
Lojda Z, Gossrau R, Schiebler T H 1979 Enzyme histochemistry. Springer-Verlag, Berlin
Mann J R, Simpson J S, McEvoy M W, Stuart J 1972 Selective in vitro cytotoxicity of a glycolytic enzyme inhibitor for immature lymphoid cells. British Journal of Haematology 23: 89
Mauer A M, Fischer V 1962 Comparison of the proliferative capacity of acute leukaemia cells in bone marrow and blood. Nature 193: 1085
Merker E R H, Castoldi G L, Linde A H 1965 Cholindehydrogenase in den Phagocyten des Blutes und des entzündlichen Exsudaten. Schweizerische Medizinische Wochenschrift 95: 1446
Morellini M, Colonna-Romano S, Meloni T, Battistuzzi G, Gandini E 1985 Glucose-6-phosphate dehydrogenase of leukocyte subpopulations in normal and enzyme deficient individuals. Haematologica 70: 390
Novikoff A B, Shin W Y, Drucker J 1960 Cold acetone fixation for enzyme localization in frozen sections. Journal of Histochemistry and Cytochemistry 8: 37
Pearse A G E 1957 Intracellular localization of dehydrogenase system using monotetrazolium salt and metal chelation of their formazan. Journal of Histochemistry and Cytochemistry 5: 515
Pearse A G E 1960 Histochemistry, theoretical and applied. Churchill, London
Pease D C 1956 An electron microscopic study of red bone marrow. Blood 11: 501
Quaglino D 1961 Aspects of dehydrogenase cytochemistry. In Proc. of the 8th Congress of the European Society of Haematology, Wien
Quaglino D, Hayhoe F G J 1960 Acetone fixation for the cytochemical demonstration of the dehydrogenases in blood and bone marrow cells. Nature 187: 85
Quaglino D, Hayhoe F G J, Flemans R J 1962 Cytochemical observations on the effect of phytohaemagglutinin in short-term tissue cultures. Nature 196: 338

Raap A K, Van Duijn P 1983 Studies on the phenazine methosulphate-tetrazolium capture reaction in NAD(P)$^+$-dependent dehydrogenase cytochemistry. II. A novel hypothesis for the mode of action of PMS and a study of the properties of reduced PMS. Histochemical Journal 15: 881
Rabinowitz J, Dietz A 1968 Effect of phytohaemagglutinin in cultures on the lactate dehydrogenases of lymphocytes from chronic lymphatic leukemia. Blood 31: 166
Roth G J, Porte D Jr 1970 Chronic lactic acidosis and acute leukemia. Archives of Internal Medicine 125: 317
Rozenszajn L A, Joseph D 1974 The isozyme patterns of glucose-6-phosphate dehydrogenase in blood cells, bone marrow and other human tissues. Biochimica et Biophysica Acta 364: 38
Rozenszajn L, Shoham D 1967 The demonstration of dehydrogenases and diaphorases in cells of peripheral blood and bone marrow. Blood 29: 737
Rozenszajn L A, Shoham D, Menashi T 1972 Evaluation of glucose-6-phosphate dehydrogenase in single erythrocytes in human blood smears. Acta Haematologica 47: 303
Rutenburg A M, Gofstein R, Seligman A M 1950 Preparation of a new tetrazolium salt which yields a blue pigment on reduction and its use in the demonstration of enzymes in normal and neoplastic tissues. Cancer Research 10: 113
Seligman A M, Gofstein R, Rutenburg A M 1949 Preparation of a radioactive iodotetrazolium salt and its distribution in mice. Cancer Research 9: 366
Smith L H, Baker F A 1959 Pyrimidine metabolism in man. I. The biosynthesis of orotic acid. Journal of Clinical Investigation 38: 1044
Starkweather W H, Cousineau L, Schoch H K, Zarafonetis C J 1965 Alterations of erythrocyte lactate dehydrogenase in man. Blood 26: 63
Sternberg W H, Farber E, Dunlap C E 1956 Histochemical localization of specific oxidative enzymes. II. Localization of DPN and TPN diaphorases and the succindehydrogenase system in the kidney. Journal of Histochemistry and Cytochemistry 4: 266
Stuart J, Simpson J S 1970 Dehydrogenase enzyme cytochemistry of unfixed leucocytes. Journal of Clinical Pathology. 23: 517
Stuart J, Simpson J S, Mann J R 1970 Intracellular hydrogen transport systems in acute leukaemia. British Journal of Haematology 19: 739
Stuart J, Simpson J S, Wallin J M 1972 Glycolytic and respiratory enzyme activity in relation to leucocyte maturation. British Journal of Haematology 23: 79
Stuart J, Skowron P N 1968 A cytochemical study of marrow enzymes in megaloblastic anaemia. British Journal of Haematology 15: 443
Stuart J, Gordon P A, Lee T R 1975 Enzyme cytochemistry of blood and marrow cells. Histochemical Journal 7: 1975
Tsou K C, Cheng C S, Nachlas M M, Seligman A M 1956 Synthesis of some p-nitrophenyl substituted tetrazolium salts as electron acceptors for the demonstration of dehydrogenases. Journal of the American Chemical Society 78: 6139
Van Noorden C J F, Butcher R G 1984 Histochemical localization of NADP-dependent dehydrogenase activity with four different tetrazolium salts. Journal of Histochemistry and Cytochemistry 32: 998
Van Noorden C J F, Tas J 1982 The role of exogenous electron carriers in NAD(P)-dependent dehydrogenase cytochemistry studied in vitro and with a model system of polyacrylamide films. Journal of Histochemistry and Cytochemistry 30: 12
Van Noorden C J F, Vogels I M C 1985 A sensitive cytochemical staining method for glucose-6-phosphate dehydrogenase activity in individual erythrocytes. British Journal of Haematology 60: 57
Van Noorden C J F, Vogels I M C, James J, Tas J 1982 A sensitive cytochemical staining method for glucose-6-phosphate dehydrogenase activity in individual erythrocytes. Histochemistry 75: 293
Vyalushkina M D 1973 Changes in the spectrum of leukemic cell dehydrogenase in children with acute leukemia at the beginning of treatment. Pediatriya 8: 51
Wachstein M 1950 Histochemical demonstration of reducing activity in normal and leukemic blood and bone marrow cells. Proceedings of the Society for Experimental Biology 73: 306
Willighagen R G J, Planteydt H T 1959 Aminopeptidase activity in cancer cells. Nature 183: 263
Wulff H R 1963 Histochemical studies of leukocytes from an inflammatory exudate. Acta Haematologica 29: 208
Zimmermann H, Pearse A G E 1959 Limitations in the histochemical demonstration of pyridine nucleotide-linked dehydrogenases ('Nothing dehydrogenase'). Journal of Histochemistry and Cytochemistry 7: 271
Zurwehme D, Mitzkat H J, Dommes B 1971 Enzyme cytochemistry of the energy metabolism in the cells of normal human bone marrow. Acta Haematologica 46: 24

CHAPTER 10

Oxidases, peroxidases and catalases

Oxidases are enzymes which catalyze oxidation by molecular oxygen, while peroxidases perform the same function by hydrogen peroxide.

Oxidases

Cytochrome oxidase is the best characterized of the three major inner mitochondrial-membrane-bound enzyme complexes in the pathway from NADH to oxygen. The first complex (NADH dehydrogenase complex) transfers electrons from NADH to ubiquinone. The b-c_1 complex, which contains two cytochromes, subsequently transports electrons from ubiquinone to cytochrome c.

The cytochrome oxidase complex, which comprises the cytochromes a and a_3 and is sensitive to cyanide inhibition, accepts four electrons from cytochrome c and ultimately transfers them to a single-bound O_2 molecule, thereby forming two water molecules.

The components of this system are widely distributed in the cells of various organisms, often in quantities too small to be detected by histochemical methods. The cytochrome oxidase reaction is estimated to account for 90% of the total oxygen uptake in most cells.

However, as far as haemic cells are concerned, the results of recent investigations suggest the existence of another oxidase system, which is dependent on NADH or NADPH as substrate, is present on the plasma membrane of neutrophils and monocytes (Briggs *et al.*, 1975; Morel and Vignais, 1984) and involves a b-type cytochrome. This cytochrome b is incorporated into the membrane of phagocytic vacuoles and is absent, or present in an aberrant form, in the neutrophils and monocytes from some patients with chronic granulomatous disease (Andrews and Babior, 1983). The NAD(P)H oxidase activity has been interpreted as the terminal enzyme of the cyanide resistant respiration of phagocytosing neutrophils and monocytes. In fact, during the 'respiratory burst' of phagocytic cells, a superoxide (O_2^-) generating oxidase system is activated that plays a central role in the aerobic microbicidal activity of these cells (Babior, 1978; Gabig and Babior, 1979; Badwey and Karnovsky, 1980). The conversion of oxygen to superoxide anion is accompanied by H_2O_2 production and increased glucose consumption via the hexose-monophosphate shunt (Iyer *et al.*, 1961).

Oxidase activity is conspicuously present in human neutrophils and monocytes stimulated with phorbol myristate acetate (PMA), while in the dimethyl sulphoxide-induced granulocytic differentiation of the human HL-60 leukaemia cell line there is a correlation between the capacity for showing a respiratory burst in response to PMA stimulation and the appearance of cytochrome b (Roberts *et al.*, 1982). O_2 generating activity has been found present in particulate preparations from granulocytes and monocytes that had been previously stimulated with PMA, but was low or absent in control preparations from unstimulated cells and in stimulated monocytes from a patient with chronic granulomatous disease (Chaudhry *et al.*, 1982). In close similarity to the O_2^- generating oxidase system from neutrophils, the superoxide generation in the preparations from stimulated normal monocytes was linearly proportional to cell protein concentration, insensitive to inhibition by azide and dependent on NADPH as substrate. However, since the oxidase activity of monocytes could not be revealed by NADH and showed a lower affinity for NADPH in comparison with the activity of granulocytes, structural differences between the two oxidase systems cannot be ruled out.

The results of all these investigations point to the existence of a fairly complex oxidase system operating in some haemic cells, part of which appears to vary considerably in relation to functional activity, cell activation and extent of differentiation.

All these facts, taken together, might explain some of the difficulties, which have been encountered in the past in detecting cytochemically oxidase activities, and why very little progress has been made until a few years ago in defining their exact distribution within cells and tissues, despite the antiquity of some of the techniques used for their demonstration (Pearse, 1960).

Cytochemical methods for oxidases

A considerable number of studies carried out on malignant and precancerous tissues (Schneider and Potter, 1943a, b; Carruthers and Suntzeff, 1947, Gallico, 1947) have indicated a gradual increase in one or other of the components of the succinic oxidase system with the onset of malignant changes, hence the considerable interest inherent in the study of enzymatic variations in neoplastic tissues, as well as in leukaemias and lymphomas.

Cytochrome oxidase (EC 1.9.3.1) has also been designated as indophenol oxidase. It has long been known (Ehrlich, 1885) that an enzyme was present in the majority of tissues which would catalyze the oxidative reaction between α-naphthol and dimethyl-p-phenylenediamine to form indophenol blue (the 'Nadi' reaction, so called because it involves a naphthol with a diamine). The name indophenol oxidase was introduced for the endogenous oxidising enzyme present in most fresh tissues, and the reaction which it catalysed was referred to as the G-Nadi reaction, after the German word Gewebe, tissue. This reaction was inhibited by acid, alkali, formalin and heat and required the presence of oxygen. A parallel but possibly non-enzymatic production of indophenol was also observed and studied especially in myeloid cells, hence being called the M-Nadi reaction. This was not inhibited by formalin, acetone, alcohol, or alkalis, was relatively heat-stable, and had an optimum pH of 12.5 to 13. Gomori (1952) took the view that the M-Nadi reaction could be adequately explained by the presence of fat peroxides in myeloid granules but Lillie and Burtner (1953) produced evidence linking the M-Nadi reaction with benzidine peroxidase (see below). As for the G-Nadi reaction, Keilin and Hartree (1938) showed that indophenol oxidase would catalyze the oxidation of reduced cytochrome, and that cytochrome oxidase would produce indophenol blue from naphthol and diamine. They concluded that indophenol oxidase, Atmungsferment and cytochrome oxidase were one and the same enzyme. It was therefore clear that the G-Nadi reaction indicated the activity of the cytochrome c-cytochrome oxidase system.

Cyanide, sodium azide and hydrogen sulphide inhibit cytochrome oxidase and prevent the formation of indophenol blue. Since endogenous cytochrome c present in tissues is not always sufficient for maximum oxidation rate, addition of cytochrome c to the incubation mixture may increase the rate of dye formation.

However, the instability of the reaction, the rapid fading of the preparations, the lipid solubility of the final reaction dye product, the coarse granular or crystalline nature of the dye, the enzyme toxicity of dimethyl-p-phenylenediamine and finally the tendency to spontaneous oxidation in the reaction, have limited its successful application, even with the most appropriate modifications (Barka and Anderson, 1963).

Crawford and Nachlas (1958) and Nachlas *et al.* (1958) have shown that substitution of the original reagent by 4-amino-1-NN-dimethylnaphthylamine considerably improves the technique, although the rate of pigment formation is slower than with dimethyl-p-phenylenediamine. The oxidative coupling of α-naphthol and 4-amino-1-NN-dimethylnaphthylamine results in the formation of indophenol purple, which is deposited as small, well localized intracellular granules. Indophenol purple is soluble in lipids and has low substantivity for proteins. The reaction mixture is relatively stable and shows no auto-oxidation within one or more hours after preparation.

In 1959 Burstone, in an endeavour to improve the original Nadi reaction, introduced a group of complex naphthols (such as 1-hydroxy-2-aceto-naphthone) and open chain and cyclic methylene compounds in conjunction with p-aminodiphenylenediamine or 2 methoxy-N-phenyl-p-phenylenediamine.

In 1960, in a further attempt to improve cytochrome oxidase localization, Burstone employed a series of coloured compounds, related to indoaniline, azomethine and indamine-azine, which are subsequently chelated to a metal in order to obtain better preservation of the preparations.

The subsequent search for still other compounds with desirable properties of chromogeneity and substantivity (Burstone, 1961) showed that the combination of p-aminodiphenylamine with 8-amino-1,2,3,4, tetrahydroquinoline or 8-hydroxy-1-4 naphthoquinone provided satisfactory localization of cytochrome oxidase.

Substantial progress in cytochrome oxidase cytochemistry was obtained by the introduction of a procedure based on the oxidation of 3,3′ diaminobenzidine (DAB), which allows detection of the enzyme at the ultrastructural level (Seligman *et al.*, 1968). Because DAB reacts with the oxidized form of cytochrome c, the reduced form of which can then be reoxidized by cytochrome oxidase and oxygen, the DAB reaction product localizes cytochrome c. Because of the involvement of cytochrome oxidase, it can also be said that the reaction demonstrates cytochrome oxidase activity (Roels, 1974).

More recently (Bedetti, 1985) an immunocytochemical method has been devised for the demonstration of cytochrome c oxidase. The immunoperoxidase method described depends on the use of a specific rabbit antiserum against bovine heart cytochrome c oxidase, following which mitochondria are specifically stained in sections of formalin-fixed, paraffin-embedded human tissues. The intensity of the immunostaining in different human tissues correlates mainly with the number of mitochondria and abundance of mitochrondrial cristae, i.e. surface area of the inner mitochondrial membrane. Mitochondrion-rich cells, such as myocardial and gastric parietal cells have very strong cytoplasmic staining. According to the author the technique may be especially useful for the identification of oncocytes, characterized by an unusually large number of mitochondria.

Briggs *et al.* in 1975 introduced a new ultrastructural cytochemical technique, based on the use of cerium, to study the cytochemical distribution of NADH oxidase in human neutrophil granulocytes which were phagocytically stimulated. Cerium ions react with H_2O_2 and precipitate as cerium perhydroxide. The electron-dense cerium reaction product was found on the cell surface and on the inner surface of the phagosome membrane. The results obtained suggested that NADH oxidase is associated with the plasmalemma of human neutrophil granulocytes and that the enzyme enters the phagosome along with the plasmalemma surrounding the particle.

Attempts were made in 1976 by Curnutte *et al.* to develop a cytochemical method for localizing sites of superoxide anion production, following stimulation of oxidative metabolism. The method, a modified DAB reaction, utilizes the oxidative activity of Mn^+ to oxidize DAB, forming an insoluble reaction product: endogenous peroxidase activity is blocked by including sodium azide in the medium. In PMA-stimulated neutrophil granulocytes reaction product for superoxide anion was consistently found within surface-derived vesicles, and in some cases on the cell surface.

Using the cerium technique, Robinson *et al.* (1978) studied the cytochemical activity of D-amino acid oxidase (D-AAO, EC 1.4.3.3). Enzyme activity was restricted to the plasmalemma and phagosome membrane of cells which had phagocytosed latex beads and were then incubated with exogenous D-amino acids.

D-AAO activity was also localized by ferricyanide reduction reaction (see below) which gave essentially the same localization as did the cerium method.

In addition to NADH oxidase and D-AAO, aldehyde oxidase (Robinson *et al.*, 1979) has been detected cytochemically in stimulated guinea-pig neutrophil granulocytes but not in human neutrophils, using the ferricyanide-type reaction, originally introduced in cytochemistry to study cholinesterase activity (Karnovsky and Roots, 1964). In principle the reaction works as follows

1. substrate + ferricyanide $\xrightarrow{\text{enzyme}}$ oxidized product + ferrocyanide
2. Cu^{++} (chelated) + ferrocyanide $\longrightarrow$ Cu^{++} − ferrocyanide

DL-dihydroxyphenyl alanine employed as substrate reveals another oxidase activity (EC 1.14.18.1), which is capable of oxidising mono- and diphenols, thus producing a blackish-brown stained quinone. The enzyme takes part in the synthesis of melanin, and therefore its histochemical demonstration has been mainly confined for a long time to dermatological investigations (Becker *et al.*, 1935). Only recently a considerable degree of phenoloxidase positivity has been described also in circulating haemic cells by Schmidt (1981). According to this author the different sensitivity to methanol fixation distinguishes two types of enzyme activity, which are unevenly distributed in different blood cell types. However, these findings still await further confirmation.

Distribution of oxidase activity in haemic cells

Attempts to demonstrate cytochrome oxidase in haemic cells have been hampered in the past by the capriciousness of the 'Nadi' reaction, by the fact that much oxidase activity is lost in the preparation by fixation of slides and by the rapid diffusion of the dye.

Notwithstanding these drawbacks, Hoffman *et al.* (1951) carried out an extensive study on cytochrome oxidase activity in lymphoid and myeloid cells and circumvented the difficulties inherent in the demonstration of the enzyme by applying the reaction to free cells in wet smears.

Hoffman and his colleagues studied cells from normal and diseased lymph nodes, cells from normal spleens and spleens of patients with Hodgkin's disease, non-Hodgkin's lymphoma and myeloid metaplasia and also cells from the blood and bone marrow of normal subjects and patients with different types of leukaemia.

The number of stained granules varied considerably in the different cell types. Cells of cardiac muscle, liver and kidney were loaded with granules, while the various cells of blood and lymphoid tissue contained comparatively few granules. Lymphocytes from normal lymph nodes and blood and from the nodes and blood of patients with Hodgkin's disease, leukaemia, non-Hodgkin's lymphoma and tuberculous lymphadenitis contained from 5 to 20 round, deep blue cytoplasmic granules and no appreciable quantitative difference in the staining reaction could be detected between normal and pathological cells.

Myelocytes showed considerably more activity than mature neutrophil polymorphs and no difference was observed between leukaemic granuloblasts and their normal counterparts. Monocytes from blood, lymph nodes and spleen contained numerous granules, the activity being more intense than that in lymphocytes. Megakaryocytes and platelets did not show any activity. The cytochrome oxidase activity in Reed-Sternberg cells from lymph nodes varied considerably. In some there was no staining at all, while in others the cytoplasm was loaded with granules.

Another study on cytochrome oxidase activity in haemic cells was carried out by Quaglino *et al.* (1965) using a modification of Burstone's technique (1961), based on the use of the substrate 8-amino-1,2,3,4 tetrahydroquinoline and 60% acetone fixation.

The number, size and distribution of the dye granules in the cytoplasm were similar to those described by Hoffman and his colleagues and would be in keeping with an intramitochondrial localization of the enzyme.

In the granulocytic series the reaction showed a decreasing intensity with progression of maturation, the reaction being characterized in neutrophil polymorphs by granules less coarse than in primitive myeloid cells. Lymphocytes showed moderate positivity, while monocytes showed the strongest reaction. Numerous coarse granules were present in normal and myelomatous plasma cells (Figs 10.1 and 10.2).

No substantial difference could be observed between normal and leukaemic cells with the exception of some untreated cases of chronic myeloid leukaemia, which showed intense cytoplasmic positivity. This behaviour is similar to that of peroxidases (see below) and is certainly not due to peroxidase interference, because of the inclusion in the incubation mixture of catalase and because of the entirely different distribution of the two enzymes. A considerable decrease in cytochrome oxidase activity was observed in some cases of chronic myeloid leukaemia, treated with 6-mercaptopurine. This observation may find some support in the reported decrease of oxygen consumption of human leukaemic cells, following treatment with 6-mercaptopurine (Vaccari and Traldi, 1956) and by the inhibition of oxygen uptake by mouse leukaemic cells treated by purine antimetabolites (Woodliff, 1963).

In conclusion, leukaemic cells and those from lymph nodes affected by different pathological conditions contain about the same degree of cytochrome oxidase activity as their normal counterparts. The low enzyme activity in haemic cells, which seems to be further depressed by the use of purine antimetabolites in leukaemia, may indicate that sources of energy other than that derived from the oxidation of carbohydrates are available in these cells.

Cytochrome oxidase activity, employing Burstone's technique based on the use of the substrates naphthol and N,N-diphenyl-1,4-phenylendiamine, has been applied by Ciplea and Mayer (1981) to assess in mice potential bone marrow damage from therapeutic substances. The authors used a special technique consisting of bone marrow extracts from experimental mice embedded in the kidney of control animals. Microphotometric determination of cytochrome oxidase activity showed strong reduction after cyclophosphamide treatment and moderate, but still evident reduction, following the use of chlorambucil, as compared to controls.

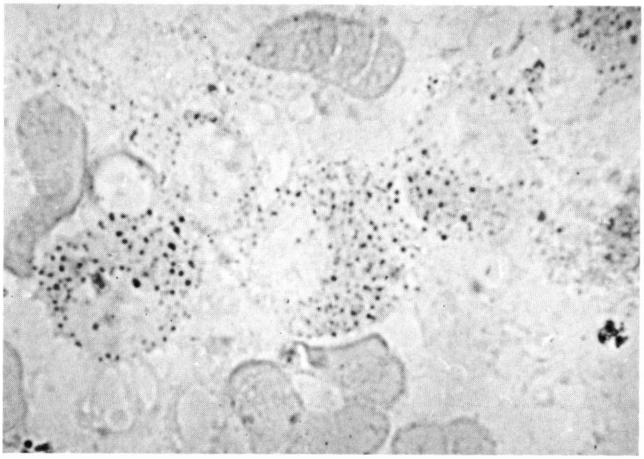

Fig. 10.1 Multiple myeloma bone marrow. Cytochrome oxidase. Plasma cells contain numerous coarse granules.

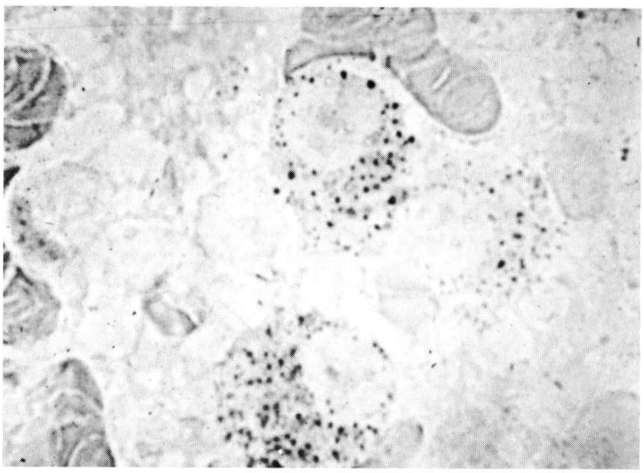

Fig. 10.2 The same preparation. Another field showing positive myeloma cells with various surrounding marrow cells of uncertain identity but weak or negative oxidase reaction.

Peroxidases (EC 1.11.1.7.)

These enzymes are particularly characteristic of plants and bacteria, although they are present in moderate amounts in the mammary gland and in the granules of leucocytes of the myeloid series and occasionally in the cytoplasm of monocytes.

Peroxidases are haemoprotein enzymes, since they consist of a protein moiety attached to a prosthetic group, formed by an iron porphyrin complex. From the functional point of view, peroxidases are enzymes which, in the presence of hydrogen peroxide, catalyze the oxidation of different specific substrates, including phenols, several amino acids and some aromatic acids, with a reaction of the type: $AH_2 + H_2O_2 \rightarrow A + 2H_2O$. Thus the peroxide peroxidase system causes the rapid oxidation of substances which peroxide alone oxidises extremely slowly (Pearse, 1960).

A peroxidase enzyme was isolated by Agner in 1941 from leucocytes. He called the enzyme myeloperoxidase or verdoperoxidase. Lillie and Burtner (1953) showed that this enzyme was identical with M-Nadi oxidase.

Sehrt (1927), Lison (1936), Gomori (1951) suggested that the peroxidase reaction was attributable to fatty acid peroxides and not to enzymatic activity. Similar conclusions were put forward by Girelli and Panagia (1954), who stated that the peroxidase reaction of neutrophil polymorphs was only partly due to Agner's verdoperoxidase, while most was caused by a non-catalytic activity of 'pseudo peroxidases'. These they identified as iron porphyrin complexes and inorganic iron salts, which would also give a positive M-Nadi oxidase reaction.

These conflicting views prompted Lillie and Burtner (1953) to undertake a careful study of the relationship between sudanophilia and peroxidase and oxidase activity. As a result of their extensive experiments, these authors concluded that the indophenol oxidase and benzidine peroxidase were essentially similar and they confirmed Agner's contention that the indophenol reaction was accomplished by the myeloperoxidase isolated from leucocytes. Lillie and Burtner also claimed that the indophenol oxidase and the benzidine peroxidase behaved dissimilarly from the sudanophilic substance in leucocytes. This, in the opinion of the authors, indicated that it was unlikely that this latter substance was concerned in the indophenol synthesis.

Several cytochemical methods are available for the demonstration of peroxidases. The standard peroxidase procedure, used in haematology since the early years of the century, involved the use of benzidine as substrate (Plate 22A,B). Following the demonstration of its carcinogenic effects, other substrates were introduced, such as o-tolidine (Quaglino and Flemans, 1958) (Plate 22C), which proved equally carcinogenic, and others such as 3,3′ diaminobenzidine (DAB) (Graham and Karnovsky, 1966) (Plate 22D) and 3-amino-9-ethyl carbazole (3AC) (Kaplow, 1975) (Plate 22E) which might represent a potential hazard.

Various other benzidine substitutes have been from time to time proposed by different groups of authors (Inagaki *et al.*, 1975; Kaplow, 1975; Papadimitriou *et al.*, 1976; Hanker *et al.*, 1977; Elias, 1980; Lippi, *et al* 1983). 3,3′ Diaminobenzidine (Graham and Karnovsky, 1966; Novikoff and Goldfischer, 1969) and 4-chloro-1-naphthol (Elias, 1980; Lippi and Cappelletti, 1982a) (Plate 22F) are at present the most widely employed substrates, for reasons which will be stated below. The patterns of reactivity observed with the various reactions are not identical, nor is the stability of the reaction products.

Another source of difficulty in interpreting the distribution of peroxidase positivity is the different reactivity of the various types of haemic cells, which might reflect the activity of several isoenzymes. The difference between neutrophil and monocyte peroxidase and that of eosinophils (and probably basophils) was established long ago on the basis of the different sensitivity to potassium cyanide inhibition (Lojda *et al.*, 1979) and of the occurrence of congenital peroxidase deficiencies involving selectively either one or other of the different cell lines (Undritz, 1966; Presentey, 1969).

Even myeloperoxidase activity related to the granulocyte-monocyte cell line, as a result of biochemical (Felberg and Schultz, 1972) and cytochemical investigations (Lippi and Cappelletti, 1982b) appears to depend upon a number of isoenzymes (Pember *et al.*, 1983). In this context, Koeffler *et al.* (1985) carried out a complex investigation in order to define the structure and expression of myeloperoxidase (MPO) in relation to myeloid differentiation. The authors isolated poly(A) containing RNA from HL-60 granulocytes, HL-60 macrophages, HL-60 blasts and normal human granulocytes. The mRNA was translated in a reticulocyte lysate system in the presence of ^{35}S-methionine. The MPO was precipitated from the lysate with rabbit IgG antiserum to human MPO. The resulting precipitate from HL-60 blasts gave a major band of radioactivity of approximately 77 000 daltons and another band at approximately 46 000 daltons on sodium dodecyl sulphate-polyacrylamide gel electrophoresis. The data obtained by the authors, combined with the results of other studies, suggest that MPO is a 135 000 dalton protein containing two heavy and two light chains of 55 000 and 15 000 daltons respectively, and that an approximate 75 000 to 77 000 dalton polypeptide is the direct precursor of the 55 000 and 15 000 dalton polypeptides. According to the authors, synthesis of the 75 000 to 77 000 dalton polypeptide precursor of MPO begins at the myeloblast stage of development, continues during the promyelocyte (HL-60) stage and ceases as cells differentiate into granulocytes and macrophages.

Following this line of investigation, Morishita *et al.* (1986) developed three monoclonal antibodies (McAbs), MA1, MA3 and MB1, which react with human MPO. In MPO-positive cultured cell lines, HL60 and NKM1, analysis by pulse-chase experiments followed by immunoprecipitation with these McAbs revealed three MPO subunits of 59 KD, 18 KD (probably an *in vitro* degradation product) and 14.0 KD, which were presumably derived from an 86 KD precursor polypeptide. MA1 and MB1 reacted with both the precursor and the mature forms of MPO, whereas MA3 reacted with only the mature forms. This fact may indicate that during the synthesis of mature MPO a new antigenic site appears. Furthermore, MA1 McAb, which detects an antigenic site present on both the precursor and the mature forms of MPO, also inhibits the enzymatic activity suggesting that the antigenic site defined by MA1 might be closely related to the enzymatic activity of MPO. The three McAbs seem therefore to recognize three different antigenic determinants. Very weak reactivity to these McAbs was shown by eosinophils, in contrast to a strong reactivity of monocytes, indicating that monocyte peroxidase exhibits the same antigenicity as neutrophil MPO.

Other anti-MPO McAbs, capable of recognizing both subunits of the mature MPO, as well as its proenzyme form, have been developed by Tetteroo *et al.* (1987). According to these authors, the anti-MPO McAbs were able to identify leukaemic myeloblasts with a fairly high degree of

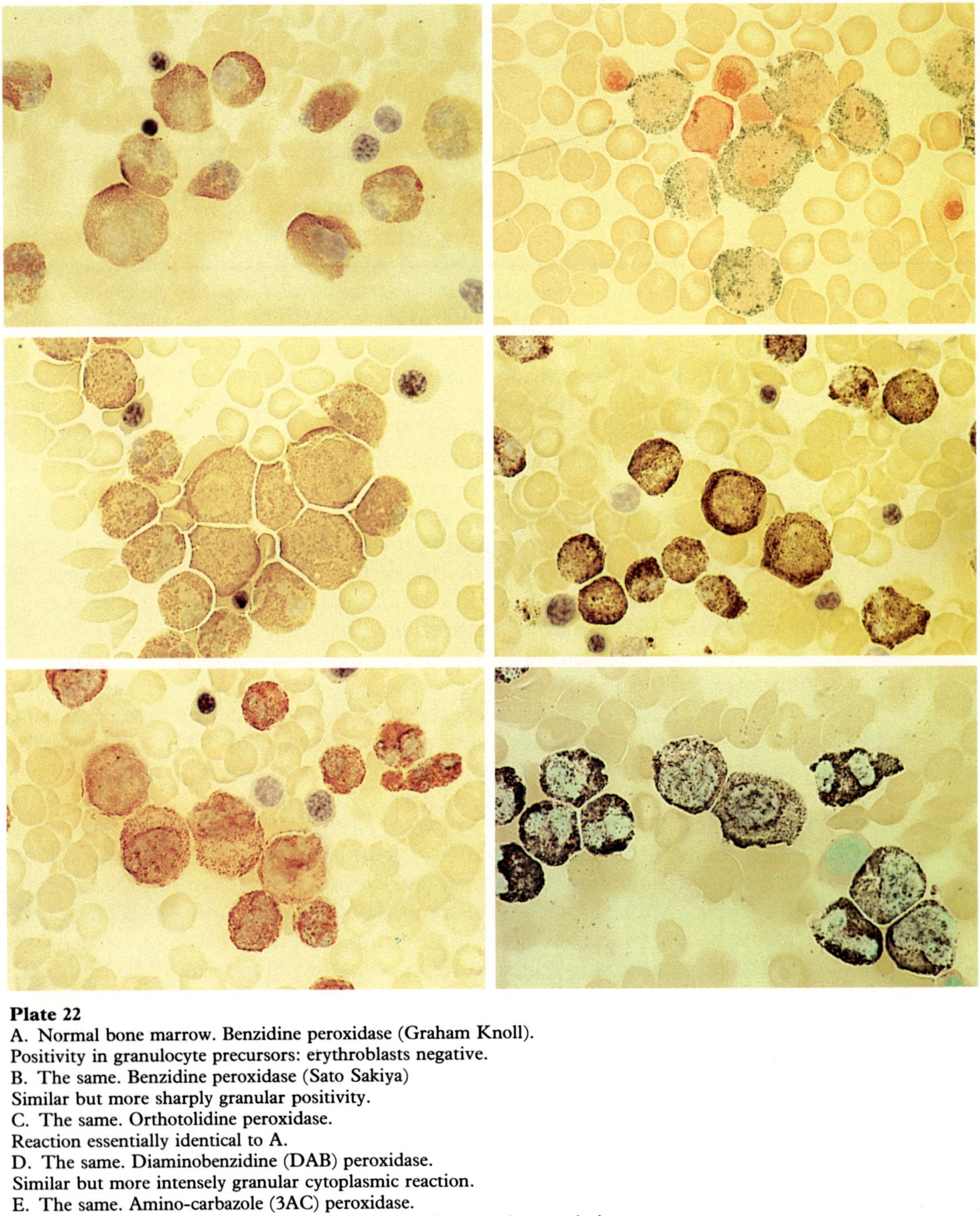

Plate 22

A. Normal bone marrow. Benzidine peroxidase (Graham Knoll).
Positivity in granulocyte precursors: erythroblasts negative.
B. The same. Benzidine peroxidase (Sato Sakiya)
Similar but more sharply granular positivity.
C. The same. Orthotolidine peroxidase.
Reaction essentially identical to A.
D. The same. Diaminobenzidine (DAB) peroxidase.
Similar but more intensely granular cytoplasmic reaction.
E. The same. Amino-carbazole (3AC) peroxidase.
Reaction comparable to orthotolidine; perhaps some increase in granularity.
F. The same. Chloro-naphthol (4CN) peroxidase.
Similar reaction to diaminobenzidine, with sharp granularity, here contrasting well with the nuclear counterstain.

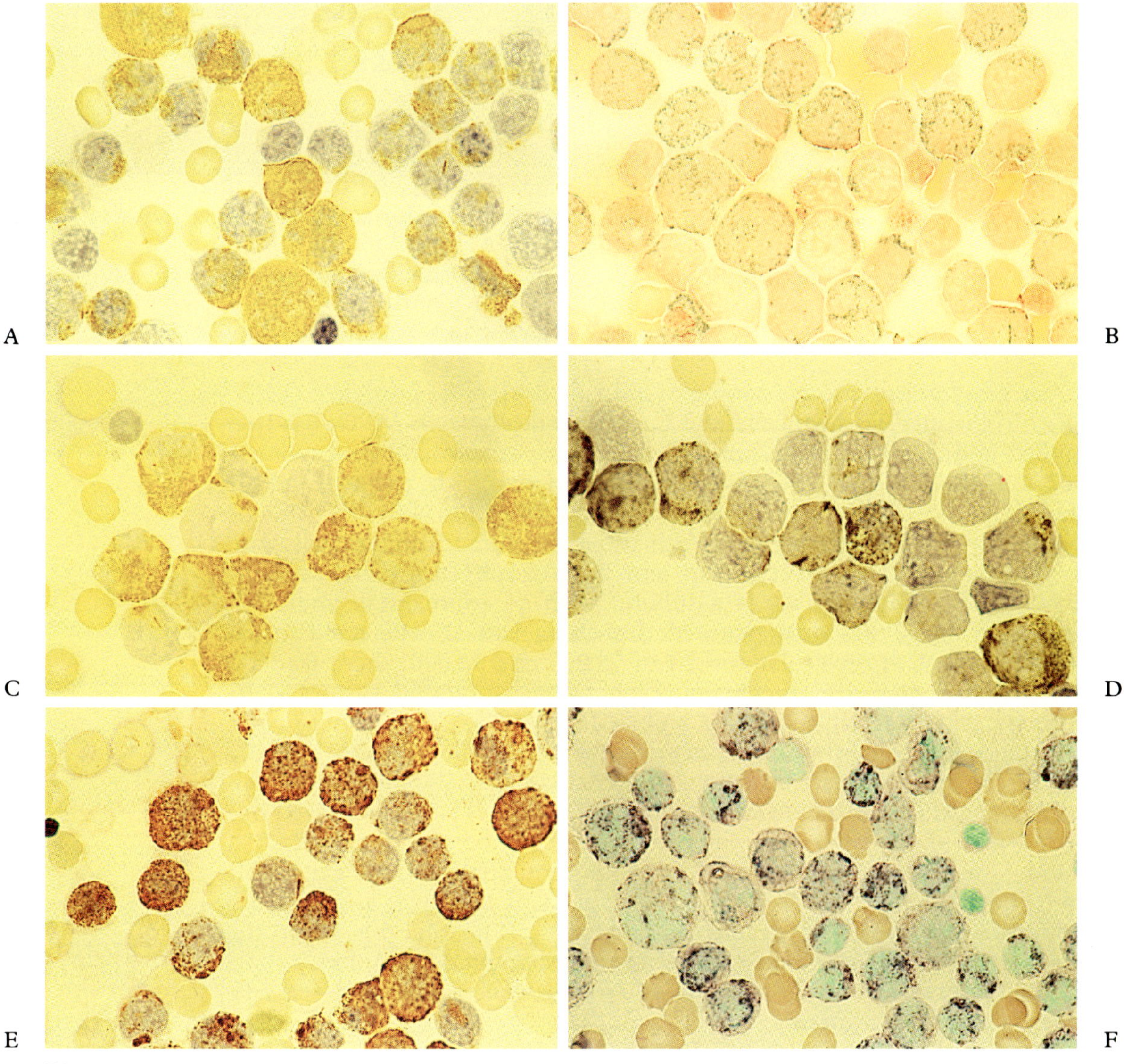

Plate 23

A. Acute granulocytic leukaemia. Benzidine peroxidase (Graham-Knoll)
A range from weak to heavy positivity is seen in the leukaemic cells, with several Auer rods.
B. The same. Benzidine peroxidase (Sato-Sakiya).
Some weaker reaction generally, with less clearly visible Auer rods.
C. The same. Orthotolidine peroxidase.
Reaction essentially identical to A.
D. The same. DAB peroxidase.
Less strong reaction than A and C, but more sharply granular, with well shown positive Auer rods. Many blast cells show little or no reaction.
E. The same. 3AC peroxidase.
Reaction comparable to orthotolidine, but even stronger and more granular with well shown Auer rods.
F. The same. 4CN peroxidase. Virtually all cells show moderate to strong positivity. Auer rods are again clearly shown.

sensitivity, superior to that shown by most McAbs directed towards myeloidspecific membrane antigens.

The human myeloperoxidase gene has been shown to be localised on chromosome 17 at 17q22–24 (Chang *et al.*, 1987), but there may perhaps also be a functional link with chromosome 1, in that a variant HL60 line showing striking atypical large azurophilic granules deficient in myeloperoxidase had a consistent chromosome 1q + abnormality which was not found in any of the parental cells examined (Sullivan *et al.*, 1986). Nevertheless the illustrative karyotype of cells from this HL60–A7 line shows not only the 1q + marker, but also multiple other defects including the loss of a chromosome 17.

The biosynthesis and processing of human eosinophil peroxidase (EPO), which is distinct biochemically and immunologically from MPO, has been studied and reported to be similar to the biosynthesis and processing of MPO. Olsson *et al.* (1985) studied in patients with eosinophilia the biosynthesis of EPO in marrow cells, using labelling with (^{14}C)leucine, followed by immunoprecipitation with anti-EPO, sodium dodecyl sulphate-polyacrylamide gel electrophoresis and fluorography for visualization of labelled EPO. Biosynthesis of a subunit with an approximate molecular weight of 50 000 was demonstrated. Labelling was also detected in two polypeptides with molecular weight of 78 000 and 72 000: these forms of EPO seem to represent precursor polypeptides, subjected to proteolytic processing in a similar manner as reported for MPO.

Diaminobenzidine appears for many authors to be the substrate of choice for the cytochemical demonstration of peroxidase activity, both at the light microscopical and at the ultrastructural level, since it is characterized by a high degree of solubility, a speedy reaction and a stable reaction product. 4-chloro-1-naphthol is not suspected of being potentially carcinogenic. The peroxidase reaction based on its use is rapid and sensitive, and is widely employed by automated differential analyzers. The blue-black reaction product is also easily recognizable at the microscopical level, but its high lipid solubility prevents the use of this substrate when a long-term record of the specimens is required.

Because of its long history of usage, the distribution of peroxidase positivity obtained with benzidine (or o-tolidine) will be used as a basis for the description of the reactivity pattern in haemic cells and will serve as a standard of comparison with regard to the other benzidine substitutes more commonly employed. Plate 23 illustrates the findings obtained in a comparative staining trial with the more common peroxidase methods in a case of acute myeloid leukaemia with a mixed myeloblastic and promyelocytic population. Differences are described in the caption.

A further comparison, using the o-tolidine, DAB and 3AC peroxidase methods together with Sudan black on normal buffy coat, and both weakly and strongly sudanophilic AML preparations, is shown in Plates 24 and 25. The reactions for myeloperoxidase and Sudan black have a general similarity in positivity and in many haematology laboratories they are used as alternatives. The comparison in Plates 24 and 25 shows that Sudan black is superior to all the peroxidase methods in sensitivity, especially with respect to monocytes and their precursors and also to weakly reacting myeloblasts; DAB provides more precisely localized (and also more stable) reaction product than either the orthotolidine (benzidine) or 3AC methods. From our comparison we conclude that if Sudan black and peroxidase are regarded as alternatives, Sudan black is preferable, but that if both reactions are to be used, as may be advantageous in the demonstration of Auer rods, for example in cases of t(8; 21) AML, then DAB is the substrate of choice.

Distribution of peroxidase positivity in normal haemic cells at light and electron microscope levels

In haemic cells studied at light microscope level peroxidase activity is present in cells of the granulocytic series, both neutrophils and eosinophils, with positive granules filling the cytoplasm, from the promyelocyte stage onwards to the segmented forms. In the basophil series there is generally strong peroxidase positivity at the promyelocyte and myelocyte stages but the reaction becomes weaker in metamyelocytes and is almost negative in fully mature basophils (Parwaresch, 1976). Positive reactions in monocytes occur in a variable proportion of cells and are usually weak with relatively few scattered granules. Red cells, lymphocytes and megakaryocytes are negative.

The results of ultrastructural cytochemistry (Behnke, 1969) are in conformity with the existence of two different types of peroxidase, a neutrophil peroxidase and an eosinophil peroxidase. The granules of mature and immature neutrophils show in fact a weaker staining reaction than those of eosinophilic granules and this might reflect the difference between the two different peroxidases and indicate that neutrophil peroxidase is more sensitive towards fixation than eosinophilic peroxidase.

The 3,3′ diaminobenzidine (DAB) reaction, introduced by Graham and Karnovsky (1966) and subsequently modified by Novikoff and Goldfischer (1969) enables the staining for both light and electron microscopy in a wide variety of plant and animal tissues of microbodies (peroxisomes), which are cellular organelles defined biochemically as containing catalase and one or more hydrogen peroxide oxidases.

As Cawley and Hayhoe (1973) point out in their monograph on the Ultrastructure of Haemic Cells, the fine localization of peroxidase activity in cells of the neutrophil series has stimulated particular interest, because much of the activity is found in well defined organelles. Indeed, peroxidase positivity has been regarded by some workers as a marker of the primary neutrophil granule in man (Dunn *et al.*, 1968; Breton-Gorius and Guichard, 1969; Bainton *et al.*, 1971; Ackerman and Clark, 1971a) and in certain animal species (Dunn *et al.*, 1968; Bainton and Farquhar, 1968b; Ackerman, 1968). However, some earlier work considered peroxidase activity to be present in secondary granules or in both the primary and secondary granules (Yamada, 1966). Work in our own laboratories described and fully illustrated by Cawley and Hayhoe (1973) has suggested that two distinct peroxidatic activities are demonstrable in the primary and secondary granules respectively. One of these two forms of peroxidase activity is

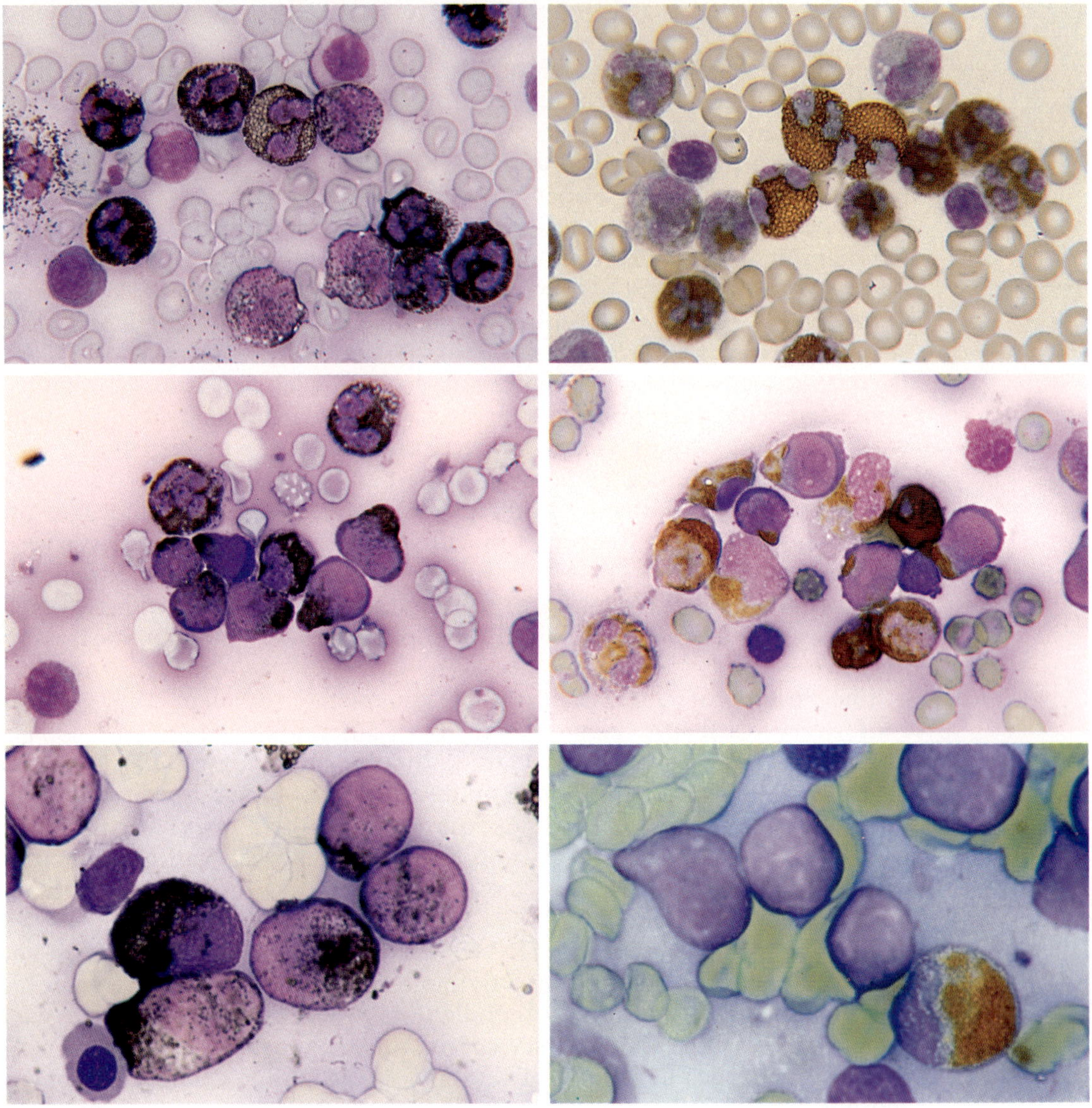

A B

C D

E F

Plate 24

Plates 24A, B and 25A, B

A sequence of Sudan black (SB), benzidine peroxidase, diaminobenzidine peroxidase (DAB) and 3-amino-9-ethyl carbazole peroxidase (3AC) in normal buffy coat leucocytes. Monocyte granules are shown best in SB, the benzidine peroxidase is heavy but diffuse in neutrophils and weak diffuse in monocytes, DAB gives a crisp granular pattern, most like SB but weaker in monocytes, and 3AC gives strong neutrophil reactions with local diffusion, weaker reactions in monocytes but again with some diffuse spread, and a notable reaction in lymphocytes and around a platelet clump.

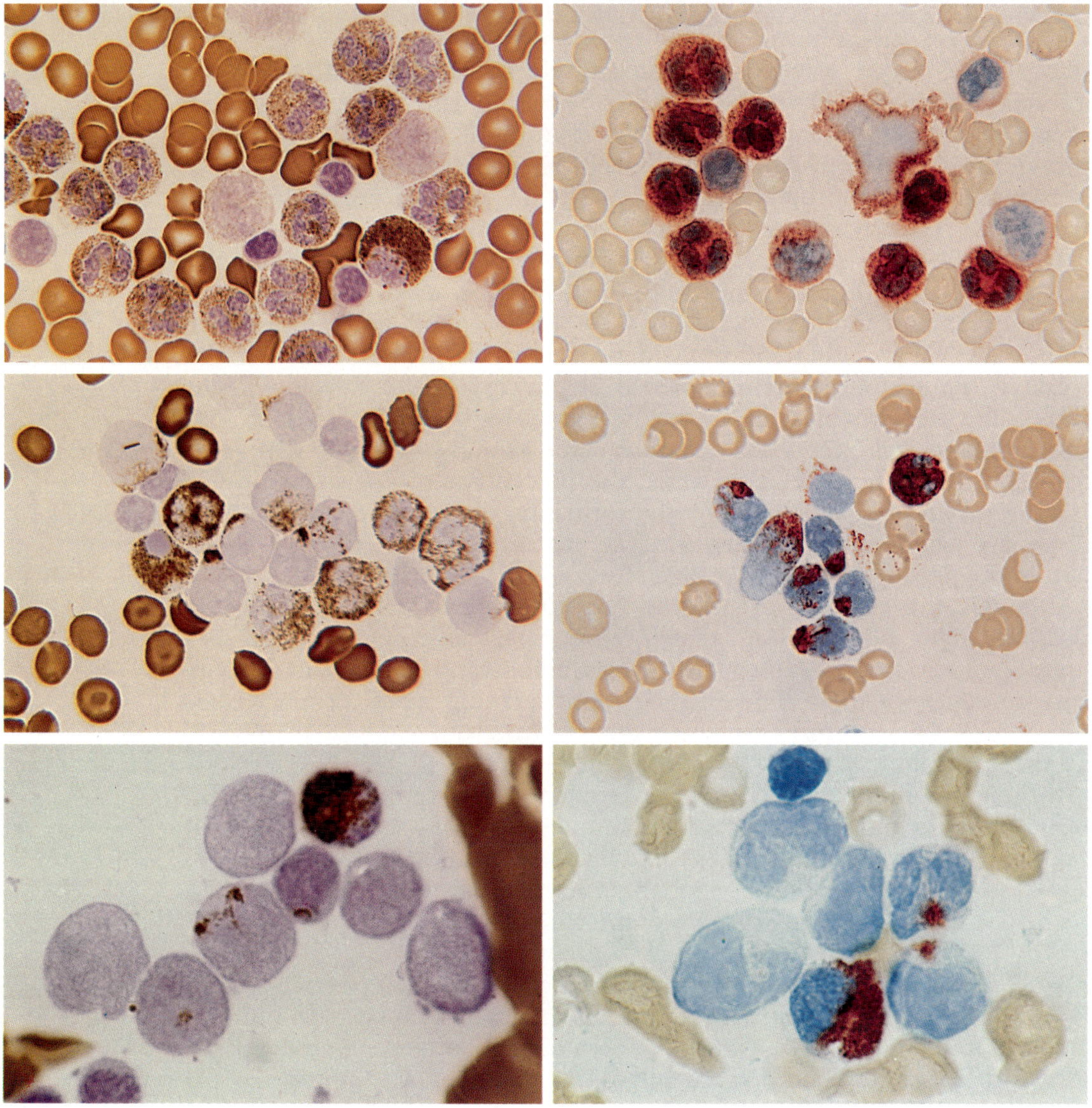

A B C D E F

Plate 25

Plates 24C, D and 25C, D

A similar sequence in a strongly reacting AML (with 8;21 translocation). All reactions are strong, SB the heaviest and DAB and 3AC both show Auer rods.

Plates 24E, F and 25E, F

A third similar sequence in an AMML with numerous monoblasts; the SB shows strong reactions with some localised positivity and scattered granules in 2 monoblasts and parts of 2 more at the edge, and heavy positivity in an early eosinophil. The 3 peroxidase reactions each show a similarly heavy reaction in an eosinophil precursor, but negative or only weak and localized reactions in the monoblasts.

demonstrated by the original Graham and Karnovsky technique using a saturated solution of DAB free base as substrate and is located in the primary neutrophil granules. The distribution of positivity with this technique thus closely resembles that of acid phosphatase-positive granules; reactive granules make up the entire population of the promyelocyte but become progressively reduced in number, with the result that in the mature neutrophil they constitute less than 10% of the total granule population.

In contrast, if a higher concentration of DAB is used, particularly at alkaline pH, as under the conditions of the Novikoff and Goldfischer technique (1969), a different peroxidatic activity is demonstrable in the secondary neutrophil granules formed at the myelocyte developmental stage, while the peroxidase in the primary granules is also detected, so that both primary and secondary granules are strongly reactive with this technique. After the myelocyte stage, the percentage of positive granules becomes progressively reduced by the formation of increasing numbers of peroxidase negative tertiary granules, so that in mature cells only about 50% of the granules show reactivity.

Although the nature of these two distinct peroxidatic activities is uncertain, their existence and distribution, taken together with the distribution of acid phosphatase positivity, provide strong evidence in support of the existence of three different neutrophil granule types, primary, secondary and tertiary, formed respectively at the promyelocyte, myelocyte and metamyelocyte stages of development.

In the eosinophil series, however, peroxidase activity, demonstrable by both ultrastructural cytochemical techniques is found in the eosinophil granules of all stages (Dunn *et al.*, 1968; Breton-Gorius and Guichard 1969; Cawley and Hayhoe, 1973). Probably, in man at least, the so-called secondary granules with crystalline internal ultrastructure are derived directly from the homogeneous primary granules as suggested by Hudson (1966), Wetzel *et al.* (1967), Scott and Horn (1970), and Breton-Gorius (1970), rather than being formed *de novo* at the myelocyte stage as Bainton and Farquhar (1967, 1970) claim. The last named authors worked principally with rabbit cells, however, and in that species the morphological evidence implicating the Golgi apparatus in secondary granule formation in the eosinophil leucocytes is more convincing than in man.

The granules of human basophils have also been shown to contain peroxidase, demonstrable at both light microscope level, as previously mentioned (Parwaresch 1976) and ultrastructurally (Ackerman and Clark, 1971b).

Cramer *et al.* (1985) recently employed gold as a marker for immunoelectron microscopy to localize lactoferrin (LF) and myeloperoxidase (MPO) in human peripheral blood neutrophils. MPO was detected in the large azurophilic granules, whereas LF was consistently localized in the matrix of another population of morphologically distinct granules, smaller and more numerous than azurophilic granules. When cytochemical detection of MPO was coupled with immunological detection of LF, LF was observed in the population of MPO-negative granules, which were identified as specific. No other neutrophil organelles displayed labelling for LF;

other blood cells were unreactive for LF. In the bone-marrow, myeloblast and promyelocyte granulations were not stained and LF-containing granules appeared at the myelocyte stage.

MPO was observed in azurophil granules by both cytochemical and immunocytochemical methods. This latter approach is of considerable practical interest since it allows identification of the cell lineage, according to the different immunologic reactivity of different apoproteins with the same cytochemical peroxidase activity (e.g. eosinophil vs. neutrophil peroxidase).

Oxidative action on the substrate 3,3′ diaminobenzidine has been demonstrated in the human platelet dense bodies (a unique organelle of the mature platelet containing serotonin) and in numerous vesicles (Parmley *et al.*, 1974). The DAB oxidative activity in the platelet dense body implies that this organelle may be formed in the platelet dense tubular system (DTS), which shows similar reactivity (Breton-Gorius and Guichard, 1972; White, 1972), rather than deriving from the alpha-granule, which contains hydrolytic and not oxidative enzymes (Murata *et al.*, 1973). There are in fact many transitional elements between the DTS and the dense bodies and in many respects all these structures may represent microperoxisomes, which may be considered dilatations of the smooth endoplasmic reticulum (Novikoff and Shin, 1964, De Duve, 1973).

The enzyme, with peroxidatic activity, present in the dense tubular system of platelets, and designated as platelet peroxidase (PPO) is also present in normal megakaryocyte precursors (Breton-Gorius, 1980). It differs from myeloperoxidase in several characteristics (it is present, for instance, in congenital myeloperoxidase deficiencies) and from platelet catalase, which is another enzyme found in granules of platelets and promegakaryoblasts (Breton-Gorius *et al.*, 1981), in its localization and cytochemical characteristics. In normal bone marrow, PPO is present in about one small cell per 10000 cells. These cells, considered as promegakaryoblasts, exhibit a lymphoid morphology and do not have recognizable organelles of megakaryocytes. PPO is present exclusively in the endoplasmic reticulum, including the perinuclear space; the Golgi apparatus is always unreactive.

Since PPO is not preserved by fixation with glutaraldehyde in phosphate buffer, several methods have been proposed to obtain satisfactory and reproducible demonstration of the enzyme activity. Breton-Gorius *et al.* (1984) have recently described a new method for detecting PPO, employing Gey's basic salts as a fixation buffer, thus obtaining reproducible cytochemical demonstration of enzyme activity, as well as excellent preservation of cell morphology and membrane antigenicity.

The importance of mantaining membrane antigenicity lies in the possibility of revealing, in addition to the cytochemical properties of megakaryocyte precursors, their phenotypic characteristics, as revealed by the use of monospecific antibodies against platelet glycoproteins (Gp) Ib, IIb, IIIa, as well as Factor VIII related antigen and platelet factor 4 (Rabellino *et al.*, 1981).

The simultaneous analysis of cytoplasmic PPO and cell surface GpIb and

GpIIIa using monoclonal antibodies (McMichael *et al.*, 1981; Tetteroo *et al.*, 1983) conjugated to colloidal gold (Breton-Gorius *et al.*, 1984) shows that only PPO-positive cells exhibit a labelling with anti-Gp IIIa and anti-GpIb and that the number of binding sites for GpIIIa is higher in more immature promegakaryocytes than for GpIb. The labelling with anti-GpIIIa is however less intense in promegakaryoblasts than in platelets. In acute leukaemias involving the megakaryocytic lineage, promegakaryoblasts may be arrested at various stages of maturation.

The very early promegakaryoblasts are characterized only by their PPO content. These blasts often have large bundles of intermediate filaments (Breton-Gorius *et al.*, 1983), but they do not synthesize glycoproteins. In slightly more mature promegakaryoblasts GpIIIa is synthesized; later Factor VIII related antigen, platelet factor 4 and platelet derived growth factor appear with a diffuse pattern and then with a granular pattern, as revealed by immunofluorescence methods. At this stage also GpIb may be detected. Therefore the simultaneous demonstration of platelet peroxidase activity and of specific antigens on the cell membrane permits the detection of two markers of one cell lineage at the level of the single cell and provides information of the level of differentiation reached by the promegakaryoblasts in the different myeloproliferative conditions.

Among myeloid cells, ultrastructurally demonstrable PPO activity, especially in the perinuclear cisternae, has been regarded for some years as a fairly specific marker of normal and malignant megakaryocytes. However, a similar pattern of PPO positivity could be detected by Koike *et al.* (1986) in the circulating erythroblasts of a patient with acute myelodysplasia and myelofibrosis in a terminal phase. The authors suggested that this unusual PPO positivity might have been the result of deranged gene expression, since perinuclear PPO activity had not been described in normal erythroblasts. Breton-Gorius *et al.* (1987), however, have reported a very interesting peroxidase study of 9 primitive-celled erythroleukaemias, 2 positive for glycophorin A and 7 of more immature phenotype but all labelling with the erythroblast-specific monoclonal antibody FA6–152, by immunogold electron microscopy. Many early erythroblasts from the more immature cases showed PPO-like activity, together with theta granules in the Golgi apparatus, and ferritin molecules in the absence of rhopheocytosis. Studies of normal bone marrow cells enriched by panning for CFU-E and depleted of non-erythroid precursors showed the presence of similar proerythroblasts with PPO positivity, theta granules and ferritin. Clearly the PPO reaction is demonstrable ultrastructurally in both normal and leukaemic early erythroid blast cells.

Khalaf and Hayhoe (1987) have recently developed a modification of the ultracytochemical method of Anderson *et al.* (1975) which allows the demonstration of PPO in platelets and megakaryocytes at light microscope level. The method uses a tannic acid-aldehyde mixture as fixative and incubation for a total of 20 hours. As with the ultrastructural method for PPO, myeloperoxidase in granulocytes is demonstrated, but positivity is also seen in normal platelets and megakaryocytes (Fig. 10.3) and in normal lymphocytes. Blast cells of acute myeloid leukaemia including megakaryo-

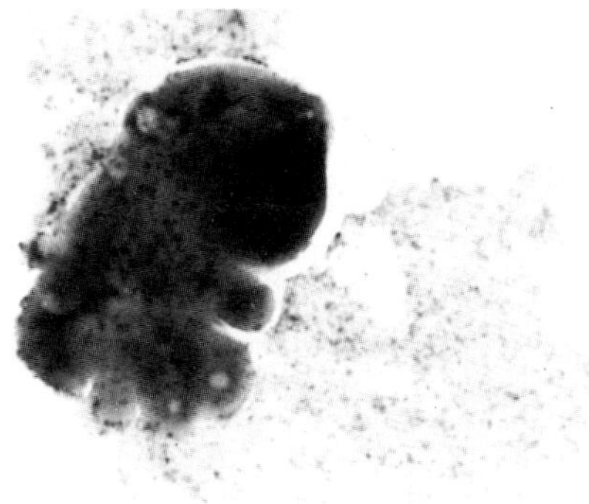

Fig. 10.3 Normal bone marrow. PPO method at light microscope level. A budding megakaryocyte and platelets show scattered granular positivity.

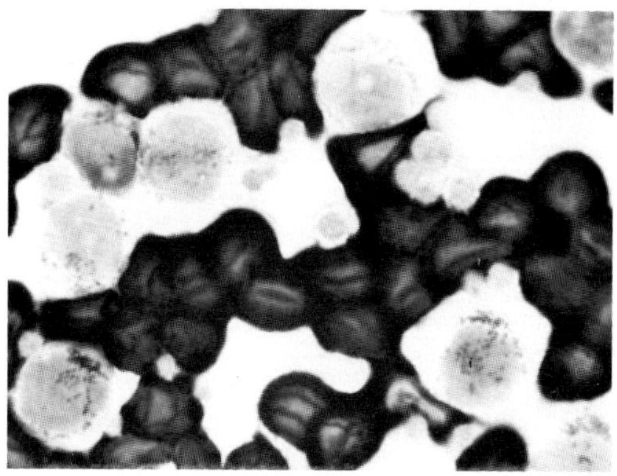

Fig. 10.4 Hairy cell leukaemia; buffy coat. PPO method shows moderate granular cytoplasmic reaction in hairy cells. Erythrocyte haemoglobin is also reactive, giving strong diffuse staining.

blasts of acute megakaryoblastic leukaemia show positivity, as do hairy cells of hairy cell leukaemia (Fig. 10.4), centroblasts in centroblastic lymphoma and plasma cells in plasma cell leukaemia. Enzyme activity was also demonstrated in mast cells in systemic mastocytosis. It thus appears that most haemic cells possess some form of peroxidatic activity demonstrable by this method, although a full range of leukaemias and related states, including lymphoblastic leukaemias, has not yet been studied.

Peroxidase in haematological disorders

The peroxidase reaction has been used in clinical haematology almost exclusively because it has been regarded of value in differentiating myeloblasts and monocyte precursors from lymphoblasts, since the former often show peroxidase granules (Plate 26B,C,D) and myeloblasts may contain peroxidase positive Auer rods or other inclusions (Figs 10.5 and 10.6),

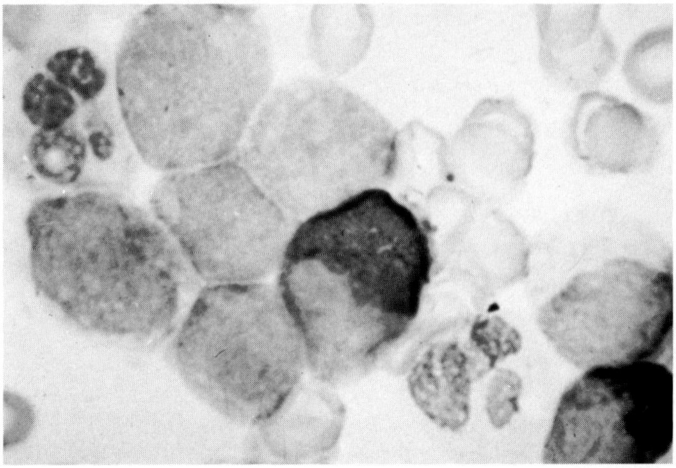

Fig. 10.5 AML bone marrow. Peroxidase reaction. Five negative blast cells, two strongly positive and one with weak localized positivity. There are two negative polymorphs, an abnormality sometimes found in myeloid leukaemias.

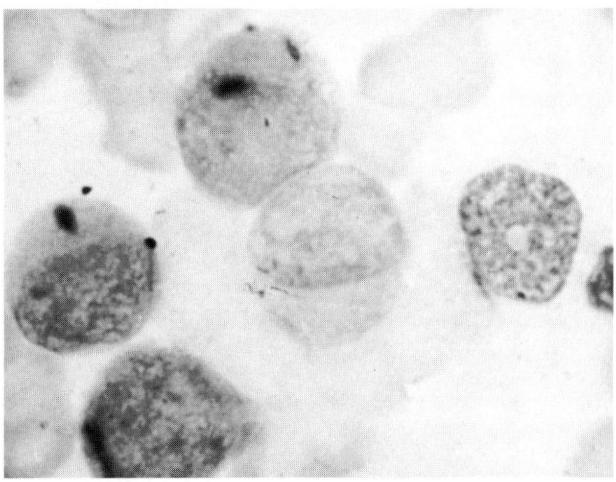

Fig. 10.6 AML bone marrow. Peroxidase reaction. Three myeloblasts contain peroxidase positive Auer rods but the cells are otherwise negative.

whereas lymphoblasts are uniformly negative (Plate 26F). Hayhoe *et al.* (1964) employed the peroxidase reaction, together with other cytochemical stains, in an attempt to characterize objectively the different nature of the primitive cells pertaining to the various cytological varieties of acute leukaemia. Further details of this study will be given in Chapter 14.

In 1961 and 1964 Quaglino carried out a semiquantitative study on the variations in peroxidase activity in neutrophil granulocytes, both in normal and pathological conditions. Peroxidase activity was found decreased in leucocytosis of infection, in Hodgkin's disease and in certain metastatic

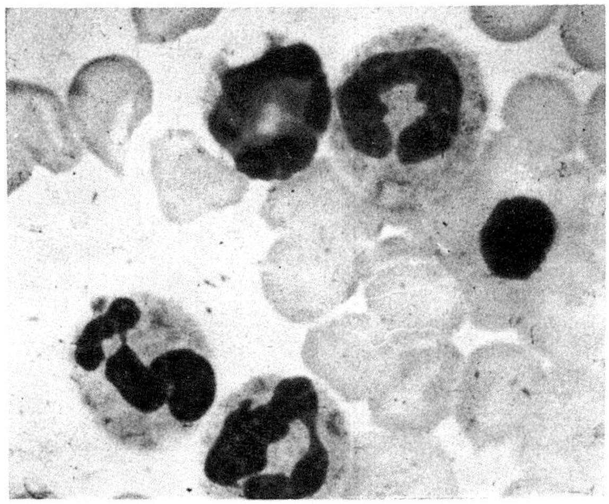

Fig. 10.7 Blood smear from a case of metastatic gastric carcinoma. Peroxidase reaction. Four neutrophils show very weak reactions.

tumours (Fig. 10.7). In the latter case the extremely low peroxidase positivity in neutrophil granulocytes might be due to an inhibitory effect of the tumour, as suggested by Hargreaves and Deutsch (1952), who showed the presence of an *in vitro* inhibition of catalase and other iron-porphyrin enzymes, including peroxidase, by tumour extracts. This material appeared to function by combination with the enzyme through the iron-porphyrin group. This inhibitory effect was essentially limited to iron-porphyrin enzymes and did not involve any other enzymatic system, including lactic dehydrogenase, ascorbic acid oxidase, succinic dehydrogenase, cholinesterase and arginase. The peroxidase enzyme used by the authors was extracted from horseradish. In infectious leucocytosis and in Hodgkin's disease the low peroxidase activity of mature granulocytes was associated with increased alkaline phosphatase reactivity in the same cells. The existence of these simultaneous but opposite variations in peroxidase and alkaline phosphatase activity might be related to the effect of ACTH on both enzyme systems. The intramuscular administration of purified corticotrophin gel in substantial doses (Valentine *et al.*, 1954) every eight hours over a 72 hour period results in an average 3–4-fold increase in leucocyte alkaline phosphatase. In contrast the injection of 25 mg of ACTH in normal subjects (Mauro and Prato, 1954) may induce a significant decrease in the number and size of peroxidase granules present in mature polymorphonuclear cells. This decrease in the intensity of the peroxidase reaction is most marked towards the fourth hour after injection, after which enzyme activity tends to return to normal. The relationship of myeloperoxidase levels and the maturity of neutrophils was further investigated by Kaplow (1986) using flow cytometric analysis. A statistically significant correlation emerged between increased peroxidase levels and features of immaturity or defective maturation, including band-cell counts, Schilling index and toxic granulation.

In striking contrast to the low peroxidase activity in infectious leucocytosis, Hodgkin's disease and metastatic tumours, the mature polymorphs in untreated cases of chronic myeloid leukaemia (Plate 26A) displayed a reaction of particular intensity (Quaglino, 1961, 1964). This finding, initially of uncertain significance, may now be related to the reported cytoplasmic immaturity of neutrophil polymorphs from cases of chronic myeloid leukaemia. This condition of biological abnormality has been revealed by ultrastructural studies (Bessis, 1968; Kakefuda, 1968), which have shown that, despite a normal morphological appearance in the light microscope, the cytoplasm of polymorphs is often abnormal, being devoid of granules in some cells and containing predominantly azurophilic with very few neutrophilic granules in others. Also endoplasmic reticulum which is usually absent or very much reduced in normal polymorphs may be instead well represented, thus providing an image corresponding to an earlier maturation stage of the granulocytic series.

Cytological and cytochemical ultrastructural investigations (Hirsh and Cohn, 1964; Enomoto and Kitani, 1966; Ackerman, 1968; Bainton and Farquhar, 1968a,b) have correlated the above mentioned cytological picture with the cytochemical pattern of leukaemic granulocytes, which appears to be characterized by an increased activity of those enzymes which are predominantly localized on the primary azurophilic granulations (or lysosomes), i.e. peroxidases and acid phosphatases and a decrease of those enzymes, such as specific alkaline phosphatases, which are associated with secondary or tertiary granules or other organelles. This imbalance between nuclear and cytoplasmic maturation may also be responsible for the poor phagocytic activity observed in leukaemic granulocytes (Pedersen and Hayhoe, 1971), since both azurophilic and specific granules may actively participate in the phenomenon of phagocytosis by producing a phosphatase positive coating to the enclosed bacteria (Horn *et al.*, 1964; Cotran and Litt, 1969; Spicer and Hardin, 1969).

Another abnormality concerning peroxidase activity was reported by Hayhoe *et al.* (1964) in mature neutrophil granulocytes from cases of acute myeloid leukaemia. These authors found that in this cytological type of acute leukaemia a variable but usually small proportion of mature granulocytes were completely peroxidase negative, in addition to their low glycogen content, their reduced sudanophilia and absent alkaline phosphatase activity.

This observation was later confirmed by Catovsky *et al.* (1972) who reported the presence of polymorphonuclear neutrophils lacking peroxidase in 12 out of 28 cases of acute myeloid leukaemia and in 2 out of 7 cases of blast cell transformation of chronic myeloid leukaemia. In acute myeloid leukaemia this phenomenon occurred much more frequently in the myeloblastic and erythroleukaemic varieties than in those with predominantly monocytic differentiation and high serum and urinary lysozyme concentrations. Auer rods were 3.3 times more common in the blast cells of acute myeloid leukaemia cases with peroxidase negative polymorphs. Low neutrophil alkaline phosphatase was also frequently associated with absence of peroxidase. The authors suggested that these abnormalities could result

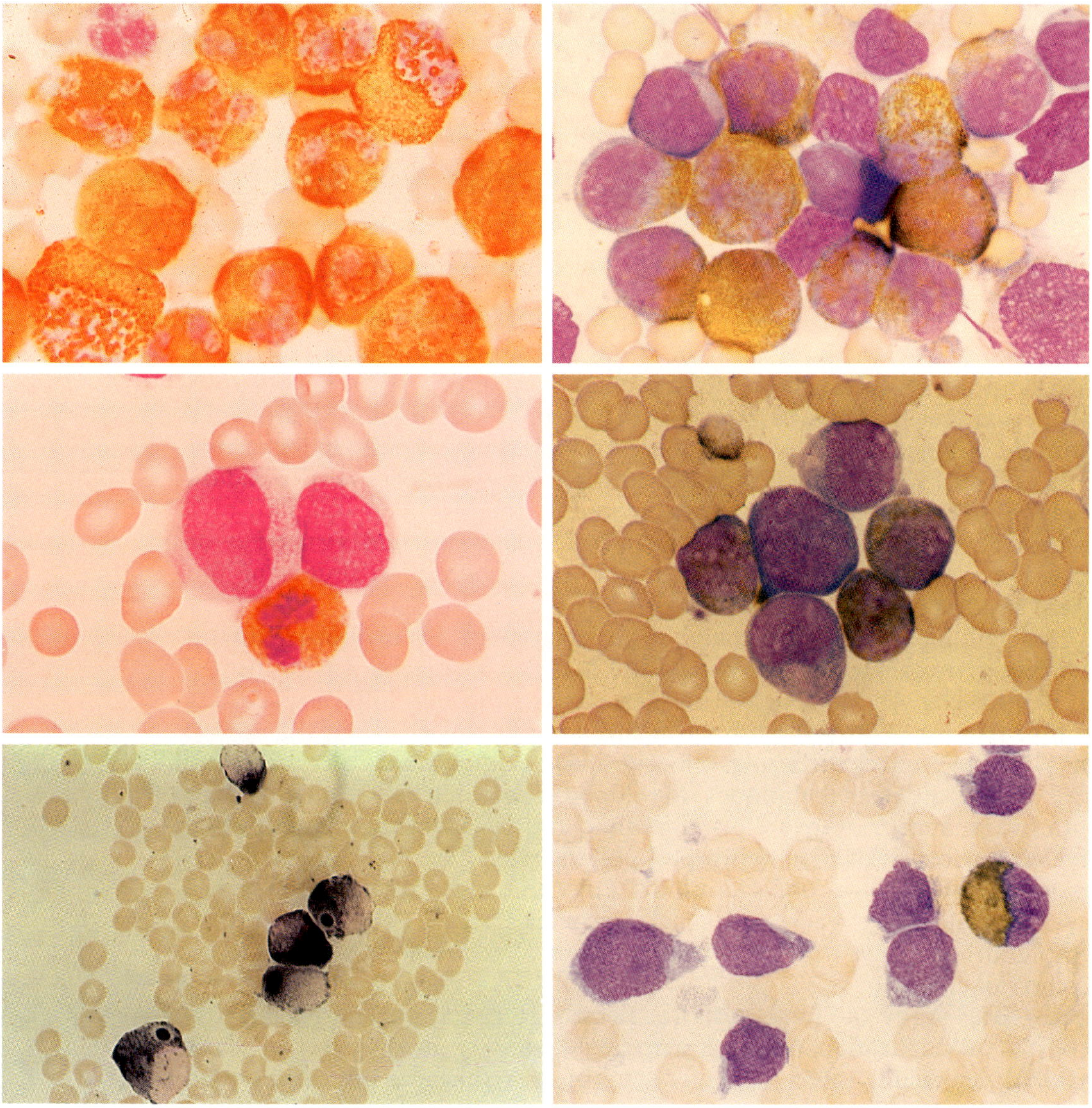

A B C D E F

Plate 26

A. CML bone marrow. Peroxidase reaction with o-tolidine. Very strong activity present in granulocytes of neutrophil and eosinophil lines.

B. Acute myeloblastic-promyelocytic leukaemia bone marrow. Peroxidase reaction with o-tolidine. Variable positivity, generally increasing in more mature cells of promyelocyte and myelocyte stages.

C. AMonL bone marrow. Peroxidase reaction with o-tolidine. A positive neutrophil and two negative monocyte precursors.

D. AMML bone marrow. Peroxidase reaction with o-tolidine. Weak to moderate cytoplasmic reaction in three of six leukaemic cells.

E. AML bone marrow. Peroxidase reaction with 4CN showing positivity in pseudo-Chediak-Higashi inclusions.

F. ALL bone marrow. Peroxidase reaction with o-tolidine. Lymphoblasts are uniformly negative. An eosinophil myelocyte shows normal positivity.

from a disturbance in the function of the Golgi complex in cells derived from leukaemic stem cells.

That a derangement in the function of the Golgi apparatus may be responsible for abnormal peroxidase reactivity is made possible by the fact that peroxidase synthesis, according to Bessis and Maigné (1970), occurs in immature cells of blood and bone marrow in the perinuclear space, the endoplasmic reticulum and the Golgi apparatus. A positive peroxidase reaction may therefore be observed in cells even in the absence of azurophil granules, in which, as we have seen, ultrastructural cytochemistry (Bainton and Farquhar, 1968a,b; Ackerman, 1968; Breton-Gorius and Guichard, 1969) and ultracentrifugation studies (Baggiolini *et al.*, 1970) have shown the peroxidase activity demonstrable by the Graham and Karnovsky method to be localized. According to Bessis and Maigné (1970) in leukaemia there are blast cells of comparable morphology, which are peroxidase negative and the reason for this behaviour is due, in their opinion, to a process of 'simplification' of the specialised functions and structures of certain clones of leukaemic cells during the progressive course of the disease.

The finding of an abnormally high percentage of myeloperoxidase-deficient granulocytes in untreated acute myeloid leukaemia was confirmed by Bendix-Hansen and Nielsen (1983a), employing two different staining methods, one based on the use of benzidine hydrochloride and the second on 3-amino-9-ethyl carbazole. No significant differences were observed with the two different techniques. Myeloperoxidase deficiency was analyzed both by a semiquantitative scoring method and by counting the number of myeloperoxidase-deficient neutrophil polymorphs. The authors found peroxidase negative neutrophil granulocytes also in 6 out of 10 cases of chronic myeloid leukaemia, whereas normal peroxidase scores were observed in lymphoid leukaemias, both acute and chronic, as previously reported by Cech *et al.*, (1982).

In a later study the same authors (Bendix-Hansen and Nielsen, 1983b) reported normalization of myeloperoxidase activity in 35 cases of acute myeloid leukaemia, who achieved complete remission after induction therapy; however no differences in pretreatment scores of myeloperoxidase activity were found when comparing the values in patients who died without achieving complete remission with the values in patients who went into complete remission after therapy.

Decreased myeloperoxidase activity of granulocytes has also been reported by several authors (Lehrer *et al.*, 1972; Breton-Gorius *et al.*, 1976; Bendix-Hansen and Kerndrup, 1985) in cases of myelodysplastic syndrome. The highest incidence of myeloperoxidase deficiency (40%) was found by Bendix-Hansen and Kerndrup in cases of chronic myelomonocytic leukaemia and the lowest (10%) in refractory anaemia with excess of blasts (RAEB). No significant correlations were observed with leucocyte alkaline phosphatase activity, white cell number and platelet counts, although in a later review (Bendix-Hansen, 1987) both myeloperoxidase-deficient and alkaline phosphatase-deficient neutrophils were noted to occur especially in cases of CML and AML with granulocytic predominance.

There is a distinction here between the overall peroxidase or LAP scores, which may show respectively high or low trends, and the proportion of totally negative individual cells, which perhaps offers a better index of neutrophil derivation from leukaemic precursors.

Reduced myeloperoxidase activity in neutrophil polymorphs may account for increased susceptibility to infections in patients with acute myeloid leukaemia (Nielsen and Bendix-Hansen, 1984), and may represent an unfavourable prognostic factor (Bendix-Hansen, 1987).

The relationship between bactericidal effects and myeloperoxidase content was shown by Lehrer and Cline (1969) in a patient with congenital myeloperoxidase deficiency and disseminated candidiasis. *In vitro* investigations of the patient's neutrophil polymorphs revealed failure in killing candida species, and *Staphylococcus aureus* and *Serratia marcescens* were killed at a slower rate than normal. Breton-Gorius *et al.* (1976) suggest that defective bactericidal activity could be related to either myeloperoxidase deficiency of azurophilic granules or failure in degranulation of azurophilic granules with normal myeloperoxidase content. The latter hypothesis might explain why increased bacterial challenge and consequent increased stimulation to degranulation of azurophilic granules can improve bactericidal effects (Repine *et al.* 1976).

A role for myeloperoxidase in the regulation of degranulation has been recently suggested by Dri *et al.* (1985), who claim that myeloperoxidase deficient neutrophil granulocytes degranulate more than normal neutrophil polymorphs. Turgeon-O'Brien *et al.* (1985) noted a reduction in polymorphonuclear myeloperoxidase, measured both biochemically and by cytochemical scoring, in cases of iron deficiency anaemia and a lesser reduction, not clearly significant, in the anaemia of chronic disorders.

Reyes *et al.* (1978) have reported the presence of a peroxidase activity, located in the endoplasmic reticulum of cells from hairy cell leukaemia. The study was carried out employing several methods based on the incubation of fixed and unfixed slides in the presence of DAB and H_2O_2. Hairy cells had been previously considered as peroxidase negative on the basis of optical (Catovsky *et al.*, 1974; Nanba *et al.* 1977) and ultrastructural cytochemistry (Daniel and Flandrin, 1974). In these studies the methods employed were suitable for the detection of myeloperoxidase, which is highly resistant to fixatives (Roels *et al.*, 1975) and is commonly detected in the maturing granulocyte and monocyte series after glutaraldehyde fixation of cells. In contrast hairy cells have a peroxidase activity sensitive to aldehyde fixation, as shown by comparing the results of the three methods used in the study of Reyes *et al.* (1978): in method A, no hairy cells were found to be reactive under conditions of myeloperoxidase detection, while in method B reactivity was present in cells of a few patients, employing a low-dose aldehyde fixation. Finally, in method C, all samples were fully reactive when cells were treated with DAB without prior fixation. Hairy cell peroxidase was also characteristic in that its localization was found to be confined to the perinuclear space and other endoplasmic reticulum strands, a distribution substantially different from that of myeloperoxidase, which in addition to cytoplasmic reticulum, always involves the Golgi apparatus and granules of the granulocytic and monocytic precursors (Breton-Gorius and Reyes, 1976).

On the basis of its sensitivity to fixatives and inhibitors and of its localization restricted to the endoplasmic reticulum, hairy cell peroxidase appears to be similar to peroxidases of other tissues (Roels *et al.*, 1975) as well as to the peroxidase activity of the megakaryocyte series (Breton-Gorius and Guichard, 1972; Breton-Gorius *et al.*, 1978). A peroxidase activity similar to that of hairy cells was not detected by Reyes and his colleagues in the various T and B cell samples obtained from patients with proliferative disorders (including cultured samples) and in immunoblasts obtained *in vitro* after culturing in the presence of phytohaemagglutinin (PHA), concanavalin A (Con A), pokeweed mitogen (PWM) and in a continuous B cell line.

In conclusion the results of this cytochemical investigation raise the possibility that hairy cell leukaemia may be composed of cells expressing a hybrid phenotype, with peroxidase expression not otherwise found in lymphocytes but with B-cell capacity to synthesize SIg (Fu *et al.*, 1974; Utsinger *et al.*, 1977) and to produce Ig *in vitro* (Golde *et al.*, 1977). This hypothesis gains some support from individual reports of hybrid characteristics in the hairy cells from some cases, such as the possession of both T- and B-cell membrane markers (Burns *et al.*, 1977).

This survey of the nature, distribution and significance of the peroxidase reaction indicates that a considerable wealth of information has accumulated in recent years on its variations in different disease states, on its probable relationship with hormonal influences, especially ACTH, and on its dependence on the presence and integrity of certain structures such as the endoplasmic reticulum, the Golgi apparatus, and the primary and secondary granules of granulocyte and monocyte precursors.

Thus the study of this enzyme, which hitherto has been restricted to the identification and differentiation of lymphoblasts and myeloblasts, has become of considerable importance for the better understanding of the metabolism and function of normal cells and the derangements which occur in various pathological conditions and in the course of leukaemia.

Catalase

Catalase (EC 1.11.1.6) has been shown to be a marker enzyme for 'peroxisomes'. The enzyme is present in different vertebrate (particularly liver and kidney cells) and invertebrate tissues and has been described in plant cells, mainly yeast.

Catalase utilizes the H_2O_2 generated by other enzymes to oxidize a variety of substrates including phenols, formic acid, formaldehyde and alcohol. In addition, catalase will convert H_2O_2 to H_2O. This latter reaction is sometimes considered a safety device that prevents dangerous accumulation of the strong oxidizing agent H_2O_2 in the absence of a sufficient supply of hydrogen donors.

The large 'peroxisomes' in liver and kidney cells are thought to be important in detoxifying various molecules. In addition to these reactions, most peroxisomes catalyze the breakdown of fatty acids to acetyl CoA using a special H_2O_2 producing enzyme. The acetyl CoA produced can be transported via the cytosol to mitochondria to feed the citric acid cycle, or it can be used for biosynthetic reactions elsewhere. It is difficult to give precise indications regarding peroxisome metabolism because the composition of this organelle varies so widely. Thus the 'peroxisomes' in different cells of the same organism often contain very different sets of enzymes. Moreover, in some cells the number and content of peroxisomes vary, depending on the conditions.

The localization of catalase in 'peroxisomal' granules of leucocytes has been reported by Szmigielski (1972), using histochemical procedures and light microscopy. The histochemical reaction was based on the use of 3,3′ diaminobenzidine tetrahydrochloride. The author reported that catalase in mature neutrophils was localized as fine granules, diffusely distributed in the cytoplasm, whereas myeloperoxidase was apparent chiefly in larger granules. This distinction was based on estimates of relative size of the granules under light microscopy.

This work, although of considerable interest, has been subjected to some criticisms because estimation of granule size under the light microscope is considered unreliable, since aggregation and clumping of granules may occur. Furthermore, when simple blood films are used, overlapping of granules may also add to the confusion.

Another point of controversy concerns the presumed optimal and specific reactivity of catalase and peroxidase at pH 9.0 and 7.0 respectively, but no data have been provided to support the enzyme specificity of the reactions at these pH values. The author employed 3-amino-1,2,4-triazole (AT) as a specific inhibitor of catalase, but the question of the specificity of AT inhibition is still a matter of uncertainty.

In 1976 Nishimura *et al.* reported ultrastructural localization of peroxidatic catalase in the cytoplasmic granules of normal and acatalasemic leucocytes from human subjects. The authors employed a modification of the cytochemical procedure of Novikoff and Goldfischer (1969) and of Fahimi (1969) to improve the specificity of the alkaline diaminobenzidine reaction for catalase. Diaminobenzidine-positive reaction for peroxidative catalase was observed in large and medium-sized granules in the cytoplasm of normal neutrophils, but a striking and notable absence of reactivity was found in acatalasemic neutrophils. The test for myeloperoxidase, with the diaminobenzidine reaction performed at neutrality, disclosed positively stained granules in both normal and acatalasemic neutrophils. Similarities in size and configuration of the positively stained granules for these enzymes suggest that catalase is localized in organelles which may be primary or azurophilic granules.

Since myeloperoxidase has also been shown to be localized in the primary granules, it is possible that catalase and myeloperoxidase are present together or separately in these granules. The staining of eosinophilic granules, in both control and acatalasemic samples, was equally intense at pH 9 as at pH 7, and in this respect was very different from the reaction of the granules in acatalasemic neutrophils. Moreover, eosinophilic granules were very resistant to AT inhibition of peroxidatic reaction. Eosinophilic peroxidase, thus, has distinctive chemical properties which are strikingly different from either catalase or myeloperoxidase.

Lymphocytes showed no evidence of diaminobenzidine-positive response at either pH 7 or 9 in control and acatalasemic individuals, thus suggesting that lymphocytes normally are not endowed with peroxidase or catalase activity.

Whereas platelets conventionally fixed and stained with uranyl acetate and lead showed numerous electron dense granules which superficially resembled those in the cytoplasm of neutrophils, according to the authors the diaminobenzidine-H_2O_2 reaction did not stain the platelet granules in control and acatalasemic individuals either at pH 7 or 9. Platelets, therefore, apparently have neither myeloperoxidase nor catalase. The authors concluded that the specificity of the diaminobenzidine-H_2O_2 reaction at pH 9 for peroxidatic catalase of neutrophils is improved significantly if post-fixation with OsO_4 is avoided, secondly that myeloperoxidase of neutrophils may be identified with the same reaction carried out at pH 7, and finally that normal neutrophilic granules are readily inhibited by AT and the inhibition is effective for diaminobenzidine-H_2O_2 reactions at pH 7 and 9, thus indicating inactivation of myeloperoxidase as well as of peroxidatic catalase.

TECHNICAL METHODS

Reaction for cytochrome oxidase

Modified from Burstone, 1961

1. Fix fresh air-dried smears of blood or bone marrow in 60% acetone at +4°C for 30 seconds
2. Prepare incubation solution as follows: 15 mg of p-amino-diphenylamine and 1 drop of 8 amino-1,2,3,4-tetrahydroquinoline are dissolved in 0.5 ml reagent ethanol and placed in a 50 ml Erlenmeyer flask. 35 ml distilled water are added, followed by 15 ml 0.2 mol/l Tris buffer pH 7.4. Following addition of the buffer, the solution is shaken and filtered through folded filter into a Coplin jar
3. Catalase (2 mg per Coplin jar) is added to the incubation solution; also cytochrome c (20 mg) is added and dissolved by mild agitation of the solution with a glass rod
4. Immerse slides in the Coplin jar and incubate for 2–3 hours
5. Following incubation, slides are transferred directly to a Coplin jar containing 10% of cobaltous acetate in 10% formalin for 1 hour
6. Slides are washed in running tap water for 15 minutes and mounted in glycerol-gelatin.

Peroxidase reaction

1. Benzidine/orthotolidine method

After Graham-Knoll, modified by Quaglino & Flemans, 1958

1. Fix air-dried smears of bone marrow or peripheral blood for 30 seconds in a solution of 10 ml of 40% formalin and 90 ml of ethanol
2. Wash with tap water and air-dry
3. In a measuring cylinder 250 mg of o-tolidine are dissolved in 6 ml of ethanol and the solution is diluted with 4 ml of distilled water. 0.02 ml of hydrogen peroxide (20 vol.) is added. No filtration is necessary because o-tolidine does not cause any heavy precipitation
4. Apply the solution to the smears for 7 minutes
5. Wash in running tap water for 2 minutes and air-dry
6. Counterstain with Leishman or May-Grünwald-Giemsa.

2. 4-chloro-1-naphthol method

Elias, 1980

1. Fix, wash and air dry smears of peripheral blood or bone marrow as in Graham-Knoll's method
2. Incubate 10 minutes at room temperature in the reaction medium prepared as follows: 2 ml of a 0.8% solution (w/v) of 4-chloro-1-naphthol in absolute ethanol are mixed with 38 ml of 0.05 M Tris-HCl buffer pH 7.6. Filter the substrate solution, then add 1 ml of freshly made 0.12% H_2O_2 in distilled water.
3. Wash in several changes of distilled water and air dry
4. Counterstain with methyl green.

3. Diaminobenzidine (DAB) method

After Novikuff and Goldfischer, 1969

Reagents

(a) Fixative: buffered formol acetone (BFA) Na_2HPO_4 40 mg, KH_2PO_4 200 mg, acetone 90 ml, concentrated formalin 50 ml; distilled water 60 ml. Store at 4°C and use cold for fixation.

(b) Stock phosphate buffer: Dulbecco A, pH 7.3. NaCl 40 g. KCl 1.0 g. Na_2HPO_4 anhydrous 5.75 g, KH_2PO_4 1.0 g.
Distilled water 1 l.

(c) Working buffer: dilute stock buffer 1 l with 4 l distilled water.

(d) 3,3′ diaminobenzidine (Sigma).

(e) H_2O_2 (100 vol).

(f) Working incubation solution: Dissolve 30 mg DAB in 60 ml working buffer and add 120 μl of 100 vol H_2O_2: use immediately.

Technique

1. Fix air dried smears in BFA for 45 secs
2. Rinse with distilled water and drain dry
3. Incubate in working substrate solution for 10 mins
4. Rinse with distilled water
5. Counterstain with Carazzi's (or other water-soluble) haematoxylin 1 min
6. Rinse with distilled water and air dry.

4. A light microscope method for 'platelet' peroxidase

Khalaf and Hayhoe, 1987

Reagents

(a) Fixative: tannic acid-aldehyde mixture prepared freshly as follows: 0.1 mol/l phosphate buffer, pH 7.2 117.6 ml, tannic acid 1.2 g, paraformaldehyde 2.4 g, 25% glutaraldehyde 2.4 ml. The mixture is filtered, and stored at 4°C.

(b) Incubation medium:

DAB·4HCl (3,3′-diaminobenzidine tetrahydrochloride)	120 mg
Tris HCl buffer 0.05 mol/l, pH 7.6	58.5 ml
Ficoll 400	2 g
When the Ficoll is completely dissolved add 3% Hydrogen peroxide	60 µl

(prepared freshly by diluting 30% solution 1 in 10).
Mix well: adjust pH to 7.6 with 1 mol/l NaOH and filter if necessary.

(c) Potassium cyanide solution (78.5 mg KCN in 60 ml of tris HCl buffer) for use in control reaction.

Technique

1. Prepare film preferably from nucleated cell concentrates to minimize interference from peroxidative activity of erythrocyte haemoglobin
2. Fix films in tannic acid-aldehyde mixture at 4°C for 4 min
3. Wash in 0.1 ml/l phosphate buffer, pH 7.2 for a few seconds and retain slides in buffer at 4°C for 10 mins. Rinse in distilled water
4. Incubate in incubation medium at 4°C for 3 h, then at room temperature for 17 h in a dark place
5. Rinse thoroughly in distilled water
6. Counterstain with Carazzi's haematoxylin 10 min
7. Rinse briefly in alkaline tap water and then distilled water and air dry.

Control films are placed in the KCN solution for 30 mins before being incubated, and 78.5 mg KCN are also added to the incubation medium in which the control slides are incubated for 20 hours.

Ultrastructural cytochemistry of peroxidases

Reagents

(a) Incubation mixture for Graham and Karnovsky method (1966). Prepare a saturated solution of the free base of 3,3′-diaminobenzidine (DAB) in 0.05 mol/l tris-HCl buffer (pH 7.6) containing 0.01 % hydrogen peroxide. Filter immediately before use. The solution should be freshly prepared
(b) Incubation mixture for Novikoff and Goldfischer method (1969). Prepare a solution of 20 mg of DAB tetrahydrochloride in 10 ml of 0.05 mol/l propanediol buffer (pH 9) containing 0.2 ml of freshly diluted 1% hydrogen peroxide. Filter immediately before use. The solution should be freshly prepared

Method

Carry out procedure as described for immunoperoxidase reaction for use in electron microscopy (p. 59), replacing step 3 by incubation for 10 minutes in the appropriate medium, either (a) or (b) above.

Fixation with 1.5% glutaraldehyde is satisfactory for the demonstration of the myeloperoxidase present in granulocytes and some monocytes, but platelet and megakaryocyte peroxidase and the peroxidase activity of hairy cells are inhibited by this aldehyde fixation and an alternative fixative procedure is required. The following fixative, recommended by Anderson *et al.* (1975), gives satisfactory results: 1% tannic acid, 2% paraformaldehyde, and 0.5% glutaraldehyde in 0.1 mol/l phosphate buffer at pH 7.2. The fixative is prepared just prior to use and chilled at 4°C. The method by Novikoff and Goldfischer, with slight modifications as described by Nashimura *et al.* (1976), may be employed for the ultrastructural demonstration of leucocyte catalase.

REFERENCES

Ackerman G A 1968 Ultrastructure and cytochemistry of the developing neutrophil. Laboratory Investigation 19: 290
Ackerman G A, Clark M A 1971a Ultrastructural localization of peroxidase activity in normal human bone marrow cells. Zeitschrift für Zellforschung und Mikroskopische Anatomie 117: 463
Ackerman, G A, Clark M A 1971b Ultrastructural localization of peroxidase activity in human basophil leucocytes. Acta Haematologica (Basel), 45: 280
Agner K 1941 Verdoperoxidase. A ferment isolated from leucocytes. Acta Physiologica Scandinavica, 2, Suppl. 8
Anderson W A, Trantalis J, Kang Y-H 1975 Ultrastructural localization of endogenous mammary gland peroxidase during lactogenesis in the rat. Results after tannic acid-formaldehyde-glutaraldehyde fixation. Journal of Histochemistry and Cytochemistry 23: 295
Andrews P C, Babior B M 1983 Endogenous protein phosphorylation by resting and activated human neutrophils. Blood 61: 333
Babior B M 1978 Oxygen-dependent microbial killing by phagocytes. New England Journal of Medicine 298: 659
Badwey J A, Karnovsky M L 1980 Active oxygen species and the functions of phagocytic

leukocytes. Annual Review of Biochemistry 49: 695
Baggiolini M, Hirsch J G, de Duve C 1970 Further biochemical and morphological studies of granule fractions from rabbit heterophil leukocytes. Journal of Cell Biology 45: 586
Bainton D F, Farquhar M G 1967 Segregation and packaging of granule enzymes in eosinophils. Journal of Cell Biology 35: 6A
Bainton D F, Farquhar M G 1968a Differences in enzyme content of azurophil and specific granules of polymorphonuclear leukocytes. I. Histochemical staining of bone marrow smears. Journal of Cell Biology 39: 286
Bainton D F, Farquhar M G 1968b Differences in enzyme content of azurophil and specific granules of polymorphonuclear leukocytes. II. Cytochemistry and electron microscopy of bone marrow cells. Journal of Cell Biology 39: 299
Bainton D F, Farquhar M G 1970 Segregation and packaging of granule enzymes in eosinophil leukocytes. Journal of Cell Biology 45: 54
Bainton D F, Ullyot J L, Farquhar M G 1971 The development of neutrophil polymorphonuclear leukocytes in human bone marrow. Origin and contents of azurophilic and specific granules. Journal of Experimental Medicine 134: 907
Barka T, Anderson P J 1963 Histochemistry, Theory, Practice and Bibliography. Hoeber Medical Division, Harper & Row, New York
Becker S W, Praver L L, Thatcher H 1935 Improved (paraffin section) method for DOPA reaction: with considerations of DOPA-positive cell, as studied by this method. Archives of Dermatology and Syphilology 31: 190
Bedetti C D 1985 Immunocytochemical demonstration of cytochrome c oxidase with an immunoperoxidase method. Journal of Histochemistry and Cytochemistry 33: 446
Behnke O 1969 Demonstration of endogenous peroxidase activity in the electron microscope. Journal of Histochemistry and Cytochemistry 17: 62
Bendix-Hansen K 1987 Enzyme cytochemistry of neutrophil granulocytes. British Journal of Haematology 65: 127
Bendix-Hansen K, Kerndrup G 1985 Myeloperoxidase-deficient polymorphonuclear leucocytes (V). Relation to FAB classification and neutrophil alkaline phosphatase activity in primary myelodysplastic syndromes. Scandinavian Journal of Haematology 35: 197
Bendix-Hansen K, Nielsen H K 1983a Myeloperoxidase-deficient polymorphonuclear leucocytes (I). Incidence in untreated myeloid leukaemia, lymphoid leukaemia and normal humans. Scandinavian Journal of Haematology 30: 415
Bendix-Hansen K, Nielsen H K 1983b Myeloperoxidase-deficient polymorphonuclear leucocytes (II). Longitudinal study in acute myeloid leukaemia, untreated, in remission and in relapse. Scandinavian Journal of Haematology 31: 5
Bessis M 1968 Ultrastructure of normal and leukemic granulocytes. In Zarafonetis CJD (ed) Proceedings of the International Conference on Leukemia-Lymphoma. Lea & Febiger, Philadelphia, p 281
Bessis M, Maigné J 1970: Le diagnostic des variétés de leucémies aiguës par la réaction des peroxydases au microscope électronique. Son intérêt et ses limites. Revue Europeene d'Etudes Cliniques et Biologiques 15: 691
Breton-Gorius J 1970 Morphological aspects of granulopoiesis. Pathologie Biologie 18: 433
Breton-Gorius J 1980 The value of cytochemical peroxidase reactions at the ultrastructural level in haematology. Histochemical Journal 12: 127
Breton-Gorius J, Guichard J 1969 Etude au microscope électronique de la localisation des peroxydases dans les cellules de la moelle osseuse humaine. Nouvelle Revue Francaise d'Hematologie 9: 678
Breton-Gorius J, Guichard J 1972 Ultrastructural localization of peroxidase activity in human platelets and megakaryocytes. American Journal of Pathology 66: 277
Breton-Gorius J, Reyes F 1976 Ultrastructure of human bone marrow cell maturation. International Review of Cytology 46: 251
Breton-Gorius J Coquin Y, Vilde J L, Dreyfus B 1976 Cytochemical and ultrastructural studies on aberrant granules in the neutrophils of two patients with myeloperoxidase deficiency during a preleukaemic state. Blood cells 2: 187
Breton-Gorius J, Reyes F, Duhamel G, Najman A, Gorin N C 1978 Megakaryoblastic acute leukemia. Identification by the ultra-structural demonstration of platelet-peroxidase. Blood 51: 45
Breton-Gorius J, Gourdin M F, Reyes F 1981 Ultrastructure of the leukemic cell. In: Catovsky D (ed) The Leukemic Cell. Churchill-Livingstone, Edinburgh, p 87
Breton-Gorius J, Tabilio A, Vainchenker W, Vinci G, Van Haeke D, Henri A, Guichard J, Canizo M C 1983 Diagnosis of megakaryoblastic leukaemias. Verhandlungen der Deutschen Gesellschaft fur Pathologie 67: 166
Breton-Gorius J, Van Haeke D, Pryzwansky K B, Guichard J, Tabilio A, Vainchenker W, Carmel R

1984 Simultaneous detection of membrane markers with monoclonal antibodies and peroxidatic activities in leukaemia: ultrastructural analysis using a new method of fixation preserving the platelet peroxidase. British Journal of Haematology 58: 447

Breton-Gorius J, Villeval J L, Mitjavila M T, Vinci G, Guichard J, Rochart H, Flandrin G, Vainchenker W 1987 Ultrastructural and cytochemical characterization of blasts from early erythroblastic leukemias. Leukemia 1: 173

Briggs R T, Drath D B, Karnovsky M L, Karnovsky M J 1975 Localization of NADH oxidase on the surface of human polymorphonuclear leukocytes by a new cytochemical method. Journal of Cell Biology 67: 566

Burns G F, Nash A A, Worman C P, Barker C R, Hayhoe F G J, Cawley J C 1977 A human leukaemic cell expressing hybrid membrane phenotypes. Nature, 268: 243

Burstone M S 1959 New histochemical techniques for the demonstration of tissue oxidase (Cytochrome oxidase). Journal of Histochemistry and Cytochemistry 7: 112

Burstone M S 1960 Histochemical demonstration of cytochrome oxidase with new amine reagents. Journal of Histochemistry and Cytochemistry 8: 63

Burstone M S 1961 Modifications of histochemical techniques for the demonstration of cytochrome oxidase. Journal of Histochemistry and Cytochemistry 9: 59

Carruthers C, Suntzeff V 1947 Succinic dehydrogenase and cytochrome oxidase in epidermal carcinogenesis induced by methylcholanthrene in mice. Cancer Research 7: 9

Catovsky D, Galton D A G, Robinson J 1972 Myeloperoxidase deficient neutrophils in acute myeloid leukaemia. Scandinavian Journal of Haematology 9: 142

Catovsky D, Pettit J E, Galton D A G, Spiers A S D, Harrison C V 1974 Leukemic reticuloendotheliosis ('hairy' cell leukemia): a distinct clinico-pathological entity. British Journal of Haematology 26: 9

Cawley J C, Hayhoe F G J 1973 Ultrastructure of Haemic Cells. Saunders, London.

Cech P, Schneider P, Bachmann F 1982 Partial myeloperoxidase deficiency. Acta Haematologica 67: 180

Chang KS, Schroeder W, Siciliano MJ, Thompson LH, McCredie K, Beran M, Freireich EJ, Liang JC, Trujillo JM, Stass SA 1987 The localization of the human myeloperoxidase gene is in close proximity to the translocation breakpoint in acute promyelocytic leukemia. Leukemia 1:458

Chaundhry A N, Santinga J T, Gabig T G 1982 The subcellular particulate NADPH-dependent O_2^- generating oxidase from human blood monocytes: comparison to the neutrophil system. Blood, 60: 979

Ciplea A G, Mayer D 1981 A new histochemical method to assess and evaluate potential bone marrow damage from therapeutic substances. Histochemistry, 71: 481

Cotran R S, Litt M 1969 The entry of granule-associated peroxidase into the phagocytic vacuoles of eosinophils. Journal of Experimental Medicine 129: 1291

Cramer E, Pryzwansky K B, Villeval J-L, Testa U, Breton-Gorius J 1985 Ultrastructural localization of lactoferrin and myeloperoxidase in human neutrophils by immunogold. Blood, 65: 423

Crawford D F, Nachlas M M 1958 Evaluation of indophenol derivatives as reagents for demonstrating cytochrome oxidase in tissue sections. Journal of Histochemistry and Cytochemistry 6: 438

Curnutte J T, Karnovsky M L, Babior B M 1976 Manganese-dependent NADPH oxidation by granulocyte particles. Journal of Clinical Investigation 57: 1059

Daniel M T, Flandrin G 1974 Fine structure of abnormal cells in hairy cell (tricholeukocytic) leukemia, with special reference to their in vitro phagocytic capacity. Laboratory Investigation 30: 1

De Duve C 1973 Biochemical studies on the occurrence, biogenesis and life history of mammalian peroxidase. Journal of Histochemistry and Cytochemistry 21: 941

Dri P, Cramer R, Menegazzi R, Patriarca P 1985 Increased degranulation of human myeloperoxidase-deficient polymorphonuclear leucocytes. British Journal of Haematology 59: 115

Dunn W B, Hardin J H, Spicer S S 1968 Ultrastructural localization of myeloperoxidase in human neutrophil and rabbit heterophil and eosinophil leukocytes. Blood, 32: 935

Ehrlich P 1885 Das Sauerstoffbedürfnis des Organismus. Hirschwald, Berlin

Elias J M 1980 A rapid, sensitive myeloperoxidase stain using 4-chloro-1-naphthol. American Journal of Clinical Pathology 73: 797

Enomoto T, Kitani T 1966 Electron microscopic studies on peroxidase and acid phosphatase reaction in human leukocytes. In normal and leukemic cells and on the phagocytosis. Acta Haematologica Japonica, 29: 554

Fahimi H D 1969 Cytochemical localization of peroxidatic activity of catalase in rat hepatic microbodies (peroxisomes). Journal of Cell Biology 43: 275

Felberg N T, Schultz J 1972 Evidence that myeloperoxidase is composed of isoenzymes. Archives of Biochemistry and Biophysics 148: 407
Fu S M, Winchester R J, Rai K R, Kunkel H G 1974 Hairy cell leukemia: proliferation of a cell with phagocytic and B-lymphocyte properties. Scandinavian Journal of Immunology 3: 847
Gabig T G, Babior B M 1979 The O_2 forming oxidase responsible for the respiratory burst of the human neutrophil: properties of the solubilized enzyme. Journal of Biological Chemistry 254: 9070
Gallico E 1947 Contributo allo studio delle deidrogenasi nei tessuti tumorali con il metodo del bleu di metilene. Tumori 21: 139
Girelli M, Panagia G 1954 Significato biochimico di alcune reazioni citoenzimatiche impiegate in ematologia. Minerva Medica, 1: 407
Golde D W, Stevens R H, Quan S G, Saxon A 1977 Immunoglobulin synthesis in hairy cell leukaemia. British Journal of Haematology 35: 359
Gomori G 1951 Histochemical staining methods. In: Visscher MB (ed) Methods in Medical Research. The Year Book Publishers, Chicago, p 14
Gomori G 1952 Microscopic Histochemistry. Principles and Practice. Univ. of Chicago Press, Chicago
Graham R C Jr, Karnovsky M J 1966 The early stages of absorption of injected horse-radish peroxidase in the proximal tubules of mouse kidney: ultrastructural cytochemistry by a new technique. Journal of Histochemistry and Cytochemistry 14: 291
Hanker J S, Yates P E, Metz C B, Rustioni A 1977 A new specific, sensitive and non-carcinogenic reagent for the demonstration of horseradish peroxidase. Histochemical Journal 9: 789
Hargreaves A B, Deutsch H F 1952 The 'in vitro' inhibition of catalase by a tumour factor. Cancer Research 12: 720
Hayhoe F G J, Quaglino D, Doll R 1964 The Cytology and Cytochemistry of Acute Leukaemias. MRC Special Report Series, No. 304, HMSO, London
Hirsch J G, Cohn A Z 1964 Digestive and autolytic functions of lysosomes in phagocytic cells. Federation Proceedings 23: 1023
Hoffmann G T, Rottino A, Stern K G 1951 Demonstration by the Nadi reaction of cytochrome oxidase activity in cells of the lymphoid and myeloid series obtained from normal individuals and patients suffering from Hodgkin's and other diseases. Blood, 6: 1051
Horn R G, Spicer S S, Wetzel B K 1964 Phagocytosis of bacteria by heterophil leukocytes. American Journal of Pathology 45: 327
Hudson G 1966 Eosinophil granules and phosphotungstic acid: an electron microscope study of guinea pig bone marrow. Experimental Cell Research 41: 265
Inagaki A, Uno S, Yoneda M, Ohkawa K 1975 2,7 fluorenediamine and 2,5 fluorenediamine as peroxidase reagents for blood smears. Journal of Laboratory and Clinical Medicine 88: 334
Iyer G Y N, Islam M F, Quastel J M 1961 Biochemical aspects of phagocytosis. Nature 192: 535
Kakefuda T 1968 Electron microscopy of normal and leukemic cells. In: Amromin G D (ed) Pathology of Leukemia. Harper & Row, New York, p 82
Kaplow L S 1975 Substitute for benzidine in myeloperoxidase stains. American Journal of Clinical Pathology 63: 451
Kaplow L S 1986 The relationship of myeloperoxidase activity to neutrophil maturity and other haematologic indicators of infection. Blood Cells 12: 153
Karnovsky M J, Roots L 1964 A "direct-coloring" thiocholine method for cholinesterases. Journal of Histochemistry and Cytochemistry 12: 219
Keilin D, Hartree E F 1938 Cytochrome oxidase. Proceedings of the Royal Society of London (Series B) 125: 171
Khalaf M R, Hayhoe F G J 1987 Cytochemical demonstration of 'platelet' peroxidase at light microscope level. European Journal of Haematology 39: 128
Koeffler H P, Ranyard J, Pertcheck M 1985 Myeloperoxidase: its structure and expression during myeloid differentiation. Blood, 65: 484
Koike T, Nagayama R, Takahashi M, Hattori A, Shibata A 1986 Perinuclear peroxidatic activity in erythroblasts. British Journal of Haematology 64: 621
Léhrer R I, Cline M J 1969 Leucocyte myeloperoxidase deficiency and disseminated candidiasis. The role of myeloperoxidase in resistance to candida infection. Journal of Clinical Investigation 48: 1478
Lehrer R I, Goldberg L S, Apple M A, Rosenthal N P 1972 Refractory megaloblastic anaemia with myeloperoxidase-deficient neutrophils. Annals of Internal Medicine 76: 447
Lillie R D, Burtner H J 1953 Stable sudanophilia of human neutrophil leucocytes in relation to peroxidase and oxidase. Journal of Histochemistry and Cytochemistry 1: 8
Lippi U, Cappelletti P 1982a Myeloperoxidase staining by the reagents of an automated differential analyzer (Hemalog D). American Journal of Clinical Pathology 78: 257
Lippi U, Cappelletti P 1982b Cytochemical evidence of multiple molecular forms of

myeloperoxidase (MPO). Haematologica, 67: 807
Lippi U, Cappelletti P, Schinella M 1983 I deficit di mieloperossidasi. Piccin Editore, Padova
Lison L 1936 Sur la recherche histochimique des oxydases par la reaction du bleu d'indophenol. Le cas des lipides. Bulletin de la Societe de Chimie Biologique 18: 185
Lojda Z, Gossrau R, Schiebler T H 1979 Enzyme histochemistry. Springer-Verlag, Berlin
Mauro G, Prato V 1954 Ricerche sul comportamento delle perossidasi leucocitarie dopo somministrazione di ACTH. Archivo per le Scienze Mediche 98: 1
McMichael A J, Rust N A, Pilch U R, Sochynsky R, Morton J, Mason D Y, Ruan C, Tobelem G, Caen J 1981 Monoclonal antibody to human glycoprotein I. I. Immunological studies. British Journal of Haematology 49: 501
Morel F, Vignais P V 1984 Examination of the oxidase function of the b-type cytochrome in human polymorphonuclear leucocytes. Biochimica et Biophysica Acta, 764: 213
Morishita Y, Morishima Y, Ogura M, Nagay Y, Ohno R 1986 Biochemical characterization of human myeloperoxidase using three specific monoclonal antibodies. British Journal of Haematology 63: 435
Murata F, Hardin J H, Spicer S S 1973 Coexistence of acid phosphatase and acid mucosubstance in the nucleoid of human blood platelet granules. Histochemie, 35: 319
Nachlas M M, Crawford D T, Goldstein T P, Seligman A 1958 The histochemical demonstration of cytochrome oxidase with a new reagent for the Nadi reaction. Journal of Histochemistry and Cytochemistry 6: 445
Nanba K, Jaffe E S, Soban E J, Braylan R C, Berard C W 1977 Hairy cell leukemia. Enzyme histochemical characterization with special reference to splenic stromal changes. Cancer 39: 2323
Nielsen H K, Bendix-Hansen K 1984 Myeloperoxidase-deficient polymorphonuclear leucocytes (III). Relation to incidence of infection in acute myeloid leukaemia. Scandinavian Journal of Haematology 33: 75
Nishimura E T, Whest G M, Yang H-Y 1976 Ultrastructural localization of peroxidatic catalase in human peripheral blood leucocytes. Laboratory Investigation 34: 60
Novikoff A B, Goldfischer S 1969 Visualization of peroxisomes (microbodies) and mitochondria with diaminobenzidine. Journal of Histochemistry and Cytochemistry 17: 675
Novikoff A B, Shin W Y 1964 The endoplasmic reticulum in the Golgi zone and its relations to microbodies, Golgi apparatus and autophagic vacuoles in rat liver cells. Journal de Microscopie 3: 187
Olsson I, Persson A-M, Strömberg K, Winqvist I, Tai P-C, Spry C J F 1985 Purification of eosinophil peroxidase and studies of biosynthesis and processing in human marrow cells. Blood 66: 1143
Papadimitriou J M, Van Duiyn P, Brederoo P, Streepkerk J D 1976 A new method for the cytochemical demonstration of peroxidase for light, fluorescence and electron microscopy. Journal of Histochemistry and Cytochemistry 24: 82
Parmley R T, Spicer S S, Wright N J 1974 Diaminobenzidine activity in the platelet dense body. Journal of Histochemistry and Cytochemistry 22: 1063
Parwaresch M R 1976 The Human Blood Basophil. Springer-Verlag, Berlin
Pearse A G E 1960 Histochemistry, theoretical and applied. Churchill, London
Pedersen B, Hayhoe F G J 1971 Relation between phagocytic activity and alkaline phosphatase content of neutrophils in chronic myeloid leukaemia. British Journal of Haematology 21: 257
Pember S, Shapira R, Kinkade J 1983 Multiple forms of myeloperoxidase from human neutrophilic granulocytes: evidence for differences in compartmentalization, enzymatic activity, and subunit structure. Archives of Biochemistry and Biophysics 221: 391
Presentey B 1969 Morphologic observations and genetic follow-up of a familial anomaly of eosinophils. American Journal of Clinical Pathology 51: 458
Quaglino D 1961 Cytochemistry of blood and bone marrow cells with particular reference to leukaemia and allied disease. Ph. D. Thesis, Cambridge University.
Quaglino D 1964 Ricerche sulle variazioni quantitative della fosfatasi alcalina e della perossidasi nei granulociti neutrofili della leucemia mieloide cronica e di alcune reazioni di tipo leucemoide. Bollettino della Società di Medicina e Chirurgia di Modena 44: 5
Quaglino D, Flemans R 1958 Peroxidase staining in leucocytes. Lancet ii: 1020
Quaglino D, Emilia G, Artusi T, Ferrari G 1965 La reazione della citocromo-ossidasi in alcuni tipi di cellule ematiche normali e patologiche. Bollettino della Società Italiana Biologia Sperimentale 41: 636
Rabellino E, Levene R B, Leung L L K, Nachman R L 1981 Human megakaryocytes. II. Expression of platelet proteins in early marrow megakaryocytes. Journal of Experimental Medicine 154: 88
Repine J E, Clawson C C, Brunning R D 1976 Abnormal pattern of bactericidal activity of

neutrophils deficient in granules, myeloperoxidase and alkaline phosphatase. Journal of Laboratory and Clinical Medicine 88: 788
Reyes F, Gourdin M F, Farcet J P, Dreyfus B, Breton-Gorius J 1978 Synthesis of a peroxidase activity by cells of hairy cell leukemia: a study by ultrastructural cytochemistry. Blood 52: 537
Roberts P J, Cross A R, Jones O T G, Segal A W 1982 Development of cytochrome b and an active oxidase system in association with maturation of a human promyelocytic (HL-60) cell line. Journal of Cell Biology 95: 720
Robinson J M, Briggs R T, Karnovsky M J 1978 Localization of D-amino acid oxidase on the cell surface of human polymorphonuclear leukocytes. Journal of Cell Biology 77: 59
Robinson J M, Badwey J, Karnovsky M L, Karnovsky M J 1979 Biochemical and cytochemical characterization of an aldehyde oxidase from guinea pig polymorphonuclear leukocytes (PMNS) Journal of Cell Biology 83: 123
Roels F 1974 Cytochrome C and cytochrome oxidase in diaminobenzidine staining of mitochondria. Journal of Histochemistry and Cytochemistry 22: 442
Roels F, Wisse E, De Prest B, Van Der Meulen 1975 Cytochemical discrimination between catalases and peroxidases using diaminobenzidine. Histochemistry 41: 281
Schmidt H 1981 Zum histochemischen Nachweis von Phenoloxidase in menschlichen Blutzellen. Folia Haematologica 108: 1981
Schneider W C, Potter V R 1943a The assay of animal tissues for respiratory enzymes. II. Succinic dehydrogenase and cytochrome oxidase. Journal of Biological Chemistry 149: 217
Schneider W C, Potter V R 1943b Succinic dehydrogenase and cytochrome oxidase. Cancer Research 3: 353
Scott R B, Horn R G 1970 Fine structural features of eosinophil granulocyte development in humans. Laboratory Investigation 23: 292
Sehrt E 1927 Die histologische Darstellung der Lipoide der weissen Blutzellen und die Beziehung dieser Lipoide zur Oxydasereaktion. Münchener Medizinische Wochenschrift 4: 139
Seligman A M, Karnovsky M J, Wasserkrug H L, Hanker J S 1968 Nondroplet ultrastructural demonstration of cytochrome oxidase activity with a polymerizing osmiophilic reagent diaminobenzidine. Journal of Cell Biology 38: 1
Spicer S S, Hardin J H 1969 Ultrastructure, cytochemistry and function of neutrophil leukocyte granules. A review. Laboratory Investigation 20: 488
Sullivan AK, Amatruda TT, Fitz-Gibbon L, Koeffler HP, Peyman J, Rowden G, Shematek G, Shihab-El-Deen A 1987 Deficiency of myeloperoxidase and abnormal chromosome 1 occurs in variant (HL60) promyelocytes. Leukemia Research 10: 501
Szmigielski S 1972 Peroxisomal enzymes in human granulocytes. I. Catalase and peroxidase. Folia Histochemica et Cytochemica 10: 47
Tetteroo P A T, Landsdorp P M, Leeksma O C, von dem Borne A E G Kr 1983 Monoclonal antibodies against human glycoprotein IIIa. British Journal of Haematology 55: 509
Tetteroo P A T, van der Schoot C E, von dem Borne A E G Kr 1987 New developments in the phenotypic analysis of (im)mature myeloid leukemia: clinical relevance. Proc. Int. Meeting on genotypic, phenotypic and functional aspects of haematopoiesis. Assisi, 6th–8th April 1987, p 13A
Turgeon-O'Brien H, Amiot J, Lemieux L, Dillon J C 1985 Myeloperoxidase activity of polymorphonuclear leucocytes in iron deficiency anaemia and anaemia of chronic disorders. Acta Haematologica 74: 151
Undritz E 1966 Die Alius-Grignaschi-Anomalie: der erblich-konstitutionelle Peroxidase-Defekt der Neutrophilen und Monozyten. Blut 14: 129
Utsinger P D, Yount W J, Fuller C R, Logue G J, Orringer E P 1977 Hairy cell leukemia: B lymphocyte and phagocytic properties. Blood 49: 19
Vaccari F, Traldi A 1956 Glucose consumption, glycolysis and cellular respiration of leukaemic leukocytes during antimitotic therapy. Schweizerische Medizinische Wochenschrift 51/52: 1466
Valentine WN, Follette JH, Hardin EB, Beck WS, Lawrence JS 1954 Studies on leucocyte alkaline phosphatase activity. Relation to 'stress' and pituitary-adrenal activity. Journal of Laboratory and Clinical Medicine 44: 219
Wetzel B K, Horn R G, Spicer S S 1967 Fine structural studies on the development of heterophil, eosinophil and basophil granulocytes in rabbits. Laboratory Investigation 16: 349
White J G 1972 Interaction of membrane systems in blood platelets. American Journal of Pathology 66: 295
Woodliff H J 1963 The effect of antimetabolites on the respiration of leukaemic mouse cells. Blood, 22: 199
Yamada E 1966 Electron microscopy of the peroxidase in granular leucocytes of rat bone marrow. Archivium Histologicum Japonicum 27: 131

CHAPTER 11

Other enzymes

SPECIFIC PHOSPHATASES, PEPTIDASES, β-GLUCURONIDASE γ-GLUTAMYL TRANSFERASE, β-GLUCOSAMINIDASE, PHOSPHORYLASES, ADENOSINE DEAMINASE, SULPHATASES, DIHYDROFOLATE REDUCTASE, TETRAHYDROFOLATE CYCLODEHYDRASE AND CYCLOHYDROLASE

Among a number of phosphatases with particular substrate specificity, those with potential relevance to haematology include 5-nucleotidase, glucose-6-phosphatase, adenosine triphosphatase and phosphamidase.

5′-Nucleotidase (5′N)

This enzyme (EC 3.1.3.5) is a specific phosphatase which splits phosphate esters on the 5-carbon of nucleosides but does not hydrolyse nucleoside phosphate esters on other carbon sites or phosphates of glycerol, phenols or sugars. Thus the enzyme is active against substrates such as adenosine-5-phosphate but not against β-glycerophosphate or p-phenylphosphate, whereas non-specific alkaline phosphatase hydrolyses all these substrates.

Demonstration of the enzyme cytochemically requires a method which prevents, so far as possible, non-specific alkaline phosphatase activity from contributing to hydrolysis of the specific substrate. This can be partially achieved by carrying out the reaction at lower pH values than usual for alkaline phosphatase, since 5-nucleotidases appear to retain considerable activity at pH levels as low as 6.5, even though their pH optima in different tissues are generally between 7.5 and 8.5 (Hardonk and de Boer, 1968). Thus Gomori (1952) and Pearse and Reis (1952) used methods similar to Gomori's calcium method for alkaline phosphatase, but with adenosine-5-phosphate as substrate and incubation at pH 8.3 and 7.5 respectively. The positivity found could be compared with that in control preparations exposed to β-glycerophosphate as substrate to allow for a component of reaction due to non-specific alkaline phosphatase. Naidoo and Pratt (1954) used a direct lead method at pH 6.5, where a similar control comparison with non-specific acid phosphatase is desirable.

With methods of this kind positivity has been found in various tissue and cellular locations differing to some extent from but often overlapping considerably with the sites of acid or alkaline phosphatase positivity. Until recent years there had been few cytochemical studies reported on the distribution of 5′-nucleotidase in haemic cells, perhaps because the methods available did not offer a clear enough specific reaction, but there has been considerable interest in the enzyme as measured biochemically because of the demonstration (Emmelot and Bos, 1965; Fleischer *et al.*, 1969 and de Pierre and Karnovsky, 1974) that it appears to provide a marker for plasma membrane activity. Enzyme assays by biochemical testing of plasma membrane extracts of lymphocytes or by surface immunofluorescence have shown low levels in chronic lymphocytic leukaemia and some other lymphoproliferative states (Lopes *et al.*, 1972; Quagliata *et al.*, 1974; La Mantia *et al.*, 1975; Kramers *et al.*, 1976), while low levels have also been reported in cord blood lymphocytes (Kramers *et al.*, 1977). Differences have not been demonstrated in the T and B subpopulations of lymphocytes, but the possibility has been raised that the enzyme expression is related to cell maturity, low levels being found in lymphocytes, whether normal or neoplastic, at an early stage in their developmental pathway. The biochemically detected differences in enzyme levels, for example, between normal and CLL lymphocytes, were sufficiently striking as to suggest that cytochemical tests, which are potentially more quantitative than immunofluorescence, should be capable of demonstrating them, and observations of this kind led to further cytochemical studies.

Silber *et al.* (1975) used a modified form of the Wachstein and Meisel lead method (1957), with the addition of tartrate and of excess β-glycerophosphate to inhibit respectively acid and alkaline phosphatase reactivity and to retain lead in solution by chelation. A prolonged incubation period of 20 hours was used and normal lymphocytes were found to be either positive or negative for the enzyme, with little variation in the extent of positivity when present. Low levels of enzyme were found in CLL and in HCL. Similar findings were reported by El-Mohandes and Hayhoe (1983a) using a more sensitive technique which enabled a semiquantitative score of 5′N positivity to be assessed in lymphocytes from 8 normal subjects and from 60 patients with various types of leukaemia or lymphoma. With this method the normal score ranged from 51 to 109 with a median of 82; moderate reductions occurred in CML and AML, and more marked reductions in ALL, HCL, and 16 of 18 cases of CLL and also in both Hodgkin's disease and non-Hodgkin's lymphoma. Higher than normal scores occurred in 2 of 18 cases of CLL.

These observations are in accordance with earlier biochemical reports (Quagliata *et al.*, 1974; Kanter *et al.*, 1979), and with the cytochemical findings of Silber *et al.* (1975), and Pangalis *et al.* (1984). The explanation is not entirely clear. Since CLL is usually a B-cell neoplasm, the decrease of 5′N activity in lymphocytes of most CLL patients could be explained on the basis of an altered B:T cell ratio if B cells were generally less positive than T cells (Kramers *et al.*, 1976; Barton and Goldschneider, 1978). But Matamoros *et al.* (1979) and Rowe *et al.* (1979) found that B lymphocytes had about four times the 5′N activity of T cells. Moreover El-Mohandes and Hayhoe (1983a) found no relationship in 9 cases of lymphoma or CLL between 5′N levels and the percentages of B and T subpopulations.

An alternative suggestion is that 5′N activity reflects cell maturity, with the CLL population consisting of less mature cells than most of those in normal peripheral blood (Conklyn and Silber, 1982). The occasional cases of CLL with normal or high scores might represent B-cell neoplasms at a somewhat more mature stage along the B-cell pathway than is usually the case. In this connection it is relevant to note the combined immunological and biochemical study of Ho *et al.* (1986) which showed that all B-CLL cases with high levels of 5′N manifested cytoplasmic immunoglobulins, were classifiable as immunocytomas, and showed features of increasing cell maturity cytologically and biochemically. Conversely, the reduced 5′N activity observed in patients with primary immunodeficiency and hypogammaglobulinaemia (Matamoros *et al.*, 1979) may reflect lack of immune developmental maturation in the B lymphocyte line. Murray *et al.* (1984) found similarly decreased 5′N activity in lymphocytes from patients with acquired immune deficiency syndrome (AIDS) and asymptomatic but sexually active homosexual men. Although the populations of helper T cells (T4) were reduced there was no correlation between 5′N activity and percentage of either T4 or suppressor T cells (T8), nor with T4/T8 ratios, and both T4 and T8-enriched samples of lymphocytes showed decreased 5′N values. Here the decreases in 5′N appear to be a generalized phenomenon affecting all peripheral blood lymphocytes.

That 5′N may be expressed at particular phases of maturation or activation of lymphocytes is further supported by observations of lymphocytes at different sites. Uusitalo (1981) compared the 5′N cytochemical reactivity of mouse lymphocytes in the thymus and spleen. In the thymus all were negative whereas some 20 to 30% of lymphocytes in the spleen showed positivity.

Perhaps also pertinent to this issue are the findings of El-Mohandes and Hayhoe (1983b) on 5′N activity in myeloma cells. They studied cytochemically the plasma cells from bone marrow aspirates from 11 patients with multiple myeloma and from 12 patients with essentially normal marrows but adequate numbers of plasma cells. About 90% of the myelomatous plasma cells showed moderately strong positivity, although little or no reactivity was found in normal plasma cells. In later studies in our laboratories Khalaf (1986) and Khalaf and Hayhoe (1987a), using a different cytochemical method, with uridine-5′-monophosphate as substrate and with an extended incubation period, confirmed these findings in general, with 4 of 6 further myeloma cases showing variable degrees of positivity in the plasma cells (Plate 27D), but also found positive reactions in a few normal plasma cells from control marrows. Macrophages from bone marrow were generally strongly positive (Plate 27B), but monocytes either in bone marrow or in peripheral blood were negative or showed at most only faintly detectable 5′N activity.

Normally megakaryocytes show little or no reaction, but immature forms may be strongly positive, both in idiopathic autoimmune thrombocytopenic purpura and in at least some cases of acute megakaryoblastic leukaemia (El-Mohandes and Hayhoe, 1983c, d), although Oliveira and her colleagues (1987) found positive reactions in only 2 or 7 cases of acute megakaryoblastic leukaemia treated, and in none of 6 megakaryoblastic transformations of CML.

There has been dispute over 5′N activity in neutrophils. Strauss and Burrows (1975) and Shirley and colleagues (1975) failed to find the enzyme in human neutrophils, whereas Rustin and Peters (1979) and Wilson *et al* (1981), by ultrastructural cytochemistry found fine granules of 5′N on the outer surface of normal granulocyte plasma membranes and also some more variable reactivity in CML granulocytes. Bainton and Farquhar (1968) reported positive reactions for 5′N in rabbit neutrophils and Robinson and Karnovsky (1983) using cerium as capture agent for EM cytochemistry of 5′N found the enzyme present at the plasma membranes as an ectoenzyme in guinea pig neutrophils. There may obviously be difficulties in distinguishing between positivity in neutrophils due to non-specific acid or alkaline phosphatases and that due to 5′N when cytochemical methods are used which do not incorporate effective inhibition of both acid and alkaline phosphatases, even if the pH of the incubation is kept lower than that usually employed for alkaline phosphatase cytochemistry and higher than that used for acid phosphatase. With the addition of levamisole and tartrate inhibitors of alkaline and acid phosphatases respectively, Khalaf & Hayhoe (1987a) found detectable but weak granular positivity in a few normal human neutrophils, and also in platelets, although neutrophils were usually negative and the main reactivity was found in lymphocytes (Plate 27A) as in most previous studies.

Positivity of blast cells in leukaemias for 5′N has proved too variable in cytochemical studies to offer a useful diagnostic supplement. Biochemically, 5′N activity has been found to be low in lymphoblasts from T-ALL cases, normal in null-ALL (pre-B) blasts, and raised in cALL positive blasts. Reduction in 5′N activity appeared to be correlated with increasing maturity in T lymphoblasts and with loss of cALL antigen in B lymphoblasts (Reaman *et al*, 1979; Gutensohn and Thiel, 1981; Sylwestrowicz *et al.*, 1982; Ma *et al.*, 1983). In B-cell ALL notably low values have also been reported (Reaman *et al.*, 1981). In AML cases generally low or normal levels have been found (Gutensohn and Thiel, 1981; Sylwestrowicz *et al.*, 1982). These biochemical findings have not been reflected in cytochemical studies in any consistent way, and there are few reports on the subject in the published literature. Veerman *et al* (1985), using the cytochemical method of Silber *et al* (1975) reported positivity in more than 30% of the blast cells from 10 of 36 cases of cALL in childhood, the remaining 26 cases being negative, as were 27 cases of T-ALL, 2 of B-ALL and 14 of null ALL. Positive 5′N reactions in cALL in this series were associated with reduced probability of continued complete remission, although there appeared to be no other significant clinical or haematological features distinguishing the 5′N positive and negative subgroups of cALL. We have ourselves carried out 5′N cytochemical staining on more than 100 acute leukaemic marrow smears and found mostly negative results in the blast cells, although positivity in some myeloblasts and monoblasts was encountered in about 10% of AML cases and in the lymphoblasts of two of 20 ALL cases, both cALL (Plate 26F). Most cALL cases like other varieties of ALL appeared cytochemically negative. With the UMP-based, long incubation, technique, Khalaf (1986) found three of 39 AML cases to show positivity in blast cells

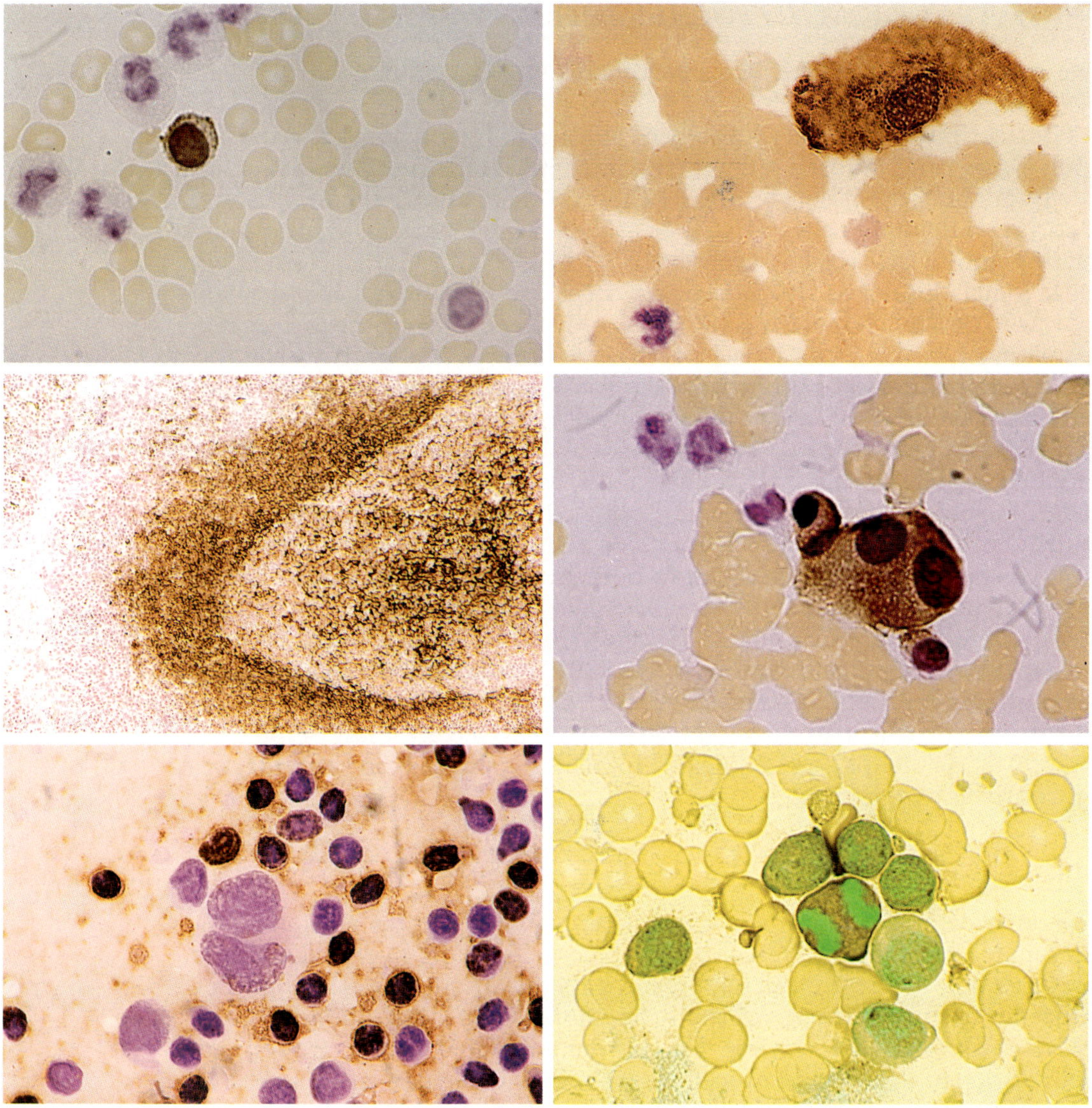

A B

C D

E F

Plate 27

A. Normal buffy coat. 5′N reaction. A strongly positive lymphocyte with overall reaction; 3 neutrophils and another lymphocyte are negative.

B. Normal bone marrow. 5′N reaction. A strongly positive macrophage and a negative neutrophil.

C. Freeze-dried cryostat section of reactive tonsil. 5′N reaction. Lymphocytes of the mantle and dendritic reticulum cells of the germinal centre are strongly positive. Paracortical lymphocytes are negative.

D. Myeloma bone marrow. 5′N reaction. Myeloma cells, one binucleated, show strong activity; three neutrophils are negative.

E. Lymph node imprint, Hodgkin's disease. 5′N reaction. An RS cell appears negative; some lymphocytes show strong positivity.

F. cALL bone marrow, 5′N reaction. Most blast cells show some membrane associated positivity, one in mitosis having strong cytoplasmic reaction.

(one myeloblastic case and two myelomonocytic cases) while all of eight ALLs were negative.

The distribution of 5′N has also been studied in lymph nodes and tonsils, using freeze-dried cryostat sections and imprints. Here the enzyme is found chiefly in the B-lymphocyte areas, where both mantle lymphocytes and dendritic reticulum cells (DRC) of the germinal centres show strong staining and T-cell regions are negative (Plate 27C) (Muller-Hermelink, 1974; Jondal and Pross, 1975; Lennert, 1978), so that in lymph nodes 5′N appears to be a much more specific marker for B cells than is the case among circulating lymphocytes. This pattern again supports the view that this enzyme expression in lymphocytes varies at different stages of cell development and perhaps at different phases of cell functional activity.

Variability seems also to be a cytochemical feature of 5′N activity in lymphomas. The circulating lymphocytes in cases of Hodgkin's disease and non-Hodgkin's lymphomas usually showed low enzyme content (El-Mohandes and Hayhoe, 1983a) but occasionally a minority of strongly positive lymphocytes may be seen in both peripheral blood and in lymph node imprints from Hodgkin's disease (Khalaf and Hayhoe, 1987a). Reed-Sternberg cells and mononuclear Hodgkin's cells in imprints and cryostat sections show negative reactions (Plate 27E), although the positivity of macrophages and DRC is conspicuous. As Crocker (1983) has pointed out, 5′N is a more effective marker of DRC than is acetate esterase, and its demonstration helps to delineate follicular patterns in centrocytic-centroblastic lymphomas.

Glucose-6-phosphatase

This enzyme (EC 3.1.3.9) is a specific phosphatase acting on glucose-6-phosphate, with pH optimum 6. It occurs chiefly in liver, kidney and intestine but is present in all organs capable of glycolysis. A genetic deficiency of this enzyme gives rise to hepatorenal glycogen storage disease.

Cytochemically, the enzyme can be demonstrated by an appropriate modification of the Gomori procedure for acid phosphatase, as proposed by Wachstein and Meisel (1956), the substrate being glucose-6-phosphate in acetate buffer at pH 6.5, with lead nitrate as capture agent for the released phosphate. Both non-specific acid and alkaline phosphatases will hydrolyse glucose-6-phosphate, so that control preparations in which the substrate is replaced by β-glycerophosphate, or with glucose-6-phosphate as substrate but with the added specific inhibitor 1,5-sorbitan-6-phosphate, must be used for comparison to provide evidence of cytochemical specificity.

Ultrastructural cytochemistry of glucose-6-phosphatase (G-6-Pase) has been performed by both lead-based (Tice and Barnett, 1962; Hugon *et al.*, 1970; Leskes *et al.*, 1971) and cerium-based methods (Robinson and Karnovsky, 1982) and the enzyme has been localized to the endoplasmic reticulum and the nuclear envelope in a variety of tissues. Relatively few studies have been directed specifically towards haemic cells but Daimon and Gotoh (1982) found the enzyme to be located in the dense tubular system of murine platelets, thus supporting the hypothesis that this system is derived from the endoplasmic reticulum of the megakaryocyte. Nichols *et al.* (1984) studied leucocytes and platelets freshly isolated from normal human peripheral blood, using a modified lead-based procedure at electron microscope level. Reaction product was found in the endoplasmic reticulum and perinuclear cisternae of all five types of leucocyte and in the dense tubular system or endoplasmic reticulum of platelets. It was not found in other intracellular membrane-limited cisternae such as the smooth pinocytic tubules of monocytes or the canalicular system of platelets.

Adenosine triphosphatase

The cytochemistry of adenosine triphosphatases (ATPases: EC 3.6.1.3) has been reviewed in detail by Firth (1978, 1980), Ernst and Hootman (1981) and Borgers and Verheyen (1985). With increasing appreciation of the complexity of the ATPase systems have come new classifications and considerable scepticism over the validity of cytochemical claims as to specific localization of particular ATPases. Firth (1978) divided the specific ATPases (excluding non-specific phosphatases such as alkaline phosphatase for which ATP is one of many possible substrates) into three main classes: proton-translocating ATPases, other ion-transporting ATPases, and cytoplasmic ATPases associated with contractile systems.

The first group are concentrated especially on the inner mitochondrial membrane and physiologically run in reverse, catalysing the synthesis of ATP. They have been demonstrated by variations of the Wachstein and Meisel (1957) method in many studies, mostly of non-haemic tissues, and although there has been considerable argument over the extent to which non-enzymatic hydrolysis of ATP may be brought about by the lead itself in certain concentrations (Rosenthal *et al.*, 1969; Jacobsen and Jørgensen, 1969; Rosenthal *et al*, 1970), it seems likely that the method can provide adequate enzyme localization if appropriate controls are used. There have been only few studies of mitochondrial ATPase in haemic cells. The granules of myeloid cells in the normal chicken show strong positivity (Haddad *et al.*, 1960) but marrow cells in man were reported by Weinstein *et al.* (1962) to show no more than a faint diffuse reaction in some neutrophil granulocytes, except for eosinophils, the granules of which were strongly positive. Omission of substrate or substitution of glycerophosphate for ATP as substrate gave negative reactions. Rinneberg *et al.* (1961) studied the distribution of ATPase in lymph nodes; lymphocytes showed some patchy positivity but strong reactions were found in plasma cells and in reticuloendothelial macrophages.

The second group of ATPases normally hydrolyse ATP and use the energy released for active transport of ions across plasmalemmal surfaces against their concentration gradients. There are Na^+ K^+ ATPase, Ca^{++}-ATPase and a carbonate-ATPase. All require sulphydryl groups for active function and are therefore inhibited by p-chloromercuricbenzoate. They also need Mg^{++}. Cytochemical demonstration of these enzymes using modified lead methods has been widely attempted, but the enzymes are extremely sensitive to lead and the resulting localizations of reaction product are of doubtful specificity (Firth, 1980). Other methods have been developed for localizing Na^{+} K^{+}-ATPase, notably a strontium capture method (Ernst, 1972) using the second phosphorylation step of this enzyme system, with p-nitrophenyl phosphate as substrate. Adequate criteria for validation of this method, which excludes reactivity of the mitochondrial and cytoplasmic ATPases, appear to exist (Ernst and Hootman, 1981), and the localization achieved, mostly at the cytoplasmic surface of the plasma membrane, seems physiologically appropriate. Further cytochemical approaches using specific activators and inhibitors include those of Mazurkiewicz *et al.* (1978) and Mayahara *et al.* (1981), while tritiated ouabain as a specific inhibitor has also been used to provide autoradiographic localization (Stirling, 1972, 1976; Ernst and Hootman, 1981). Immunocytochemical techniques using ferritin-labelled antibodies (Kyte, 1974, 1976a) or peroxidase methods (Wood *et al.*, 1977; Schwartz *et al.*, 1980, 1981) have been employed, with sharp localization of reaction.

All these studies have been performed on solid tissues and few appear to have been directly applied to haemic cells. Attempts in our own laboratories to use various modifications of the lead capture procedure, with added theophylline and levamisole to inactivate phosphodiesterase and alkaline phosphatase and with controls including ouabain and heat-treatment have shown some reaction in most haemic cells, both normal and pathological, when unfixed, but the reactions, though inhibited by heat, were not inhibited by ouabain, and could not be attributed with certainty to this ATPase system (Khalaf, 1986; Khalaf and Hayhoe, 1987b). An ATPase dependent on Ca^{++} and Mg^{++} activation has been studied by Cuschieri (1982), using fresh unfixed slides made from suspensions of lymphocytes before and after activation by phytohaemagglutinin (PHA). Unactivated lymphocytes showed little activity in tiny cytoplasmic granules, but in PHA-transformed cells this ATPase was intensely positive and localized in a crescentic perinuclear area of cytoplasm: in mitotic cells there were scattered discrete granules or elongated mitochondria-like structures with positive reactions. In ultrastructural studies Cuschieri *et al* (1982) showed the Ca^{++} Mg^{++} activated ATPase reaction product to be localized in endoplasmic reticulum, nuclear membrane, Golgi apparatus and mitochondria and on membranes surrounding large electron-dense bodies. A similar but much weaker enzyme distribution was seen in unstimulated cells, where there was much less endoplasmic reticulum. These findings, and numerous similar reports relating to solid tissues, are not attributable with certainty to enzymes of any specific ATPase subgroup because of lack of inhibitors of clear specificity and the unreliability of ion-dependence as a differentiating criterion (Borgers and Verheyen, 1985).

Finally the third group, cytoplasmic ATPases, appear to be inhibited by Mg^{++} but stimulated by Ca^{++} and have been demonstrated ultrastructurally in myofibrils (Somogyi and Sotonyi, 1969) but not in haemic cells.

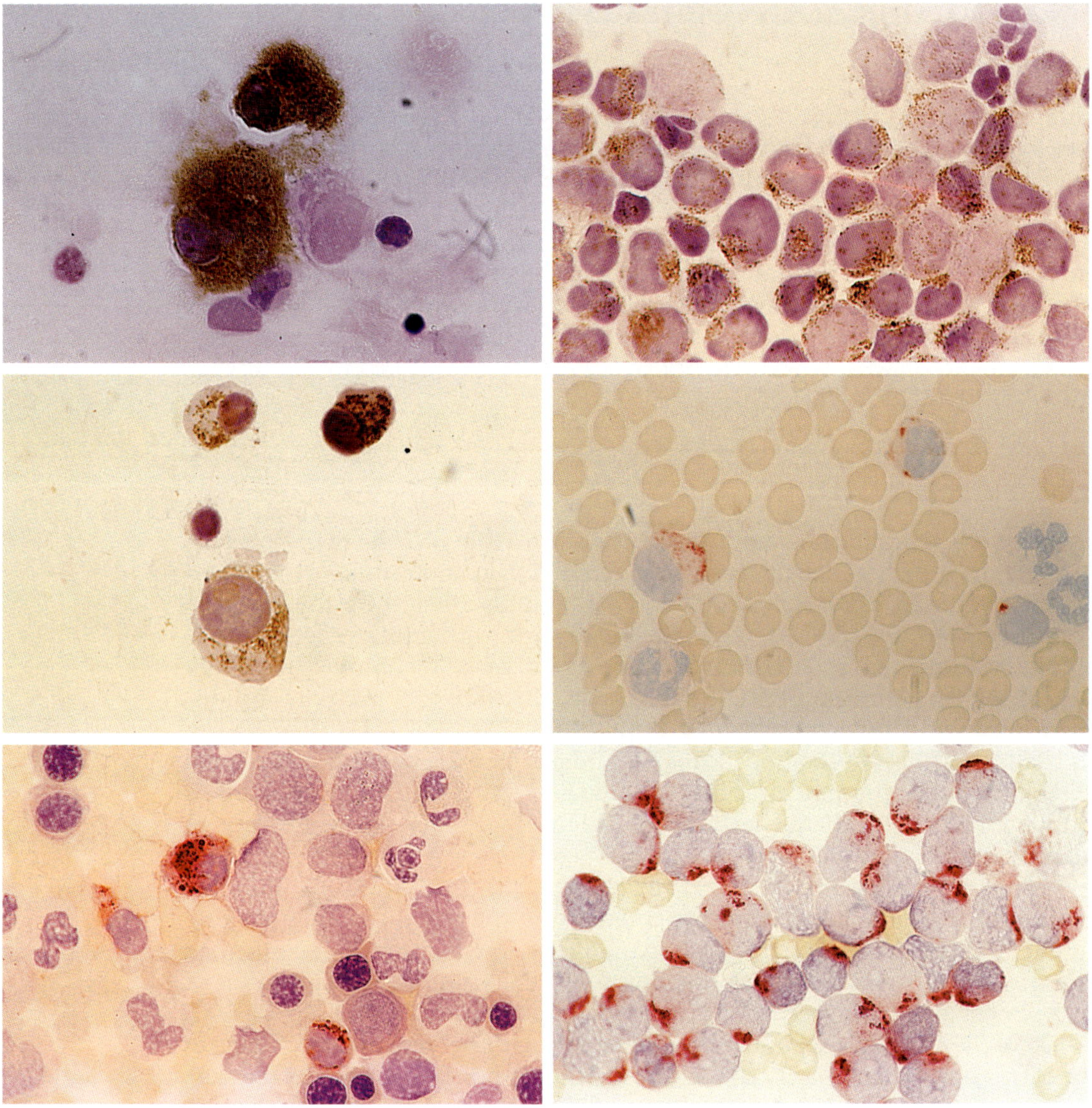

A B

C D

E F

Plate 28

A. Normal bone marrow. Adenylate cyclase. Two strongly reacting macrophages and two negative lymphocytes.
B. AMML buffy coat. Adenylate cyclase. Moderately strong cytoplasmic positivity is present in all cells.
C. Myeloma bone marrow. Adenylate cyclase. Strong granular cytoplasmic positivity in 3 myeloma cells; a lymphocyte appears negative.
D. Normal buffy coat. DAP II. Variable granular positivity in 3 lymphocytes. A monocyte and a neutrophil are negative.
E. Normal bone marrow. DAP II. Positivity in a lymphocyte, a plasma cell and a macrophage; granulocytes and erythroblasts are negative.
F. AMML bone marrow. DAP II. Strong localized reaction in all blast cells.

Adenylate cyclase

This enzyme (AC., EC 4.6.1.1, adenylyl cyclase, ATP pyrophosphate-lyase) is involved in the synthesis of cyclic adenosine monophosphate. It was first demonstrated at EM level in rat liver by Reik *et al.* (1970), and in blood and bone marrow cells at light microscope level by Khalaf (1986) and Khalaf and Hayhoe (1987d). Difficulties arose from enzyme inhibition by all fixatives tried, but partially satisfactory enzyme demonstration was achieved in unfixed haemic cells where most normal cells and their neoplastic counterparts showed some activity. Examples of reaction patterns in macrophages, leukaemic myeloid precursors and myeloma cells are shown in Plate 28 A, B, C.

Phosphamidase

Enzymes which split phospho-amide bonds probably include many proteolytic enzymes such as pepsin, trypsin, chymotrypsin and rennin (Holter and Li, 1950) as well as a more specific phosphamidase listed by the Enzyme Commission as EC 3.9.1.1. Cytochemically, phosphamidase activity has been demonstrated by various modifications of a technique originally introduced by Gomori (1952), in which the substrate was p-chloroanilidophosphonic acid and lead nitrate at acid pH was used as capture agent as in his acid phosphatase method. Phosphamidase, however, unlike lysosomal acid phosphatase, is not inhibited by fluoride. Although Gomori himself found difficulties with the technique, especially in lack of uniformity in staining, he claimed that moderate amounts of the enzyme were to be found in practically all animal tissues but very high concentrations in most malignant epithelial neoplasms, the intensity of the reaction being roughly parallel to the degree of histological malignancy. Burstone (1962) confirmed this finding but also noted a tendency for indiscriminate precipitation of lead phosphate to occur during the cytochemical reaction, presumably as a result of lability of the phospho-amide bond. The technical difficulties appear to have been largely overcome by the use of polyvinyl alcohol as a substrate stabiliser and the omission of fixative procedures which may reduce enzyme activity and lead to non-specific absorption of lead (Chayen *et al.*, 1973).

The distribution of phosphamidase activity in haemic cells has been little studied.

Peptidases

Peptidases are hydrolases which split the peptide bonds in peptides of various molecular weights. They may be simply divided into endopeptidases and exopeptidases, the former catalysing cleavage of peptide bonds within the protein molecule and the latter splitting of terminal peptides at either amino or carboxyl ends of the chain. Endopeptidases such as the major proteolytic enzymes like trypsin and cathepsins B, D and E have not been widely studied cytochemically, although cathepsin B has been demonstrated in macrophages both in sections of tonsils and in HL60 cell-line cultures by an immunohistological technique (Burnett *et al.*, 1983; Crocker *et al.*, 1984). Several exopeptidases, however, have proved of interest to haematological cytochemists. These include especially the aminopeptidases and dipeptidyl amino peptidases.

Aminopeptidases

Historically, the former enzymes were first studied cytochemically with the use of leucine-β-naphthylamine and alanyl-β-naphthylamine as substrates (Gomori, 1954; Burstone and Folk, 1956; Nachlas *et al.*, 1957, 1962). The naphthol released by aminopeptidase action is captured by a diazonium salt such as Fast blue B or Fast Garnet GBC. Using a method of this type, Ackerman (1960, 1963) studied the distribution of aminopeptidase in cells of the blood and bone marrow of man. Normally positivity was localized to the cytoplasm of neutrophils, monocytes and eosinophils from the late myelocyte stage onwards: lymphocytes, basophils, myeloblasts, promyelocytes and early myelocytes, and erythrocyte precursors were non-reactive. Wide-ranging studies were also carried out in pathological states and the relative activity of the enzyme compared with normal. Increased aminopeptidase activity was found in myelocytes and mature neutrophils in chronic myeloid leukaemia, with a return to normal or decreased levels after anti-leukaemic therapy. Similar increases in enzyme activity at the myelocyte and monocyte stages were seen in acute leukaemias of myeloid origin where these later cell stages were present: myeloblasts, promyelocytes and monoblasts in acute leukaemia remained negative.

No changes from the normal pattern of enzyme distribution and intensity were seen in CLL, lymphomas of Hodgkin's or non-Hodgkin's type, polycythaemia, haemolytic and iron deficiency anaemias, thrombocytopenic purpura, pancytopenia or infectious mononucleosis. Increased levels of aminopeptidase were, however, noted in the neutrophils of patients with pyogenic infections.

We have used the reaction in a comparable series of studies in our own laboratories, with generally similar findings to those reported by Ackerman, but have not found the differences in positivity in pathological states to be sufficiently striking to have much value in differential diagnosis.

Dipeptidyl aminopeptidases

With dipeptidyl aminopeptidases more distinctive patterns of positivity have emerged. Although there are four known dipeptidyl aminopeptidases (DAP I, II, III and IV) only DAP II, (EC 3.4.14.2) and DAP IV (EC 3.4.14.4) have been profitably studied cytochemically. The development and usefulness of histochemical methods for these enzymes have been reviewed by Lojda (1981).

DAP II (dipeptidyl aminopeptidase II, dipeptidyl peptidase II, dipeptidyl arylamidase II, carboxytripeptidase, lysyl-alanyl-2-naphthylamidase) catalyses the hydrolysis of the second peptide link from the amino-terminal

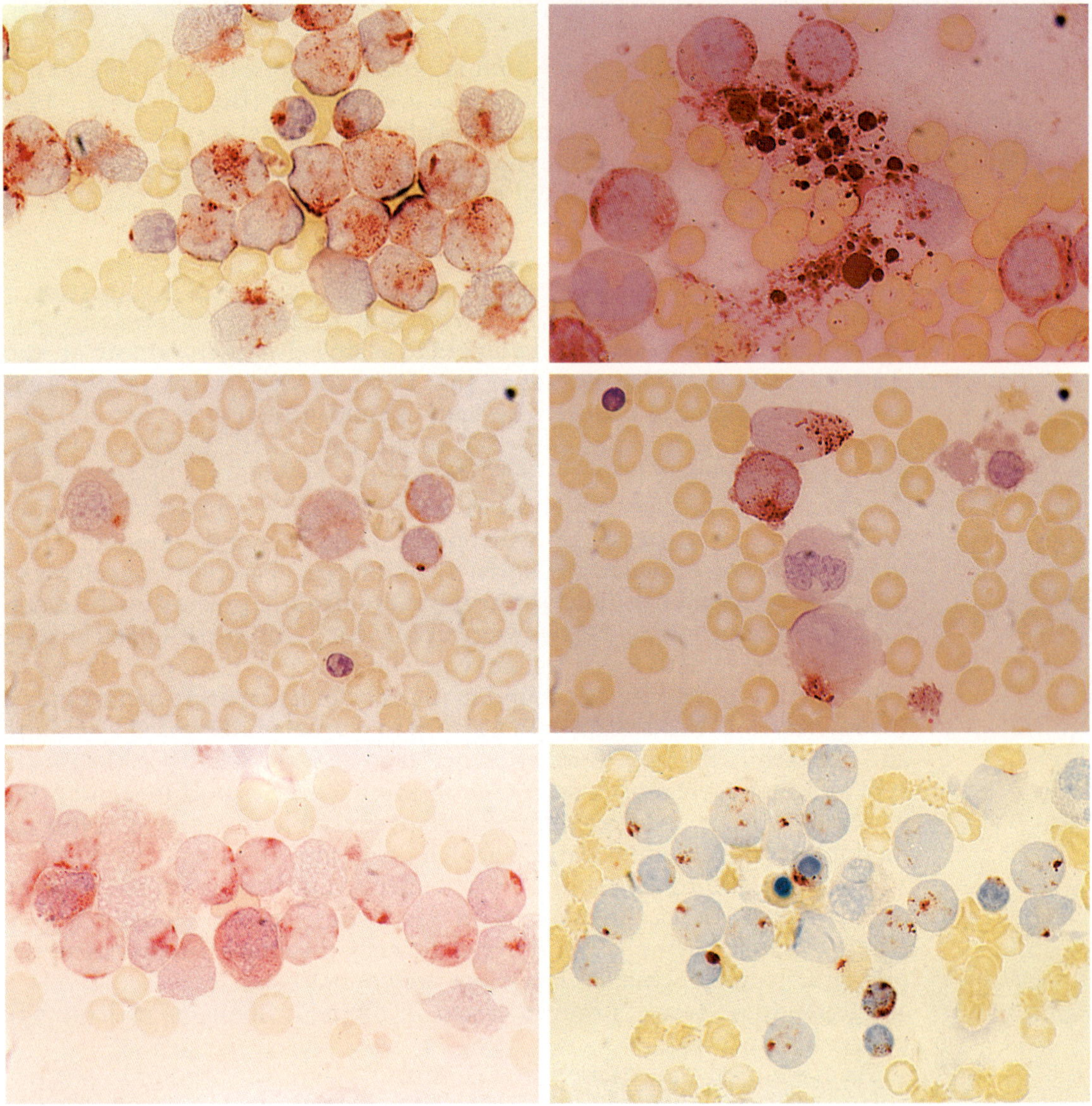

Plate 29

A. AMML buffy coat. DAP II. Blast cells show scattered granular reaction, including possible Auer rod positivity; three lymphocytes show focal reaction.

B. AMonL bone marrow. DAP II. Granular cytoplasmic positivity in four of five monoblasts; an RE cell shows very strong reaction.

C. Erythroleukaemia: bone marrow. DAP II. Dot-like positivity in one erythroblast and in two lymphocytes.

D. AMegL buffy coat. DAP II. Strong positivity in three megakaryoblasts and varying activity in platelets.

E. Mixed myeloid transformation of CML, buffy coat; DAP II. Various granulocyte, monocyte and erythroid precursors show positive reactions.

F. cALL, buffy coat. DAP II. Most blast cells show focal positivity. One late erythroblast shows localized positivity and another is negative.

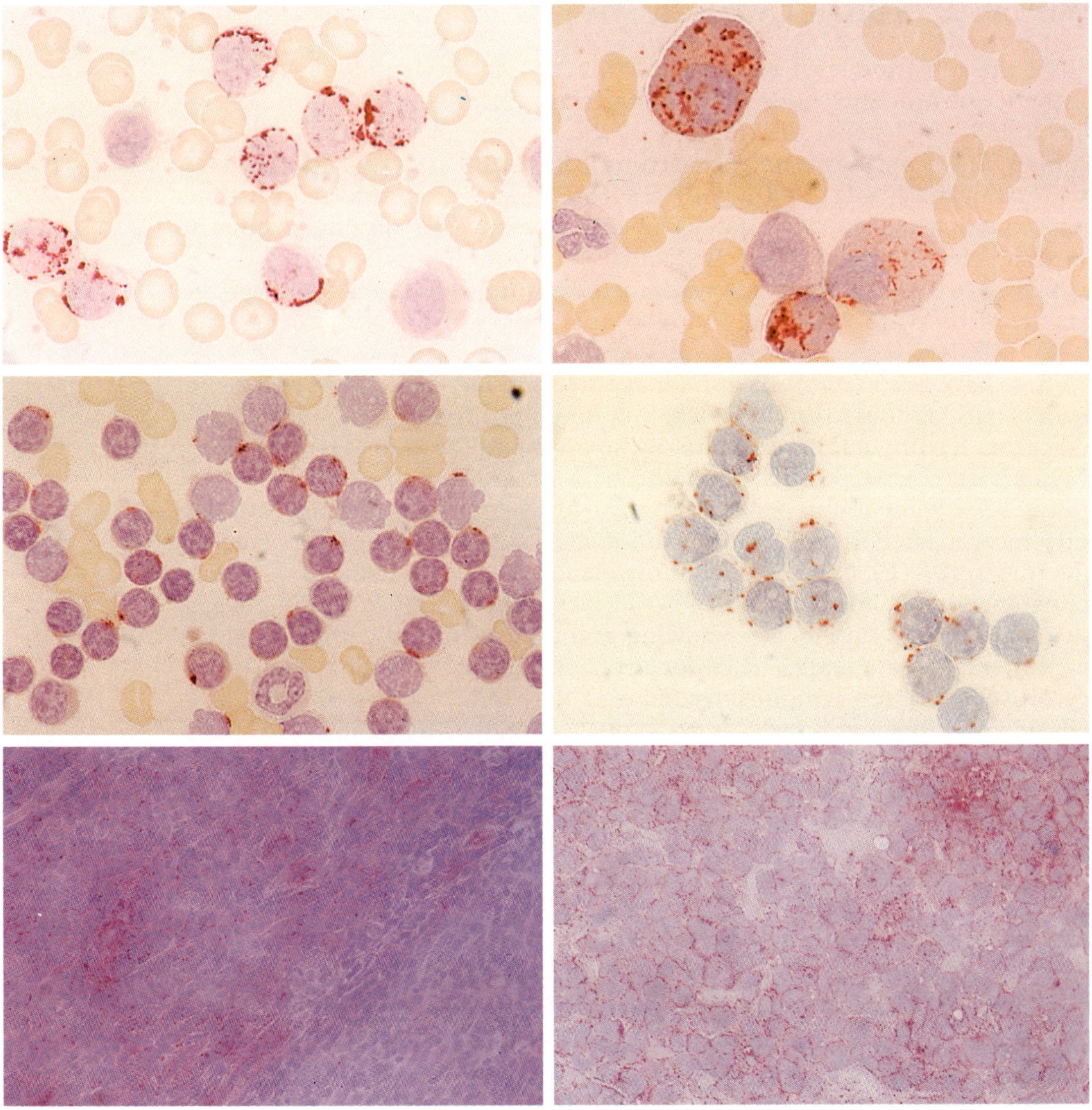

Plate 30

A. HCL buffy coat. DAP II. Strong cytoplasmic positivity in seven of nine HCs.

B. Myeloma bone marrow. DAP II. Coarse granular positivity in three myeloma cells. One large lymphocyte is negative.

C. CLL buffy coat. DAP II. Single dot positivity in most lymphocytes; a neutrophil is negative.

D. Sezary syndrome, buffy coat. DAP II. Scattered coarsely granular reaction in all Sezary cells.

E. Freeze-dried cryostat section of reactive lymph node. DAP II. Most lymphocytes of the paracortical area (T cell) show positivity, and cells of the germinal centre (B cells) are negative.

F. Freeze-dried cryostat section of lymph node from T-cell lymphoma. DAP II. Membrane associated positivity is apparent in the neoplastic cells.

end of the substrate lysyl-alanyl-4-methoxy-2-naphthylamide (LAMN), a synthetic substrate first used for peptidase histochemistry by Smith (1969), who found a lysosomal distribution of the enzyme in the cells of kidney tubules. The same substrate was used by McDonald *et al.* (1971) and later by Gossrau and Lojda (1980), with hexazotised pararosaniline or Fast Blue B as coupling agents, in studies of various organs, including pituitary, thyroid, liver, and gut. The lysosomal localization of the enzyme was confirmed by these light microscopy techniques and again by ultra-cytochemistry with essentially similar methods (Smith and van Frank, 1975).

There have been few studies so far reported on DAP II cytochemistry in haemic cells, although Lojda (1981) first noted positivity in blood lymphocytes from healthy adults. At pH 5.5, the optimum for DAP II, some 60% of lymphocytes gave positive reactions, with a scattering of small and larger granules and some diffuse staining in the cytoplasm. More recently, studies of haemic cells in both normal and pathological states have been reported by Invernizzi *et al.* (1985) and from our own laboratories (Khalaf *et al.*, 1986a). Using incubation for 2 hours at 37°C we found about 80% of normal peripheral blood lymphocytes showing granular positivity, either localized or scattered (Plate 28D), occasionally with diffuse staining. Tests on erythrocyte-rosetted lymphocytes showed that DAP II was present in T cells and also in some B cells, as previously suggested by Lojda (1981) and Invernizzi *et al.* (1985). Nevertheless a proportion of B cells were negative and the lower percentage of bone marrow lymphocytes showing DAP II reactivity probably reflects the higher percentage of B cells at this site. We also found DAP II positivity in normal marrow plasma cells and macrophages and in an occasional normal myeloblast, but later granulocytes, monocytes and megakaryocytes were invariably negative in 11 normal bone marrows studied (Plate 28E).

In 52 cases of AML studied, DAP II positivity was found to a variable extent in leukaemic myeloblasts, monoblasts, erythroblasts and megakaryoblasts, and later granulocytes and monocytes also showed weak positivity in cases of mixed myeloid leukaemia and atypical Ph^1 negative CML, although typical CML usually showed negative later granulocytes (Plate 28F, 29A–E).

In seven cases of ALL a wide variation of DAP II activity was found, one case being entirely negative and others showing from 5 to 95% of positive blast cells (Plate 29F). There was no clear correlation with surface marker phenotype.

Hairy cells in ten cases of HCL were usually positive (Plate 30A) as were virtually all the plasma cells in five cases of multiple myeloma (Plate 30B) and two of plasma cell leukaemia. These findings differ from those of Invernizzi *et al.* (1985) who found hairy cells negative and myeloma cells only occasionally positive, but the discrepancy may well be due to the relatively long fixation time used by these authors which is likely to have inhibited enzyme activity, and perhaps also to their use of Fast Garnet GBC as coupling agent, as we have found this less efficient than Fast Blue B.

In other B-cell neoplastic states variable positivity for DAP II was found;

in 14 cases of B-CLL (Plate 30C) and three of non-Hodgkin's lymphomas the percentage of positive cells was generally lower than in normal blood lymphocytes, but showed a wide range, although the prolymphocytes in one case of B-PLL were completely negative.

In a case of Sézary's syndrome around 90% of the helper T cells were positive for DAP II (Plate 30D) and in a case of suppressor T-cell CLL only 40% of the cells were positive.

Khalaf *et al.* (1986b) also applied the reaction to freeze-dried cryostat sections of normal or reactive tonsils, lymph nodes and spleens, where the paracortical zones, predominantly T-cell, showed scattered and variable positivity in a proportion of lymphocytes and strong positivity in branching macrophages. In germinal centres and medulla only macrophages showed reaction (Plate 30E). In lymph node imprints from Hodgkin's disease, Reed Sternberg cells were negative, but some lymphocytes showed positivity, while in 12 cases of B-cell lymphoma the B lymphocytes were entirely negative. In 13 cases of T-cell lymphoma discrepant results were found; in five cases all cells were negative and in the other eight they were strongly positive for DAP II (Plate 30F). Apparently DAP II is absent from both normal and pathological B cells from nodes, although present in a proportion of circulating B cells. In this respect it resembles several other lysosomal enzymes which are found in circulating lymphocytes but may be absent from either B or T cells, or both, in tonsils and lymph nodes, such as acid phosphatase, beta-glucuronidase and N-acetyl-beta-glucosaminidase, largely restricted to T-lymphocyte regions, and 5′N, confined to B-lymphocyte areas.

DAP IV (Dipeptidyl aminopeptidase IV, dipeptidyl peptidase IV, glycyl-prolyl-2-naphthylamidase) catalyses the hydrolysis of the second peptide link from the amino-terminal end of various synthetic substrates such as glycyl-proline-2-naphthylamide, first introduced by Hopsu-Havu and Ekfors (1969) and developed by McDonald *et al.* (1971) for the cytochemical demonstration of this enzyme at both light and electron microscope (EM) levels. The first studies of haemic cells were carried out by Lojda, who introduced glycyl-prolyl-4-methoxy-2-naphthylamide (GPMN), as a more specific and effective substrate for DAP IV, with Fast Blue B as the capture agent (Lojda, 1977; Lojda *et al.*, 1979a). The method was later adapted for EM work by Lojda *et al.* (1979b) and Lojda (1981). Reaction product was found in 39% of lymphocytes in normal human peripheral blood with ultrastructural localization to lysosomes and cell membranes. Positivity appeared to be restricted to E-rosette-forming T lymphocytes, and the numbers of positive cells were not decreased in patients with myelofibrosis, myeloma, CML or HCL, but were much reduced in CLL and in advanced stages of malignant lymphomas. Subsequent studies have generally confirmed the restriction of DAP IV positivity to T lymphocytes (Feller and Parwaresch, 1981; Feller, 1982, Feller *et al.*, 1982, 1984; Wirthmüller *et al.*, 1983). Hirt *et al.* (1984) found 52.2 $\pm$ 15.8% of Leu-3a (T4) positive T cells to show DAP IV reactivity, while only 29.1 $\pm$ 10.9% of Leu-2a (T8) positive T cells showed positivity. A greater proportion of positively reacting T4 (helper) lymphocytes was also found by Crockard *et al.* (1984)

with 72% showing single or several discrete DAP IV positive granules, whereas only 41% of T8 (suppressor) cells reacted.

In pathological states, Crockard *et al.* (1984) reported the DAP IV reaction to be strongest in the T cells of T-lymphoblastic lymphoma and T-PLL, but absent or in few cells in the more immature blast cells of T-ALL and also absent from the neoplastic T cells of a suppressor-cell (T8) T-CLL and cases of adult T-cell leukaemia-lymphoma and Sézary's syndrome. Chilosi *et al.* (1984) found a correlation between cellular phenotype, clinical behaviour and DAP IV positivity in eight cases of T-CLL. Aggressive disease with T4 helper phenotype was associated with DAP IV expression, whereas a more favourable clinical course with large granular lymphocyte cytology and T8 suppressor phenotype showed negative DAP IV reaction in the neoplastic cells.

Khalaf *et al.* (1986a, b) studied DAP IV positivity in a wide range of haematological states. In normal buffy coat preparations neutrophils, monocytes and B cells were negative; T cells showed localized positivity (Plate 31A). The absence of positivity from B cells and the marked reduction in numbers of positive lymphocytes in B-CLL, representing the residual T-cell population, was confirmed. Marked reduction of DAP IV activity was also found in a case of suppressor T-CLL as earlier reported by Crockard *et al.* (1984) and Chilosi *et al.* (1984). Hairy cells, myeloma cells and all B-cell neoplastic states studied, including B-PLL, B-CLL, and a range of B-cell lymphomas, showed negative reactions. Positivity was, however, found in a small percentage of lymphoblasts in three of five cases of cALL and in one of two T-cell lymphomas. Sézary cells were negative. Non-lymphoid haemic cells were almost invariably negative, but a few megakaryoblasts from an acute megakaryoblastic leukaemia showed scattered granular postivity (Plate 31C) and some weak reaction was observed in neutrophils and monocytes in an atypical Ph^1 negative CML. In lymph nodes some macrophages in both T and B zones showed strong DAP IV positivity, but otherwise positivity was confined to T-cell areas (Plate 31B)

The general pattern of DAP IV positivity confined almost entirely to T cells of all but the least mature stages, with more frequent and conspicuous reactions in the helper (T4) subset in both normal and neoplastic states, is only occasionally contravened. In Sézary's syndrome (SS) the T4 neoplastic cells have usually been negative (Wirthmüller *et al.*, 1983; Crockard *et al.*, 1984; Khalaf *et al.*, 1986) but Feller *et al.* (1983) found one positive case among eight studied, suggesting a functional phenotypic heterogeneity in this disease. The usual DAP IV negativity of helper T cells in SS contrasts with the positivity of helper T lymphocytes in T-CLL (Chilosi *et al.*, 1984; Invernizzi *et al.*, 1985), and perhaps reflects a difference in stage of differentiation.

Srivastava and Bhargava (1986) studied DAP IV reactions in cells from leukaemic patients and in 33 normal or leukaemic cell lines. No correlation was found between DAP IV reactivity and OKT4 positivity or the maturation stage in T cells. DAP IV was unexpectedly found to be expressed in 4 B-cell lines and one histiocytic lymphoma cell line, indicating either lack of specificity or infidelity of gene expression in culture.

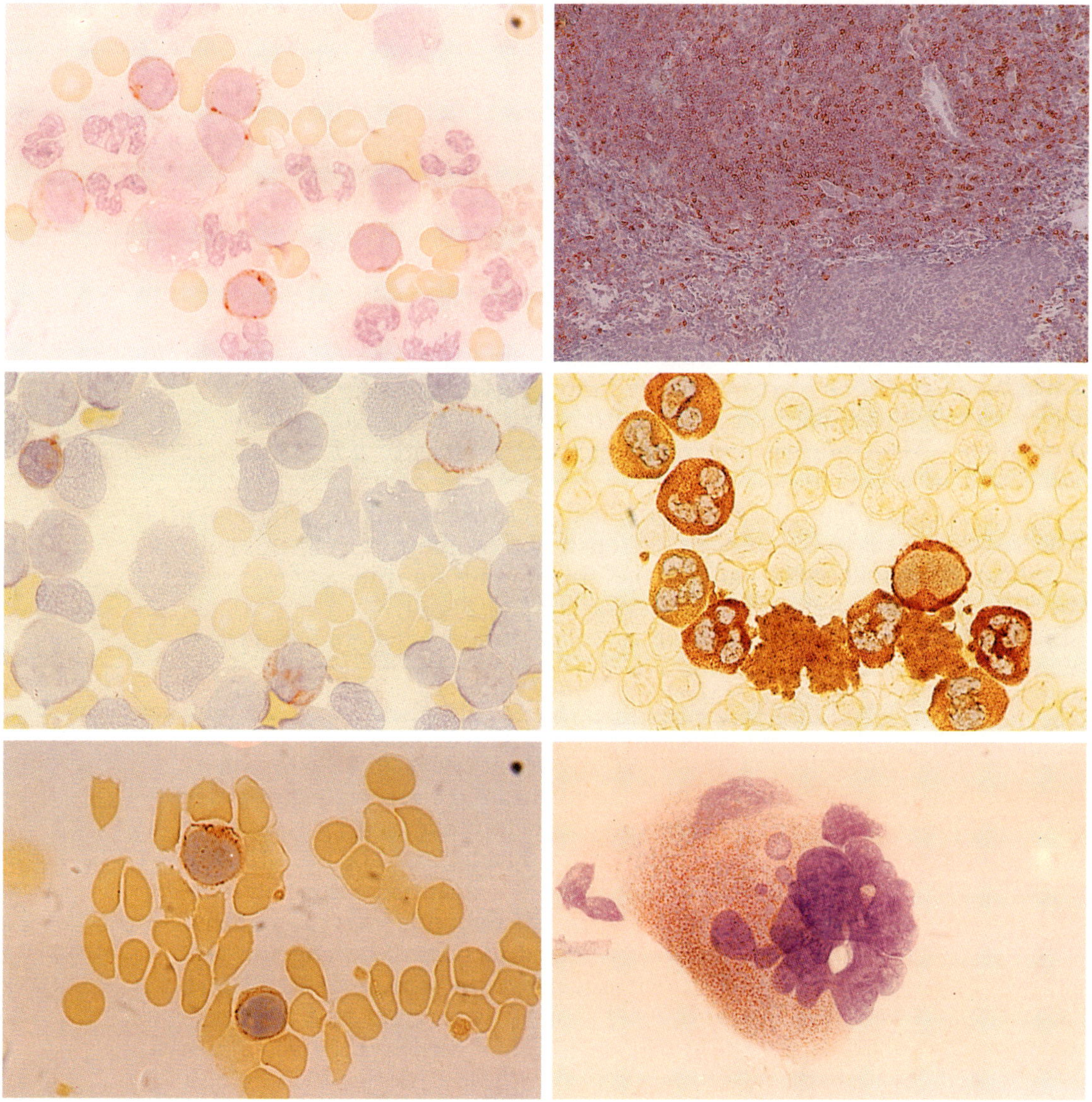

Plate 31

A. Normal buffy coat. DAP IV. Four presumed T lymphocytes show dot-like focal positivity and 2 others show weak membrane reactions. All remaining cells, monocytes, neutrophils and presumed B lymphocytes are negative.

B. Freeze-dried cryostat section of reactive lymph node. DAP IV. Positivity is essentially confined to the paracortical T-cell zone.

C. AMegL, buffy coat. DAP IV. Two megakaryoblasts show cytoplasmic reactions; a single probable T lymphocyte shows dot-like positivity.

D. Normal buffy coat. γGT. Neutrophils, monocytes and platelets are strongly positive.

E. Normal buffy coat. γGT. Moderate granular positivity in two lymphocytes.

F. Normal bone marrow. γGT. Diffuse granular reaction in a megakaryocyte.

γ-Glutamyl transferase

γ-GT (EC 2.3.2.2, γ-glutamyl transpeptidase) is an enzyme that catalyses the transfer of a γ-glutamyl group from peptides and other compounds that contain it to various acceptors. One of the main functions of this enzyme is to facilitate the uptake of amino acids from the extracellular fluid to the cytoplasm (Meister 1973).

The first histochemical demonstrations of γGT in animal and human tissues were by Albert *et al.* (1961) and Glenner *et al.* (1962). Szmigielski *et al.* (1965) first reported the cytochemical distribution of the enzyme in human blood and bone marrow cells and more recently Girino *et al.* (1985) reported a cytochemical study in 24 normal and 84 abnormal haemic cell preparations. Most normal leucocytes showed positivity, but low levels were found in malignant states except for three cases of T-CLL and ten cases of myeloma where the neoplastic cells showed strong positivity. At ultrastructural level γGT has been demonstrated on the surface membrane of human lymphoid cells and in the membranes of the endoplasmic reticulum and the Golgi apparatus (Marathe *et al.*, 1980).

Biochemical studies of γGT have revealed differences between normal haemic cells and their leukaemic counterparts (Kramers and Catovsky, 1978; Hultberg and Sjogren, 1980).

An extensive study of the distribution of γGT using a more sensitive modification of the technique described by Rutenberg *et al.* (1969),with γ-glutamyl-4-methoxy-2-naphthylamide as substrate, N-glycylglycine as activator and Fast Garnet GBC as diazonium salt coupling agent, was recently carried out in our laboratories (Khalaf, 1986; Khalaf and Hayhoe, 1987c). Thirty normal buffy coat preparations and 20 essentially normal bone marrows provided a background for comparison with over a hundred pathological specimens from haematological disorders. The enzyme was found to be widely distributed cytochemically, with some parallel to acid phosphatase; it was demonstrable in almost all leucocytes and in megakaryocytes and platelets as Girino *et al.* (1985) had earlier reported (Plates 31D–F, 32A). In conformity with biochemical studies showing γGT to be a cell surface marker of lymphoid cells (Novogrodsky, 1976), positivity was found at the cell membrane, but also, consistent with the ultrastructural findings of Marathe *et al.* (1980), within the cytoplasm.

In AML the enzyme activity varied widely among blast cells, although most monoblasts in AMonL consistently showed strong activity (Plate 32B), contrary to the finding of Girino *et al.* (1985) that only cases of acute promyelocytic leukaemia showed any strong positivity.

Among cells of the lymphoid series γGT activity in tonsillar imprints proved exceptionally strong in plasma cells, with marked localization at the cell surface membrane, while tonsillar lymphocytes showed generally stronger reactions than did peripheral blood lymphocytes (Plate 32A). This confirms cytochemically the observation previously made at a biochemical level by Kramers and Catovsky (1978) who suggested that maturation into plasma cells rather than mere lymphocyte activation may be the main factor associated with the increase of γGT activity. Macrophages were generally strongly positive (Plate 32C).

In all the lymphoproliferative disorders studied (ALL, B-CLL, HCL, Sezary's syndrome (SS), and non-Hodgkin's lymphoma (NHL)) the lymphoid cells showed reduced activity of γGT compared to normal lymphocytes (Plate 32C, D), suggesting a lower rate of uptake of amino acids from the extracellular fluid by these neoplastic cells, and offering some parallel with the reduced activity of dipeptidyl aminopeptidase IV and 5′nucleotidase in these disorders. These findings are in agreement with the biochemical studies of Kramers and Catovsky (1978) except in SS, where these authors found normal or increased enzyme levels in two cases. A single case of T-CLL showed only weak positivity, contrary to the findings of Girino *et al.* (1985) in three cases of T-CLL where intense finely granular positivity was observed, with a tendency to polar distribution.

In two cases of Hodgkin's disease, nodal imprints showed normal or increased lymphocyte activity for γGT and very strong positivity in Reed-Sternberg cells. (Plate 32E).

Strong, fine or coarse, granular positivity was found in the plasma cells of all of eight cases of multiple myeloma (Plate 32F) and two of plasma cell leukaemia, contrary to the earlier findings of Szmigielski *et al.* (1965) who reported decreased activity of γGT in neoplastic plasma cells of myeloma, but in conformity with the more recent cytochemical finding of Girino *et al.* (1985) and the biochemical findings of Novogrodsky *et al.* (1976) and Kramers and Catovsky (1978). These raised levels may well be associated with the differentiation and secretory activity of the malignant plasma cells.

β-Glucuronidase

This enzyme is present in most mammalian tissues, its intracellular location being chiefly in lysosomal and microsomal fractions, very much in parallel with acid phosphatase (de Duve *et al.*, 1955). β-glucuronidase (EC 3.2.1.31) hydrolyses a variety of glucuronides, and the successful use of naphthol compounds as substrates for the cytochemical demonstration of other hydrolases led to the introduction of naphthol AS-LC glucuronide as a substrate for β-glucuronidase cytochemistry (Pugh and Walker, 1961). Fast Garnet GBC was used as the diazonium coupling agent, but Hayashi *et al.* (1964) substituted hexazonium pararosanilin as coupler, with naphthol AS-BI glucuronide as substrate, and obtained reproducible, precisely localized, staining with negligible artefacts, the positivity being confined to cytoplasmic sites or organelles. Lorbacher *et al.* (1967) adapted this technique for use on blood and bone marrow smears, with fixation in methanol formalin followed by a period of 40 to 60 min freezing, and then incubation at 37°C at pH 5.4. The staining patterns in mature granulocytes and in their precursors back to the early myelocyte or promyelocyte stage were generally diffuse but with some granular component, increasing in strength with increasing maturity, and more intense in eosinophils than in other granulocytes. Weak non-granular reactions occurred also in

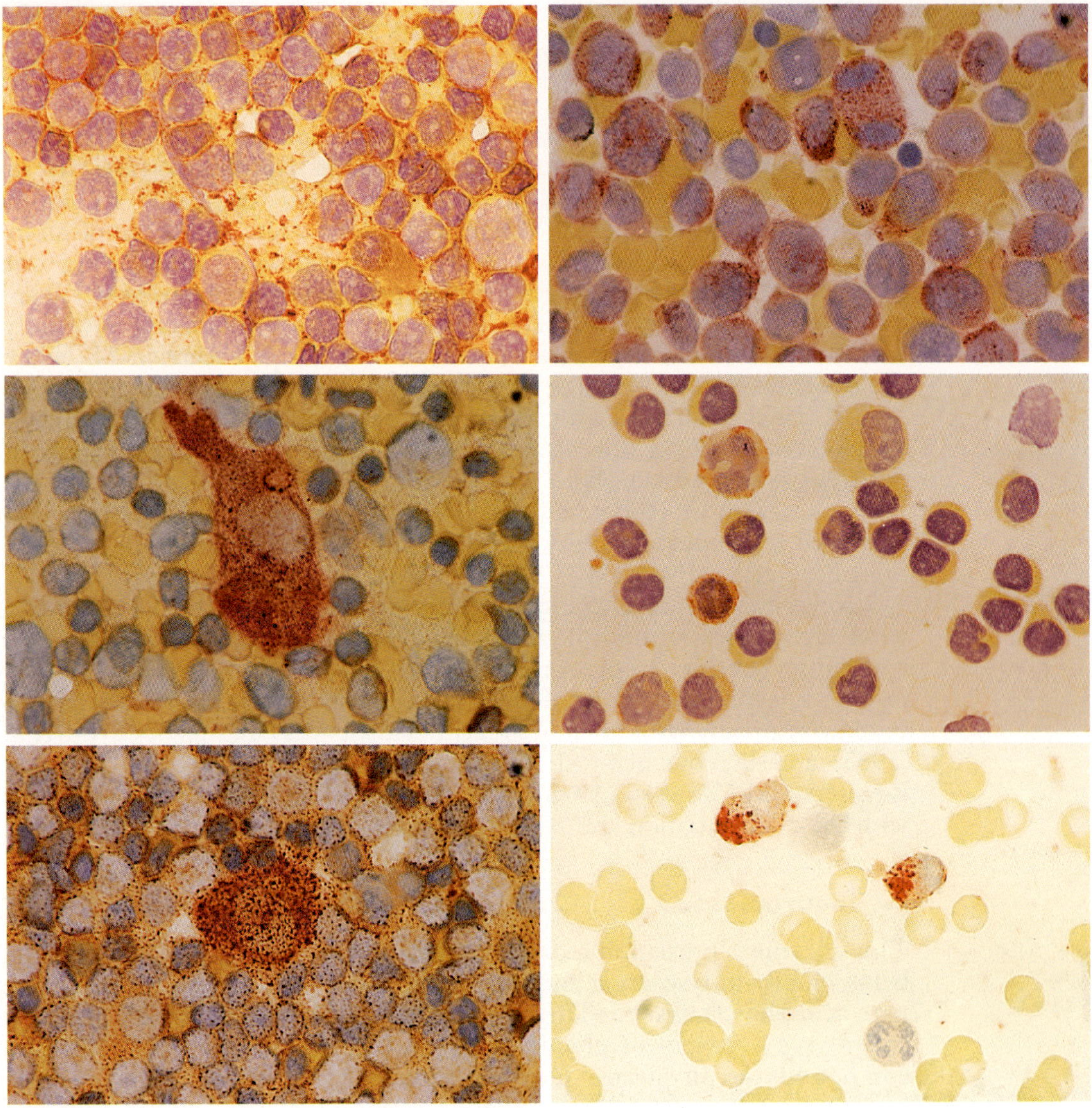

Plate 32

A. Imprint of reactive tonsil, γGT. Moderate to strong positivity in virtually all lymphocytes and in a conspicuous plasma cell.

B. AMonL, buffy coat. γGT. Variable positivity in monoblasts, including one in mitosis.

C. Imprint of lymph node with NHL. γGT. Most neoplastic lymphocytes are negative, in contrast to a strongly positive macrophage.

D. CLL buffy coat. γGT. Lymphocytes are all negative except for one; a monocyte also shows weak positivity.

E. Imprint of lymph node with HD. γGT. Scattered granular activity in most lymphocytes and strong positivity in an R-S cell.

F. Myeloma bone marrow. γGT. Two myeloma cells show strong positivity. A neutrophil is negative.

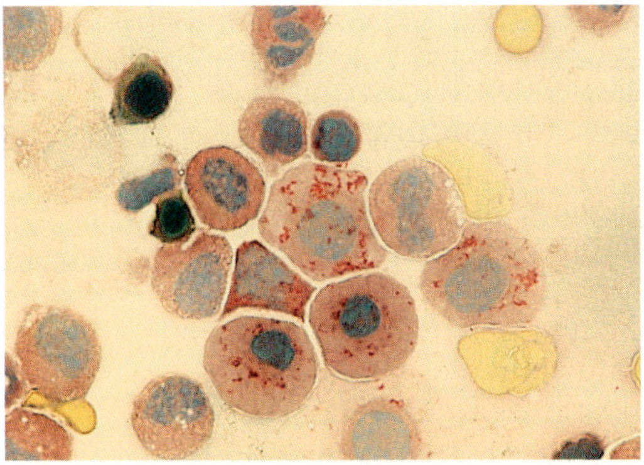
A

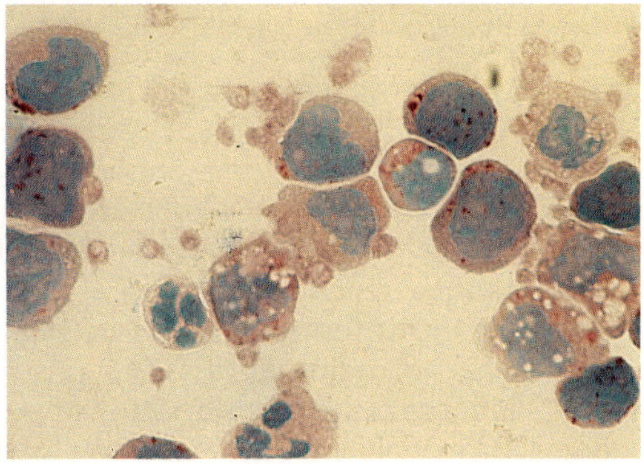
B

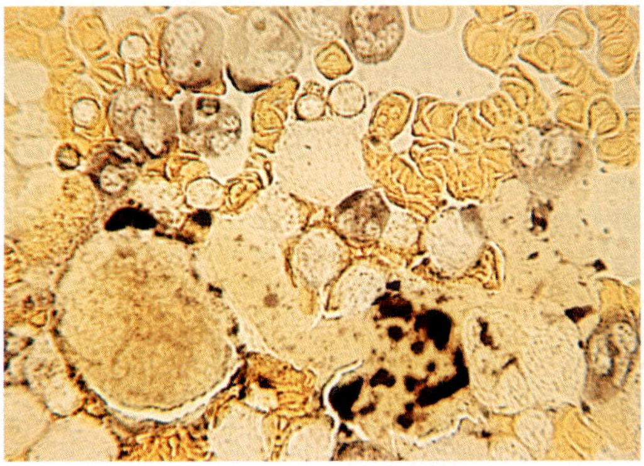
C

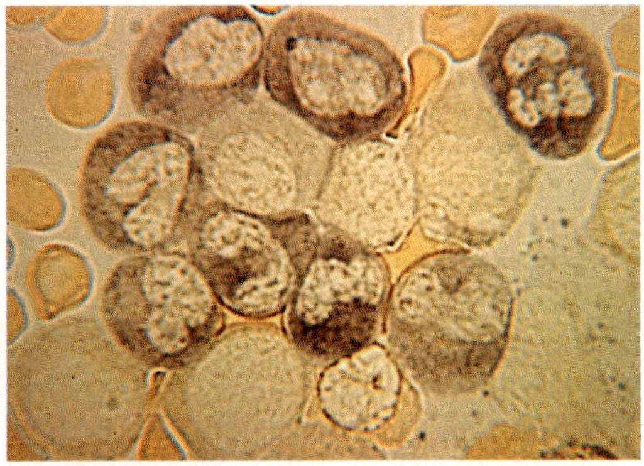
D

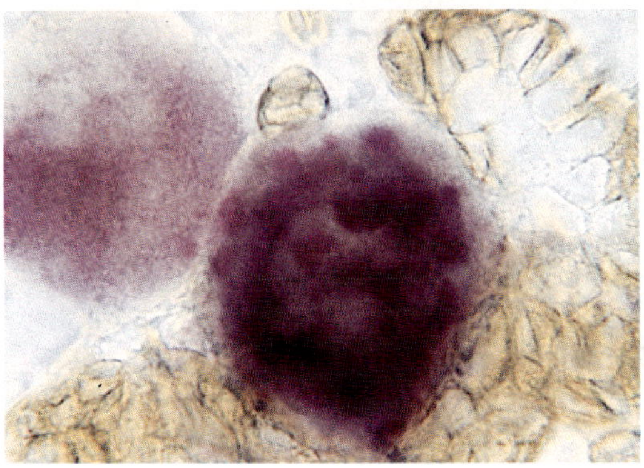
E

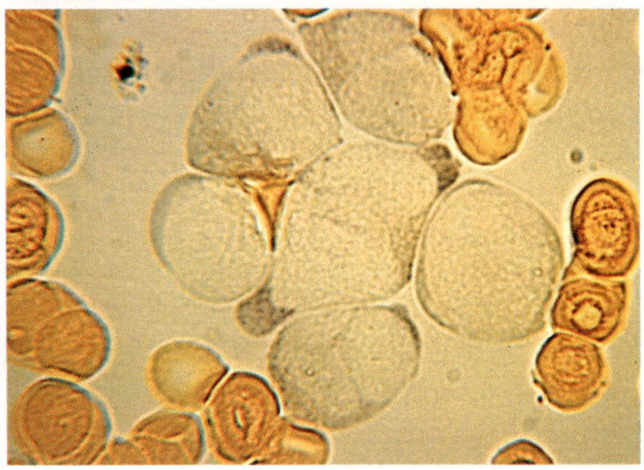
F

Plate 33

A. Myeloma bone marrow. β-glucuronidase. Diffuse and granular positivity in myeloma cells.

B. Sézary syndrome, Buffy coat. β-glucuronidase. Diffuse and granular positivity in most Sézary cells.

C. Normal bone marrow. Phosphorylase reaction. Variable, sometimes coarse positivity in megakaryocytes, negative reaction in early myeloid precursors and diffuse positive reaction in later granulocytes.

D. Normal bone marrow. Phosphorylase reaction. Increasing positivity with increasing cell maturity in the granulocyte developmental sequence.

E. Normal bone marrow. Phosphorylase reaction. Two megakaryocytes, showing strong and moderate diffuse and finely granular positivity.

F. Acute myeloblastic leukaemia bone marrow. Phosphorylase reaction. Weak tinge of cytoplasmic positivity, paralleling a similar reaction with PAS.

monocytes, megakaryocytes and early erythroblasts. Two patterns of positivity were found in lymphocytes, the commonest, present in many small lymphocytes, showing a few coarse granules at one pole of the cytoplasm, while large numbers of granules scattered throughout the cytoplasm were more characteristic of larger lymphocytes. Plasma cells showed strong diffuse and granular reactions (Plate 33A).

The existence of differing patterns of lymphocyte positivity raised the question whether content of this enzyme might provide a marker for immunologically separable lymphocyte subpopulations. Results have been contradictory; Tamaoki and Essner (1969) found both acid phosphatase and β-glucuronidase in thymus dependent areas of lymph nodes, whereas Mueller *et al.* (1975) did not find either of these enzymes to be differentially active in thymus-dependent and independent areas. Barr and Perry (1976) studied both acid phosphatase and β-glucuronidase levels in separated T, B and null cells from normal peripheral blood. They found a single large block of cytoplasmic β-glucuronidase positivity in about 45% of the T cells, multiple small granules being present in another 45%, and some 10% being negative. Among B cells and null cells from 10 to 76% showed the multiple granule pattern and single blocks were rarely seen.

Davey *et al.* (1977) used a range of cytochemical reactions including β-glucuronidase on T and B cell populations from both normal and lymphomatous states. Positivity to β-glucuronidase was more frequent in T than in B cell populations normally, and in 17 of 19 B-cell lymphomas only few cells showed cytochemical positivity while in three of four T-cell lymphomas many cells were positive. De Waele and his colleagues (1983) incubated human peripheral blood leucocyte suspensions with the mouse monoclonal antibodies OKT3, T4, T8 and M1, and then with goat anti-mouse antibodies labelled with colloidal gold. The cells were fixed and cytocentrifuge preparations made, followed by a range of cytochemical reactions including one for β-glucuronidase. By this elegant technique the T subsets could be recognised by their gold labelling and their individual cytochemistry noted. T lymphocytes frequently showed dot-like activity for β-glucuronidase, as also for acid phosphatase and α-naphthyl acetate esterase, but no difference could be seen in the cytochemical profiles of any of these three enzymes between OKT4-positive helper cells and OKT8-positive suppressor cells. Nevertheless, Orye *et al.* (1984), in a study of Tμ (T4) and Tγ (T8) helper and suppressor subsets, claimed that the former showed chiefly dot-like positivity for β-glucuronidase and the latter a more scattered granular pattern. B cells showed no reaction.

The reaction is clearly not sufficiently discriminating among normal lymphocytes to allow for more than the recognition of a proportion of T cells, and in pathological conditions the patterns do no more than provide some help in separating T and B cell neoplastic states. Among acute leukaemias the myeloid group show diffuse positivity of variable intensity (Flandrin and Bernard, 1975), while lymphoblastic cases show cellular reactions ranging from negative to moderately strongly positive, with a granular pattern, but without any characteristic differences between T-cell derived cases and the common non-T non-B cases (Brouet *et al.*, 1976). The levels of β-glucuronidase, like those of acid phosphatase, are usually reduced in the circulating lymphocytes of chronic lymphocytic leukaemia (Yam and Mitus, 1968; Douglas *et al.*, 1973) and are high in Sézary cells (Plate 33B) (Flandrin and Daniel, 1974). Studies of lymph node imprints from malignant lymphomas (Pangalis *et al.*, 1977) showed very low levels in the lymphocytes in diffuse non-Hodgkin's lymphomas, whether the cells were well or poorly differentiated, but levels similar to those in control node imprints occurred in Hodgkin's disease and in reactive hyperplasia, tuberculosis and sarcoidosis. In chronic lymphocytic leukaemia three of seven cases showed low levels and the other four high normal scores. Parallel semiquantitative scoring assessment of β-glucuronidase positivity showed scores in peripheral blood lymphocytes in the controls and in patients with Hodgkin's disease to be three times higher than those in the lymph node cells. The authors suggested that the differences were most likely to reflect differing functional activities or states of maturation between lymphocytes in the two sites.

In a more recent extensive study of the cytochemistry of acid hydrolases, including β-glucuronidase, in chronic B and T cell leukaemias, Crockard and his colleagues (1982) reported findings in 50 patients. There was strong positivity for β-glucuronidase in the neoplastic cells from five cases of T-CLL but weak or negative reactions in 15 of B-CLL and in 10 of B-PLL. Reactions were variable in the cells from 11 cases of T-PLL, strong in a case of Sézary's syndrome, and negative in the cells from six cases of HCL. In a further report, Crockard *et al.* (1985) sought correlations between immunological variants of B-CLL involving Ig expression, rosetting characteristics and acid hydrolase expression. None was found, and β-glucuronidase reactions were noted to be generally similar to those for acid phosphatase. Indeed, the International Committee for Standardization in Haematology (ICSH) in a recent set of cytochemical recommendations (1985), concluded that β-glucuronidase staining provided little information not already available from acid phosphatase staining.

β-N-acetyl-D-glucosaminidase

This lysosomal acid hydrolase (EC 3.2.1.30, β-hexosaminidase) has been studied cytochemically mostly in parallel with acid phosphatase and β-glucuronidase, whose distribution it generally follows (Basso *et al.*, 1980, 1981; Dempsey *et al.*, 1980; Crockard *et al.*, 1982). Invernizzi and Perugini (1987) looked especially at the cytochemistry of this enzyme in normal and leukaemic T cells. Normally, most cells reacting with T4 monoclonal antibody had focal enzyme activity. In chronic T-lymphoproliferative diseases such as T-CLL and T-PLL the reaction product in T4 cells consisted of a single coarse granule, while T8 cases showed many scattered granules. In acute T-cell malignancies (T-ALL and T-lymphoblastic lymphoma) all blast cells were positive with either a single coarse granule or many coarse and small granules at one pole of the cell. The parallels between these findings and earlier reports dealing with other acid hydrolases are again apparent.

While the cytochemical demonstration of this enzyme, using the specific substrate naphthol β-N-acetyl-D-glucosamide with azo dye coupling, thus offers no substantial diagnostic advantage, it may eventually prove possible to exploit cytochemically the rather specific isoenzyme patterns demonstrable by analytical isoelectric focusing or disc electrophoresis among acute leukaemic cells. The literature on this subject is reviewed by Drexler *et al.* (1984). Four isoenzyme bands are recognized, A, B, I and C (Drexler and Gaedicke, 1983). In all cases of T-ALL the β-hexosaminidase forms A and B were found, as in normal lymphocytes; in most cases of cALL, pre B-ALL, null-ALL, undifferentiated leukaemias (AUL) and some myeloblastic cases the intermediate component (Hex I) was found, but in T-ALL, B-ALL and acute monocytic (AMonL) or myelomonocytic (AMML) cases it was not present. During remission this enzyme marker disappeared. More complex isoenzyme patterns within the immunological and morphological phenotypes of acute leukaemia occur, which may indicate different maturation and differentiation status of cells expressing the same immunological markers, but certainly the intermediate band Hex I appears to be a very early marker of lympho-haemopoietic development, common to most less differentiated ALLs but expressed also, perhaps transiently, at an early stage of granulopoietic differentiation.

Phosphorylases

The main phosphorylase enzyme, glycogen or α-glucan phosphorylase (EC 2.4.1.1) effectively extends the glycogen chain by catalysing the attachment to it of glucose residues from glucose-1-phosphate. The enzyme is found chiefly in the supernatant fraction on centrifugation (Dixon and Webb, 1964). In 1955 Takeuchi and Kuriaki described a histochemical method for the demonstration of glycogen phosphorylase in animal tissues, based on the specific iodine staining reaction of newly synthesised polysaccharide derived from the addition of glucose units to a glycogen primer as a result of enzyme action. Takeuchi and Kinoshita (1956) studied the distribution of the enzyme in blood cells, using frozen sections, and blood or bone marrow smears fixed for 60 sec in methanol. Phosphorylase activity was found to be greatly enhanced by the addition of adenosine-5-phosphate. In preparations of human blood and bone marrow cells enzyme was found in mature neutrophils and eosinophilic granulocytes, in metamyelocytes, myelocytes and promyelocytes, but not in erythrocytes, monocytes, lymphocytes or basophils. Megakaryocytes showed only a weakly positive reaction.

Quaglino and Hayhoe (1962) carried out further studies of the distribution of phosphorylase in haemic cells using a modified substrate containing polyvinylpyrrolidone and a high concentration of glucose-1-phosphate as suggested by Eränkö and Palkama (1961). Smears give better final preservation and enzyme reactivity if left to dry in air for 3–4 hr before being fixed in 70% acetone for 30 sec. After incubation in the substrate solution for 2 hr the slides were exposed to iodine, when a dark blue or violet blue colour showed the sites of phosphorylase activity. Positive reactions were found in all cells of the granulocyte series from myeloblasts onwards, with increasing intensity as maturation progressed (Plate 33C,D). Basophils had coarsely positive granules and eosinophils showed reaction in the cytoplasm around the specific granules. Monocytes and lymphocytes were sometimes negative as described by Takeuchi and Kinoshita, but not infrequently showed diffuse or granular positivity of varying intensity. Erythroblasts were normally negative, while megakaryocytes and platelets showed diffuse positive staining with occasional coarse granules (Plate 33E). Quaglino and Hayhoe noted that this pattern of positivity followed very closely that of preformed glycogen as demonstrated by the periodic acid-Schiff reaction, controlled by salivary amylase digestion, and observed a similar close parallel in studying phosphorylase cytochemistry in pathological states (Plate 33F and Fig. 11.1). In short, where PAS postivitiy was found, phosphorylase activity was also found, generally with similar diffuse or granular distribution. This applied in, for example, both acute and chronic leukaemias. An especially striking example of this concordance was described by Ackerman (1964) in a case of eosinophilic leukaemia, where the neoplastic cells showed intense cytoplasmic accumulation of glycogen and markedly elevated phosphorylase activity between the granules, notably in the frequent multilobed, poorly granular cells. Eosinophils in secondary or reactive eosinophilia did not show this feature, which Ackerman regarded as of differential diagnostic value.

Hayhoe *et al.* (1964) noted that phosphorylase positivity occurred in the erythroblasts of erythraemic myelosis in parallel with PAS positivity, and this observation was confirmed and extended by Kass and Hadi (1975) who reported phosphorylase positivity in the cytoplasm of erythroblasts in six of seven patients with chronic erythraemic myelosis, in proerythroblasts and megaloblasts from three patients with pernicious anaemia and two with severe folate deficiency, and in neoplastic lymphocytes from acute lymphoblastic leukaemia and lymphoma patients. In all these cases the cells showing phosphorylase reactivity had contained increased amounts of glycogen demonstrated by PAS positivity, whereas PAS negative erythroblasts and lymphocytes from normal blood and marrow and some other types of anaemia showed no phosphorylase reactions.

Quaglino and Hayhoe (1962) argued that the phosphorylase reaction could not be staining pre-existing glycogen, because negative results were obtained when glucose-1-phosphate was omitted from the substrate. However, the possibility was raised that the existence of preformed glycogen in the cell might help to dictate the apparent cytochemical localization of phosphorylase activity by acting as an in-built primer, supplementary to the primer included in the substrate mixture. That this is indeed the case was demonstrated 10 years later by Martin and Engel (1972) in an elegant series of experiments using phosphorylase staining and biochemical assay before and after stimulation of gastrocnemius muscle to the point of severe glycogen depletion. Absence of glycogen led to a sharp fall in histochemically demonstrable phosphorylase activity, but no change in the biochemi-

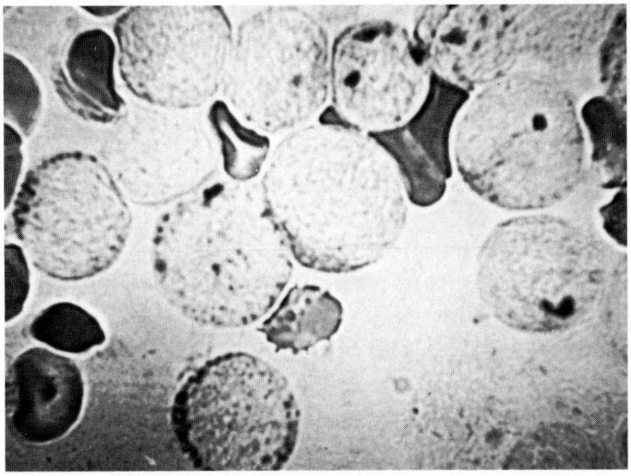

Fig. 11.1 ALL bone marrow. Phosphorylase reaction. Positivity in the form of coarse granules and blocks resembles that found in similar cases with the PAS reaction.

cally determined content, thus strongly suggesting that cytochemical staining of phosphorylase was dependent upon an adequate amount of intrinsic glycogen.

The action of phosphorylase would lead one to expect its distribution in cells and tissues to be similar to that of glycogen, and although the similarity observed in haemic cells is very close the distributions of the two are not identical. Polymorphs show greater variability in phosphorylase content than in PAS positivity, and blast cells in some acute leukaemic cases show weak diffuse phosphorylase staining when the PAS reaction is entirely negative.

With this technique the iodine staining gives a blue colour for the most part, but a violet or brownish tinge may be seen occasionally, especially at sites where coarse glycogen granules or blocks probably exist, as in some leukaemic lymphoblasts. This variation in colour probably relates to the chain-length of the polysaccharide produced by phosphorylase action at that site and the possible effect of branching enzyme on chains above a critical length (Takeuchi, 1958).

The validity of the phosphorylase reaction as described by Takeuchi and his colleagues has been challenged especially by Eckner *et al.* (1969), and the difficulty of distinguishing 'native' and 'newly formed' glycogen emphasised (Eckner, 1971). Nevertheless this distribution is further validated by electron histochemical observations (Takeuchi and Sasaki, 1970) which show new polyglucose particles synthesised by phosphorylase to be much finer and more amorphous than particles of native glycogen, and staining less densely with lead. Probably the differences in interpretation can largely be explained by the special role of intrinsic glycogen as primer for the reaction.

Evidence for the existence of at least five isozymes of phosphorylase in mammalian organs has been reported by Yonezawa and Hori (1975), following earlier studies showing differences in activity, immunological properties and amino acid composition of liver and muscle phosphorylase (Henion and Sutherland, 1957; Sevilla and Fischer, 1969; Schliselfelt, 1973). The five isozymes were designated L, LI, I, II and III, and a degree of organ-specificity was found, isozymes L and LI being richly concentrated in spleen and leucocytes of adult rats and pigs. The previous reports of liver-type phosphorylases in human leucocytes (Yunis and Arimura, 1964, 1968) seem likely to indicate the predominance of phosphorylase L in these cells.

Purine nucleoside phosphorylase (PNP)

This enzyme (EC 2.4.2.1) catalyses the reversible reaction:

purine nucleoside + phosphate $\rightleftharpoons$ α-D-ribose-1-phosphate + purine base

A cytochemical method for demonstrating the enzyme was devised by Rubio *et al.* (1972), using hypoxanthine as substrate, and this method has been applied to haemic cells by Borgers and his associates (1977, 1978). Cells separated from peripheral blood are fixed in glutaraldehyde and frozen and thawed to encourage penetration of the incubation medium. This consists of 4 mmol/l α-D-ribose-1-phosphate, 10 mmol/l hypoxanthine and 220 mmol/l sucrose in tris maleate buffer at pH 7.2–7.4, with 3 mmol/l lead nitrate as phosphate capture agent. After 30 minutes incubation the cells are centrifuged, washed and exposed to ammonium sulphide, and smears subsequently made and counterstained with haematoxylin. A similar procedure may be used for ultrastructural demonstration of PNP. The distribution of positivity is widespread, but it appears that among lymphocytes positivity is scanty and weak in B cells and strong and extensive throughout the cell in T cells (Borgers *et al.*, 1977). In CLL the lymphocytes are generally negative, in conformity with their common B cell origin, but in a single case of T cell CLL studied the lymphocytes were strongly positive (Borgers *et al.*, 1978). The reaction clearly merits further study in ALL and other lympho-proliferative states.

Biochemical studies have shown significant reduction in PNP levels in T-ALL cells as compared with normal lymphocytes, and perhaps also in B-ALL cells, but other ALL cases show normal levels, with some decrease in null-ALL and pre-B cases as compared with the cALL antigen-positive group (Demeocq *et al.*, 1982; Sylwestrowicz *et al.*, 1982; Drexler *et al.*, 1984).

In AML the PNP levels were generally somewhat higher than in ALL, although little above that of normal control cells, but AML cases with a monocytic or monoblastic component had higher levels than those without (Sylwestrowicz *et al.*, 1982).

Adenosine deaminase (ADA)

ADA (EC 3.5.4.4), a further enzyme concerned with the purine pathway, catalyses the irreversible conversion of adenosine and deoxyadenosine to inosine and deoxyinosine respectively. It is a cytoplasmic enzyme, essential for differentiation of lymphoid cells, especially T lymphocytes. Congenital absence of this enzyme is associated with severe combined immunodeficiency. Biochemical studies, reviewed by Drexler *et al.* (1984), have shown variability in levels of ADA in leukaemias, with most but not all reports suggesting higher levels in T-ALL than in other ALL variants, and with mostly high levels in AML, especially in cases with a monocytic or monoblastic component. ADA is of particular clinical interest because this enzyme is the target of a specific inhibitor, 2′-deoxycoformycin, which has been the subject of trials in chemotherapy of acute leukaemias, especially in T-ALL (Prentice *et al.*, 1980; Sylwestrowicz *et al.*, 1982).

Satisfactory cytochemical methods have not been devised for use on haemic cell preparations, but an immunohistochemical procedure has been applied by Chechik *et al.* (1984) to study the localization of ADA in benign reactions of human extrathymic lymphoid tissues and in B-cell lymphomas. With immune peroxidase and fluorescence methods reactive lymph nodes showed ADA reactions in the cell nuclei in small cleaved centrocytes and weak nuclear and/or cytoplasmic staining in large lymphoid cells (centroblasts) in the germinal centres. Cells in the mantle zone were negative. Plasma cells showed mainly cytoplasmic staining. In B-cell lymphomas a similar pattern was found to that of normal B cells, with heterogeneity perhaps associated with differentiation or proliferative activity.

Sulphatases

Arylsulphatases (EC 3.1.6.1) hydrolyze arylsulphates; they are widely distributed and occur in at least three main forms—A, B and C—differing in pH optima, behaviour towards inhibitors, and substrate specificity and in their intracellular localization, A and B being lysosomal and C microsomal (Dodgson and Spencer, 1957; Milsom *et al.*, 1968). Cytochemical methods using various naphthyl sulphates as substrates were devised by Seligman and his colleagues (Seligman *et al.*, 1950; Rutenburg *et al.*, 1952), the most satisfactory substrate being potassium 6-benzoyl-2-naphthyl sulphate. A modified version of this technique was described as suitable for use in smears of blood and marrow cells and organ imprints by Austin and Bischel (1961), who found granular positivity in the cytoplasm of eosinophils, megakaryocytes and neutrophils in descending order of density. Tanaka *et al.* (1962) used biochemical assays on leucocyte extracts and found the enzymic activity to vary markedly with cell type rather than with specific disease states. With an arbitrary score of one for average neutrophil activity, basophils had a value of roughly 2 and eosinophils about 8, whether from normal subjects or from patients with various diseases, including leukaemias and polycythaemia. Lymphocytes and lymphoblasts appeared to be devoid of biochemically detectable activity. The leucocytic enzymes had properties consistent with arylsulphatases A and B of Dodgson and Spencer (1957), being active against nitrocatechol sulphate but not against p-nitrophenyl sulphate, and being strongly inhibited by fluoride and phosphate. The intensity of arylsulphatase activity in eosinophils might suggest a role for the enzyme in allergic responses, reactions to parasitic invasion or in modifying the inflammatory response, but the specific mechanism of any such function and the identity of naturally occurring substrates remain speculative.

In the course of a study of arylsulphatase activity in acute leukaemic states, Lawrinson and Gross (1964) noted that, while primitive cells showed little cytoplasmic reaction, those of the lymphoid line had diffuse nuclear positivity while myeloid blasts had unstained nuclei (Figs 11.2, 11.3). With this method the common type of acute leukaemia of childhood, then sometimes described as 'stem cell' or 'undifferentiated', showed the positive nuclear reaction of lymphoblasts. The finding was confirmed in general by Ekert and Denett (1966), but others have found positive nuclear reactions in myeloblasts and monoblasts (Boysen, 1969; Smutka and Brunning, 1969) and the differential value of nuclear positivity is clearly less than initially suggested.

Ultrastructural studies have provided more accurate localizations for aryl sulphatase activity. Most earlier studies, reviewed by Hopsu-Havu and Helminen (1974), used either nitrocatechol sulphate or hydroxyquinoline sulphate as substrate and the heavy metals lead and barium as capturing agents, demonstrating mostly aryl sulphatases A and B. A more recent alternative (Chang and Moudgil, 1984) used p-nitrocatechol sulphate as substrate, but the released nitrocatechol was coupled in the presence of cupric ferricyanide to produce Hatchett's brown which reacts with di-aminobenzidine to give an osmiophilic polymer. Using this method and inhibiting aryl sulphatase C by low pH and B by addition of pyrophosphate, Chang and Moudgil found reaction product of aryl sulphatase A in fibroblasts to be located mostly in large periplasmic vesicles near the cell surface, and not in lysosomes. Most earlier studies, however, have found the enzyme activity in lysosomes of many different tissues, with less consistent activity in Golgi cisternae and GERL and occasionally in endoplasmic reticulum and nuclear envelope (Bainton and Farquhar, 1968, 1970; Bainton *et al.*, 1971). Azurophilic granules in human granulocytes have shown positivity (Bainton *et al.*, 1971) as have the α-granules of rabbit blood platelets (Murata *et al.*, 1975).

Biochemical studies of lysosomal aryl sulphatases A and B of peripheral blood leucocytes from acute and chronic leukaemias compared with healthy controls have been reported by Uehara *et al.* (1983). An increase in phosphorylated forms of these lysosomal hydrolases was found in CML and also in AML, perhaps representing a feature of proliferating neoplastic granulocytes. A decrease was found in ALL and CLL.

Dihydrofolate reductase (DHFR)

This enzyme (EC 1.5.1.3) plays an important part in the conversion of dietary folate to the active coenzymatic form, tetrahydrofolate, and again during the synthesis of thymidylate from deoxyuridylate, when tetrahydro-folate is oxidized to dihydrofolate and the reductase re-converts dihydro- to tetrahydrofolate. *In vitro* the enzyme can be viewed as an isomerase,

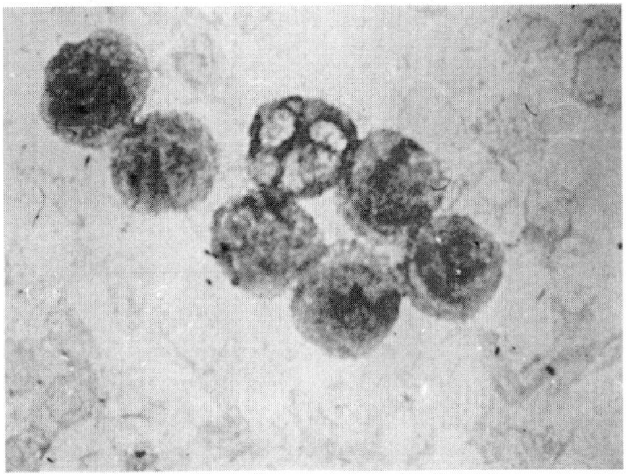

Fig. 11.2 ALL bone marrow. Aryl sulphatase reaction. The lymphocyte precursors have nuclear positivity, while the single polymorph present has a negative nucleus.

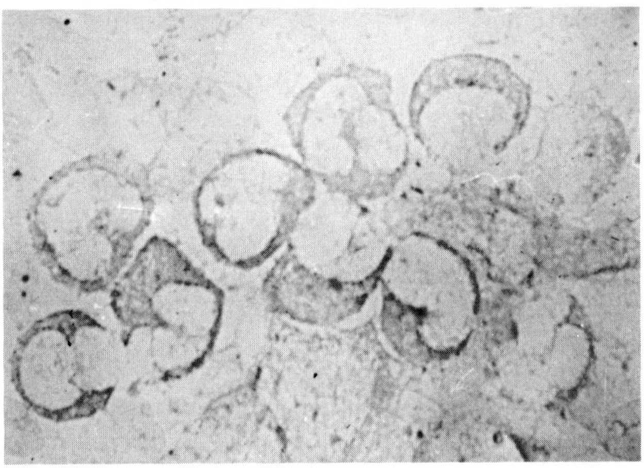

Fig. 11.3 CML bone marrow. Aryl sulphatase reaction. The granulocyte precursors have unstained nuclei.

establishing an equilibrium between tetrahydrofolate (FH_4), dihydrofolate (FH_2) and folate, and it is also named FH_4 dehydrogenase, although the equilibrium favours the tetrahydro form. The reversible reaction is NADPH (or NADP) dependent.

$$\text{Folate} + \text{NADPH}^+ \xrightleftharpoons{\text{DHFR}} \text{FH}_2 + \text{NADP}^+ \qquad \text{(Step 1)}$$

This is a very slow reaction.

$$\text{FH}_2 + \text{NADPH}^+ \xrightleftharpoons{\text{DHFR}} \text{FH}_4 + \text{NADP}^+ \qquad \text{(Step 2)}$$

This is a fast reaction.

The enzyme has been much studied biochemically and has come to be regarded as the prime target enzyme for methotrexate (MTX) action

(Futterman, 1957; Osborn *et al.*, 1958; Peters and Greenberg, 1958; Roberts and Wodinsky, 1968). Cellular resistance to inhibitory analogues of folic acid may be associated with the appearance of increased intracellular levels of DHFR (Bertino *et al.*, 1962; Schrecker *et al.*, 1971; Gundersen *et al.*, 1972). Study of the interaction of MTX and DHFR would be facilitated if reliable techniques for the cytochemical demonstration of DHFR were available and a number of techniques have been devised.

Gerzeli and de Piceis Polver (1969, 1972) used FH_4 as substrate, NADP as coenzyme and nitro blue tetrazolium (NBT) as hydrogen acceptor. They blocked diaphorase action with p-chloromercuric benzoate (PCMB) and inhibited interference from cytochrome oxidase by blocking this enzyme with azide. They attempted to prevent the spontaneous oxidation of FH_4 by dissolving it in 2-mercaptoethanol and carrying out the incubation in an atmosphere of nitrogen. Despite these precautions they encountered considerable non-specific precipitation of formazan extracellularly and presumably intracellularly also, almost certainly due to the persistence of some non-enzymatic reduction of NBT by FH_4. Moreover, the enzymic contribution to positivity is likely to be small with FH_4 as substrate, as the reaction $FH_2 \xrightleftharpoons{\text{DHFR}} FH_4$ strongly favours the tetrahydro form. Onicescu and colleagues (1970, 1972a, b, and 1973) approached the reaction from the reverse direction by using folic acid as substrate, adding both NAD and NADP, but the reaction $\text{FA} \xrightleftharpoons{\text{DHFR}} FH_2$ is very slow and the positivity they encountered may well have been attributable to diaphorases acting with endogenous NADPH as tetrazolium reductases.

Tzortzatou and Hayhoe (1974a) devised a method which uses FH_2 as substrate, blocks diaphorase and cytochrome oxidase activity, and permits the non-enzymatic reduction of NBT by FH_4 formed by the action of DHFR on the substrate, thus revealing intracellular sites of enzyme activity. The method appears to be specific and shows only intracellular localization of the blue formazan deposits of reduced tetrazolium. Control films treated with the same reaction mixture but without FH_2 were negative. In human peripheral blood (Figs 11.4 and 11.5) a very strong granular reaction was found in eosinophils and a strong granular positivity together with a weak diffuse reaction in neutrophils. The size of the granules varied and they were often too numerous to count. Monocytes had fewer granules and more restricted areas of diffuse positivity in the paranuclear region. Basophils showed moderate granular positivity. Lymphocytes had only few cytoplasmic granules or none. Erythrocytes were negative. In bone marrow smears (Fig. 11.6), the cells of the granulocyte series were all positive, with weakest reaction in myeloblasts and increasing positivity with increasing cell maturity. Proerythroblasts had moderate numbers of granules and later normoblasts showed weak or negative reactions. Plasma cells and megakaryocytes were negative or showed only a few granules.

The cytochemically demonstrable activity of DHFR could not be inhibited by pre-incubation of leucocytes or fibroblasts with folate antagonists, even at very high concentrations (Tzortzatou and Hayhoe, 1974b). Inhibition was obtained only when the antagonists were included during incubation with the cytochemical reaction mixture in very high

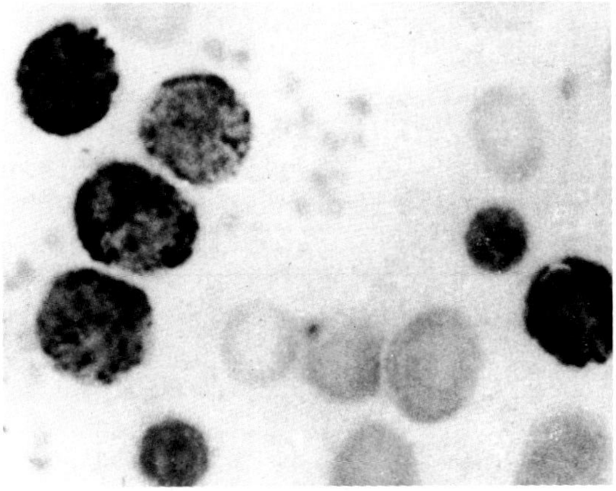

Fig. 11.4 DHFR activity in normal buffy coat cells. Three neutrophils show strongly positive granules, two monocytes have fewer scattered granules and two lymphocytes are almost negative.

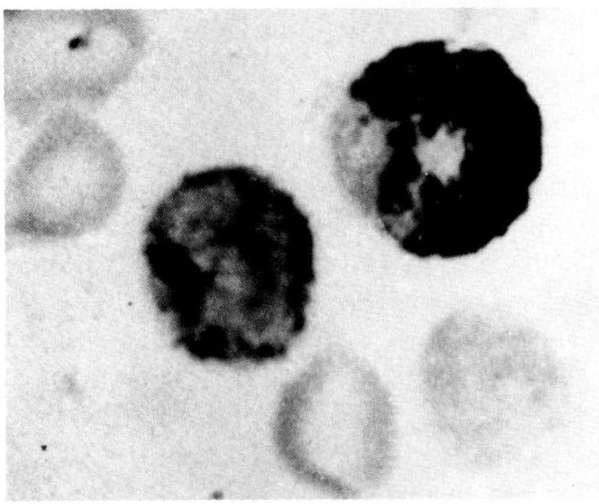

Fig. 11.5 DHFR activity in an eosinophil (very strongly positive) and a neutrophil.

concentrations (20 mmol/l). Injection of antagonists *in vivo* into AKR mice, even at highly toxic and eventually lethal concentrations, failed to produce any detectable inhibition of DHFR activity in cells of bone marrow, spleen or other tissues. These results suggest that, contrary to generally accepted belief, the binding of folate antagonists to DHFR in the intact cell is weak and reversible, that the antagonists are readily replaced by the physiological substrate, dihydrofolate, and that some mechanism other than inhibition of DHFR may have a part to play in the therapeutic action of folate antagonists.

Studies of DHFR activity in cytochemical preparations from various haematological diseases were reported by Tzortzatou (1975, 1977a). The variation between normal controls and cells from Hodgkin's disease, myeloma, polycythaemia, chronic lymphocytic and chronic myeloid leukaemias was not significant. In acute myeloid and lymphoblastic leukaemias

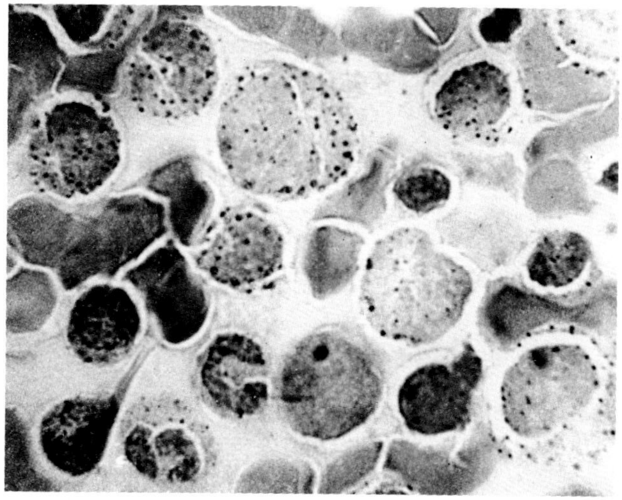

Fig. 11.6 DHFR activity in cells of normal bone marrow. Positivity in granulocyte series is less strong than in peripheral blood.

the blast cells were only weakly positive, or negative. In patients treated with MTX, increased DHFR was found cytochemically in the granulocytes.

Using a further technical modification Nano *et al.* (1982) reported finding intense polar positivity in leukaemic lymphoblasts, which contrasted with scattered granular positivity in myeloblasts. These authors also found some discriminant staining patterns for DHFR in CLL as between T and B-cell variants (Invernizzi *et al.*, 1983). In another study (Nano *et al.*, 1987) DHFR activity was studied in bone marrow erythroblasts of normal subjects and of 24 untreated cases of myelodysplastic syndromes, including all the cytological subtypes. In contrast to the pattern of positivity found in normal erythroblasts and in those of RA and RAS, where the reaction product was mostly shown as a perinuclear ring of granules or, less frequently, was asymmetrically localized in a paranuclear area, a predominance of the paranuclear staining pattern was reported in RAEB, RAEB-t and CMML.

5-Formyl tetrahydrofolate cyclodehydrase and 5,10-methenyl tetrahydrofolate cyclohydrolase activity

The enzyme 5-formyl FH_4 cyclodehydrase (folinic isomerase) catalyses the conversion of 5-formyl-FH_4 (leucovorin) to 5,10-methenyl-FH_4 in the presence of ATP and NADPH. A second enzyme, 5,10-methenyl-FH_4 cyclohydrolase, catalyses the hydrolysis of 5,10-methenyl-FH_4 to 10-formyl-FH_4. The products of each of these reactions play an important part in the biosynthesis of purines.

Using a modification of the method previously developed for the cytochemical demonstration of DHFR (Tzortzatou and Hayhoe, 1974a),

Tzortzatou (1977b) succeeded in revealing intracellular sites of activity of both the cyclodehydrase and the cyclohydrolase. The respective substrates of 5-formyl-FH_4 and 5,10-methenyl-FH_4 were used, possible interference by other enzymes was blocked, and the non-enzymatic reduction of NBT by the reaction products produced blue deposits of formazan. Strong positivity for both enzymes was found in granulocytes, moderate positivity in monocytes, and weak or negative reactions in lymphocytes and red cells. In both blood and marrow cells the general distribution of these two enzymes appears similar to that of DHFR.

TECHNICAL METHODS

5′-nucleotidase

El-Mohandes and Hayhoe (1983) after Wachstein and Meisel (1957)

Reagents

Fixative

Buffered acetone solution, pH 4.2:
0.03 M sodium citrate, 64 ml
0.03 M citric acid, 336 ml
absolute acetone, 600 ml

Incubation medium

0.1 N barbiturate buffer, pH 7.5*	26 ml
(0.1 M sodium barbiturate, 15 ml; 0.1 N hydrochloric acid, 10 ml; distilled water, 5 ml)	
Adenosine-5′-monophosphoric acid†	30 mg
0.1 M manganous acetate‡	10 ml
1% L-tartaric acid§	1.35 ml
2% lead acetate¶	6 ml
Distilled water	10 ml

Control medium

AMP is replaced by β-glycerophosphate, a substrate split by alkaline phosphatase but not by 5′-nucleotidase.

* pH is not critical. 5′-Nucleotidase is active between 6.5 and 8.5, with pH optimum towards the alkaline end, but at pH 7.5–8.5 nonspecific alkaline phosphatase is more active. A pH between 6.5 and 7.5 is, therefore, best for demonstrating 5′-nucleotidase specifically.

† AMP: Type II sodium salt from equine muscle, from Sigma

‡ Manganous, as distinct from manganic, acetate, is used—$(CH_3COO)_2\ Mn.4H_2O$; from BDH Ltd.

§ L-tartaric acid obtained from Sigma

¶ Either basic lead acetate ($(CH_3COO)_2Pb.Pb(OH)_2$) or neutral lead acetate ($(CH_3COO)_2Pb$) can be used. The former is less soluble than the latter, and gives a final pH in the incubation medium of approximately 7.2, while the neutral salt gives a final pH of 6.7. Either solution can be filtered before being added drop by drop to the medium with continuous stirring, or even the whole incubation medium can be filtered before use without significantly affecting the reaction. Analar reagents from BDH Ltd. are used.

Technique

1. Fix air-dried smears in buffered acetone solution, pH 4.2, for 15 s
2. Wash in distilled water
3. Incubate air-dried smears in Coplin jars for 6 h at 37°C. The demonstration of 5′-nucleotidase was carried out at pH 7.2 at which the hydrolysis of AMP and β-glycerophosphate by nonspecific alkaline phosphatase is low
4. Wash thoroughly in distilled water
5. Fix in formalin vapour for 5 min
6. Wash in distilled water
7. Immerse in 2% ammonium sulphide solution for 2 min
8. Wash in distilled water
9. Counterstain with 2% Methyl Green for 10 min
10. Wash in distilled water and examine.

Khalaf (1986) after Wachstein and Meisel (1957)

Reagents

1. Fixative: Citrate buffered acetone (CBA) (Kaplow and Burstone, 1963)
2. Substrate: uridine-5′-monophosphate-disodium salt. H_2O.
3. Buffer: 0.2 mol/l tris-maleate buffer pH 7.2
4. Activator: 0.02 mol/l manganese sulphate tetrahydrate
5. Inhibitors of AlP and AcP: Levamisole and Na K-tartrate.
6. Coupler: 0.02 mol/l lead nitrate solution
7. Post-incubation fixative: formaldehyde vapour (FV)
8. Reducing agent: 2% ammonium polysulphide solution
9. Counterstain: Carazzi's haematoxylin
10. Blueing agent: Scott's tap water substitute
11. Mountant: DPX.

Preparation of the incubation medium (IM)

Prepare the IM freshly before use with the aid of a magnetic stirrer.

0.2 mol/l tris-maleate buffer pH 7.2	24 ml
Distilled water (DW)	18 ml
Uridine-5′-monophosphate-disodium salt. H_2O	46 mg
0.02 mol/l manganese sulphate tetrahydrate	12 ml
Levamisole	29 mg
Na K tartrate	169 mg
Sucrose	1.8 mg
Ficoll	0.6 g
0.02 mol/l lead nitrate	5.4 ml

(Add very slowly, the tip of the pipette should be under the surface of IM)

Check pH, adjust to 7.2 if necessary with 0.5 mol/l of either HCl or NaOH and filter using Whatman filter paper no. 1.

Technique

1. Fix air-dried blood films in CBA pH 4.2 at room temperature (RT) for 15 s
2. Incubate films in a Coplin jar at RT for 16 h and fixed freeze-dried cryostat sections for only 1 h
3. Wash briefly in DW × 2
4. Post fix in FV for 5 min
5. Wash briefly in DW × 3
6. Immerse in 2% ammonium polysulphide for 2 min
7. Wash briefly in DW × 3
8. Counterstain films in Carazzi's haematoxylin at RT for 10 min (sections for 5 min)
9. Mount the sections in DPX mountant.

Glucose-6-phosphatase

After Wachstein and Meisel (1956)

Reagents

Incubation medium

Glucose-6-phosphate-Na salt	26 mg
0.1 mol/l acetate buffer, pH 6.5	40 ml
0.1 mol/l lead nitrate	1 ml

Control medium

Replace the glucose-6-phosphate with 25 mg sodium β-glycerophosphate.

Technique

1. Fix air-dried smears in formalin vapour 5 minutes
2. Wash briefly in tap water
3. Immerse in freshly prepared incubation medium at 37°C for 30 minutes
4. Wash thoroughly in distilled water
5. Immerse in 5% (v/v) ammonium sulphide solution for 2 minutes
6. Wash in distilled water and examine.

Adenosine triphosphatase

Khalaf (1986) after Wachstein and Meisel (1957)

Reagents

1. Substrate: adenosine-5′-triphosphate-disodium salt
2. Buffer: 0.2 mol/l tris-maleate buffer pH 7.2
3. Activator: 0.02 mol/l magnesium sulphate heptahydrate
4. Inhibitors of interfering phosphatases:
   a. 0.02 mol/l theophylline
   b. 0.05 mol/l levamisole
   c. 0.3 mol/l alloxan
5. Stabilizer: Dextran T500.
6. Coupling agent: 0.02 mol/l lead nitrate
7. Rinsing solution: 0.9% sodium chloride
8. Post-incubation fixative: formaldehyde vapour
9. Reducing agent: 2% ammonium polysulphide
10. Counterstain: Carazzi's haematoxylin
11. Blueing agent: Scott's tap water substitute

Incubation medium (IM)

Prepare the following IM freshly before use:

0.2 mol/l tris-maleate buffer pH 7.2	24 ml
Distilled water	20 ml
0.02 mol/l theophylline	6 ml
ATP-disodium salt	18 mg
0.1 mol/l magnesium sulphate.7 H_2O	2.4 ml
0.05 mol/l levamisole	0.6 ml
0.3 mol/l alloxan	2 ml
Dextran T500	2.4 g
0.02 mol/l lead nitrate	5 ml

Mix well after each addition. Adjust pH to 7.2 with 1 mol/l HCl. Filter through Whatman filter paper no. 4 if necessary.

Technique

1. Leave the unfixed buffy coat films covered on the bench or in deep freeze at −25°C for about 18–24 h
2. Incubate the films in Coplin or Hellendahl jars at 37°C for 45 min on a slow-moving shaker
3. Wash films briefly in 0.2 mol/l tris-maleate buffer pH 7.2
4. Rinse briefly × 1 in 0.9% sodium chloride
5. Fix in formaldehyde vapour at room temperature (RT) for 8 min
6. Wash thoroughly in distilled water (DW) for about 30 s × 3
7. Immerse in 0.2% ammonium polysulphide for 2 min
8. Wash thoroughly in DW for about 30 s × 3
9. Counterstain in Carazzi's haematoxylin for 10 min
10. Wash briefly in alkaline tap water or immerse in Scott's tap water substitute for 30 s then rinse briefly in DW.

Phosphamidase

After Gomori (1952), Chayen *et al.* (1973)

Reagents

Incubation medium

Solution A:	0.05 mol/l acetate buffer, pH 5.4	50 ml
	0.1 mol/l (3.3 g/100 ml) lead nitrate	2 ml
	Sodium chloride	0.1 g
	Polyvinyl pyrrolidone (PVP)	2 g
Solution B:	(prepare immediately before use)	
	p-chloroanilidophosphonic acid	0.15 g
	1 mol/l sodium hydroxide	0.5 ml

Add solutions A and B.

Control solution

1. Omit solution B
2. Add 0.01 mol/l sodium fluoride to medium: phosphamidase activity is *not* inhibited.

Technique

1. Fix air-dried smears in formalin vapour 5 minutes
2. Wash briefly in tap water
3. Immerse in freshly prepared incubation medium at 37°C for 30 minutes
4. Wash thoroughly in distilled water
5. Immerse in 5% (v/v) ammonium sulphide solution for 2 minutes
6. Wash in distilled water and examine.

Aminopeptidases

After Ackerman (1960)

Reagents

Incubation medium

Alanyl β-naphthylamide 4 mg in 5 ml distilled water

0.1 mol/l phosphate buffer pH 6.7	5 ml
Magnesium sulphate	1 mg
Fast Garnet GBC salt	5 mg

Technique

1. Fix air-dried smears in formalin vapour 5 minutes
2. Wash in tap water

3. Immerse in incubation medium for 4–6 hours at room temperature
4. Wash in tap water and blot dry
5. Counterstain with aqueous haematoxylin 5 minutes.

Dipeptidylaminopeptidases II and IV (DAP II and IV)

After Lojda *et al.* (1979a) modified by Khalaf *et al.* (1986a)

Reagents

Incubation medium for DAP II

Lysyl-alanyl-4-methoxy-2-naphthylamide (LAMN)	4 mg
dissolved in	
Dimethyl formamide (DMF)	0.25 ml
and added to	
cacodylate buffer 0.1 mol/l pH 5.2	4.75 ml
Fast blue B salt	5 mg

Adjust pH to 5.2 if necessary by 0.2 mol/l HCl and filter.

Incubation medium for DAP IV

Glycyl-prolyl-4-methoxy-2-naphthylamide (GPMN)	4 mg
dissolved in DMF	0.2 ml
and added to	
Phosphate buffer 0.1 mol/l pH 7.2	4.6 ml
Fast blue B salt	5 mg
dissolved in DMF	0.2 ml.

Adjust pH to 7.2 if necessary by 0.2 mol/l HCl and filter.

Technique

1. Fix air-dried smears for 2 min in formalin vapour at room temperature
2. In a humid chamber: apply the incubation medium to smears using a 200 µl pipette (200 µl/smear) and cover each smear with a 22 × 64 mm coverslip spreading the mixture over the smear
3. Incubate for 2 h at 37°C (DAP II) or for 45 min at room temperature (DAP IV)
4. Remove the coverslips carefully and wash in distilled water
5. Counterstain in Carazzi's haematoxylin for 5–10 min
6. Wash briefly in alkaline tap water or immerse in Scott's tap water substitute for 30 s and rinse in distilled water
7. Mount the smears using Apathy's mounting medium.

Comments

1. The substrates are expensive so preparation of large amounts of incubation medium for use in Coplin jars or even smaller plastic containers

is not usually justified. The use of a humid chamber is recommended.

2. Because of their reported stability (McDonald and Schwabe, 1977) it is feasible to study the DAP II and IV activities in smears up to several months old.

3. Mounting in Apathy's mounting medium is used to preserve the azo dye reaction product as it dissolves in immersion oil used in microscopy. It also fades with time so that early scoring is preferable.

γ-Glutamyl transferase (γGT)

Khalaf (1986) after Rutenberg *et al.* (1969)

Reagents

1. Fixatives:
   a. Phosphate buffered acetone formalin (PBAF) mixture pH 6.6
   b. Acetone
   c. 50% Acetone
2. Substrate: gamma-glutamyl-4-methoxy-2-naphthylamide (GGMN)
3. Solvent for the substrate: dimethyl sulphoxide (DMSO) and 1 mol/l sodium hydroxide (NaOH)
4. Buffer: 0.1 mol/l tris-HCl buffer pH 7.2
5. Activator/acceptor: N-glycylglycine
6. Coupling agent: Fast Garnet GBC (Gurr/BDH)
7. Rinsing buffer: veronal acetate pH 9.2
8. Counterstain: Carazzi's haematoxylin
9. Blueing agent: Scott's tap water substitute
10. Mounting medium: Apathy's mounting medium.

Preparation of the substrate solution

Dissolve 10 mg of GGMN in 0.2 ml DMSO and 0.2 ml of 1 mol/l NaOH; add 3.6 ml distilled water (DW).

Incubation medium

Prepare the following IM freshly before use:

0.1 mol/l tris-HCl buffer pH 7.2	68ml
GGMN solution (2.5 mg/ml)	4 ml
Distilled water	8 ml
Glycylglycine	60 mg
Fast Garnet GBC	28 mg

Mix well after each addition, if necessary adjust pH to 7.2 with 1 mol/l HCl and filter with Whatman filter paper no. 1.

Technique

1. Fix air-dried buffy coat films in PBAF at 4–10°C for 15 s, then wash briefly × 3 in distilled water. Fix bone marrow films in acetone at 4–10°C

for 15 s followed immediately by 50% acetone at 4–10°C for 30 s

2. Incubate films in Coplin jars or small plastic containers at room temperature for 1 h 45 min on a slow-moving shaker
3. Wash the films briefly × 3 in veronal acetate buffer pH 9.2
4. Rinse briefly × 1 in distilled water
5. Counterstain in Carazzi's haematoxylin for 5–10 min
6. Wash briefly in alkaline tap water or immerse in Scott's tap water substitute for 30 s and rinse in distilled water
7. Mount the films in Apathy's mounting medium.

β-Glucuronidase

After Hayashi *et al.* (1964)

Reagents

Substrate stock solution

(may be stored at room temperature for 4–6 weeks)

Dissolve naphthol AS-BI glucuronide	28 mg
In 0.05 mol/l sodium bicarbonate	1.2 ml

Add 0.2 mol/l acetate buffer pH 5 to 100 ml.

Pararosaniline solution

Dissolve 1 g pararosaniline hydrochloride in 20 ml distilled water plus 5 ml concentrated HCl, using gentle heat. Cool, filter and store at room temperature.

Sodium nitrite solution

4% sodium nitrite in distilled water. Prepare freshly each week.

Incubation medium

Prepare immediately before use. Add 0.3 ml pararosaniline solution to 0.3 ml of sodium nitrite solution; after 1 minute, add 10 ml of substrate stock solution. Adjust pH to 5.2 with NaOH solution, and make up final volume to 20 ml with distilled water. Filter before use.

Technique

1. Fix air-dried smears in formalin vapour 5 minutes
2. Wash briefly in tap water
3. Immerse in freshly prepared incubation medium at 37°C for 30 minutes
4. Rinse well with distilled water
5. Counterstain with methyl green for 15 minutes or with aqueous haematoxylin for 5 minutes.

Phosphorylase

After Takeuchi and Kinoshita (1956) and Quaglino and Hayhoe (1962)

Reagents

Incubation medium

Glucose-1-phosphate, dipotassium salt	400 mg
Adenosine-5-phosphate	40 mg
Glycogen	8 mg
Sodium fluoride	720 mg
PVP	3.6 g
Ethanol	8 ml
Acetate buffer 0.1 mol/l, pH 5.9	40 ml
Insulin 40 i.u./ml	0.1 ml

The solution is filtered and remains satisfactory for 4–5 days.

Grams iodine solution

Iodine	1 g
Potassium iodide	2 g
Sucrose solution 0.32 mol/l	300 ml

Technique

1. Fix air-dried smears 3–4 hours after preparation by covering with 70% acetone for 30 seconds at room temperature
2. Tip off fixative and blot dry without washing
3. Immerse in incubation medium for 2 hours at 37°C
4. Transfer to 40% ethanol for 2 minutes
5. Allow to dry and immerse in 0.32 mol/l sucrose solution for 5 minutes
6. Transfer to Gram's iodine solution in sucrose for 5 minutes
7. Mount in iodine-glycerine (equal parts of glycerine and Gram's iodine).

Arylsulphatases

After Austin and Bischel (1961)

Reagents

Incubation medium

Dissolve 25 mg potassium 6-benzoyl-2-naphthyl sulphate in 80 ml 0.85% NaCl at 95–99°C. Cool, and add 20 ml of 0.5 mol/l acetate buffer pH 6.1 and 2.6 g NaCl.

This remains stable at 4°C for 4 weeks. Immediately before use, add 35 mg Fast blue RR salt and filter in the dark into a Coplin jar.

Technique

1. Unfixed slides are immersed consecutively in three graded tonicities of NaCl, 0.85, 1.0 and 2.0%, with 3 to 5 minutes in each solution.
2. Transfer to the filtered incubation medium and incubate for 48 hours in the dark at 37°C
3. Pass slides through a reverse sequence of saline solutions, from 2.0 to 0.85 per cent, 3 to 5 minutes in each
4. Wash in tap water and air dry.

Dihydrofolate reductase

After Tzortzatou and Hayhoe (1974)

Reagents

Incubation medium prepared immediately before use.

Ficoll	0.3 mmol/l
NaCN	10 mmol/l
NBT	1 mmol/l (filtered before adding)
NADPH	1 mmol/l
Dihydrofolic acid	6 mmol/l (added last, already dissolved in a small volume of buffer).

These reagents are made up in 0.18 mol/l tris-HCl buffer, pH 7.4.

Technique

1. Fix smears, air-dried for 1 hour, in formalin vapour for 5–7 minutes
2. Air-dry for a further 30 minutes and incubate with PCMB (para-chloromercuric benzoic acid), 3 mmol/l in 0.18 mol/l tris-HC1 buffer, pH 7.4, for 15 minutes at 37°C
3. Air-dry and immerse in incubation medium for 60–90 minutes at 37°C in an atmosphere of nitrogen and in darkness
4. Wash briefly in tap water and counterstain with methyl green for 15 minutes or nuclear Fast Red for 20 minutes.

Cyclodehydrase and cyclohydrolase

After Tzortzatou (1977)

Reagents

Incubation medium

1. *for cyclodehydrase*

NBT	3 mg
NADPH	3 mg
Leucovorin	3 mg (1 ml)

Dissolve in 0.2 mol/l citric phosphate buffer pH 7.0, to give a final volume of 2 ml.

2. *for cyclohydrolase*

NaCN	1.0 mg
NBT	3 mg
5,10-methenyl-FH_4	1.5 ml

Dissolve in 0.15 mol/l tris-HCl buffer pH 7.2, to give a final volume of 2 ml.

For both reactions, prepare immediately before use, and apply to slide surface within a wax ring and with coverslip.

Technique

1. Films are air-dried for 1 hour and fixed in formalin vapour for 7 minutes
2. Air-dry for a further hour and incubate with 3 mmol/l PCMB in 0.18 mol/l tris-HCl buffer, pH 7.4, for 15 minutes at 37°C
3. Air-dry and apply a thin ring of wax on the slide to contain the incubation medium
4. Add the appropriate incubation medium, coverslip, and incubate for 60–90 minutes for cyclodehydrase and 60 minutes for cyclohydrolase, in the dark and in an atmosphere of nitrogen
5. Wash briefly in tap water, remove the wax ring and coverslip and counterstain with nuclear Fast Red for 20 minutes or methyl green for 15 minutes.

REFERENCES

Ackerman G A 1960 Histochemical demonstration of aminopeptidase activity in the leucocytes of blood and bone marrow. Journal of Histochemistry and Cytochemistry 8: 36

Ackerman G A 1963 Histochemical demonstration of aminopeptidase activity in the cells of the blood and bone marrow from various haematological disorders. Nature 197: 189

Ackerman G A 1964 Eosinophilic leukaemia—a morphological and histochemical study. Blood 24: 372

Albert Z, Orlowski M, Szewczuk A 1961 Histochemical demonstration of gamma-glutamyl transpeptidase. Nature 191: 767

Austin, J H and Bischel M 1961 A histochemical method for sulfatase activity in hemic cells and organ imprints. Blood, 17: 216

Bainton D F, Farquhar M G 1968 Differences in enzyme content of azurophil and specific granules of polymorphonuclear leukocytes. Journal of Cell Biology 39: 299

Bainton D F, Farquhar M G 1970 Segregation and packaging of granule enzymes in eosinophilic leukocytes. Journal of Cell Biology 45: 54

Bainton D F, Ullyot J L, Farquhar M G 1971 The development of neutrophilic polymorphonuclear leukocytes in human bone marrow: origin and content of azurophil and specific granules. Journal of Experimental Medicine 134: 907

Barr R D, Perry S 1976 Lysosomal acid hydrolases in human lymphocyte subpopulations. British Journal of Haematology 32: 565

Barton R W, Goldschneider I 1978 5′-Nucleotidase activity in subpopulations of rat lymphocytes. Journal of Immunology 121: 2329

Basso G, Cocito MG, Semenzato G, Pezzutto A, Zanesco L 1980 Cytochemical study of thymocytes and T lymphocytes. British Journal of Haematology 44: 577

Basso G, Cocito M G, Pezzutto A, Semenzato G 1981 N-acetyl-β-D-glucosaminidase activity in T chronic lymphocytic leukaemia. Acta Haematologica 66: 69

Bertino J R, Donohue D R, Gabris B W, Silber R, Alenty A, Meyer M, Huennekens F M 1962 Increased level of dihydrofolic reductase in leucocytes of patients treated with amethopterin. Nature 193: 140

Borgers M, Verheyen A 1985 Enzyme Cytochemistry. International Review of Cytology 95: 163

Borgers M, Verhaegen H, De Brabander M, Thone F, Geuens G 1977 Purine nucleoside

phosphorylase, a possible histochemical marker for T cells in man. Journal of Immunological Methods 16: 101
Borgers M, Verhaegen H, De Brabander M, De Cree J, De Cock W, Thone F, Geuens G 1978 Purine nucleoside phosphorylase in chronic lymphatic leukaemia (CLL). Blood 52: 886
Boysen G 1969 An evaluation of aryl sulphatase activity in leukaemic cells. Scandinavian Journal of Haematology 6: 246
Brouet J C, Valensi F, Daniel M-T, Flandrin G, Preud'Homme J L and Seligmann M 1976 Immunological classification of acute lymphoblastic leukaemias: evaluation of its clinical significance in a hundred patients. British Journal of Haematology 33: 319
Burnett D, Crocker J, Vaughan A 1983 Synthesis of cathepsin B by cells derived from the HL60 promyelocytic leukaemia cell lines. Journal of Cell Physiology 115: 249
Burstone M S 1962 Enzyme Histochemistry and its Application in the Study of Neoplasms. Academic Press, New York
Burstone M S, Folk J E 1956 Histochemical demonstration of aminopeptidase. Journal of Histochemistry and Cytochemistry 4: 217
Chang P L, Moudgil G 1984 A specific ultrastructural stain for arylsulphatase A activity in human cultured fibroblasts. Journal of Histochemistry and Cytochemistry 32: 617
Chayen J, Bitensky L, Butcher R G 1973 Practical Histochemistry. Wiley, London, p 129
Chechik B E, Schrader W P, Perets A, Fernandes B 1984 Immunohistochemical localization of adenosine deaminase in human benign extrathymic lymphoid tissues and B-cell lymphomas. Cancer 53: 86
Chilosi M, Pizzolo G, Semenzato G, de Rossi G, Pandolfi F 1984 Heterogeneous expression of dipeptidyl-amino-peptidase (DAP IV) in T-cell chronic lymphocytic leukemia. Acta Haematologica 71: 277
Conklyn M, Silber R 1982 Properties of lymphocyte 5′-nucleotidase in normal subjects and patients with chronic lymphocytic leukaemia. Leukemia Research 6: 203
Crockard A D, Chalmers D, Matutes E, Catovsky D 1982 Cytochemistry of acid hydrolases in chronic B and T cell leukemias. American Journal of Clinical Pathology 78: 437
Crockard A D, MacFarlane E, Andrews C, Bridges J M, Catovsky D 1984 Dipeptidylaminopeptidase IV activity in normal and leukemic T-cell subpopulations. American Journal of Clinical Pathology 82: 294
Crockard A D, MacFarlane E, Markey G M, McConnell R E, Alexander H D, Agnew A N D, Morris T C M, Bridges J M 1985 Acid hydrolase activities in B cell chronic lymphocytic leukaemia lymphocytes. Correlation of cytochemical reactions with immunological phenotype. Scandinavian Journal of Haematology 34: 242
Crocker J 1983 The Enzyme Histochemistry of Lymphomas. Dissertation for the MD degree. University of Cambridge, p 218
Crocker J, Burnett D, Jones E L 1984 Immunohistochemical demonstration of cathepsin B in the macrophages of benign and malignant lymphoid tissues. Journal of Pathology 142: 87
Cuschieri A 1982 Cytochemical demonstration of increased adenosine triphosphatase activity in lymphocytes activated in vitro by PHA or by the mixed lymphocyte reaction. Histochemical Journal 14: 139
Cuschieri A, Mughal S, Kharbat B A 1982 Ultrastructural localization of adenosine triphosphatase activity in lymphocytes activated in vitro by phytohaemagglutinin. Histochemical Journal 14: 593
Daimon T, Gotoh Y 1982 Cytochemical evidence of the origin of the dense tubular system in the mouse platelet. Histochemistry 76: 189
Davey F R, Huntingdon S J, McCallum J, MacMath J M 1977 Cytochemical reactions of normal and neoplastic lymphocytes. Journal of Clinical Pathology 30: 653
de Duve C, Pressman B C, Gianetto R, Wattiauz R, Appelmans F 1955 Tissue fractionation studies. 6. Intracellular distribution patterns of enzymes in rat liver tissue. Biochemical Journal 60: 604
de Pierre J W, Karnovsky M L 1974 Ectoenzymes of the guinea pig polymorphonuclear leukocyte. II. Properties and suitability as markers for the plasma membrane. Journal of Biological Chemistry 249: 7121
De Waele M, DeMey J, Moeremans M et al 1983 Cytochemical profile of immunoregulatory T-lymphocyte subsets defined by monoclonal antibodies. Journal of Histochemistry and Cytochemistry 31: 471
Demeocq F, Viallard J L, Bounsell L 1982 The correlation of adenosine deaminase and purine nucleoside phosphorylase activities in human lymphocyte subpopulations and in various lymphoid malignancies. Leukemia Research 6: 211
Dempsey S L, Crockard A D, Bridges J M 1980 An estimation of beta-glucuronidase and N-acetyl-beta-glucosaminidase activity in normal and chronic lymphocytic leukaemia lymphocytes. Acta Haematologica 64: 141
Dixon M, Webb E C 1964 Enzymes. 2nd edn. Longmans, London

Dodgson K S, Spencer B 1957 Assay of sulfatases. In: Glick D (ed) Methods of Biochemical Analysis. Vol. 4 Interscience Publishers, New York, p 211
Douglas S D, Cohnen G, König E, Brittinger G 1973 Lymphocyte lysosomes and lysosomal enzymes in chronic lymphocytic leukaemia. Blood 41: 511
Drexler H G, Gaedicke G 1983 Isoenzyme studies in human leukemia-III. β-hexosaminidase (EC 3.2.1.30) Leukemia Research 7: 611
Drexler H G, Gaedicke G, Minowada J 1984 Enzyme markers in acute leukemias: advances during the last decade. Journal of the National Cancer Institute 72: 1283
Eckner F A O 1971 Demonstration of phosphorylase and uridine diphosphate glucose-glycosyl transferase activities. Journal of Histochemistry and Cytochemistry 19: 133
Eckner F A O, Riebe B H, Moulder P V, Blackstone E H 1969 Polysaccharide synthesis in tissue sections. Evaluation of methods as an example of quality control in histochemistry. Histochemie 19: 340
Ekert H, Denett X 1966 An evaluation of nuclear aryl sulphatase activity as an aid to the cytological diagnosis of acute leukaemia. Australian Annals of Medicine 15: 152
El-Mohandes E, Hayhoe F G J 1983a 5′Nucleotidase in lymphocytes from various Clinical disorders. Leukemia Research 7: 57
El-Mohandes E, Hayhoe F G J 1983b 5′Nucleotidase activity in myeloma cells. Clinical and Laboratory Haematology 5: 265
El-Mohandes E, Hayhoe F G J 1983c 5′-Nucleotidase activity in megakaryocytes of idiopathic thrombocytopenic purpura. Haematologica 68: 328
El-Mohandes E, Hayhoe F G J 1983d 5′-Nucleotidase acitivity of megakaryoblasts in a case of acute megakaryoblastic leukaemia. British Journal of Haematology 53: 523
Emmelot P, Bos C J 1965 Differential effect of neuraminidase on the magnesium ATP-ase, sodium-potassium-magnesium ATP-ase and 5-nucleotidase of isolated plasma membranes. Biochimica et Biophysica Acta 99: 578
Eränkö O, Palkama A 1961 Improved localization of phosphorylase by the use of polyvinyl pyrrolidone and high substrate concentration. Journal of Histochemistry and Cytochemistry 9: 585
Ernst S A 1972 Transport adenosine triphosphatase cytochemistry. 1. Biochemical characterization of a cytochemical medium for the ultrastructural localization of ouabain-sensitive, potassium-dependent phosphatase activity in the avian salt gland. Journal of Histochemistry and Cytochemistry 20: 13
Ernst S A, Hootman S R 1981 Microscopical methods for the localization of Na^+, K^+-ATPase. Histochemical Journal 13: 397
Feller A C 1982 Cytochemical reactivity of Tμ lymphocytes in human lymphatic tissue for dipeptidylaminopeptidase IV. Histochemical Journal 14: 889
Feller A C, Parwaresch M R 1981 Specificity and polymorphism of diaminopeptidase IV in normal and neoplastic T lymphocytes. Journal of Cancer Research and Clinical Oncology 101: 59
Feller A C, Heijnen C J, Bellieux R E, Parwaresch M R 1982 Enzyme histochemical staining of T lymphocytes for glycyl-proline-4-methoxy-beta-naphthylamide-peptidase (DAP) IV. British Journal of Haematology 51: 227
Feller A C, Ziegler A, Sterry W, Goos M, Parwaresch M R 1983 Phenotypic heterogeneity of leukaemic Sezary cells. Blut 47: 333
Feller A C, Parwaresch M R, Lennert K 1984 Cytochemical distribution of dipeptidylaminopeptidase IV (DAP IV: EC 3.4.14.5) in T-lymphoblastic lymphoma/leukaemia characterised with monoclonal antibodies. Leukemia Research 8: 397
Firth J A 1978 Cytochemical approaches to the localization of specific adenosine triphosphatases. Histochemical Journal 10: 253
Firth J A 1980 Reliability and specificity of membrane adenosine triphosphatase localization. Journal of Histochemistry and Cytochemistry 28: 69
Flandrin G, Bernard J 1975 Cytological classification of acute leukaemias. A survey of 1400 cases. Blood Cells 1: 7
Flandrin G, Daniel M T 1974 β-glucuronidase activity in Sezary cells. Scandinavian Journal of Haematology 12: 23
Fleischer B, Fleischer S, Ozawa H 1969 Isolation and characterization of Golgi membranes from bovine liver. Journal of Cell Biology 43: 59
Futterman S 1957 Enzymatic reduction of folic acid and dihydrofolic acid to tetrahydrofolic acid. Journal of Biological Chemistry 288: 1031
Gerzeli G, de Piceis Polver P 1969 Un nuovo metodo citochimico per lo studio dell'attivita tetraidrofolicoreduttasica in Plasmodium Berghei: Analisi dell'effetto di inibitori specifici. Rivista Parassitologia 30: 19
Gerzeli G, de Piceis Polver P 1972 A critical examination of some histochemical methods for demonstrating tetrahydrofolate dehydrogenase. Histochemical Journal 4: 79

Girino M, Invernizzi R, Perseghin P, Michienzi M 1985 Cytochemical study of gamma glutamyl transpeptidase activity in normal blood cells and in blood malignancies. Haematologica 70: 266

Glenner G G, Folk J E, McMillan P J 1962 Histochemical demonstration of gamma-glutamyl transpeptidase-like activity. Journal of Histochemistry and Cytochemistry 10: 481

Gomori G 1952 Microscopic Histochemistry. Univ. Chicago Press, Chicago

Gomori G 1954 Enzymatic hydrolysis of acyl naphthylamines. Proceedings of the Society for Experimental Biology (New York) 85: 570

Gossrau R, Lojda Z 1980 Study on dipeptidylpeptidase II (DPP II). Histochemistry 70: 53

Gundersen L E, Dunlap R B, Harding N G L, Freisheim J H, Otting F, Huennekens F M 1972 Dihydrofolate reductase from amethopterin-resistant Lactobacillus casei. Biochemistry 11: 1018

Gutensohn W, Thiel E 1981 High levels of 5′nucleotidase activity in blastic chronic myelogenous leukaemia with common ALL antigen. Leukemia Research 5: 505

Haddad N, Weinstein D, Bonar R A, Beaudreau G S, Beard D, Beard J W 1960 Virus of avian myeloblastosis. XV. Structural loci of virus synthesis and adenosinetriphosphatase activity by electron and light microscopy of myeloblasts from tissue culture. Journal of the National Cancer Institute 24: 971

Hardonk M J, Boer H G A de 1968 5-nucleotidase. III. Determinations of 5-nucleotidase isoenzymes in tissues of rat and mouse. Histochemie 12: 29

Hayashi M, Nakajima Y, Fishman W H 1964 The cytological demonstration of β-glucuronidase employing naphthol AS-BI glucuronide and hexazonium pararosanilin: a preliminary report. Journal of Histochemistry and Cytochemistry 12: 293

Hayhoe F G J, Quaglino D, Doll R 1964 The Cytology and Cytochemistry of Acute Leukaemias. MRC Spec. Report Series No. 304, HMSO: London

Henion H F, Sutherland E W 1957 Immunological differences of phosphorylases. Journal of Biological Chemistry 224: 477

Hirt A, Auner H L, Wagner H P 1984 Cytochemistry of human T lymphocyte subpopulations in peripheral blood. Scandinavian Journal of Haematology 32: 190

Ho A D, Dorken B, Ma D F, Pezzutto A, Hunstein W, Hoffbrand A V 1986 Purine degradative enzymes and immunological phenotypes in chronic B-lymphocytic leukaemia: indications that leukaemic immunocytoma is a separate entity. British Journal of Haematology 62: 545

Holter H, Li S O 1950 Phosphamidase activity of some proteolytic enzymes and rennin. Acta Chemica Scandinavica 4: 1321

Hopsu-Havu V K, Ekfors T O 1969 Distribution of a dipeptide naphthylamidase in rat tissues and its localization by using diazo coupling and labeled antibody techniques. Histochemie 17: 30

Hopsu-Havu VK, Helminen H 1974 In: Hayat MA (ed) 'Electron Microscopy of Enzymes. Principles and Methods Vol II. Van Nostrand-Reinhold, Princeton, New Jersey, pp. 90–109

Hugon J S, Borgers M, Maestracci D 1970 Glucose-6-phosphatase and thiamine pyrophosphatase activities in the jejunal epithelium of the mouse. Journal of Histochemistry and Cytochemistry 18:361

Hultberg B, Sjogren U 1980 L-γ-glutamyl transpeptidase activity in normal and leukaemic leukocytes. Acta Haematologica 63: 132

International Committee for Standardization in Haematology (ICSH) (1985) Recommended methods for cytological procedures in haematology. Clinical and Laboratory Haematology 7: 55

Invernizzi R, Perugini O 1987 Cytochemistry of N-acetyl-beta-glucosaminidase in normal and leukaemic T-cells. Acta Haematologica 77: 6

Invernizzi R, Nano R, Ippoliti G, Girino M, Gerzeli G 1983 Cytochemical study of tetrahydrofolate dehydrogenase in chronic lymphocytic leukemia cells. Haematologica 68: 742

Invernizzi R, Bertolino G, Girino M, Perseghin P, Michienzi M, Nano R 1985 Cytochemistry of dipeptidyl-amino-peptidase IV (DAP IV) and II (DAP II) in normal and neoplastic lymphoid cells. Blut 50: 277

Jacobsen N O, Jørgensen P L 1969 A quantitative biochemical and histochemical study of the lead method for localization of adenosine triphosphate-hydrolyzing enzymes. Journal of Histochemistry and Cytochemistry 17: 443

Jondal M, Pross H 1975 Surface markers on human B and T lymphocytes. VI. Cytotoxicity against cell lines as a functional marker for lymphocyte subpopulations. International Journal of Cancer 15: 596

Kanter R J, Freiberger I A, Rai K R, Sawitsky A 1979 Lymphocyte populations with 5′nucleotidase in chronic lymphocytic leukaemia. Clinical Immunology and Immunopathology 12: 351

Kaplow LS, Burstone MS 1963 Acid-buffer acetone as a fixative for enzyme cytochemistry. Nature 200: 690

Kass L, Hadi M Z 1975 Phosphorylase activity in chronic erythraemic myelosis. American Journal of Clinical Pathology 64: 503

Khalaf M R 1986 Enzyme Cytochemical Studies in Haemic Cells. Ph. D. Dissertation, Cambridge University.
Khalaf M R, Hayhoe F G J 1987a Improved cytochemistry of a specific 5′nucleotidase in haemic cells. Histochemical Journal (in press)
Khalaf M R, Hayhoe F G J 1987b A cytochemical method for demonstrating magnesium-activated adenosine triphosphatase in haemic cells. Acta Haematologica (in press).
Khalaf M R, Hayhoe F G J 1987c Cytochemistry of gamma-glutamyl transferase in haemic cells and malignancies. Histochemical Journal (in press)
Khalaf M R, Hayhoe F G J 1987d The cytochemical demonstration of adenylate cyclase in normal and neoplastic haemic cells. Journal of Histochemistry and Cytochemistry (in press)
Khalaf M R, Bevan P C, Hayhoe F G J 1986a Comparative cytochemical study of dipeptidyl aminopeptidase (DAP) II and IV in normal and malignant haemic cells. Journal of Clinical Pathology 39: 891
Khalaf M R, Aqel N, Hayhoe F G J 1986b Histochemistry of dipeptidyl aminopeptidase (DAP) II and IV in reactive lymphoid tissues and malignant lymphomas. Journal of Clinical Pathology 40: 480
Kramers M T C, Catovsky D 1978 Cell membrane enzymes: L-γ-glutamyl transpeptidase, leucine aminopeptidase, maltase and trehalase in normal and leukaemic lymphocytes. British Journal of Haematology 38: 453
Kramers M T C, Catovsky D, Foa R, Cherchi M, Galton D A G 1976 5′Nucleotidase activity in leukaemic lymphocytes. Biomedicine 25: 363
Kramers M T C, Catovsky D, Cherchi M, Galton D A G 1977 5′-nucleotidase in cord blood lymphocytes. Leukaemia Research 1: 279
Kyte J 1974 The reactions of sodium and potassium ion-activated adenosine triphosphatase with specific antibodies. Journal of Biological Chemistry 249: 3652
Kyte J 1976a Immunoferritin determination of distribution of (Na^+ + K^+) ATPase over the plasma membranes of renal convoluted tubules. I. Distal segment . Journal of Cell Biology 68: 287
Kyte J 1976b Immunoferritin determination of the distribution of (Na^+ + K^+) ATPase over the plasma membranes of renal convoluted tubules . II. Proximal segment. Journal of Cell Biology 68: 304
La Mantia K, Conklyn M, Quagliata F, Silber R 1975 Immunologic studies of 5′nucleotidase in normal and chronic lymphocytic leukaemia patients. Blood 46: 1042
Lawrinson W, Gross S 1964 Nuclear aryl-sulphatase activity in primitive haemic cells. Laboratory Investigation 13: 1612
Lennert K 1978 Malignant Lymphomas other than Hodgkin's Disease. Springer-Verlag, Berlin
Leskes A, Siekevitz P, Palade G E 1971 Differentiation of endoplasmic reticulum in hepatocytes. 1. Glucose-6-phosphatase distribution in situ. Journal of Cell Biology 49: 264
Lojda Z 1977 Studies on glycyl-proline naphthylamidase. Histochemistry 54: 299
Lojda Z 1981 Proteinases in pathology. Usefulness of histochemical methods. Journal of Histochemistry and Cytochemistry 29: 481
Lojda Z, Gossrau R, Schiebler T H 1979a Enzyme Histochemistry. A Laboratory Manual. Springer-Verlag, Berlin p 199
Lojda Z, Hermansky F, Benesova E, Salkova J, Lodrova V 1979b Dipeptidyl aminopeptidase IV in lymphocytes of peripheral blood of man and the importance of its evidence in malignant lymphoma. Sbornik Lekarsky 81: 200
Lopes J, Zucker Franklin D, Silber R 1972 Heterogeneity of 5′nucleotidase activity in lymphocytes in chronic lymphocytic leukaemia. Journal of Clinical Investigation 52: 1297
Lorbacher P, Yam L T, Mitus W J 1967 Cytochemical demonstration of β-glucuronidase activity in blood and bone marrow cells. Journal of Histochemistry and Cytochemistry 15: 680
Ma D D, Masseia M, Sylwestrowicz T A, Price G, Hoffbrand A V 1983 Comparison of purine degradative enzymes and terminal deoxynucleotidyl transferase in T-cell leukaemias and in normal thymic and post-thymic T cells. British Journal of Haematology 54: 451
Marathé C V, Damle N S, Haschemeyer R H, Tate S S 1980 Localization of γ-glutamyl transpeptidase in lymphoid cells. FEBS Letters 115 (2): 273
Martin D L, Engel W K 1972 Dependency of histochemical phosphorylase staining on amount of cellular glycogen. Journal of Histochemistry and Cytochemistry 20: 476
Matamoros N, Horwitz D A, Newton C, Asherson G L, Webster A D B 1979 Histochemical studies of 5′-nucleotidase and alpha naphthyl (non-specific) esterase in lymphocytes from patients with primary immunoglobulin deficiencies. Clinical and Experimental Immunology 36: 102
Mayahara H, Fujimoto K, Ando T, Ogawa K 1981 A new one-step method for the ultracytochemical localization of ouabain-sensitive, potassium-dependent p-nitrophenylphosphatase activity. Histochemistry 67: 125
Mazurkiewicz J E, Hossler F E, Barnett R J 1978 Cytochemical demonstration of sodium-

potassium-adenosine triphosphatase by a hemepeptide derivative of ouabain. Journal of Histochemistry and Cytochemistry 26: 1042
McDonald J K, Schwabe C 1977 Intracellular exopeptidases. In: Barrett A J (ed) Proteinases in Mammalian Cells and Tissues, 2nd Edn. North-Holland, Amsterdam
McDonald J K, Callahan P X, Stanley E, Smith R 1971 Polypeptide degradation by dipeptidyl aminopeptidase 1 (Cathepsin C and related peptidases). In: Barrett A J, Dingle J T (eds) Tissue Proteinases. North-Holland, Amsterdam, p 86
Meister A 1973 On the enzymology of amino acid transport. Transport in kidney and probably other tissues is mediated by a cycle of enzymic reactions involving glutathione. Science 180: 33
Milsom D W, Rose F A, Dodgson K S 1968 Assay of a microsomal marker enzyme: rat liver arylsulphatase C. Biochemical Journal 109: 40P
Mueller J, Brun del Re G, Buerki H, Keller H U, Hess M W, Cottier H 1975 Non-specific acid esterase activity: a criterion for differentiation of T and B lymphocytes in mouse lymph nodes. European Journal of Immunology 5: 270
Müller-Hermelink H K 1974 Characterization of the B-cell and T-cell regions of human lymphatic tissue through enzyme histochemical demonstration of ATPase and 5′nucleotidase activities. Virchows Archiv B Cell Pathology 16: 371
Murata F, Nagata T, Spicer S S 1975 Fine structural localization of arylsulfatase B activity in the rabbit blood platelets. Histochemistry 44: 307
Murray J L, Reuben J M, Munn G C et al 1984 Decreased 5′nucleotidase activity in lymphocytes from asymptomatic sexually active homosexual men and patients with AIDS. Blood 64: 1016
Nachlas M M, Crawford D T, Seligman A M 1957 The histochemical demonstration of leucine aminopeptidase. Journal of Histochemistry and Cytochemistry 5: 264
Nachlas M M, Goldstein T P, Seligman A M 1962 An evaluation of amino-peptidase specificity with seven chromogenic substrates. Archives of Biochemistry and Biophysics 97: 223
Naidoo D, Pratt O E 1954 The activity and localization of adenosine 5-phosphatase in the thiamine deficient chicken brain. Biochimica et Biophysica Acta 15: 291
Nano R, Gervaso M V, Invernizzi R, Gerzeli G 1982 Tetrahydrofolate dehydrogenase cytochemistry and classification of acute leukaemia. Acta Haematologica 67: 160
Nano R, Invernizzi R, Rezzani R, Arisi C, Gerzeli G 1987 Cytochemical study of tetrahydrofolate dehydrogenase in myelodysplastic syndromes. Haematologica 72: 11
Nichols B A, Setzer P Y, Bainton D F 1984 Glucose-6-phosphatase as a cytochemical marker of endoplasmic reticulum in human leukocytes and platelets. Journal of Histochemistry and Cytochemistry 32: 165
Novogrodsky A, Tate S S, Meister A 1976 γ-glutamyl transpeptidase, a lymphoid cell-surface marker: relationship to blastogenesis, differentiation and neoplasia. Proceedings of the National Academy of Sciences USA 73: 2414
Oliveira M S P de, Gregory C, Matutes E, Parreira A, Catovsky D 1987 Cytochemical profile of megakaryo blastic leukaemia: a study with cytochemical methods, monoclonal antibodies and ultrastructural cytochemistry. Journal of Clinical Pathology 40: 663
Onicescu D, Tasca L, Trutza E 1970 A histochemical method for dihydrofolate reductase. Histochemical Journal 2: 289
Onicescu D, Tasca L, Mischiu L 1972a On some histoenzymological and DNA variations induced by folate and methotrexate in normal and malignant cells in vitro. Acta Histochemica et Cytochemica 5: 33
Onicescu D, Popescu M, Tasca L, Stoica I 1972b On the specificity of a histochemical method for localizing dihydrofolate dehydrogenase. Histochemical Journal 4: 87
Onicescu D, Stoica I, Tasca L 1973 A histochemical study on the dihydrofolate reductase activity in leukaemia. Acta Histochemica 45: 214
Orye E, Plum J, De Smedt M 1984 β-glucuronidase activity in human T and B lymphocytes and the Tμ and Tγ subpopulations. Histochemistry 81: 287
Osborn M J, Freeman M, Huennekens F M 1958 Inhibition of dihydrofolic reductase by aminopterin and amethopterin. Proceedings of the Society for Experimental Biology and Medicine 97: 429
Pangalis G A, Yataganas X, Fessas Ph 1977 β-glucuronidase activity of lymph node imprints from malignant lymphomas and chronic lymphocytic leukaemia. Journal of Clinical Pathology 30: 812
Pangalis G A, Roussou P P, Tsavaris N B 1984 Cytochemical distribution of 5′nucleotidase activity in human leukemic cells. In: Polliack A (ed) Human Leukaemias: Cytochemical and Ultrastructural Techniques in Diagnosis and Research. Martinus Nijhoff Publishing, Boston, p 15
Pearse A G E, Reis J L 1952 The histochemical demonstration of a specific phosphatase (5-nucleotidase). Biochemical Journal 50: 534
Peters J M, Greenberg D M 1958 Dihydrofolic acid reductase. Journal of the American Chemical Society 80: 6679

Prentice H G, Smyth J F, Ganeshaguru K 1980 Remission induction with adenosine deaminase inhibitor 2'deoxy-coformycin in Thy—lymphoblastic leukaemia. Lancet ii: 170
Pugh D, Walker P G 1961 Histochemical localization of β-glucuronidase and N-acetyl-β-glucosaminidase. Journal of Histochemistry and Cytochemistry 9: 105
Quagliata F, Faig D, Conklyn M, Silber R 1974 Studies on the lymphocyte 5′ nucleotidase in chronic lymphocytic leukaemia, infectious mononucleosis, normal subpopulations and phytohaemagglutinin stimulated cells. Cancer Research 34: 3197
Quaglino D, Hayhoe F G J 1962 Phosphorylase activity in haemic cells. Nature, 194: 929
Reaman G H, Blatt J, Poplack D G 1981 Lymphoblast purine pathway enzymes in B-cell acute lymphoblastic leukemia. Blood 58: 330
Reaman G H, Levin N, Muchmore A, Holiman B J, Poplack D G 1979 Diminished lymphoblast 5′N activity in acute lymphoblastic leukemia with T-cell characteristics. New England Journal of Medicine 300: 1374
Reik L, Petzold G L, Higgins J A, Greengard P, Barnett R J 1970 Hormone-sensitive adenyl cyclase: cytochemical localization in rat liver. Science 168: 382
Rinneberg H, Schubert J C F, Lennert K 1961 Der Histochemische Nachweis der Adenosintriphosphatase, 5-nucleotidase und Glycerophosphatase in menschlichen Lymphknoten und Tonsillen. Proc. 8th Congr. Europ. Soc. Haemat., S. Karger: Basel, New York. 1: Art. 21
Roberts D, Wodinsky I 1968 On the poor correlation between the inhibition by methotrexate of dihydrofolate reductase and of deoxynucleoside incorporation into DNA. Cancer Research 28: 1955
Robinson J M, Karnovsky M J 1982 Cerium as an ultrastructural cytochemical probe for phosphatases. Journal of Cell Biology 95: 462a
Robinson J M, Karnovsky M J 1983 Ultrastructural localization of 5-nucleotidase in guinea pig neutrophils based upon the use of cerium as capturing agent. Journal of Histochemistry and Cytochemistry 31: 1190
Rosenthal A S, Moses H L, Tice L, Ganote C E 1969 Lead ion and phosphatase histochemistry. III. The effect of lead and adenosine triphosphate concentration on the incorporation of phosphate into fixed tissue. Journal of Histochemistry and Cytochemistry 17: 608
Rosenthal A S, Moses H L, Ganote C E 1970 Interpretation of phosphatase cytochemical data. Journal of Histochemistry and Cytochemistry 18: 915
Rowe M, De Gast C G, Platts-Mills T A E, Asherson G L, Webster A D B, Johnson S M 1979 5′-Nucleotidase of B and T lymphocytes isolated from human peripheral blood. Clinical and Experimental Immunology 36: 97
Rubio R V, Wiedmeier T, Berne R M 1972 Nucleoside phosphorylase: localization and role in the myocardial distribution of purines. American Journal of Physiology 222: 550
Rustin G J S, Peters T J 1979 Studies on the subcellular organelles of neutrophils in chronic granulocytic leukaemia with special reference to alkaline phosphatase. British Journal of Haematology 41: 533
Rutenburg A M, Cohen R B, Seligman A M 1952 Histochemical demonstration of arylsulfatase. Science, 116: 539
Rutenberg A M, Kim H, Fischbein J W, Hanker J S, Wasserkrug H L, Seligman A M 1969 Histochemical and ultrastructural demonstration of gamma-glutamyl transpeptidase activity. Journal of Histochemistry and Cytochemistry 17(8): 517
Schliselfelt L H 1973 Comparative studies of phosphorylase isozymes from the rabbit. Annals of the New York Academy of Sciences 210: 181
Schrecker A W, Mead J A R, Greenberg N H, Goldin A 1971 Dihydrofolate reductase activity of leukaemia L1210 during development of methotrexate resistance. Biochemical Pharmacology 20: 716
Schwartz M, Seigel G J, Chen N, Agranoff B W 1980 Goldfish brain (Na^+, K^+)-ATPase: purification of the catalytic polypeptide and production of specific antibodies. Journal of Neurochemistry 34: 1745
Schwartz M, Ernst S A, Siegel G J, Agranoff B W 1981 Immunocytochemical localization of (Na^+, K^+)-ATPase in the goldfish optic nerve. Journal of Neurochemistry 36: 107
Seligman A M, Nachlas M M, Cohen R B 1950 Histochemical demonstration of β-glucuronidase and sulfatase. Cancer Research 10: 240
Sevilla C L, Fischer E H 1969 The purification and properties of rat muscle glycogen phosphorylase. Biochemistry 8: 2161
Shirley P S, Wang P, DeChatelet L R, White M 1975 Absence of the membrane marker enzyme 5′-nucleotidase in human polymorphonuclear leukocytes. Biochemical Medicine 15: 289
Silber R, Conklyn M, Grusky G, Zucker-Franklin D 1975 Human lymphocytes: 5′nucleotidase-positive and -negative subpopulations. Journal of Clinical Investigation 56: 1324
Smith R E (1969) Subcellular localization of peptidases hydrolysing aminoacyl and dipeptidyl

derivatives of 4-methoxy-betanaphthylamine. Journal of Cell Biology 43: 135a
Smith R E, van Frank R M 1975 The use of amino acid derivatives of 4-methoxy-2-naphthylamine for the assay and subcellular localization of tissue proteinases. In: Dingle J T, Den R J (eds) Lysosomes in Biology and Pathology: Vol. IV. North-Holland, Amsterdam, p 218
Smutka P, Brunning R D 1969 An evaluation of nuclear aryl sulphatase activity in acute leukaemias. Acta Haematologica 41: 290
Somogyi E, Sotonyi P 1969 Data to the electron microscopic demonstration of ATP-ase. Acta Histochemica 34: 70
Srivastava M D, Bhargava A K 1986 Expression of dipeptidyl-aminopeptidase-IV in some human T and B cell lines. Acta Haematologica 76: 25
Strauss R G, Burrows S E 1975 Letter to the Editor. Blood 46: 655
Stirling C E 1972 Radioautographic localization of sodium pump sites in rabbit intestine. Journal of Cell Biology 53: 704
Stirling C E 1976 High resolution autoradiography of ^{3}H-ouabain binding in salt transporting epithelia. Journal of Microscopy 106: 145
Sylwestrowicz T A, Ma D D, Murphy P P, Prentice H G, Hoffbrand A V, Greaves M F 1982 5′Nucleotidase, adenosine deaminase and purine nucleoside phosphorylase activities in acute leukemia. Leukemia Research 6: 475
Sylwestrowicz T, Piga A, Murphy P 1982 The effects of deoxycoformycin and deoxyadenosine on deoxyribonucleotide concentrations in leukaemic cells. British Journal of Haematology 51: 623
Szmigielski S, Litwin J, Zupanska B 1965 Histochemical demonstration of gamma glutamyl transpeptidase activity in bone marrow and blood cells. Journal of Clinical Pathology 18: 244
Takeuchi T 1958 Histochemical demonstration of branching enzyme (amylo-1, 4-1,6-transglucosidase) in animal tissues. Journal of Histochemistry and Cytochemistry 6: 208
Takeuchi T, Kinoshita K 1956 Histochemical demonstration of phosphorylase in blood and bone marrow cells. Blood, 11: 375
Takeuchi T, Kuriaki H 1955 Histochemical detection of phosphorylase in animal tissues. Journal of Histochemistry and Cytochemistry 3: 153
Takeuchi T, Sasaki M 1970 Validity of the histochemical phosphorylase and uridine-diphosphate-glycogen transglucosylase methods. Journal of Histochemistry and Cytochemistry 18: 761
Tamaoki N, Essner E 1969 Distribution of acid phosphatase, β-glucuronidase and N-acetyl-β-glucosaminidase activities in lymphocytes of lymphatic tissues of man and rodents. Journal of Histochemistry and Cytochemistry 17: 238
Tanaka K R, Valentine W N, Fredricks R E 1960 Aryl sulfatase activity of human leukocytes: distinctive pattern in eosinophils. Journal of Clinical Investigation 39: 1034
Tanaka K R, Valentine W N, Fredricks R E 1962 Human leucocyte arylsulphatase activity. British Journal of Haematology 8: 86
Tice L W, Barnett R J 1962 The fine structural localization of glucose-6-phosphatase in rat liver. Journal of Histochemistry and Cytochemistry 10: 754
Tzortzatou F 1975 Dihydrofolate reductase and the action of folate antagonists. PhD dissertation, University of Cambridge.
Tzortzatou F 1977a Cytochemical demonstration of dihydrofolate reductase in leukaemia and other haematological diseases. Scandinavian Journal of Haematology 19: 5
Tzortzatou F 1977b Cytochemical demonstration of 5-formyl tetrahydrofolate cyclodehydrase and 5,10-methenyl tetrahydrofolate cyclohydrolase activity. Journal of Histochemistry and Cytochemistry 25: 349
Tzortzatou F, Hayhoe F G J 1974a Cytochemical demonstration of dihydrofolate reductase activity. British Journal of Haematology 28: 199
Tzortzatou F, Hayhoe F G J 1974b The effect of folate antagonists on dihydrofolate reductase activity demonstrated cytochemically. British Journal of Haematology 28: 209
Uehara Y, Gasa S, Makita A, Sakurada K, Miyazaki T 1983 Lysosomal arylsulfatases of human leukocytes: increment of phosphorylated B variants in chronic myelogenous leukemia. Cancer Research 43: 5618
Uusitalo R J 1981 5′-nucleotidase activity in lymphocytes. Histochemical Journal 13: 525
Veerman A J P, Hogeman P H G, van Zantwijk C H, Bezemer P D 1985 Prognostic value of 5′-nucleotidase in acute lymphoblastic leukemia with the common-ALL phenotype. Leukemia Research 9: 1227
Wachstein M, Meisel E 1956 Histochemistry of substrate specific phosphatases at a physiological pH. Journal of Histochemistry and Cytochemistry 4: 424
Wachstein M, Meisel E 1957 Histochemistry of hepatic phosphatase at a physiological pH. American Journal of Clinical Pathology 27: 13
Weinstein D, Sommer J R, Beard J W 1962 Cytochemical localization of an adenosine triphosphate dephosphorylating enzyme in the granules of human eosinophils. Blood, 19: 612

Wilson P D, Rustin G J S, Smith G P, Peters T J 1981 Electron microscopic cytochemical localization of nucleoside phosphatases in normal and chronic granulocytic leukaemic human neutrophils. Histochemical Journal 13: 73
Wirthmüller R, Dennig D, Oertel J, Gerhartz H 1983 Dipeptidylaminopeptidase IV (DAP IV) activity in normal and malignant T cell subsets as defined by monoclonal antibodies. Scandinavian Journal of Haematology 31: 197
Wood J G, Jean D H, Whitaker J N, McLaughlin B J, Albers R W 1977 Immunocytochemical localization of sodium, potassium activated ATPase in knifefish brain. Journal of Neurocytology 6: 571
Yam L T, Mitus W J 1968 The lymphocyte β-glucuronidase activity in lymphoproliferative disorders. Blood 31: 48
Yonezawa S, Hori S H 1975 Electrophoretic studies on the phosphorylase isozymes. Journal of Histochemistry and Cytochemistry 23: 745
Yunis A A, Arimura G K 1964 Enzymes of glycogen metabolism in white blood cells. l. Glycogen phosphorylase in normal and leukaemic human leucocytes. Cancer Research 24: 489
Yunis A A, Arimura G K 1968 Isozymes of glycogen phosphorylase in human leucocytes and platelets: relation to muscle phosphorylase. Biochemical and Biophysical Research Communications 33: 119

CHAPTER 12

Cytochemical findings in cultured haemic cells

Studies in PHA-transformed lymphocytes

Soon after Nowell (1960) had shown that phytohaemagglutinin (PHA) was capable of inducing certain peripheral blood lymphocytes to transform into large basophilic blast cells capable of mitotic activity, we carried out cytochemical studies of this change (Quaglino *et al.*, 1962; Quaglino and Hayhoe, 1965). Apart from standard Romanowsky stains we studied the levels of acid and alkaline phosphatase, peroxidase, non-specific esterases, succinic and lactic dehydrogenases, and carried out Sudan black and PAS reactions. The results showed that stimulated lymphocytes, prior to their transformation to blast cells, accumulated strong PAS positivity (Fig. 12.1), but that the blast cells themselves were mostly devoid of similar positivity, though rich in dehydrogenases (Fig. 12.2) and acid phosphatase (Fig. 12.3). This sequence suggested that accumulation of glycogen in stimulated lymphocytes represented a source of energy for the transformation process, the glycogen being progressively utilized in the course of cellular activation, and its removal being associated with an increase of dehydrogenating enzymes active in the glycolytic process and of acid phosphatase. The latter enzyme is notably strong in certain other metabolically and phagocytically active cells such as plasma cells and macrophages.

These studies were extended by the use of labelled DNA and RNA precursors. After exposure to tritiated thymidine and with PAS counterstaining of the autoradiographs it was found that the small lymphocytes which showed increase in PAS positivity did not take up any label, but that intermediate nucleolated lymphocytes, which started to appear in culture between 24 and 48 hours and which often had even coarser PAS positivity than the small lymphocytes, showed considerable uptake (Figs 12.1 and 12.4). High grain counts were found in the PAS negative blast cells, which became increasingly numerous at 72 hours.

Studies on RNA synthesis were carried out using uridine tritiated in the 5-position, which generally loses its tritium label on conversion to thymidine and therefore acts as a largely specific label for RNA (Hayhoe and Quaglino, 1965). The localization and time sequence of appearance of uridine-labelled RNA differed from that of DNA, with 60% of the mature-looking lymphocytes taking up the RNA label within the first hour of culture, and over 90% from 24 hours onwards (Fig. 12.5). None of these cells showed thymidine uptake at any stage. Intermediate nucleolated

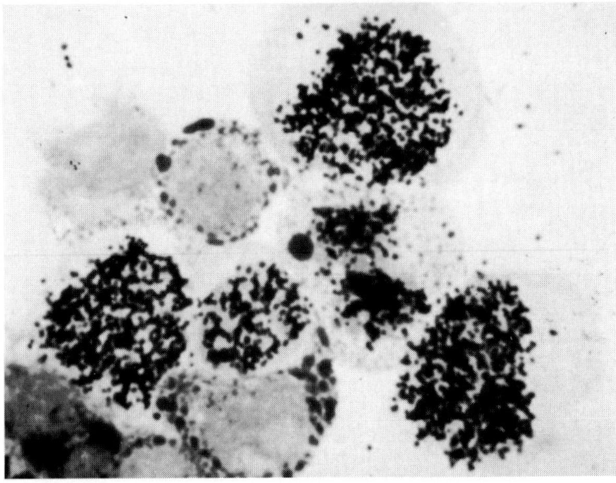

Fig. 12.1 PHA stimulated lymphocytes. PAS reaction and ^{3}HTdR autoradiography. Strong PAS positivity with coarse granules and blocks is seen in the earlier stages of transformation. Transformed blast cells, once in mitosis, show little PAS positivity, but strong DNA labelling.

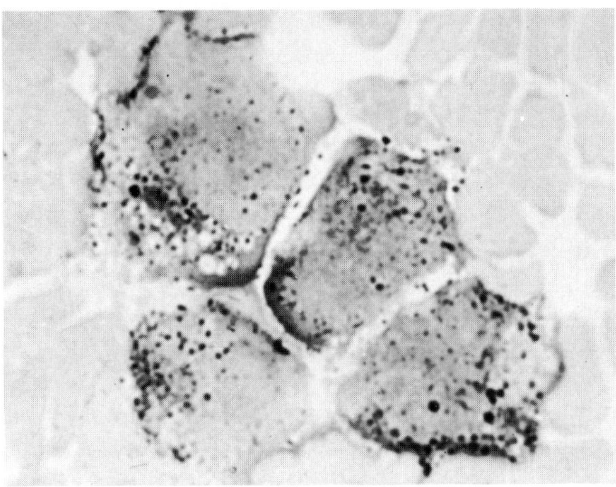

Fig. 12.2 PHA transformed cells. Lactic dehydrogenase. The reaction in blast cells is strongly positive.

lymphocytes always showed strong uridine uptake, while at 48 and 72 hours 70% failed to incorporate thymidine. Similarly blast cells were always heavily labelled after incubation with uridine, whereas between 20 and 30% showed no thymidine uptake. When cells were transferred to non-radioactive medium the RNA label, initially exclusively nuclear, shifted progressively from nucleus to cytoplasm, whereas DNA label showed no such movement (Fig. 12.6).

The combination of cytochemical and autoradiographic findings in normal cultures is summarised in Figure 12.7.

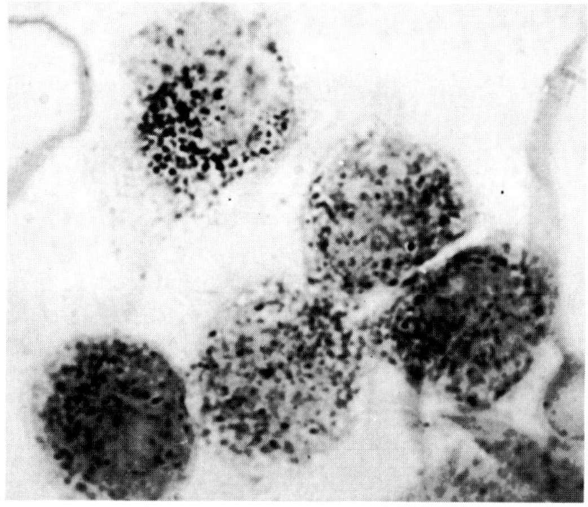

Fig. 12.3 The same preparation. Acid phosphatase. The blast cells show strong positivity.

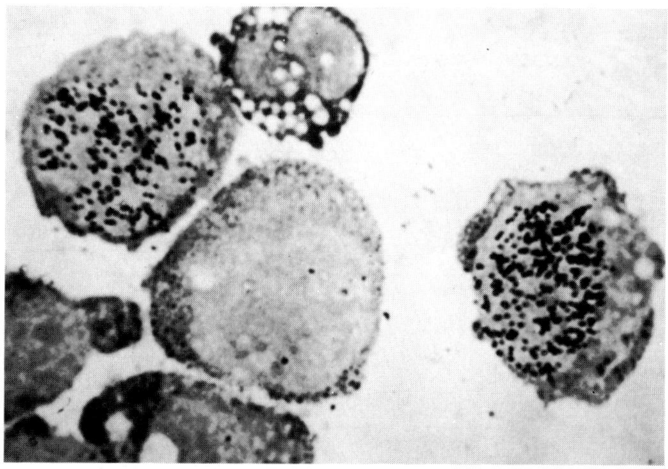

Fig. 12.4 PHA stimulated lymphocytes. PAS reaction and ^{3}HTdR autoradiography. Positivity to PAS is present in all cells, especially strong in early and intermediate phases of transformation. DNA labelling is seen in two intermediate lymphocytes, but not in a single transformed blast cell or in the unconverted lymphocytes.

Studies with leukaemic peripheral blood cells

Studies of the cytochemical changes occurring in populations of cells from CLL and non-Hodgkin's lymphoma when cultured with PHA were reported by Quaglino and Cowling (1964). Parallel with the minimal response to PHA was a failure of the neoplastic cells to show increased accumulation of glycogen. Hayhoe *et al.* (1967) showed that in short-term cultures of CLL lymphocytes stimulated by PHA there existed a non-responsive neoplastic population and a responsive 'normal' population, usually not far outside the

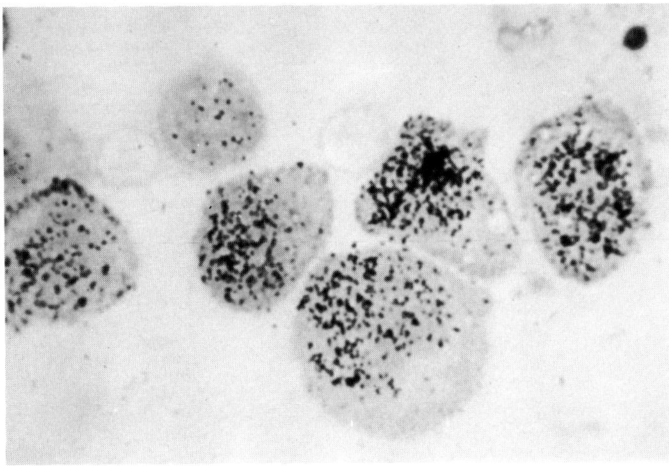

Fig. 12.5 PHA stimulated lymphoctytes. PAS reaction and 5-[^{3}H] uridine autoradiography. RNA labelling occurs in all cells, but is confined to nuclear RNA.

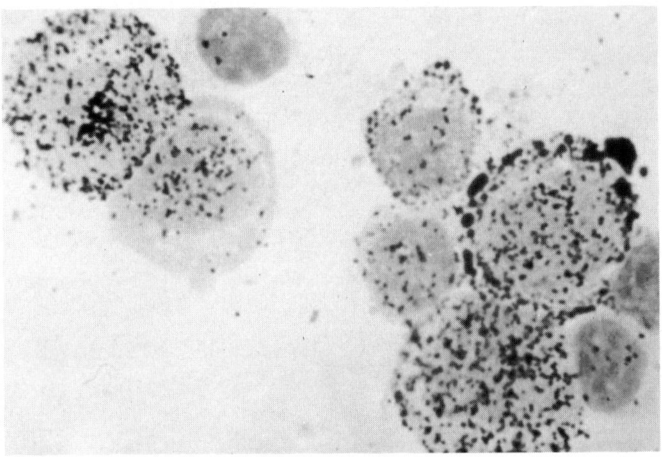

Fig. 12.6 The same preparation after a period in non-radioactive medium: the RNA label shows movement from nucleus to cytoplasm. One intermediate cell to the right of the field shows also coarse PAS positivity in the cytoplasm.

normal range of lymphocyte numbers, although sometimes considerably above these limits. These early findings preceded the recognition of the usual B cell nature of the neoplastic component of CLL, but were later confirmed by the observation (Catovsky *et al.*, 1974) that PHA responsive T cells were present and sometimes increased in numbers in B-CLL. Using combined cytochemical and autoradiographic studies in short-term cultures of cells from patients with acute or chronic myeloid leukaemias, Quaglino *et al.* (1964) found increased cellular glycogen during culture and a moderate increase in DNA-synthetic activity as measured by uptake of tritiated thymidine, but there was no evidence of increased mitotic activity. Cells

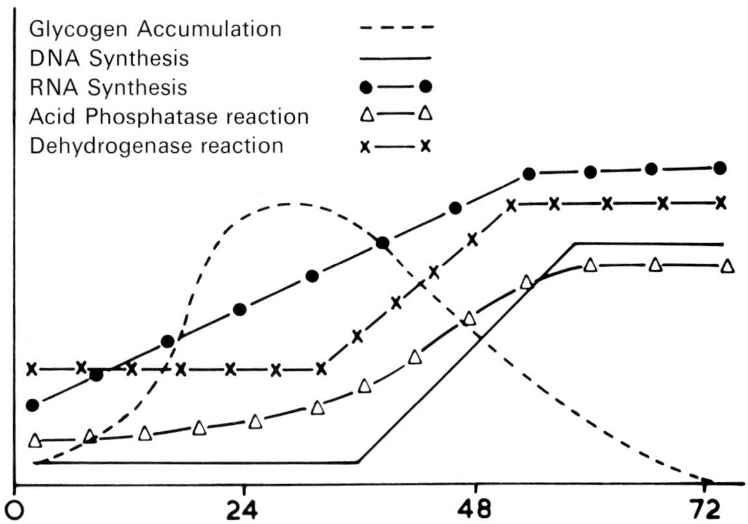

Fig. 12.7 Summary of cytochemical and autoradiographic findings in normal PHA-stimulated lymphocyte cultures.

from subacute and chronic myeloid leukaemia showed a tendency to mature in culture, although with the methods used cells could be maintained for only 72 to 96 hours.

Yoda *et al.* (1984) used peroxidase activity and ultrastructure to evaluate monocyte differentiation in short term cultures of 20 peripheral blood samples from childhood ALL as compared with 15 normal control samples. In the controls peroxidase activity, limited to the monocyte granules, decreased steadily during a period of culture for 4 days, the decrease being clearly associated with change to macrophages. In the ALL cases studied before treatment monocytes showed minimal changes in peroxidase activity and in morphology during the 4 day culture period, but when remission had been induced the pattern reverted to normal. It seems therefore that the capacity of monocytes to differentiate to macrophages is impaired in childhood ALL.

A combined morphological and isoenzymatic study of B-CLL induced to differentiate by TPA (phorbol ester, 12-0-tetradecanoyl phorbol 13-acetate) was reported by Drexler *et al.* (1986). Morphological changes in the B-CLL cells included cytoplasmic and nuclear enlargement, vacuolisation, protrusion of cytoplasmic processes, eccentric localization of nuclei, development of perinuclear clear zones and production of oval or elongated cell forms. Isoenzyme profile analysis by isoelectric focusing on polyacrylamide gel was performed for esterase, acid phosphatase, hexosaminidase and lactic dehydrogenase. All four showed increase in number and intensity of isoenzyme bands, indicating maturation along the B-cell pathway. New expression of tartrate resistant acid phosphatase (TRAP) was triggered by TPA, and also of the hexosaminidase intermediate isoenzyme (Hex I), not present in B-CLL but conspicuous in multiple myeloma cells. The overall picture was of induced phenotypic changes suggestive of differentiation to B-cell stages corresponding to hairy cells or to pre-plasma cells.

Cultures of hairy cells from 11 patients with HCL were studied by Divine *et al.* (1984). Fresh samples showed ultrastructural peroxidase and a predominantly B-cell phenotype, with SIg^+, $B1^+$, $OKT3^-$, $OKT11^-$ profile, but after culture in the presence of PHA and irradiated T cells the phenotype converted to SIg^-, $B1^+$, $OKT3^-$, $OKT11^+$, while peroxidase positivity remained unchanged. Some hairy cells also reacted with OKM1, a granulocyte-monocyte marker, but did not give clear positivity with Mo2, a more specifically monocytic monoclonal marker.

Reactivity of cultured hairy cells with the monoclonal antibody S-HCL3, which reacts normally with histiocytes, monocytes and granulocytes, has been reported by Falini *et al.* (1985), further suggesting that hairy cells may manifest features in common with monocytes, but Hsu and Ho (1986) showed that other B-cell lines after differentiation induced by TPA may also express S-HCL3 antigen.

Studies of bone marrow-derived cells in culture

These studies fall into two categories, those carried out on short-term liquid suspension cultures set up directly from marrow aspirates, and those on established cell lines cultured for long periods and studied either by sampling at intervals from the long-term culture or by transfer to a new environment in which maturation may be encouraged, such as diffusion chambers implanted in irradiated host animals.

Cytochemical studies in short-term cultures

In the first group, periods of culture up to about 14 days are possible using, for example, *in vitro* diffusion chambers (Golde and Cline, 1973), although towards the end of this period mononuclear phagocyte precursors come to predominate heavily over other haemopoietic precursor cells of granulocytic and erythroid lines. Cytochemical studies have proved helpful in distinguishing the changing relationships of marrow cell lines in these circumstances. Willcox *et al.* (1976) harvested cells at intervals from such cultures and stained them with a Romanowsky stain, methyl green pyronin, PAS, and with cytochemical techniques for peroxidase, naphthol AS-D chloroacetate esterase, α-naphthyl butyrate esterase and acid phosphatase.

After 6 days in culture monocytes and macrophages had increased from about 2% to 40%, more mature granulocytes (myelocytes to polymorphs) were much unchanged at 45%, while erythroblasts had fallen sharply from 25% to about 5%. Blast cells, promyelocytes and promonocytes remained much the same at around 5%, and lymphocytes also remained unaltered at 10%. At this time peroxidase activity in mature neutrophils was decreased and many cells were negative; at day 14 most mature granulocytes were negative, but promyelocytes and early myelocytes still gave strong reactions. Macrophages were largely peroxidase negative at 6 and 14 days, apart from some phagocytosed material in occasional cells. Chloroacetate esterase reactivity persisted strongly in all granulocytic cells, mature and immature, throughout the 14 days of culture, but was negative in monocytes and lymphocytes and showed only occasional positivity, possibly following ingestion of dead neutrophils, in macrophages. Naphthyl butyrate esterase, initially found only in monocytes—apart from a small dot of positivity seen in lymphocytes throughout the culture period, presumably in T cells (Higgy *et al.*, 1977)—was strongly positive in macrophages as these cells appeared. This enzyme was regarded as a most helpful marker of the mononuclear phagocyte series in cultured cells, and together with chloroacetate esterase enabled the monocyte-macrophage cells, all with butyrate esterase positivity inhibited by fluoride, but little or no chloroacetate esterase reactivity, to be clearly separated from granulocytes and their precursors back to promyelocytes, all with chloroacetate esterase positivity and virtually no butyrate esterase reactivity.

Acid phosphatase was present in most cell types, being exceptionally strong in macrophages, and persisted throughout the 14 days of culture. RNA staining with the methyl green pyronin reaction showed the expected initial reactions, strong in the cytoplasm of blast cells, weaker in cells of intermediate maturity and in lymphocytes and absent in mature granulocytes and monocytes. The pattern persisted, with some increase in intensity of positive cells during culture. The emerging macrophages showed strong positive staining. PAS reactivity also showed typical patterns at day 0, with increasing positivity in the granulocyte series and most lymphocytes negative. During culture the granulocytic reactions tended to lessen in intensity and the macrophages showed variable positivity, from moderate to strong. In general, acid phosphatase, RNA staining and PAS reactions were not sufficiently specific to be of much help in identifying cell types in cultures.

With the use of differentiation-inducing agents signs of cell maturation can be found even in short term fluid cultures. Mastrangelo *et al.* (1987), for example, reported their findings on cytochemical analysis of AML cells cultured with PHA-lymphocyte conditioned medium. Bone marrow cells were cultured from two cases with apparently pure granulocytic precursor involvement on morphological and cytochemical criteria, without a monocytic component demonstrable by alpha-naphthyl butyrate esterase staining. After 6 days in culture the cells showed complete loss of myeloperoxidase activity, retention of chloroacetate esterase positivity, but with some 40–50% of cells now also showing butyrate esterase positivity. The simultaneous expression of 'monocytic' and 'granulocytic' esterases emerging in short term culture was thought to be due to selection of CFU-GM (granulocyte-macrophage colony forming units) from among the early myeloid precursors. The finding is closely similar to reports on the differentiation capacity of established cell lines such as HL60, reviewed later in this Chapter, and suggests the probable dual granulocyte/monocyte lineage involvement in most apparently single lineage AMLs.

Cytochemical studies in established haemic cell lines

Cytochemical studies formed part of a wide-ranging cytological, cytogenetic and immunological survey of continuous long-term cultured lines established from bone marrow or peripheral blood of 28 patients with haematological malignancies by Karpas *et al.* (1977). Properties were studied after 6 months to 2 years growth in liquid culture. Apart from a single T-cell line, all lines tested in this report were positive for the Epstein–Barr virus nuclear antigen (EBNA) and all but the T-cell line and two others showed cell surface characteristics similar to those of blastoid lines of lymphoid origin. The cells also often showed PAS positivity (Fig. 12.8). Nevertheless, many of the cell lines showed cytochemical positivity for alkaline phosphatase (Fig. 12.9), not normally demonstrable in lymphoid cells of blood or marrow, the extent of esterase (Fig. 12. 10) and acid phosphatase positivity (Fig. 12.11) with different substrates and incubation periods was often more than would be expected with blastic cells of lymphoid origin, and some cell lines produced lysozyme. Whether these combined surface marker and cytochemical findings indicate that the EBNA positive lines are all lymphoblastoid, but with acquired myeloid enzymatic-functional characteristics, or whether, alternatively, some of the lines may be of myeloid origin but with susceptibility to EBV and with capacity for production of some surface membrane immunoglobulin (SmIg) is still uncertain.

A further study (Karpas *et al.,* 1978) of an EBNA-positive but SmIg negative haemic cell line derived from a patient with acute myeloid leukaemia has established that the alkaline phosphatase present in the cells of this line is of the myeloid type (Neumann *et al.,* 1976), characterised by capacity to catalyse the hydrolysis of both O-substituted monoesters of orthophosphoric acid and S-substituted monoesters of thiophosphoric acid with equal efficiency. This finding supports the suggestion that human myeloid cells may in some circumstances be susceptible to infection by EBV.

Additional approaches to confirm whether EBNA positive SmIg positive cells of certain lines might also show synthesis of myeloid enzymes were made by Greenberger *et al.* (1978). They treated cells of these established lines with butyric acid, which induces erythroid cells to differentiate and some mouse leukaemic cells to produce lysozyme (Leder and Leder, 1975; Reuben *et al.,* 1976). Several lines showed increasing absolute numbers of alkaline phosphatase (LAP) and chloroacetate esterase positive cells, with a direct dose-response relationship, after exposure *in vitro* to butyric acid and subsequent transfer of cells to diffusion chambers in irradiated rats. Enzyme induction could not be attributed to growth alteration or selection of a subpopulation, since multiple cloned sublines showed a quantitatively similar effect.

One of the EBNA positive SmIg positive lines tested for the nature of its LAP activity again showed the myeloid enzyme, while other lines showed LAP of lymphoid type.

These findings in long-term cell lines are of great interest, though their

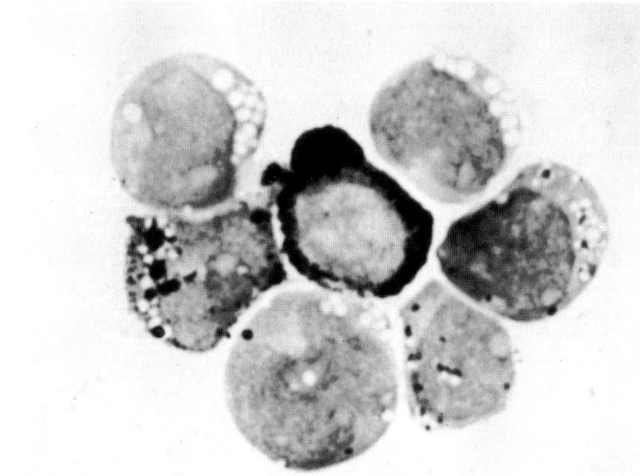

Fig. 12.8 Haemic cell line after six months culture. PAS reaction. Strong positivity in one cell and scattered blocks in three others.

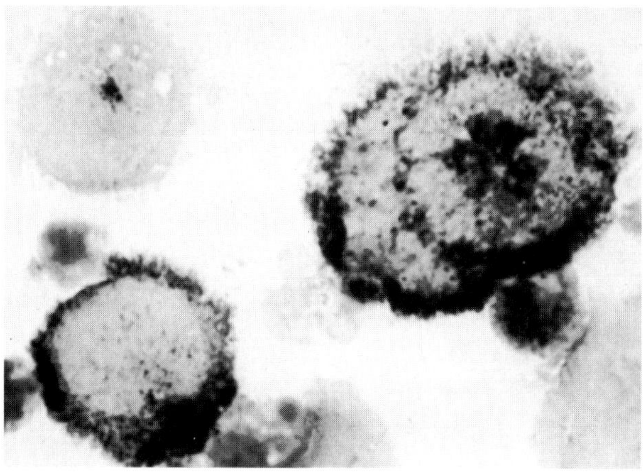

Fig. 12.9 The same preparation. Alkaline phosphatase, azo-dye method. Strong positivity present in some cultured cells, one of which is in mitosis.

interpretation remains uncertain. The possibilities reviewed by Greenberger *et al.* (1978) expand those earlier suggested by Karpas *et al.* (1977), namely, that the findings may represent the expression in a myeloid cell line of the lymphocyte associated surface receptor for EB virus that led to its infection and transformation *in vitro*, the expression in EB virus-transformed lymphoid cells of myeloid cellular gene products, or perhaps the existence of undifferentiated cell lines expressing both myeloid and lymphoid genes.

There is certainly increasing evidence that long-term cultured cell lines may be derived from human and animal granulocyte-macrophage malignancies, and that such lines may express, spontaneously or after induction, a

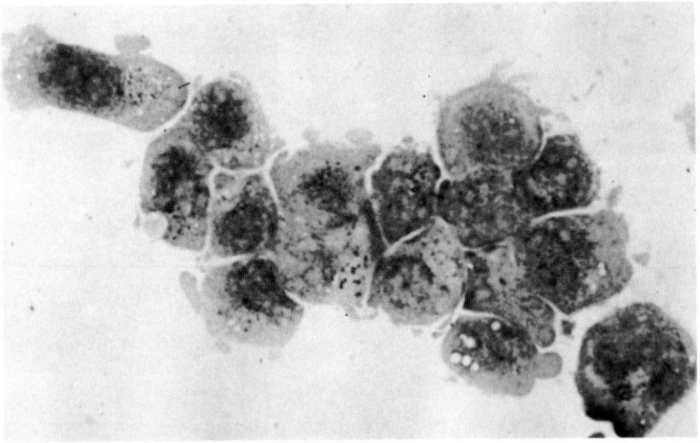

Fig. 12.10 The same preparation. Chloroacetate esterase. Scattered granular positivity in many cells.

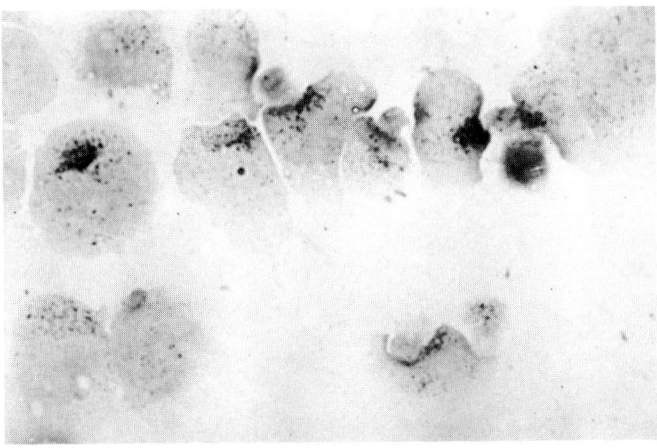

Fig. 12.11 The same preparation. Acid phosphatase. Most cells show scattered granular positivity, and some strong localized paranuclear reactions.

variety of myeloid differentiation markers, cytochemical and functional (Greenberger *et al.*, 1978).

Studies of the cytochemical profile of cells biopsied from lymphocytic leukaemias, myelomas and lymphomas were carried out by Sundström and Nilsson (1977) and compared with the findings in a panel of established cell lines derived from similar diseases and of lymphoblastoid cell lines of presumed non-neoplastic derivation. The staining methods used included Oil Red O, the PAS reaction, peroxidase, alkaline and acid phosphatases, the usual range of esterases and β-glucuronidase.

In general the reactions found in the cultured cell lines did not differ in any marked respect from those of the corresponding fresh biopsy cells, thus indicating an unaltered qualitative expression of enzyme production in lymphoid lines after prolonged *in vitro* growth. Apart from occasional positivity to Oil Red O, B-cell lines including non-neoplastic lymphoblastoid lines and lymphoma lines gave weak or negative reactions to all stains as did cells of CLL and fresh lymphoma biopsies. T-cell lines showed in three of four cases strong PAS positivity with coarse granules or blocks, and moderately strong α-naphthyl acetate esterase reactions, but with negative acid phosphatase and β-glucuronidase reactions. Myeloma lines, like fresh myeloma cells, were strongly β-glucuronidase positive, but unlike fresh cells showed strong positivity for α-naphthyl acetate and AS-D acetate esterases and only weak acid phosphatase. A single histiocytic lymphoma line was strongly positive with AS-D acetate esterase, inhibited by fluoride, in the monocytic pattern, but three fresh biopsies of supposed histiocytic lymphomas reacted as B-cell tumours and were probably of lymphoid origin. One myeloid line was examined which appeared to show more monocytic than granulocytic reactions, being negative to peroxidase and with more positivity to α-naphthyl acetate and AS-D acetate esterases than to chloroacetate esterase. Moreover the acid phosphatase reaction was very strong, as might be expected in cells of monocytic origin.

In a further expansion of this work, Sundström *et al.* (1982) compared the activity and isoenzyme patterns for acid esterase and acid phosphatase, using isoelectric focusing, in two T-ALL lines (J.M. and Molt-4), in a B-ALL line (Ball-1), in one non-T non-B ALL line (KM3) and in the APL line HL60. Acid esterase was expressed as in the native cell types involved in each line, but acid phosphatase showed additional isoenzymes and/or reductions in intensity of the typical expected isoenzymes in the T-ALL and B-ALL lines, although regular expression was found in the HL60 and KM3 lines. The authors concluded that long term culture conditions could repress phenotypic properties and/or derepress gene activities.

Srivastava *et al.* (1983) carried out an extensive study of 47 human leukaemia/lymphoma cell lines of myelocytic, monocytic, non-T, non-B lymphoblastic, T-cell and B-cell lineages, representing various levels of maturation, using immunological markers in parallel with cytochemical reactions for acid and alkaline phosphatases, α-naphthyl acetate and butyrate esterases (at pH 5.8 and 8.0), naphthol AS-D acetate (non-specific) esterase, chloroacetate esterase, chymotrypsin-like protease, deoxyribonuclease II, β-glucuronidase, and Sudan black and periodic acid Schiff reactions. The results were compared with findings in fresh cells from normal and leukaemic subjects.

Strong reactions to Sudan black, and for non-specific esterase and chloroacetate esterase occurred only in myelocytic and monocytic cell lines, with reaction intensity generally increasing progressively in more mature cells. Focal acid phosphatase reactions were found in all T-ALL cell lines, whereas myeloid and monocytic lines showed semicircular cytoplasmic distribution and B-cell lines cytoplasmic scattered granules. Acid phosphatase activity appeared to decrease with maturation along myeloid and also T-cell lines. There was high activity of acetate and butyrate esterases at both pH 5.8 and 8.0, and also of β-glucuronidase, in myeloid/monocytoid lines. Some esterase positivity was also detected in both B- and T-cell lines, in contrast to the negativity of peripheral blood B cells. Butyrate esterase activity declined with increasing cell maturation along the myeloid lineages.

Alkaline phosphatase was not detected in any cell lines except weakly in two B-cell lines. Monocyte esterase activity was inhibited by sodium fluoride, and acid phosphatase by L-tartrate except for the enzyme in a hairy cell leukaemia line, which was tartrate-resistant.

PAS reactions were not clearly distinctive in myeloid, T-, B- or non-T, non-B cell lines, but gave characteristic patterns with many perinuclear coarse granules in an erythroblastic leukaemia cell line and also showed a general increase in intensity in more mature cells of the myeloid series. Deoxyribonuclease II and chymotrypsin-like protease staining were not discriminatory.

In summary, the results of this study showed that cytochemical staining characteristics of the various leukaemia and lymphoma cell lines were comparable to those of corresponding cells from patients, and that the intensity and pattern of expression of these activities were related to cell type and degree of cell maturation.

In contrast to these findings, Srivastava and Bhargava (1986) in a study of cells from 33 normal or leukaemic cell lines found the enzyme dipeptidyl aminopeptidase IV (DAP IV), normally found specifically in T cells, to be expressed unexpectedly by four B-cell lines and one histiocytic lymphoma line, indicating a possible infidelity of gene expression in cultured cells. Among T-cell lines there appeared to be no correlation between DAP IV activity and OKT4 positivity or maturation stage.

A study of ectoenzyme activities in leukaemic cells and in a range of established haematopoietic cell lines was carried out by Morell *et al.* (1986). They confirmed earlier reports that 5′-nucleotidase (5′N) was present in greater amounts in cALL than in T-ALL or null-ALL, that cells from AML and AMML had greater activity of γ-glutamyl transferase (γGT) than those from ALL, and that normal B lymphocytes differed from T cells and monocytes chiefly in 5′N, whereas in monocytes γGT was more active than in other normal blood cells. Distinct patterns of ectoenzyme activity were found in the cell lines studied. Most B-cell lines including Raji, RPMI 1788 and 8226, induced by EB virus, had high 5′N and Na-K-Mg adenosine triphosphatase (ATPase), whereas myeloid lines had high γ-GT values. In lymphoid stem cells and T lymphoblast lines, most ectoenzymes had low activity.

Among other recent studies of *in vitro* cultures of leukaemic cells with potential for further cytochemical exploitation are several involving unusual variants of leukaemia. Saito *et al.* (1985) succeeded in establishing a new line, EoL, from the blood of a patient with Ph^1 negative eosinophilic leukaemia. Cells in the line expressed Ia (HLA-DR), the myeloid antigens IF10 and MY9 and receptors for IL2. A small percentage of the cells retained typical eosinophil features with eosinophil granules and positivity for Luxol-fast-blue and peroxidase. In the presence of an alkaline medium or after the addition of DMSO to induce maturation the cells with features typical of eosinophils increased to around 40%. The cultured cells and the original leukaemic cells had similar chromosomal anomalies. Ogura and his colleagues (1985) reported the establishment of a novel human megakaryoblastic leukaemia cell line, MEG-01, Ph^1 positive, derived from bone marrow cells of a case of Ph[1] positive CML in blastic transformation. The cells were positive for PAS, acid phosphatase and α-naphthyl acetate esterase, and for ultrastructural platelet peroxidase (PPO), but were negative for MPO, α-naphthyl butyrate esterase, naphthol AS-D chloroacetate esterase and alkaline phosphatase. Cytoplasmic factor VIII related antigen was weakly positive in the larger MEG-01 cells by immunofluorescence and immunoperoxidase with appropriate antibodies and platelet glycoprotein could also be demonstrated on most cell surfaces with antiplatelet GIIb and IIIa monoclonal antibodies, and with GIb on the surface of larger MEG-01 cells only. There were no detectable B- or T-cell or myeloid markers. The chromosomes were hyperdiploid with 56-58 XY Ph[1]+ constitution.

Strong and Kersey (1985) used a methyl cellulose colony assay system, with semi-solid conditioned medium containing T-cell growth factor, and a feeder cell layer of T cells, to obtain satisfactory growth *in vitro* of cells from two cases of acute leukaemia with the (4;11) translocation and with more than 90% of blasts expressing TdT. The cultured cells expressed the CD24 and BA2 B-cell markers but had colony growth requirements, with low oxygen need, like cells of AML rather than ALL. The cell line showed commitment to lymphoid lineage in expressing Ig heavy and light chain gene rearrangement and with some cells expressing cytoplasmic Ig, but there were also some myeloid features including reaction with the myeloid monoclonals CD15 and IG10. After treatment with the differentiation inducer PTA the cells developed phagocytic capacity and expressed the myelomonocytic markers CD11/OKM1 and CD13/MCS2 (Strong *et al.*, 1985).

The induction of differentiation in cell lines, especially of early myeloid cell constitution, provides an opportunity for study of cytochemical differentiation markers. Patterns of differentiation and the inducing agents which have proved effective in different cell lines were reviewed by Koeffler (1983). The early myeloblast line KG-1a and the myeloblast line KG-1 differentiated to macrophages under the influence of phorbol diesters, and showed some response to growth stimulatory factors. The promyelocytic line HL60 showed maturation and differentiation to granulocytes and macrophages after exposure to many different inducers. Lines ML-1 and ML-3, myelomonoblastic in cytology, responded to phorbol diesters and to teliocidins chiefly by production of macrophages, as did the monocytic line 0 937 and the monoblastic line THP1. Early erythroblast lines K562 and HEL responded to haemin and butyrate induction to show erythroid differentiation. Few detailed cytochemical studies appear to have been carried out in the work summarized in this review, although clearly an opportunity exists for analysis of the early stages of emergence of individual cytochemical maturation markers.

Cytochemical esterase markers were indeed included in a more recent study of phorbol diester-induced myeloid differentiation in acute undifferentiated leukaemia. Gregg *et al.* (1984) exposed cells from 32 cases of acute leukaemia to culture with TPA, and found changes in cell surface antigens, notably the emergence of MCS-2 positivity, a marker of granulocytic and monocytic differentiation, in cells from primary acute myeloblastic or acute undifferentiated leukaemias. No such induced changes occurred in cells from ALL. Undifferentiated blast cells could also be induced to express the non-specific leucocyte differentiation markers BA-2 and TA-1, and to produce non-specific (monocytic) esterase and sometimes chloroacetate esterase, while losing TdT positivity.

The interesting study by Nauseef (1986) of myeloperoxidase (MPO) biosynthesis by the HL60 promyelocytic cell line illustrates the potential value of using cell lines for this purpose. The cells of this line produce MPO, a glycoprotein of 150 000 MW composed of two $\alpha_2\beta_2$ heavy-light protomers with subunits of 59 000 and 13 500 MW. The primary translation product of MPO mRNA isolated from HL60 cells was a single polypeptide of 80 000 MW. In HL60 cells labelled with ^{35}S-methionine, labelled MPO was produced with a MW of 89 000, susceptible to endoglycosidase H, but with no detectable peroxidatic activity. This pro-MPO contained the α and β subunits of MPO and was presumed to be a glycosylated product of the primary translation polypeptide. Further post-translational modification of pro-MPO leads to localization at lysosomes and proteolytic maturation to active MPO. Studies with the HL60 line suggest that hereditary MPO deficiency, where the neutrophil polymorphs contain an 89 000 MW protein, may result from defective post-translational processing of pro-MPO.

In a further study of granulocyte maturation and enzyme expression Sullivan *et al* (1986) found a variant line of HL60, called HL60–A7, derived by growth in actinomycin D. This line contained large azurophilic granules deficient in MPO and the products of *in vitro* translation of A7 RNA contained less than 5% of the immunoreactive MPO found in the parent line. Karyotypic analysis showed a consistent chromosome 1q + abnormality not found in any of the parent HL60 cells. The human MPO gene appears to be located on chromosome 17 at q22–24 (Chang *et al.*, 1987) and although the association between MPO deficiency and the abnormality of 1q shown in the A7 line suggests that expression of the enzyme may be controlled in some way by genes involved in that structural defect, it is noteworthy that the representative karyotype of A7 cells illustrated in the report of Sullivan *et al.* (1986) shows not only the 1q + marker but also an absent chromosome 17 and multiple other numerical and structural defects.

In a very extensive series of studies in human leukaemia-lymphoma cell lines, Drexler *et al.* (1985a, b, c, d) compared the patterns of carboxylic esterase, acid phosphatase, β-hexosaminidase and lactate dehydrogenase isoenzyme distribution in large numbers of permanently established cell lines which had been classified by a full immunological analysis using poly- and monoclonal antibodies. The isoenzymes were separated by IEF on polyacrylamide gels and visualised by histocytochemical techniques.

Among 74 leukaemia-lymphoma cell lines (26T-, 34B-, 6 myelomonocytic, 8 non-T non-B lines) and 12 'normal' Epstein–Barr virus transformed B-lymphoblastoid lines no leukaemia-specific or, except for monocytes, blood cell type-specific isoenzyme or isoenzyme pattern of carboxylic esterase could be detected. The monocytic element in some cell lines showed a strong isoenzyme band selectively and completely inhibited by sodium fluoride. In general the number and staining intensity of esterase isoenzymes increased with differentiation in both T- and B-cell lines, with mature B-cell lines showing similar isoenzyme patterns to the 'normal' B-lymphoblastoid lines.

For acid phosphatase isoenzymes a similar pattern of progressive increase in number and staining intensity of bands in parallel with advancing stages of differentiation along their developmental pathways was found for T and B-cell lines. Again, no leukaemia-lymphoma specific isoenzyme was found, nor was any isoenzyme or pattern of isoenzymes characteristic of a particular cell lineage. The tartrate resistant acid phosphatase isoenzyme was detected in nine of the 74 malignant cell lines (4T-, 2B-, 1 myelomonocytic, 2 non-T, non-B lines) and in 10 of the 12 'normal' B-lymphoblastoid cell lines. Curiously enough, the single HCL-derived line (JOK-1) studied did not express this isoenzyme.

A somewhat comparable pattern of sequential β-hexosaminidase (Hex) isoenzyme expression paralleling differentiation stages in T- and B-cell lines was found. Very immature T-cells appear to have no Hex isoenzyme, somewhat more differentiated T-blasts have at first bands Hex A and I, and then lose the I band, while mature T-cells show again both A and I strongly. Among B-cells, pre-B blasts express Hex I strongly plus Hex B, at the next separable stage no isoenzymes can be detected, but with further maturation Hex A and B and finally Hex I appear. Plasma cells show all three bands strongly. Myelomonocytic lines displayed strongly stained bands of all three isoenzymes.

Finally lactic dehydrogenase (LDH) isoenzyme patterns once again showed increase in complexity and staining intensity as successive developmental stages were studied among the T- and B-cell lines and in the myelomonocytic lines.

The isoenzyme patterns associated with increasing differentiation, generally more complex and more intense than those found in less mature cell lines, can be induced to appear by the use of various agents such as phorbol esters (TPA) dimethylsulphoxide (DMSO), dimethylformamide (DMF), retinoic acid and 1,25 dihydroxyvitamin D3. A more natural agent, differentiation-inducing factor (DIF), released from mitogen-stimulated T lymphocytes, and possessing capacity to trigger differentiation in leukaemia cell lines has been reported by Drexler *et al* (1987) to induce similar mature-cell expression of isoenzymes in all of 15 leukaemic cell lines tested. In particular, lines HL60 (promyelocytic) and HEL (erythroleukaemic) displayed a newly developed monocyte-specific isoenzyme, while several myelomonocytic and erythroleukaemic cell lines expressed a tartrate resistant acid phosphatase isoenzyme. Lymphoid cell lines showed more limited differentiation responses.

A similar combined study of two cell lines, HDLM-2 and L-428, derived from pleural effusions of patients with Hodgkin's disease, showed TPA-induced isoenzyme profiles resembling those of myelomonocytic leukaemia cell lines (Drexler *et al* 1986). Both cell lines were negative fro immunological markers of the T- and B-cell lineages but reacted with several markers associated with the myelomonocytic cell lineages. The overall results suggested that Hodgkin and Reed-Sternberg cells constituted a unique cell type which might be derived from cells of the myelomonocytic cell lineage. This question is further discussed in Chapter 15.

These comprehensive studies show that the use of multiple isoenzyme analysis has a valuable contribution to make to cell phenotyping in leukaemic and lymphomatous states. For most of the cell lines studied the patterns of isoenzymatic differentiation developed in close parallel to the maturation sequence on which the immunological classification of these lines was based.

It is therefore already abundantly clear that with the increasing availability of stable cell lines of differing cell lineages and stages of maturation and with differing properties and patterns of gene expression, cytochemically, isoenzymatically, immunologically and otherwise determinable, we have excellent experimental models for the analysis of both genetic and environmental factors influencing the development and behaviour of haemic cells of all kinds.

REFERENCES

Catovsky D, Miliani E, Okos A, Galton D A G 1974 Clinical significance of T-cells in chronic lymphocytic leukaemia. Lancet ii: 751

Chang K S, Schroeder W, Siciliano M J, Thompason L H, McCredie K, Beran M, Freireich E J, Liang J C, Trujillo J M, Stass SA 1987 The localization of the human myeloperoxidase gene is in close proximity to the translocation breakpoint in acute promyelocytic leukemia. Leukemia 1: 458

Divine M, Farcet J P, Gourdin M F, Tabilio A, Vasconcelos A, Andre C, Jouault H, Bouguet J, Reyes F 1984 Phenotype study of fresh and cultured hairy cells with the use of immunologic markers and electron microscopy. Blood 64: 547

Drexler H G, Gaedicke G, Minowada J 1985a Isoenzyme studies in human leukaemia-lymphoma cell lines. I. Carboxylic esterase. Leukemia Research 9: 209

Drexler H G, Gaedicke G, Minowada J 1985b Isoenzyme studies in human leukaemia-lymphoma cell lines. II. Acid phosphatase. Leukemia Research 9: 537

Drexler H G, Gaedicke G, Minowada J 1985c Isoenzyme studies in human leukaemia-lymphoma cell lines. III. Beta-hexosaminidase. Leukemia Research 9: 549

Drexler H G, Gaedicke G, Minowada J 1985d Isoenzyme studies in human leukaemia-lymphoma cell lines. IV. Lactate dehydrogenases. Leukemia Research 9: 561

Drexler H G, Klein M, Bhoopalan N, Gaedicke G, Minowada J 1986 Morphological and isoenzymatic differentiation of B-chronic lymphocytic leukaemia cells induced by phorbolester. British Journal of Cancer 53: 181

Drexler H G, Gaedicke G, Lok M S, Diehl V, Minowada J 1986 Hodgkin's disease derived cell lines HDLM-2 and L-428: comparison of morphology, immunological and isoenzyme profiles. Leukemia Research 10: 487

Drexler H G, Notani J, Otsuka K, Gaedicke G, Minowada J 1987 Isoenzyme studies in human leukemia-lymphoma cell lines—V. Induction of differentiation by T-cell derived differentiation-inducing activity. Leukemia Research 11: 85

Falini B, Schwarting R, Erber W et al 1985 The differential diagnosis of hairy cell leukemia with a panel of monoclonal antibodies. American Journal of Clinical Pathology 83: 289

Golde D W, Cline M J 1973 Growth of human bone marrow in liquid culture. Blood 41: 45

Greenberger J S, Karpas A, Gans P J, Neumann H, Moloney W C 1978 Increased myeloid-associated enzymes in Epstein-Barr virus nuclear antigen-positive human cell lines exposed to butyric acid in vitro. Blood 51: 1073

Greenberger J S, Newburger P E, Karpas A, Moloney W C 1978 Constitutive and inducible granulocyte-macrophage functions in mouse, rat and human myeloid-leukaemia-derived continuous tissue culture lines. Cancer Research 38: 3340

Gregg S L, Gajl-Peczalska K J, LeBien T W, Bloomfield C D, Brunning R, Segawa K 1984 Monoclonal antibody MCS-2 as the marker of phorbol diester-induced myeloid differentiation in acute undifferentiated leukemia. Cancer Research 44: 2724

Hayhoe F G J, Quaglino D 1965 Autoradiographic studies on RNA and DNA metabolism of human leucocytes cultured with phytohaemagglutinin: uridine-5-T as a specific precursor of RNA. Nature 205: 151
Hayhoe F G J, Sinks L F, Flemans R J 1967 Studies on the transformation in vitro of lymphocytes from chronic lymphocytic leukaemia. In: Yoffey J M (ed) The Lymphocyte. Immunology and Haemopoiesis. Edward Arnold, Bristol, p 66
Higgy K E, Burns G F, Hayhoe F G J 1977 Discrimination of B, T and null lymphocytes by esterase cytochemistry. Scandinavian Journal of Haematology 18: 437
Hsu S M, Ho Y S 1986 Hairy cell leukemia still a B cell neoplasm. American Journal of Clinical Pathology 85: 252
Karpas A, Hayhoe F G J, Greenberger J S, Barker C R, Cawley J C, Lowenthal R M, Moloney W C 1977 The establishment and cytological, cytochemical and immunological characterization of human haemic cell lines: evidence for heterogeneity. Leukemia Research 1: 35
Karpas A, Hayhoe F G J, Greenberger J, Neumann H 1978 Alkaline phosphatase in Epstein-Barr viral nuclear antigen-positive cell lines. Science 202: 318
Koeffler H P 1983 Induction of differentiation of human acute myelogenous leukemia cells: therapeutic implications. Blood 62: 709
Leder A, Leder P 1975 Butyric acid, a potent inducer of erythroid differentiation in cultured erythroleukaemia cells. Cell 5: 319
Mastrangelo D, Orrico A, Pasqui AL, Capecci PL, Ceccatelli L, Laghi Pasini F. 1987 Double esterase (DE) positivity in AML cells. British Journal of Haematology 65: 119
Morell A, Losa G, Carrel S, Heumann D, Fliedner V E von 1986 Determination of ectoenzyme activities in leukemic cells and in established hematopoietic cell lines. American Journal of Haematology 21: 289
Nauseef W M 1986 Myeloperoxidase biosynthesis by a human promyelocytic leukemia cell line: insight into myeloperoxidase deficiency. Blood 67: 865
Neumann H, Klein E, Hanck-Granoth R, Yachmin S, Ben-hassat H 1976 Comparative study of alkaline phosphatase activity in lymphocytes, mitogen-induced blasts, lymphoblastoid cell lines, acute myeloid leukaemia and chronic lymphatic leukaemia cells. Proceedings of the National Academy of Sciences USA 73: 1432
Nowell P C 1960 Phytohaemagglutinin: an initiator of mitosis in culture of normal human leucocytes. Cancer Research 20: 462
Ogura M, Morishima Y, Ohno R, Kato Y, Hirabayashi N, Nagura H, Saito H 1985 Establishment of a novel human megakaryoblastic leukemia cell line, MEG-01, with positive Philadelphia chromosome. Blood 66: 1384
Quaglino D, Cowling D C 1964 Cytochemical studies on cells from chronic lymphocytic leukaemia and lymphosarcoma cultured with PHA. British Journal of Haematology 10: 358
Quaglino D, Hayhoe F G J 1965 Metabolic changes in short-term 'in vitro' cultures of normal and leukaemic cells: studies with cytochemistry and autoradiography. In: Hayhoe F G J (ed) Current Research in Leukaemias. Cambridge Univ. Press, Cambridge, p 124
Quaglino D, Hayhoe F G J, Flemans R J 1962 Cytochemical observations on the effect of phytohaemagglutinin in short-term tissue cultures. Nature 196: 338
Quaglino D, Cowling D C, Hayhoe F G J 1964 Cytochemical and autoradiographic studies on normal and leukaemic cells in short-term tissue cultures. British Journal of Haematology 10: 417
Reuben R C, Wife R L, Breslow R, Rifkind R A, Marks P A 1976 A new group of potent inducers of differentiation in murine erythroleukemia cells. Proceedings of the National Academy of Sciences USA 73: 862
Saito H, Bourinbaier A, Ginsburg M, Minato K, Ceresi E, Yamada K, Machover D, Bŕead J, Mathé G 1985 Establishment and characterization of a new human eosinophilic leukemia cell line. Blood 66: 1233
Srivastava B I S, Rossowski W, Minowada J 1983 Cytochemical comparison of immunologically characterized human leukaemia/lymphoma cell lines representing different levels of maturation. British Journal of Cancer 47: 771
Srivastava M D, Bhargava A K 1986 Expression of dipeptidyl-aminopeptidase-IV in some human T and B cell lines. Acta Haematologica 76: 25
Strong R C, Kersey J H 1985 In vitro culture of leukemic cells in t(4;11) acute leukemia. Blood 66: 439
Strong R C , Korsmeyer S T, Parkin J L, Arthur D C, Kersey J H 1985 Human acute leukemia cell line with the t(4;11) chromosomal rearrangement exhibits B-lineage and monocytic characteristics. Blood 65: 21
Sullivan A K, Amatruda T T, Fitz-Gibbon L, Koeffler H P, Peyman J, Rowden G, Shematek G, Shihab-El-Deen A 1986 Deficiency of myeloperoxidase and abnormal chromosome 1 occurs in variant (HL60) promyelocytes. Leukemia Research 10: 501

Sundstrom C, Nilsson K 1977 Cytochemical profile of human haematopoietic biopsy cells and derived cell lines. British Journal of Haematology 37: 489
Sundstrom C, Radzun H J, Parwaresch M R 1982 Isozymes of acid esterase and acid phosphatase in permanent human hematopoietic cell lines. Journal of the National Cancer Institute 69: 1255
Willcox M B, Golde D W, Cline M J 1976 Cytochemical reactions in human haemopoietic cells in liquid culture. Journal of Histochemistry and Cytochemistry 24: 979
Yoda S, Tsakada M, Hara Y, Komiyama A, Akabane T 1984 Impaired monocyte differentiation in children with acute lymphoblastic leukaemia; its evaluation by peroxidase activity and ultrastructure. Scandinavian Journal of Haematology 32: 525

CHAPTER 13

Major cytochemical findings in non-neoplastic pathological haemopoiesis

This Chapter will be mainly concerned with non neoplastic haematological disorders involving either the red or white cell series or cells of the monocyte-macrophage cell line. The first part of the Chapter will therefore be devoted to the different types of anaemia, the second to disorders related to non neoplastic alterations of leucocytes and the third to lysosomal storage diseases.

Cytochemistry of the red cell series

The contribution of cytochemistry towards a better understanding of the pathogenesis underlying many disorders of the red cell series has certainly been less remarkable than that of biochemistry and especially molecular biology; these methods of study have in fact provided a considerable wealth of information on the enzyme changes and alterations in globin structure, present in cases of haemolytic anaemia and in haemoglobinopathies.

The reason for this imbalance is that for many metabolic processes there are no suitable cytochemical methods and furthermore that many of the cytochemical reactions used in the study of the erythropoietic tissue are not entirely reliable from the point of view of their sensitivity and specificity.

Nevertheless cytochemistry may in many instances provide accurate information on the localization of enzymes and substrates, so that this type of investigation may be valuable in providing fundamental information on cell pathology, especially when cytochemical reaction patterns show striking differences from normal, and may be of help in the differential diagnosis of superficially similar conditions affecting the erythropoietic tissue.

The transformation of the proerythroblast, first stage in the red cell developmental line, into a mature red cell devoid of nucleus and cytoplasmic organelles, is associated with a number of cytochemical changes, some of which may be measured in fairly accurate quantitative manner, while others, and they are the majority, may be assessed only semiquantitatively or described only on the basis of subjective criteria.

With regard to normal erythroblasts, quantitative variations in nuclear DNA between 2n and 4n values have been reported in the proliferative compartment of the erythropoietic tissue, in relation to DNA synthetic activity. Quantitative changes in RNA and proteins (including haemoglobin), in the course of the maturation process, have been examined with great accuracy. On the basis of the results obtained with quantitative cytochemical methods (see also Chapter 2) erythroblasts may be divided into four groups: in the first group (proerythroblasts) the concentration of RNA is maximal, in the second RNA progressively decreases and basic cellular proteins such as globin are synthesized, in the third haemoglobin is produced in great quantities and finally in the fourth group (late erythroblasts) there is a conspicuous reduction in all metabolic activities.

Autoradiographic studies with radioactive precursors of DNA, RNA and proteins have fully confirmed the metabolic sequence outlined above (Mauri, 1961; Storti and Torelli, 1965).

As for the topographical localization of protein constituents in erythroblasts, histones may be found in the nucleus while the various reactions for proteins give a diffuse positivity both in the nucleus and cytoplasm of all cells, with the sole exception of protein-bound sulphydryl groups, which are associated with the granules of neutrophil and eosinophil polymorphs and are present in the cytoplasm of erythroblasts (Mauri *et al.*, 1954; Hertl, 1959).

Of the various iron-protein compounds, only traces of free iron may be observed with cytochemical techniques in 30–40% of intermediate and late erythroblasts.

In normal erythroblasts and erythrocytes it is not possible to demonstrate cytochemically the presence of glycogen.

With regard to the cytochemistry of lipids, red cells become fluorescent with Berg's 3–4 benzpyrene technique and are faintly stained by Sudan black B. The considerable number of investigations carried out (Ackerman, 1952; Mauri, 1963) indicate that red cells contain phospholipids and sphingomyelins. In addition to nuclear positivity, visible after 'unmasking procedures' (Mauri and Silingardi, 1963), erythroblasts in uncounterstained preparations show sudanophilic granules in the form of a faint perinuclear crown, due to the presence of phospholipids, in connection with chondriosomes. The cytochemical study of lipids in erythroblasts, like that of proteins, has contributed relatively little information on the origin, synthetic pathways and significance of these substances in the metabolism of the cell, and on their variations in pathological processes.

Megaloblastic anaemias

Various cytochemical observations suggest that the fundamental abnormality in megaloblastic anaemias is one involving nucleic acid metabolism. With regard to DNA, early cytophotometric studies on the intensity of the Feulgen reaction (Marinone, 1953) have shown that the mean DNA content of megaloblasts is consistently increased compared with that of normal erythroblasts. Mauri *et al.* (1960), studying the incorporation of tritiated thymidine in megaloblasts, reported a decreased rate of DNA synthesis, in agreement with the observations of Reisner (1958). However, other investigators do not believe that the megaloblast is characterized by any derangement or decrease in the rate of DNA synthesis. Lajtha and

Kumatori (1957) and Schmid *et al.* (1964) did not observe any difference in the rate of incorporation of either formate ^{14}C or tritiated thymidine in megaloblasts and normal erythroblasts and tend to exclude the existence in megaloblastic anaemia of alterations in DNA synthesis, distinguishable by autoradiographic methods.

A clearer understanding of the distribution of proliferating megaloblasts in the various stages of interphase using a combination of Feulgen microdensitometry and tritiated thymidine autoradiography has been obtained following the investigations of several workers (Menzies *et al.*, 1966; Wickramasinghe *et al.*, 1967 and 1968; Yoshida *et al.*, 1968). Such studies have revealed two important cytokinetic disturbances in vitamin B_{12} deficient megaloblasts. These consist in (a) an increased frequency of cells in the G_2 phase of the cell cycle and (b) the presence of cells with DNA contents between the 2n and 4n values, which were not in DNA synthesis, as judged by their failure to incorporate tritiated thymidine *in vitro* (U cells).

The presence of U cells and an accumulation of cells in G_2 have also been described in the bone marrow of two cases with a mild megaloblastic anaemia due to folate deficiency (Menzies *et al.*, 1966). However, as identification of cell types was impossible, there is no information concerning differences in the severity of these abnormalities in basophilic and early polychromatic cells of folate deficient patients.

Both the pile-up of cells in G_2 and the presence of U cells in vitamin B_{12}-deficient bone marrows have been confirmed using gallocyanin chromalum as an alternative staining reaction for the quantitative cytochemical estimation of DNA (Wickramasinghe *et al.*, 1969). Unlike the Feulgen method which depends on a partial hydrolysis of the DNA molecule and the formation of aldehyde groups on the deoxyribose moiety, the gallocyanin chromalum complex binds directly with the DNA phosphate groups (see also Chapter 2).

The abnormalities in the distribution of erythroblasts in interphase cannot ed detected in all patients with vitamin B_{12} deficiency. Thus the two cases with the highest haemoglobin levels studied by Wickramasinghe *et al.* (1968) showed no pile-up of cells in G_2 and no abnormality in the S/G_2 ratio.

The explanation for the finding that the cell cycle abnormalities, when present, are maximal in the more mature proliferating stages (i.e. the early polychromatic erythroblasts) is uncertain. It may be related to the fact that the uptake of vitamin B_{12} during erythropoiesis is largely confined to the basophilic erythropoietic cells (Wickramasinghe and Carmel, 1972). If suboptimal amounts of vitamin B_{12} are accumulated by the promegaloblasts and the more mature cells take up little of this vitamin, the last dividing cell class might be maximally affected by the consequences of vitamin B_{12} deficiency. An alternative explanation could be that early polychromatic erythroblasts have less thymidine kinase activity than basophilic erythropoietic cells (Wickramasinghe, 1975).

In this context, it is perhaps appropriate to recall that in mammalian tissues there are only two vitamin B_{12} dependent reactions, which have been identified with certainty (Wintrobe, 1981). The first of these is the conversion of methylmalonyl CoA to succinyl CoA, which represents a step in the catabolism of propionate. The second reaction, namely the synthesis of the amino acid methionine from homocysteine, appears to be of special interest since it requires not only methylcobalamin but also a folate coenzyme, N^5-methyl-tetrahydrofolate:

$$\text{Homocysteine} + N^5\text{-methyl-}FH_4 \xrightarrow[\text{vitamin } B_{12}]{\text{methyl transferase}} \text{methionine} + FH_4$$

Although the relationships of the coenzyme functions of vitamin B_{12} to the metabolic consequences of B_{12} deficiency have not yet been clearly established, according to certain lines of thought most of the effects of vitamin B_{12} on DNA synthesis are mediated by folate (Kass, 1979). In particular, the disturbance of DNA synthesis in megaloblastic anaemias could be the result of a reduced activity of the enzyme methyl-tetrahydrofolate homocysteine cobalamin methyl transferase, which leads to decreased tetrahydrofolate available for conversion to various folate derivatives. Of these 5,10-methylene-tetrahydrofolate plays a key role, since it is the cofactor for thymidylate synthetase. The lack of tetrahydrofolate and folate intermediates, which function as cofactors for enzymes involved in DNA synthesis, forms the basis for the so-called 'methyl-tetrahydrofolate trap' hypothesis (Metz *et al.*, 1968; Waxman *et al.*, 1969; Nixon and Bertino, 1972).

Some of the metabolic derangements related to nucleoprotein metabolism and to reduced activity of methyl-tetrahydrofolate homocysteine cobalamin methyl transferase, may be revealed by cytochemical reactions (Kass, 1979).

For instance, the ammoniacal silver stain indicates the presence of abundant lysine-rich histones in the nuclei and the virtual absence of arginine-rich histones. The Sakaguchi reaction for arginine tends to confirm these findings, since promegaloblasts show the weakest colour reaction, whereas erythraemic proerythroblasts show a particularly intense nuclear colour reaction. Normal proerythroblasts appear to exhibit a staining intensity intermediate between promegaloblasts and erythraemic proerythroblasts. Promegaloblasts also show less total histones stainable by Fast green than either normal or erythraemic proerythroblasts.

Differences in histones in the two types of erythroid precursors may be a factor in the pathogenesis of those morphological changes termed megaloblastic and megaloblastoid (Kass, 1972). The finely dispersed thin chromatin strands of the promegaloblast may result in part from the observation that lysine-rich histones are more weakly linked to DNA than the other histones (Busch, 1965). This phenomenon may cause DNA strands to be dispersed and widely separated in the megaloblast, resulting in a fenestrated appearance of its nucleus. On the other hand, it is possible that the increased calibre of chromatin strands found in the erythraemic erythroblast, compared to megaloblasts, may be partly the result of scattered punctate areas of arginine-rich histones, binding these strands together into larger coils of chromatin.

As previously mentioned, in megaloblastic anaemia decreased amounts of tetrahydrofolate are found because of decreased activity of methyl transferase. Tetrahydrofolate can be demonstrated in the cytoplasm of cells with the use of *molybdenum dichloride dioxide,* a substance that interacts with tetrahydrofolate, thus forming an insoluble black precipitate. Using this reaction, tetrahydrofolate appears to be markedly reduced and in some instances absent from the cytoplasm of polymorphonuclear leucocytes from patients with untreated megaloblastic anaemia.

Employing *nickel chloride,* which forms an insoluble yellow-brown precipitate with homocysteine, increased amounts of homocysteine are found in the cytoplasm of megaloblasts as a result of decreased activity of methyl transferase and subsequent reduced conversion of homocysteine to methionine. Reduction in methionine may also be detected with another cytochemical test, involving the use of a bacterial overlay technique employing *Leuconostoc mesenteroides,* a microrganism that has an absolute growth requirement for methionine. In megaloblastic anaemia, decreased activity of methyl transférase leads to decreased generation of methionine, which is reflected in decreased growth of the methionine-requiring bacteria in contact with megaloblasts.

Alizarin red S stains the cytoplasm of megaloblasts a pink colour, which is believed due in part to accumulation of methylmalonyl CoA, as a result of decreased activity of coenzyme B_{12}-dependent methylmalonyl CoA mutase, an enzyme that catalyzes the conversion of methylmalonyl CoA to succinyl CoA.

The cytophotometrics studies of Thorell (1947) suggest that RNA metabolism is also disturbed in megaloblastic anaemia, since nuclear and cytoplasmic RNA persist to a later stage in the line of differentiation than in normal erythropoietic cells. In contrast, Wickramasinghe (1970), using Azur B as a metachromatic stain for the quantitative cytochemical estimation of RNA, found a progressive reduction in the total RNA content of the megaloblast with increasing maturity, the fractional decrease in mean RNA content from one cell class to the next being similar to that during normal erythropoiesis. The explanation for this discrepancy is uncertain. It is possible that the small sample of megaloblasts studied by Thorell was rich in G_2 cells, which are now known to be present in large numbers in vitamin B_{12} deficient marrows. Such G_2 cells would be expected to have high cytoplasmic RNA concentrations due to previous RNA synthesis during the G_1 and S phases. Alternatively megaloblasts might contain large amounts of soluble mono- and oligo-nucleotides which absorb ultraviolet light intensely (Gordon and Nurnberger, 1955) and are lost during fixation and subsequent staining.

A study of the PAS reaction in the megaloblasts of megaloblastic anaemia has shown that in most cases this staining reaction is completely negative (Plate 34A), while in a few cases moderate or fairly strong PAS positivity has been observed, mainly confined to early megaloblasts (Plate 34B) (Quaglino and Hayhoe, 1960)

Free iron granules are reported in megaloblasts rather more frequently than in normal red cell precursors (Plate 34C), although the degree of sideroblastosis never reaches the intensity found in refractory sideroblastic

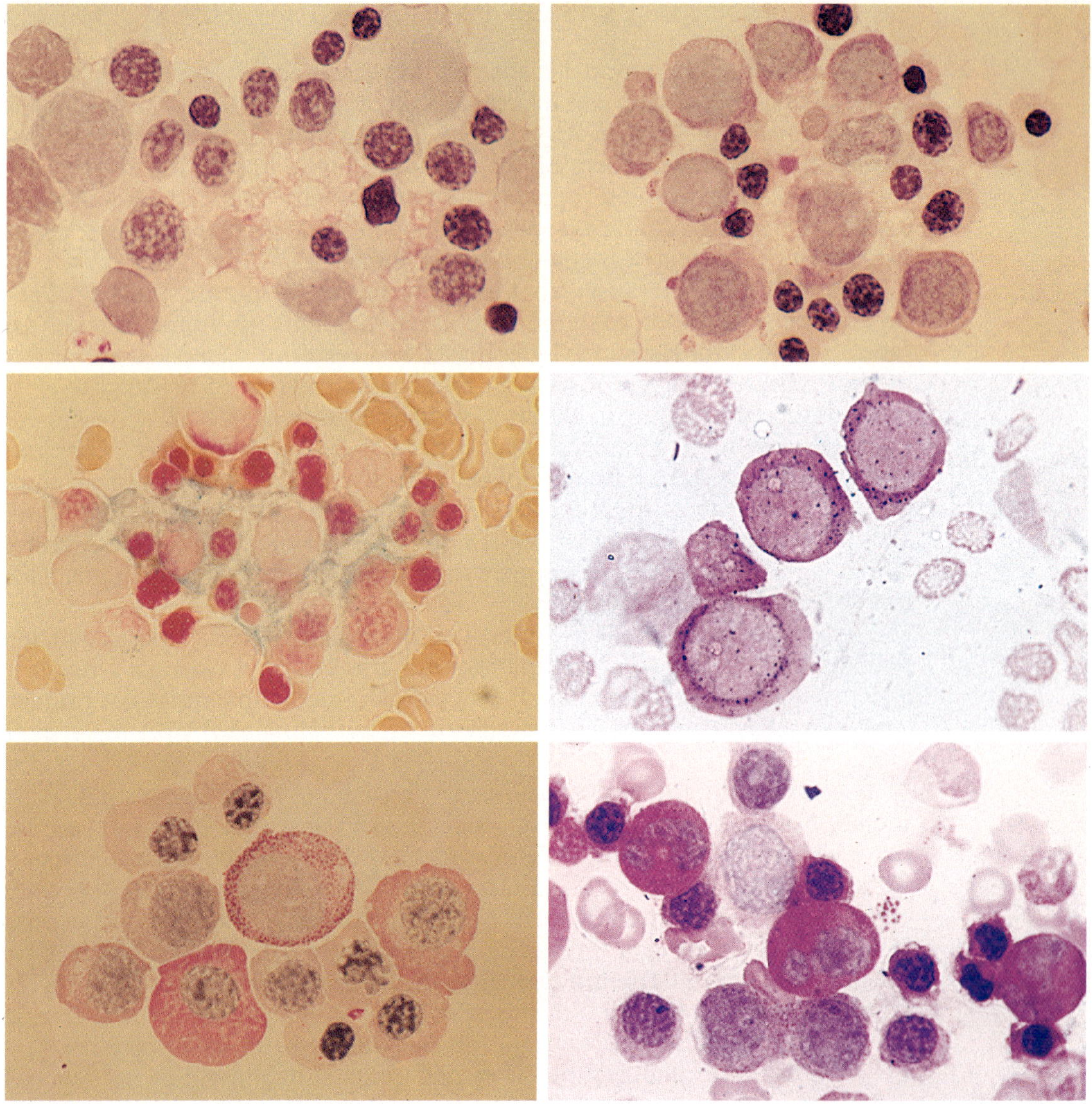

A B C D E F

Plate 34

A. Megaloblastic bone marrow. PAS stain. There is weak positivity in a macrophage, but all the surrounding megaloblasts at every stage of maturity are negative.

B. Megaloblastic bone marrow. PAS stain. In this case there is weak diffuse and granular positivity in early megaloblasts while later erythroblasts are negative.

C. Megaloblastic bone marrow. Prussian blue stain. Megaloblasts around a central macrophage show occasional free iron granules but no gross sideroblastosis.

D. Megaloblastic bone marrow. Lactic dehydrogenase. A group of megaloblasts show moderately intense granular positivity.

E. Strong PAS positivity in erythroblasts from β-thalassaemia.

F. Iron deficiency anaemia bone marrow. PAS stain. Moderate to strong PAS positivity in the small and ragged normoblasts surrounding normally positive granulocyte precursors.

anaemia. However, a few patients with megaloblastic changes caused by malnutrition or malabsorption have been reported to show many ringed sideroblasts (MacGibbon and Mollin, 1965; Dacie and Mollin, 1966).

Using the substrate α-naphthyl acetate Rozenszajn *et al.* (1968) have reported an increase in non-specific esterase activity in the megaloblasts of pernicious anaemia as well as in those of Di Guglielmo's disease and aplastic anaemia, and in conditions with high serum levels of vitamin B_{12}. The results suggested that increased α-naphthyl acetate esterase activity was the consequence of a derangement in nucleic acid metabolism. This derangement may produce a disturbance in the regulation of protein synthesis, a sign of which may be an enhanced esterase activity. By correcting the metabolism of nucleic acids with vitamin B_{12} treatment, the regulation system of protein synthesis returns to normal. This results in the disappearance of megaloblastosis and reduction of the strong α-naphthyl acetate esterase activity. Another explanation of the increased esterase activity may be that in normal erythroblasts vitamin B_{12} and folic acid act as repressors or inhibitors of α-naphthyl acetate esterase.

Stuart and Skowron (1968), using cytochemical methods, observed in megaloblasts an increase in the activity of representative enzymes of the Embden-Meyerhof pathway, pentose phosphate shunt and tricarboxyl cycle. The enzymes studied by Stuart and Skowron included lactate dehydrogenase, malate dehydrogenase, succinate dehydrogenase, α-glycerophosphate (soluble NAD-dependent) dehydrogenase, α-glycerophosphate (mitochondrial NAD-independent) dehydrogenase, glucose-6-phosphate dehydrogenase and 6-phosphogluconate dehydrogenase. Increased serum lactate dehydrogenase levels in patients with megaloblastic anaemia appear to be derived from intramedullary destruction of large numbers of megaloblasts containing an increased amount of this enzyme (Plate 34D).

As mentioned in an earlier chapter (7), megaloblasts occasionally show a positive alkaline phosphatase reaction, in the form of dots and rods, the significance of which is still uncertain (Quaglino, 1964). This finding has been recently confirmed by Woessner *et al.* (1984), who found 24% alkaline phosphatase positive erythroblasts in a case of pernicious anaemia. However the authors also reported the presence of alkaline phosphatase positivity in a variable proportion (1 to 36%, mean 9.4%) of erythroblasts from normal controls. Alkaline phosphatase activity was also observed in a variety of pathological conditions, characterized by dyserythropoiesis, ranging from iron-deficiency anaemia, haemolytic anaemia to cirrhosis of the liver and chronic myeloproliferative disorders. The same authors also studied the acid phosphatase reaction of erythroblasts of normal controls and different pathological conditions, displaying features of acquired dyserythropoiesis. A considerable number of red cell precursors from normal controls (60 to 92%, mean 74%) showed a granular centrosomic positivity, which was sensitive to tartrate inhibition. The most outstanding feature in the pathological material examined was the markedly strong positivity in the megaloblasts in a case of pernicious anaemia and in the erythroid precursors of a case of erythraemic myelosis. Finally, strong N-acetyl-β-D-glucosaminidase activity was found in a case of pernicious anaemia, where 23% of megaloblasts were positive, and in a case of acute erythraemic myelosis (62% positive erythroblasts), in contrast with the low percentage of positive cells (mean 3%) in normal controls.

Beta-thalassaemia

The anaemia in patients with homozygous beta-thalassaemia is mainly caused by a combination of shortened red cell life-span and ineffective erythropoiesis. The shortening of red cell survival is largely due to an intrinsic cell defect, which consists in a failure or in a severe reduction in the rate of beta-chain synthesis. Hypersplenism may be a contributory factor in the causation of anaemia.

Examination of the bone marrow shows a marked increase in the number of erythroid cells. Erythropoiesis is normoblastic, but early and late polychromatic cells show a rather pale-staining cytoplasm, suggestive of reduced haemoglobin synthesis.

Many of the erythroblasts show a strong PAS positive reaction (Plate 34E), first reported by Astaldi *et al.* (1954) in thalassaemia major and by Greig and Metz (1957) in thalassaemia minor. A series of cytochemical investigations led Astaldi and his co-workers to believe that the substance responsible for the PAS reaction might be a gluco- or a mucoprotein or a neutral polysaccharide. However the results of other investigators (Stegagno and Pollitzer, 1953; Perugini and Soldati, 1955; Sansone and Bertolotti, 1958; Mattioli and Del Piano, 1961) indicate that the PAS positive material in thalassaemic erythroblasts is glycogen. In particular Perugini and Soldati (1955) carried out extensive cytochemical investigations in a variety of pathological conditions, including cases of thalassaemia major, in order to establish the nature and significance of PAS positivity in erythroblasts and erythrocytes. Prior treatment of the smears with amylase or diastase caused the disappearance of the PAS positive granules while the diffuse cytoplasmic tinge was unaffected. Incubation with lysozyme and hyaluronidase, as well as prior pyridine extraction or blocking of aldehyde groups did not modify in any way PAS reactivity. Acetylation and deacetylation were performed in all cases. Sudan black B staining and reactions for metachromasia were always negative. The Feulgen reaction was applied to the material under study, because of the report of Rondanelli and coworkers (1955), who observed in two cases of chronic granulocytic leukaemia the presence of PAS positive granules which reacted positively to the Feulgen stain. In all the erythroblasts and erythrocytes of thalassaemic patients examined by Perugini and Soldati PAS positive granules were consistently Feulgen negative. The authors therefore concluded that the PAS positive granules in the erythroblasts of the pathological conditions examined, including cases of thalassaemia major, most probably consisted of glycogen, while the diffuse cytoplasmic tinge present in both erythroblasts and erythrocytes was due to muco- or glycoproteins or neutral mucopolysaccharides.

Detailed investigations by Fessas and Papayannopoulou (1965) confirmed that in thalassaemic erythroblasts PAS positivity was dependent on the presence of glycogen. PAS positivity was demonstrable mainly in polychromatic and to a smaller extent in orthochromatic erythroblasts, while the degree of positivity was in relation to the severity of the disease and was more conspicuous in younger patients. Finally there did not appear to be any direct relationship between PAS positivity and the level of fetal haemoglobin.

As Fessas and Papayannopoulou point out, glycogen is a major component serving as an energy source in the animal kingdom. Normally erythroblasts do not contain glycogen, at least not in amounts histochemically demonstrable, all energy being used up or stored otherwise. Its demonstration in erythroblasts under certain circumstances proves that in these cells the enzymatic pathways for glycogen synthesis are present and potentially operative. It is therefore suggested that in thalassaemia glycogen tends to accumulate because of the existence of a considerable surplus of energy due to the great reduction or complete absence of beta-chain synthesis. During the normal maturation process protein synthesis rapidly increases because of intense haemoglobin synthesis. Any interference with this energy-requiring process leads to the presence of energy in excess of needs, with the result that this surplus is channeled into the normally latent pathway of glycogen synthesis.

The presence of PAS positive material in thalassaemic erythroblasts has also been demonstrated by a very sensitive fluorescent PAS reaction (Yataganas *et al.*, 1973). The intracytoplasmic fluorescent material may either be diffusely distributed or may appear as granules of variable size. Treatment with amylase completely abolishes the fluorescent staining reaction, indicating that the PAS positive material consists of glycogen. Yataganas and coworkers also observed that the proportion of PAS positive erythroblasts in patients with beta-thalassaemia increased from 0.5–9.5% in basophilic erythroblasts to 0.9–45% in early polychromatic cells. Yataganas and his associates also determined by microfluorimetry the average glycogen content per PAS-positive early and late polychromatic erythroblasts and reported that in patients with thalassaemia major the glycogen content was higher than in patients with thalassaemia intermedia. The same authors have shown that the majority of the glycogen-containing basophilic and early polychromatic erythroblasts in beta-thalassaemia are in the G_1 phase and that the remainder are in the S phase. These findings suggest that the increased amounts of glycogen observed in the G_1 cells may be a reflection of decreased energy requirements in cells which are either arrested in G_1 or progressing through a prolonged G_1 period. The fraction of fluorescent PAS-positive G_1 early polychromatic cells is higher in clinically severe cases as compared to less severe cases.

In patients with beta-thalassaemia a proportion of erythroblasts contain intracytoplasmic inclusions, consisting of precipitated alpha-chains. Precipitation occurs because of an imbalance between beta- and alpha-chain synthesis, the former being severely depressed or absent. These cytoplasmic inclusions may be demonstrated by staining erythroid cells supravitally with methyl violet (Yataganas and Fessas, 1969).

Increased sideroblastosis has been observed in beta-thalassaemia (Kaplan *et al.*, 1954), although the reaction is not so intense as in refractory sideroblastic anaemia and the formation of ringed sideroblasts is not a typical feature of the disease.

Of particular interest in beta-thalassaemia is the study of the number of erythrocytes containing fetal haemoglobin, which may be detected by means of the acid elution technique of Kleihauer and Betke (1963); with this method it is possible to distinguish a dual cell population, a finding which differs from that occurring in another haemoglobinopathy, 'hereditary persistence of fetal haemoglobin', in which HbF is present in the entire erythroid population.

Electron microscopy has confirmed the presence in thalassaemic erythroblasts of multiple abnormalities (Polliack and Rachmilewitz, 1973; Polliack *et al.*, 1974). The ultrastructural alterations consist in the presence of nuclear and intracytoplasmic Heinz bodies, which correspond to sites of precipitated alpha chains, in the presence of granular deposits of ferritin, autophagic vacuoles, accumulation of glycogen and various membranous structures. Nuclei show considerable abnormalities in addition to the presence of intranuclear Heinz bodies.

In beta-thalassaemia, as in other pathological conditions, cell proliferation has been studied by the combination of Feulgen microdensitometry and autoradiography with tritiated thymidine. The results obtained by different authors (Wickramasinghe *et al.*, 1970; Kesse-Elias *et al.*, 1970; Queisser *et al.*, 1971a) indicate that there is a slight increase in the proportion of basophilic erythroblasts in the S phase and a reduction of the same cells in the G_1 phase. This behaviour has been related to a shortening of the cell cycle in response to high erythropoietin levels. In contrast polychromatic erythroblasts are characterized by a marked decrease in the proportion of cells in the S phase and a simultaneous increase in the percentage of cells in G_1. This finding may be attributed to a substantial prolongation of G_1 or a failure of G_1 cells to enter the S phase. Both hypotheses are in keeping with a depression of cell proliferation at this stage. Early polychromatic cells also show a low S/G_2 ratio due to a relative excess of G_2 cells, despite the small numbers of cells in the S phase, indicating a failure or delay of G_2 cells to enter mitosis.

Wickramasinghe *et al.* (1970) have postulated that a high death rate of early polychromatic erythroblasts, which accumulate in G_1 , may be partly responsible for the ineffectiveness of erythropoiesis in homozygous beta-thalassaemia. Cellular damage followed by cell death may be the result of intracellular alpha chain precipitates (Fessas, 1963; Yataganas and Fessas, 1969; Wickramasinghe *et al.*, 1970). The latter authors investigated this hypothesis by correlating the extent of intracytoplasmic alpha chain precipitates, determined by supravital staining with methyl violet, with *in vitro* protein and DNA synthetic activity in individual cells, studied by autoradiography. It was found that none of the early polychromatic erythroblasts with moderate or large amounts of alpha chain precipitates incorporated tritiated thymidine and that over 80% of late polychromatic cells with large amounts of alpha chain precipitates did not incorporate tritiated leucine or tritiated phenylalanine. However the labelling index with tritiated thymidine of early polychromatic erythroblasts devoid of intracytoplasmic alpha chain precipitates is subnormal and 10–23% of late polychromatic cells with no precipitates do not incorporate radioactive aminoacids. A possible explanation for this finding may lie in the fact that the derangement in cell proliferation and the reduction in the rate of protein synthesis may be related to the accumulation of free alpha chains rather than to the presence of alpha chain precipitates (Wickramasinghe *et al.,* 1973).

Haemoglobinopathies

An important investigation, although not strictly cytochemical, is the demonstration in these conditions of intraerythrocyte inclusion bodies. The presence of HbH causes the intracellular precipitation of denatured haemoglobin in the form of fine granules (Fig. 13.1); the phenomenon is particularly pronounced when a drop of blood is incubated in isotonic solution for two hours at 37°C with brilliant cresyl-blue.

Heinz bodies may be demonstrated in erythrocytes containing unstable haemoglobins by supravital staining, using either 0.5% methyl violet in saline or a saline solution of brilliant cresyl-blue.

Iron-deficiency anaemia

In iron-deficiency anaemia erythropoiesis is characterized by the presence of hypochromic microcytic erythrocytes. In addition to their small size late polychromatic cells present a ragged outline and an irregularly stained cytoplasm.

According to Stohlman and co-workers (1963) the microcytosis of iron-deficiency anaemia is due to an abnormally slow rate of haemoglobin synthesis which allows one or more additional erythroblast divisions before reaching the critical haemoglobin level at which cell division ceases. Another possibility is that microcytosis occurs because erythroblasts and erythrocytes contain less haemoglobin (Wickramasinghe, 1975).

The most relevant cytochemical findings in iron-deficiency anaemia are obtained with the Prussian blue reaction and the PAS stain.

Douglas and Dacie (1953), Kaplan *et al.* (1954), Björkman (1956) found that sideroblasts and siderocytes were almost completely absent in the bone marrows of the cases examined and their reappearance was associated with the administration of iron or the addition of iron to bone marrow cultures *in vitro*.

Marrow aspirates from patients with iron-deficiency anaemia also show a complete absence of stainable iron in the reticulum cells of the marrow fragments (Plate 10E) (Ellis *et al.,* 1964; Nixon and Olson, 1968).

Quaglino and Hayhoe (1960) observed in iron-deficiency anaemia a variable number (9–94%) of erythroblasts showing slight to strong PAS positivity (Plate 34F). The highest PAS scores were found in cases with very low MCHC, while the weakest reactions were seen in patients presenting only a mild degree of hypochromic anaemia or in those who

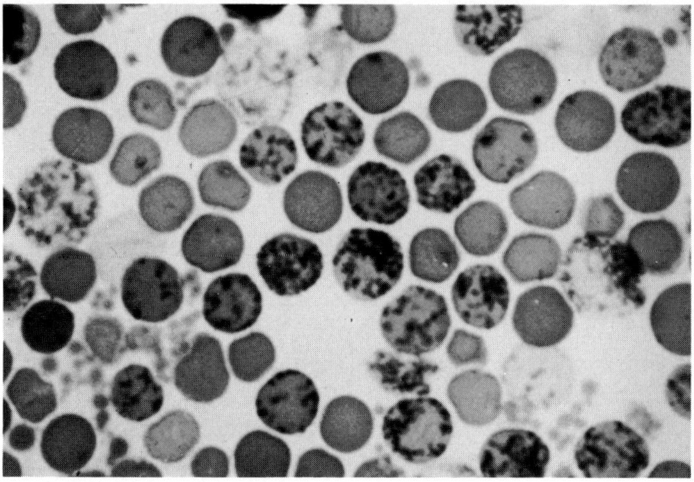

Fig. 13.1 Red cells showing Hb inclusion bodies.

showed signs of partial recovery under iron therapy. A positive reaction, generally diffuse but occasionally granular, was present in a small proportion of mature red cells.

The study of cell proliferation in iron-deficiency anaemia, employing the combination of Feulgen microdensitometry and tritiated thymidine autoradiography has shown no abnormality in the distribution of early polychromatic erythroblasts in the different phases of the cell cycle, excepting an increase in the percentage of cells in G_1 (Wickramasinghe, 1975). In one case studied by this author a slight increase of U cells was observed. From these investigations it seems unlikely that there is any significant degree of cell death during the proliferative phase of iron-deficient erythropoiesis. Late polychromatic erythroblasts often showed nuclear abnormalities, the DNA content being mostly diploid, but occasionally tetraploid or intermediate. It is therefore probable that any ineffective erythropoiesis which may occur in iron-deficiency anaemia may be largely confined to the late polychromatic cell compartment, in which nuclear abnormalities are more frequent. Finally the average rate of DNA synthesis in the haemopoietic cells of patients with iron-deficiency anaemia appears to be normal in the majority of cells. However Dörmer (1973), employing quantitative autoradiography, reported that the S phase of erythroblasts in one patient with iron-deficiency anaemia was significantly prolonged.

Haemolytic anaemias

In these conditions the Prussian blue and the PAS reactions may reveal distinctive cytochemical abnormalities. The former commonly shows an increase in stainable iron in erythroblasts and erythrocytes (Plate 35A, B); ringed sideroblasts are not generally found, but siderocytes appear numerous especially after splenectomy when increased numbers of reticulocytes are present in the circulation (Douglas and Dacie, 1953).

The behaviour of the PAS reaction in some cases of haemolytic anaemia has been investigated by Perugini and Soldati (1955) and by Quaglino and Hayhoe (1960). The latter observed that in acquired haemolytic anaemia and in hereditary spherocytosis the intensity of the reaction was slight and the percentage of positive cells rather low. The strongest positivity in the form of coarse blocks of glycogen in proerythroblasts and of a diffuse tinge in late erythroblasts was seen in two cases of severe foetal haemolytic disease.

In haemolytic anaemias due to glucose-6-phosphate dehydrogenase deficiency, where exaggerated Heinz body formation occurs on exposure of red cells to phenyl hydrazine (Figs 13.2 and 13.3), it is also possible to differentiate cytochemically between normal erythrocytes and those with reduced or absent glucose-6-phosphate dehydrogenase. In the technique described by Tönz and Rossi (1964) red cells of normal, hemizygous and homozygous individuals, as well as heterozygous females, were treated with sodium nitrite to convert the haemoglobin to methaemoglobin. They were then incubated with glucose and methylene blue for three hours at 37°C. During this period methaemoglobin is reduced by glucose-6-phosphate dehydrogenase to oxyhaemoglobin. If any methaemoglobin remains, it is converted to cyanmethaemoglobin by the addition of potassium cyanide. Since oxyhaemoglobin is active as a peroxidase and cyanmethaemoglobin has no such activity, cells with glucose-6-phosphate dehydrogenase activity can be distinguished from those without enzyme activity. For observation the red cells were spread as a film on a slide, dried and the cyanmet-haemoglobin eluted by a hydrogen peroxide solution. Finally, the slides were stained using haematoxylin-erythrosin. In normal individuals all erythrocytes are stained. In hemizygous men and in one homozygous female the red cells appeared as non-stained ghosts. In heterozygous females two distinctly different types of cells appeared, one behaving as in normals and one as in hemizygous men with favism. No intermediary changes were observed.

In Chapter 9 details were given of a recent, highly sensitive cytochemical technique for the demonstration of glucose-6-phosphate dehydrogenase activity in erythrocytes (Van Noorden and Vogels, 1985), capable of detecting in normal individuals subtle variations of enzyme activity, as well as of discriminating between heterozygous and homo or hemizygous patients with glucose-6-phosphate dehydrogenase deficiency.

Finally the technique of Kleihauer and Betke (1963) for demonstrating HbF may be useful for detecting the presence of fetal erythrocytes in the maternal circulation in the course of isoimmunization of the pregnant mother (Fig. 13.4).

Congenital dyserythropoietic anaemia

In this rare group of congenital disorders of unknown aetiology cytochemical findings appear to be similar in all varieties, with some weak diffuse or finely granular PAS positivity (Plate 35C), an increase in free iron in erythroblasts, but no ringed sideroblast formation (Plate 35D), an absence

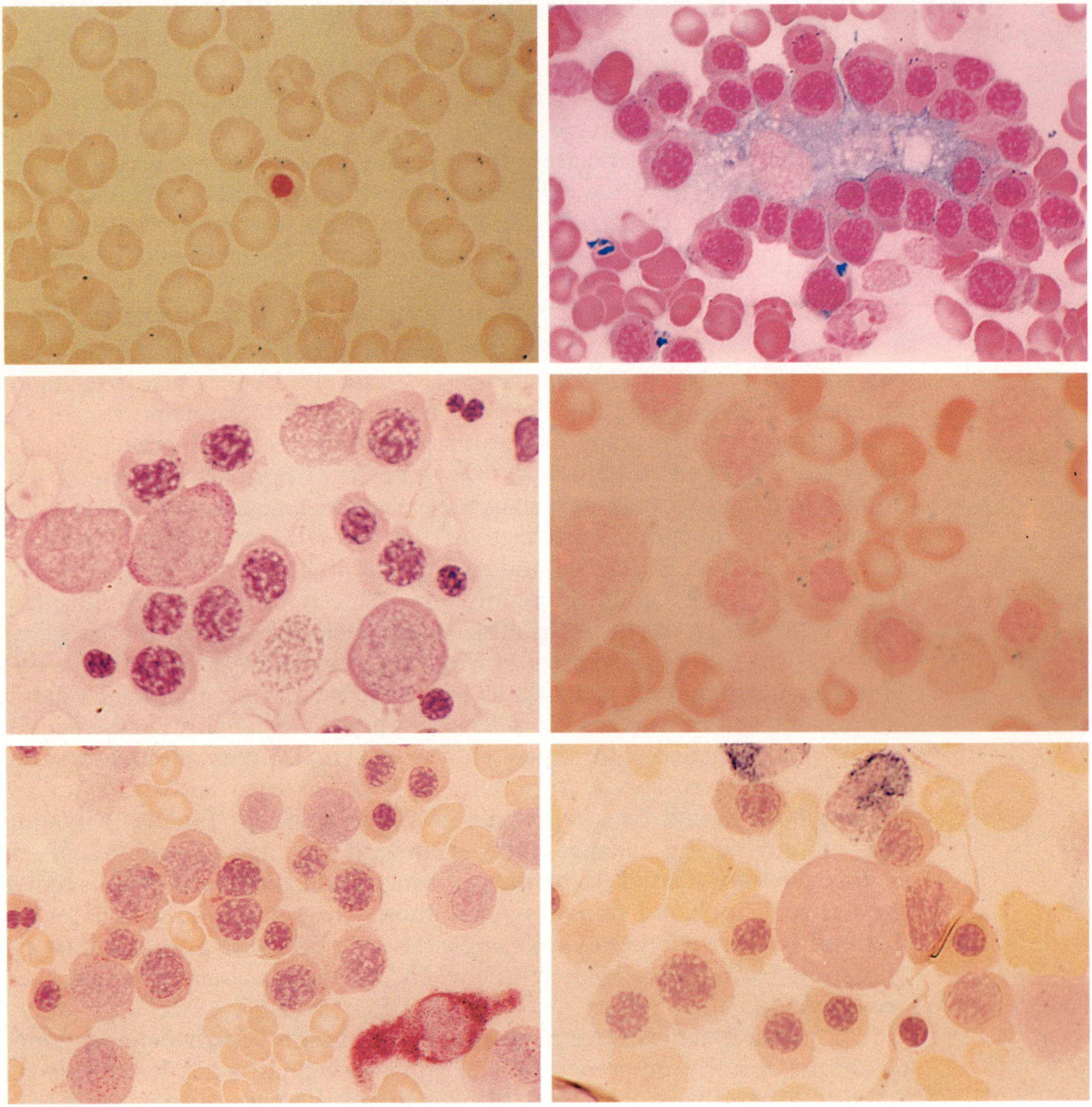

A B C D E F

Plate 35

A. Peripheral blood from pyruvate kinase (PK) deficiency haemolytic anaemia. Prussian blue stain showing a sideroblast and numerous siderocytes.

B. Bone marrow from PK deficiency anaemia. Prussian blue stain showing a central strongly positive RE cell with a group of surrounding normoblasts some of which show siderotic granules.

C. Congenital dyserythropoietic anaemia (CDA) Type 1 bone marrow. PAS reaction. Some weak diffuse and finely granular PAS positivity in erythroblasts, more prominent in the earlier stages.

D. The same case. Prussian blue reaction, showing some increase in free iron in erythroblasts but no ringed forms.

E. The same case. Acid phosphatase reaction. Strong reaction in an RE cell but erythroblasts are negative.

F. The same case. Dual esterase reaction. Normal CE positivity in neutrophils while all erythroblasts are both CE and BE negative.

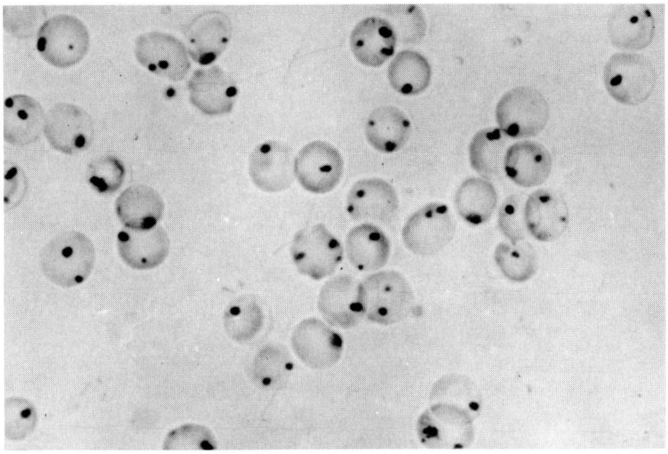

Fig. 13.2 Heinz bodies in normal red cells exposed to phenyl hydrazine.

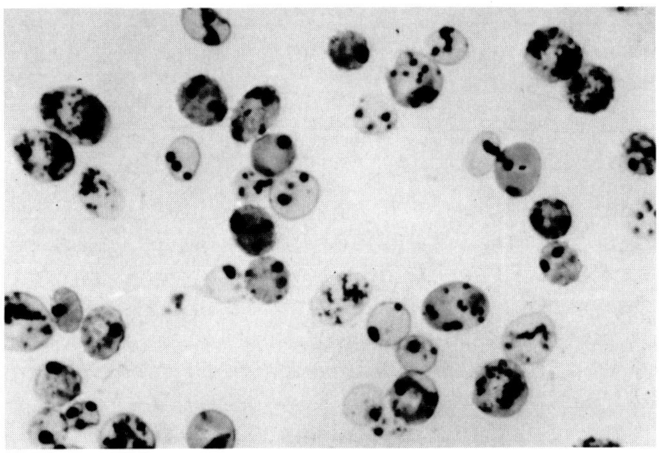

Fig. 13.3 Heinz bodies in G-6-PD deficiency. Exaggerated response to phenyl hydrazine.

of acid phosphatase in contrast to erythroleukaemia (Plate 35E) and an absence of both butyrate and chloroacetate esterases (Plate 35F). Abnormalities of cell proliferation, detectable by microdensitometry and autoradiography, have been studied in three different forms of the disease. In CDA type I the major abnormality found by Queisser *et al.* (1971b) was a severe reduction in the proportion of mono- and bi-nucleate polychromatic and orthochromatic erythroblasts taking up a DNA label. In type II a somewhat similar defect in DNA synthesis was found, also most prominently seen at the polychromatic erythroblast stage (Queisser *et al.*, 1971b). In type III CDA, again, ineffective erythropoiesis and the emergence of giant mononucleate and bi- and multi-nucleate erythroblasts with high death rates appears to follow an arrest of DNA synthesis in normal-

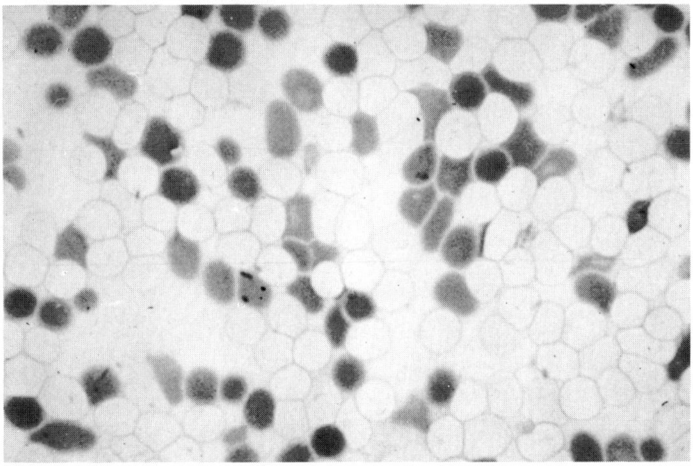

Fig. 13.4 Fetal erythrocytes reacting positively by the Kleihauer method, amongst negatively reacting maternal erythrocytes.

sized early polychromatic and orthochromatic erythroblasts (Wickramasinghe and Goudsmit, 1979).

Non-neoplastic leucocytoses

Mention has already been made in Chapter 7 (p. 130) of the increased levels of alkaline phosphatase activity in inflammatory leucocytosis, observed with both the Gomori-Takamatsu and azo-dye methods. However the high levels of leucocyte alkaline phosphatase in the neutrophils of inflammatory states are not directly related to the appearance of immature stab cells, since these have lower phosphatase ratings than do mature segmented cells (Hayhoe and Quaglino, 1958).

Increased levels of alkaline phosphatase activity have also been reported in pregnancy, a condition which is well known to be associated with an increase in the white cell count in the peripheral blood and principally in granulocytes and monocytes (Fleming, 1975). The increase in alkaline phosphatase activity may be the result of hormonal influence since differences in the activity of the enzyme have been reported in man and woman (Rosner and Lee, 1965; Ray and Pinkerton, 1969), and in the latter during ovulation (Polishuk *et al.,* 1970; Elder *et al.,* 1971) and under the influence of oestrogens (Rosner and Lee, 1965).

Neutrophil peroxidase activity has also been reported to be increased during pregnancy (Sulowicz *et al.,* 1981). The activity was highest before delivery and gradually fell during the puerperium. Also the increase in peroxidase activity appears to be hormone dependent. Jacobs *et al.* (1973) reported that oestrogens stimulate the bactericidal activity of myeloperoxidase-containing leucocytic fractions. However, despite the activation of the neutrophil system, in pregnancy there is an increased risk of infection, especially of the urinary tract.

Whereas in pregnancy there is a parallel increase in both alkaline phosphatase and peroxidase activity, an inverse relation between peroxidase and alkaline phosphatase levels may be often seen in CML, where peroxidase is high and alkaline phosphatase low, and in some leukaemoid reactions, secondary to neoplastic states, in Hodgkin's disease and in leucocytosis of infection (see also Chapter 10). In all these conditions peroxidase positivity and sudanophilia of polymorphs do not run strictly in parallel, since, despite their low peroxidase levels, granulocytes in infective states usually have normal or high sudanophilia (Quaglino, 1961).

The problem of the variations in myeloperoxidase activity in inflammatory states has also been tackled employing automated differential leucocyte counting devices, which identify neutrophils on the basis of their size and peroxidase staining. By this method, Peacock *et al.* (1982) investigated the correlation between the presence of high peroxidase (HPX) neutrophils and that of circulating band neutrophils, employing leucocytes harvested during haemodialysis. In comparing predialysis (control) samples with 60-minute samples (the latter containing a band neutrophil subpopulation enriched by 700%), no predictable increase in HPX cell concentration was observed. In fact, the majority of patients exhibited decreased concentration of HPX neutrophils in samples which contained markedly increased numbers of band forms.

The authors also evaluated the myeloperoxidase concentration of the predialysis and 60-minute samples and found that the mean enzyme content of the neutrophil populations was not significantly different. It would thus appear that band neutrophils (presumably recruited from the marginal leucocyte pool) are not identifiable as HPX cells, nor do they exhibit increased MPO activity. On the contrary, a positive correlation was found between an increased concentration of HPX cells and the presence of toxic neutrophils. The latter, according to the studies of McCall *et al.* (1969, 1973), are characterized by the presence of toxic granules, Dohle bodies, cytoplasmic vacuolization, increased affinity for Romanowsky dyes, increased lysosomal activity and increased oxidative metabolism.

However, when trying to relate the data provided by automated differential analyzers with the results of conventional cytochemical studies one has to bear in mind that, according to Peacock *et al.* (1982), an increase in HPX cells, as revealed by automated devices, does not correspond to an increase of cellular myeloperoxidase concentration, but seems to be rather attributable to an alteration in enzyme distribution within the neutrophil. In any case, the suggestions put forward disclose a new field of investigation, which in due course may explain the ample variations in peroxidase activity observed in different physiological and pathological conditions.

Lysosomal storage diseases

The lysosomal storage diseases include most of the lipid storage disorders, the mucopolysaccharidoses, the mucolipidoses, glycoprotein storage diseases and many others.

The concept of lysosomal storage diseases arose from the studies of type

II (Pompe's) glycogen storage disease. The demonstration of lysosomal accumulation of glycogen as the result of α-glucosidase deficiency and data from other disorders led Hers and Van Hoof (1973) to define an inborn lysosomal disease as one in which a single lysosomal enzyme is deficient and abnormal deposits of substrate lie within vacuoles related to lysosomes. This definition was later modified to encompass single gene defects, affecting one or more lysosomal enzymes and thus include disorders such as mucolipidoses and multiple sulphatase deficiency.

The role of cytochemical techniques in lysosomal storage diseases appears to be twofold: they may in fact be employed both for demonstrating specific enzyme deficiencies, and for defining more precisely the nature of the material stored.

With regard to specific enzyme deficiencies, reduced activity of acid α-naphthyl acetate esterase is cytochemically demonstrable in cholesteryl ester storage disease and in Wolman's disease. In the latter, according to the data of Lake (1971), only 0–14% of lymphocytes (presumably T) display the typical dot-like acid esterase positivity, in contrast with 70–96% of positive lymphocytes in control subjects.

In G_{M1} gangliosidosis the deficiency of β-galactosidase activity may be detected cytochemically employing the azoindoxyl procedure of Gossrau (1977), which normally provides good lysosomal localization.

Absence of β-glucuronidase activity is observed in mucopolysaccharidosis type VII (Peterson *et al.*, 1982).

In ceroid lipofuscinosis (Kuf's disease) an abnormal proportion of atypical insoluble peroxidase appears to be responsible for apparent enzyme deficiency in neutrophils and eosinophils (Bozdek et al., 1980).

A considerable number of cytochemical investigations have been devoted to the study of stored materials (Hann *et al.*, 1983). For instance, in the infantile, type A form of Niemann-Pick disease, characterized by sphingomyelin lipidosis, the typical foamy cells stain weakly with Sudan black, and may often also show a red birefringence in polarized light.

In the B/E adult type of Niemann-Pick disease, foamy cells are few,while sea-blue histiocytes predominate. The latter cells are packed with blue-staining granules which are also PAS positive, are stained by Sudan black and show marked autofluorescence.

In ophthalmoplegic lipidosis (also known as Niemann-Pick disease, types C or D), the foamy cells are usually unstained with Sudan black, although some may show a greyish granular cytoplasm. There is no birefringence in polarized light. The cells are PAS positive and exhibit strong acid phosphatase activity.

In Gaucher's disease, there is no difference in the morphology of the cells from infants, juveniles or adults with the disease, although the abnormal cells are much more numerous in the infantile form. Gaucher cells are often multinuclear and their striped cytoplasm is well demonstrated by both the PAS and Sudan black stains. The cells also show strong acid phosphatase activity, occasional cytoplasmic free iron, and strong acetate or butyrate esterase positivity (Plate 36). In Wolman's disease and cholesteryl ester storage disease the large foamy histiocytes, in contrast with other storage

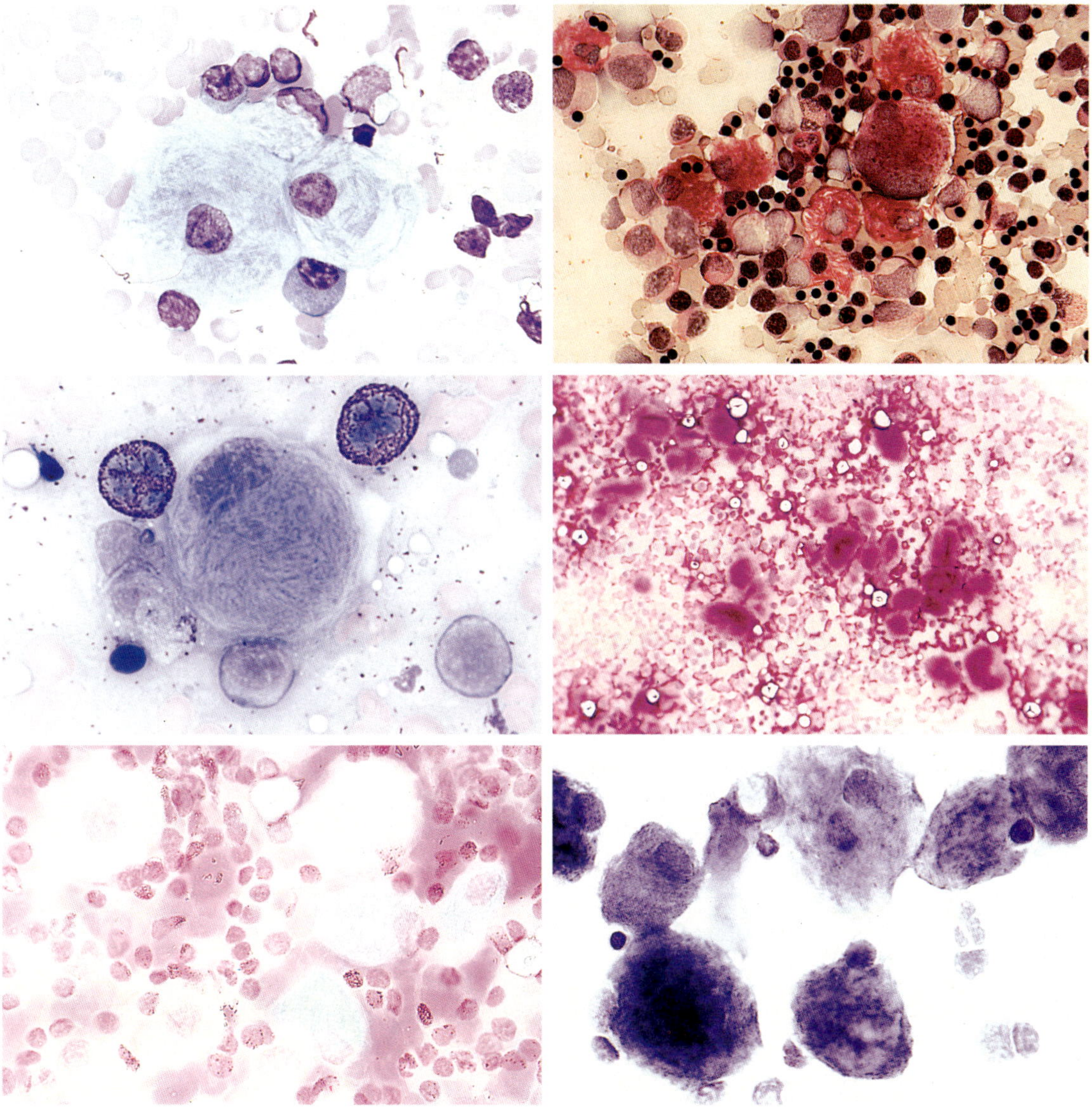

Plate 36

A. Spleen imprint in Gaucher's disease. Leishman stain, showing two typical lipid-laden macrophages.

B. Bone marrow aspirate in Gaucher's disease. PAS reaction. Strong positivity in several Gaucher cells and normal PAS reaction in a megakaryocyte and other haemic cells.

C. Spleen imprint in Gaucher's disease. SB stain. Two neutrophils are normally positive but the Gaucher cell is SB negative, although the laminated cytoplasmic structure is well shown by the counterstain.

D. The same case. Acid phosphatase stain. The numerous Gaucher cells in this low power view show strong positivity.

E. The same case. Prussian blue stain. Some Gaucher cells are negative for free iron, but others show a weak diffuse positive reaction.

F. The same case. Butyrate esterase reaction, showing intense positivity in all Gaucher cells.

disorders, contain large amounts of sudanophilic material, including free fatty acids, in addition to neutral fat, since they stain deep blue/purple with Nile blue stain.

In Pompe's disease (glycogen storage disease II) lymphocytes show small discrete vacuoles, containing PAS positive, water soluble, glycogen. This appearance is particularly prominent and frequent in the infantile form, while in adults fewer lymphocytes display these morphological and cytochemical features.

In G_{M1} gangliosidosis type II, scattered storage-cells with a sky-blue cytoplasm may be observed in the bone marrow. These cells superficially resemble Gaucher cells, but they may be distinguished by their characteristic blue colour, as well as by the deficiency of β-galactosidase activity.

In all types of mucopolysaccharidosis, except type IV (Morquio syndrome), lymphocytes, when stained with toluidine blue, show the presence of pink metachromatic inclusions. The proportion of lymphocytes affected is highest (more than 20%) in the Sanfilippo syndrome (type III), while in Hurler (type I H) and Hunter (type II) syndromes less than 20% lymphocytes contain inclusions. In the Sanfilippo syndrome, only few storage cells may be seen, but occasional scattered histiocytes may be found, containing basophilic granules which stain metachromatically with toluidine blue.

In mucopolysaccharidosis VI (Maroteaux-Lamy syndrome) all neutrophils show Alder granules, which differ from toxic granulations because of their metachromasia and birefringence. In the bone marrow, storage cells, when stained with toluidine blue, show a very striking densely-basophilic granular cytoplasm.

In sialidosis I, foamy cells with bluish-staining vacuoles are present, but rare. They stain intensely with the PAS reaction and show strong acid phosphatase activity at the periphery of the vacuoles. Finally in cystinosis, histiocytes, in preparations stained with alcoholic basic fuchsin, show the presence of water-soluble cystine crystals. Also in wet preparations of anticoagulated bone marrow aspirate, histiocytes show, when viewed in polarized light, cystine crystals with their characteristic brick shape.

REFERENCES

Ackerman G A 1952 A modification of the Sudan black B technique for the possible cytochemical demonstration of masked lipids. Science 115: 629

Astaldi G, Rondanelli E G, Bernardelli E, Strosselli E 1954 An abnormal substance present in the erythroblasts of thalassaemia major. Cytochemical investigations. Acta Haematologica 12: 145

Björkman S E 1956 Chronic refractory anemia with sideroblastic bone marrow. A study of four cases. Blood 11: 250

Bozdek M J, Bainton D F, Mustacchi P 1980 Partial myeloperoxidase deficiency in neutrophils and eosinophils associated with neurologic disease. American Journal of Clinical Pathology 73: 409

Bush H 1965 Histones and other nuclear proteins. Academic Press, New York

Dacie J V, Mollin D L 1966 Siderocytes, sideroblasts and sideroblastic anaemia. Acta Medica Scandinavica (Suppl) 445: 237

Dormer P 1973 Kinetics of erythropoietic cell proliferation in normal and anemic man. A new approach using quantitative ^{14}C-autoradiography. Progress in Histochemistry and Cytochemistry 6: 1

Douglas A S, Dacie J V 1953 The incidence and significance of iron containing granules in human erythrocytes and their precursors. Journal of Clinical Pathology 6: 307
Elder M G, Bonello F, Elkil J 1971 Neutrophil alkaline phosphatase in pregnancy and its relationship to urinary estrogen excretion and serum heat stable alkaline phosphatase levels. American Journal of Obstetrics and Gynecology 111: 319
Ellis L D, Jensen W N, Westerman M P 1964 Needle biopsy of bone and marrow. An experience with 1445 biopsies. Archives of Internal Medicine 114: 213
Fessas Ph 1963 Inclusions of hemoglobin in erythroblasts and erythrocytes of thalassemia. Blood 21: 21
Fessas Ph, Papayannopoulou T 1965 Cytochemical observations on beta-thalassaemia. I. The PAS positive substance of erythroblasts. Acta Haematologica 34: 1
Fleming A 1975 Haematological changes in pregnancy. Clinics in Obstertrics and Gynaecology 2: 269
Gordon H W, Nurnberger J I 1955 Acute cold-induced changes in cellular protein and nucleic acids in rat liver and brain. A description and critique of biochemical and cytochemical stains. Experimental Cell Research 8: 279
Gossrau R 1977 Azoindoxylverfahren zum Hydrolasennachweis. II. Biochemische und histochemische Untersuchung der sauren β-galactosidase. Histochemistry 51: 219
Greig H B W, Metz J 1957 The periodic acid-Schiff reaction as a diagnostic aid in thalassaemia. South African Journal of Medical Science 22: 7
Hann I A, Rankin A, Lake B D, Pritchard J 1983 Colour atlas of paediatric haematology. Oxford University Press, Oxford
Hayhoe F G J, Quaglino D 1958 Cytochemical demonstration and measurement of leucocyte alkaline phosphatase activity in normal and pathological states by a modified azo-dye coupling technique. British Journal of Haematology 4: 375
Hers H G, Van Hoof F (eds) 1973 Lysosomes and storage diseases. Academic Press, New York
Hertl M 1959 Zytochemische Untersuchungen an Knochenmarkzellen. Folia Haematologica (NF) 3: 269
Jacobs A A , Selvaraj R J, Straus R R, Paul B B, Mitchel G W, Sbarra A J Jr 1973 The role of the phagocyte in host-parasite interactions. XXXIX. Stimulation of bactericidal activity of myeloperoxidase-containing leukocytic fractions by estrogens. American Journal of Obstetrics and Gynecology 117: 671
Kaplan E, Zuelzer W W, Mouriquand C 1954 Sideroblasts: a study of stainable non-hemoglobin iron in marrow normoblasts Blood 9: 203
Kass L 1972 Demonstration of histones in proerythroblasts in pernicious anemia and the Di Guglielmo syndrome. Journal of Histochemistry and Cytochemistry 20: 817
Kass L 1979 Bone marrow interpretation. J B Lippincott Company, Philadelphia
Kesse-Elias M, Yataganas X, Fessas Ph 1970 Erythropoiesis in thalassaemia: a cytochemical and autoradiographic study. Proc. XIIIth Congr. Int. Soc. Haemat., Munich
Kleihauer E, Betke K 1963 Die farberische Darstellung von ihre praktischen Anwendungsgebiete. In: Merker H (ed) Zyto- und Histochemie in der Hämatologie. Springer, Berlin, p 160
Lajtha L G, Kumatori T 1957 Nucleic acid metabolism in megaloblastic marrows in vitro. Nature 180: 991
Lake B D 1971 Histochemical detection of enzyme deficiency in blood films in Wolman's disease. Journal of Clinical Pathology 24: 617
MacGibbon B H, Mollin D L 1965 Sideroblastic anaemia in man: observations on seventy cases. British Journal of Haematology 11: 59
Marinone G 1953 Gli Acidi Nucleici in Ematologia. Biblioteca Haematologica, Pavia
Mattioli G, Del Piano E 1961 Su due casi di talassemia minima a monoespressivita genica. Haematologica 46: 83
Mauri C 1961 Die DNS-Synthese in den Knochenmarkererythroblaste; autoradiographische Untersuchungen mittels ^{3}H-Thymidin. Folia Haematologica (NF) 6: 239
Mauri C 1963 Über Lipide und ihre Bedeutung in den Zellen der blutbildenden Systeme. In: Merker H (ed) Zyto-und Histochemie in der Hamatologie. Springer, Berlin, p 433
Mauri C, Silingardi V 1963 Histochemical localisation of lipids in the nuclei of human blood and hemopoietic cells. Experientia 19: 622
Mauri C, Vaccari F, Sabotto B 1954 Ricerche istochimiche sui gruppi tiolici delle cellule del sangue e degli organi ematopoietici dell'uomo normale. Medicina Clinica e Sperimentale 4: 321
Mauri C, Torelli U, Grossi G 1960 Studi sulla dinamica della sintesi dell'ADN nelle cellule ematiche. II. L'incorporazione della timidina marcata con tritio nei megaloblasti dell'anemia perniciosa biermeriana. Bollettino della Societa di Medicina e Chirurgia di Modena 60: 703
McCall C E, Katayama I, Cotran R A, Finland M 1969 Lysosomal and ultrastructural changes in human toxic neutrophils during bacterial infection. Journal of Experimental Medicine 129: 267
McCall C E, De Chatelet L R, Cooper M R, Shannon C 1973 Human toxic neutrophils. III.

Metabolic characteristics. Journal of Infectious Diseases 127: 26
Menzies R C, Crossen P E, Fitzgerald P H, Gunz F W 1966 Cytogenetic and cytochemical studies on marrow cells in B_{12} and folate deficiency. Blood 28: 581
Metz J, Kelly A, Swett V C, Waxman S, Herbert V 1968 Deranged DNA synthesis by bone-marrow from vitamin B_{12}-deficient humans. British Journal of Haematology 14: 575
Nixon P F, Bertino J R 1972 Interrelationships of vitamin B_{12} and folate in man. Journal of Clinical Investigation 51: 1431
Nixon R K, Olson P J 1968 Diagnostic value of marrow haemosiderin patterns. Annals of Internal Medicine 69: 1249
Peacock J E, Ross D W, Cohen M S 1982 Automated cytochemical staining and inflammation. Further assessment of the left shift. American Journal of Clinical Pathology 78: 445
Perugini S, Soldati M 1955 Modificazioni citochimiche degli eritroblasti e degli eritrociti in condizioni patologiche. Proc. XIII Congr. Soc. It. Emat., Emes, Roma, p. 397
Peterson L, Parkin J, Nelson A 1982 Mucopolysaccharidosis type VII. A morphologic, cytochemical and ultrastructural study of the blood and bone marrow. American Journal of Clinical Pathology 78: 544
Polishuk W Z, Diamant Y Z, Zuckerman H, Sadovsky E 1970 Leukocyte alkaline phosphatase in pregnancy and the puerperium. American Journal of Obstetrics and Gynecology 107: 604
Polliack A, Rachmilewitz E A 1973 Ultrastructural studies in beta-thalassaemia major. British Journal of Haematology 24: 319
Polliack A, Yataganas X, Thorell B, Rachmilewitz E A 1974 An electron microscopic study of the nuclear abnormalities in erythroblasts in beta-thalassaemia major. British Journal of Haematology 26: 201
Quaglino D 1961 Cytochemical methods in haematology. PhD Thesis, Cambridge University
Quaglino D 1964 Alkaline phosphatase positivity in erythroblasts. Haematologica Latina 7: 263
Quaglino D, Hayhoe F G J 1960 Periodic acid-Schiff positivity in erythroblasts with special reference to Di Guglielmo's disease. British Journal of Haematology 6: 26
Queisser W, Betzler M, Heimpel H, Kleihauer E 1971a Erythropoietic cell proliferation in different clinical states of beta-thalassaemia. Acta Haematologica 45: 303
Queisser W, Spiertz E, Jost E, Heimpel H 1971b Proliferation disturbances of erythroblasts in congenital dyserythropoietic anaemia type I and II. Acta Haematologica 45: 65
Ray P K, Pinkerton P H 1969 Leukocyte alkaline phosphatase. The effect of age and sex. Acta Haematologica 42: 18
Reisner E H 1958 The nature and significance of megaloblastic blood formation. Blood 13: 313
Rondanelli E G, Strosselli E, Gorini P 1955 Anomalia citochimica degli eritroblasti in casi di leucemia mieloide. Progresso Medico 11: 230
Rosner F, Lee S L 1965 Endocrine relationships of leukocyte alkaline phosphatase. Blood 25: 356
Rozenszajn L, Leibovich M, Shoham D, Epstein J 1968 The esterase activity in megaloblasts, leukaemic and normal haemopoietic cells. British Journal of Haematology 14: 605
Sansone G, Bertolotti E 1958 Ricerche citochimiche sugli eritroblasti di prematuri anemici. Atti VII Congr. Soc. Int. Emat., vol. 2. Pensiero Scientifico, Roma
Schmid J R, Moeschlin S, Haegi V 1964 Pernicious anaemia: an erythrokinetic and autoradiographic study, using ^{3}H-thymidine, ^{3}H-uridine and ^{3}H-cytidine. Acta Haematologica 32: 65
Stegagno G A and Pollitzer C 1953 Ricerche citochimiche sui polisaccaridi e lipidi delle cellule ematiche e midollari di bambini affetti da Talassemia major (morbo di Cooley). Archivio Italiano di Pediatria e Puericoltura 16: 247
Stohlman F Jr, Howard D, Beland A 1963 Humoral regulation of erythropoiesis. XIII. Effect of erythropoietin and iron on cell size in iron deficiency. Proceedings of the Society for Experimental Biology and Medicine 113: 986
Storti E, Torelli U 1965 Blood cell studies with Radioautography. In: Szirmai E (ed) Nuclear Hematology. Academic Press, New York, p 171
Stuart J, Skowron P N 1968 A cytochemical study of marrow enzymes in megaloblastic anaemia. British Journal of Haematology 15: 443
Sulowicz W, Strzalka H, Jakobiec M, Dawidiuk W, Szczepaniec M 1981 Neutrophil peroxidase activity in women during the perinatal period. Acta Haematologica 65: 122
Thorell B 1947 Studies on the formation of cellular substances during blood cell production. Acta Medica Scandinavica (Suppl) 200
Tönz O, Rossi E 1964 Morphological demonstration of two red cell populations in human female heterozygous for glucose-6-phosphate dehydrogenase deficiency. Nature 202: 606
Van Noorden C J F, Vogels I M C 1985 A sensitive cytochemical staining method for glucose-6-phosphate dehydrogenase activity in individual erythrocytes. British Journal of Haematology 60: 57
Waxman S, Metz J, Herbert Y 1969 Defective DNA synthesis in human megaloblastic bone-

marrow. Effects of homocysteine and methionine. Journal of Clinical Investigation 48: 284
Wickramasinghe S N 1970 RNA content of megaloblasts in pernicious anaemia. Scandinavian Journal of Haematology 7: 398
Wickramasinghe S N 1975 Human Bone Marrow. Blackwell Scientific Publications, Oxford
Wickramasinghe S N, Carmel R 1972 Uptake of ^{57}Co vitamin B_{12} by human haemopoietic cells 'in vitro'. British Journal of Haematology 23: 307
Wickramasinghe S N, Goudsmit R 1979 Some aspects of the biology of multinucleate and giant mononucleate erythroblasts in a patient with CDA, type III. British Journal of Haematology 41: 485
Wickramasinghe S N, Chalmers D G, Cooper E H 1967 Disturbed proliferation of erythropoietic cells in pernicious anaemia. Nature 215: 189
Wickramasinghe S N, Cooper E H, Chalmers D G 1968 A study of erythropoiesis by combined morphologic, quantitative, cytochemical and autoradiographic methods. Normal human bone marrow, vitamin B_{12}deficiency and iron deficiency anaemia. Blood 31: 304
Wickramasinghe S N, Chalmers D G, Cooper E H 1969 Arrest of cell proliferation and protein synthesis in megaloblasts of pernicious anaemia. Acta Haematologica 41: 65
Wickramasinghe S N, McElwain T J, Cooper E H, Hardisty R M 1970 Proliferation of erythroblasts in beta-thalassaemia. British Journal of Haematology 19: 719
Wickramasinghe S N, Letsky E, Moffat B 1973 Effect of alpha-chain precipitates on bone marrow function in homozygous beta-thalassaemia. British Journal of Haematology 25: 123
Wintrobe M M 1981 Clinical hematology (8th edn). Lea and Febiger, Philadelphia
Woessner S, Lafuente R, Florensa L, Sans-Sabrafen J 1984 Cytochemical detection of erythroblastic enzymes in acquired dyserythropoiesis. Acta Haematologica 72: 303
Yataganas X, Fessas Ph 1969 The pattern of hemoglobin precipitation in thalassaemia and its significance. Annals of the New York Academy of Sciences 165: 270
Yataganas X, Gahrton G, Fessas Ph, Kesse-Elias M, Thorell B 1973 Proliferative activity and glycogen accumulation of erythroblasts in beta-thalassaemia. British Journal of Haematology 24: 651
Yoshida Y, Todo A, Shirakawa S, Wakisaka G, Uchino H 1968 Proliferation of megaloblasts in pernicious anaemia as observed from nucleic acid metabolism. Blood 31: 292

CHAPTER 14

Major cytochemical findings in leukaemias, preleukaemic myelodysplasias and related states

In earlier chapters we have described the general distribution and intensity of cytochemical reactions of many kinds in normal haemic cells and the striking departures from normal which sometimes occur in pathological states. In this chapter are brought together the main cytochemical features of the leukaemias and related states with emphasis on those findings or combinations of findings that have particular value as aids to differential diagnosis.

The leukaemias and preleukaemic states were generally classified on primarily cytomorphological criteria, with some additional cytochemical discrimination, from the early nineteen sixties (Hayhoe *et al.*, 1964) until the mid-seventies (Mathé and Rappaport, 1976). The commonly agreed cytological subgroups among the acute leukaemias at that time were codified and given numerals in the French-American-British (FAB) classification (Bennett *et al.*, 1976), while the same authors later offered a classification of the preleukaemic or myelodysplastic states (Bennett *et al.*, 1982) and at the same time revised the earlier criteria for acute leukaemia grouping. Additional revisions of the FAB groups of ALL were presented in 1981 and of AML in 1985. Essentially these revisions do no more than refine the basic cytological criteria, with minimal cytochemical support, for the respective lymphoblastic groups L1, L2 and L3, and the myeloid groups M1 to M6. The main features of the FAB groupings are incorporated in Table 14.1, where they are set in the context of a more general classification scheme, less precise in cytomorphological criteria of distinction between groups, but giving greater weight to newer discriminative criteria, especially surface antigenic and biochemical markers of cell lineage and stages of maturity, and the emerging cytogenetic groupings.

The development of immunocytochemical techniques, including the use of fluorescent dyes and gold particles as labels for antibodies, and the refinement of immunoperoxidase and immuno-alkaline phosphatase methods, has brought cell surface immunology within the purview of the haematological cytochemist. Avidin-biotin immunoperoxidase methods (Fishleder *et al.*, 1984), and combinations of immunogold and myeloperoxidase techniques at ultrastructural level (Polli *et al.*, 1984) exemplify this advancing immunocytochemical approach. The principles and the main technical applications relevant to haematology have been described in Chapter 3, where the important impact of monoclonal antibody (McAb) technology together with these immunocytochemical techniques in the study of haemic cell maturation and differentiation is briefly summarized.

The use of McAbs in characterizing leukaemic cells has expanded dramatically in the past decade; among innumerable reports in this research field especially relevant reviews so far as the acute leukaemias are concerned include those of Janossy (1981) which analyses much of the earlier literature, Janossy and Campana (1985) and Chan *et al.* (1985), which provide more recent assessments of the availability and specificity of McAbs with the best discriminative capacity among the varieties of acute leukaemia. A broad immunological phenotyping panel of McAbs would include examples such as the following, with the specification described, for a first diagnostic round on an acute leukaemia.

1. Stem cell: CD34 (3C5, My10), Class II anti-HLA-DR.
2. Myeloid: mixed—CD13 (My7), CD11 (OKM1)
   granulocyte—CD15 (VIM-D5, FMC10)
   monocyte—CD14 (MO2, UCHM1)
3. Pre-B and cALL: CD19 (B4, Leu-12)
   CD10 (J5, cALL)
4. T-cell ALL: CD7 (3A1, WT1, Leu-9)
5. Erythroid: anti-glycophorin A
6. Megakaryocytic: CDw41 (J15, GPIIb/IIIa)

To these would normally be added anti-SIg reagents and a test for terminal deoxynucleotidyl transferase, TdT.

A second diagnostic round may be used to subdivide further the ALLs, although lack of specificity among myeloid McAbs and the frequent existence of mixed cytology has continued so far to hinder more precise immunological classifications in the myeloid group. Among ALLs, however, several antigens allow subdivision, for example, in T-ALLs, CD3 (OKT3, UCHT1, Leu 4), CD6 (T12, RFT12, MBG6), CD4 (OKT4, Leu 3a), CD8 (OKT8, RFT8, Leu 2), CD1 (OKT6, NA 1/34) and CD25 (TAC, anti-IL-2 receptor). The increasingly complex surface antigen phenotypes distinguishable among the ALLs have already exceeded clinical requirements in terms of prognostic significance, but the broad groupings separated by the first round reagents have some clinical importance, and have frequent parallels with enzyme markers and cytochemical findings, although they are less clearly related to the detailed cytomorphological subdivisions of the FAB classification.

Thus, in a study of 61 cases of acute leukaemia classified by morphological means into the FAB groupings and also screened by monoclonal antibodies (McAbs) for lineage specific surface markers, Avnstrøm *et al.* (1985) found concordance only in the separation of lymphoblastic from myeloid cases. There was no apparent correlation between c-ALL, other pre-B-ALL, T-ALL and the FAB types L1 and L2, while FAB L3 morphology did not correlate with B-cell ALL. Among the acute myeloid cases no correlation was found between FAB subtypes and the expression of myeloid antigens SA1, MY9 and MO2. These findings are typical of many, most authors showing an absence of correlation for the lymphoblastic subsets severally defined (van Eys *et al.*, 1986), and broad overlapping between the surface antigen expression in different myeloid subgroups, with correlation only detectable with difficulty or by computer analysis (Reijden *et al.*, 1983). This is perhaps not surprising since mixed lineage cytology among the myeloid leukaemias is so common.

Again, the biochemical enzyme markers and their isoenzyme patterns available for discrimination of acute leukaemias, notably acid phosphatase (AcP), adenosine deaminase (ADA), carboxyl esterase (Est), β-glucosaminidase (hexosaminidase (Hex)), 5′-nucleotidase (5′-N),purine-nucleoside phosphorylase (PNP) and terminal deoxynucleotidyl transferase (TdT), generally show quantitative or qualitative differences of diagnostic value relating closely to immunological analyses rather than to FAB subtypes, among cases of ALL. In AML enzyme markers reflect mixed myeloid populations and in particular help to detect substantial components of monocytic lineage through, for example, characteristic Est isoenzyme patterns and high levels of Hex or PNP (Drexler *et al.*, 1984).

Terminal deoxynucleotidyl transferase (TdT), the nuclear enzyme able to catalyse the polymerisation of deoxynucleotides in the absence of a template, is measurable biochemically and detectable in individual cells immunocytochemically. It was thought at one time to be present in lymphoid but not in myeloid precursors, and so to provide a simple biochemical distinction between ALL and AML (McCaffrey *et al.*, 1975b; Coleman and Hutton, 1981). This is no longer a tenable view; some 10 to 20% of cases with AML cytology, cytochemistry and immunological marker reactions show TdT positivity, and among ALLs B-cell cases are almost always negative. The extensive literature is reviewed by Drexler *et al.* (1984) who share the conclusion of Hutton *et al.* (1982) that 'measurement of TdT alone is not helpful in differential diagnosis and does not seem to be able to guide therapy'. Nevertheless, in combination with other markers, TdT may be useful. Heumann *et al.* (1985), for example, used combined measurements of TdT and plasma membrane-associated γ-glutamyl transpeptidase to characterise undifferentiated acute leukaemias (AUL). In 37 AMLs only 4 had less than 5 units of γ-GT and in 37 ALLs none had more than 5 units. In 16 cases of AUL there were 8 with high TdT levels and 8 with normal levels; 6 of the 8 with high TdT levels were nearly devoid of γ-GT, showed completely sudanophobic blasts and were presumably ALL cases; 6 of the 8 with low TdT levels had high γ-GT values; 4 of them showed 3 to 5% of Sudan black positive blasts and were presumably AML cases. Two cases had no increase in either TdT or γ-GT and had Sudan black negative blasts and 2 had increases in both, with 2.5% Sudan black positive blasts.

While combined studies of this kind may allow more precise allocation of poorly differentiated cases to lymphoid or myeloid classes, there remains a minority of acute leukaemias with equivocal findings. We have referred in Chapter 5 to the possible significance of sudanophilia in blast cells with cALL markers and positive TdT reactions. Cuttner *et al.* (1984) described

13 acute leukaemias with mature or early monocyte lineage antigens, intense Ia (HLA-DR) expression, TdT activity and monocytic cytochemistry, including characteristic Sudan black and PAS reactions and monocytic esterase. The authors thought these cases might represent a malignant proliferation arising at a less mature stage in monocyte development than usual in either monocytic or myelomonocytic leukaemias. Lanham *et al.* (1984) reported 3 cases of acute leukaemia with respectively 55, 57 and 83% of TdT positive cells and 89, 25 and 10% of peroxidase positive cells, and with clear dual positivity in 8, 1 and 23% of cells, yet with a single cytogenetic leukaemic clone in each case. The question of mixed acute leukaemias, with involvement of both lymphoid and myeloid lineages, and the possibility of dual phenotypic expression in the same cell appears highly relevant to these anomalous case reports. Janossy and Campana (1985) summarize a collection of some 20 cases with probable 'double' phenotypes and 13 with probable 'mixed' lineages, in the sense of expressing either in the same cells or in different cells of the same case both myeloid markers and TdT. They argue that most of these 'double' or 'mixed' phenotypes probably represent malignancies arising from the human pluripotential stem cell, and suggest that such findings are likely to be encountered more often with the increasing use of double staining methods. A further extensive review of examples of apparent lineage infidelity led Greaves and his associates (1986) to conclude that truly aberrant gene expressions probably occurred with 'insufficient regularity to be of general significance, many reports being suspect on technical grounds'. When true lineage infidelity occurred these authors thought it less likely to be due to genetic misprogramming than to the reflection of a normally transient phase of multiple gene expression occurring in bipotential or multipotential progenitor cells. This view is essentially in agreement with that expressed by Janossy and Campana. Until the cytological and immunological characteristics of the putative pluripotential stem cell stages with multiple gene expression are defined, this allocation must remain speculative, and we have not attempted to separate a category of mixed leukaemias in Table 14.1, although reference is made there to their existence and possible frequency. They do not, of course, fit within the FAB morphological grouping.

Cytogenetically, translocations, inversions and deletions such as t(8;14) in B-ALL, t(1;19) in pre-B ALL, t(8;21) in granulocytic leukaemias either of M1 or M2 FAB subclass, inv(16) or del(16) probably in either M2 or M4 subclasses, t(15;17) in the promyelocytic (M3) subclass, and 3q rearrangements in mixed AML with abnormal thrombopoiesis, all seem to define entities which, though sometimes loosely related to the cytological FAB classes in the myeloid if not the lymphoid group, are not co-extensive with them and sometimes overlap between them.

We have therefore compromised in Table 14.1 by offering a scheme of classification which incorporates findings from all these related disciplines, together with a few salient cytochemical data. Subsequent more detailed consideration of specifically cytochemical features among this group of diseases follows this broad scheme.

For acute leukaemias, a very much simplified classification which never-

Table 14.1 Classification and major cytological features of leukaemias

1. Acute leukaemias

Predominantly primitive cells of stem cell to early differentiated blastic and intermediate forms. Generally reflect a stage of normal early differentiation, within either the myeloid (marrow-derived) series, or the lymphoid series of cells. Mixed myeloid-lymphoid cases either with dual phenotypic expression in the same blast cells, or with a mixture of cell lineages, perhaps arising from a pluripotential stem cell, may comprise from 10 to 20% of all cases. Immature cells of either series react with 3C5 and anti-HLA-DR McAbs and are TdT positive.

(a) *Acute lymphoblastic leukaemias (ALL)*
Cases frequently show at least some and often many cells with coarse Periodic acid-Schiff (PAS) positivity—granules or blocks—against a negative background. Sudan black (SB) and peroxidase reactions are negative. Most cases show TdT positivity.

Null cell (stem cell, non B non T, but mostly pre-B)—react with CD19 McAbs. Ph^1 chromosome t(9;22) positive in about a third of cases, especially in adults. Cases with cIg may show t(1;19).

Common ALL—probably all early B-cell cases, showing cALL surface marker and equivalent CD10 McAb reactivity; show high ADA and 5′N levels.

B-ALL—uncommon cases with surface membrane Ig (SmIg). May show t(8;14). 5′N and PNP levels low. No Hex I.

T-ALL—cells react with CD7 McAbs and possibly CD1,3,4, 6 or 8. Acid phosphatase positivity is often strongly localised. ADA levels high, 5′N and PNP low. No Hex I. May show t(11;14).

Mixed types of ALL occur rarely if at all. Chromosome anomalies such as 8q−, 14q+ and 6q− and hyperdiploidy or pseudodiploidy do not show specific links with ALL subvariants. t(4;11) may occur in null or cALL, and may be associated with monocytic transformation.

FAB groups L1 (small cells, high N/C ratio, round nuclei, inconspicuous nucleoli); L2 (large cells, moderate N/C ratio, irregular nuclei, one or two large nucleoli); and L3 (large cells, moderate N/C ratio, round nuclei, one or two prominent nucleoli, very deep cytoplasmic basophilia and often prominent cytoplasmic vacuoles) correlate poorly with above phenotypes.

(b) *Acute myeloid leukaemia (AML)*

Myeloblastic; myeloblastic myelocytic (AMyL)—localized SB/peroxidase positivity, chloroacetate esterase (CE) +, PAS weak, react with CD13, CD11 and CD15 McAbs. A subset shows t(8;21), another t(6;9) and a third 9q−. Inv(16) may occur, with eosinophilia. FAB groups M1 (without maturation) and M2 (with more than 10% maturing granulocytes).

Promyelocytic (APL)—frequent Auer rods; bilobed nuclei, may have coarse granules. SB/peroxidase positivity strong. CE + +. PAS +. React with CD13, CD11 and CD15. Often show t(15;17). FAB group M3.

Myelomonocytic (AMML)—some twisted nuclei. Mixed localized and scattered SB, mixed CE and butyrate esterase (BE) reactions in the respective granulocyte and monocyte components. Hex level high, but no Hex I band. Show mixed granulocyte/monocyte McAb reactivity. May show 11q−. FAB group M4. A subset with abnormal marrow eosinophils shows Inv(16) or del(16) (FAB group M4e).

Monoblastic (AMonL)—BE reaction usually predominant. SB shows scattered granules. Acid phosphatase may be strong. Hex level high, but no Hex I band. React with CD13, CD11 and CD14 McAbs. PAS ranges from negative to coarse granules. May show 11q− or t(4;11). FAB group M5, a without and b with maturation.

Erythroleukaemia (AEL)—SB/peroxidase negative. PAS usually strongly positive; localized acid phosphatase, moderate BE or CE reactions or mixture. Gross aneuploidy and major karyotypic anomalies common. Cells react with antiglycophorin antibodies. FAB group M6 (if > 50% erythroblasts).

Megakaryocytic (AMegL)—SB negative, platelet peroxidase reaction positive at EM level. PAS usually strongly positive; CE and BE negative but acetate esterase (AE) positive. 5′-N may be positive. Cells react with an anti-platelet antibody (J15 McAb).

Mixed myeloid leukaemias are common; the presence of any PAS positivity in erythroblasts probably indicates involvement of the red cell series and is associated with poorer prognosis. Mixed AML with high platelet counts and/or dysmegakaryocytopoiesis may show inv. or transl. involving 3q21–26. Common chromosome abnormalities, not associated with specific subtypes of AML, include trisomy 8, monosomy 7 or 7q−. Defects of 7 including especially t(1;7) and 5q− are very common in AML developing secondary to earlier treatment with ionising radiation and/or alkylating agents. Some 20% of cases show TdT positivity.

(c) *Blastic transformation* of chronic myeloid leukaemia (CML-Bl)—may show any of the variants listed above except (8;21). Ph^1, t(9;22) common

Simple classification of AL

ALL- Pre-B (null), $CD19^+$; early B (cALL), $CD10^+$; B, $SmIg^+$; and T, $CD7^+$.
Cytogenetic prognostic groups: hyperdiploid—good; diploid—intermediate; structural rearrangements—poor.

AML- *Type Ia*
Granulocyte (G) and/or Monocyte (M) lineage only; well differentiated (includes t(8;21), t(15;17), Inv 16.
Type Ib
G and/or M lineage only; poorly differentiated (<50% cells show cytochemical positivity; includes most t(v;11)).
Type II
Multiple lineage involvement G and/or M + E (dysplastic/PAS + ve erythroblasts) and/or Me (dysplastic megakaryocytes/precursors) (includes most 5 or 7 monosomies or 5q−, 7q−, t(1;7) and trisomy 8, whether in secondary or apparently de novo cases and most 9q−, t(6;9), and 3q rearrangements).

Mixed ALL/AML—includes t(4;11).

2. Preleukaemic states

These conditions tend to transform sooner or later into acute myeloid leukaemias; they are usually associated with pancytopenia in the peripheral blood and with striking cytological dysplasia involving various myeloid cell lines on marrow examination. The picture may show chiefly erythroid dysplasia as in sideroblastic anaemia or Di Guglielmo's syndrome (FAB—Refractory Anaemia (RA) and refractory anaemia with sideroblasts (RAS)), but more commonly shows some left shift in granulocytic and megakaryocytic series, with blast cells increased from around 5 to 20% (FAB—refractory anaemia with excess blasts (RAEB)), or from 20 to 30%, or showing Auer rods (RAEB in transformation). There may be a relative or absolute monocytosis in both blood and marrow, with raised lysozyme production, a preleukaemic variant sometimes called chronic myelomonocytic leukaemia (FAB—CMML). Chromosome anomalies in all these myelodysplastic states are generally similar to those in AML except for the t(8;21), t(15;17) and t(9;22) which are not found. A 5q− refractory anaemia with little excess of blasts but with frequent megakaryocytic anomalies, and male-to-female ratio of 1:5 is especially notable in elderly patients, as are refractory anaemia with excess of blasts (RAEB) or smouldering leukaemia.

3. Chronic leukaemias

Predominance of more mature cells of either myeloid or lymphoid series.

(a) Chronic myeloid leukaemia (CML)—Ph^1 chromosome t(9;22) present in 85% of cases: LAP score very low.
(b) Chronic lymphocytic leukaemia (CLL)—B-cell variant—usual; shows monoclonal SmIg and Fc and C3 receptors.

Table 14.1 Continued

T-cell variant—uncommon; reacts with T-cell monoclonals. Usually suppressor subset.

(c) Hairy-cell leukaemia (HCL)—an uncommon disease with typical 'hairy' mononuclear cells of uncertain origin; they show tartrate-resistant acid phosphatase and are probably B-cell derived.

(d) Prolymphocytic leukaemia (PLL)—a variant of CLL with larger more primitive cells, often nucleolated.
B-cell variant—usual.
T-cell variant—uncommon; perhaps more often of helper than suppressor subset.

(e) Sézary syndrome (SS)—leukaemic phase of mycosis fungoides (MF)—T-cell origin, helper subset.

(f) Adult T-cell leukaemias (ATL)—an endemic disease in South-West Japan with demonstrated viral aetiology. A parallel occurs in Caribbean patients of African origin.

theless takes account of the most important features with biological and prognostic significance and in the case of AML recognizes the importance of variable lineage involvement, extensively discussed later in this chapter, is given as an inset to Table 14.1.

This classification is simple to use as a primary form of subdivision, and can of course be elaborated with whatever additional features emerge in any individual case, including immunological and cytogenetic data as well as the relevant FAB allocation.

The subdivision of AML cases into 3 broad groups—cases with definite or possible erythroid or megakaryocytic components being recorded as Type II while the remainder are divided into well differentiated, Type Ia, and poorly differentiated, Type Ib, cases—has been applied retrospectively to 715 cases from a recent Medical Research Council trial of treatment in AML for whom adequate data on relevant aspects of marrow morphology and cytochemistry had been recorded (Swirsky *et al.*, 1986). Of 183 Type Ia cases 135 (74%) entered complete remission, of 232 Type Ib cases 151 (65%) did so, and of 300 Type II cases 172 (57%) did so. The differences were clearly significant as were the differences in overall survival among the three groups.

The cytochemical details and discussion given below are arranged so as to be helpful whichever classification of the acute leukaemias is preferred, and indeed, as has been emphasized above, classification schemes are not mutually exclusive but may helpfully supplement one another.

Acute leukaemias

Cytochemical methods of every kind have been employed in the study and differentiation of acute leukaemias, and it is in this group of diseases that some of the most important diagnostic contributions of cytochemistry occur. The results of extensive investigations fall into two categories: those common to all forms of acute leukaemia and those showing differences between the various cytological subdivisions of acute leukaemia and indeed helping to define these subdivisions.

Cytochemical findings common to all forms of acute leukaemia

These include particularly studies of cell cycle activities by autoradiographic and densitometric methods. That a considerable, if variable, fraction of the leukaemic blast cell population is in an inactive non-proliferating phase was suggested many years ago by the observed infrequency of mitotic figures. Astaldi and Mauri (1954) measured the percentage of cells entering mitosis in short-term culture during a given period of time—the stathmokinetic index—and found very low values for leukaemic blast cells as compared with those for myeloblasts or later granulocytes from normal or chronic leukaemic marrows. Labelling experiments with tritiated thymidine (^{3}HTdR) confirmed the diminished proliferative activity; Gavosto *et al.* (1959, 1960) first showed by autoradiography that relatively few leukaemic primitive cells incorporated ^{3}HTdR as compared with precursor cells from normal marrows, most of which became labelled. The mean grain counts in labelled primitive cells, whether leukaemic or normal, were the same, implying that the rate of DNA synthesis during the S phase of the cell cycle was essentially similar, but that a higher proportion of leukaemic cells were in the resting G_1 and G_2 phases or in a G_0 phase, out of cycle.

Measurements of DNA content of acute leukaemic cells from blood or bone marrow were carried out using an integrating microdensitometer by Hale and Wilson (1959, 1961), who found a considerable proportion of cells to contain a tetraploid (4n) quantity of DNA and therefore presumably to be in the G_2 premitotic phase.

Numerous subsequent studies, both *in vivo* and *in vitro*, have confirmed the low labelling index of leukaemic cell populations, especially at first diagnosis or in full relapse, and suggested the existence of a relatively small proliferating component and a larger non-proliferating or dormant component (Mauer and Fisher, 1962, 1966; Gavosto *et al.*, 1964). The proportion of marrow cells labelled ranged from 1 to around 20%, with generally lower figures in peripheral blood samples, and a two cell population could often be recognized morphologically, with large blast cells showing a high, normal, labelling index, while small blasts were almost always unlabelled and were therefore at least temporarily out of cycle (Fig. 14.1). The large blasts appear usually to give rise to small ones and the latter are released preferentially into the peripheral blood (Killman *et al.*, 1963; Mauer and Fisher, 1966; Gavosto *et al.*, 1967; Lin and Bouroncle, 1967). The small blasts may re-enter the cell cycle, especially when the proliferating pool is diminished by chemotherapy or other means, and at this time the labelling index may rise sharply (Saunders and Mauer, 1969; Chan and Hayhoe, 1971). Similarly, when early relapse first reappears after remission the proliferating fraction is usually higher than at diagnosis, presumably because the total mass of leukaemic cells is smaller and the tumour doubling time shorter than in full relapse (Foadi *et al.*, 1967, 1968; Saunders *et al.*, 1967; Clarkson, 1969).

Calculations based on *in vitro* ^{3}HTdR labelling give figures for the 'proliferating fraction' of marrow blast cells at the time of diagnosis of

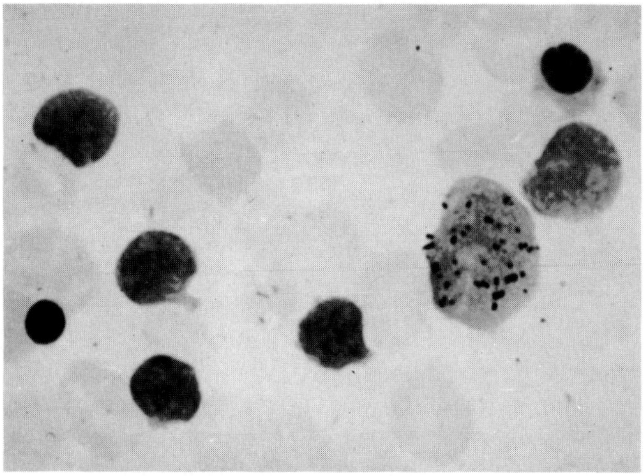

Fig. 14.1 ALL bone marrow. ^{3}HTdR autoradiography. A large blast cell shows positive labelling while five smaller blast cells and two normoblasts are unlabelled.

between 15 and 35% of the total blast cell population (Killmann, 1972). Continuous or repeated *in vivo* administration of ^{3}HTdR gives an increased fraction of labelled cells,with 90 to 99% entering the S-phase and acquiring a label by 21 days (Clarkson, 1969), but this method will include in the proliferating fraction any cells moving into it from the resting fraction during the period of study. Using a technique, modified from that devised by Nelson and Schiffer (1973), for the autoradiographic detection of labelled thymidine-5-triphosphate (^{3}H-TTP) incorporated into nuclear DNA by cells containing DNA-polymerase and exposed to the labelled triphosphate together with the other three nucleotide triphosphates, Wantzin *et al.* (1976) found that from 47.3 to 82.5% of leukaemic cells in six patients studied took up the label. If, as seems likely, TTP labelling reflects the potential for DNA replication, these findings imply that the growth fraction of leukaemic cells may be considerably higher than the earlier studies with ^{3}HTdR *in vitro* had suggested, and that the distinction between proliferating and quiescent fractions may be arbitrary.

The contribution of autoradiographic methods to understanding of leukaemic cell kinetics has chiefly arisen from studies of DNA synthesis, but there has also been parallel work summarized in Chapter 2 on the synthesis and movement of RNA which lends itself to a generally similar interpretation in regard to the varying activities and interchangeability of growth and quiescent fractions. Since the modes of action of chemotherapeutic agents can often be related to phases of the cell cycle, and these drugs are sometimes classified in particular as cell cycle stage specific (CCSS) or cell cycle stage non-specific (CCSN), it might be expected that increased knowledge of patterns of leukaemic cell kinetics would allow a more rational and effective design of drug usage. To an extent this may be true—at least theoretical explanations may now be forthcoming for the effectiveness of drug combinations established empirically—but in general the complexity of cell kinetic patterns in leukaemia, as in other neoplastic states, has been such that predicted effective dose schedules based on experimental measurements of changing cell kinetic patterns at diagnosis and during response to different chemotherapeutic agents are still probably no better than dose schedules designed by trial and error (van Putten, 1974; Arlin *et al.*, 1978).

This disappointing conclusion probably arises chiefly from the remarkable variability found from patient to patient in the cell kinetic patterns, which makes it difficult to design treatments on other than individual case data (Mauer and Look, 1985), especially in patients with AML or relapsed ALL whose prognosis with existing empirical forms of therapy is relatively poor and for whom improved treatment regimens are urgently needed.

Smets *et al.* (1983) have attempted to apply individually designed timing of high dose cytosine arabinoside in children with AML using schedules calculated from initial labelling indices. The results suggest that improved remission expectation may be achieved by such patient-specific treatment scheduling, although the basic investigations necessary may restrict the approach to research centres.

There may be promise for the future in the establishment of relationships between cell proliferative activity in acute leukaemias and the surface marker phenotype or cytogenetic structure. Look *et al.* (1982), for example, showed that proliferative activity among ALLs was greatest in surface immunoglobulin positive B-cell ALL, intermediate in T-cell ALLs and least in cALLA positive (pre-B) ALLs. This hierarchy of proliferative activity is paralleled by clinical features of tumour bulk and height of initial white cell count at diagnosis and is inversely related to remission duration in patients who relapse (Mauer and Look, 1985). Moreover, the large growth fraction in SmIg positive B-ALL is associated with the 8;14 translocation, suggesting that the high proliferative rate characteristic of this form of ALL may be a result of the translocation of the c-myc gene into the region of the Ig heavy chain locus and its consequent activation. Similarly cytoplasmic Ig positive ALL cells with a 1;19 translocation appear also to have increased proliferative activity, whereas T-ALLs with an 11;14 translocation have decreased percentages of cells in S-phase and growth inhibition (Williams *et al.*, 1984; Mauer and Look, 1985).

DNA flow cytometry has been used to detect hyperdiploidy in childhood ALL (Smets *et al.*, 1987). Among 107 evaluable patients (from a total of 114 studied) 39 showed a single aneuploid hyperdiploid cell line with mean DNA index 1.22 (range 1.1–1.4). Those 39 cases did not differ from diploid cases in age, sex, FAB classification, immunophenotype or height of leucocyte count, but they showed significantly longer disease-free survivals.

Similar studies using flow cytometry with simultaneous analysis of DNA and protein content to evaluate cell cycle activity in AML were reported by Ffrench *et al.* (1986). Sixty-seven cases were studied before the start of treatment, and the proportion of cells in S phase, in S + G2 + M, and the low protein content (LPC) fraction and high protein content (HPC) fraction of G1 (marking early and late phases of G1 respectively) were determined. LPC fractions were significantly lower in patients over 50 who failed to achieve complete remission (CR) as compared with remitters in general and also younger non-remitters. Better survivals were found in all age groups in patients with LPC fractions above 26% and HPC below 70%. A high LPC fraction implies a high proportion of blast cells in a quiescent phase or in early stages of G1, whereas a high HPC fraction (and also high S + G2 + M fraction) represents actively proliferating cells. The findings in this study therefore indicate that patients older than 50 with proliferative forms of AML have a notably poorer chance of remitting, while in all age groups proliferative disease was associated with poorer survival. The data did not include a breakdown related to cytological or cytogenetic subtypes of AML, but further studies of this kind seem likely to provide new information in relation to these variants and may lead to important advances in understanding gene regulatory activities.

Quantitative flow cytometric techniques may also be used to assess differences in surface antigen reactivity with monoclonal antibodies (McAbs) on the basis of fluorescence intensity (FI). In a study of immunostained lymphoblasts using the HLe-1 McAb, for example, Caldwell *et al.* (1987) correlated FI with prognosis—patients with dimmest blasts responding better to treatment than those with brighter cells. The difference may have reflected maturation stages, less mature cells giving dimmer FI, since HLe-1 appears to be differentiation linked.

Variable cytochemical characteristics

Among the studies which have revealed sharp differences between individual cases of acute leukaemia, and the existence of patterns helpful in classification, are those concerning the peroxidase reaction, Sudan black B staining, the PAS reaction, various esterase reactions and the acid phosphatase, β-glucuronidase, β-glucosaminidase and dipeptidyl aminopeptidase IV reactions in blast cells, and the level of alkaline phosphatase in mature neutrophils. Details of these individual reactions are described in earlier chapters, and their value in differential diagnosis is summarised here in the form of a series of tables expanded and modified from those previously published by Hayhoe *et al.* (1964), Hayhoe and Cawley (1972) and Hayhoe (1984). Certain other features relevant to classification are also incorporated in some of these tables.

Acute lymphoblastic leukaemia (ALL)

Apart from the main discriminative features of this group of leukaemias given in Table 14.1, several additional characteristics frequently present in ALL or helpful in distinguishing the stem cell or common pre-B forms, from the less common T-cell form and perhaps from the rare B-cell form are given in Table 14.2.

Among the cytochemical features listed in the table and illustrated in Plates 37 and 38 are several deserving of further amplification. The very occasional sudanophilic reaction in the blast cells in some cases of ALL has been discussed in detail in Chapter 5 and a case showing this unusual finding is illustrated in Plate 38 A, B, C. The PAS reaction is especially

Table 14.2 Cytochemical features of acute lymphoblastic leukaemias (ALL)

The Sudan black and peroxidase reactions are almost always negative. Atypical coarse sudanophilia has rarely been recorded.

The PAS reaction in primitive cells is not diffuse, but nearly always present as coarse granules or blocks against a negative cytoplasmic background in at least a few and often many cells.

Phosphorylase positivity generally follows a pattern similar to the PAS reaction.

Purine nucleoside phosphorylase biochemically low in T and B-ALL, normal in other ALL cases.

Adenosine deaminase, biochemically, raised in T-ALL, not in other forms.

Acid phosphatase shows a localized cytoplasmic reaction often in T-cell but rarely in other forms of ALL: T-ALL with strong focal AcP cytochemistry shows intense characteristic isoenzyme bands.

Biochemically 5′-Nucleotidase is very low in B-ALL and weak in T-ALL blasts, moderate in null-ALL and raised in cALL, and these findings may be erratically reflected in 5′N cytochemistry. Gamma-glutamyl transpeptidase weak or negative.

DAP II variably positive with no correlation with morphology or surface marker phenotype, DAP IV negative in T-ALL blasts, and B-ALL; few reactive cells in cALL.

Beta-glucuronidase activity is negative to moderately strong with granular patterns in both T and null cell cases.

Beta-glucosaminidase (Hex); biochemically isoenzyme Hex I found in null, pre-B and cALL but not in T or B-ALL. Cytochemical positivity resembles beta-glucuronidase.

Esterase activities against naphthol AS-D acetate and alpha-naphthyl acetate and butyrate are negative or weakly positive with scattered granules in null cell ALL; localized coarse granules may be seen in T-cell ALL. The activities are fluoride resistant. A characteristic T-group isoenzyme occurs in T-ALL.

Extramitochondrial dehydrogenases are active (e.g. lactate, malate).

Intramitochondrial dehydrogenase activities are depressed (e.g. succinate, glutamate).

The leucocyte alkaline phosphatase (LAP) score is normal or high.

Erythroblasts are PAS negative.

often misinterpreted in acute leukaemias. Coarse positivity in primitive cells has sometimes been regarded as necessary for the diagnosis of ALL and as confined to this group of acute leukaemias. Neither belief is true. Lymphoblasts may all be PAS negative in some cases of ALL, although in our experience this finding is rare, occurring in less than 5% of ALL cases, whereas most cases show, as in the table, coarse granules or blocks of PAS positivity in at least a few and often many cells. The 'few' may be less than 1%, but the occurrence of a typical pattern of coarsely granular or block positivity against a clear negative background in blast cells recognisably similar in general morphology to the predominating Sudan black negative blast cell population, and the absence of finely granular and diffuse positivity in similar cells, makes the diagnosis of ALL very likely. A series of typical PAS reaction patterns in cases of ALL is shown in Figures 14.2 to 14.7. Equally coarse positivity may be seen in myeloid leukaemias, especially in some myelomonocytic variants with predominance of monocyte precursors and in the erythroblasts of erythroleukaemia, but the coarse granules or blocks of PAS positivity are commonly found against a background of finer granules or diffuse staining in most of the positively reacting primitive cells or erythroblasts in both these conditions. The recognition of these differences of pattern is of key importance.

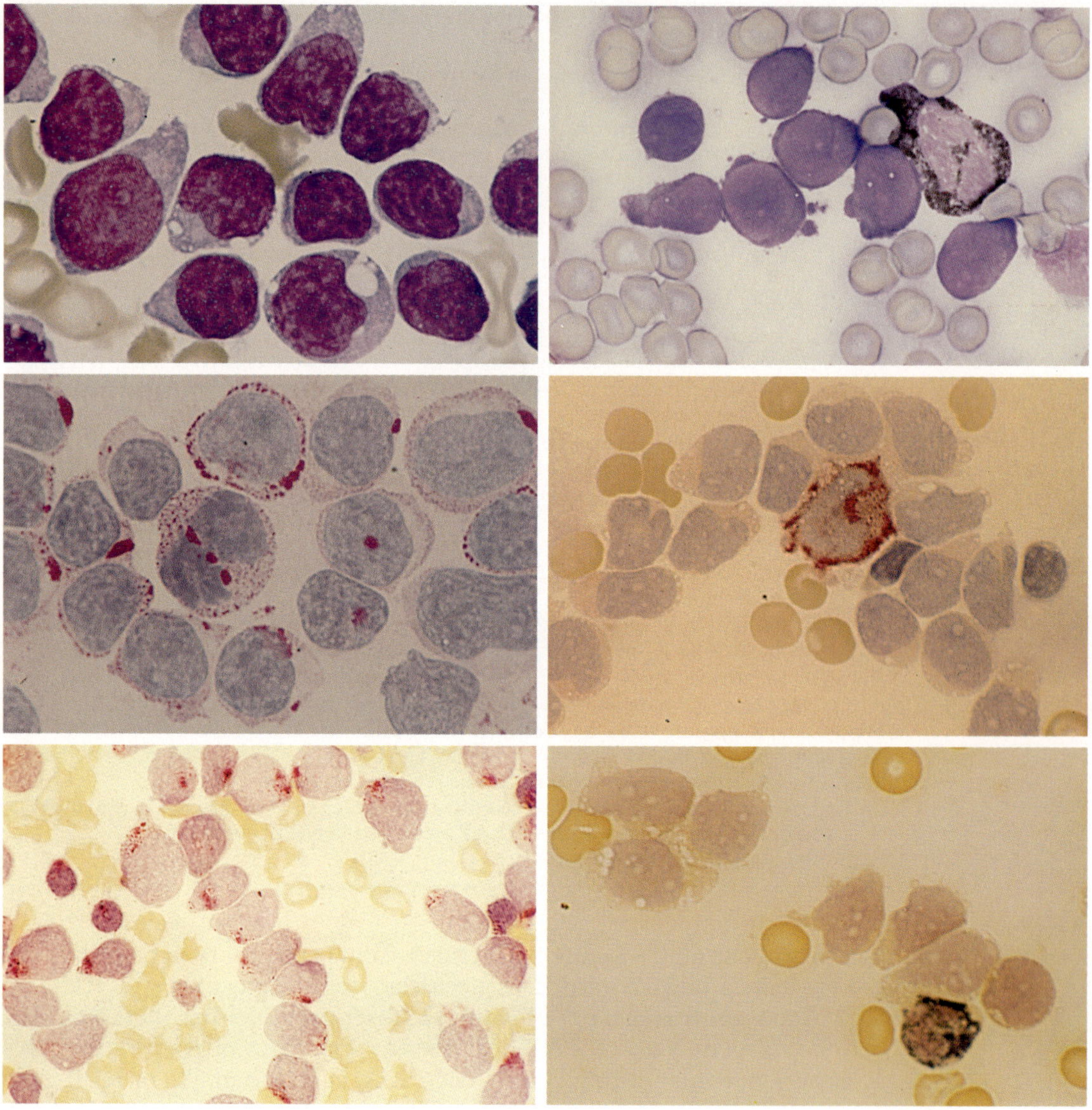

Plate 37

A. cALL bone marrow. MGG stain.
B. The same case. SB stain shows negative blasts with a strongly positive myelocyte.
C. The same case. PAS stain shows heavy and coarse granularity against a negative background in all blast cells.
D. The same case. Acid phosphatase stain, showing a positive metamyelocyte and uniformly negative blast cells.
E. T-cell ALL bone marrow. Acid phosphatase stain showing localized polar reactivity of blast cells.
F. cALL bone marrow. Dual esterase reaction, showing one CE positive neutrophil and uniformly negative blast cells.

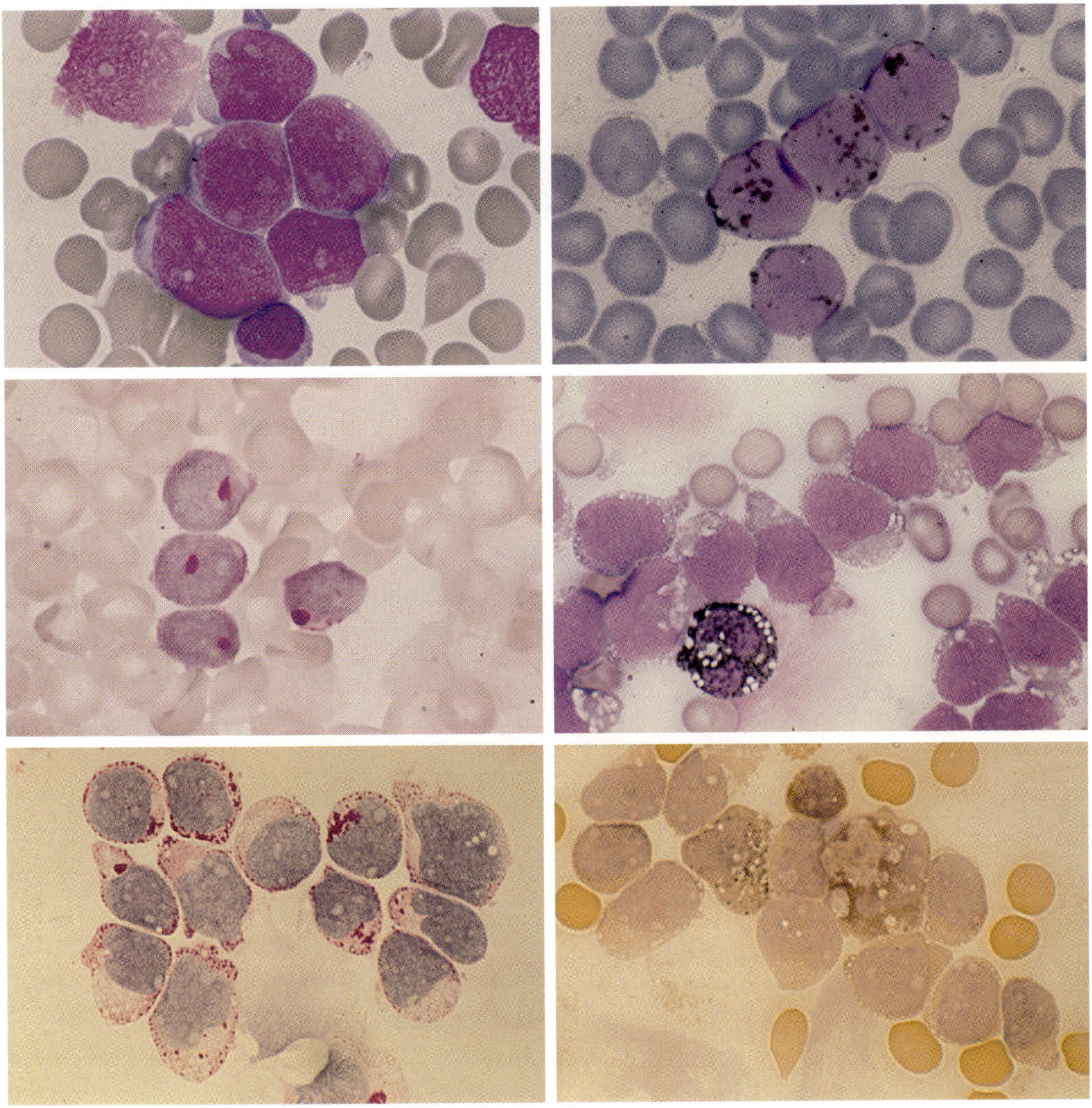

A B C D E F

Plate 38
A. cALL bone marrow. MGG stain. A group of lymphoblasts.
B. The same case. SB stain showing the rare finding of sudanophilia in lymphoblasts, the positivity showing a coarsely globular or almost rod-like disposition.
C. The same case. PAS stain, showing typical block positivity in all cells: one has also a few scattered granules. The picture is typical of ALL.
D. cALL cells in pleural effusion. SB stain showing negative reaction in blast cells, with a single normally positive neutrophil.
E. The same case. PAS stain showing strong positivity in all blast cells, with blocks and coarse granules against a clear background.
F. The same case. Dual esterase stain showing completely negative blast cells, with BE positivity in accompanying monocytes.

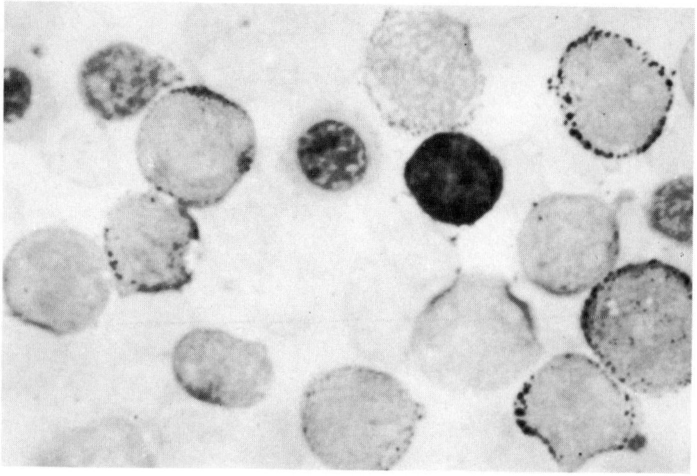

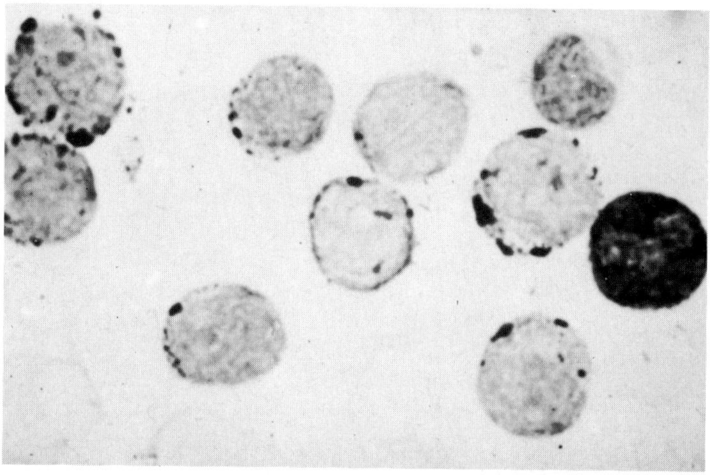

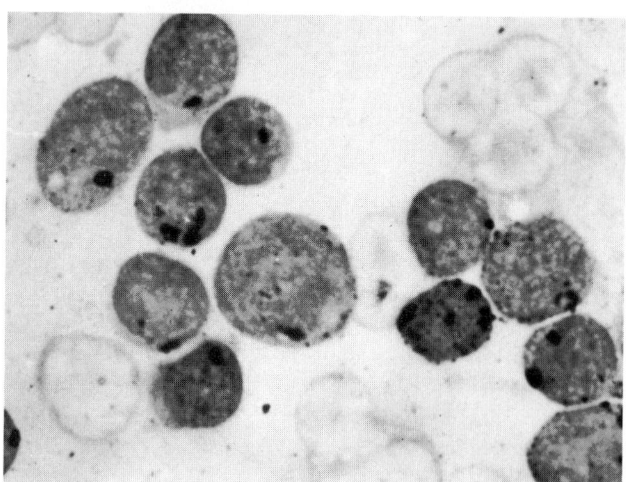

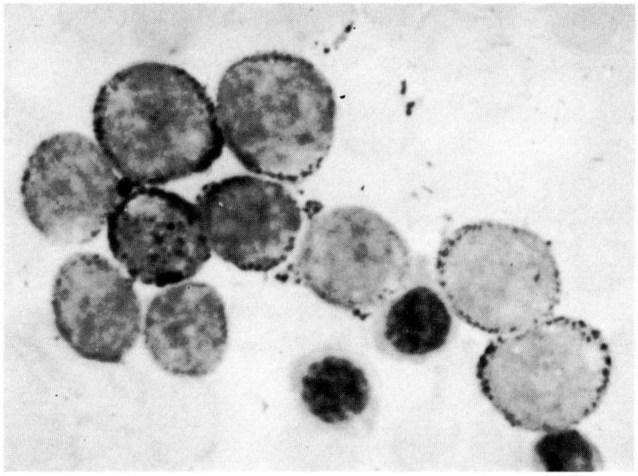

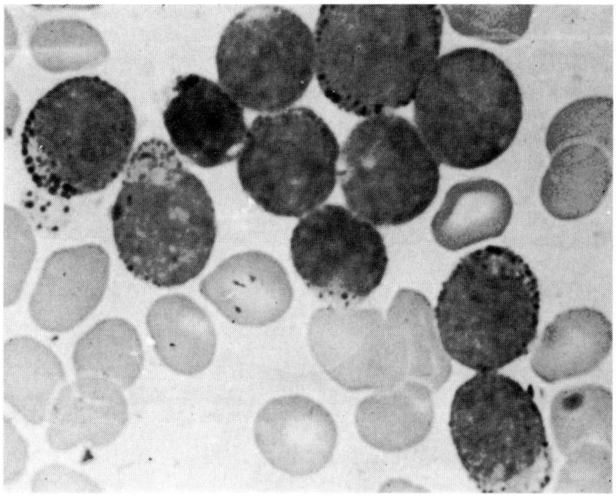

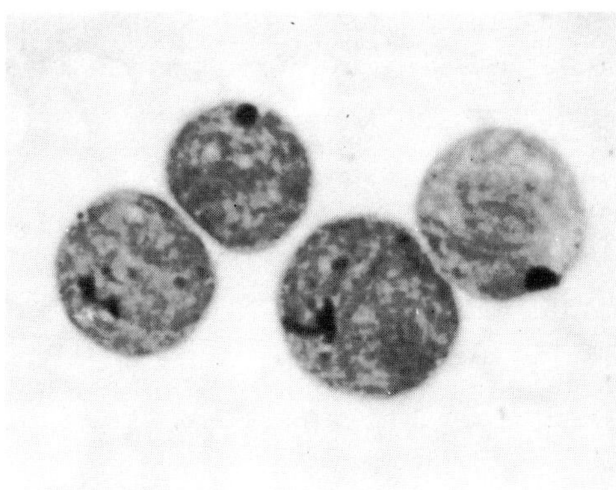

Fig. 14.2–14.7 Various ALL bone marrow smears. PAS reaction. Representative positive reactions range from single blocks to fine or moderately coarse granular positivity in the cytoplasmic rim. Some lymphoblasts are entirely negative, and such cells may predominate. The neutrophil polymorphs occasionally present show normally strong positivity.

Laurie (1968) first reported a correlation between extent of PAS positivity and survival, frequent PAS positivity in the blast cells being associated with better prognosis. This finding has since been supported by many independent studies (Vowels and Willoughby, 1973; Feldges *et al.*, 1974; Ascari *et al.*, 1975; Paolucci *et al.*, 1977) but denied by others (Bennett and Henderson, 1969; Berrebi *et al.*, 1973; Humphrey *et al.* , 1974; Shaw *et al.*, 1977). The differences are not easy to explain, but a study by Lilleyman and his associates (1979) suggests that the distribution of other features carrying poor prognosis, such as age over 14 years or high presenting leucocyte count, may have influenced the comparative figures. In their series, actuarial survival studies showed that patients with more than 20% PAS positive blast cells survived longer than those with less, but that the difference was statistically significant only after the exclusion of other adverse factors.

Alkaline phosphatase has not hitherto been demonstrable cytochemically by standard techniques in lymphoblasts of ALL, although biochemically they may contain the lymphoid type of enzyme, N-APase (see Chapter 6) (Neumann *et al.*, 1979). The LAP score in neutrophils, however, is normal or high.

Acid phosphatase has been observed to give localized cytoplasmic reactions more frequently and strongly in T-cell ALL (Plate 37E) than in the common pre-B cell type (Plate 37D), but the difference is not entirely specific and cannot be relied upon as diagnostic in an individual case. Thus, McKenna *et al.* (1979) found focal acid phosphatase positivity in most blast cells from 8 of 11 patients with E-rosette positive lymphoblasts (T-ALL) as compared with only 3 of 48 non-T, non-B cases. The finding of focal reactivity is thus suggestive of T-ALL rather than conclusive, and this is in accordance with our own experience, although some authors (Savage *et al.*, 1981; Telek *et al.*, 1983; Boesen *et al.*, 1984) seem to have found strong reactions confined to T lymphoblasts with only faint granular reactions in some non-T non-B cells.

An interesting association between azurophil granularity and punctate granular acid phosphatase positivity, often also with PAS positivity, has been observed in ALL cases by a number of authors (Grogan *et al.*, 1981; Stein *et al.*, 1983; Rodriguez-Fuch *et al.*, 1984). Darbyshire and Lilleyman (1987), in a substantial study, checked 320 ALL marrows for azurophil granules in the lymphoblast cytoplasm and found 20 (7%) to have more than 5% positive cells, 19 of them having the cALL phenotype. The granules stained strongly for acid phosphatase. The finding was more common in males and in cases with predominantly large cell (L2) morphology. PAS positivity was also often positive, although less consistently so than in the otherwise comparable smaller series of patients earlier reported by Stein *et al.* (1983). The finding does not appear to have obvious prognostic significance.

Biochemical studies of acid phosphatase isoenzymes using polyacrylamide gel electrophoresis (Li *et al.*, 1970a, b) and later isoelectric focusing (Gaedicke *et al.*, 1979; Drexler *et al.*, 1984) revealed complex patterns with marked heterogeneity even within such immunological subgroups as cALL and T-ALL. The subject is discussed in Chapter 7. It should be noted in the present context that no leukaemia-specific isoenzymes could be detected and that the general patterns of isoenzyme distribution were broadly similar in AML and ALL. Cases of T-ALL (and also of T-cell lymphomas) with strong cytochemically demonstrable focal acid phosphatase positivity showed more complete band patterns than other ALL cases (or B-cell lymphomas), with strong group IV and V bands, in the nomenclature of Gaedicke *et al.* (1979), as well as the main group I-III bands shown by other ALLs. Even less reliable in this regard is the beta-glucuronidase reaction, although here too coarsely granular positivity is more often seen in T-cell cases, but with similar increases in some non-T, non-B ALLs (Telek *et al.*, 1983).

The distribution of esterases in different leukaemic cell types varies considerably according not only to the different available substrates but also to the precise methods used for the cytochemical reactions. In lymphoblasts of all kinds chloroacetate esterase reactions are negative provided prolonged incubation in excess of 30 minutes is avoided. With naphthol AS-D acetate, α-naphthyl acetate or α-naphthyl butyrate as substrates a pattern of granular positivity may be seen. In our hands and with the butyrate as substrate the common T-cell pattern of a few coarse granules to one side of the nucleus, perhaps characteristic of the Tμ subpopulation (p. 000), has occasionally been found in the blast cells in cases of T-cell ALL, whereas cases of cALL have shown no more than a scattered fine granular reaction in the majority of cells, comparable to the pattern seen in normal null-cells from the peripheral blood and are often negative (Plate 37D). Any positive reactions are generally much weaker than those often seen in monocytes and their precursors, and unlike the reactions in monocytes are not inhibited by fluoride.

Using a long incubation acid esterase method, however, Pieters and Veerman (1987) have found esterase positivity to be present in some cases of childhood ALL and to be associated with significantly lower continued complete remission rates among 32 cALL cases than esterase negativity. In 61 null-ALLs the prognostic difference between positively and negatively reacting cases was not statistically significant. In neither group was reactivity for acid phosphatase prognostically linked.

Extensive isoenzyme studies of esterases in leukaemias have been reported and are reviewed in chapter 8. Here it may be noted that Drexler and Gaedicke (1983) found distinct isoenzyme groups with different patterns in AML and ALL respectively and that among the ALLs those of T-cell origin had a characteristic 'T group' isoenzyme complex, which nevertheless did not correlate with the 'acid esterase' reaction described by Mueller *et al.* (1975) in later T cells and, indeed, showed a pH-optimum between 7 and 10. Cases of c-ALL showed considerable heterogeneity suggesting that at least three differentiation stages could be separated within this immunological subgroup (Drexler *et al.*, 1984). One of these stages, like null-ALL and undifferentiated leukaemias, showed no esterase isoenzymes.

Veerman *et al.* (1983) attempted to relate the immunological phenotype to acetate esterase and acid phosphatase reactions in childhood ALL. Among 83 cases studied were 31 T-ALLs, 30 of them negative for acid alpha-naphthyl acetate esterase but 29 positive for acid phosphatase, 38 cALLs with variable positivity for either enzyme, 12 null ALLs which were all acid phosphatase negative although four showed esterase positivity, and two B-ALL, negative for both enzymes. Boesen *et al.* (1984) also found T lymphoblasts mostly negative or weakly reacting for esterase, like cALL blasts, whereas they usually showed strong granular acid phosphatase positivity, unlike the faint reactions in c-ALL.

Gaedicke and Drexler (1984) further extended these studies by combined analysis of the isoenzyme band patterns for acid phosphatase, esterases, and hexosaminidase in childhood ALLs. The various immunological subtypes of ALL showed different patterns, but no specifically leukaemic bands, not found in normal cells, were detected and none specific to any particular leukaemic subtype. They again confirmed that the immunologically based subclassification of ALL could be further extended by subdivision on the basis of isoenzyme patterns representing particular points on the normal haemopoietic cell differentiation sequence at which the leukaemic cell population appeared to be arrested in 'frozen differentiation'. Such detailed subclassification remains beyond the reach of cytochemical or immunocytochemical techniques at present, but as Drexler *et al.* (1984) point out may prove to have prognostic importance or even to have therapeutic implications if enzyme inhibitors such as deoxycoformycin come into more general use.

From the prognostic point of view Basso *et al.* (1984) in an interesting study of 120 childhood non-T, non-B ALLs, correlated the duration of first remission with initial cytochemical positivity of the blast cells for four cytochemical reactions, acid phosphatase, alpha-naphthyl acetate esterase, beta-glucuronidase and N-acetyl-beta-glucosamidase. The percentage in complete remission up to 6 years was always higher for children whose blasts lacked these enzymes, although the differences were statistically significant only between those positive or negative for beta-glucuronidase. At 6 years 89% of ALLs with no cytochemical positivity for the 4 enzymes studied were in complete remission, as were 59% of those with one positive reaction, but less than 39% of those with two or more positive reactions. The probable interpretation of this finding is that the least mature cell types among the non-B non-T ALL variants, with least marker enzyme expression, respond best to therapy.

Both 5′-nucleotidase (5′N) and adenosine deaminase (ADA) appear to have discriminative value among the immunological variants of ALL, when measured biochemically. 5′N shows a range from low to high levels in the sequence B-ALL, T-ALL, null-ALL and cALL, and ADA generally shows highest activity in T-ALLs, with variable levels in B-ALL and moderate increases in both null ALL and cALL over the values in normal lymphocytes (Drexler *et al.*, 1984). The cytochemical methods available for 5′N reflect the biochemical findings only irregularly (see Chapter 11), although Veerman *et al.* (1985) found 10 of 36 cALL cases to show 5′N positive reactions, with associated lower probability of continuous complete remission than in the 5′N negative cases. Among 27 T-ALLs 14 null-ALLs and 2 B-ALLs little or no positivity for 5′N was found. ADA is not yet demonstrable cytochemically. For beta-glucosaminidase (beta-hexosaminidase, Hex), characteristic isoenzyme patterns are demonstrable biochemically in the variants of ALL, with the intermediate band Hex I absent from T-ALL but present in other forms, while cytochemically this discriminant feature cannot be observed, the pattern of positivity being generally similar to that of acid phosphatase and beta-glucuronidase (see Chapter 11). Purine nucleoside phosphorylase (PNP) also shows biochemical variability among ALL cases, with low levels in T-ALL and B-ALL, and normal levels (those of normal lymphocytes) in null-ALL, and pre-B cases with or without cALL antigen positivity (Drexler *et al.*, 1984). The cytochemical distribution of PNP has been studied in normal blood and in CLL (Borgers *et al.*, 1977, 1978) but not hitherto in ALL.

The dipeptidyl aminopeptidases, DAP II and DAP IV (reviewed in Chapter 11) have been studied cytochemically in acute leukaemias (Khalaf *et al.*, 1986). Variable positivity occurred with DAP II in ALL, not related clearly to immunological phenotype, and although most blast cells in ALL were negative for DAP IV, occasional positive cells could be found in some cALL cases. The general restriction of DAP IV positivity to T cells is not clearly manifest among acute leukaemias; Crockard *et al.* (1984) found DAP IV to be absent from or detectable in only few cells in T-ALL cases.

Another peptidase, gamma-glutamyl transferase, shows low levels in ALL biochemically (Kramers and Catovsky, 1978) and is absent or very weak cytochemically (Khalaf and Hayhoe, 1987).

The dehydrogenase findings in ALL, with relatively high levels of extra-mitochondrial glycolytic enzyme activity and low intramitochondrial and α-glycerophosphate shuttle enzyme activity (Stuart *et al.*, 1970) may reflect the deficiency of mitochondria in lymphoblasts, demonstrable ultrastructurally.

Acute myeloid leukaemia (AML)

Under this broad heading is incorporated a range of acute leukaemias arising in marrow-derived cells. Whereas in Table 14.1 the ALLs have been subdivided primarily on immunological (surface antigen) criteria, this is not at present appropriate for AMLs, because pure forms of AML with a single cell lineage are the exception rather than the rule, as is the case in ALLs, and because McAbs against myeloid cell surface antigens frequently show cross-lineage overlap. The morphological and cytochemical criteria basically used for subdividing AMLs are concerned chiefly with (a) defining the predominant cell type and determining its position in the maturation sequence of that cell lineage, and (b) assessing the extent of involvement of other myeloid cell lineages in the leukaemic process. The classifications resulting from process (a) gave myeloblastic, myeloblastic-promyelocytic, and promyelocytic variants when granulocyte precursors predominated, and monoblastic, monoblastic-promonocytic, and monocytic variants when the monocyte line predominated, with less common erythroid or megakaryocytic lineage dominance. When process (b) is added a high proportion of cases show some admixture of cell lines, granulocytic and monocytic most often, but also quite frequently involvement of erythroid or megakaryocytic lines. Obviously an almost infinite range of combinations may exist and the broad classification given in Table 14.1 does no more than recognise conventionally accepted but not sharply defined groupings, with imprecise and overlapping boundaries, except where some 'all or none' feature such as a structural cytogenetic anomaly exists, as in the t(15;17) of APL or the t(8;21) of one myeloblastic variant. Attempts to achieve a simplified and reproducible classification by drawing arbitrary lines at selected levels among these variations have been made by, for example, the FAB group, whose subdivisions are shown in Table 14.1.

The FAB nomenclature is widely used, although the recommended differential counting and morphological criteria on which the divisions are based between M1 and M2, M2 and M4, M4 and M5a or b, M5a and M1, and M6 and any other variant with a mixed erythroid component, are sufficiently elaborate and esoteric as rarely to be practised. Whether the divisions of the FAB classification are sensible or meaningful is another matter. There seems no good reason why the granulocytic lineage AMLs other than APL should be separated into two groups, M1 and M2, with respectively less or more than 10% of maturing cells beyond myeloblastic stage, rather than with less or more than 5 or 15 or 30%; these cases show a spectrum of maturation which might equally well be divided into multiple subsets, or not divided at all, on the criterion of percentages of promyelocytes or myelocytes present. Other evidence of lineage commitment and cytological maturation, such as sudanophilia, Auer rods and other inclusions may be equally important, as discussed below. Similarly with the maturation aspect of monocytic lineage AMLs, where again two groups, labelled in this instance M5a and b, are arbitrarily divided within the spectrum of variable maturation ranging from 100% monoblasts to 100% promonocytes or monocytes. Again, mixed granulocyte-monocyte lineage cases in which the minor cell population makes up less than 20% of the whole are classified as unmixed by the FAB group, as are leukaemias with an erythroid component of less than 50%. This cannot be biologically sensible.

Perhaps the increasing potentialities of cytogenetic analysis, with high resolution banding, will eventually allow for a classification based on aetiology and pathogenesis, related to particular oncogene activation. Certainly the separation of several subtypes of AML associated with structural chromosome anomalies and having characteristic cytology and cytochemistry seems justifiable, as for example t(15;17), t(8;21), t(6;9), perhaps 9q-, and inv(16). But there still remain many AML cases without easily detectable cytogenetic abnormalities and others with chromosomal changes unassociated with specific cytological pictures. For the present perhaps the most effective classification of AML cases would be one which provided assessments of maturation and of lineage admixture, as proposed in the inset to Table 14.1.

An attempt to determine the prognostic relevance of cytological and cytochemical findings has been made in recent years in our laboratories (Swirsky *et al.*, 1985; Hayhoe, 1985; Swirsky *et al.*, 1986). The study involved over 600 patients with *de novo* AML, entered into the Medical

Research Council's 8th trial of therapy. The patients were all treated with the same remission induction regimen and pretreatment variables comprising age, clinical status (performance index), haematological parameters (haemoglobin, leucocyte total and differential counts, and platelet count) and a detailed cytological and cytochemical assessment of bone marrow aspirates were analysed. As has repeatedly been found in earlier large-scale studies, the most important single prognostic feature was age, with significantly poorer remission rates in patients over 60 years old. Other poor prognostic features at presentation included low performance index, and platelet counts below $25 \times 10^9/1$. Account had to be taken of these features, especially age, in comparing cytological and cytochemical findings in different prognostic groups. This study showed that cases of AML with cytological maturation along either granulocytic or monocytic lines, or both, evidenced by granules, Auer rods, more than 5% Sudan black positive blast cells and morphological and cytochemical abnormalities of neutrophils were associated with a good response to treatment. Marrow eosinophilia was also a good prognostic feature. Nuclear features of immaturity such as increasing numbers and prominence of nucleoli had poor prognostic significance. The proportion of erythroblasts in marrow aspirates showed a relation to remission rates, a rising percentage of erythroblasts being associated with a falling remission rate. Abnormalities of the erythroid series, especially PAS positivity, present in over 20% of the cases in this study, carried a poor prognostic significance. The contribution of simple cytochemical methods to this analysis was of importance in several areas, further discussed under relevant headings below.

Auer rods

Sudan black staining greatly facilitated the recognition of Auer rods, which were found in 67% of specimens on the Sudan black stain as compared with 32% on the Romanowsky Stain. Auer rods are virtually pathognomonic of an acute myeloid leukaemia with a granulocytic component; they are probably derived largely from the incorporation of primary granules into autophagic vacuoles (Cawley and Hayhoe, 1973) and their crystalline content, with birefringence faintly visible in polarized light, reflects the internal ultrastructure of the primary granules forming them. Thus the promyelocytic primary granules in most cases of AML have an internal structure showing either homogeneity or parallel lamellae with periodicity between 6 nm and 13 nm, and the same substructure, either homogeneous or with an identical periodicity, is found in the Auer rods in these cases.

Recent work in our laboratories (Pearson, 1986) has shown that Auer rods from AMyL, even when apparently amorphous in sections, have in fact a truly three dimensional crystal structure when observed in the goniometer stage of the electron microscope, showing periodicity in at least one direction when suitably tilted. Using an optical diffractometer illuminated by laser, Pearson derived measurements consistent with a monoclinic unit cell, with edge lengths 6.6 nm, 8.6 nm and 9.6 nm respectively. This structure was consistent with periodicities previously recorded in Auer rods from AMyL (McDuffie, 1967; Bessis and Breton Gorius, 1969; Tulliez and Breton-Gorius, 1979).

In APL (Fig. 14.8), with recognizable coarse azurophilic granules or multiple Auer rods visible in Romanowsky preparations, the ultrastructure of the granules and of the Auer rods is made up of tubules in a hexagonal array with a periodicity of 25 nm (Fig. 14.9). This structure is quite different from that of the Auer bodies in AMyL, and granules or rods showing it are commonly, though not invariably, found in the cells of APL (Breton-Gorius and Houssay, 1973; Tulliez and Breton-Gorius, 1979; McKenna *et al.*, 1982; Pearson *et al.*, 1986). Both types of granule and Auer rod are peroxidase and Sudan black positive and react for chloroacetate esterase, although the former reactions are weaker than usual and the latter stronger in the Auer rods of APL, where the reactions involve the rod envelope and leave a negative central core (Plates 41E,F, 42A). Other cytochemical reactions shown by Auer rods are entirely in conformity with the derivation of these structures from the primary lysosomal granules of early granulocytes. Thus they are usually, though not always, oxidase (M-Nadi reaction) and peroxidase positive at both light microscope and ultrastructural level (Ackerman, 1950; Kondo *et al.*, 1966), and may stain weakly with the PAS reaction (Ackerman, 1950). They contain acid but not alkaline phosphatase (Harada, 1951; Goldberg, 1964) the former being also demonstrable at EM level (White, 1967; Huhn *et al.*, 1971).

Inclusions other than Auer rods, ranging in size from giant granules 1–3 μm in diameter to inclusions up to 8 μm in diameter occur from time to time in AMyL. In t(8;21), blasts and later granulocytes often contain giant granules, resembling those of the Chediak-Steinbrinck-Higashi anomaly, and there may also be glossy pink-staining inclusions in Romanowsky preparations. The cytochemistry of these giant granules and inclusions is uncertain, but they are probably positive for Sudan black, peroxidase (Plate 42F) and chloroacetate esterase (Plate 19D), like Auer rods. In a detailed cytochemical and ultrastructural study of an unusual case of AMyL with giant inclusions, 2 to 8 μm in diameter, Senelar *et al.* (1984) noted that they were mostly round or oval but occasionally rectangular with a crystalline appearance; they showed strong positivity for Sudan black, peroxidase, acid phosphatase and chloroacetate esterase, but were butyrate and ASD-acetate esterase negative. Most showed PAS positivity. On electron microscopy a rim of myeloperoxidase positivity surrounded the inclusions, which showed weakly positive centres. Some massive inclusions had a homogeneous matrix containing strongly myeloperoxidase positive granulations like azurophilic lysosomal primary granules. No clearly crystalline structure was observed, but the cells were not studied in the goniometer stage with tilt. There is no information given on the cytogenetics in this case.

Some cases with deletions of the long arm of chromosome 9, 9q-, show swollen granular inclusions somewhat like early stages in Auer rod formation, with marked sudanophilia but negative PAS and virtually negative chloroacetate esterase reaction (Plate 44 A–F) (Hoyle *et al.*, 1987).

In all these cases there is need for further study of the three dimensional structure of the inclusions, which may prove to have characteristic features.

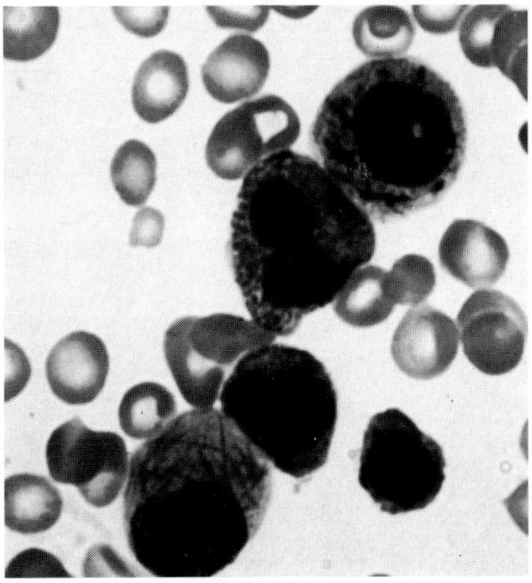

Fig. 14.8 APL bone marrow. Leishman stain. Coarse azurophilic granules and multiple Auer rods are present.

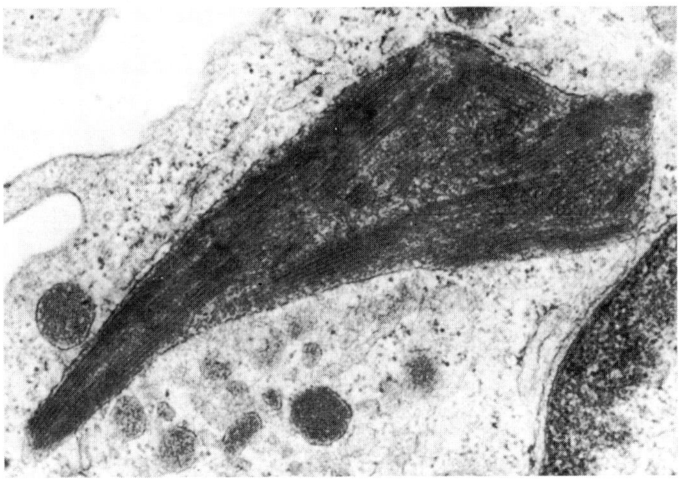

Fig. 14.9 APL bone marrow. EM preparation showing an Auer rod and primary granules with tubular substructure and periodicity of about 25 nm.

A third type of Auer rod, differing from the common AMyL and the APL variants, has in fact been proposed by Tulliez and Breton-Gorius (1979) on the basis of the ultrastructure visible after peroxidase cytochemistry in a case of AMyL with giant granules, although the precise structure is not clearly defined.

Other cytochemical features of AMyL and APL

Other important cytochemical features commonly found in AMLs with predominantly or exclusively granulocytic components are summarized in

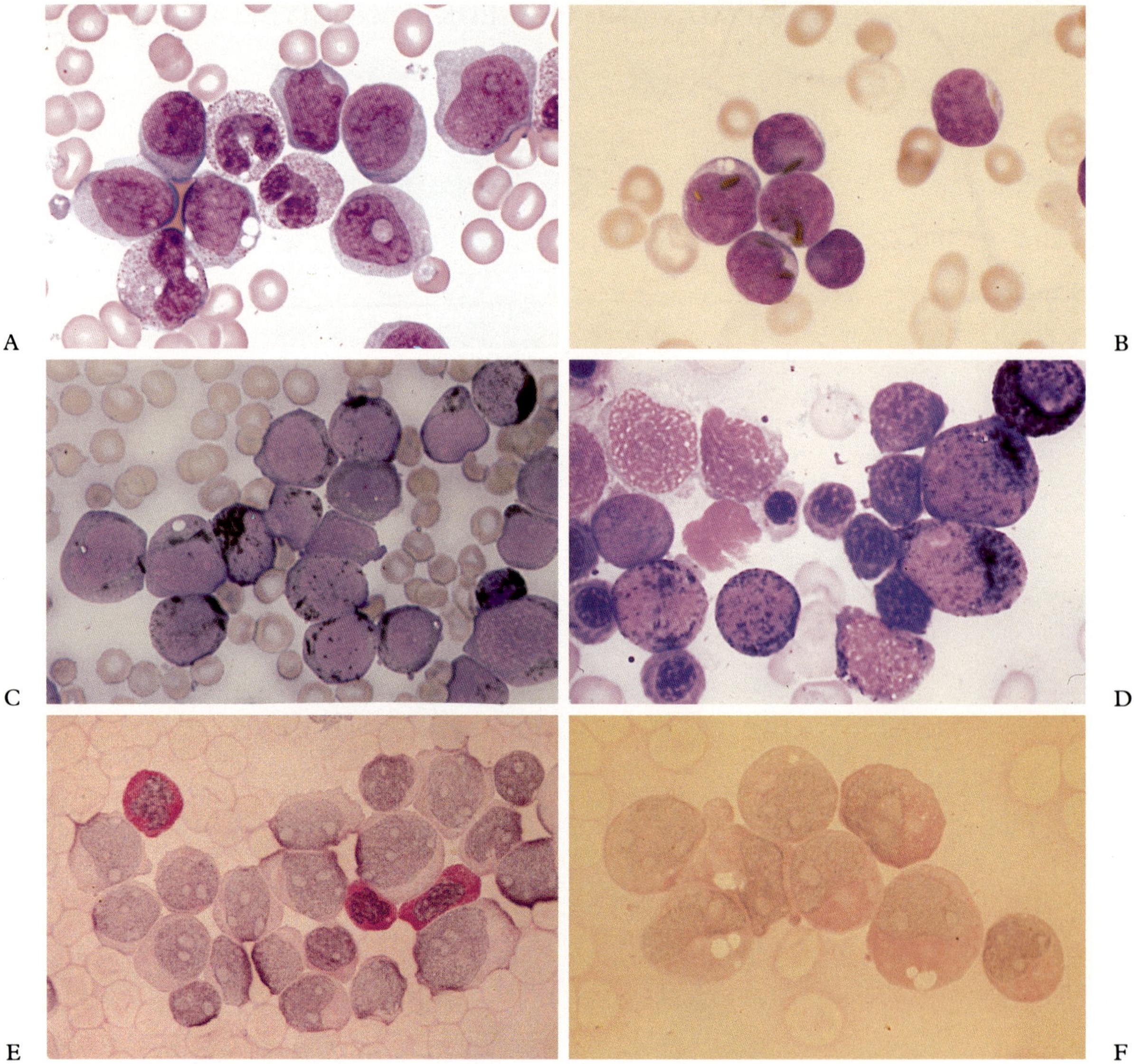

Plate 39

A. AMyL bone marrow. Myeloblasts predominate but there are later granulocyte stages also present.

B. AMyL bone marrow. Peroxidase reaction (o-tolidine) showing positive Auer rods in most cells.

C. AMyL bone marrow. SB stain. Most blast cells show positivity, either Auer rods, weak or moderately strong localized cytoplasmic positivity or some combination.

D. AMyL bone marrow in emerging remission. SB stain. Granulocyte precursors show sudanophilia like that in normal bone marrow.

E. AMyL bone marrow. PAS stain, showing negative myeloblasts and three normally positive neutrophils.

F. AMyL bone marrow. PAS stain. Eight myeloblasts show weak diffuse positive cytoplasmic staining.

Table 14.3 Cytochemical features of acute myeloblastic leukaemia (AMyL)

The leucocyte alkaline phosphatase (LAP) score is usually low.
Many cells are Sudan black-positive and peroxidase-positive, as are Auer rods.
The PAS reaction is negative in many cells, but some show a diffuse tinge with or without superimposed fine granules.
PAS positive erythroblasts may be seen in AMyL with an erythroid component.
Acid phosphatase shows only weak scattered positivity, as do beta-glucuronidase and beta-glucosaminidase, but biochemically isoenzyme Hex I may be found.
Naphthol AS-D chloroacetate esterase (CE) activity is positive (increasing positivity with increasing maturity) but alpha-naphthyl acetate and butyrate esterase (BE) activities are usually negative.
Auer rods in most AMyL cases are CE negative or only weakly positive.
Naphthol AS-D acetate esterase reaction is positive (particularly in promyelocytes and myelocytes) but less so than in monocytoid cells and is not inhibited by sodium fluoride (cf. monocytic cells).
Other cytochemical methods are as yet of little value. 5′-Nucleotidase is generally weak or negative, DAP II variably positive, but DAP IV uniformly negative, gamma-glutamyl transpeptidase variable but stronger than in ALL.

Table 14.4 Cytochemical features of APL

In addition to the general features of AMyL, stronger Sudan black and peroxidase positivity is usual.
Auer rods, whether single or multiple, may show hollow sudanophobic cores.
PAS reaction generally shows a strong diffuse tinge, often with fine granules: Auer rods may show weak positivity. PAS positive erythroblasts uncommon.
Naphthol AS-D chloroacetate esterase (CE) is usually strongly positive; Auer rods may appear as CE-positive cylinders (or rings, in early stages) with unstained hollow centres or occasionally with alpha-naphthyl butyrate esterase (BE)-positive centres.
Mixed CE and BE positivity in cytoplasm occurs in about a third of cases, with dual positivity often in the same cell. Very rarely the whole granular content of APL leukaemic cells may be BE positive and CE negative.

Table 14.3, and additional features in APL and in the 8;21 translocation are shown in Tables 14.4 and 14.5. All these appearances are illustrated in Plates 39 to 44.

The cytochemistry of neutrophils in acute granulocytic leukaemias deserves some comment. A low LAP score is often found (Hayhoe *et al.*, 1964), perhaps representing a deficiency of the tertiary specific granules in which alkaline phosphatase is probably located (Cawley and Hayhoe, 1973), or possibly an isolated defect in the production of the enzyme. Certainly hypogranular and agranular, though otherwise mature, neutrophils are frequently to be found in this group of AMLs (Leinonen, 1970) and they may also have reduced amounts of peroxidase and Sudan black positivity and of chloroacetate esterase (Schmalzl *et al.*, 1970).

Although in the subset of AMyL associated with the 8;21 chromosomal translocation Sudan black and peroxidase-negative neutrophils are uncommon, the neutrophils frequently show evidence of abnormality with pseudo Pelger-Huet anomaly and defective granularity, and the LAP score seems invariably to be low (Kamada *et al.*, 1981). As we have seen earlier, such abnormalities are linked with other cytological signs of maturation in

Table 14.5 Cytochemical features of cases with chromosomal translocation 8;21

In addition to the general features of AMyL, the LAP score is invariably low, and the Sudan black reaction shows heavy localized cytoplasmic positivity. Auer rods may have hollow sudanophobic cores.

Peroxidase reaction is weaker than Sudan black but Auer rods are more conspicuous (less obscured by coarse positivity) and sometimes revealed in both neutrophil and eosinophil polymorphs.

Sudan black- and peroxidase-negative neutrophils are uncommon, but agranular and PAS negative neutrophils occur.

PAS usually diffusely positive in blast cells, with occasional coarse granules and rare blocks. Intragranular positivity in eosinophils. PAS positive erythroblasts absent.

The activity of chloroacetate esterase is high in many blasts, with occasional positive Auer rods.

No butyrate or acetate esterase activity. 5′Nucleotidase negative; DAP II variable, DAP IV negative.

Other features as in Table 14.3.

the leukaemic cell compartment as associated with improved prospects of remission on treatment. On the other hand, neutrophils with myeloperoxidase (MPO) deficiency appear to be less effective in suppressing infection in AML. Nielsen and Bendix-Hansen (1984) studied 74 cases of AML among which 36 had more than 4% of MPO deficient neutrophils. In this group there were significantly more febrile attacks, more infective foci and more bacteraemias or other causative agents detected than in the 38 patients with normal MPO levels in neutrophils, irrespective of the height of the neutrophil count.

The presence in AMyL of more than 5% and usually more than 85% of primitive cells showing Sudan black positivity had emerged clearly from the study of Hayhoe *et al.* (1964) but the type of positivity requires emphasis. In leukaemic granulocyte precursors the sudanophilia usually shows a strong localized or heavy overall reaction (Plates 39C,D; 41A), distinguishable from the more finely granular and widely scattered pattern of discrete positivity found in monocytes and their precursors (Plate 45A,B). Since, as we have seen in Chapter 5, the precise nature of the material stained by Sudan black in still uncertain and in particular the relation to peroxidase activity is not as clearcut as is sometimes supposed, it is worth drawing attention to our own repeated observations that Sudan black positivity is usually more marked than peroxidase in acute granulocytic leukaemia and may sometimes be present when peroxidase activity appears absent by orthodox cytochemical techniques. Moreover, the distinction between patterns of positivity in granulocyte and monocyte precursors seen with the Sudan black stain is rarely detectable with peroxidase. There is at present no satisfactory means of demonstrating sudanophilia at ultrastructural level, but the myeloperoxidase of neutrophil precursors can be stained by ultrastructural cytochemistry (Chapter 10) and is seen to be located in the primary azurophilic granules and to some extent in the perinuclear space, the endoplasmic reticulum and the Golgi apparatus (Bessis and Maigné, 1970). The latter, non-granular, dispositions may account for the peroxidase positivity (and perhaps the peroxidase-related sudanophilia) of some leukaemic primitive cells with no granules visible in Romanowsky preparations. Positivity to both Sudan black and peroxidase stains is generally very strong in APL, as might be expected from the heavy azurophilic granulation of the predominant cells.

Even in the poorly granular or hyperbasophilic microgranular variants of APL the strong sudanophilia and peroxidase positivity are usually found (Plate 40F, 41A).

Chang *et al.* (1987) have recently shown by Southern blot hybridization of myeloperoxidase (MPO) cDNA to DNA from a somatic cell hybrid clone panel, that the human MPO gene cosegregates with chromosome 17. In situ hybridization mapped the MPO gene to 17q22–24, close to the t(15;17) (q22;q21–22) breakpoint of APL. However, DNA samples from 4 APL patients did not reveal MPO gene rearrangement, although APL patients with t(15;17) expressed high levels of MPO mRNA. The authors speculate that the high levels of MPO gene expression in APL might be due to arrest of granulocyte differentiation at the promyelocyte stage because of the translocation of genes localized at the 17q21–24 region which encode functional proteins possibly related to cell differentiation. Alternatively, the MPO gene may itself be translocated to chromosome 15 in APL and so acquire enhanced expression, or, if it remains on chromosome 17, be further activated by a translocated gene from chromosome 15. In one APL patient with t(15;17) (q22;q21) the MPO gene has in fact been shown by in situ hybridization to be translocated to chromosome 15q + (Chang *et al.*, 1987, addendum).

Both peroxidase and Sudan black reactions are of particular interest in relation to t(8;21) cases. In a study from our own laboratories of 30 patients with this translocation (Swirsky *et al.*, 1984), strong sudanophilia was invariably found, with 20 to 100% of leukaemic cells showing positivity, mostly as a localized densely-staining clump of deposit to one side or in a cleft of the nucleus, often so coarse as to obscure Auer rods buried within it which are therefore only occasionally visible, though often, when seen, showing hollow centres (Plate 42D,E). In Romanowsky preparations Auer rods in t(8;21) cases are commonly seen but are usually single, long and slender; in Sudan black stains, multiple rods are seen rather more frequently. Peroxidase staining shows patterns generally similar to Sudan black, but, especially with the most discriminative peroxidase technique, that using diaminobenzidine (DAB), the localized reaction could be more easily resolved into granules and Auer rods were easier to discern, even occurring in maturing eosinophils. All these features are illustrated in Plate 42C–F and Plate 43A,B.

The PAS reaction in AMyL shows a range from negativity to diffuse cytoplasmic staining with or without fine superimposed granules (Hayhoe *et al.*, 1964) (Plate 39E,F), the reaction being generally stronger in cases of APL (Plate 41B,C). Coarse granules or blocks of positivity are not common in AMyL or in APL, but may occasionally be seen in blast cells from cases of t(8;21), although here the more usual pattern is one of diffuse cytoplasmic reaction with some localized clumping of moderate positivity and some scattered granules (Plate 43A). Very coarse granules or blocks of PAS

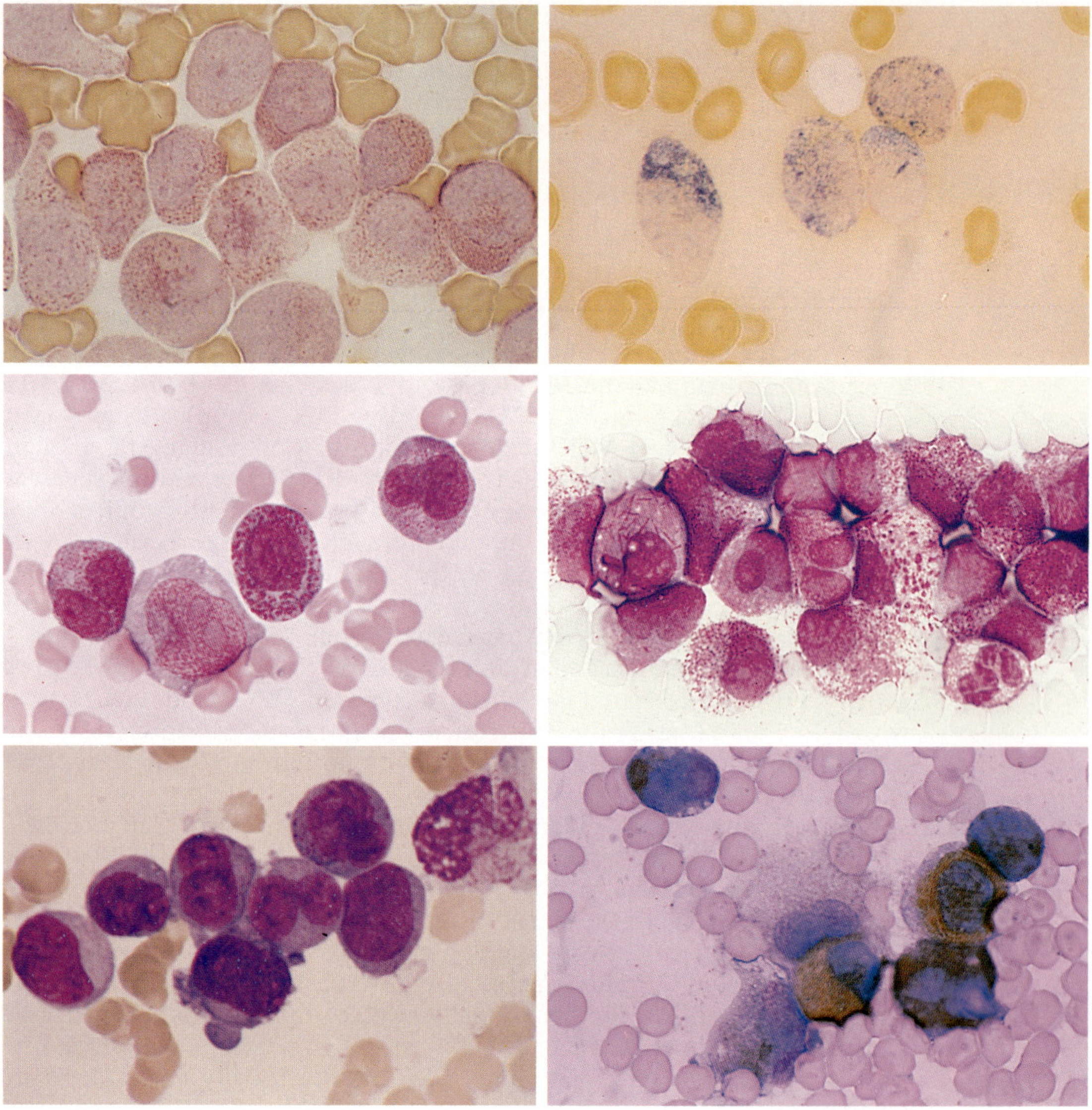

Plate 40

A. AMyL bone marrow. Acid phosphatase reaction showing scattered granules of positivity in the myeloblasts.

B. The same case. Dual esterase reaction. Four blast cells with variably strong CE positivity, including a single probable Auer rod.

C. APL bone marrow. MGG stain. One of 4 promyelocytes shows coarse granularity, the other 3 have less conspicuous granules or none, but the nuclear configuration is again that of APL.

D. Another APL bone marrow. MGG stain. Leukaemic promyelocytes predominate, showing very coarse azurophil granules, multiple Auer rods in some cells, and the characteristic nuclear lobulation or twinning often seen in APL.

E. A third APL bone marrow. MGG stain. In this case the cells are agranular, but show the typical nuclear appearance of APL.

F. APL bone marrow. Peroxidase reaction (o-tolidine).

Coarse positivity shown in most promyelocytes; the Auer rods appear to have a negative core with positive outline.

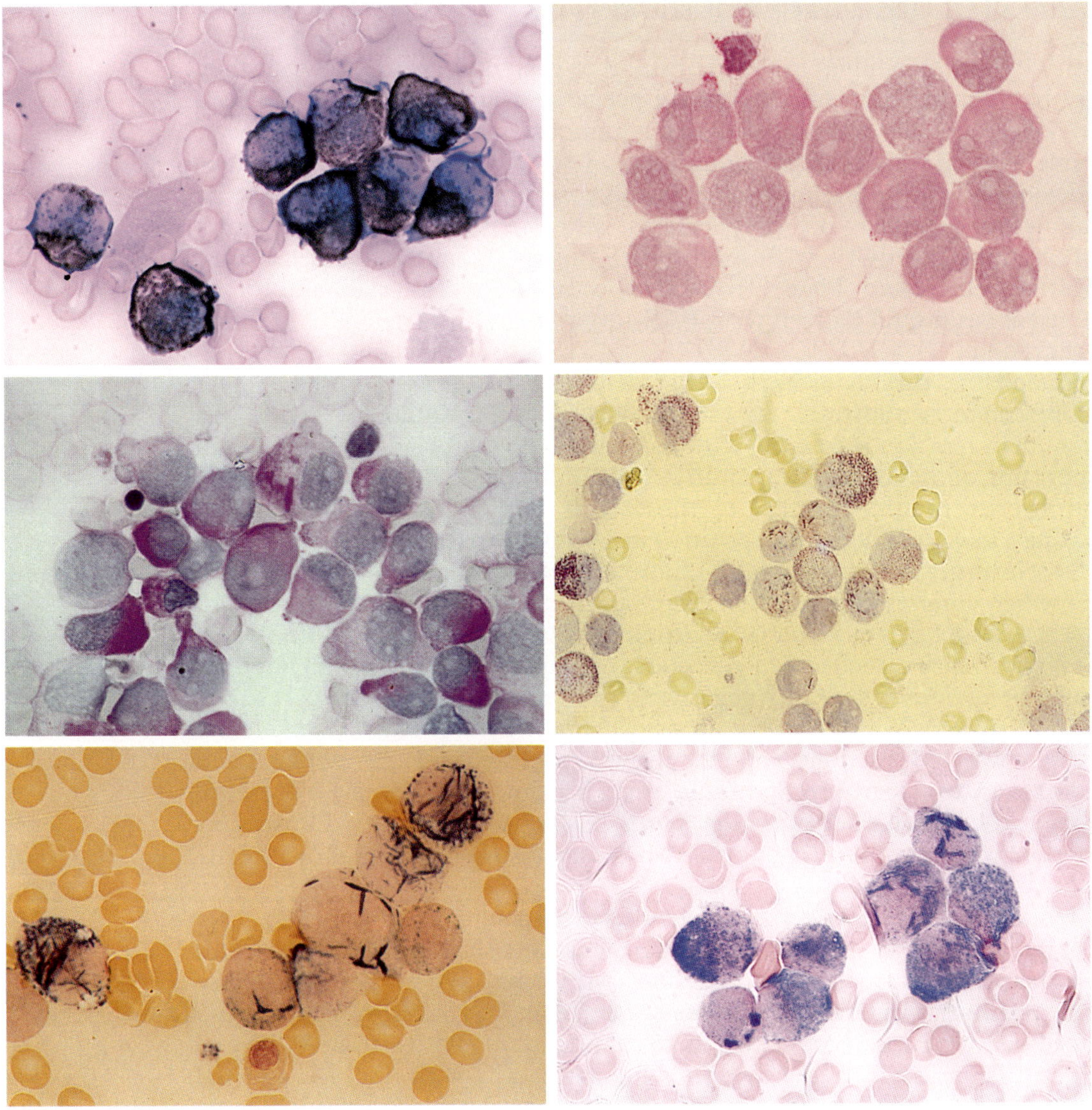

A B C D E F

Plate 41

A. APL bone marrow. SB stain. The reaction is even more strongly positive than the peroxidase reaction, but Auer rods again show a negative core.

B. APL bone marrow. PAS stain. Intense diffuse positivity in promyelocytes.

C. APL bone marrow. MGG stain. One of 4 promyelocytes shows coarse granularity, the other 3 have less conspicuous granules or none, but the nuclear configuration is again that of APL.

D. Another APL bone marrow. MGG stain. Leukaemic promyelocytes predominate, showing very coarse azurophil granules, multiple Auer rods in some cells, and the characteristic nuclear lobulation or twinning often seen in APL.

E. APL bone marrow. Dual esterase. Strongly CE-positive Auer rods are conspicuous.

F. APL bone marrow. Dual esterase. Strong CE positivity in multiple Auer rods and coarse inclusion material.

positivity are uncommon in the neutrophil or eosinophil cell series, but they may be seen in leukaemic monocyte precursors and in the basophil cell series and are not restricted to the lymphoblastic leukaemias where, of course, they are much more common.

Eosinophils in t(8;21) may show mixed eosinophil and basophil staining granules in Romanowsky preparations, and the PAS stain often shows intragranular positivity, with solidly PAS positive granules against a more negative background (Plate 43E) as also occurs in Inv(16) cases, and unlike the usual eosinophil pattern with PAS-negative granules against a positive background.

Among the various esterases, that demonstrated with naphthol AS-D chloroacetate as substrate shows positivity in the granulocyte series, the strength of reaction increasing with increasing cell maturity. In most cases a proportion of the leukaemic primitive cells show a positive reaction, and the findings are roughly parallel to those with Sudan black or peroxidase, although in general less often positive in undifferentiated or agranular blast cells. Nevertheless, occasionally chloroacetate esterase positivity may be seen when other cytochemical markers of the granulocyte series are absent, and the reaction is a valuable addition to the discriminating complex of cytological and cytochemical methods. α-Naphthyl acetate and butyrate esterase reactions are usually negative in granulocyte precursors, but this is not always true. We have seen already (Plate 17C) that normal precursors may show mixed reactions, and in leukaemic proliferations, myeloblasts and promyelocytes, clearly recognisable as such from other criteria—presence of Auer rods, coarse sudanophilia, peroxidase positivity, weak diffuse PAS reaction, positive chloroacetate esterase—may also sometimes show acetate or butyrate esterase positivity.

Tomonaga *et al.* (1985) made a detailed cytochemical analysis of 25 cases of APL. Sixteen (8 with the 15;17 translocation) showed pure neutrophil cytochemistry, with strong peroxidase (MPO) and chloroacetate esterase (CE) positivity, but no butyrate esterase (BE); 7 (5 with t(15;17)) had a mixed neutrophil/monocyte cytochemical phenotype with strong MPO and strong BE positivity, every case showing some cells with dual BE and CE positivity. The last two cases fitted neither group; one had intense lysozyme production but no BE and the other was negative for MPO and showed toluidine-blue metachromasia, suggestive of basophilic granulation. In Plates 41E, F and 42B are illustrated examples of APLs with the CE positive neutrophil pattern, and a mixed CE/BE reactivity—a rare example of predominantly BE activity. Presumably the esterase isoenzyme patterns usually confined to granulocyte and monocyte cell lines respectively may occasionally both be activated in the same cell line as a normally occurring variant or as a consequence of a leukaemic defect in the mechanism of cell metabolic control.

Unlike APL, with its frequent (15;17) translocation, cases of t(8;21) never showed BE positivity in the leukaemic cells (Swirsky *et al.*, 1984), but were always CE positive (Plate 43B), with CE-positive Auer rods detectable in about 25% of cases. CE-positive granules have been reported in eosinophils involved in the leukaemic process (Flandrin and Daniel, 1981), but we have not seen them in the 8;21 translocation. Most cases of AMyL with defects of chromosome 9q that we have studied show cytochemical features much resembling those of t(8;21) cases (Plate 44 A–F).

The naphthol AS-D acetate esterase reaction gives granular positivity, less strong than is usual in leukaemias with a monocytic component, and not inhibited by fluoride. Once again there are exemptions to this generalisation, where at least partial fluoride inhibition occurs, presumably also due to the presence of both granulocytic and monocytic esterase isoenzymes.

Cytochemical features of AML with mixed lineages, including monocytic, erythroid and megakaryocytic involvement

Apart from AMyL and APL the remaining morphological variants of AML listed in Table 14.1 usually involve some mixture of cell lineages; AMML by definition, and AMonL, AEL and AMegL either as a detectable feature at diagnosis or as a matter of natural developmental history during the course of the respective diseases. Indeed, studies with cells cultured in vitro suggest that even apparently pure granulocyte precursor AMLs may have a minimal monocytic component capable of emerging on stimulation with differentiation-inducing agents (Mastrangelo *et al.*, 1987). The subject is further discussed in Chapter 12.

Acute myelomonocytic leukaemia (AMML) and monoblastic leukaemia (AMonL)

Classification of acute leukaemias involving monocyte precursors has been a subject of controversy for many years, the difficulties reflecting uncertainty about the existence of a dual origin for monocytes. Historically the term 'pure monocytic' or 'Schilling type' leukaemias was applied to forms of the disease where there was little or no recognizable involvement of granulocyte precursors; these forms were thought to have arisen from stem cells within the reticulo-endothelial system with capacity to form monocytes and histiocytes, as postulated by Schilling (1926). In contrast, many cases with an obvious monocytic component also showed a variable but undoubted admixture of immature granulocytes and were referred to as the 'Naegeli type' of monocytic leukaemia, with presumed origin from a multipotential stem cell in the bone marrow, in conformity with Naegeli's (1923) view of the origin of monocytes from myeloblasts or early myelocytes.

Our own cytological and cytochemical studies (Hayhoe *et al.*, 1964) led us to believe that this difference proposed between two types of monocytic leukaemia in accordance with their supposed sites of origin was no longer valid, but that all could most sensibly be grouped in a single category of 'myelomonocytic' leukaemia with widely varying potential proportions of monocytic to granulocytic elements, and that all were probably derived from a bone marrow stem cell. Laboratory work on the kinetics of promonocytes and monocytes in the bone marrow, and with marrow colony growth *in vitro* has supported this view by confirming that normally monocytes and macrophages are derived solely from the bone marrow

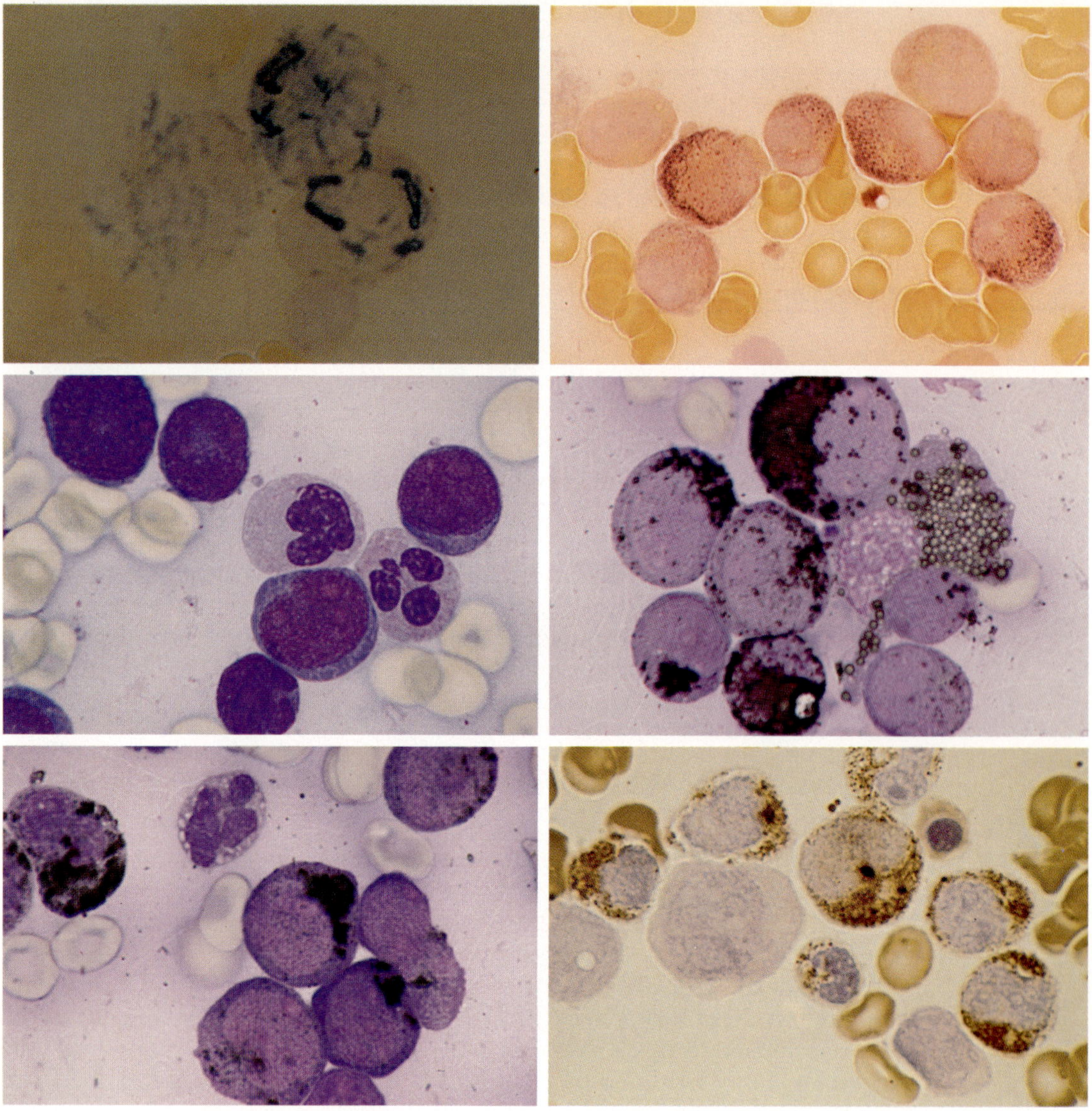
A B C D E F

Plate 42

A. APL bone marrow. Dual esterase. A group of promyelocytes with CE positivity and conspicuous Auer rods some of which show a negatively reacting core with CE positive outline.

B. APL bone marrow. Dual esterase. This case, unusually, shows a mixture of both CE and BE granular reactivity, with a predominance of BE positivity.

C. t(8;21) AMyL bone marrow. Heyl stain. Variable size myeloblasts and agranular neutrophils.

D. The same case. SB stain. Very dense, usually localized sudanophilia, probably obscuring Auer rods; an eosinophil is partially degranulated but shows typical SB reaction.

E. The same case. SB stain. Another field showing similarly strong localized reactions but with hollow Auer rods visible in two cells. A mature neutrophil is SB negative.

F. Another t(8;21) AMyL bone marrow. Peroxidase reaction (DAB) showing strong localized positivity and reactions in pseudo-Chediak Higashi inclusions.

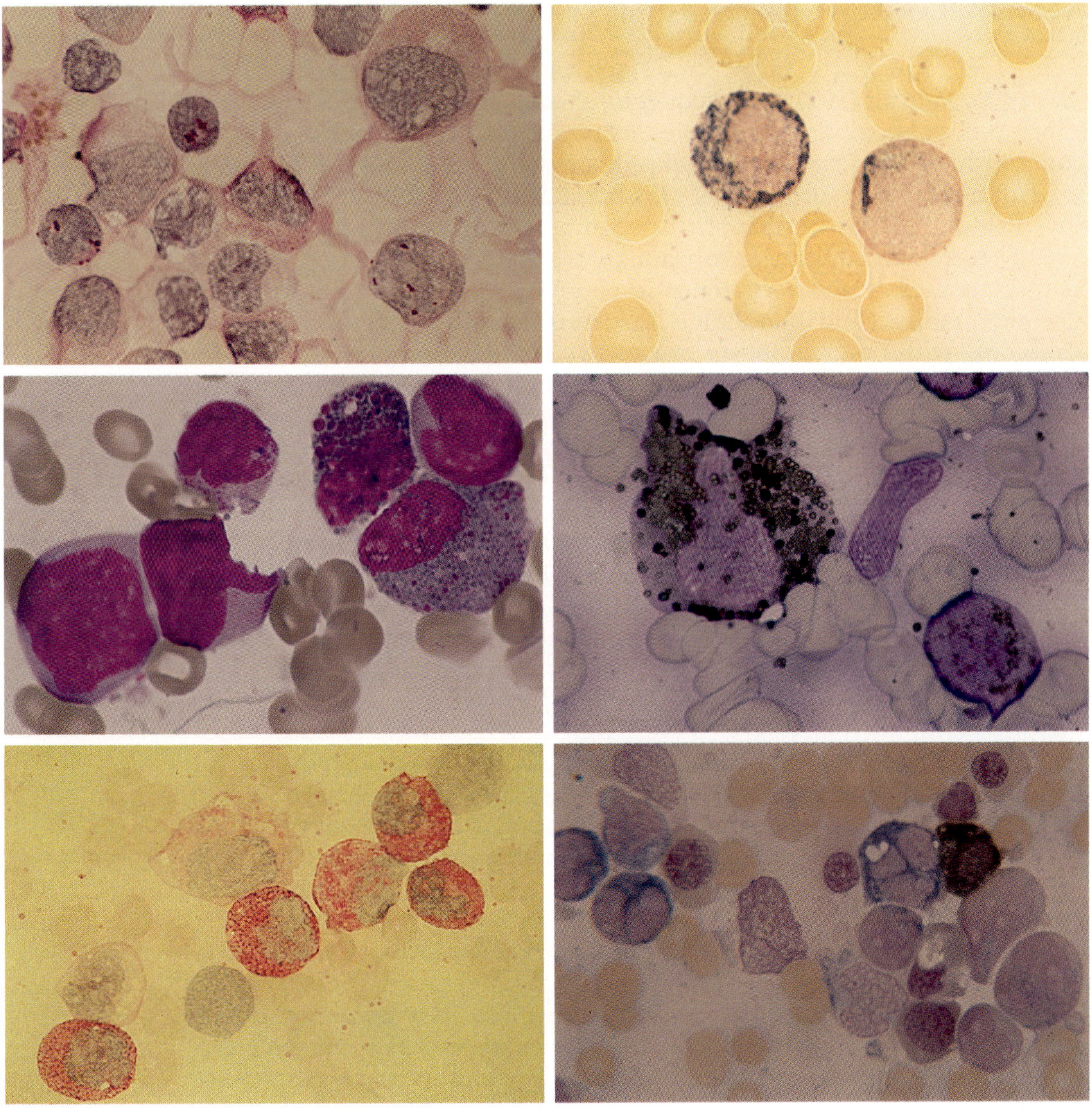

A B C D E F

Plate 43

A. t(8;21) AMyL bone marrow. PAS stain. Occasional coarse rods and blocks of positivity in some blast cells, together with a weak diffuse reaction in most leukaemic cells.

B. The same case. Dual esterase reaction showing strong CE positivity with a CE-positive Auer rod.

C. Inv(16) AMyL bone marrow. Heyl stain. Three blast cells and a promyelocyte together with two abnormal eosinophils showing variably mixed granules.

D. The same case. SB stain. Two abnormal eosinophils with variable sudanophilia of granules, some showing hollow centres and others not.

E. Another Inv(16) AMyL bone marrow. PAS stain. Variable positivity of granules in a group of atypical eosinophils, some showing hollow centres and others staining solidly. Blast cells present are negative or show weak diffuse positivity.

F. Another Inv(16) AMyL bone marrow. Dual esterase.

Erythroblasts are negative; myeloblast cells vary from negative to strongly CE positive; a single probable monoblast is strongly BE positive and 2 abnormal eosinophils, one of them disrupting, show CE positive granules.

(Leder, 1967; van Furth and Cohn, 1968; van Furth and Diesselhoff-den Dulk, 1970; Metcalf, 1971). Subsequently, however, a number of reports of clinical and cytological case studies again raised the suggestion that a distinct form of pure monocytic leukaemia may be separated (McKenna *et al.*, 1975; Shaw and Nordquist, 1975; Schiffer *et al.*, 1975; Brynes *et al.*, 1976). Features characteristic of this category are said to include clinically a tendency to extramedullary involvement, frequent lymphadenopathy and hepatosplenomegaly and occurrence in young age groups, cytologically an exclusively monocytoid population as determined by light and electron microscopy, with no granulocyte precursors and an absence of Auer rods, and cytochemically negative peroxidase, Sudan black and chloroacetate esterase reactions, but variably positive PAS reactions, with granular positivity, and fluoride inhibited α-naphthyl acetate, butyrate and AS-D acetate esterase reactions.

This collection of features contains nothing, other than the absence of granulocyte precursors, not found from time to time in cases of obviously mixed myelomonocytic cytology. The monocytes and promonocytes in such mixed leukaemias have characteristic features by light and electron microscopy differing in no way from those of the so-called 'pure' or Schilling monocytes; Auer rods probably never occur in monocytes; the monocytes and their precursors from mixed cases are often predominantly and sometimes completely negative to both Sudan black and peroxidase; they too may be negative for chloroacetate esterase and are usually positive for other non-specific esterases, with fluoride inhibition, although some mixed reactions may be found when double esterase staining is used; and they too show variably positive PAS reactions. There seems no reason from cytological or cytochemical considerations to believe the monocytic component of cases with 20%, 50% or 80% of monocytes to be essentially different in kind from the component in the occasional cases with 90 to 100% monocytes.

A combined cytological, ultrastructural, cytochemical and immunological study of the cells of a 'pure monocytic' acute leukaemia as compared with six cases of clearly mixed myelomonocytic leukaemias, with percentages of monocytic cells varying from about 15 to 80, was carried out by Cawley *et al.* (1978). The lack of differential specificity of morphological and cytochemical criteria was confirmed, and the leukaemic monocytes in both pure and mixed cases were found to express a receptor for the Fc of IgG which probably acted as a phagocytosis receptor, since IgG coated particles were preferentially ingested. A variable but lower percentage of leukaemic monocytes had a receptor for mouse C3, but all cells uniformly lacked surface immunoglobulin and receptors for the Fc of IgM and for unsensitized mouse erythrocytes. Neither the immunological nor the cytochemical or ultrastructural studies lent support to the concept that distinct types of leukaemic monocytes existed.

The advent of monoclonal antibodies (McAbs) reacting with cell surface antigens of various specificities including those with monocyte lineage specificity, such as CD14 (MO2, UCHM1) has not provided evidence to support such a distinction, although it has brought to light more heterogeneity among cases of AMonL than expected from simple cytological criteria, and thrown further doubt on the reliability of 'monocyte' esterase cytochemistry as a marker of monocyte precursors. Milligan *et al.* (1984), for example, applied the McAbs MO2, 63D3 and LeuM3, each with a degree of monocyte lineage specificity, to the leukaemic cells from 12 cases of 'morphologically unequivocal' AMonL. Six were esterase positive and McAb positive, 3 esterase positive and McAb negative, and 3 esterase negative and McAb positive.

Finally the advances during the last decade in enzyme markers in acute leukaemias (Drexler *et al.*, 1984) while confirming the high levels of acid phosphatase, beta-glucuronidase and beta-glucosaminidase, and also that of lysozyme in monocyte precursors, have not suggested that separable types of monocyte, different in origin or lineage, exist.

For the present therefore, and in the absence of evidence to the contrary, we continue to regard all forms of monocytic leukaemia, whether the monocyte precursors are obviously mixed with granulocyte precursors (AMML) or appear relatively unmixed (AMonL) as only very broadly separable groupings within a continuous spectrum. The defining features of these groups, shown in Table 14.1, require no elaboration, and the additional features of diagnostic help, listed in Table 14.6, have already been partially discussed above. In general, the cytochemistry shows a mixed pattern which helps to assess the involvement of granulocytic and monocytic cell lines. Sudanophilia and peroxidase positivity are conspicuous in the granulocyte precursors whereas most leukaemic monocytes show negative reactions to peroxidase and only occasionally weak discrete scattered granule type positivity to Sudan black (Plate 45 A,B,C). On the other hand the acid phosphatase reaction and the β-glucuronidase and β-glucosaminidase reactions are usually strong in monocyte precursors (Fig. 14.10, Plate 46A).

The variability of the PAS reaction in acute myelomonocytic leukaemia also partly reflects the mixed population of cells present. Some blast cells are entirely negative, others show a reaction like that found in myeloblastic leukaemia, with weak tinging of the cytoplasm and perhaps fine granulation, while others give the PAS reaction typical of many normal monocytes, with fine to moderately coarse granulation against a background of diffuse cytoplasmic staining (Plate 45D). Rarely, much stronger positivity may be encountered in leukaemic monocyte precursors, with coarse granules and even blocks of glycogen found (Plate 45E,F). When this type of positivity is present it can nearly always be readily distinguished from that of lymphoblasts because monoblasts and promonocytes show a background of diffuse positivity in most of the reacting cells, whereas lymphoblasts do not.

The esterase reactions have come to be regarded as especially helpful in recognising leukaemic monocyte precursors, because these cells often contain strongly reacting non-specific esterase demonstrable with butyrate, acetate or AS-D acetate (Plate 46B,C,D) as the commonly used substrates, and sharply inhibited by fluoride. Chloroacetate esterase, present in the granulocytic component of mixed myelomonocytic leukaemias, is typically absent from monocyte presursors. These generalisations, while offering

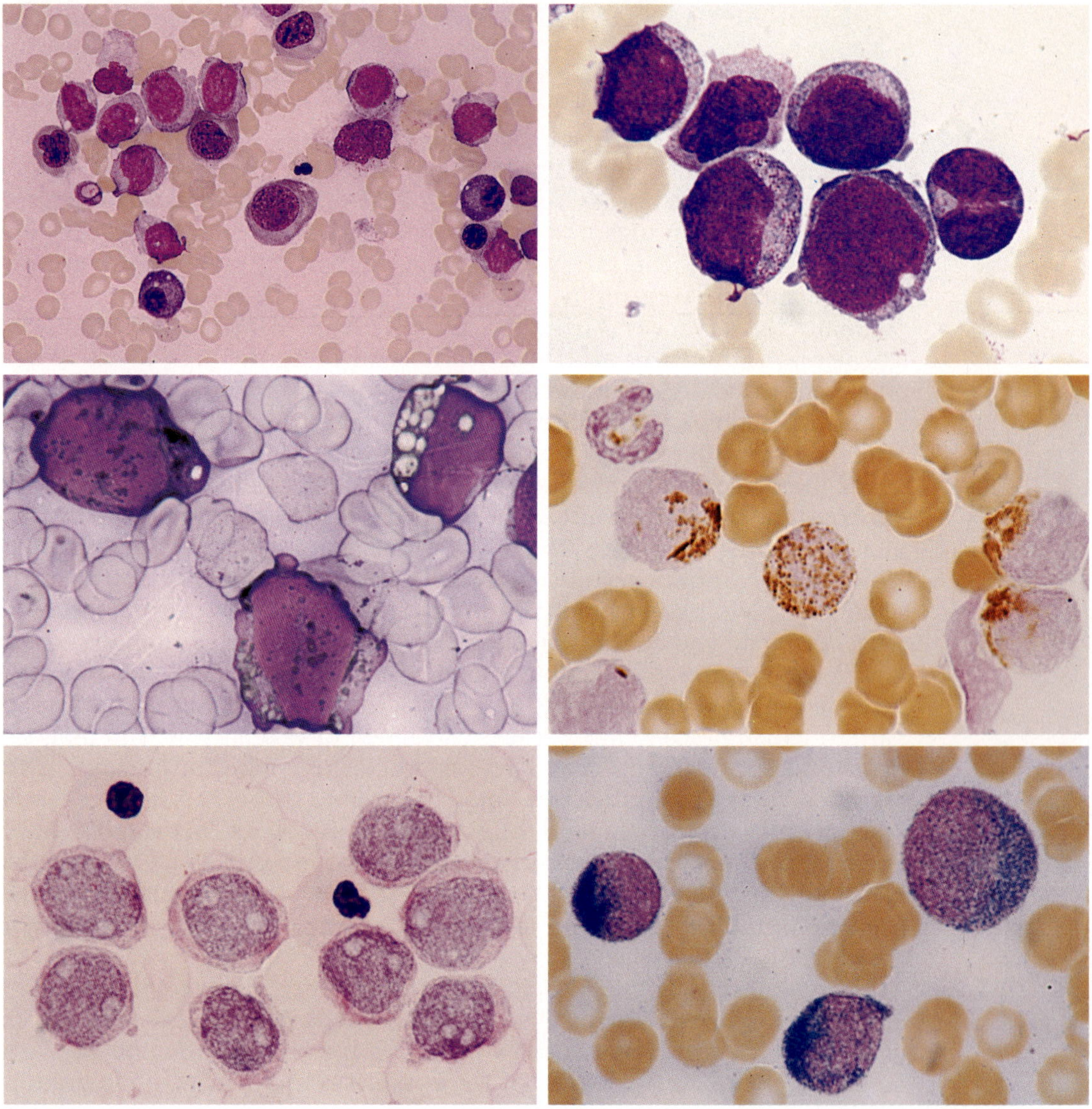

A B C D E F

Plate 44

A. 9q- AMyL bone marrow; Heyl stain. Low power view showing nine myeloblasts with often irregular nuclear outlines and ample cytoplasm, sometimes vacuolated, six erythroblasts, some with dysplastic features, and two plasma cells.

B. The same case. Heyl stain. A high power view of 6 blast cells, one with 'cottage-loaf' nucleus, one monocytoid and one with promyelocytic granules.

C. Another 9q- AMyL bone marrow. SB stain. Three blast cells show a range of sudanophilic inclusions from faintly staining vacuolar structures to needle-like Auer rods.

D. The same case as in A and B. Peroxidase stain (DAB). A weakly reacting stab, three blast cells with localised positivity and various Auer rods and inclusion material, and a probable abnormal eosinophil with positive granules.

E. The same case as in C. PAS stain. Eight blast cells show only weak diffuse cytoplasmic positivity; two negative erythroblasts are present.

F. The same case as in A, B and D. Dual esterase. Three myeloblasts show strong CE positivity.

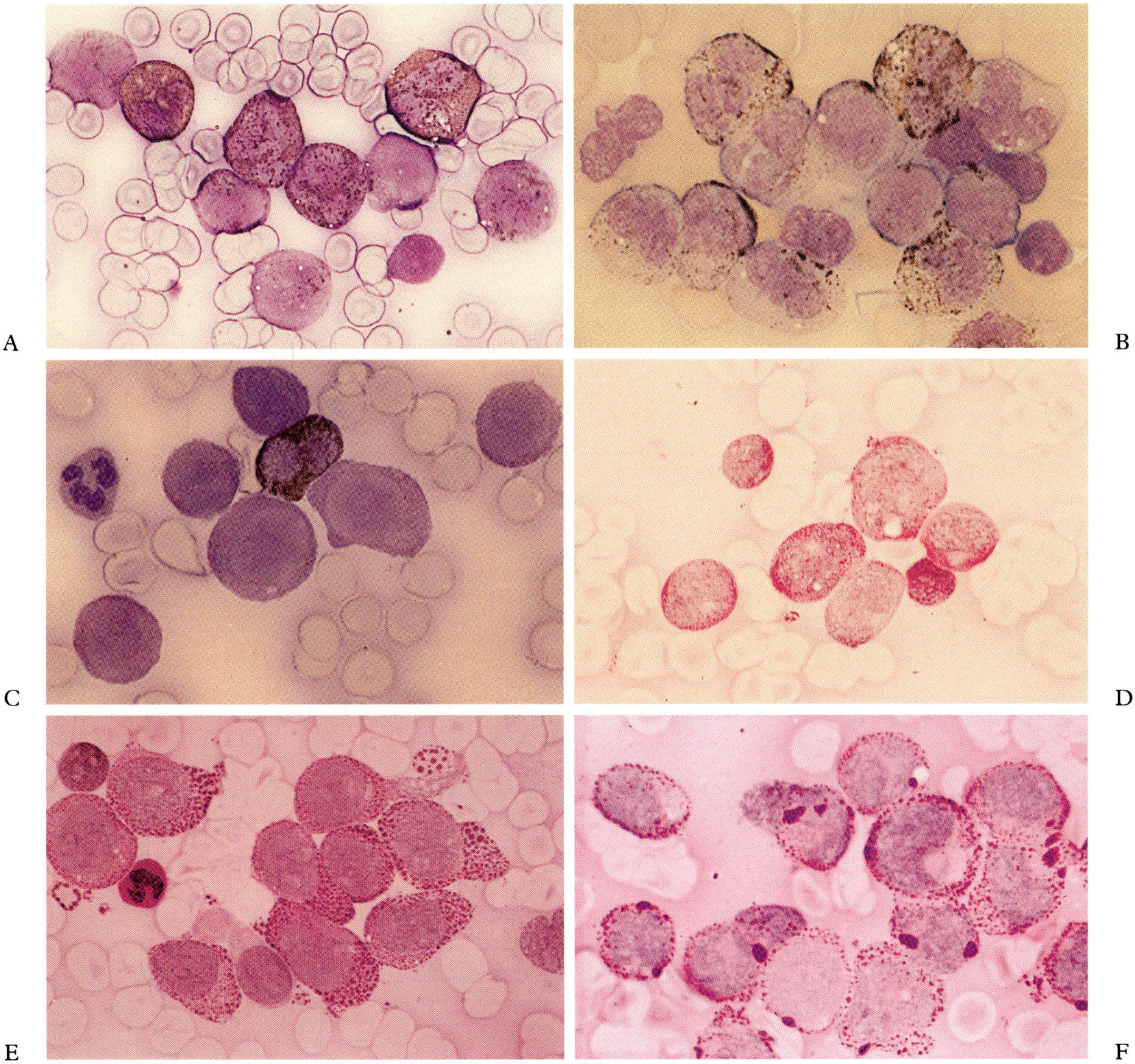

Plate 45

A. AMML bone marrow. SB stain. There is scattered discrete granular positivity in most leukaemic cells of monocytoid lineage with heavier overall positivity in granulocyte precursors and a positive polymorph.

B. AMML bone marrow. SB stain. A more predominantly monocytic sample with fine scattered granules in most cells.

C. AMML bone marrow. SB stain. In this case immature monoblasts predominate, showing negative reaction to SB. A negative neutrophil and a positive metamyelocyte are also seen.

D. Same case as in A. PAS stain. Fine granular and diffuse cytoplasmic positivity in all cells.

E. Same case as in B, PAS stain. One normally positive neutrophil, a negative micromyeloblast and coarse granular and diffuse positivity in the predominating monocytoid blasts.

F. Same case as in C. PAS stain. Coarsely granular positivity with frequent blocks against a background of diffuse positivity. This is a very characteristic, though relatively uncommon, PAS pattern in AMML with monocytic predominance (AMonL).

Table 14.6 Cytochemical features of acute myelomonocytic leukaemia (AMML) and acute monocytic leukaemia (AMonL)

The leucocyte alkaline phosphatase (LAP) score is usually normal or high.

Blast cells often show Sudan black positivity, granulocyte precursors localized or dense overall reactions, monocyte precursors may be negative or may show discrete scattered granules.

Peroxidase reactions of monocyte precursors are usually negative.

PAS reaction is variable; usually ranges from negative to diffuse positivity with fine granules superimposed. Coarse granules or even blocks sometimes seen, but usually on a background of diffuse positivity. PAS positive erythroblasts may be seen when there is an erythroid component.

Acid phosphatase strongly positive, especially in monocyte precursors, as is beta-glucuronidase and beta-glucosaminidase. Isoenzyme Hex I is weak or absent.

Esterase activities against alpha-naphthyl acetate and butyrate are positive in monocytic cells, varying from weak to intense.

Naphthol AS-D acetate esterase is more positive in monocytic cells than in granulocytes and is inhibited by sodium fluoride, unlike the positivity in the granulocytic series.

Naphthol AS-D chloroacetate esterase (CE) is usually negative in monocytic cells but positive in the granulocytic component; mixed reactions may occur. Abnormal eosinophils in a subset with Inv(16) or del(16) and eosinophilia may show CE-positive granules.

5′nucleotidase generally negative, DAP II usually positive but DAP IV negative, gamma-glutamyl transpeptidase strongly positive in most monocyte precursors, variable in granulocyte precursors.

valuable guide lines to the interpretation of esterase cytochemistry in the acute leukaemias, do not invariably hold true. Double esterase reactions, for example, will not infrequently show positivity to both chloroacetate (granulocytic) and butyrate or acetate (monocytic) esterases in the same cell (Plate 46D), while a variable proportion, sometimes high, of the leukaemic monocyte precursors in some cases of myelomonocytic leukaemia may show weak or even negative reactions for monocytic esterase.

Huhn and Twardzik (1983) retrospectively studied 39 cases of myelomonocytic or monocytic leukaemias, classified as FAB M4 or M5, by the use of myeloperoxidase (MPO), acetate esterase (AE) and chloroacetate esterase (CE) stains. In 8 cases almost all the leukaemic blast cells showed dual granulocytic/monocytic enzyme expression, with both MPO and CE positivity, but also AE positivity. On the basis of MPO and CE reactions these 8 cases might be regarded as AMyL (M1 or M2), while on the basis of the AE reaction they would be regarded as M5. Similarly marked cytochemical heterogeneity especially in esterase expression was reported by Scott and his colleagues (1985) in a study of 39 cases classified as M4 (AMML). They found more consistent positivity in lysozyme production and in the expression of characteristic monocytic esterase isoenzyme patterns demonstrable by isoelectric focusing. Immunological studies with a panel of McAbs also showed more consistent reactions than did cytochemistry in marking monocyte precursors, although expression of monocyte-associated antigenic determinants also showed considerable variability. The occurrence of mixed reactions was again confirmed by Li *et al.* (1986), who

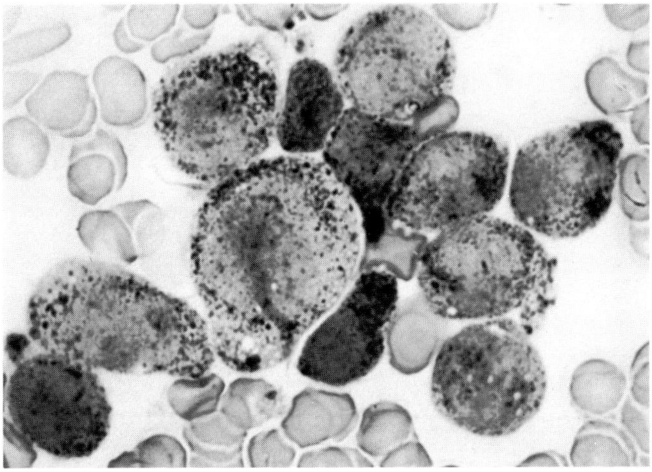

Fig. 14.10. AMML. Acid phosphatase. Strong reaction in monocyte precursors.

reported five cases of AMML, all men, two with a preceding preleukaemic phase, all of whom showed many leukaemic cells positive for both CE and BE.

Even when only acetate esterase reactions are studied, the positivity may vary in depth and pattern very considerably. Wrotnowski *et al.* (1987) found, for example, that 6 of 14 cases of acute monocytic leukaemia classified as M5 had finely granular diffuse cytoplasmic positivity for acetate esterase, whereas 8 had dense focal staining against a weak cytoplasmic background reaction. All of the first group had typical hypercellular acute leukaemic bone marrows, as did 4 of the second group, but the remaining 4 of that group were hypocellular, although no consistent differences were found among the cases in clinical features or in the course of the disease.

The technical methods available for demonstrating surface antigen expression have been discussed in some detail in Chapter 3, but it is relevant in the present context to note that different methods may not always give the same results. Plate 46 E, F shows clear positivity against McAb OKM1 manifest by acute leukaemic blast cells when the immuno-alkaline phosphatase (APAAP) method was used, but negative reaction with the same antibody and the same blast cells when the immunoperoxidase (PAP) was used. The difference in this case may well arise from the previous use of methanol and hydrogen peroxide treatment in the latter technique to inhibit endogenous peroxidase, but methodological variation is already as important in immunochemistry as in cytochemistry. These variations from the usual pattern deprive the esterase reactions of complete specificity; nevertheless they are among the most useful of current cytochemical techniques when their capacity is not overestimated and the possibility of mixed phenotypic expression of differentiation genes usually restricted to separate lineages is kept in mind. An even more remarkable and unusual example of such mixed expression was reported by Tomonaga *et al.* (1984) in a child with AMonL whose leukaemic monocytes contained a subpopulation with

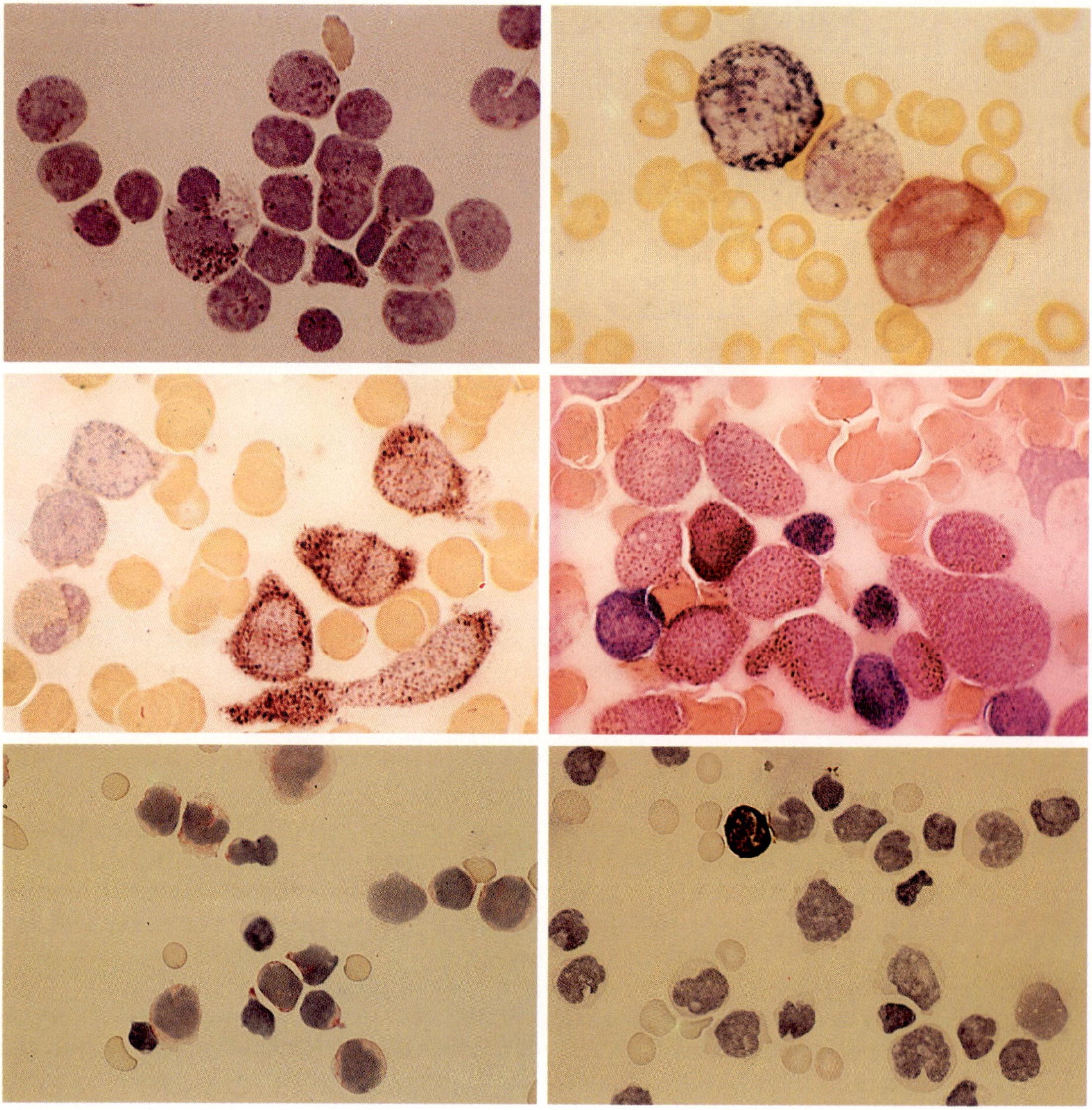

Plate 46

A. AMML bone marrow. Acid phosphatase, showing strong granular positivity.

B. Same case as in 45A and D. Dual esterase, with strong CE positivity in a granulocyte precursor, strong BE positivity in a monocyte precursor and weaker mixed reactivity in a probable myelocyte.

C. Same case as in 45B and E. Dual esterase, with strong BE positivity in four monocytoid blasts and weaker CE reaction in two myeloblasts. An eosinophil is esterase negative.

D. Same case as in 45C and F. Dual esterase, with strong BE positivity in eleven monoblasts, strong CE reaction in two myeloblasts and two neutrophils and a mixed reaction in a cell of indeterminate lineage.

E. AMML bone marrow. Immunoalkaline phosphatase (APAAP) reaction with McAb OKMl, showing positive reaction in most cells.

F. Same case as in E. Immunoperoxidase (PAP) reaction, with same McAb, showing positivity in only one cell.

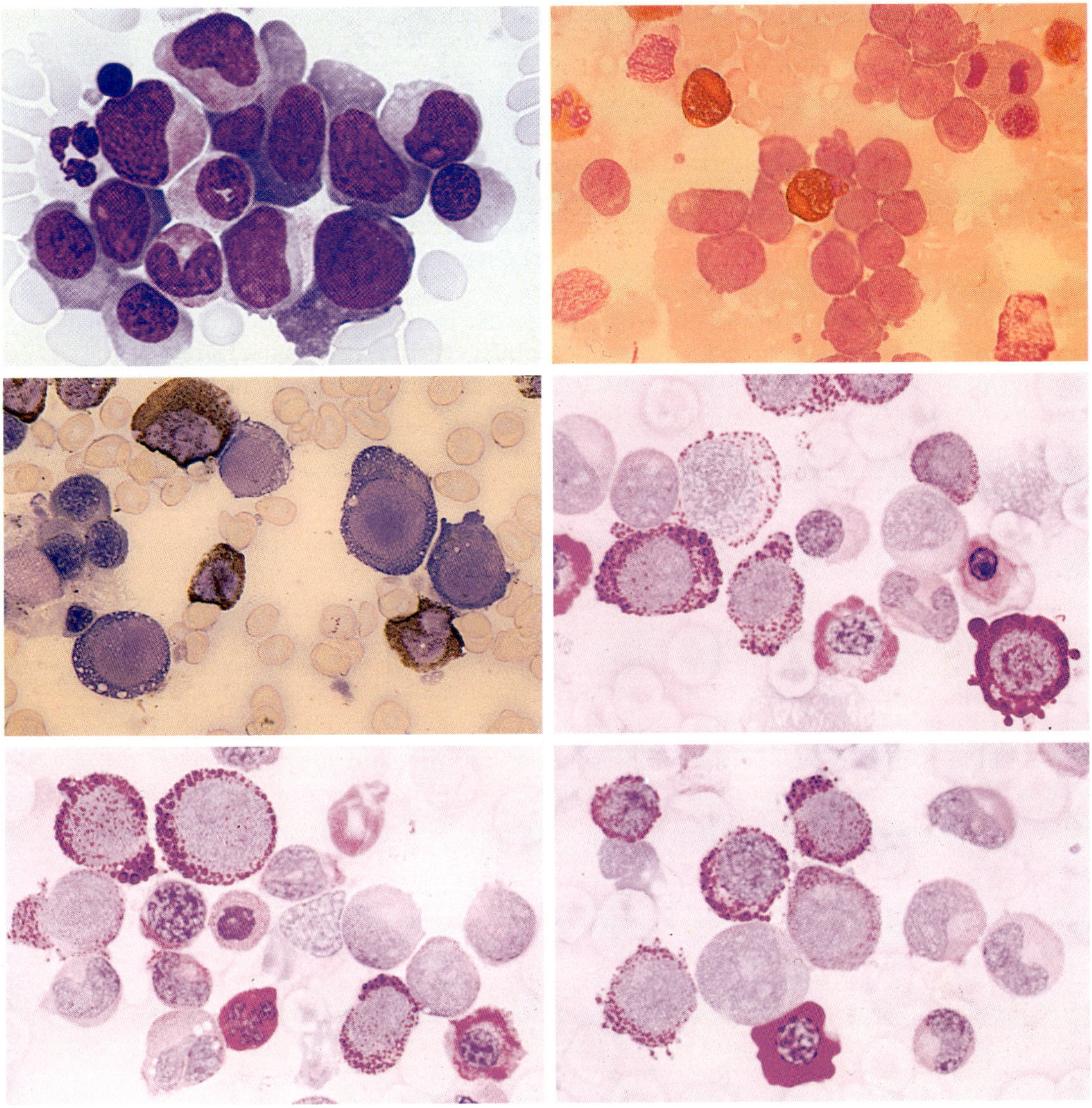

Plate 47

A. Erythroleukaemia. Bone marrow. Leishman stain. A mixture of granulocyte and erythrocyte precursors, one of the proerythroblasts being binucleated.

B. The same case. Peroxidase reaction. The few granulocyte precursors in this field show positive reactions: the predominating proerythroblasts are peroxidase negative.

C. The same case. SB reaction. The erythroid precursors are uniformly negative. Normal sudanophilia is shown by the few granulocytes present.

D. The same case. PAS reaction. Myeloblasts show a weakly positive cytoplasmic tinge, while proerythroblasts and normoblasts at all stages of maturity show intensely positive diffuse and coarsely granular reactions.

E. Erythroleukaemia. Another case. PAS reaction showing intense positivity from proerythroblast to late normoblast and diffuse positivity in a mature erythrocyte. Granulocyte precursors from myeloblast onwards show weaker reactions but increasing positivity with increasing cell maturity.

F. Another field from the same preparation as E. The erythroblastic and myeloblastic components of the neoplastic state are clearly separable with intensely positive PAS reaction in the former.

alkaline phosphatase activity and 'neutrophilic' granules despite being strongly positive for monocytic esterase and active in phagocytosis of opsonised zymosans.

A variant of AML with abnormal marrow eosinophils associated with an inversion or deletion of chromosome 16 involving a breakpoint at 16q22 generally shows mixed myelomonocytic cytology (Arthur and Bloomfield, 1983; Le Beau *et al.*, 1983). Bitter *et al.* (1984) reported a morphological and cytochemical study of 19 patients with inv(16) (p13q22). All had myelomonocytic cytology, 15 of the 19 showed an increase in marrow eosinophils and 18 of the 19 showed abnormal eosinophils, with large basophilic granules. These granules were frequently PAS positive and showed chloroacetate esterase positivity (Plate 43 C–F).

Erythroleukaemia

Although the term Di Guglielmo's disease or erythraemic myelosis has been used to refer to cases of neoplastic proliferation of erythroblasts, there is virtually always evidence that such proliferations are accompanied by parallel involvement of other myeloid precursors (Scott *et al.*, 1964; Hayhoe *et al.*, 1964). Just as a continuous spectrum of mixed proportions of granulocytic and monocytic cells occurs in myelomonocytic leukaemias, so in erythroleukaemia there is a continuous spectrum from a small to a preponderant component of erythroblasts in a mixed myeloid proliferation where either granulocyte or monocyte precursors, or both, make up the remainder of the neoplastic population.

The additional features of value in the diagnosis of erythroleukaemia listed in Table 14.7 reflect this combination of cell series. When granulocyte precursors are involved promyelocytes may be recognised, Auer rods may be seen, and cytochemical staining with Sudan black, peroxidase, PAS and chloroacetate esterase show the granulocyte type of reaction. When monocyte precursors, or a mixture of monocyte and granulocyte precursors are present, the cytochemistry shows the mixed reactions to be expected in these cell lines. The neoplastic erythroblasts themselves often show abnormal morphology and mitotic irregularities, and cytochemically have commonly but not always, strong PAS positivity (Plate 47 D,E,F) (Baldini *et al.*, 1959; Quaglino and Hayhoe, 1960).The intensity of the reaction and the percentage of erythroblasts showing positivity vary considerably in different patients and also in the same patient during the course of the illness. The strongest reactions are usually associated with bizarre cytological and mitotic abnormalities, but high PAS levels occur also when erythropoiesis shows a more normal pattern. The positive material is mostly removed by amylase and is therefore most probably glycogen. In contrast with iron-deficiency anaemia, in which a rough correlation existed between the degree of hypochromia and the intensity of PAS positivity, no such relationship was observed between the degree of PAS positivity and hypochromia in erythroleukaemia and erythraemic myelosis. A parallel increase in phosphorylase positivity in erythraemic erythroblasts was noted by Hayhoe *et al.* (1964) and by Kass and Hadi (1975). They also usually

Table 14.7 Cytochemical features of erythroleukaemia (EL)

Strong PAS positivity, granular or diffuse, is present in red cell series at all stages of maturity. Abnormal myeloid precursors present show PAS positivity resembling that in granulocytic or myelomonocytic leukaemia.
Sudan black and MPO reactions are negative in erythroblasts, but with typical reactions in involved leucocyte precursors, which also show the esterase reactions expected for granulocytic or monocytic cell lines. PPO-like activity present in erythroblasts.
Alpha-naphthyl butyrate and acetate esterases are strongly positive in erythroblasts.
Strong paranuclear unipolar acid phosphatase activity present in erythroblasts.

show positive reactions for α-naphthyl acetate and butyrate esterase (Plate 48D, E) and a marked localized paranuclear acid phosphatase activity (Plate 48C) (Leder, 1967). Cytoplasmic basophilia and associated pyroninophilia are often a conspicuous feature in neoplastic erythroblasts (Plate 48F).

In an interesting recent study of the ultrastructural and cytochemical characterization of blasts from early erythroblastic leukaemias, Breton-Gorius and her colleagues (1987) examined cells from 9 cases, 2 with positivity to glycophorin A and 7 with a more immature phenotype, labelling with FA6-152 McAb by an ultrastructural immunogold technique. Many early erythroblasts showed platelet peroxidase (PPO)-like activity, had theta granules in the Golgi apparatus and contained ferritin molecules in the absence of rhopheocytosis. Parallel studies of the same specimens identified occasional promegakaryoblasts by immunogold labelling with C17, a McAb to platelet glycoprotein IIIa, and showed that while these cells also manifested strong PPO activity they lacked theta granules and ferritin. A similar PPO-like activity and ferritin molecules were also detected in immature erythroblasts from normal bone marrow.

Erythroleukaemia as defined in the FAB classification (M6) with erythroblasts making up 50% or more of the nucleated cells in the bone marrow, is an uncommon form of AML. In our study of 619 AML cases only 15 met this criterion. (Swirsky *et al.*, 1986). All of these showed granular or diffuse PAS positivity of erythroid precursors. However, erythroblast positivity was found in a further 118 cases not classifiable as erythroleukaemia on FAB criteria, and was associated with a poor response rate. Moreover, a rising percentage of erythroid precursors irrespective of morphological and cytochemical features was also associated with a falling remission rate. It seems unlikely, therefore, that most red cell precursors in AML marrows represent residual normal marrow cells, and more probable that increasing numbers of erythroblasts represent an increasing likelihood of erythroid line involvement in the neoplastic process. Erythroleukaemia in the sense of leukaemia with an erythroid component, not necessarily dominant, appears to be relatively common, thus underlining the fact that 'predominant cell' classification systems are to a greater or less extent arbitrary creations which may encourage standardization of terminology partly at the expense of biological relevance.

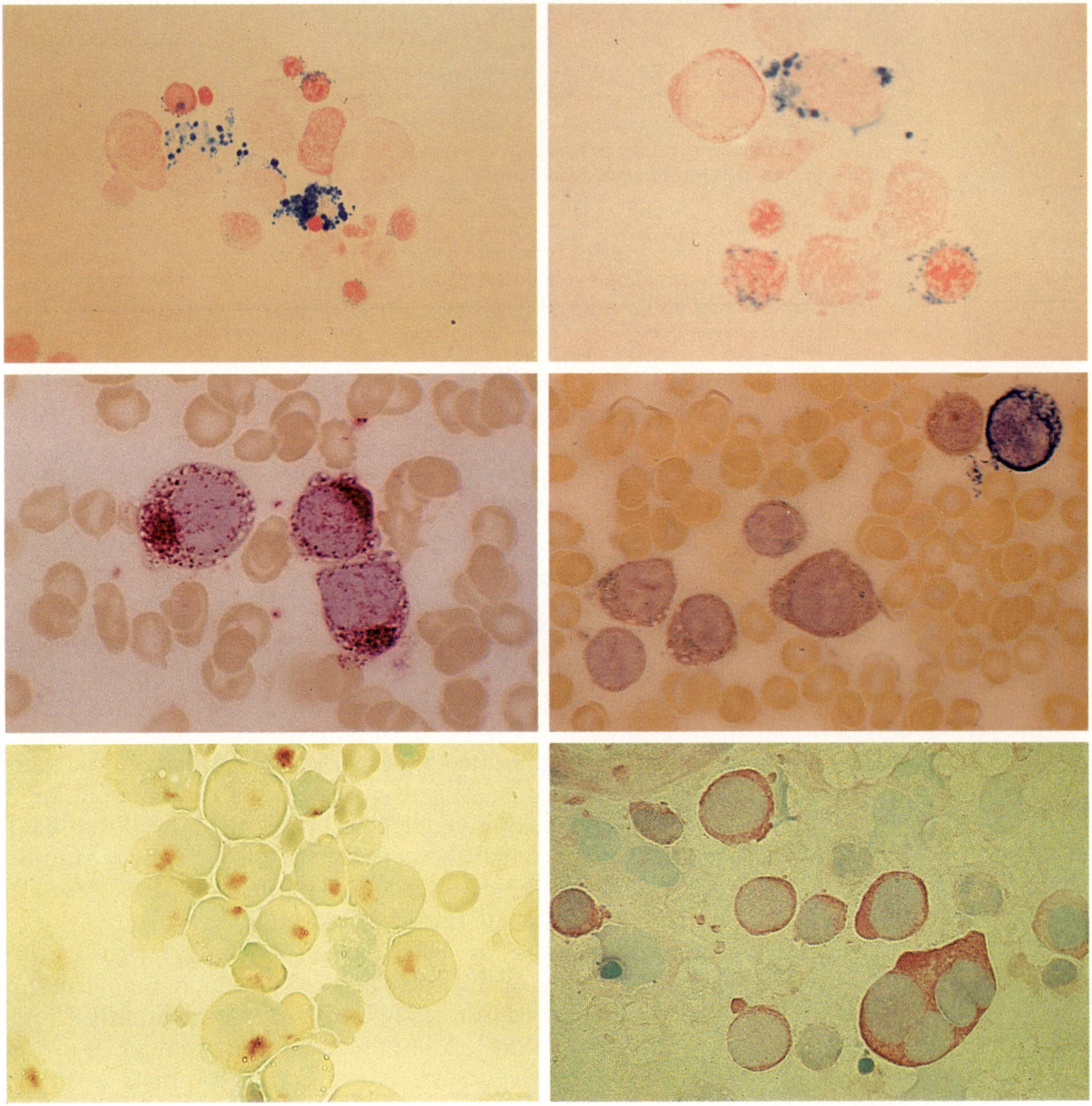

A B C D E F

Plate 48

A. Erythroleukaemia bone marrow. Prussian blue stain. Scattered iron granules from a disrupted macrophage, and positive reaction in later erythroblasts.

B. Another field from the same preparation as in Plate 48A. An iron laden macrophage and free iron particles in two intermediate erythroblasts. Other cells are negative.

C. Erythroleukaemia bone marrow. Acid phosphatase. Strong focal positivity in three early erythroblasts.

D. Erythroleukaemia bone marrow. Dual esterase. A single strongly CE positive granulocyte precursor and six erythroblasts showing weaker mixed esterase reactions, BE predominating.

E. Erythroleukaemia bone marrow. Acetate esterase (pararosaniline as receptor), showing strong focal positivity in the erythroid precursors, chiefly proerythroblasts.

F. Erythroleukaemia bone marrow. Methyl green-pyronine stain, showing strong cytoplasmic pyroninophilia, indicative of high RNA content.

Acute megakaryocytic leukaemia (AMegL)

AML cases with a megakaryoblastic-megakaryocytic component, or even predominance, have come increasingly to be recognized with the use of ultrastructural platelet peroxidase and immunological markers to supplement the PAS reaction, differential esterase activities and possible 5′ nucleotidase 5′N reactivity. The main diagnostic features are summarized briefly in Table 14.1 and the cytochemical findings are listed more fully in Table 14.8 and illustrated in Plate 49.

In the course of a report on three cases of this disease, den Ottolander and his colleagues (1979) drew attention to the differential value of non-specific esterase reactions in recognising megakaryoblasts which, as earlier reported by Li *et al.* (1973) and confirmed by Grusovin and Castoldi (1976), may show a positive reaction with α-naphthyl acetate as substrate but negative reaction with α-naphthyl butyrate. In our experience this discrimination, although highly suggestive, is not always clearcut, and the characteristic pattern of strong peripheral PAS positivity shown by many cells of the megakaryocyte series is more often helpful, but both reactions are of value in doubtful cases.

Ultrastructural evidence of bull's eye granules and demarcation membrane system may not always be visible in early megakaryoblasts, but the specific platelet peroxidase pattern may be demonstrable (Breton-Gorius *et al.*, 1978).

In studies of 5′N cytochemistry in the blast cells from over 50 cases of AML in our laboratories most were found to be negative with only slight reactivity in one AMyL and two AMML cases, whereas a single clear case of AMegL with predominance of megakaryoblasts showed strong positivity and a second mixed case showed lesser positivity in a few blast cells; an acute megakaryoblastic transformation of CML, however, gave negative 5′N reactions (El Mohandes and Hayhoe, 1983; Khalaf and Hayhoe, 1987).

Oliveira *et al.* (1987) found the 5′N reaction positive in only 2 of 7 cases of AMegL, however, and in none of 6 cases of megakaryoblastic transformation of CML. They confirmed the presence of acetate esterase positivity as scattered granules mostly in the Golgi region and the weaker or negative butyrate esterase reaction shown by megakaryoblasts, and also noted acid phosphatase positivity in most cells, predominantly localised in the Golgi zone.

The peroxidase demonstrable by ultrastructural cytochemistry in the dense tubular system of platelets (Breton-Gorius and Guichard, 1972; White, 1972), designated platelet peroxidase (PPO), which appears early in megakaryoblast maturation before platelet glycoproteins, and is probably the most sensitive marker of promegakaryoblasts (Breton-Gorius *et al.*, 1984) has been discussed already in Chapter 10. Its application to acute leukaemias suggests that the frequency of 'promegakaryoblastic' leukaemias is much higher than has been thought especially in transforming CML and in patients with secondary leukaemias. Breton-Gorius *et al.* (1985) collected 27 cases with from 10 to 95% of PPO-positive blasts; 3 appeared to be primary leukaemias, 2 showed marked myelofibrosis and could be classified

Table 14.8 Cytochemical features of acute megakaryocytic leukaemia (AMegL)

The frequently mixed myeloid character of cases with predominating megakaryoblasts gives a mixed cytochemistry, but the salient features of the megakaryoblast are Sudan black and peroxidase negativity at the light microscope level (there is usually peroxidase positivity of the platelet peroxidase type visible at the electron microscope level).

Acid phosphatase usually strongly positive.

PAS reaction shows diffuse and granular positivity in many blasts, with coarse peripheral granules and sometimes blocks conspicuous.

Chloroacetate esterase activity is negative, alpha-naphthyl butyrate esterase is usually weak or negative, but alpha-naphthyl acetate esterase is positive.

5′-Nucleotidase occasionally positive. DAP II usually positive; occasional rare positivity for DAP IV. Gamma-glutamyl transpeptidase moderately to strongly positive.

as 'malignant myelosclerosis' or 'acute myelofibrosis', 8 represented the megakaryoblastic crisis of CML and 6 were in patients previously treated with alkylating agents and/or radiotherapy. Other cases have occurred with widely variable background; in Down's syndrome; following preleukaemic states and in other myeloproliferative disorders. There has been a frequent association with erythroid involvement, so that mixtures of several lineages are usual when promegakaryoblasts are present.

Koike (1984) studied 3 patients with acute leukaemia whose blast cells showed ultrastructural PPO positivity in more than 70% of cells; acid phosphatase was detected in the nuclear envelope and in endoplasmic reticulum but not in the Golgi cisternae. Other cytochemical and immunocytochemical studies suggested the following sequence of emergence of maturation and differentiation markers in the megakaryocyte cell line: HLA-DR, PPO, acid phosphatase, acetate esterase, platelet glycoproteins (PG) II and III, PGI, Fc receptors, alpha granules, Factor VIII antigen and finally PAS positivity. Expression of these markers in AMegL may be erratic, however, with variable positivity, not necessarily reflecting particular stages in the postulated developmental sequence. Thus Huang *et al.* (1984) studied 12 patients with positive reactions for either HP1-1D or anti-factor VIII McAbs, or both, whose cells were negative for esterases and other cytochemical reactions, except for diffuse acid phosphatase and granular PAS positivity. On the other hand, Breton-Gorius *et al.* (1985) recognised AMegL cells to show maturation blocking at some 5 different levels, representing sequential stages in marker expression along rather similar lines to those described by Koike.

Chromosome studies in AMegL have not shown any constant or commonly recurrent abnormality, although at least one example of a rearrangement involving chromosome 3 with a break at 3q21 has been reported in a childhood case (Cairney *et al.*, 1986), with a clear similarity to the rearrangements earlier reported to be associated with abnormal thrombopoiesis in AML (Bernstein *et al.*, 1982). Rearrangements of chromosome 3 involving bands 3q21 and 3q26 seem clearly to represent a further cytogenetic variant among AML cases generally with abnormalities of megakaryocytic, erythroid and granulocytic lineages. Bitter *et al.* (1985) described 14 patients with abnormalities of 3q, 8 of whom had bands 3q21 and 3q26 involved. Seven of these 8 had platelet counts over 100 × 10^9/l (5 normal or high, and 4 very high, from 600 to 1731 × 10^9/l). All 8 had marked abnormalities of megakaryocytopoiesis with increased numbers of megakaryocytes and numerous micromegakaryocytes. The morphological and cytochemical features included hypogranular neutrophils and pseudo Pelger Huet anomaly in the peripheral blood of 7, absent Auer rods except in one case, PPO detectable in one case and blasts with acetate but not butyrate esterase positivity in another. Bloomfield *et al.* (1985) reported 3 cases of t(1;3)(p36;q21), also with similar findings, and suggested that this cytogenetic variant identifies a new cytologically mixed clinicopathological subtype of AML. Myeloperoxidase appears to be only rarely positive in the blast cells, but further cytochemical analysis remains to be reported.

Preleukaemic (myelodysplastic) states

This rather poorly defined group of disorders includes a wide range of conditions showing ineffective myelopoiesis involving to variable extents any or all of the myeloid cell lines. As indicated in Table 14.1 the most important members of this complex group are conditions involving an increase in granulocyte or megakaryocyte precursors, to which the name 'refractory anaemia with excess of blasts' (RAEB) was given by Dreyfus *et al.* (1970). Alternative names previously used include 'smouldering leukaemia' (Rheingold *et al.*, 1963) and oligoblastic or low-percent leukaemia (Dameshek and Gunz, 1958). RAEB, with 5–20% of marrow blast cells and RAEBt (representing a supposedly transforming variety of RAEB) with 20–30% of marrow blast cells, form two of the five subcategories of the myelodysplastic states (MDS) in a classification proposed by the FAB group (Bennett *et al.*, 1982). The other categories are refractory anaemia (RA), with less than 5% of blasts and less than 15% of the erythroblasts showing the features of ring sideroblasts, refractory anaemia with sideroblasts (RAS) where there are again less than 5% of blast cells, but 15% or more of the erythroblasts show iron rings, and finally chronic myelomonocytic leukaemia (CMML), with less than 20% of blasts but 1 or more monocytes × 10^9/l in peripheral blood. Categories with 'excess of blasts' and probably CMML, as well as at least some of the refractory anaemias, are almost certainly clonal disorders affecting the myeloid stem cell, closely comparable with acute myeloid leukaemias (AML). Indeed there is considerable evidence both from morphology and studies of G6PD isoenzyme distribution that the preleukaemic myelodysplasias, and probably also certain of the AMLs apparently arising *de novo* but with dysplastic changes in erythroblasts and megakaryocytes, arise from a pluripotential stem cell. The evidence is reviewed by Tricot *et al.* (1986). The incidence of cytogenetic abnormalities and their nature have considerable similarities and common features in the two disorders.

The position with RA and RAS is complex; many variants of sideroblastic anaemia occur, hereditary, primary and secondary acquired, and pyridoxine-responsive. Perhaps only the 'primary acquired' cases are truly 'preleukaemic' clonal disorders. Others respond to withdrawal of inducing

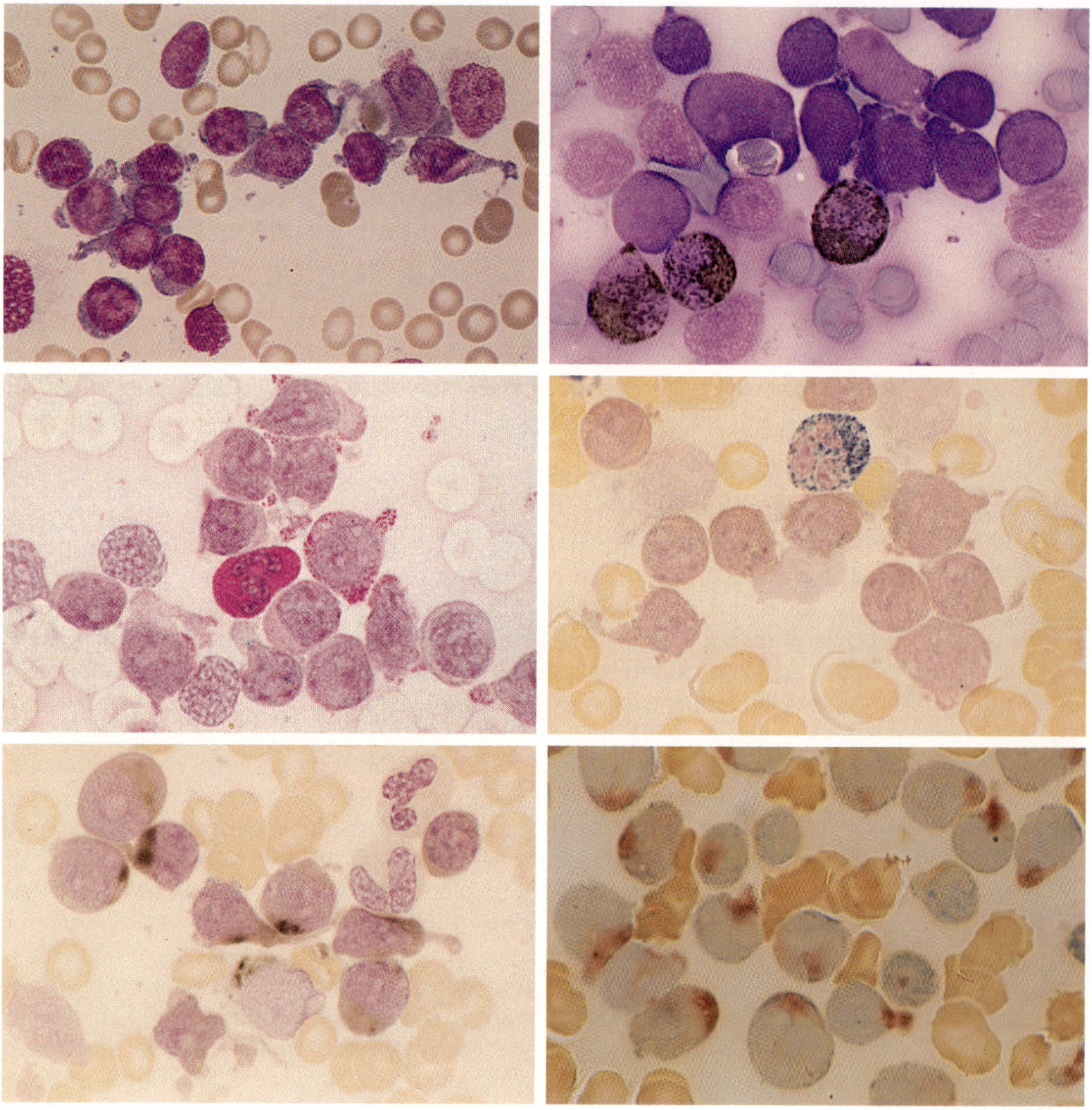

Plate 49

A. AMegL. Leishman stain. Megakaryoblasts show somewhat irregular outlines with conspicuous cytoplasmic budding, and one contains a phagocytosed erythrocyte.

B. The same case. SB stain. Megakaryoblasts are sudanophobic, while three granulocyte precursors show normal reactivity.

C. The same case. PAS stain. There is diffuse and granular positivity in the megakaryoblasts, granules being especially prominent in cytoplasmic protrusions.

D. The same case. Dual esterase. A single granulocyte shows strong CE positivity while the megakaryoblasts show no more than faint BE reactions.

E. The same case. Acetate esterase. Megakaryoblasts show variable positivity, but often much stronger than was shown with butyrate as substrate, in Plate 49D.

F. Another case of AMegL. Acetate esterase (pararosaniline as receptor). Megakaryoblasts again show quite strong acetate esterase positivity.

agents such as isoniazid, cycloserine, alcohol, chloramphenicol, or lead, and some improve with folic acid and pyridoxine. Transformation to leukaemia occurs in a minority of cases (10–20%), although most patients are refractory to treatment and die after a protracted course from the consequences of progressive marrow failure, including infection and haemorrhage.

Patients with sideroblastic anaemias show a hypercellular bone marrow with predominantly micronormoblastic but occasionally megaloblastic erythropoiesis.

Cytochemical investigations with the Prussian blue reaction have shown the presence in this disease of numerous coarse siderotic granules arranged in concentric rings around the nucleus in orthochromatic, polychromatic and occasionally basophilic erythroblasts (Björkman, 1956; Dacie *et al.*, 1959; Hayhoe and Quaglino, 1960). Siderocytes are also frequent.

The number of sideroblasts as well as the amount of stainable iron in individual cells tend to be higher in primary acquired sideroblastic anaemia than in the two other forms, i.e. hereditary sideroblastic anaemia and secondary acquired sideroblastic anaemia. In primary acquired sideroblastic anaemia ringed sideroblasts may be observed in all the differentiating stages of the erythroblastic population, including basophilic erythroblasts, while in the other two types of sideroblastic anaemia ringed sideroblasts rarely occur in basophilic erythroid cells.

In all three types of sideroblastic anaemia electron microscope studies (Heilmeyer *et al.*, 1962; Bessis and Breton-Gorius, 1962; Larizza and Orlandi, 1964; Tanaka *et al.*, 1966; Goodman and Hall, 1967; Grasso and Hines, 1969) have shown abnormal accumulation of iron within the mitochondria of a variable number of erythroblasts. Besides the mitochondrial alteration erythroblasts in sideroblastic anaemias show an increase in the number of ferritin molecules within the cytoplasm.

Semiquantitative estimations of PAS positivity (Hayhoe and Quaglino, 1960) revealed the existence in sideroblastic anaemias of three different patterns. In the first the reaction was negative at all stages of maturation of the red cell series. In the second, the earliest stages showed moderate or coarse granular positivity, while later erythroblasts were negative. In the third, moderate positivity was observed in both proerythroblasts and in intermediate and late erythroblasts. The stained material was most probably glycogen since the reaction was abolished or greatly reduced by treatment with amylase.

Vilter and his associates (1967) also observed amongst their group of patients with sideroblastic anaemia cases with fairly strong PAS positivity, which underwent a leukaemic or erythroleukaemic transformation or eventually became aplastic.

The alkaline phosphatase reaction in sideroblastic anaemias shows great variability. Dacie and his colleagues (1959) reported the absence of leucocyte alkaline phophatase activity in two cases, and normal amounts in a third. Heilmeyer *et al.* (1960) reported very low values in one case, low normal values in a second and increased levels in a third. In erythroleukaemic states alkaline phosphatase scores may be either below normal or above (Hayhoe *et al.*, 1964). Since in sideroblastic anaemias this enzyme may be either high or low, the presence of reduced alkaline phosphatase activity in some cases cannot be taken as suggestive of erythraemic transformation. On the other hand low alkaline phosphatase levels in sideroblastic anaemias may precede a transformation into a myeloblastic type of leukaemia, in which the enzyme is often absent from mature granulocytes (Hayhoe *et al.*, 1964). The possibility of such a transformation is exemplified by the occurrence of a terminal leukaemic phase in one of Bjorkman's cases of sideroblastic anaemia and by the fact that at least three of the cases listed by Heilmeyer (1959) ended with the clinical and haematological picture of an acute leukaemia.

Studies of cell proliferation with Feulgen microdensitometry and autoradiography in primary acquired sideroblastic anaemia have shown a decrease in the S/G_2 ratio of early polychromatic erythroblasts (Wickramasinghe, 1975). In three of the four cases studied there was also an increase in the proportion of U cells. These findings indicate a slowing or an arrest in the progression of early polychromatic erythroblasts through G_2 and an arrest of these cells after a period in DNA synthesis. Wickramasinghe (1975) has shown that early polychromatic erythroblasts with a complete ring of coarse siderotic granules around the nucleus do not incorporate tritiated thymidine and that only a small proportion of polychromatic cells with a ring of fine granules incorporate the radioactive precursor. It appears likely that the accumulation of siderotic granules within early polychromatic erythroblasts is associated with an arrest of DNA synthesis and of entry of G_2 cells into mitosis. This occurs probably as a result of mitochondrial damage, which, when sufficiently severe, may interfere with various cellular processes including RNA and protein synthesis, DNA synthesis and entry of cells into prophase, with consequent cellular death (Wickramasinghe *et al.*, 1968; Wickramasinghe and Hughes, 1978).

Cell proliferation derangements similar to those described in primary acquired sideroblastic anaemia were also reported in one case of secondary acquired sideroblastic anaemia presenting a clinical and morphological picture similar to subacute erythraemic myelosis (Wickramasinghe *et al.*, 1968).

In hereditary sideroblastic anaemia Wickramasinghe *et al.* (1971) found that cell cycle distribution of the early polychromatic erythroblasts of three patients was essentially normal and concluded that no cell death occurs during the proliferative phase of erythropoiesis in this form of sideroblastic anaemia. Such a behaviour appears to be related to the fact that in the hereditary form ringed sideroblasts occur predominantly in non-dividing late polychromatic erythroblasts.

Broad differences between the patients with refractory anaemia and those with excess blasts emerged from a study of DNA content and an analysis of cell cycle activity reported by Peters *et al.* (1986). They used a sensitive technique of flow cytometry with a fluorescence activated cell sorter (FACSIII), capable of detecting small differences in DNA content such as those found between male and female lymphocytes due to the difference in size of the respective Y and X chromosomes. Twenty four patients with RA and 10 with RAEB were studied. A high incidence (56%) of abnormal DNA content was found, RA cases tending to be hyperdiploid, and abnormal DNA levels were associated with poorer prognosis. These findings parallel the incidence of chromosomal abnormalities, also around 50–60%, and also associated in some series with poor prognosis (Nowell and Finan, 1978; Greenberg, 1983) though not in others (Tricot *et al.*, 1984).

Earlier studies by Montecucco *et al.* (1983a, b) had shown fewer DNA abnormalities in flow cytometry of MDS patients, and poor correlation with changes in karyotype, but both groups agreed in finding a lower proportion of cells from RA patients in G_0 or G_1 phases of the cell cycle and a higher proportion in S, G_2, or M phases than occurred in either normal controls or in RAEB patients. This suggests increased proliferative activity, or a failure to progress normally through the later phases of the cell cycle, and is in conformity with the findings of Wickramasinghe *et al.* (1968) in cases of primary acquired sideroblastic anaemia, described above. In cases with excess blasts, however, the findings were often close to normal, although some showed a reduced proportion of cells in S, G_2 and M phases, suggesting a failure to initiate DNA synthesis in G_1, characteristics similar to those found in patients with acute leukaemia (Barlogie *et al.*, 1980) and perhaps indicating malignant progression.

Cytochemical studies using the common reactions for cytoplasmic metabolites and enzymes generally show findings resembling those in the more floridly leukaemic counterparts of the various preleukaemic states. Schmalzl *et al.* (1978) summarized the main findings as falling into the following categories; (a) defective enzyme equipment of apparently mature cells, e.g. low levels of peroxidase, sudanophilia, chloroacetate esterase and sometimes alkaline phosphatase in polymorphs, and decreased non-specific esterases and acid phosphatase in monocytes, (b) accumulation of materials due to defective synthetic processes, e.g. free iron and PAS-positive material (glycogen) in erythroblasts, (c) unusual increases in some enzymes suggesting abnormal function, e.g. neutrophil alkaline phosphatase in some cases, acid phosphatase and esterases in erythroblasts and (d) defective cytoplasmic organization, e.g. patchily granular PAS positivity in some neutrophils.

All these features are certainly to be found among cases of preleukaemic myelodysplasia, and may be diagnostically helpful. They all reflect similar abnormalities occurring among the AML group with more or less prominence in different variants. Atypical expression of ‘granulocytic’ and ‘monocytic’ esterases seen in some cases of APL, AMML and AMonL especially, are also found in preleukaemic states. Scott *et al.* (1983), for example, found atypical granular reactions for both acetate and chloroacetate esterases in individual cells, thought usually to be granulocyte precursors, among various megaloblastic anaemias and in both primary and secondary preleukaemic myelodysplastic states. Dual esterase positivity appeared not to be related to the proportion of immature granulocytes and was most marked among marrow aspirates showing sideroblastic changes, 60% of which showed this feature. The authors argued that the transition from common myeloid precursor to committed granulocytic or monocytic cells is normally accompanied by differentiation-linked activation of respective granulocyte or monocyte-specific esterases, as suggested by Markert and Hunter (1959) and Yourno and Mastropaolo (1981), but that the dysmyelopoiesis present in both preleukaemic and non-malignant conditions is associated with a failure in differential gene activation and is not to be interpreted as directly indicating evolution towards AMML, as had been suggested by Gordon and Hubbard (1979).

Chronic myeloid leukaemia (CML)

The most striking established abnormality of cytochemistry in CML concerns the low alkaline phosphatase of mature neutropils (Plate 50A). This finding has been discussed at length in Chapter 6, but here it is relevant to emphasise that the characteristic is sufficiently constant to be of help in the differential diagnosis of CML from non-leukaemic myeloproliferative states where alkaline phosphatase scores in the neutrophils are generally high. Exceptions to the rule do occur, however, and the distinction cannot be regarded as an absolute one. The lowest scores in CML are found in untreated cases and those with high peripheral leucocyte counts. The low levels in full relapse tend to return towards normal as remission is established, although scores seldom rise above the lower ranges of normal even in full remission. When blastic crisis emerges the LAP score may rise sharply to normal or higher than normal levels (Plate 50B).

Other cytochemical studies in CML have revealed less consistent changes. The strong diffuse staining with the PAS reaction in later granulocytes makes it difficult to assess changes in intensity unless they are gross, but quantitative cytochemical assays have shown the levels of glycogen in the granulocytes in CML to be reduced (Gahrton, 1966; Gahrton *et al.*, 1969). Unusually strong peroxidase positivity has been encountered in our own experience, especially in cases in relapse and with high peripheral leucocyte counts (Plate 50C). Indeed, an inverse relationship between peroxidase and alkaline phosphatase levels in neutrophils could be observed (Quaglino, 1961).On the other hand, there may very occasionally be neutrophils in CML which give very little in the way of cytochemical reactions, being negative to Sudan black and peroxidase as well as to chloroacetate esterase and alkaline phosphatase.

Basophils may sometimes be numerous in CML, and their characteristic cytochemical reactions need to be recognised and distinguished from those of the neutrophil series. This is especially true in relapse and in terminal metamorphosis, when both mature and immature basophils may be seen, the latter sometimes being sufficiently primitive and poorly granular to be confused with blast cells of other series, and especially, because of their PAS reaction, with lymphoblasts. To deal with this point first, the PAS reaction in basophils at all stages of maturity may be much increased in CML and other myeloproliferative states, with coarse granularity or block type positivity (Plate 50D); the reaction is due to glycogen present between the basophil granules rather than in them, as clearly established by Kaung

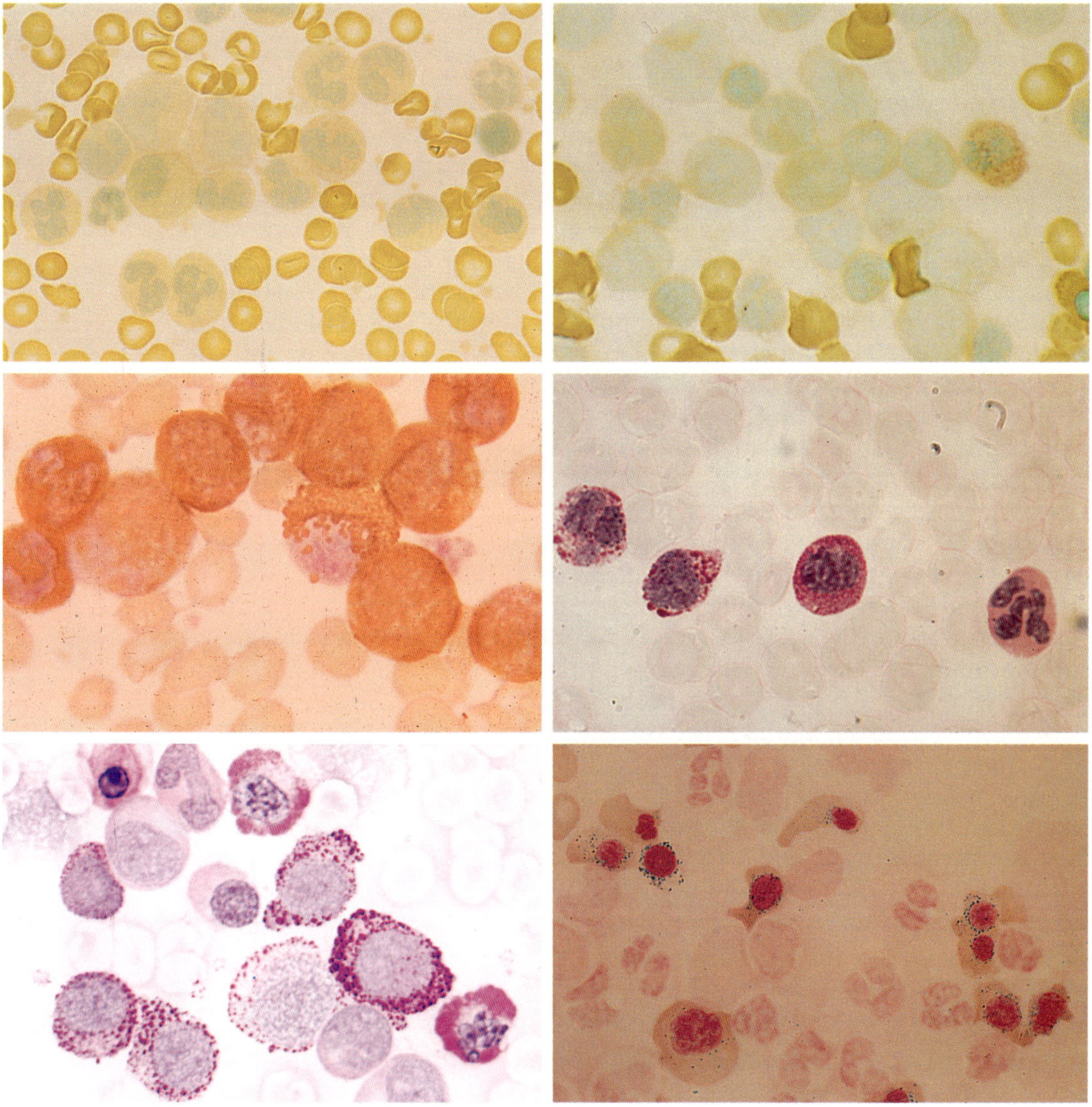

A B

C D

E F

Plate 50

A. CML. Peripheral blood. Alkaline phosphatase. Numerous neutrophil polymorphs and precursors, showing largely negative reactions. A single polymorph shows faint 1 plus positivity.

B. CML in blastic crisis. Peripheral blood. Alkaline phosphatase. The neutrophil polymorph present among the predominantly blast cell population is positive, with LAP score rating of 2 plus.

C. CML. Peripheral blood. Peroxidase reaction. Unusually strong positivity in polymorphs and myelocytes.

D. CML. Peripheral blood. PAS reaction. A neutrophil, an eosinophil and two basophils, all showing PAS positivity; the reaction in the basophils shows coarse granularity with formation of irregular pools of positivity.

E. Erythroleukaemic blastic phase of CML; bone marrow. PAS stain. Myeloblasts show a weakly positive cytoplasmic tinge and the neoplastic erythroblasts show positivity like that in 'de novo' erythroleukaemia

F. Another blastic transformation of CML with erythroid component; bone marrow. Prussian blue stain shows numerous sideroblasts including some coarsely ringed forms.

(1969), using both light and electron microscopy. The appearance is very characteristic, especially in earlier basophils, where the positivity commonly forms irregular pools or lakes of reaction product, clearly distinguishable from the more regular, round, oval or paracrystalline appearance of blocks in ALL. Another helpful cytochemical guide to the recognition of abnormal basophils, even when their granules are inconspicuous in Romanowsky preparations, is the peculiar Sudan black reaction they show (Plate 51A). This is chiefly of a reddish metachromatic hue, with only occasional black granules of normally reacting sudanophilic material. The stain is often stronger in basophil precursors than in mature cells and in the basophils of CML and other myeloproliferative states than in those of normal blood, and this is especially true of the normally reacting component, which in this respect runs parallel to the myeloperoxidase content of basophils, where strong positivity occurs in early cells but later ones are normally negative though they may be positive, sometimes strongly, in myeloproliferative states. These reactions and other aspects of the cytochemistry of blood basophils are reviewed extensively by Parwaresch (1976).

Eosinophilic leukaemia, though rare, is probably separable as a distinct disease entity from CML, AML or ALL, conditions in all of which a more or less prominent component of eosinophilia may occur as in AML with Inv(16), but the differential diagnosis of leukaemic eosinophilia from eosinophilia secondary to other disease states, especially immune disorders and collagen diseases, may sometimes be difficult. Nevertheless, cytochemical studies in occasional apparently genuine examples of eosinophilic leukaemia have suggested that coarse extragranular glycogen and phosphorylase accumulation and strongly positive acid phosphatase, together with chloroacetate esterase positivity, may be especially observed in leukaemic eosinophils (Ackerman, 1964; Schaefer *et al.*, 1973; Presentey *et al.*, 1979).

Malignant progression, metamorphosis and blastic crisis in CML

After an unpredictable period ranging from a few weeks to many years in the relatively benign and controllable chronic phase, CML undergoes a malignant progression, sometimes associated with cytogenetic changes, giving rise to a cell population no longer sensitive to chemotherapeutic control with busulphan. Whether the emerging resistant cell population arises *de novo* shortly before the development of crisis, or alternatively represents an increasing predominance of a previously existing malignant subclone, perhaps subject to progressive metabolic and cytogenetic derangement, is a matter for argument, but much evidence favours the latter view (Pedersen and Killmann, 1971; Stryckmans, 1974; Quaglino *et al.*, 1977). Such a hypothesis has received further support from studies of molecular biology (Gale and Cannani, 1985) which imply a multistep oncogene activation responsible for progression to the acute phase. In many patients this resistance is associated with a frank transformation into an acute leukaemic state, with blast cells exceeding 30% and often strongly predominating in blood and bone marrow. In a substantial minority of patients, however, variable patterns of hypoplasia, aplasia, myelofibrosis, thrombocytosis and megakaryoblastosis, thrombocytopenia, basophilia, myeloblastosis or erythroblastosis may occur, sometimes in succession, without a fully acute leukaemic picture. In such cases the appropriate and characteristic cytochemical reactions are generally manifest in the cell types involved, and may indeed be helpful in cytological diagnosis (Plate 51B–F).

In blastic transformation, however, cytochemical reactions unlike those usually found in acute myeloblastic and promyelocytic leukaemias have long been noted to occur in some cases (Hayhoe *et al.*, 1964). Apart from the common elevation of LAP scores, discussed in Chapter 7, weak or negative peroxidase and Sudan black reactions and coarse granular PAS positivity in blast cells occur in more than 50% of cases. Both cytologically and cytochemically these cases of blastic crisis may resemble acute lymphoblastic leukaemia, and indeed a response to vincristine and prednisone therapy is often observed in such cases, whereas those with more typical granuloblastic cytology and cytochemistry usually fail to respond (Marmont and Damasio, 1973; Boggs, 1976). Further support for the concept that blastic transformation might involve a lymphoid stem cell was provided by membrane marker analysis (Beard *et al.*, 1976; Janossy *et al.*, 1976a). Summarizing early experience with membrane markers, Janossy *et al.* (1976b) reported analysis of the surface phenotype of leukaemic cells from 40 patients in blastic crisis of Ph^1-positive CML. Eighteen of these showed cytological and cytochemical findings suggestive of lymphoblastic character, and among these 15 reacted with cALL antiserum. The remaining 22 showed cytological and cytochemical findings compatible with myeloid blast crisis and all but one, and that equivocal, failed to react with anti-ALL serum.

There is further evidence in favour of lymphoblastic transformation occurring in CML. Most blastic cases, including those with lymphoblastic morphology and membrane phenotype, possess the Ph^1 chromosome, and the continued presence of this clonal marker from chronic to acute phase can reasonably be taken to exclude the development of a second independent leukaemia and to confirm the concept of malignant progression within the original clone. Earlier views that CML arose from a committed myeloid stem cell seemed therefore likely to be incorrect, and the possibility that the origin might be from a pluripotential stem cell with capacity to give rise to all haemopoietic cell lines, myeloid or lymphoid, gained ground. An hypothesis reconciling the observed patterns of transformation with possible stem cell interrelationships including the possible origin of lymphoid blast cells of crisis from a Ph^1 positive pluripotential stem cell was proposed by Janossy *et al.* (1976b).

Studies with other techniques which help to discriminate between lymphoid and myeloid cell lines generally support the existence of both types of blastic crisis in CML. Thus the enzyme terminal deoxynucleotidyl transferase (TdT) is usually positive in ALL and in lymphoblast-type crises of CML but often negative in AML, and in both chronic phase and myeloid-type crises of CML (McCaffrey *et al.*, 1975a and b), while parallel findings

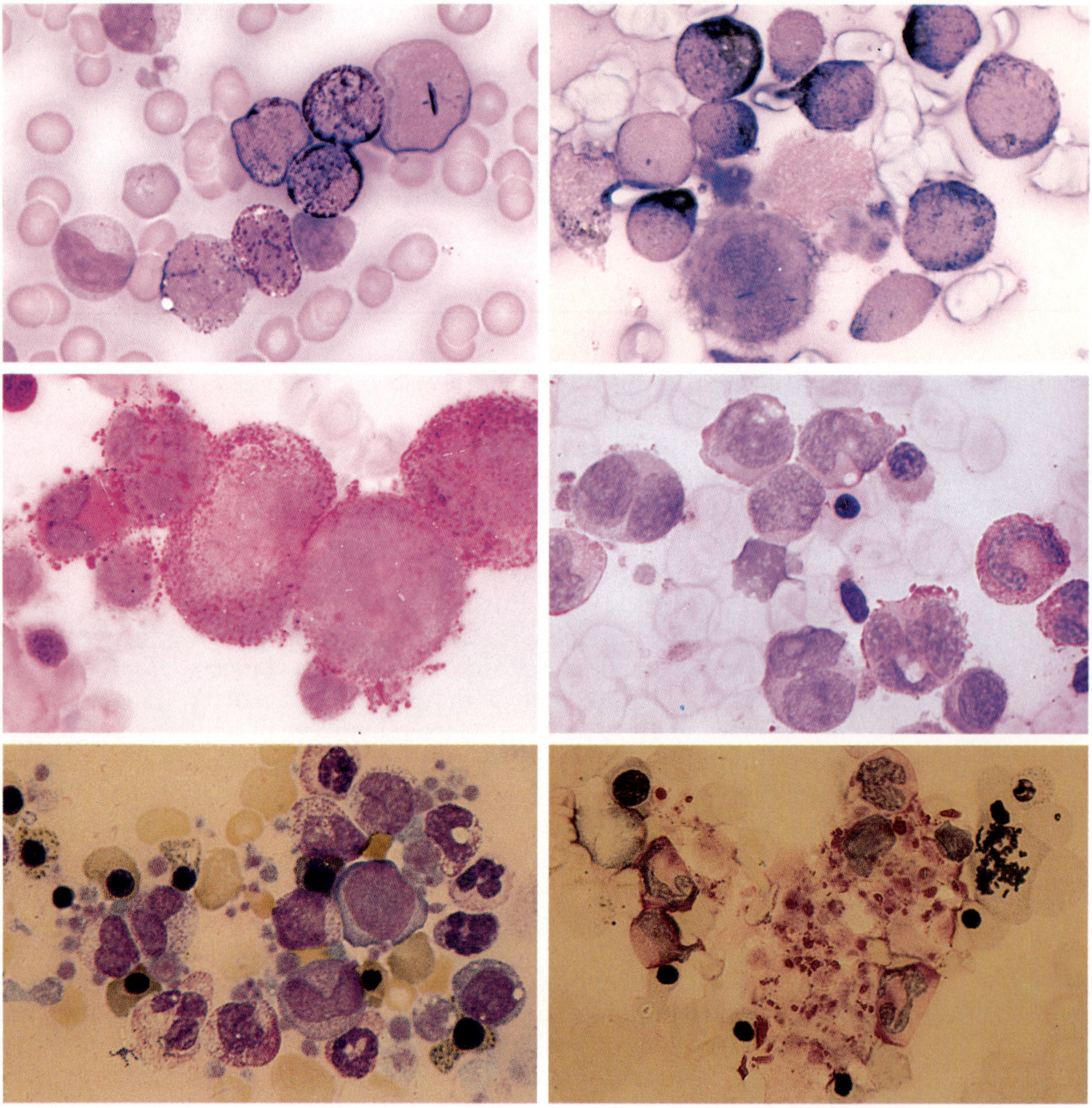

A B C D E F

Plate 51

A. CML metamorphosis to mixed myelofibrosis and blastic crisis. Buffy coat. SB stain. Sudanophilic reaction with characteristic staining pattern in neutrophil granulocyte precursors, including positivity in Auer rods; a basophil shows reddish metachromatic staining of granules.

B. CML metamorphosis to myeloblastic crisis with megakaryocytic component. SB stain. Myeloblasts show variable sudanophilia and a megakaryocyte precursor shows a possible Auer rod.

C. CML metamorphosis to megakaryocytic myelosis. Bone marrow. PAS reaction. The characteristic mixed positive reactions shown by megakaryocytes are seen here in both clearly recognisable megakaryocytes and smaller atypical cells, whose lineage is thus seen also to be megakaryocytic.

D. CML metamorphosis to megakaryoblastosis. Bone marrow. PAS reaction. Bizarre primitive cells show positive reactions with diffuse cytoplasmic stain and in places peripheral granular reaction suggesting their megakaryoblastic nature.

E. Buffy coat from mixed metamorphosis in CML. Heyl stain. Highly pleomorphic mixed cell population with giant platelets.

F. The same specimen as in E. PAS stain. Bizarre blocks of PAS positivity in platelets. Granulocytes are positive but erythroblasts are negative.

have been reported on the occurrence of the lymphoid and myeloid types of leucocyte alkaline phosphatase, which are found respectively in the TdT positive lymphoblasts and TdT negative myeloid types of crisis (Neumann *et al.*, 1979).

Nevertheless, as in the *de novo* acute leukaemias, the TdT levels do not always correlate closely with lymphoblastic morphology, and absence of this marker may occur in cases responding to vincristine and prednisone (Srivastava *et al.*, 1977; Marks *et al.*, 1978; Barton and Conrad, 1978). Moreover, blast cells which appear of undifferentiated or lymphoblastic appearance by light microscopy, using standard cytological and cytochemical stains, and even show lack of differentiation in electron microscope preparations, may be recognisable as of the myeloid series—myeloblastic, monoblastic, megakaryoblastic or erythroblastic—by the ultrastructural demonstration of peroxidase and acid phosphatase activity having identical characteristics with that of normal precursors of these cell lines (Breton-Gorius *et al.*, 1978, 1985). Even the patients whose cells, studied by Janossy *et al.* (1976b), reacted with cALL antiserum showed heterogeneous clinical and mixed cytological features. Five of them presented with supposed 'lymphoid' blast crisis, but went into remission with minimal or no myeloid abnormality, so that these may really have been examples of Ph^1 positive ALL *ab initio* and not transformations of CML, in which remission would have led to a return to the Ph^1 positive chronic myeloid phase (Rowley, 1978). Another three presented in blast crisis, but with a mixture of myeloid and lymphoid blasts which must have led to considerable difficulties of interpretation.

Later investigations have shown that not only granulocytes, monocytes, erythroblasts and megakaryocytes but also B lymphocytes and some T cells carry the Ph^1 chromosome and share the same glucose-6-phosphate dehydrogenase isoenzyme, indicating a common clonal origin (Bernheim *et al.*, 1981; Fialkow *et al.*, 1981; Nogueira-Costa *et al.*, 1985). Moreover, enzymatic and immunological methods including the application of monoclonal antibodies to the analysis of membrane phenotypes in blast crisis of CML have clearly demonstrated that the Ph^1 positive clone can give rise to blasts with characteristics of any myeloid or lymphoid lineage (Greaves *et al.*, 1979, Hernandez *et al.*, 1982; Griffin *et al.*, 1983). A recent study (Bettelheim *et al.*, 1985) of 45 cases of blastic crisis using a panel of 26 McAbs showed 28 with myeloid blasts, frequently having admixtures of granulocyte, monocyte, erythroid and megakaryocyte lineages, 14 with lymphoid blasts, resembling the variants found among ALLs, 2 with mixed lymphoid/myeloid lineage and one unclassifiable. Three of 16 patients serially tested showed phenotypic switches of lineage involvement.

These immunological findings are broadly compatible with cytochemical experience, in that the features manifest among the variants of acute leukaemias of all kinds, as already described, may be recognised in blastic crises of CML. The contribution of cytochemical techniques to elucidating the nature of blastic transformation lacks the specificity of immunocytochemistry with McAbs and the basic importance in relation to the molecular biology of the disease of cytogenetic evolution, but nevertheless retains importance, with light microscopic morphology, as the simplest and most widely available tool for primary diagnostic analysis.

Chronic lymphocytic leukaemia (CLL)

The lymphocytes in this disease show a much richer content of glycogen than do normal lymphocytes (Plate 52B), but the pattern is not peculiar to CLL, since a general increase in glycogen in lymphocytes of various stages of maturity occurs also in Hodgkin's disease, non-Hodgkin's lymphoma and in other non-leukaemic lymphoproliferative diseases (Quaglino and Hayhoe, 1959). The significance of these changes is discussed in Chapter 4, but it may be noted here that in CLL a close relationship appears to exist between the level of PAS scores and the effect of treatment. When suitable therapy induces a progressive fall in the leucocyte count, decreasing levels of glycogen are found in the lymphocytes, while leukaemic relapse is associated with a rise to high levels of glycogen once more. However, the occasional transformation to an immunoblastic proliferation may be associated with weak PAS positivity (Plate 52E).

We have seen in earlier chapters that B and T cells may be distinguished by cytochemical features as well as by functional and immunological ones, and that the distinctions frequently hold true in lymphoproliferative states. Thus T-cell proliferations, including the relatively uncommon T-cell variant of CLL, generally show strong localised positivity for one or more of the hydrolases, acid phosphatase, butyrate or acetate esterase, beta-glucuronidase, beta-glucosaminidase and dipeptidylaminopeptidase (DAP IV) while in B-cell diseases, including the great majority of CLL cases, these hydrolases are negative or show only weak scattered reactivity (Plate 52C) (Cawley and Burns, 1980; Feller *et al.*, 1982; Crockard, 1984; Pangalis, 1984).

Most cases of CLL show the expected non-specific esterase negativity of B cells, but a few show different patterns, some the crescentic positivity usually seen in hairy cells (Fig. 14.11) and others the null cell pattern of scattered granules as described and discussed in Chapter 8. Other cytochemical features of CLL lymphocytes include low levels of the plasma membrane enzyme 5′-nucleotidase, and, at least in circulating lymphocytes, reduced levels of both acid phosphatase and β-glucuronidase. Phosphorylase activity is elevated in parallel with the PAS reaction.

Crockard *et al.* (1985), noting that the expression of acid hydrolases may be a marker of cell maturity especially in T cells (Basso *et al.*, 1980), examined hydrolase cytochemistry in 30 patients with B-CLL to determine whether, despite the generally lower levels in B cells, variations of importance might exist. Counting cells showing any clear reaction, however weak, as positive, they found 5 cases with less than 50% cells positive for both acid phosphatase (AcP) and acetate esterase (AE), 11 with more than 50% cells AcP positive but less than 50% AE positive, 7 with the reverse of that finding, and 7 with more than 50% positive for both enzymes. Beta-glucuronidase was generally similar to AcP, whereas beta-glucosaminidase

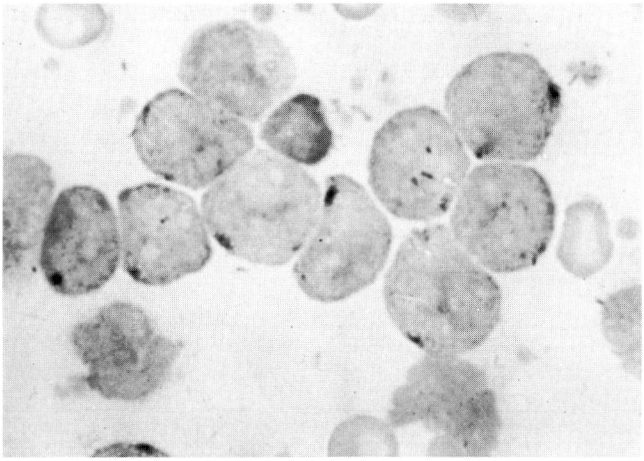

Fig. 14.11. CLL. Dual esterase. A pattern of localized crescentic BE positivity, somewhat like that seen in hairy cells, is present in this case.

was found in most lymphocytes of all patients. DAP IV was detected in less than 20% of cells in all cases, presumably in residual normal T cells. There was no relation between the cytochemical groupings and either clinical or other haematological findings, including height of peripheral leucocyte count, E or mouse rosette values and light or heavy immunoglobulin chain expression. The lack of correlation between acid hydrolase expression and Ig heavy chain isotype, a probable marker of immunological maturation (Ling, 1983), must be set against the finding of Grossi *et al.* (1982) that AP and AE positivity are associated with ultrastructural features of increasing B-cell maturity in CLL. If hydrolase positivity is indeed a marker of B-cell maturation, as seems likely towards the plasma cell stage of the B-cell chain, perhaps B-CLL involves too narrow a band between pre-B cells and mature B lymphocytes for this component of maturation to be manifest.

In T-CLL, which probably accounts for no more than 5% of all CLL cases, the acid hydrolases generally show much more marked positivity (Yang *et al.*, 1982; Crockard *et al.* 1982; Boesen *et al.*, 1984), with a strong correlation between localized dot-like or globular reactions for both AcP and AE and the neoplastic T-cell components. Huhn (1984) distinguished three variants of T-CLL, lymphocytic, Sézary-cell like, and with large granular lymphocytes, the first two having functional properties of T-helper lymphocytes, although not always showing Fcμ receptor or OKT4 positivity and the third representing Fcγ, OKT8 positive suppressor T cells. The first two types showed scattered weak AcP and AE positivity and the third a strong localized paranuclear AcP reaction but weak AE. PAS reactions were finely granular.

Mature neutrophils in CLL show normal or increased amounts of alkaline phosphatase, scores usually being higher in untreated cases or those in relapse and lower in well-controlled cases (Hayhoe and Quaglino, 1958).

Markey *et al.* (1987) recently drew attention to an apparent link between

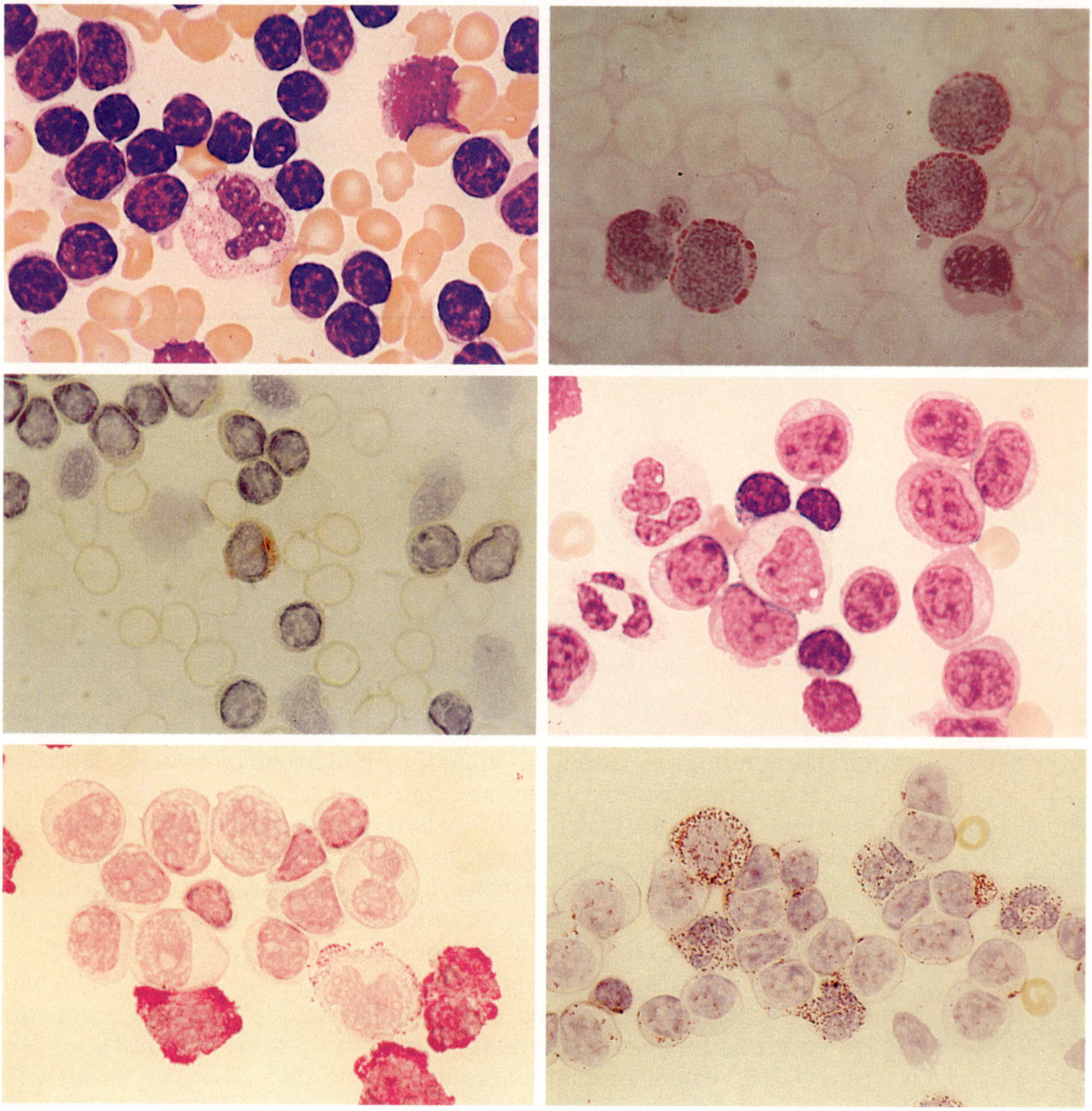

A B C D E F

Plate 52

A. B-CLL peripheral blood. Leishman stain. Typical cytology.

B. B-CLL bone marrow. PAS reaction. Strongly positive reaction in three of four lymphocytes. A normoblast is PAS-negative.

C. B-CLL peripheral blood. DAP IV reaction, showing focal positivity in a single residual T lymphocyte. The neoplastic B cells are negative.

D. B-CLL in transformation to immunoblastic form. Buffy coat. Leishman stain.

E. The same specimen as in D. PAS stain.

The neoplastic cells are mostly PAS-negative; one cell shows a fringe of positive granules. Three neutrophils are strongly positive.

F. The same specimen as in D and E. Acid phosphatase. Lymphocytes and immunoblasts show only a few coarse granules; neutrophils and a monocyte are more strongly positive.

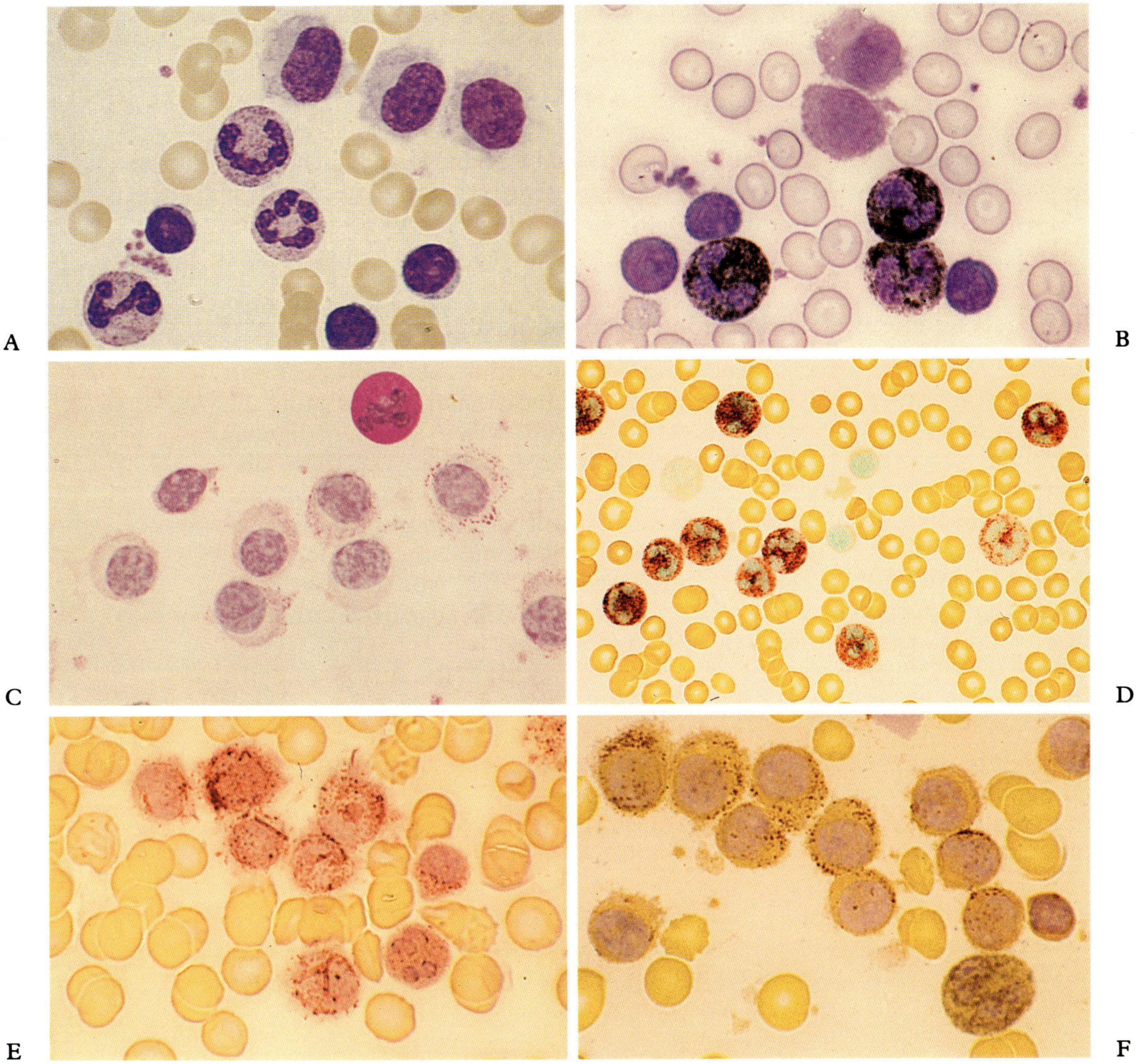

Plate 53

A. HCL. Buffy coat. Leishman stain. Three neutrophils, three lymphocytes and two hairy cells with characteristic morphology.

B. HCL. Buffy coat. SB stain. Three positive neutrophils, three negative lymphocytes and two negative hairy cells.

C. HCL. Another case. Buffy coat. PAS reaction. Apart from one strongly positive neutrophil all cells in this field are leukaemic hairy cells with positive diffuse and granular reactions, quite unlike the reactions usually shown by either normal or neoplastic lymphocytes. This pattern seems to be peculiar to hairy cells.

D. HCL. Buffy coat. Alkaline phosphatase. Most neutrophil polymorphs have strong positivity, with individual cell ratings of 2 or 3 plus, giving a high LAP score.

E. HCL. Buffy coat. Acid phosphatase. Seven hairy cells show strong positivity including rods which may represent ribosome-lamellar complexes.

F. HCL. Another case. Buffy coat. Dual esterase. Apart from one CE positive stab cell and a single T cell, all cells in this field are leukaemic hairy cells showing the range of BE positivity typical of those cells, with from a few to many positive granules and a strong tendency for the positive reaction to show a crescentic disposition at one or more peripheral sites.

monocytic esterase deficiency and lymphoproliferative states, notably non-Hodgkin's lymphoma and CLL. Among NHL-CLL patients 7.1% showed the deficiency, whereas only 1.7% of other patients and normal subjects were esterase deficient. In a study of relatives of NHL or CLL patients, esterase deficiency was found to be a familial characteristic, suggesting that the anomaly predisposes or indicates a predisposition to the development of NHL or CLL.

Hairy cell leukaemia (leukaemic reticuloendotheliosis)

Hairy cell leukaemia (HCL) has a number of distinctive cytological and cytochemical features which may be used as diagnostic aids in separating the condition from CLL or chronic myelofibrosis or even ALL or malignant lymphoma, since confusion may sometimes arise with any of these disorders. These features of HCL include the presence of tartrate-resistant acid phosphatase (Plate 53E) (isoenzyme 5) in a proportion of hairy cells (Yam *et al.*, 1971); distinctive cytoplasmic ribosome-lamellar inclusions (Katayama *et al.*, 1972; Daniel and Flandrin, 1974) and characteristic cytoplasmic projections (Plate 53A) particularly well visualised either by phase contrast or transmission electron microscopy (Katayama *et al.*, 1972; Catovsky, 1977); the presence of a receptor for IgM on the surface of hairy cells (Burns *et al.*, 1977a) and the absence of complement receptors (Burns *et al.*, 1977b); the finding of a high leucocyte alkaline phosphatase score (Plate 53D) (Hayhoe *et al.*, 1977); and the existence of a distinctive pattern of α-naphthyl butyrate esterase positivity in hairy cells (Plate 53F) (Higgy *et al.*, 1978). Hairy cells are Sudan black negative (Plate 53B) and usually show quite strong PAS positivity, with both diffuse and granular reactions (Plate 53C). A peroxidase can be demonstrated at EM level (p. 261). The cells are DAP II positive (Plate 30A), but DAP IV negative (Khalaf *et al.*, 1986). While the accumulation of several of these features lends strong support to the diagnosis of HCL, none of them individually is entirely pathognomonic since each may be absent in any particular case and any may occur individually in other pathological states.

Tartrate-resistant acid phosphatase (TRAP) is usually present in the cells of HCL, but in our own experience (Higgy *et al.*, 1978) and that of others (Catovsky, 1977) activity may be weak or absent in a small proportion of cases. The characteristic pattern of positivity to α-naphthyl butyrate esterase (described in Chapter 8 and here illustrated in Plate 53F) was, however, invariably found in a high proportion of the cells of HCL, and appears to offer not only a simple and reliable diagnostic aid but also a means of estimating the numbers of circulating leukaemic hairy cells. The particular hairy cell pattern of crescentic positivity is only very rarely encountered in normal lymphocytes, and among other lymphoproliferative states we have hitherto found it only in a few examples of CLL.

Katayama and Aiba (1980) compared the reliability of the butyrate esterase (BE) test with two techniques for TRAP, one using fast Garnet GBC as capture agent and the other hexazotised pararosaniline. In blood smears from 27 patients with HCL the first two were positive in every case but TRAP with pararosaniline was less reliable. BE proved best for detecting HCs, but neutrophils were less easily recognised by this staining method. Although Yam *et al.* (1972) had early appreciated that TRAP positivity could be detected cytochemically and by gel electrophoresis in preparations from conditions other than HCL, finding it present, for example, in 1 of 11 normals, 2 of 15 CLL, 2 of 5 lymphomas and 6 of 9 infectious mononucleosis patients, the TRAP test remains the most widely used screening procedure in the differential diagnosis of HCL, with generally good correlation between positivity and presence of HCs (Katayama and Yang, 1977; Cawley *et al.*, 1980). Variakojis *et al.* (1980) found TRAP positivity ranging from 1 to 100% of HCs in 23 of 24 patients with HCL studied, the intensity of positivity varying widely. Esterases, both butyrate and acetate, showed weak to moderate or strong positivity in 16 (BE) and 19 (AE) of 20 cases studied, while the PAS reaction showed weak to strong granular positivity in the HCs of 22 of the 24 cases studied.

From a recent review of 15 years experience of TRAP staining Yam *et al.* (1987) concluded that, when fast Garnet was used as capture agent, and intense TRAP positivity was taken as the criterion, the test was effectively diagnostic of HCL. Negative TRAP reactions occurred in only 2 or 200 HCL cases included in the review, while positive TRAP reactions were seen in only 3 of 800 non-HCL cases.

Arguments still continue over the nature of HCs (Cawley *et al.*, 1980), and especially as to whether a normal counterpart exists, perhaps at some point in the B-cell developmental chain, since HCs can express surface immunoglobulin (SIg). In this context, Machii and Kitani (1984) observed similarities between IgG-bearing lymphocytes and hairy cells. They noted that of six patients with HCL studied five showed SIgG. Among normal lymphocytes some showed SIgG and some SIgM; 76% of the former showed cytoplasmic processes under phase contrast microscopy and 65% of them had TRAP, whereas most of the SIgM-positive lymphocytes had smooth surfaces and only 19% had TRAP. When SIgM positive cells that also had SIgG positivity were excluded, only 10% of the remainder had surface processes and 12% possessed TRAP. These findings have some parallel with the observations on induced differentiation of B-CLL cells towards a point in the chain of B-cell development corresponding to the hairy cell phenotype, reviewed in Chapter 12.

Immunocytochemical approaches confirm the existence of common features yet also distinctive ones between hairy cells and other neoplastic B cells. Using a range of McAbs, Falini *et al.* (1985) showed that hairy cells expressed HLA-DR, leucocyte common antigen, the B-cell antigens reacting with TO15 and B1, and reacted with the 'anti-hairy cell' McAbs S-HCL3, HC1 and HC2. Other neoplastic B cells also expressed HLA-DR, leucocyte common antigen and B-cell antigens, but not S-HCL3.

The raised LAP score in HCL, first noted by Hayhoe *et al.* (1977), has generally been confirmed, although a minority of cases with low scores have since been reported. Aiba *et al.* (1980) found high LAP scores in 17 of 23 patients, but normal scores in the remaining 6, the scores appearing to show an inverse correlation with the peripheral blood neutrophil count, but no relation to age, splenectomy or other clinical features. These authors thought the high LAP score to be secondary to decreased marrow granulocyte reserves. Jansen (1981) reviewed 25 patients and found normal or low LAP scores in only 4 cases. These examples had atypical features but showed a less clear correlation between score and height of neutrophil count. Atypical clinical features, with generalized lymphadenopathy and a complete remission on CHOP chemotherapy, were recorded in an unusual HCL case with persistently low LAP scores (Carbone *et al.*, 1981).

Prolymphocytic leukaemia (PLL)

This variant of CLL is characterised clinically by substantial splenomegaly and high lymphocyte count (usually over $100 \times 10^9/l$), and cytologically by lymphocytes with a conspicuous central nucleolus and a larger size and greater amount of cytoplasm than is usual in CLL. The cytochemistry of PLL in both the common B-cell variant and the rarer T-cell variant has been described at both light and electron microscope level by Costello *et al.* (1980), Crockard *et al.* (1982) and Matutes *et al.* (1983). In general the reactions for acid hydrolases are negative or show no more than weak scattered positivity in B-PLL, as in B-CLL, while more variable positivity is found in T-PLL. Positive reactions for both beta-glucuronidase and beta-glucosaminidase were reported in T-PLL, as in T-CLL, but with strong positivity for alpha-naphthyl acetate esterase (AE) in T-PLL contrasting with weak or negative reactions in T-CLL. Acid phosphatase was also more variable in T-PLL than in T-CLL although Tsai *et al.* (1984) reported strong paranuclear AP activity in all of 7 T-PLL cases tested for this enzyme among 19 cases collected from the literature. These findings were initially thought to reflect a predominance of helper cells among T-PLLs, as against a predominance of suppressor cells among T-CLLs, but the cytochemical reactions on further study appeared to be unrelated to membrane phenotype as defined by McAbs (Crockard *et al.*, 1982). In our experience PLL cases of whatever subtype show little PAS positivity, with only occasionally fine peripheral granularity. A typical set of cytochemical findings in T-PLL is illustrated in Plate 54.

Sézary syndrome (SS)

This disease is the leukaemic phase of mycosis fungoides and is of T-cell origin, chiefly but not always of the helper subset. Cytochemical studies of Sézary cells have been reviewed by Flandrin and Daniel (1984). The PAS reaction ranges from negative to variably positive with a finely granular pattern but occasionally strong and coarse positivity may be seen. The acid hydrolases show unevenly scattered granular reactions of moderate to strong intensity but only rarely solitary large granules or dots as seen in some T-cell populations. DAP II has been found positive (Plate 30D) (Khalaf *et al.*, 1986), but DAP IV negative (Wirthmüller *et al.*, 1983; Crockard *et al.*, 1984; Khalaf *et al.*, 1986), except in one of 8 SS cases reported by Feller *et al.* (1983).

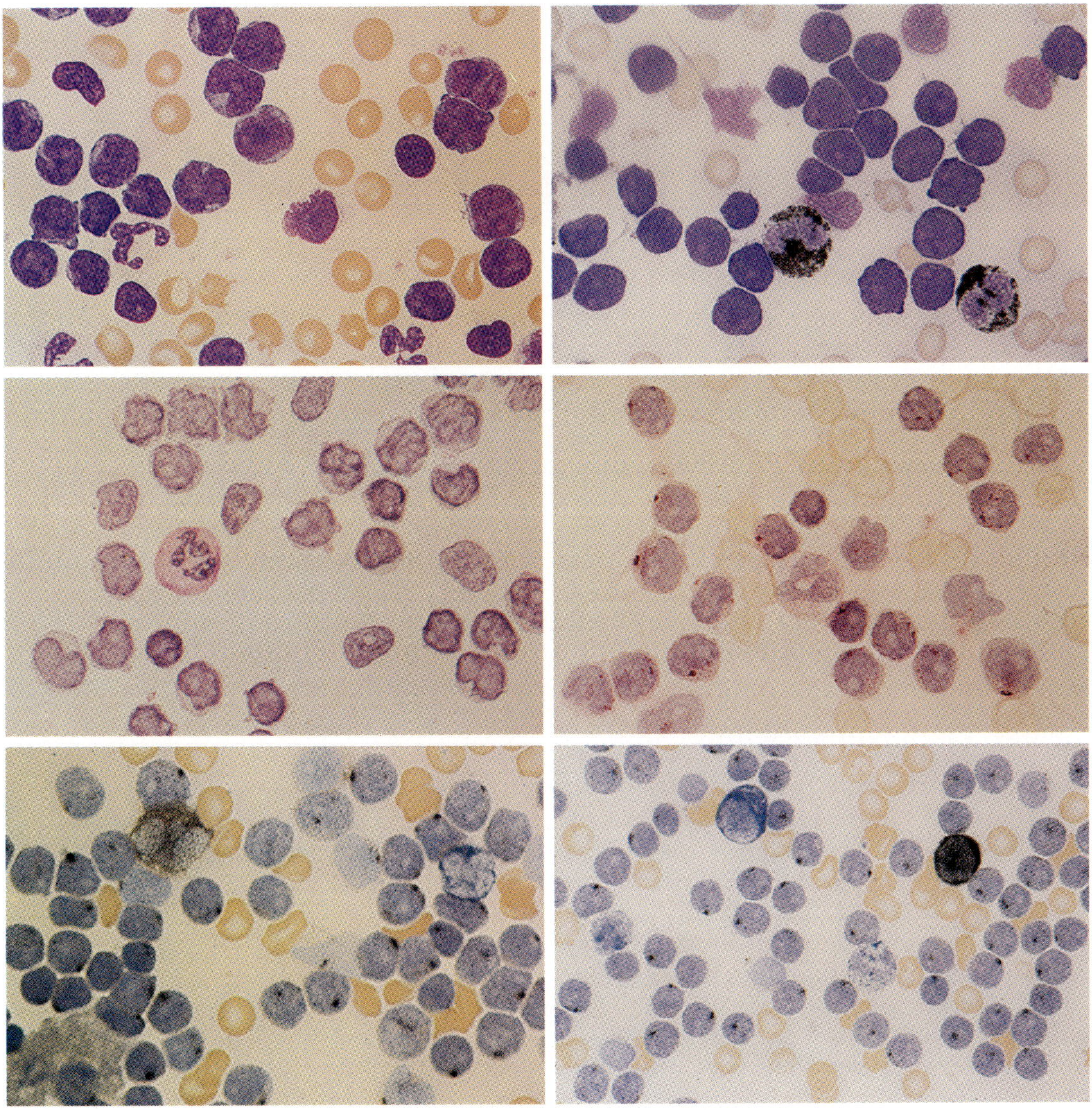

A B

C D

E F

Plate 54

A. T-PLL, buffy coat. Leishman stain. Nuclear irregularity, nucleoli and variable nuclear–cytoplasmic ratio with some cytoplasmic granularity are shown.

B. The same specimen. SB stain. The neoplastic cells are entirely negative, contrasting with two positive stab cells.

C. The same specimen. PAS stain. A single neutrophil shows weak positivity, but the T prolymphocytes are negative.

D. The same specimen. Acid phosphatase. Most cells show dot-like focal positivity, occasionally with additional scattered granules.

E. The same specimen. Dual esterase. A granulocyte is CE positive, a monocyte BE positive, while the T cells show mostly a dot-like focal positivity with BE.

F. A second field in the same preparation as E. A CE-positive myelocyte and a BE positive monocyte are again seen, with a weakly CE positive basophil, and the great majority of neoplastic T-prolymphocytes have focal BE positivity.

Adult T-cell leukaemia (ATL)

This disorder, extensively reviewed in a recent monograph by Hanaoka *et al.* (1982), occurs most often in South-West Japan where it is endemic, but cases of a similar, perhaps identical, clinical syndrome have been observed in black patients from the West Indies (Catovsky *et al.*, 1982). The disease is induced by a retrovirus (ATLV or HTLV1) and affects a $CD4^+$ but functionally suppressor T-cell subset. There are similarities to SS, but the skin lesions are more papular and less generalised, the onset is more acute, clonal karyotypic abnormalities, notably trisomy 7, are often found, and antiviral antibodies occur in all cases whereas they are absent from SS.

The cytochemistry of ATL cells from 3 cases was studied by Mikata *et al.* (1982); reactions were generally in keeping with those of other T-cell lymphomas, with moderately strong acid phosphatase and beta-glucuronidase, but weak or negative esterases. Nasu *et al.* (1981) found TRAP positivity in the neoplastic cells from eight patients with lymphomas, thought to be examples of preleukaemic phases of ATL, together with focal acetate esterase positivity in more mature-looking cells which the authors regarded as neoplastic, but which may perhaps have represented a residual non-neoplastic T-cell population.

Chronic T-cell lymphocytosis

This recently recognized disorder involves a chronic, possibly neoplastic, proliferation of suppressor T cells associated with cytopenias and commonly splenomegaly. In a review of 21 cases, Newland *et al.* (1984) showed that the lymphocytes involved, with peripheral counts ranging from 0.75 to $24 \times 10^9/l$, were morphologically and phenotypically homogeneous at presentation and post-splenectomy; they were $OKT3^+$ and $OKT8^+$, contained azurophil granules, stained strongly for acid phosphatase which was tartrate sensitive, but were generally negative or only weakly positive in a minority of cells for acetate esterase. PAS staining was variable but mainly negative with only occasional dot positivity. Chan *et al.* (1984) also noted strong acid phosphatase positivity in the large granular suppressor T cells of four patients with this disorder, together with multiple parallel tubular arrays on electron microscopy.

Multiple myeloma

As in the leukaemias, the mitotic and proliferative activity in this disease is low; the low proportion of cells showing thymidine incorporation and the abnormal patterns of protein synthesis have been described in earlier chapters. Cytochemically, myeloma cells do not differ sharply in their reactions from normal plasma cells. Microspectrophotometric measurements show increased levels of RNA in the cytoplasm of myeloma cells as compared with their normal counterparts (Olhagen *et al.* 1949), and Schmid *et al.* (1966) found the RNA synthesis rate elevated about threefold, but these observations may reflect no more than greater immaturity of the myeloma cells. Varying patterns of PAS positivity may be seen, but reactions are usually negative or weak and diffuse, with sometimes fine granularity (Plate 55A). Russell bodies are commonly weakly PAS positive (Plate 55B), probably because of mucoprotein contact rather than glycogen, since the material is not digested by diastase (Pearse, 1949). Intense enzymatic activity of dehydrogenases (Plate 55F), cytochrome oxidase (Figs 10.1-10.2), acid phosphatase (Plate 55C), non-specific esterase (Plate 55E) and β-glucuronidase (Plates 33A and 55D) is generally found in myeloma cells (Quaglino *et al.*, 1967; Suzuki *et al.*, 1969). Hayhoe and Neuman (1976), from a study of 195 cases of multiple myeloma divided according to immunoglobulin patterns, found no significant differences between the uniformly high acid phosphatase scores, the generally weak PAS reactions, and the normal to high LAP scores in the separated IgG, IgA or Bence-Jones variations of the disease. Nor were differences to be found in relation to the nature of the light chains.

Cassuto *et al.* (1977) compared the acid phosphatase scores in myeloma cells with those in normal plasma cells and in polyclonal plasmacytosis and found the myeloma scores to be markedly higher. In non-myelomatous monoclonal dysglobulinaemia lower, normal, scores were found in the 10 cases studied. These findings were partially confirmed by Tortatolo *et al.* (1981) who recorded a mean score for acid phosphatase (range 0–400) of 343 in 30 cases of myeloma, as compared with 204 in 21 non-myelomatous monoclonal paraproteinaemias, but found a considerable overlap between the patient groups, as did Johansen and Krogh-Jensen (1980) and Miliani *et al.* (1982). Hoffmann *et al.* (1985) used a revised scoring method, with eight subclasses instead of the conventional four, and applied this to 26 cases of multiple myelomas, 28 of monoclonal gammopathy of undetermined significance (MGUS), 11 of reactive plasmacytosis and 22 normals. Clearly significant differences without overlap were found between myeloma (350 ± 30) and both reactive plasmacytosis (302 ± 22) and normals (304 ± 22). The MGUS group (322 ± 27) did not differ significantly from the myeloma group, nor from the other groups, showing considerable overlap. The authors appreciate that their scoring system is too complex and time-consuming for routine use, but recommend that it be used as a supplementary aid in equivocal cases of plasmacytosis and to confirm a probable diagnosis of myeloma.

Alkaline phosphatase scoring in neutrophils may be more helpful in differentiating myeloma and benign monoclonal gammopathy than is the acid phosphatase score in the plasma cells. Johansen and Krogh-Jensen (1980) found the LAP score higher in the malignant group, with a positive correlation between the M band level and the LAP score in myeloma cases. Remarkably high LAP scores were found by Mukiibi and Kyobe (1982) in African patients with myeloma, with a mean of 182.5 ± 57.4 compared with a normal figure of 64.2 ± 17.9. This finding may reflect gross disease with very high M protein levels at diagnosis.

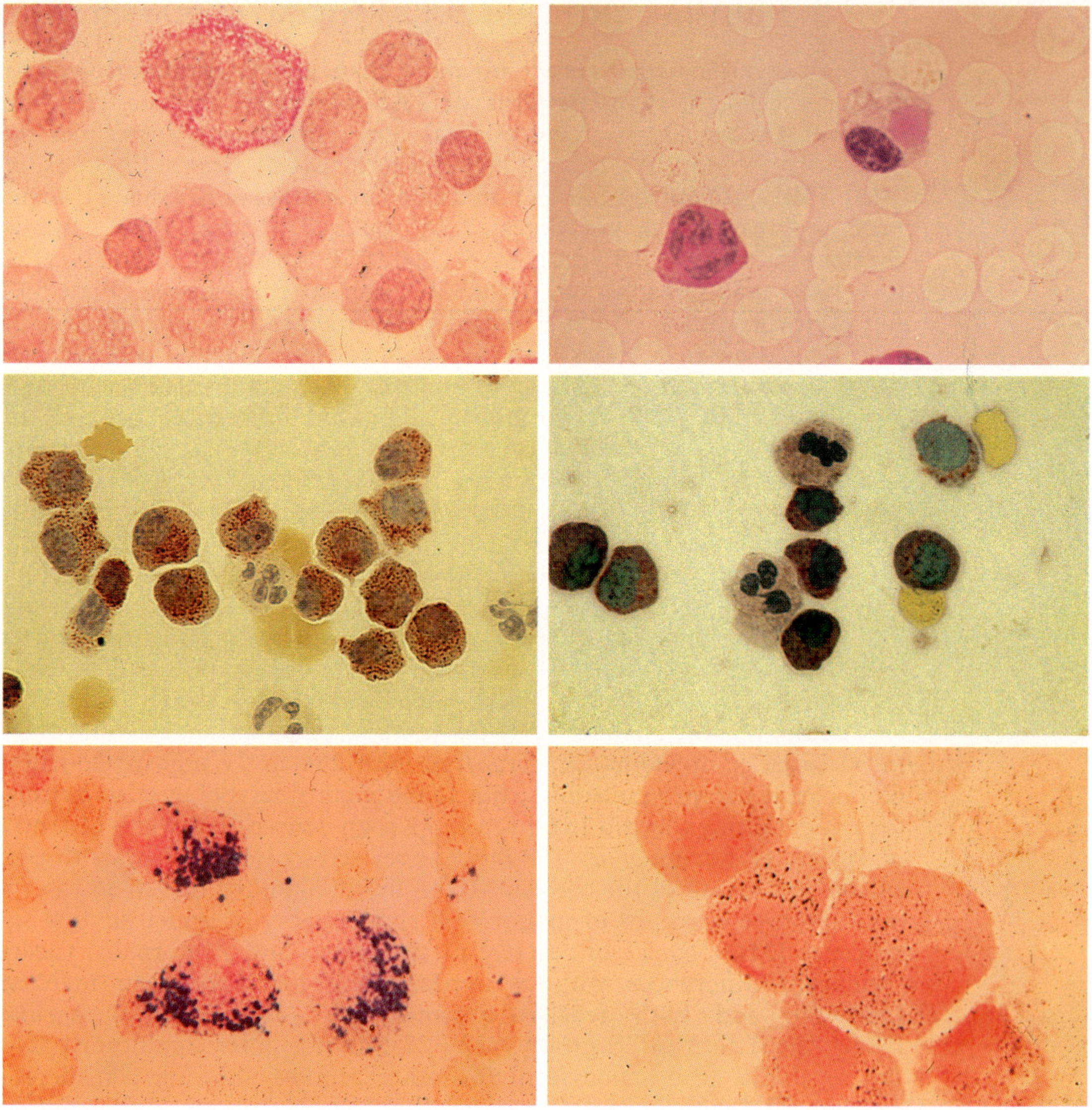

A B C D E F

Plate 55

A. Multiple myeloma. Bone marrow. PAS reaction. Strong positivity in a binucleate cell; the remaining myeloma cells are virtually negative.

B. Multiple myeloma. Bone marrow. PAS reaction. Positive reaction in a large Russell body.

C. Multiple myeloma. Bone marrow. Acid phosphatase. Strong granular cytoplasmic positivity in all plasma cells.

D. The same case. Bone marrow. β glucuronidase. All plasma cells show strong granular and diffuse cytoplasmic positivity.

E. Multiple myeloma. Bone marrow. Naphthyl acetate esterase. Strong positive reaction in three myeloma cells.

F. The same case. Bone marrow. Succinic dehydrogenase reaction. Variable, but sometimes strong reactivity in a group of myeloma cells.

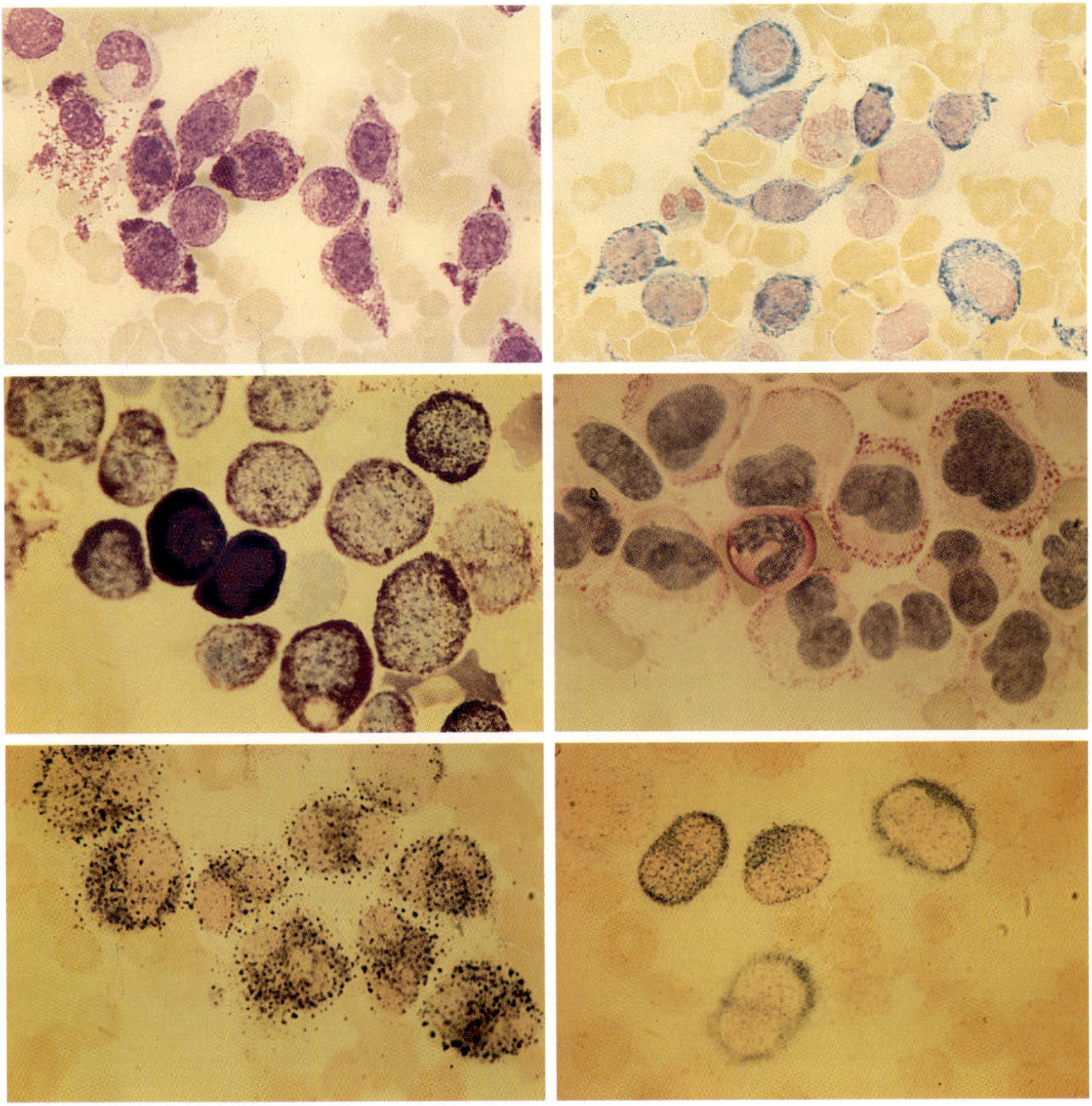

A B C D E F

Plate 56

A. Tissue mast cell leukaemia. Bone marrow. Heyl stain. The granular tissue mast cells are well shown.

B. The same specimen. Dual Esterase. Strong CE positivity in virtually all the tumour cells.

C. Tissue mast cell leukaemia. Another case. Bone marrow. Toluidine blue. The field is composed almost entirely of mast cells, which show strong to very strong positivity of their granules with toluidine blue.

D. The same case. PAS reaction. Most cells show granular positivity, commonly coarse, especially at the cell periphery.

E. The same case. Lactic dehydrogenase. Strong scattered granular positivity is present in all tissue mast cells.

F. The same case. Chloroacetate esterase. The mast cells show strong CE positivity.

Polycythaemia and myelofibrosis

These two diseases may be considered together, since they are both myeloproliferative states, and transition from the first to the second often takes place. Cytochemically they show similarities; probably the most important, and certainly the most useful in differentiating forms of the diseases with high leucocyte counts from CML, is the level of leucocyte alkaline phosphatase, which is usually considerably above normal (Wachstein, 1946; Wiltshaw and Moloney, 1955; Hayhoe and Quaglino, 1958). When treatment is successful in inducing some degree of remission there is a tendency for the high scores to diminish, but they usually remain above the normal range. Scores within the normal range have sometimes been reported in polycythaemia vera (Meislin *et al.*, 1959; Anstey *et al.*, 1963) but such cases form a small minority and between 70 and 90% of all cases have elevated scores (Berlin, 1975). Similarly in myelofibrosis, while most patients have raised LAP scores, normal or even low scores may sometimes be found, especially when there is marked cellular immaturity in the peripheral blood (Valentine *et al.*, 1952; Mitus *et al.*, 1958; Koler *et al.*, 1958; Mitus and Kiossoglou, 1968). These cases often show more acute clinical development than usual and it is possible that they provide a link between the relatively benign leucocytic proliferation of polycythaemia and myelofibrosis and the more clearly malignant proliferation of CML.

Proliferation of the megakaryocytic line is usually conspicuous in myelofibrosis and predominates in the megakaryocytic myelosis variant of the myeloproliferative states. The acute form of 'malignant myelofibrosis' may show features suggestive of megakaryoblastic leukaemia, although the blast cells are often difficult to recognise and bone marrow biopsy may be necessary, since aspiration may yield a dry tap (Demmler *et al.*, 1970).

The high levels of LAP found in polycythaemia vera are not present in the neutrophils from patients with erythrocytosis secondary to chronic pulmonary disease, congenital heart disease or primary renal lesions (Mitus *et al.*, 1959), although of course the presence of infection may elevate the score in these conditions and this reduces the diagnostic value of the test. Nevertheless, the finding of a normal LAP score in a patient with erythrocytosis strongly points to the condition being a secondary erythrocytosis and not polycythaemia vera.

Other cytochemical reactions do not differ markedly from normal in polycythaemia or myelofibrosis, although the glycogen content of the polymorphs has long been known to be high in the former (Wagner, 1947) and there is occasional PAS positivity to be seen in a small proportion of the erythroblasts in the latter (Quaglino and Hayhoe, 1960).

Systemic mastocytosis

This condition of widespread tissue infiltration with mast cells, with its rare malignant leukaemic phase, offers an opportunity to study the cytochemistry of mast cells. The characteristic granules are known to contain heparin, histamine, 5-hydroxytryptamine and hyaluronic acid (Parwaresch, 1976) and stain strongly with toluidine blue (Plate 56C). The cells also show positivity to PAS (Plate 56D) and contain lactic dehydrogenase (Plate 56E) and naphthol AS-D chloroacetate esterase (Plate 56B,F), as well as acid phosphatase.

REFERENCES

Ackerman G A 1950 Microscopic and histochemical studies on the Auer bodies in leukaemic cells. Blood 5: 847

Ackerman G A 1964 Eosinophilic leukemia. A morphologic and histochemical study. Blood 24: 372

Aiba M, Raffa P P, Katayama I 1980 Significance of leukocyte alkaline phosphatase in hairy cell leukemia. American Journal of Clinical Pathology 74: 297

Anstey L, Kemp N H, Stafford J L, Tanner R K 1963 Leucocyte alkaline phosphatase activity in polycythaemia rubra vera. British Journal of Haematology 9: 91

Arlin Z A, Fried J, Clarkson B D 1978 Therapeutic role of cell kinetics in acute leukaemia. Clinics in Haematology 7: 339

Arthur D C, Bloomfield C D 1983 Partial deletion of the long arm of chromosome 16 and bone marrow eosinophilia in acute nonlymphocytic leukemia: a new association. Blood 61: 994

Ascari E, Marini G, Invernizzi R, Ippolito G, Casirola G, Fontana G, Rizzo S C 1975 On the usefulness of PAS reaction for the prognosis of acute lymphoblastic leukemia. Haematologica 60: 300

Astaldi G, Mauri C 1954 Recherches sur l'activité proliferative de l'hémocytoblaste de la leucemie aigue. Revue Belge de Pathologie et de Medecine Experimentale 23: 59

Avnstrøm S, Ralfkier E, Clausen N, Plesner T, Hansen N E, Nissen N I 1985 Immunological markers in acute leukaemia. Scandinavian Journal of Haematology 34: 16

Baldini M, Fudenberg H M, Furutake K, Dameshek W 1959 The anaemia of the Di Guglielmo syndrome. Blood 14, 334

Barlogie B, Latrielle J, Freireich E J, Fu C T, Melland M, Andreeff M 1980 Characterisation of hematologic malignancies by flow cytometry. Blood Cells 6: 719

Barton J C, Conrad M E 1978 Current status of blastic transformation in chronic myelogenous leukaemia. American Journal of Hematology 4: 281

Basso G, Cocito M G, Semenzato G, Pezzutto A, Zanesco L 1980 Cytochemical study of thymocytes and T lymphocytes. British Journal of Haematology 44: 577

Basso G, Agostini C, Cocito M G, Pezzutto A, Destro R, Capuzzo F, Gazzola M V, Raimondi R, Zanesco L, Semenzato G 1984 Non-T, non-B childhood acute lymphoblastic leukaemia: correlation between cytochemical markers and first complete remission. Cancer 54: 981

Beard M E J, Durrant J, Catovsky D, Wiltshaw E, Amass J L, Brearley R L, Kirk B, Wrigley P F M, Janossy G, Greaves M F, Galton D A G 1976 Blast crisis of chronic myeloid leukaemia (CML). I. Presentation simulating acute lymphoid leukaemia (ALL). British Journal of Haematology 34: 167

Bennett J M, Henderson E S 1969 Lymphoblastic leukaemia (letter). British Medical Journal 2: 513

Bennett J M, Catovsky D, Daniel M-T, Flandrin G, Galton D A G, Gralnick H R, Sultan C 1976 Proposals for the classification of the acute leukaemias. British Journal of Haematology 33: 451

Bennett J M, Catovsky D, Daniel M-T, Flandrin G, Galton D A G, Gralnick H R, Sultan C. The French-American-British (FAB) Cooperative Group 1981 The morphological classification of acute lymphoblastic leukaemia: concordance among observers and clinical correlations. British Journal of Haematology 47: 553

Bennett J M, Catovsky D, Daniel M-T, Flandrin G, Galton D A G, Gralnick H R, Sultan C. The French-American-British (FAB) Cooperative Group 1982 Proposals for the classification of the myelodysplastic syndromes. British Journal of Haematology 51: 189

Bennett J M, Catovsky D, Daniel M-T, Flandrin G, Galton D A G, Gralnick H R, Sultan C 1985 Proposed revised criteria for the classification of acute myeloid leukemia. Annals of Internal Medicine 103: 626

Berlin N J 1975 Diagnosis and classification of the polycythaemias. Seminars in Hematology 12: 339

Bernheim A, Berger R, Preud'homme J L, Labaume S, Bussel A, Barot-Ciorbarn R 1981 Philadelphia chromosome positive blood B lymphocytes in chronic myelocytic leukemia. Leukemia Research 5: 331

Bernstein R, Pinto M R, Behr A, Mendelow B 1982 Chromosome 3 abnormalities in acute nonlymphocytic leukemia (ANLL) with abnormal thrombopoiesis: report of three patients with a 'new' inversion anomaly and a further case of homologous translocation. Blood 60: 613

Berrebi A, Malaskova V, Oberling F, Myer S 1973 Application des techniques cytochimiques a 47

cas de leucemies aigues: modifications cytochimiques au cours de évolution. Semaine des Hopitaux de Paris 49: 633
Bessis M, Breton-Gorius J 1962 Iron metabolism in the bone marrow as seen by electron microscopy: a critical review. Blood 19: 635
Bessis M, Breton-Gorius J 1969 Pathologies et asynchronisme de developpement des organelles cellulaires au cours des leucemies aigües granulocytaires. Etude au microscope electronique. Nouvelle Revue Francaise d'Hematologie 9: 245
Bessis M, Maigné J 1970 Le diagnostic des variétés de leucémies aigues par la reaction des peroxydases au microscope électronique. Son intérêt et ses limites. European Journal of Clinical and Biological Research 15: 691
Bettelheim P, Lutz D, Majdic O et al 1985 Cell lineage heterogeneity in blast crisis of chronic myeloid leukaemia. British Journal of Haematology 59: 395
Bitter M A, LeBeau M M, Larson R A, Rosner M C, Golomb H M, Rowley J D, Vardiman J W 1984 A morphologic and cytochemical study of acute myelomonocytic leukemia with abnormal marrow eosinophils associated with Inv (16) (p13q22). American Journal of Clinical Pathology 81: 733
Bitter M A, Neilly M E, LeBeau M M, Pearson M, Rowley J D 1985 Rearrangements of chromosome 3 involving bands 3q21 and 3q26 are associated with normal or elevated platelet counts in acute nonlymphocytic leukemia. Blood 66: 1362
Björkman S E 1956 Chronic refractory anaemia with sideroblastic bone marrow. A study of four cases. Blood 11: 250
Bloomfield C D, Garson O M, Volin L, Knuutila S, de la Chapelle A 1985 t(1;3) (p36;q21) in acute nonlymphocytic leukemia: a new cytogenetic-clinicopathological association. Blood 66: 1409
Boesen A M, Hokland P, Jensen O M 1984 Acid phosphatase and acid esterase activity in neoplastic and non-neoplastic lymphoid cells. A semiquantitative evaluation related to immunological markers in 112 cases. Scandinavian Journal of Haematology 32: 313
Boggs D R 1976 The pathogenesis and clinical patterns of blastic crisis of chronic myeloid leukaemia. Seminars in Oncology 3: 289
Borgers M, Verhaegen H, De Brabander M, Thone F, Geuens G 1977 Purine nucleoside phosphorylase, a possible histochemical marker for T cells in man. Journal of Immunological Methods 16: 101
Borgers M, Verhaegen H, De Brabander M et al 1978 Purine nucleoside phosphorylase in chronic lymphatic leukaemia (CLL). Blood 52: 886
Breton-Gorius J, Guichard J 1972 Ultrastructural localization of peroxidase activity in human platelets and megakaryocytes. American Journal of Pathology 66: 277
Breton-Gorius J, Houssay D 1973 Auer bodies in acute promyelocytic leukemia. Demonstration of their fine structure and peroxidase localization. Laboratory Investigation 28: 135
Breton-Gorius J, Reyes F, Duhamel G, Najman A, Gorin N C 1978 Megakaryoblastic acute leukemia; identification by the ultrastructural demonstration of platelet peroxidase. Blood 51: 45
Breton-Gorius J, Reyes F, Vernant J P, Tulliez M, Dreyfus B 1978 The blast crisis of chronic granulocytic leukaemia: megakaryoblastic nature of cells as revealed by the presence of platelet-peroxidase—a cytochemical ultrastructural study. British Journal of Haematology 39: 295
Breton-Gorius J, Vanhaeke D, Pryzwansky K B, Guichard J, Tabilio A, Vainchenker W, Carmel R 1984 Simultaneous detection of membrane markers with monoclonal antibodies and peroxidatic activities in leukaemia: ultrastructural analysis using a new method of fixation preserving the platelet peroxidase. British Journal of Haematology 58: 447
Breton-Gorius J, Guichard J, Tabilio A, Vainchenker W, Vinci G, Henri A 1985 Ultrastructural cytology and cytochemistry of non lymphocytic acute leukaemias. In: Quaglino D, Hayhoe FGJ (eds) The Cytobiology of Leukaemias and Lymphomas. Raven Press, New York, p 179
Breton-Gorius J, Villeval J L, Mitjavila M T, Vinci G, Guichard J, Rochart H, Flandrin G, Vainchenker W 1987 Ultrastructural and cytochemical characterization of blasts from early erythroblastic leukemias. Leukemia 1:173
Brynes R K, Golomb H M, Desser R K, Recant W, Reese C, Rowley J 1976 Acute monocytic leukaemia. Cytologic, histologic, cytochemical, ultrastructural and cytogenetic observations. American Journal of Clinical Pathology 65: 471
Burns G F, Cawley J C, Barker C R, Goldstone A H, Hayhoe F G J 1977a New evidence relating to the nature and origin of the hairy cell of leukaemic reticuloendotheliosis. British Journal of Haematology 36: 71
Burns G F, Cawley J C, Barker C R, Hayhoe F G J 1977b Absence of a receptor for fixed C3 on the hairy cells of leukaemic reticuloendotheliosis. Clinical and Experimental Immunology 29: 442
Cairney A E L, McKenna R, Arthur D C, Nesbit M E Jr, Woods W G 1986 Acute megakaryoblastic leukaemia in children. British Journal of Haematology 63: 541
Caldwell C W, Patterson W P, Hakami N 1987 Alterations of HLe-1(T200) fluorescence intensity

on acute lymphoblastic leukemia cells may relate to therapeutic outcome. Leukemia Research 11: 103

Carbone A, Volpe R, Galligioni E, Tirelli U, Veronesi A, Trovo G, Tumulo S, Grigoletto E 1981 Unusual HCL case with splenomegaly plus generalised lymphadenopathy showed persistently low LAP scores. CR on CHOP. American Journal of Clinical Pathology 76: 728

Cassuto J P, Hammou J C, Pastorelli E, Dujardin P, Massayeff R 1977 Plasma cell acid phosphatase, a discriminative test for benign and malignant monoclonal gammopathies. Biomedicine 27: 197

Catovsky D 1977 Hairy cell leukaemia and prolymphocytic leukaemias. Clinics in Haematology 6: 245

Catovsky D, Greaves M F, Rose M et al (1982) Adult T-cell lymphoma-leukaemia in blacks from the West Indies. Lancet i: 639

Cawley J C, Burns G F 1980 The cytochemistry of human lymphoreticular subpopulations. Immunology Today 1: 85

Cawley J C, Hayhoe F G J 1973 Ultrastructure of Haemic Cells. W B Saunders Co Ltd, London

Cawley J C, Burns G F, Worman C P, Flemans R J, Sibbald R, Barker C R, Roberts B E 1978 Morphological and immunological similarity of the monocytes from pure and mixed monocytic leukaemias. Scandinavian Journal of Haematology 21: 233

Cawley J C, Burns G F, Hayhoe F G J 1980 Hairy-cell Leukaemia. Springer-Verlag, Berlin

Chan L C, Pegram S M, Greaves M F 1985 Contribution of immunophenotype to classification and differential diagnosis of acute leukaemia. Lancet 1: 475

Chan B W B, Hayhoe F G J 1971 Changes in proliferative activity of marrow leukaemic cells during and after extracorporeal irradiation of blood. Blood 37: 657

Chan W C, Check I, Schick C, Brynes R K, Kateley J, Winton E F 1984 A morphologic and immunologic study of the large granular lymphocyte in neutropenia with T lymphocytosis. Blood 63: 1133

Chang K S, Schroeder W, Siciliano M J, Thompson L H, McCredie K, Beran M, Freireich E J, Liang J C, Trujillo J M, Stass S A 1987 The localization of the human myeloperoxidase gene is in close proximity to the translocation breakpoint in acute promyelocytic leukemia. Leukemia 1: 458

Clarkson B 1969 Review of recent studies of cellular proliferation in acute leukaemia. National Cancer Institute Monograph No. 30, p 81

Coleman M S, Hutton J J 1981 Terminal transferase. In: Catovsky D (ed) The Leukaemic Cell. Churchill-Livingstone, Edinburgh p 203

Costello C, Catovsky D, O'Brien M, Morilla R, Varadi S 1980 Chronic T-cell leukaemias. I. Morphology, cytochemistry and ultrastructure. Leukemia Research 4: 463

Crockard A D 1984 Cytochemistry of lymphoid cells: a review of findings in the normal and leukaemic state. Histochemical Journal 16: 1027

Crockard A, Chalmers D, Matutes E, Catovsky D 1982 Cytochemistry of acid hydrolases in chronic B- and T-cell leukaemias. American Journal of Clinical Pathology 78: 437

Crockard A D, MacFarlane E, Andrews C, Bridges J M, Catovsky D 1984 Dipeptidylaminopeptidase IV activity in normal and leukaemic T-cell subpopulations. American Journal of Clinical Pathology 82: 294

Crockard A D, MacFarlane E, Markey G M, McConnell R E, Alexander H D, Agnew A N D, Morris T C M, Bridges J M 1985 Acid hydrolase activities in B cell chronic lymphocytic leukaemia lymphocytes: correlation of cytochemical reactions with immunological phenotype. Scandinavian Journal of Haematology 34: 242

Cuttner J, Saremetis S, Najfeld V, Dimitriu-Bona A, Winchester R A 1984 TdT-positive acute leukaemia with monocytoid characteristics; clinical, cytochemical, cytogenetic and immunological findings. Blood 64: 237

Dacie J V, Smith M D, White J C, Mollin D L 1959 Refractory normoblastic anaemia: a clinical and haematological study of seven cases. British Journal of Haematology 5: 56

Dameshek W, Baldini M 1958 The Di Guglielmo Syndrome. Blood 13: 192

Dameshek W, Gunz F W 1958 Leukemia. Grune & Stratton, New York

Daniel M T, Flandrin G 1974 Fine structure of abnormal cells in hairy cell (tricholeucocytic) leukaemia with special reference to their in vitro phagocytic capacity. Laboratory Investigation 30: 1

Darbyshire P J, Lilleyman J S 1987 Granular acute lymphoblastic leukaemia of childhood: a morphological phenomenon. Journal of Clinical Pathology 40: 251

Demmler K, Burkhardt R, Prechtel K 1970 Megakaryoblastische Myelose. Klinische Wochenschrift 48: 1168

Den Ottolander G J, te Velde J, Brederoo P, Geraedts J P M, Slee P H T, Willemze R, Zwaam F E, Haak H L, Muller H P, Bieger R 1979 Megakaryoblastic leukaemia (acute myelofibrosis): a report of three cases. British Journal of Haematology 42: 9

Drexler H G, Gaedicke G 1983 Isoenzyme studies in human leukemia. II. Carboxylic esterase (EC. 3.1.1.1). Leukemia Research 7: 599

Drexler H G, Gaedicke G, Minowada J 1984 Enzyme markers in acute leukemias: advances during the last decade. Journal of the National Cancer Institute 72: 1283
Dreyfus B, Rochant H, Sultan C, Clauvel J P, Yvart J, Chesneau A M 1970 Les anemies refractaires avec exces de myeloblastes dans la moelle. Etude de ll observations. Nouvelle Presse Medicale 78: 359
El-Mohandes E, Hayhoe F G J 1983 5′-Nucleotidase activity of megakaryoblasts in a case of acute megakaryoblastic leukaemia. British Journal of Haematology 53: 523
Eys J van, Pullen J, Head D et al 1986 The French-American-British (FAB) classification of leukemia. The pediatric oncology group experience with lymphocytic leukemia. Cancer 57: 1046
Falini B, Schwarting R, Erber W, Posnett D N, Martelli M F, Grignani F, Zuccaccia M, Gatter K C, Cernetti C, Stein H, Mason D Y 1985 The differential diagnosis of hairy cell leukemia with a panel of monoclonal antibodies. American Journal of Clinical Pathology 83: 289
Feldges A J, Aur R J A, Verzosa M S, Daniels S 1974 Periodic acid-Schiff reaction, a useful index of duration of complete remission in acute childhood lymphocytic leukemia. Acta Haematologica 52: 8
Feller A C, Parwaresch M R, Bartels H, Lennert K 1982 Enzyme cytochemical heterogeneity of human chronic T-lymphocytic leukaemia as demonstrated by reactivity to dipeptidylaminopeptidase IV (DAP IV; EC.3.4.14.4). Leukemia Research 6: 801
Feller A C, Ziegler A, Sterry W, Goos M, Parwaresch M R 1983 Phenotypic heterogeneity of leukaemic Sezary cells. Blut 47: 333
Ffrench M, Bryon, P A, Fiere D et al 1986 Cell-cycle prognostic value in adult acute myeloid leukemia. The choice of the best variables. Leukemia Research 10: 51
Fialkow P L, Martin P L, Najfeld V, Penfold G K, Jacobson R J, Hansen J A 1981 Evidence for a multistep pathogenesis of chronic myelogenous leukemia. Blood 58: 158
Fishleder A J, Tubbs R R, Savage R A, Lucas F V, Norris D, Weick J K, Valenzuela R 1984 Immunophenotypic characterisation of acute leukemia by immunocytology. American Journal of Clinical Pathology 81: 611
Flandrin G, Daniel M-T 1981 Cytochemistry in the classification of leukaemias. In: Catovsky D (ed) The Leukemic Cell, Churchill Livingstone, Edinburgh, p 41
Flandrin G, Daniel M T 1984 Sézary syndrome: cytochemistry and ultrastructure. In: Polliack A (ed) Human Leukemias. Martinus Nijhoff Publ, Boston, p 327
Foadi M D, Cooper E H, Hardisty R M 1967 DNA synthesis and DNA content of leucocytes in acute leukaemia. Nature 216: 134
Foadi M D, Cooper E H, Hardisty R M 1968 Proliferative activity of leukaemic cells at various stages of acute leukaemia of childhood. British Journal of Haematology 15: 269
Gaedicke G, Drexler H G 1984 Leukemic cell differentiation in childhood leukemias. Analysis by enzyme markers. European Journal of Pediatrics 142: 157
Gaedicke G, Arndt R, Kabisch H, Landbeck G 1979 Isoenzyme studies in human leukemia. 1. Acid phosphatase. Leukemia Research 3: 417
Gahrton G 1966 Normal human neutrophil leucocytes as a reference system for the microspectrophotometrically quantitated periodic acid-Schiff reaction. Journal of Histochemistry and Cytochemistry 14: 45
Gahrton G, Brandt L, Franzen S, Norden A 1969 Cytochemical variants of neutrophil léucocyte populations in chronic myelocytic leukaemia. Scandinavian Journal of Haematology 6: 365
Gale R P, Cannani E 1985 The molecular biology of chronic myelogenous leukaemia. British Journal of Haematology 60: 395
Gavosto F, Pileri A, Maraini G 1959 Incorporazione di timidina marcata con tritio negli elementi del midollo osseo normale e leucemico—indagini autoradiografiche. Haematologica 44: 997
Gavosto F, Maraini G, Pileri A 1960a Radioautographic investigations on DNA and protein metabolism in two cases of Di Guglielmo's disease. Blood 16: 1122
Gavosto F, Pileri A, Maraini G 1960b Proliferative capacity of acute leukaemia cells. Nature 187: 611
Gavosto F, Pileri A, Vachi C, Pegoraro L 1964 Proliferation and maturation defect in acute leukaemia cells. Nature 203: 92
Gavosto F, Pileri A, Gabutti V, Masera P 1967 Cell population kinetics in human acute leukaemia. European Journal of Cancer 3: 301
Goldberg A F 1964 Acid phosphatase activity in Auer bodies. Blood 24: 305
Goodman J R, Hall S G 1967 Accumulation of iron in mitochondria of erythroblasts. British Journal of Haematology 13: 335
Gordon D S, Hubbard M R 1979 Neutrophilic nonspecific esterase: a reply (Letter). Blood 53: 340
Grasso J A, Hines J D 1969 A comparative electron microscopic study of refractory and alcoholic sideroblastic anaemia. British Journal of Haematology 17: 35
Greaves M F, Verbi W, Reeves B R, Hoffbrand A V, Drysdale H C, Jones L, Sacker L S, Samaratunga I 1979 'Pre-B' phenotypes in blast crisis of Ph[1] positive CML: evidence for a pluripotential stem cell

'target'. Leukemia Research 3: 181
Greaves M F, Chan L C, Furley A J, Watt S M, Molgaard H V 1986 Lineage promiscuity in hemopoietic differentiation and leukemia. Blood 67: 1
Greenberg P 1983 The smouldering myeloid leukemia states. Clinical and biological features. Blood 4: 1035
Griffin J D, Todd F J, Ritz J, Nadler L M, Canellos G P, Rosenthal D, Gallivan M, Beveridge R P, Weinstein H, Karp D, Schlossman S F 1983 Differentiation patterns in the blastic phase of chronic myeloid leukaemia. Blood 61: 85
Grogan T M, Insalaco S J, Savage R A, Vail M L 1981 Acute lymphocytic leukaemia with prominent azurophilic granulation and punctate acidic non-specific esterase and phosphatase activity. American Journal of Clinical Pathology 75: 716
Grossi C E, Zicca A, Leprini A, Cadoni A, Pistoia V, Ferrarini M 1982 Acid hydrolases as markers of maturation in B cell chronic lymphocytic leukemia. Blood 60: 220
Grusovin G D, Castoldi G L 1976 Characterization of blast cells in acute nonlymphoid leukemias by consecutive cytochemical reactions. Acta Haematologica 55: 338
Hale A J, Wilson S J 1959 The deoxyribonucleic acid content of leucocytes in normal and in leukaemic human blood. Journal of Pathology and Bacteriology 77: 605
Hale A J, Wilson S J 1961 The deoxyribonucleic acid content of the leucocytes in human blood, bone marrow and lymph glands. Journal of Pathology and Bacteriology 82: 483
Hanaoka M, Takatsuki K, Shimoyama M (eds) 1982 Adult T cell Leukemia and Related Diseases. Plenum Press, New York
Harada N 1951 Histochemical studies on the Auer body. Nagoya Journal of Medical Science 14: 129
Hayhoe F G J 1984 Cytochemistry of the acute leukaemias. Histochemical Journal 16: 1051
Hayhoe F G J 1985 Integrated approach to classification of AML in relation to prognosis and survival. In: Quaglino D, Hayhoe F G J (eds) The Cytobiology of Leukaemias and Lymphomas. Raven Press, New York, p 225
Hayhoe F G J, Cawley J C 1972 Acute leukaemia: cellular morphology, cytochemistry and fine structure. Clinics in Haematology 1: 49
Hayhoe F G J, Neuman Z 1976 Cytology of myeloma cells. Journal of Clinical Pathology 29: 916
Hayhoe F G J, Quaglino D 1958 Cytochemical demonstration and measurement of leucocyte alkaline phosphatase activity in normal and pathological states by a modified azo-dye coupling technique British Journal of Haematology 4: 375
Hayhoe F G J, Quaglino D 1960 Refractory sideroblastic anaemia and erythraemic myelosis: possible relationship and cytochemical observations. British Journal of Haematology 6: 381
Hayhoe F G J, Quaglino D, Doll R 1964 The Cytology and Cytochemistry of Acute Leukaemias. A study of 140 cases. MRC Special Report Series No. 304 HMSO: London
Hayhoe F G J, Flemans R J, Burns G F, Cawley J C 1977 Leucocyte alkaline phosphatase scores in hairy cell leukaemia. British Journal of Haematology 37: 158
Heilmeyer L 1959 The sideroblastic anaemias. German Medical Monthly 4: 403
Heilmeyer L, Keiderling W, Merker H, Clotten R, Schubothe H 1960 Die anaemia refractoria sideroblastica und ihre Beziehungen zur Lebersiderose und Hämochromatose. Acta Haematologica 23: 1
Heilmeyer L, Merker H, Mölbert E, Neidhart M 1962 Zur Mikromorphologie der hereditären hypochromen sideroachrestischen Anämie. Acta Haematologica 27: 78
Hernandez P, Carnot J, Cruz C 1982 Chronic myeloid leukaemia blast crisis with T-cell features. British Journal of Haematology 51: 175
Heumann D, Losa G, Barras C, Morell A, von Fliedner V 1985 Characterisation of acute undifferentiated leukemia by combined analysis of plasma membrane-associated γ-glutamyl transpeptidase and soluble terminal transferase. Blood 66: 255
Higgy K E, Burns G F, Hayhoe F G J 1978 Identification of the hairy cells of leukaemic reticuloendotheliosis by an esterase method. British Journal of Haematology 38: 99
Hoffmann J J M L, Breed W P M, Swaak-Lammers N 1985 Plasma cell acid phosphatase score in multiple myeloma and related disorders. British Journal of Haematology 59: 67
Hoyle C, Sherrington P, Hayhoe F G J 1987 Cytological features of 9q- deletions in AML. British Journal of Haematology 66: 277
Huang M-J, Li C-Y, Nichols W L, Young J-H, Katzmann J A 1984 Acute leukemia with megakaryocytic differentiation; a study of 12 cases identified immunocytochemically. Blood 64: 427
Huhn D 1984 Cytochemistry and ultrastructure of T-cell chronic lymphocytic leukemia. In: Polliack A (ed) Human Leukemias. Martinus Nijhoff Publ, Boston, p 271
Huhn D, Twardzik L 1983 Acute myelomonocytic leukaemia and the French-American-British classification. Acta Haematologica 69: 36
Huhn D, Schmalzl F, Krug U 1971 Acute myeloid leukaemia. Blut 23: 189
Humphrey G B, Nesbit M E, Brunning R D 1974 Prognostic value of the periodic acid-Schiff (PAS)

reaction in acute lymphoblastic leukemia. American Journal of Clinical Pathology 61: 393
Hutton J J, Coleman M S. Moffitt S et al 1982 Prognostic significance of terminal transferase activity in childhood acute lymphoblastic leukemia: a prospective analysis of 164 patients. Blood 60: 1267
Janossy G 1981 Membrane markers in leukaemia. In: Catovsky D (ed) The Leukemic Cell. Churchill Livingstone, Edinburgh, p 129
Janossy G, Campana D 1985 The use of monoclonal antibodies in characterizing leukaemia cells. In: Deinhardt F (ed) Proc. XII Symp. Comp. Res. Leuk.
Janossy G, Greaves M F, Revesz T, Lister T A, Roberts M, Durrant J, Kirk B, Catovsky D, Beard M E J 1976a Blast crisis of chronic myeloid leukaemia (CML). II. Cell surface marker analysis of 'lymphoid' and myeloid cases. British Journal of Haematology 34: 179
Janossy G, Roberts M, Greaves M F 1976b Target cell in chronic myeloid leukaemia and its relationship to acute lymphoid leukaemia. Lancet ii: 1058
Jansen J 1981 Significance of leucocyte alkaline phosphatase index in hairy cell leukemia. American Journal of Clinical Pathology 75: 765
Johansen P, Krogh Jensen M 1980 Enzyme cytochemistry and immunohistochemistry in monoclonal gammopathy and reactive plasmacytosis. Acta Pathologica et Microbiologica Scandinavica (Section A) 88: 377
Kamada N, Dohy H, Okada K, Oguma N, Kuramoto A, Tanaka K, Uchino H 1981 In vivo and in vitro activity of neutrophil alkaline phosphatase in acute myelocytic leukemia with 8;21 translocation. Blood 58: 1213
Kass L, Hadi M Z 1975 Phosphorylase activity in chronic erythraemic myelosis. American Journal of Clinical Pathology 64: 503
Katayama I, Aiba M 1980 Tartrate-resistant acid phosphatase reaction. American Journal of Clinical Pathology 73: 143
Katayama I, Yang J 1977 Reassessment of a cytochemical test for differential diagnosis of leukemic reticuloendotheliosis. American Journal of Clinical Pathology 68: 268
Katayama I, Li C Y, Yam L T 1972 Ultrastructural (cytochemical) demonstration of tartrate-resistant acid phosphatase isoenzyme activity in 'hairy cells' of leukaemic reticuloendotheliosis. American Journal of Pathology 69: 471
Kaung D T 1969 Periodic acid-Schiff reaction in human basophilic leucocytes. Acta Haematologica 42: 269
Khalaf M R, Hayhoe F G J 1986 Improved cytochemistry of a specific 5′nucleotidase in haemic cells. Histochemical Journal (in press).
Khalaf M R, Bevan P C, Hayhoe F G J 1986 Comparative cytochemical study of dipeptidyl aminopeptidase (DAP) II and IV in normal and malignant haemic cells. Journal of Clinical Pathology 39: 891
Killmann S A 1972 Kinetics of leukaemic blast cells in man. Clinics in Haematology 1: 95
Killmann S A, Cronkite E P, Robertson J S, Fliedner T M, Bond V P 1963 Estimation of phases of the life cycle of leukaemic cells from labelling in human beings in vivo with tritiated thymidine. Laboratory Investigation 12: 671
Koike T 1984 Megakaryoblastic leukemia: the characterisation and identification of megakaryoblasts. Blood 64: 683
Koler R D, Seaman A J, Osgood E E, Vanbellinghen P 1958 Myeloproliferative diseases: diagnostic value of the leucocyte alkaline phosphatase test. American Journal of Clinical Pathology 30: 295
Kondo K, Yoshitake J, Takemura K 1966 The fine structure of Auer bodies. Journal of Electron Microscopy 15: 237
Lanham G R, Bollum F J, Williams D L, Stass S A 1984 Simultaneous occurrence of terminal deoxynucleotidyl transferase and myeloperoxidase in individual leukemic blasts. Blood 64: 318
Larizza P, Orlandi F 1964 Electron microscopic observations on bone marrow and liver tissue in non-hereditary refractory sideroblastic anaemia. Acta Haematologica 31: 9
Laurie H C 1968 Duration of remissions in lymphoblastic leukaemia of childhood. British Medical Journal 2: 95
Le Beau M M, Larson R A, Bitter M A, Vardiman J W, Golomb H M, Rowely J D 1983 Association of an inversion of chromosome 16 with abnormal marrow eosinophils in acute myelomonocytic leukemia: a unique cytogenetic-clinicopathological association. New England Journal of Medicine 309: 630
Leder L D 1967 Die fermentcytochemische Erkennung normaler und neoplastischer Erythropoesezellen in Schnitt und Ausstrich. Blut 15: 289
Leder L D 1969 On the PAS reaction in acute paraerythroblastic hemoblastoses. Acta Haematologica 41: 328
Leinonen E A 1970 Cytochemical studies of acute leukaemias. Acta Haematologica 43: 219
Li C Y, Lam K W, Yam L T 1973 Esterases in human leucocytes. Journal of Histochemistry and Cytochemistry 21: 1

Li C-Y, Phyliky R L, Yam L T 1986 Acute myelomonocytic leukemia: an unusual variant with both granulocytic and monocytic esterases in the leukemic cells. Mayo Clinic Proceedings 61: 104
Li C-Y, Yam L T, Lam K W 1970a Acid phosphatase isoenzymes in human leucocytes in normal and pathological conditions. Journal of Histochemistry and Cytochemistry 18: 473
Li C-Y, Yam L T, Lam K W 1970b Studies of acid phosphatase isoenzymes in human leukocytes, demonstration of isoenzyme cell specificity. Journal of Histochemistry and Cytochemistry 18: 901
Lilleyman J S, Mills V, Sugden P J, Britton J A 1979 Periodic acid-Schiff reaction and prognosis in lymphoblastic leukaemia. Journal of Clinical Pathology 32: 158
Lin M S, Bouroncle B A 1967 The size and transit time of non-dividing subpool of precursor cells in acute leukaemia. Blood 29: 63
Ling N R 1983 Immunoglobulin as a differentiation and clonal marker. Journal of Immunological Methods 65: 1
Locquin M, Bessis M 1949 Sur quelques caractères des corps d'Auer et de Charcot. Compte Rendu des Seances de la Societe de Biologie 143: 609
Look A T, Melvin S L, Williams D Z, Brodeur G M, Dahl G V, Kalwinsky D K, Murphy S B, Mauer A M 1982 Aneuploidy and percentage of S-phase cells determined by flow cytometry correlate with cell phenotype in childhood acute leukemia. Blood 60: 959
Machii T, Kitani T 1984 Similarities between IgG-bearing lymphocytes and hairy cells: cytologic and cytochemical studies. Blood 64: 166
Markert C L, Hunter R L 1959 The distribution of esterases in normal tissue. Journal of Histochemistry and Cytochemistry 7: 42
Markey G M, Morris T C M, Alexander H D, Kyle A, Middleton D, Turner A, Burnside P, Drexler H G, Gaedicke G, Hartman E, Robertson J H 1987 Monocyte esterase? A factor involved in the pathogenesis of lymphoproliferative neoplasia. Leukemia 1: 236
Marks S M, Baltimore D, McCaffrey R 1978 Terminal transferase as a predictor of initial responsiveness to vincristine and prednisone in blastic chronic myelogenous leukaemia. A cooperative study. New England Journal of Medicine 298: 812
Marmont A M, Damasio E E 1973 The treatment of terminal metamorphosis of chronic granulocytic leukaemia with corticosteroids and vincristine. Acta Haematologica 50: 1
Mastrangelo D, Orrico A, Pasqui A L, Capecci P L, Ceccatelli L, Laghi Pasini F 1987 Double esterase (DE) positivity in AML cells. British Journal of Haematology 65: 119
Mathe G, Rappaport H 1976 Histological and Cytological Typing of Neoplastic Diseases of Haematopoietic and Lymphoid Tissues. WHO, Geneva
Matutes E, Crockard A D, O'Brien M, Catovsky D 1983 Ultrastructural cytochemistry of chronic T-cell leukaemias. A study with 4 acid hydrolases. Histochemical Journal 15: 895
Mauer A M, Fisher V 1962 Comparison of the proliferative capacity of acute leukaemia cells in bone marrow and blood. Nature 193: 1085
Mauer A M, Fisher V 1966 Characteristics of cell proliferation in four patients with untreated acute leukaemia. Blood 28: 428
Mauer A M, Look T 1985 Leukemic cell kinetics today. In: Quaglino D, Hayhoe F G J (eds) The Cytobiology of Leukaemias and Lymphomas. Serono Symposia Publications, Raven Press, New York, p 209
McCaffrey R, Greaves M, Harrison T A, Revesz T, Beard M, Baltimore D 1975a Biochemical and immunological evidence for lymphoblastic conversion in chronic myelogenous leukaemia (Abstract). Blood 46: 1043
McCaffrey R, Harrison T A, Parkman R, Baltimore D 1975b Terminal deoxynucleotidyl transferase activity in human leukaemic cells and in normal human thymocytes. New England Journal of Medicine 292: 775
McDuffie N G 1967 Crystalline patterns in Auer bodies and specific granules of human leucocytes. Journal of Microscopy 6: 321
McKenna R W, Bloomfield C D, Dick F, Nesbit M E, Brunning R D 1975 Acute monoblastic leukaemia: diagnosis and treatment of ten cases. Blood 46: 481
McKenna R W, Brynes R K, Nesbit M E et al 1979 Cytochemical profiles in acute lymphoblastic leukemia. American Journal of Pediatic Hematology and Oncology 1: 263
McKenna R W, Parkin J, Bloomfield C D, Sundberg R D, Brunning R D 1982 Acute promyelocytic leukaemia: a study of 39 cases with identification of a hyperbasophilic microgranular variant. British Journal of Haematology 50: 201
Meislin A G, Lee S L, Wasserman L R 1959 Leucocyte alkaline phosphatase activity in haematopoietic disorders: A correlative study by biochemical and cytochemical techniques. Cancer, 12: 760
Metcalf D 1971 Transformation of granulocytes to macrophages in bone marrow colonies in vitro. Journal of Cell Physiology 77: 277
Mikata A, Suzuki H, Suzuki S, Shimoyama M 1982 Histochemical and cytochemical observations

on normal and neoplastic T lymphocytes. In: Hanaoka M, Takatsuki K, Shimoyama M (eds). Adult T cell leukemia and related disorders. Plenum Press, New York, p 77
Miliani A, Anichini P, DiGuglielmo R et al 1982 Etude cytoenzymologique des plasmocytes medullaires des dysglobulinemies monoclonales myelomateuses et non-myelomateuses. Nouvelle Revue Francaise d'Hematologie 24: 27
Milligan D W, Roberts B E, Limbert H J, Jalihal S 1984 Cytochemical and immunological characteristics of acute monocytic leukaemia. British Journal of Haematology 58: 391
Mitus W J, Kiossoglou K A 1968 Leucocyte alkaline phosphatase in myeloproliferative syndrome. Annals of the New York Academy of Sciences 155: 976
Mitus W J, Mednicoff J B, Dameshek W 1959 Alkaline phosphatase of mature neutrophils in various polycythemias. New England Journal of Medicine 260: 1131
Mitus W J, Bergna L J, Mednicoff I B, Dameshek W 1958 Alkaline phosphatase of mature neutrophils in chronic forms of the myeloproliferative syndrome. American Journal of Clinical Pathology 30: 285
Montecucco C, Riccardi A, Traversi E, Giordano P, Mazzini G, Ascari E 1983a Proliferation activity of bone marrow cells in primary dysmyelopoietic (preleukaemic) syndromes. Cancer 52: 1190
Montecucco C, Riccardi A, Traversi E, Danova M, Ucci G, Mazzini G, Giordano P 1983b Flow cytometric DNA content in myelodysplastic syndromes. Cytometry 4: 238
Mueller J, Brun Del Re G, Buerki H, Keller H U, Hess M W, Cottier H 1975 Non-specific acid esterase activity: a criterion for differentiation of T and B lymphocytes in mouse lymph nodes. European Journal of Immunology 5: 270
Mukiibi J M, Kyobe J 1982 Leukocyte alkaline phosphatase (LAP) activity in African myeloma patients. East African Medical Journal 59: 37
Naegeli E 1923 Blutkrankheiten und Blutdiagnostik. 4th Auflage. Julius Springer, Berlin
Nasu K, Tatsumi E, Takiuchi Y, Kita K, Nishikori N, Shirakawa S, Uchino H 1981 Several marker analysis of T-lymphoma cells in eight Japanese adults. American Journal of Clinical Pathology 76: 670
Nelson J S R, Schiffer L M 1973 Autoradiographic detection of DNA polymerase containing nuclei in sarcoma 180 ascites cells. Cell and Tissue Kinetics 6: 45
Neumann H, Klein E, Russell R M, Epstein S 1979 N-alkaline phosphatase: a potential disease marker for lymphoproliferative disorders. British Journal of Haematology 41: 519
Newland A C, Catovsky D, Linch D, Cawley J C, Beverley P, SanMiguel J F, Gordon-Smith E C, Blecher T E, Shahriari S, Varadi S 1984 Chronic T cell lymphocytosis: a review of 21 cases. British Journal of Haematology 58: 433
Nielsen H K, Bendix-Hansen K 1984 Myeloperoxidase-deficient polymorphonuclear leucocytes. III. Relation to incidence of infection in acute myeloid leukaemia. Scandinavian Journal of Haematology 33: 75
Nogueira-Costa R, Spitzer G, Cork A, Trujillo J M 1985 E rosette-positive agar colonies containing the Philadelphia chromosome in chronic myeloid leukaemia. Scandinavian Journal of Haematology 34: 184
Nowell P, Finam J 1978 Chromosome studies in preleukaemic states. IV. Myeloproliferative versus cytopenic disorders. Cancer 42: 2254
Olhagen B, Thorell B, Wising P 1949 The endocellular nucleic acid distribution and plasma protein formation in myelomatosis. Scandinavian Journal of Clinical and Laboratory Investigation 1: 49
Oliveira MSP de, Gregory C, Matutes E, Parreira A, Catovsky D 1987 Cytochemical profile of megakaryoblastic leukaemia: a study with cytochemical methods, monoclonal antibodies and ultrastructural cytochemistry. Journal of Clinical Pathology 40: 663
Pangalis G A 1984 Contribution of cytochemistry in leukaemia. In: Polliack A (ed) Human Leukaemias: Cytochemical and Ultrastructural Techniques in Diagnosis and Research. Martinus Nijhoff Publ, Boston, p 3
Paolucci G, Rosito P, Mancini A, Vecchi V, Vivarelli F, Faldella G 1977 Relazione fra percentuale di blasti PAS positivi alla diagnosi e durata della prima remissione nella leucemia acuta linfoblastica del bambino. Haematologica 62: 250
Parwaresch M R 1976 The Human Blood Basophil. Springer-Verlag, Berlin
Pearse A G E 1949 The nature of Russell bodies and Kurloff bodies: observations on the cytochemistry of plasma cells and reticulum cells. Journal of Clinical Pathology 2: 81
Pearson E C 1986 Crystal structure of Auer rods in acute myeloblastic leukaemia (AMyL). Journal of Clinical Pathology 39: 569
Pearson E C, Matthews J G, Hayhoe F G J 1986 Ultrastructure and cytogenetics in seven cases of acute promyelocytic leukaemia (APL). British Journal of Haematology 63: 247
Pedersen B, Killmann S-A 1971 Leukaemic subclones in chronic myelogenous leukaemias. Differences in drug sensitivity and differentiation capacity. Acta Medica Scandinavica 190: 61

Peters S W, Clark R E, Hoy T G, Jacobs A 1986 DNA content and cell cycle analysis of bone marrow cells in myelodysplastic syndromes (MDS). British Journal of Haematology 62: 239
Pieters R, Veerman A J P 1987 Prognostic significance of acid alpha-naphthyl acetate esterase and acid phosphatase in childhood acute lymphoblastic leukaemia. Leukemia Research (in press)
Polli N, Zola H, Catovsky D 1984 Characterisation by ultrastructural cytochemistry of normal and leukaemic myeloid cells reacting with monoclonal antibodies. American Journal of Clinical Pathology 82: 389
Presentey B, Jerushalmy Z, Mintz U 1979 Eosinophilic leukaemia: morphological, cytochemical and electron microscopic studies. Journal of Clinical Pathology 32: 261
Quaglino D 1961 Cytochemistry of blood and bone marrow cells with particular reference to leukaemia and allied diseases. Ph. D. Thesis, Cambridge Univ.
Quaglino D, Hayhoe F G J 1959 Observations on the periodic acid-Schiff reaction in lymphoproliferative diseases. Journal of Pathology and Bacteriology 78: 521
Quaglino D, Hayhoe F G J 1960 Periodic acid-Schiff positivity in erythroblasts with special reference to Di Guglielmo's disease. British Journal of Haematology 6: 26
Quaglino D, Torelli U, Sauli S, Mauri C 1967 Cytochemical and autoradiographic investigations on normal and myelomatous plasma cells. Acta Haematologica 38: 79
Quaglino D, Emilia G, Piccinini L, De Pasquale A, Zanni G, Guerzoni O, Giuliani M 1977 Relationship between autoradiographic, cytochemical and cytogenetic findings in chronic myeloid leukaemia. Haematologica 62: 360
Reijden H J van der, Rhenen D J van, Lansdorp P M et al 1983 A comparison of surface marker analysis and FAB classification in acute myeloid leukaemia. Blood 61: 443
Rheingold J J, Kaufman R, Adelson E, Lear A 1963 Smoldering acute leukemia. New England Journal of Medicine 268: 812
Rodriguez-Fuch C A, Marin C A, Bezares R F, Carreras L O, De Tezanos–Pinto M 1984 Granular acute leukaemia not always of myeloid origin. Lancet i: 636
Rowley J D 1978 The cytogenetics of acute leukaemia. Clinics in Haematology 7: 385
Saunders E F, Mauer A M 1969 Re-entry of non-dividing cells into a proliferative phase in acute childhood leukemia. Journal of Clinical Investigation 48: 1299
Saunders E F, Lampkin B C, Mauer A M 1967 Variation of proliferative activity in leukemic cell populations of patients with acute leukemia. Journal of Clinical Investigation 46: 1356
Savage R A, Valenzuela R, Hoffman G C 1981 Acid phosphatase staining pattern as an indication of T-cell acute leukemia. American Journal of Clinical Pathology 76: 760
Schaefer H E, Hellriegel K P, Hennekeuser H, Hübner G, Zach J, Fischer R, Gross R 1973 Eosinophilenleukämie, eine unreifzellige Myelose mit Chloroacetatesterase-positiven Eosinophilen. Blut 26: 7
Schiffer C A, Sanel F T, Stechmiller B K, Wiernik P H 1975 Functional and morphological characteristics of the leukemic cells of a patient with acute monocytic leukemia: correlation with clinical features. Blood 46: 17
Schilling V 1926 Der Monozyt in trialistischer Auffassung und seine Bedeutung im Krankheitsbilde. Medizinische Klinische 22: 563
Schmalzl F, Lederer B, Braunsteiner H 1970 Atypical myeloblastic leukaemia with differentiation into paraneutrophils. Blut 20: 337
Schmalzl F, Konwalinka G, Michlmayr G, Abbrederis K, Braunsteiner H 1978 Detection of cytochemical and morphological abnormalities in 'preleukaemia'. Acta Haematologica 59: 1
Schmid J R, Kiely J M, Tauxe W N, Owen C A Jr 1966 In vitro DNA and RNA synthesis in human bone marrow cells: a study of 12 normal subjects and 12 patients with lymphoplasmocytic disorder. Blood 27: 310
Scott C S, Cahill A, Bynoe A G, Ainley M J, Hough D, Roberts B E 1983 Esterase cytochemistry in primary myelodysplastic syndromes and megaloblastic anaemias: demonstration of abnormal staining patterns associated with dysmyelopoiesis. British Journal of Haematology 55: 411
Scott C S, Morgan M, Limbert H J, Mackarill I D, Roberts B E 1985 Cytochemical, immunological and ANAE isoenzyme studies in acute myelomonocytic leukaemia: a study of 39 cases. Scandinavian Journal of Haematology 35: 284
Scott R B, Ellison R R, Ley A B 1964 A clinical study of twenty cases of erythroleukaemia (Di Guglielmo's syndrome). American Journal of Medicine 37: 162
Senelar R, Jaubon C, Taib J, Escola M J, Vannereau H, Vannereau A, Dubois A, Charlier D 1984 Acute myeloid leukemia with giant inclusions; cytochemical and ultrastructural study. Acta Haematologica 71: 165
Shaw M T, Nordquist R E 1975 'Pure' monocytic or histiomonocytic leukaemia: a revised concept. Cancer 35: 208
Shaw M T, Humphrey G B, Lawrence R, Fisher D B 1977 Lack of prognostic value of the periodic acid-Schiff reaction and blast cell size in childhood acute lymphocytic leukemia. American Journal of Haematology 2: 237

Smets L A, Taminian J, Hählen K, de Waal F, Behrendt H 1983 Cell kinetic responses in childhood acute nonlymphocytic leukemia during high-dose therapy with cytosine arabinoside. Blood 61: 79
Smets L A, Homan-Blok J, Hart A, de Vaan G, Behrendt H, Hählen K, de Waal F J 1987 Prognostic implication of hyperdiploidy as based on DNA flow cytometric measurement in childhood acute lymphocytic leukemia—a multi-centre study. Leukemia 1: 163
Srivastava B I S, Khan S A, Minowada J, Gomez G A, Rakowski I 1977 Terminal deoxynucleotidyl transferase activity in blastic crisis of chronic myelogenous leukemia. Cancer Research 37: 3612
Stein P, Peiper S, Butler D, Melvin S, Williams D, Strass S 1983 Granular acute lymphoblastic leukaemia. American Journal of Clinical Pathology 79: 426
Stryckmans P A 1974 Current concepts in chronic myelogenous leukaemia. Seminars in Haematology 11: 101
Stuart J, Simpson J S, Mann J R 1970 Intracellular hydrogen transport systems in acute leukaemia. British Journal of Haematology 19: 739
Suzuki A, Shibota A, Onodera S, Miura A B, Sakamoto S, Suzuki C 1969 Histochemical study on plasma cells. Tohoku Journal of Experimental Medicine 97: 1
Swirsky D M, Li Y S, Matthews J G, Flemans R J, Rees J K H, Hayhoe F G J 1984 8;21 translocation in acute granulocytic leukaemia: cytological, cytochemical and clinical features. British Journal of Haematology 56: 199
Swirsky D M, de Bastos M, Parish S, Rees J K H, Hayhoe F G J 1985 Factors influencing remission induction in acute myeloid leukaemia. In: Quaglino D, Hayhoe FGJ (ed) The Cytobiology of Leukaemias and Lymphomas. Raven Press, New York, p 271
Swirsky D M, de Bastos M, Parish S, Rees J K H, Hayhoe F G J 1986 Factors affecting outcome during remission induction of acute myeloid leukaemia in 619 adult patients. British Journal of Haematology 64: 435
Tanaka Y, Brecher G, Bull B 1966 Ferritin localization on the erythroblast cell membrane and rhopheocytosis in hypersiderotic human bone marrow. Blood 28: 758
Telek B, Kiss A, Pecze K, Rak K 1983 Differential diagnostic value of acid phosphatase and beta-glucuronidase in acute leukaemia. Blut 46: 67
Tomonaga M, Ichimaru M, Tsuji Y 1984 Alkaline phosphatase-positive leukaemic monocytes in a child with acute monocytic leukaemia. Acta Haematologica 71: 66
Tomonaga M, Yoshida Y, Tagawa M, Jinnai I, Kuriyama K, Amenomori T, Yoshioka A, Matsuo T, Nonaka H, Ichimaru M 1985 Cytochemistry of acute promyelocytic leukemia (M3): leukemic promyelocytes exhibit heterogeneous patterns in cellular differentiation. Blood 66: 350
Tortarolo M, Cantore N, Grande M, Buonanno G, Dammacco F 1981 Plasma cell acid phosphatase as an adjunct in the differential diagnosis of monoclonal immunoglobulinemias. Acta Haematologica 65: 103
Tricot G, De Wolf-Peeters C, Vlietinck R, Verwilghen R L 1984 Bone marrow histology in myelodysplastic syndromes. II. Prognostic value of abnormal localization of immature precursors in MDS. British Journal of Haematology 58: 217
Tricot G, Mecucci C, Van den Berghe H 1986 Annotation: Evolution of the myelodysplastic syndromes. British Journal of Haematology 63: 609
Tsai L-M C, Tsai C-C, Hyde T P, Thomas L A, Broun G O 1984 T-cell prolymphocytic leukemia with helper-cell phenotype and a review of the literature. Cancer 54: 463
Tulliez M, Breton-Gorius J 1979 Three types of Auer bodies in acute leukemia. Visualization of their protein by negative contrast after peroxidase cytochemistry. Laboratory Investigation 41: 419
Valentine W N, Beck W S, Follette J H, Mills H, Lawrence J S 1952 Biochemical studies in chronic myelocytic leukemia, polycythaemia vera and other idiopathic myeloproliferative disorders. Blood 7: 959
van Furth R, Cohn Z A 1968 The origin and kinetics of mononuclear phagocytes. Journal of Experimental Medicine 128: 415
van Furth R, Diesselhoff-Den Dulk M M C 1970 The kinetics of promonocytes and monocytes in the bone marrow. Journal of Experimental Medicine 132: 813
van Putten L M 1974 Are cell kinetic data relevant for the design of tumour chemotherapy schedules? Cell Tissue Kinetics 7: 493
Variakojis D, Vardiman J W, Golomb H M 1980 Cytochemistry of hairy cells. Cancer 45: 72
Veerman A J P, Huismans D R, Van Zantwijk C H 1983 Immunological phenotype related to acid alpha-naphthyl acetate esterase and acid phosphatase in childhood acute lymphoblastic leukaemia. Acta Haematologica 69: 32
Veerman A J P, Hogeman P H G, van Zantwijk C H, Bezemer P D 1985 Prognostic value of 5′nucleotidase in acute lymphoblastic leukemia with the common-ALL phenotype. Leukemia Research 9: 1227
Vilter R W, Will J J, Jarrold T 1967 Anemia refrattaria con midollo iperplastico (anemia arigenerativa). Aggiornamenti di Ematologia 4: 257

Vowels M R, Willoughby M L N 1973 Cyclic chemotherapy in acute lymphoblastic leukaemia of childhood: 5-year survivals. Archives of Diseases of Childhood 48: 436
Wachstein M 1946 Alkaline phosphatase activity in normal and abnormal human blood and bone marrow cells. Journal of Laboratory and Clinical Medicine 31: 1
Wagner R 1947 Studies on the physiology of the white blood cell: the glycogen content of leucocytes in leukemia and polycythemia. Blood 2: 235
Wantzin G L, Karle H, Killmann S A 1976 Nuclear DNA-polymerase estimation in human leukaemic myeloblasts. British Journal of Haematology 33: 329
White J G 1967 Fine structural demonstration of acid phosphatase activity in Auer bodies. Blood 29: 667
White J G 1972 Interaction of membrane systems in blood platelets. American Journal of Pathology 66: 295
Whittaker J A, Withey J, Powell D E B, Parry T E, Khurshid M 1979 Leukaemia classification: a study of the accuracy of diagnosis in 456 patients. British Journal of Haematology 41: 177
Wickramasinghe S N 1975 Human Bone Marrow. Blackwell Scientific Publications, Oxford.
Wickramasinghe S N, Hughes M 1978 Capacity of ringed sideroblasts to synthesise nucleic acid and protein in patients with primary acquired sideroblastic anaemia. British Journal of Haematology 38: 345
Wickramasinghe S N, Chalmers D G, Cooper E H 1968 A study of ineffective erythropoiesis in sideroblastic anaemia and erythraemic myelosis. Cell Tissue Kinetics 1: 43
Wickramasinghe S N, Fulker M J, Losowsky M S, Hall R 1971 Microspectrophotometric and electron microscopic studies of bone marrow in hereditary sideroblastic anaemia. Acta Haematologica 45: 236
Williams D L, Look A T, Melvin S L, Roberson P K, Dahl G, Flake T, Stass S 1984 New chromosomal translocations correlate with specific immunophenotypes of childhood acute lymphoblastic leukemia. Cell 36: 101
Wiltshaw E, Moloney W C 1955 Histochemical and biochemical studies on leucocyte alkaline phosphatase activity. Blood 10: 1120
Wirthmüller R, Dennig D, Oertel J, Gerhartz H 1983 Dipeptidylaminopeptidase IV (DAP IV) activity in normal and malignant T cell subsets as defined by monoclonal antibodies. Scandinavian Journal of Haematology 31: 197
Wrotnowski U, Innes D J, Hobson A S 1987 Nonspecific esterase staining patterns in acute monocytic leukemia. American Journal of Clinical Pathology 87: 515
Yam L T, Li C Y, Lam K W 1971 Tartrate resistant acid phosphatase isoenzyme in the reticulum cells of leukemic reticuloendotheliosis. New England Journal of Medicine 284: 357
Yam L T, Li C Y, Finkel H T 1972 Leukemic reticuloendotheliosis: the role of tartrate resistant acid phosphatase in diagnosis and splenectomy in treatment. Archives of Internal Medicine 130: 248
Yam L T, Janckila A J, Li C-Y, Lam WKW 1987 Cytochemistry of tartrate-resistant acid phosphatase: 15 years experience. Leukemia 1: 285
Yang K, Bearman R M, Pangalis G A, Zelman R J, Rappaport H 1982 Acid phosphatase and alpha-naphthyl acetate esterase in neoplastic and non-neoplastic lymphocytes. A statistical analysis. American Journal of Clinical Pathology 78: 141
Yourno J, Mastropaolo W 1981 Nonspecific esterases of the formed elements. Zymograms produced by pH 9.5 polyacrylamide gel electrophoresis. Blood 58: 939

CHAPTER 15

Major cytochemical and histochemical findings in lymphocyte subpopulations and in normal, reactive and lymphomatous lymph nodes

Normal peripheral blood lymphocytes

The search for immunological markers of lymphocytes, until a few years ago, was based mainly on the use of immunofluorescence or immunoenzymatic techniques for the detection of surface immunoglobulins (SIg), and of the sheep red blood cell (E) rosetting test, which represented the principal methods for identifying respectively B or T lymphocytes. More recently, the application of immunological techniques based on the use of monoclonal antibodies has been found of great value in defining the distribution pattern of several lymphocyte populations and subsets, and in studying their functional characteristics.

Thus, B lymphocytes can now be recognized by their reactivity with some monoclonal antibodies, such as B1 (CD20), which recognizes a pan-B surface antigen, with a molecular weight of 35 000 daltons. Similarly, within the T cell system, many monoclonal antibodies can be employed as pan-T cell markers. Some of them (OKT1, Leu 1, RFA 1 etc.) (CD5) recognize an antigen with a molecular weight of 67 000 daltons, which is also shared by a small subpopulation of B-cells (see below). In addition, it is worth mentioning the antibodies (OKT3, T3, anti-Leu 4 and others) (CD3) directed towards the T cell antigen receptor, and those (OKT11, T11, anti-Leu 5, Lyt 3 etc.) (CD2) which bind the E rosette receptor. Mature T cells, from late medullary thymocytes onwards, segregate into helper/inducer and suppressor/cytotoxic. These T cell subsets can be broadly identified by the presence of membrane receptors for the Fc fragments respectively of IgM and of IgG, as well as by their reactivity with monoclonal antibodies. OKT4, T4A, anti-Leu 3a and others (CD4) have been employed for detecting the helper/inducer T cells, whereas OKT8, T8A, anti-Leu 2a and similar antibodies (CD8) have been considered as specific markers of the suppressor/cytotoxic subset. However, as we have already discussed (see Chapter 3) only a partial overlapping exists between the expression of these markers and that of Fc receptors, and the exact relationship between membrane phenotype and functional activity is not always obvious.

In the peripheral blood of healthy individuals, a mean helper ($CD4^+$) to suppressor ($CD8^+$) cell ratio of 1.8 has been reported (Schroff *et al.*, 1983).

The use of additional monoclonal antibodies is capable of identifying a number of functionally different subsets within the two major T cell subpopulations: they will not be dealt with here, since at present possible differences in their cytochemical behaviour have been little explored.

Several monoclonal antibodies have been recently employed also in order to detect lymphocyte subsets capable of exerting natural killer (NK) activities; those most widely utilized appear to be anti-Leu 7 (Abo and Balch, 1981) and anti-Leu 11 (CD16) (Phillips and Babcock, 1983). Only partial overlapping exists between the range of positivity of these two antibodies, and the most intense NK activity is seemingly exerted by the Leu 7^-, Leu 11^+ subset (Lanier *et al.*, 1983). NK cells, as defined by monoclonal antibodies, reveal a heterogeneous membrane phenotype (Abo *et al.*, 1982), which in part of the cells corresponds to that of suppressor/cytotoxic ($CD8^+$) T lymphocytes, but in other cells points to a T helper ($CD4^+$) or non-T nature (Velardi *et al.*, 1985). From the morphological point of view, NK cells display the features of large granular lymphocytes, which represent about 15% of circulating mononuclear cells in healthy adult subjects (Abo and Balch, 1981).

A large number of investigations have been carried out, especially during the last decade, with the aim of detecting possible specific patterns of cytochemical reactivity in immunologically defined lymphocyte subsets.

As far as normal circulating B lymphocytes are concerned, a certain degree of specificity may be attributed to the alkaline phosphatase reaction, which however identifies a very small percentage of cells (see Chapter 7). Some authors (Matamoros *et al.*, 1979; Rowe *et al.*, 1979) suggested that a four times higher membrane 5′-nucleotidase activity distinguishes B from T lymphocytes: however this reaction has been more frequently exploited for the study of malignant disorders (see Chapter 11) rather than for detecting B lymphocytes in normal peripheral blood.

Cytochemical studies on normal circulating T lymphocytes and T cell subsets have been mainly concerned with the detection of hydrolases (acid phosphatase, non-specific esterases, β-glucuronidase, N-acetyl-β-glucosaminidase, dipeptidylaminopeptidases etc.). The relevant findings obtained by the individual reactions have been discussed and illustrated in some detail in the relative chapters: in general, it can be said that a higher activity of these enzymes characterizes T cells in comparison with B cells, and that a striking tendency of the reaction product to be localized in one or few paranuclear areas (dot-like positivity) is evident in a high proportion of normal circulating T cells.

Using fluorescence activated cell sorting with the monoclonal antibody UCHT1 (CD3), Armitage *et al.* (1982) have shown that approximately 80% of peripheral blood T cells display dot-like α-naphthyl acetate esterase and acid phosphatase reactions. Bernard and Dufer (1983) used a combined monoclonal antibody rosetting technique (in which monoclonal antibodies were linked to ox erythrocytes) to identify lymphocyte subpopulations.

This study revealed that 80–90% of T cells (OKT3$^+$) show α-naphthyl acetate esterase, acid phosphatase and β-glucuronidase positivity, with two thirds of esterase-reactive cells displaying dot-like reactions. Lymphocyte reactivity with monoclonal antibodies may also be revealed by the use of immunocolloidal gold methods. Using this approach, De Waele *et al.* (1983) and Crockard and Catovsky (1983) have shown that 80–90% of circulating T cells (OKT3$^+$) are positive for α-naphthyl acetate esterase, acid phosphatase, and β-glucuronidase. The majority (60–75%) of acetate esterase-positive T cells display characteristic dot-like reactivity, while approximately 15–30% of cells show scattered granular reactions. DAP IV reactivity is present in approximately 70% of OKT3$^+$ cells.

More controversial results emerged from the study of hydrolase distribution in the two major T-cell subsets. A degree of discriminating capacity has been attached to the non-specific esterase and to the DAP IV reactions. As far as the first is concerned, a dot-like positivity pattern is more frequently encountered in the CD4$^+$ subset, and a scattered granular positivity in the CD8$^+$ subset, but the opposite cytochemical behaviour may sometimes be observed in both subsets (see Chapter 8).

DAP IV reactivity has been reported as restricted to T cells bearing Fcμ receptors (Feller *et al.*, 1982), although both the majority of CD4$^+$ and a minority of CD8$^+$ T cells display enzyme activity (Chilosi *et al.*, 1982). A minority of CD8$^+$ T lymphocytes has been also described as positive for the naphthol AS-D chloroacetate esterase reaction, which according to some reports identifies also a proportion of large granular lymphocytes. The latter cell type would be also characterized by a fluoride-sensitive α-naphthyl acetate esterase activity: all these findings have been already dealt with extensively in Chapter 8.

Normal and reactive lymph nodes

The understanding of normal lymph node architecture has been much influenced in recent years by the rapid advances in the theoretical and practical aspects of immunology, and in particular by a more precise knowledge of the radical changes in lymphocyte morphology, which take place as a result of the response of the lymphocyte to antigenic stimuli.

In studying normal lymph node architecture and the morphological changes which may occur in reactive and malignant conditions, it is always necessary to search carefully for alterations in the different histological landmarks of the lymph node structure—i.e. the follicles and the paracortical tissue, the medulla and the sinuses—and more precisely to note their possible absence or presence, and in the latter case to determine whether they are normal or abnormal structurally and in cell content. However, the lymph node has only a limited repertoire of responses to a wide variety of stimuli (Robb-Smith and Taylor, 1981), so that in some cases a reactive condition may simulate a lymphoreticular neoplasm and vice versa: hence the great importance of a thorough and detailed appraisal of all relevant histological and cytological data and the great interest in the contribution of both immunological and histochemical markers.

Since cytoimmunological techniques have in many instances enabled a more functional interpretation of some morphological aspects and cytochemical patterns of hitherto uncertain significance (e.g. the presence of cleaved nuclei in germinal centre cells, the particular aspects of acid phosphatase positivity in T lymphocytes), a brief survey of some of the more recent and significant immunological findings in normal and reactive lymph nodes will precede a more detailed analysis of the histochemical and cytochemical results obtained in the same conditions.

Immunological findings

The application of immunohistochemical techniques with panels of monoclonal antibodies has been widely exploited in order to define the distribution pattern of lymphocyte populations and to study the interaction of a lymphocyte subset with other cells in frozen sections of normal and reactive human lymph nodes. One of the more thorough investigations carried out on normal human lymphoid tissues is that of Hsu *et al.* (1983a), who employed mouse monoclonal antibodies capable of recognizing differentiation antigens on the surface of distinct human B, T, NK cell or monocyte subsets. Labelling reagents included biotin conjugated horse anti-mouse IgG and avidin-biotin peroxidase complex. This study confirmed on a more solid basis previous reports (Curran and Jones, 1977; Breard *et al.*, 1980; Bhan *et al.*, 1981) regarding the general distribution and compartmentalization of T and B cells in normal human lymphoid tissues.

In peripheral lymphoid tissues, T and B cells generally are located within separate compartments. T cells are largely restricted to paracortical or interfollicular areas, B cells are concentrated in lymphoid follicles. The majority of cells within the follicles including both the mantle zone and the germinal centre, are positive for the pan-B cell marker B1. The cells of the mantle zone also express SIg: they are positive for μ and δ and can be subdivided in k^+ and λ^+ cells. On the contrary, the number of Ig positive B cells within germinal centres is highly variable, depending on the stage of differentiation or transformation (Hsu *et al.*, 1983b). IgG positive cells constitute less than 10% of the total cell population, and are scattered in the light zone of the germinal centre. 0 to 10% of cells are IgA positive, and tend to be distributed randomly either in the light or in the dark zone. IgM and IgD positive cell percentage ranges from less than 10 to 70%, with a tendency to localize in the dark zone. Moreover, an extracellular substance is present, which stains intensely with anti-μ, γ, κ, λ and B1 antibodies. In the germinal centre are also detectable lymphocytes with staining characteristics of both helper (Leu 3a$^+$) and NK (Leu 7$^+$) cells, whereas suppressor (Leu 2a$^+$) lymphocytes are very rare.

Other cell types normally present within reactive germinal centres, such as dendritic reticulum cells (Naiem *et al.*, 1983) or tissue macrophages (Dimitriu-Bona *et al.*, 1983) may also be identified by monoclonal antibodies.

In addition to their prevalence in the mantle zone, SIgM and SIgD positive B lymphocytes are scattered throughout lymphoid tissues in small numbers. However, lymphocytes bearing SIgG or SIgA are rarely observed. In the medulla, B cells are admixed to T cells with an irregular distribution.

Several monoclonal antibodies are available for the study of the development and distribution of the B lymphocyte lineage. A tentative scheme of B cell differentiation has been proposed by Caligaris-Cappio *et al.* (1982), who exploited the use of a wide range of monoclonal antibodies including the RFA series, in double marker assays with anti-IgM, IgD, Ia-like antigen and TdT. These investigations have been carried out on suspensions of peripheral blood, lymphoid tissues and bone marrow, as well as in various areas (germinal centre, lymphocyte corona, medullary cords) of lymph nodes.

In the bone marrow, pre-B cells and $SIgM^+$, $SIgD^-$ B lymphocytes were RFA $2^-, 3^-, 4^-$. In peripheral lymphoid tissues, small B lymphocytes with abundant expression of both SIgM and SIgD in the mantle zone of lymphoid follicles were RFA 2^+, 3^+ ($SIgM^+$, $SIgD^+$ bone marrow lymphocytes expressed the same phenotype). A large proportion of these peripheral B cells (60%) were also positive for RFA 4. The larger blasts in the middle of the germinal centres had a phenotype intermediate between the two major categories of bone marrow and peripheral (mantle zone) B cells: $SIgM^+$ (weakly), $SIgD^-$, RFA 2^-, 3^+,4^+. A small subset (5%) of weakly $SIgM^+$, $SIgD^+$, RFA 2^+,3^+,4^- lymphocytes, observed around the boundaries of the germinal centre, was capable of forming rosettes with mouse red blood cells and of reacting with RFA 1 (CD5). A similar immunological phenotype is expressed by the vast majority of B lymphocytes in lymph nodes of the normal human fetus at the 17th gestational week (Bofill *et al.*, 1985): however, fetal B lymphocytes do not rosette with mouse erythrocytes.

Plasma cells showed different phenotypes in normal bone marrow and peripheral lymphoid tissues. The majority of bone marrow plasma cells were RFA 2^-,3^-,4^-, while the majority of plasma cells in the tonsils and lymph nodes as well as those obtained in cultures of peripheral blood lymphocytes stimulated with pokeweed mitogen were RFA 2^+,3^+,4^+. These observations suggest a differentiation scheme for B lymphocytes in which the germinal centre exerts a pivotal role between bone marrow and peripheral lymphoid tissues.

As far as the T cell system is concerned, the results of Hsu *et al.* (1983a) indicate that Lyt 3 (CD2) appears to be the most useful pan-T cell marker to detect T cells in tissue sections, because of the constancy of staining observed both in terms of cell number and intensity. In peripheral lymphoid tissues the number of Leu 4^+, $OKT3^+$ or Lyt 3^+ cells was greater than that of Leu 1^+ cells, indicating that a small proportion of T cells are devoid of Leu 1 (CD5) antigen. The authors also identified in peripheral lymphoid tissues $OKT6^+$ lymphocytes, a finding which indicates that a small percentage of mature T cells retain the OKT6 (CD1) associated antigen, expressed by the majority of thymocytes (see below).

$CD4^+$ and $CD8^+$ T lymphocytes are present with a 2–3:1 ratio in the T dependent areas of peripheral lymphoid tissues. The proportion of $CD4^+$

cells may increase (4–5:1) in tissues with marked reactive hyperplasia.

Confirmation of T cell subset distribution was obtained by Pizzolo and Chilosi (1984) employing double immunostaining of tissue sections. The authors tested monoclonal antibodies directly conjugated with the red fluorochrome phycoerythrin, in combination with haptenated monoclonal antibodies (arsenilated or biotinated) detected by second or third layers labelled with fluorescein isothiocyanate (FITC, green), or else in combination with monoclonal antibodies of different mouse-Ig subclasses detected by FITC-conjugated sera anti-mouse-Ig subclass.

The following combinations were described as the most useful for characterizing T cell subsets:

anti-Leu 3a/anti-Leu 2a: T lymphocytes were confined manly to paracortical areas, where Leu $3a^+$ cells were predominant. Several Leu $3a^+$ cells were also found in the germinal centres, where Leu $2a^+$ lymphocytes were virtually absent.

anti-Leu 3a and anti-Leu 2a/anti-HLA-DR: HLA-DR^+ lymphocytes were stained in the germinal centre and lymphocyte corona. Strongly HLA-DR^+ interdigitating reticulum cells were detected in the paracortical area in close contact with Leu $3a^+$ lymphocytes. This combination appears to be useful in studying the relationship between T cell subsets and accessory cells (Janossy *et al.*, 1980; 1981).

anti-Leu 3a and anti-Leu 2a/anti-Leu 7: this combination demonstrated that the few Leu 7^+ cells found in lymphoid organs are confined to the germinal centre. The Leu 7^+ cells were mostly double stained with anti-Leu 3a and none with anti-Leu 2a. This finding is of particular interest in relation to the heterogeneity and functions of the Leu 7^+ population, which includes cells with K and NK function (Abo and Balch, 1981).

Most of the observations reported in normal and reactive lymph nodes hold true also for B and T dependent areas of the *spleen*. Thus, according to the data of Hsu *et al.* (1983a), the cell composition of splenic follicles closely resembles that of lymph node follicles, whereas the great majority of lymphocytes in the periarteriolar sheaths display T cell markers, with 70–90% of Leu $3a^+$ and 10–30% of Leu $2a^+$ cells.

In the marginal zone of the white pulp, T cells make up 50–70%, and B cells up to 40% of lymphocytes. The majority of B cells are stained with anti-μ, δ and B1 antibodies. Very often, a considerable number of $OKM1^+$ positive mononuclear cells (monocytes) is also detectable in the marginal zone.

Both T and B cells are randomly distributed within sinusoids or in cords of the red pulp. In contrast to the predominance of $CD4^+$ cells in the periarteriolar sheaths, in the red pulp the majority of lymphocytes (60–80%) appears to be positive for CD8. Lining cells of sinusoids are capable of reacting with both anti-Leu 3a and anti-Leu 2a monoclonal antibodies.

Histochemical findings

Enzyme histochemical techniques have been available for many years, but have been employed mainly with frozen sections and touch preparations, which do not allow precise correlations of enzymes with tissue and cell morphological characteristics. Nonetheless, enzyme histochemistry has been used to identify T cells (Seymour *et al.*, 1978; Lennert, 1978; Knowles and Halper, 1982) and B cells (Müller-Hermelink, 1974: Lennert, 1978; Beckstead *et al.*, 1982) in benign and neoplastic lymph node disorders: it has also been extremely useful in identifying cells of the mononuclear phagocyte system and the neoplasms derived from them (Carbone *et al.*, 1981; Van der Valk *et al.*, 1981). Many of the accessory cells involved in the immune response are difficult to recognize morphologically but are readily identifiable histochemically. The application of suitable enzyme techniques to the study of these cells might improve our understanding of the reactive and neoplastic changes in lymph nodes.

In an attempt to improve enzyme histochemical demonstration in normal and pathological paraffin-embedded lymphoid tissues, Chilosi *et al.* (1981) introduced a modified tissue embedding procedure, consisting principally in the use of 10% neutral calcium-formalin fixation at low temperature, tap water rinsing and acetone dehydration. The authors investigated the activity of several enzymes, namely acid and alkaline phosphatase, tartrate-resistant acid phosphatase, α-naphthyl acetate esterase, adenosine triphosphatase, myeloperoxidase and naphthol AS-D chloroacetate esterase. The results obtained by detecting these enzymes both on cryostat sections and sections from paraffin-embedded tissues, processed with the technique described by the authors, were quite comparable, with the sole exception of ATPase. The authors could distinguish B from T lymphocytes. Myeloid cells, capillary endothelium, fibroblastic reticulum cells and various subpopulations of macrophages could also be identified.

In a further effort to increase the discriminating ability of enzyme reactions, Beckstead (1983) developed a plastic-embedding technique, as a result of which the morphological appearance of the lymph node tissue was strikingly better than in paraffin sections and the enzyme localization was extremely satisfactory. The method involved fixation in phosphate-buffered paraformaldehyde solution, washing in buffered 3% sucrose, and acetone dehydration, prior to glycol methacrylate embedding under vacuum. The whole procedure was carried out at 4°C.

The author investigated the following histochemical reactions: ATPase, 5′-nucleotidase, acid and alkaline phosphatase, non-specific esterase, chloroacetate esterase, β-glucuronidase and peroxidase. Positive reactions could be identified easily in individual cells, and frequently were well localized to a part of the cell. In Beckstead's experience, enzymes varied considerably in their resistance to inhibition by fixation and processing: chloroacetate esterase was extremely stable, peroxidase and non-specific esterase were very resistant, alkaline and acid phosphatase were intermediate in their resistance, whereas 5′-nucleotidase, ATPase and β-glucuronidase were more sensitive to inhibition.

On the basis of Beckstead's observations and of those of other authors (Leder, 1967; Müller-Hermelink, 1974; Lennert, 1978; Chilosi *et al.*, 1981) the major histochemical findings observed in the different cell types present in normal and reactive human lymph nodes may be summarized as follows:

Lymphoid cells

Within secondary follicles, most of the mantle zone lymphocytes show a positive ATPase reaction, sensitive to overfixation, while up to 50% of the same cells show weak to moderate plasma membrane alkaline phosphatase activity, and approximately 50% show moderate to strong membrane positivity for 5′-nucleotidase (Plate 27C). Germinal centre cells display negative enzyme reactions, except for weak to moderate, chiefly granular acid phosphatase activity in large blast cells (centroblasts and immunoblasts), which in imprints may also appear occasionally PAS positive.

Plasma cells show strong granular cytoplasmic staining with acid phosphatase and β-glucuronidase, and proportionally less intense α-naphthyl acetate esterase positivity: the latter has been reported as sensitive to fluoride inhibition (Dufer *et al.*, 1984). A strong plasma membrane ATPase activity is also detectable in most plasma cells.

The cytochemical positivity pattern of lymphocytes in T dependent areas of peripheral lymphoid tissues closely resembles that of circulating T cells: thus, a considerable percentage of cells in the paracortex of lymph nodes and in the periarteriolar sheaths of the spleen show punctate (dot-like) positivity with the acid phosphatase, β-glucuronidase, non-specific esterase and both DAP II and DAP IV reactions (Plates 30E, 31B).

Reticulum cells and mononuclear phagocyte system

With the term reticulum cell, it is possible nowadays to identify four main types of non-lymphoid structural cells, belonging to the stationary compartment of the lymph node, which can be distinguished by their different cytochemical, ultrastructural and immunological characteristics: the histiocytic reticulum cell, the fibroblastic reticulum cell, the dendritic reticulum cell and the interdigitating reticulum cell.

Histiocytic reticulum cells (synonymous with macrophages) are regarded as immigrant monocytes (Lennert, 1978); interdigitating reticulum cells probably derive from circulating precursors (Veldman and Kaiserling, 1980). The relatively more stable fibroblastic reticulum cells and dendritic reticulum cells seem to originate from a local primitive mesenchymal cell, even though the intermediate stages are not yet clear (Müller-Hermelink *et al.*, 1981). The main cytochemical features of these four different cell types may be described as follows:

Histiocytic reticulum cells contain the greatest amount of acid phosphatase and non-specific esterase known in the lymphatic tissue. The PAS reaction is generally negative, but following phagocytosis these cells may show globular PAS positive intracytoplasmic deposits. They may also contain Sudan black-positive material, haemosiderin and other substances (Plate 61

B–F). A granular cytoplasmic staining for 5′-nucleotidase and a membrane positivity for ATPase are evident in most histiocytic reticulum cells.

Fibroblastic reticulum cells contain very small amounts of acid phosphatase and non-specific esterase: they reveal instead large amounts of membrane-bound alkaline phosphatase, which can be shown by light and electron microscopy. The demonstration of alkaline phosphatase on fibroblastic reticulum cells represents a criterion of functional activity.

Dendritic reticulum cells, which are localized within lymphoid follicles, show a very strong membrane and cytoplasmic positivity with the 5′-nucleotidase reaction, whereas non-specific esterases, acid phosphatase and ATPase display a rather weak reactivity. The PAS reaction is negative.

Interdigitating reticulum cells show, in contrast to histiocytic reticulum cells, scarce non-specific esterase activity. The acid phosphatase reaction is weakly positive and usually present in a localized area of the cytoplasm at the site of nuclear indentations, probably in or close to the Golgi area. 5′-nucleotidase activity is not detectable. On the other hand, interdigitating reticulum cells show a strong membrane-bound reaction with ATPase. The alkaline phosphatase reaction is negative, while the PAS reaction frequently reveals very fine granules. These granules, according to Kaiserling and Lennert (1974) may be better visualized by means of the periodic acid-silver methenamine method on electron microscopy, which is based on the same chemical reactions as the PAS reaction (De Martino and Zamboni, 1967).

In normal and reactive lymph nodes, some additional cell types are recognizable, which together with histiocytic reticulum cells and interdigitating reticulum cells form the so-called mononuclear phagocyte system.

Sinus histiocytes show some degree of membrane positivity with ATPase, whereas *monocytes* are negative. Both show strong cytoplasmic positivity for non-specific esterases, monocytes being the more active of the two cell types. As far as other enzymes are concerned, sinus histiocytes react positively with 5′-nucleotidase, acid phosphatase and to a lesser degree with β-glucuronidase, while monocytes only show moderate acid phosphatase and myeloperoxidase activity.

Epithelioid cells are special monocyte-derived evolutionary forms of macrophages. They are found chiefly in the T-dependent regions of lymphoid tissue, but may also be frequently observed in germinal centres and sinuses. Epithelioid cells exhibit high metabolic activity, as suggested by high enzymatic activities of the pentose cycle, the Krebs cycle, anaerobic glycolysis, enzymes of the respiratory chain and hydrolytic enzymes (Williams *et al.*, 1969; Palva *et al.*, 1973). The non-specific esterase and acid phosphatase histochemical reactions are weakly to strongly positive, but not often in the same degree as in macrophages. Moreover, the acid phosphatase activity of epithelioid cells, in contrast to that of macrophages, appears to be resistant to inhibition by L-tartaric acid (Chilosi *et al.*, 1981). A granular PAS positivity is easily seen, especially in imprints. 'Foreign body' giant cells or Langhans cells, usually found in association with epithelioid cells in reactive or chronic inflammatory states, show very similar cytochemical features as illustrated in Plate 57.

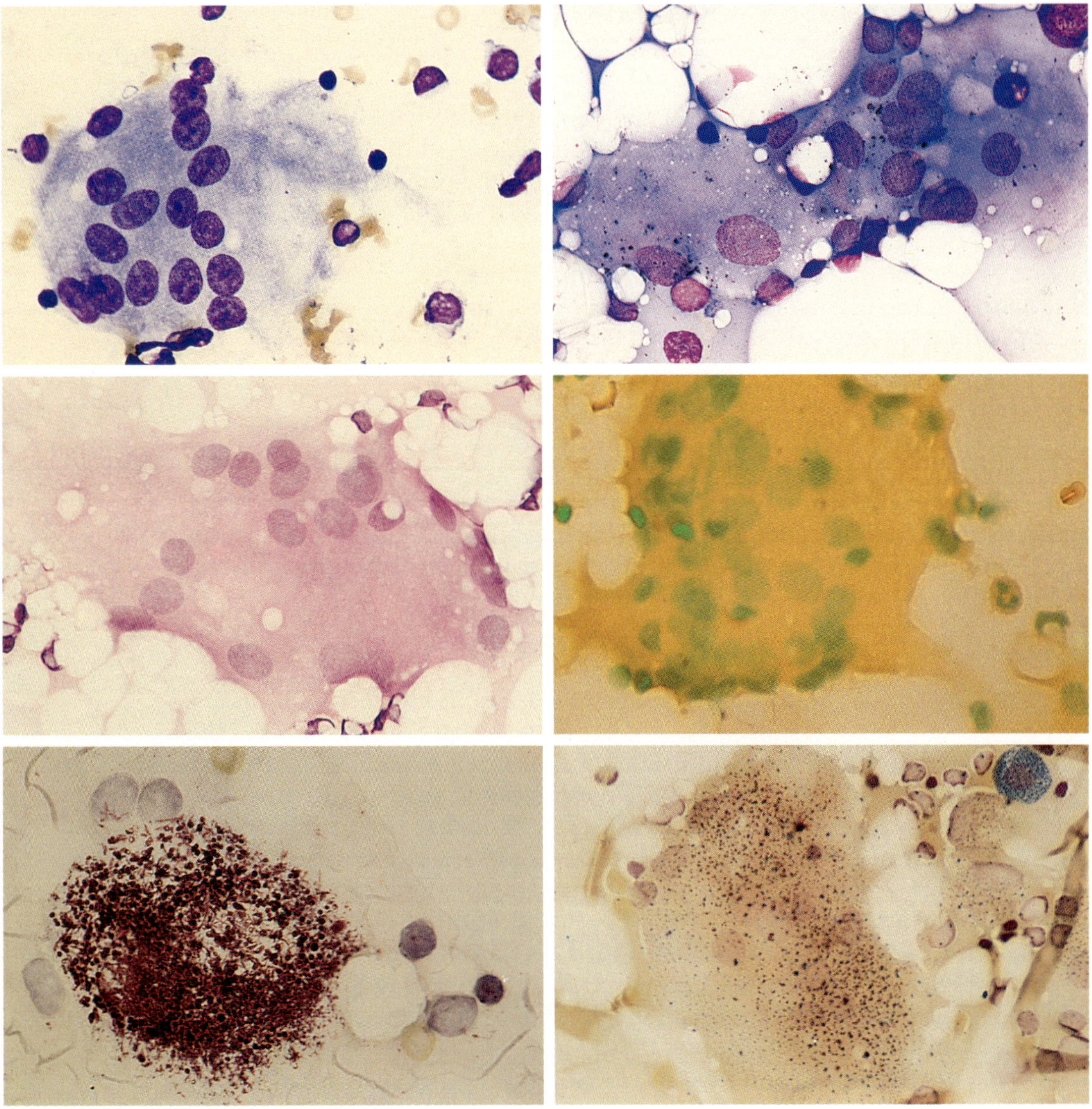

A B C D E F

Plate 57

A. Langhan's giant cell in lymph node imprint. Leishman stain. Typical multinucleated cytology.
B. The same preparation. SB stain. Two Langhan's cells show only a few scattered sudanophilic granules.
C. The same preparation. PAS stain. A weak diffuse and very finely granular positivity is seen.
D. The same preparation. Alkaline phosphatase. Langhan's cells, unlike many macrophages, are negative for this enzyme.
E. The same preparation. Acid phosphatase. A Langhan's cell shows exceptionally strong positivity for this enzyme.
F. The same preparation. Dual esterase. A Langhan's cell shows moderately strong BE positivity. A single neutrophil is CE positive.

Vascular endothelial cells

The ATPase reaction stains the plasma membrane of virtually all vascular endothelial cells found in the human lymphoid tissues, with the exception of the lymphatic endothelial (sinus lining) cells: on the other hand, the latter react strongly with 5′-nucleotidase, which instead is only weakly or moderately positive in all other endothelial cells. The alkaline phosphatase reaction is present in a variable degree in the different vascular endothelial cells, bound to the plasma membrane, except in lymphatic endothelial cells where the reaction is negative. All other enzyme ractions in endothelial cells appear to be negative.

Other cell types

Neutrophil granulocytes and mast cells may be easily identified by their reactivity for chloroacetate esterase. Neutrophils and eosinophils react strongly for myeloperoxidase, which is occasionally seen also in monocytes.

Thymus

One of the first extensive studies based on the use of monoclonal antibodies (Reinherz *et al.*, 1980) explored from the immunological point of view the different stages of intrathymic differentiation of human T cells. According to the results of this and subsequent investigations, three main stages of development may be recognized in thymic cells. The first stage is represented by early cortical thymocytes, which express on the cell membrane only the T11 (CD2) antigen. In the subsequent stage, corresponding to the majority of cortical thymocytes, the cells, in addition to T11, express on their surface T6 (CD1), T4 (CD4) and T8 (CD8) antigens. The third intrathymic stage of differentiation is represented by medullary thymocytes, which do not express T6, but display the T3 (CD3) antigen, associated with the T-cell antigen receptor, along with T11, T4 and T8, A 'mature' phenotype, corresponding to that of circulating T cells, is achieved by segregation into $T4^{+}$, $T8^{-}$ and $T4^{-}$, $T8^{+}$ subsets. For a correct interpretation of immunohistological findings, one has to bear in mind that the antigenic structures responsible for cell reactivity with monoclonal antibodies may be present within the cytoplasm prior to their expression on the cell membrane. For this reason, the majority of cortical thymocytes appear to be positive for T3 on frozen sections investigated by immunohistochemical techniques (Hsu *et al.*, 1983a).

To a certain extent, discrete stages of intrathymic cell differentiation may be recognized also by histochemical and cytochemical investigations. Thus, Basso *et al.* (1980) studied cells from fetal and postnatal thymuses, in order to determine the sequence of expression of some cytochemically detectable hydrolases during T cell differentiation. According to the results presented by the authors, a focal positivity for the acid phosphatase reaction is already present at an early stage of differentiation. In subsequent stages, β-glucuronidase and acid α-naphthyl acetate esterase activities are sequentially expressed. Only at a relatively late stage of differentiation would T cells be positive for N-acetyl-β-glucosaminidase, α-naphthyl butyrate and neutral α-naphthyl acetate esterase reactions.

According to Crockard (1984), focal distribution of acid phosphatase reaction product is a feature of thymocytes, but not of post-thymic cells, where scattered granular reactions are more usual. The distribution of β-glucuronidase activity broadly parallels that of acid phosphatase, but the highest levels of enzyme activity are observed in late thymocytes, and focal reactivity is not prominent. N-acetyl-β-glucosaminidase activity is detectable at all stages of T cell development, being proportionally higher in post-thymic cells. Acid α-naphthyl acetate esterase activity is absent or very weak in early thymocytes, whereas it becomes prominent in late thymocytes. Similarly, the highest levels of DAP IV activity are found in thymocytes at a relatively late stage of differentiation.

NON-HODGKIN'S LYMPHOID MALIGNANCIES

The parallel application of cytoimmunological and cytochemical techniques to the study of malignant lymphoproliferative disorders has resulted in the identification in the neoplastic cells of an increasing number of membrane and/or cytoplasmic immunological markers, enzyme activities and cellular structures, many of which appear to be more or less specific for certain lymphoid cell populations or subsets.

The first practical consequence resulting from this combined approach has been the acquisition of the concept that cells of lymphoproliferative disorders are in many ways quite similar to their physiological counterpart, since neoplastic cells often retain, at least in part, the structural and functional properties of the normal lymphoid cells to which they are related. This line of thought has led to the development of the Kiel (Gérard Marchant *et al.*, 1974) and Lukes and Collins (1974, 1975) classifications of lymphoid neoplasms, based on the recognition of the physiological counterpart of malignant cells by means of morphological, cytochemical and cytoimmunological techniques, and to the various attempts at identifying the different cytological and clinical varieties of acute lymphoblastic leukaemias.

Another fundamental concept which has emerged from the cytobiological study of lymphoid malignancies is represented by the fact that they are monoclonal cell proliferations, i.e. that they derive from the neoplastic transformation of a single progenitor cell. Several findings strongly support this hypothesis. Immunological studies of the cell phenotype show that in B-cell malignancies the surface and/or cytoplasmic immunoglobulins synthesized by neoplastic cells display a single type of light chain (Plate 60 A,B) and identical idiotypic determinants. The results of karyotypic analysis demonstrate in many cases of lymphoid malignancies typical non-random chromosomal alterations, which are mostly identical within a single neoplastic cell population. Even more important in this respect have been, in recent years, the data obtained by techniques of molecular genetics, capable of investigating the rearrangement and the expression of genes coding for immunoglobulins or for T-cell antigen receptors (Korsmeyer *et al.*, 1983; O'Connor *et al.*, 1985). These techniques not only have been useful in identifying discrete stages of early differentiation, respectively along the B and T cell line, but can also assess, through the identical pattern of gene rearrangement, the monoclonal nature of a cell proliferation.

Although the general statement that neoplastic cells resemble in many ways their normal counterparts is valid for the majority of cases, it is possible that in some instances, as a result of neoplastic derangement or cell anaplasia, the functional properties or phenotype characteristics may be abnormally expressed, both in term of quality and quantity, by malignant cells, so that in these circumstances it may be impossible to detect those features which are linked to the cells of a particular normal lymphoid population or subpopulation. Nonetheless, the B or T cell derivation of neoplastic cells is easily assessable in most cases, thus defining the two main categories of non-Hodgkin's lymphoid malignancies.

B cell malignancies

The malignant cell population of this large group of neoplasms is principally characterized, from the functional point of view, by the capacity of synthesizing and expressing monoclonal surface and/or cytoplasmic immunoglobulins.

In a first group of B-cell disorders, the neoplastic cells show a fully expressed functional behaviour: they are able to synthesize, and in most instances to excrete, monoclonal immunoglobulins. In this group are included plasmacytomas, Waldenström's disease, the rare cases of plasmacytic malignant lymphomas, and lymphoplasmacytoid immunocytoma. The ability of synthesizing cytoplasmic immunoglobulins is often expressed also by a proportion of cells in B-immunoblastic lymphomas (Plate 60A, B). According to Caligaris Cappio *et al.* (1982), the reactivity with monoclonal antibodies of the RFA series indicates that the cells of multiple myelomas are the malignant counterpart of normal bone-marrow plasma cells, whereas those in Waldenström's disease, solitary plasmacytomas, plasmacytic lymphomas and immunocytomas correspond to plasma cells of peripheral lymphoid tissues.

In multiple myeloma, comparison of SIg on peripheral blood lymphocytes with SIg and CIg present on or within tumour cells and with secreted monoclonal Ig shows that in many cases the SIg of a proportion of circulating lymphocytes and the CIg of malignant plasma cells show the same idiotypic activity, indicating that both types of cells belong to the same neoplastic clone (Lennert, 1978). Moreover, especially in the more aggressive forms of plasmacytoma and in plasma cell leukaemias, a considerable percentage of bone-marrow or circulating lymphocytes may express the cALL antigen (Bergui *et al.*, 1985).

In a number of cases, the synthesis of CIg is not followed by their secretion. This behaviour could be caused either by a real deficiency of the secreting mechanism, as for instance in non-secreting plasmacytomas and in the majority of cases of lymphoplasmacytoid immunocytomas, or by the fact that neoplastic cells are the malignant counterpart of normal non-secretory plasma cell precursors, such as plasmablasts and immunoblasts.

Especially when an intense Ig synthesis is not followed by an adequate secreting activity, Russell bodies are formed and become visible in the cytoplasm or as intranuclear inclusions. On the contrary, when Ig synthesis takes place at a reduced rate and is confined to a minority of malignant cells, such as in B-immunoblastic lymphomas, Russell bodies are infrequent and in the absence of cytoimmunological or cytochemical investigations it may be almost impossible to differentiate these types of lymphoma from certain undifferentiated carcinomas (e.g. nasopharyngeal) and from true reticulum cell malignancies (Lennert, 1978).

Cytochemically, when intracytoplasmic immunoglobulins are present in considerable amounts and give rise to the presence of Russell bodies, a diastase resistant PAS positivity becomes apparent, either in the form of coarse granules or inclusions (Plate 56B) or even of a diffuse cytoplasmic tinge. Hitherto PAS-positive inclusions have been thought to represent retained immunoglobulins, and this suggestion was supported by the results of many studies. As previously stated, the inclusions are diastase-resistant, and in addition they do not diminish during processing of tissues for paraffin-embedding. On electron microscopy they correspond to protein precipitates (Stein *et al.*, 1973). By means of immunofluorescent techniques, they have also been identified as immunoglobulins (Diebold *et al.*, 1974). Employing Ladewig's modification of Mallory's connective tissue stain and the modified Masson trichrome stain according to Goldner, Lennert (1978) reported that the PAS positive inclusions in lymphoplasmacytoid immunocytomas represented mostly IgM. In fact, with the Ladewig and Goldner methods IgM stained gray blue or green respectively, while deposits of IgG and IgA usually stained orange-red to bright crimson with both methods. Retained light chains and other incomplete components may take on the same colours, but in such cases the PAS reaction is negative.

On the other hand, more recent studies, based both on cytochemical and immunocytochemical findings, suggest that the PAS-positive substance of Russell bodies is not represented by the immunoglobulin itself, but by a condensed by-product from the processing of immunoglobulins (Hsu *et al.*, 1982; Khaukin, 1984). PAS-positive intranuclear inclusions are by no means specific for immunocytomas and plasmacytoma. They can be found also in other types of malignant lymphoma (Lennert *et al.*, 1975), such as centroblastic/centrocytic malignant lymphoma, and in plasma cells of non-malignant diseases. The biological significance of the intranuclear inclusions is generally thought to lie in a disturbance of secretion. Diebold *et al.* (1974) believe that an initial, abnormally rapid secretion might cause the disturbance.

Active Ig synthesis is often associated with a strong positive reaction for hydrolases: usually acid phosphatases are more intensely positive than non-specific esterases, however with ample variations in individual cases. In particular, in cases of plasmacytoma, detailed investigations seem to exclude any direct relationship between the degree of enzyme activity in plasma cells and the class of immunoglobulin produced (Hayhoe and Neuman, 1976).

Since the acid phosphatase activity of malignant plasma cells is usually more intense than that of normal plasma cells (Plate 55C), Cassuto *et al.* (1977) suggested that the reaction could be usefully employed to discriminate between malignant plasma cell proliferations and 'benign' monoclonal gammapathies. Studies by other authors have alternatively confirmed (Tortarolo *et al.*, 1981) or challenged (Johansen and Krogh Jensen, 1980) the results of Cassuto *et al.* More recently, there has been a growing tendency to accept the possibility of differentiation between malignant and benign plasmacytic proliferations by means of the acid phosphatase reaction, provided appropriate and sufficiently elaborate scoring methods are employed (Geniram *et al.*, 1984; Hoffman *et al.*, 1985) (See also Chapters 7 and 14).

Myeloma cells also show strong β-glucuronidase (Plate 55D) and non-specific esterase reactions (Plate 55E) (Sundstrom *et al.*, 1978; Bataille *et al.*, 1980). Paranuclear localization of α-naphthyl acetate esterase reaction has been described by Halper *et al.* (1982).

According to Miliani *et al.* (1979) the ATPase reaction might be considered of value in differential diagnosis, since enzyme activity appears to be decreased in the malignant plasma cells and normal in benign gammapathies.

When the cytomorphological features are not sufficiently discriminating, the presence of a diastase resistant PAS positivity in circulating lymphocytes and of a marked hydrolase activity may favour the diagnosis of a non secretory lymphoplasmacytoid immunocytoma in a leukaemic phase rather than that of other forms of chronic lymphocytic leukaemia. Even more reliable results in this respect may be achieved using immunocytochemical techniques (Papadimitriou *et al.*, 1979).

In the majority of non-Hodgkin's B-cell lymphomas, the cytological features closely resemble those of one or the other type of germinal centre cells—centroblasts and centrocytes according to Lennert (1978)—usually mixed in variable proportions. The predominant cell type and the growth pattern, whether follicular or diffuse, are the main features which are responsible for the identification of different histological forms, which according to the Kiel classification may be designated as centroblastic, centrocytic, or centroblastic–centrocytic malignant lymphomas, each variety being capable of a follicular or diffuse type of growth pattern. Centrocytes and centroblasts are non-secretory cells, so that in most instances the search for cytoplasmic immunoglobulins is negative. However, in some borderline cases, which are presumably progressing towards immunoglobulin synthesizing forms, a variable proportion of cells may show the presence of cytoplasmic Ig, sometimes in the form of Russell bodies.

It is now generally accepted that Burkitt's lymphoma and the Burkitt-like lymphoblastic forms (including B-ALL with Burkitt-like cells) are closely related to the lymphoproliferative disorders of the germinal centre, because, although the physiological counterpart of the malignant cells is hardly identifiable, Burkitt's lymphoma in its first stages of development is characterized by a growth pattern of the follicular type (Lukes and Collins, 1974). From the cytoimmunological point of view, the most relevant feature consists in the presence of surface Ig, usually IgM, with a considerable density and a variable tendency to form 'capping' (Lennert, 1978).

Some varieties of malignant lymphomas, which had not been recognized by the Kiel classification, appear to represent the neoplastic counterpart of the lymphocytes of the mantle zone of secondary follicles (Weisenburger *et al.*, 1982), or in rare cases of the B-cells of the marginal zone of spleen follicles (De Pasquale *et al.*, 1986). The boundaries between these forms and germinal centre cell lymphomas appear to be fairly hazy, and many borderline pictures may be envisaged within the group of so-called 'intermediate lymphocytic' lymphomas (Weisenburger *et al.*, 1981).

Cytochemically, germinal centre cell lymphomas and related forms usually show a weak to moderate granular diastase-sensitive PAS reaction in a small number of cells (Plate 59A–D). The percentage of PAS positive cells may however become considerably elevated in the leukaemic phase of these disorders (Plate 58C and 59E). A conspicuous proportion of malignant cells displayed a marked diastase sensitive PAS positivity, in the form of blocks or granules, also in two cases of supposed 'marginal zone lymphoma' (De Pasquale *et al.*, 1986).

The presence of immunoglobulin synthesizing cells is associated instead with a diastase-resistant PAS positive reaction and indicates a tendency to evolve towards a lymphoplasmacytoid immunocytoma. When diastase resistant PAS positivity is observed in centrocytes, the question arises of whether the step from centrocyte to plasma cell takes place invariably via large basophilic forms (immunoblasts and/or plasmablasts), or whether a direct transition from centrocytes to plasma cells is possible.

Similarly to PAS positivity, the extreme variability of the acid phosphatase and non-specific esterase reactions, in most cases weak or absent (Plate 58D, E, F), may be related to a different degree of progression towards a post-germinal centre stage. However, since a considerable degree of acid phosphatase activity is frequently detectable in the cells of highly malignant histological varieties (centroblastic (Plate 60D), immunoblastic (Plate 60E), and Burkitt-like forms (Plate 60F)), this metabolic behaviour is probably related also to the degree of proliferative activity. In contrast, macrophages give strongly positive reactions for both acid phosphatase and non-specific esterases.

The reaction for 5′-nucleotidase is only weakly positive.

Of considerable interest in this group of lymphomas is the existence of cases showing a positive membrane alkaline phosphatase reaction. This finding has been reported in germinal centre cell lymphomas, including Burkitt's lymphoma (Nanba *et al.*, 1977), in mantle zone lymphoma (Palutke *et al.*, 1982) and in marginal zone lymphoma (De Pasquale *et al.*, 1986). A similar cytochemical behaviour may also be observed in some cases of prolymphocytic leukaemia (Leder, 1978), indicating the possible existence of close links with the aforementioned conditions (see below).

The Sudan black B reaction may be positive in cells of the African type of Burkitt's lymphoma (O'Conor, 1961), in relation to a high content of

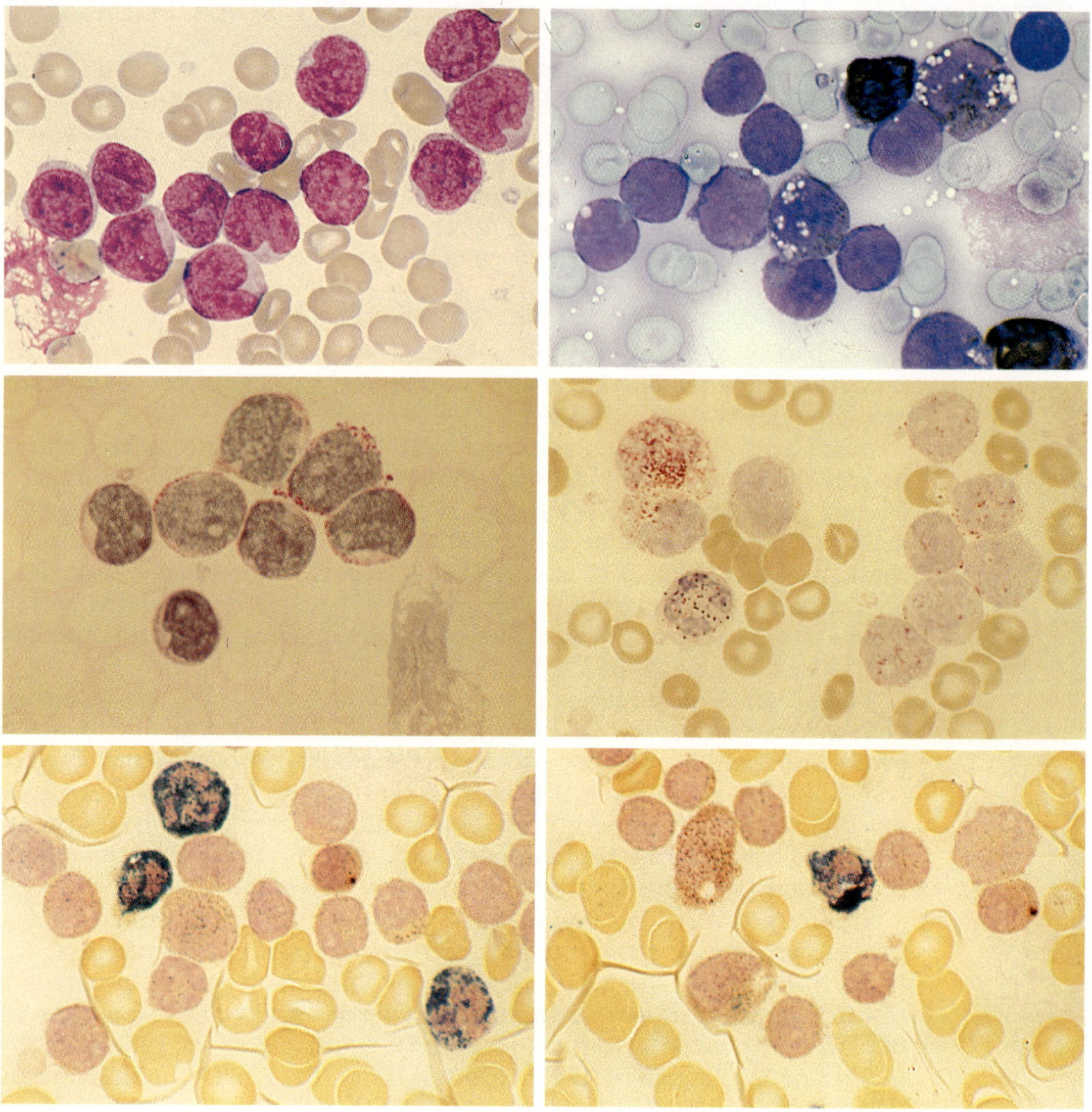

Plate 58

A. Centrocytic/centroblastic lymphoma in leukaemic phase. Buffy coat. Leishman stain. The lymphoma cells show characteristic nuclear notching.

B. The same preparation. SB stain. The lymphoma cells are negative, while typical Sudanophilia is shown in two vacuolated monocytes and two neutrophils.

C. The same preparation. PAS stain. Three lymphoma cells show some weak to moderately strong granular reaction and three are PAS negative.

D. The same preparation. Acid phosphatase. A neutrophil and a monocyte show typical positivity, while eight lymphoma cells show only weak granular reaction.

E. The same preparation. Dual esterase. There are three CE positive neutrophils, a T cell with dot-like BE positivity, but the remaining lymphoma cells are essentially negative with only minimal fine granular mixed reactivity.

F. The same preparation. Dual esterase. A neutrophil is CE positive and a monocyte BE positive, while a single T cell shows a dot of BE positivity. The lymphoma cells are again virtually negative.

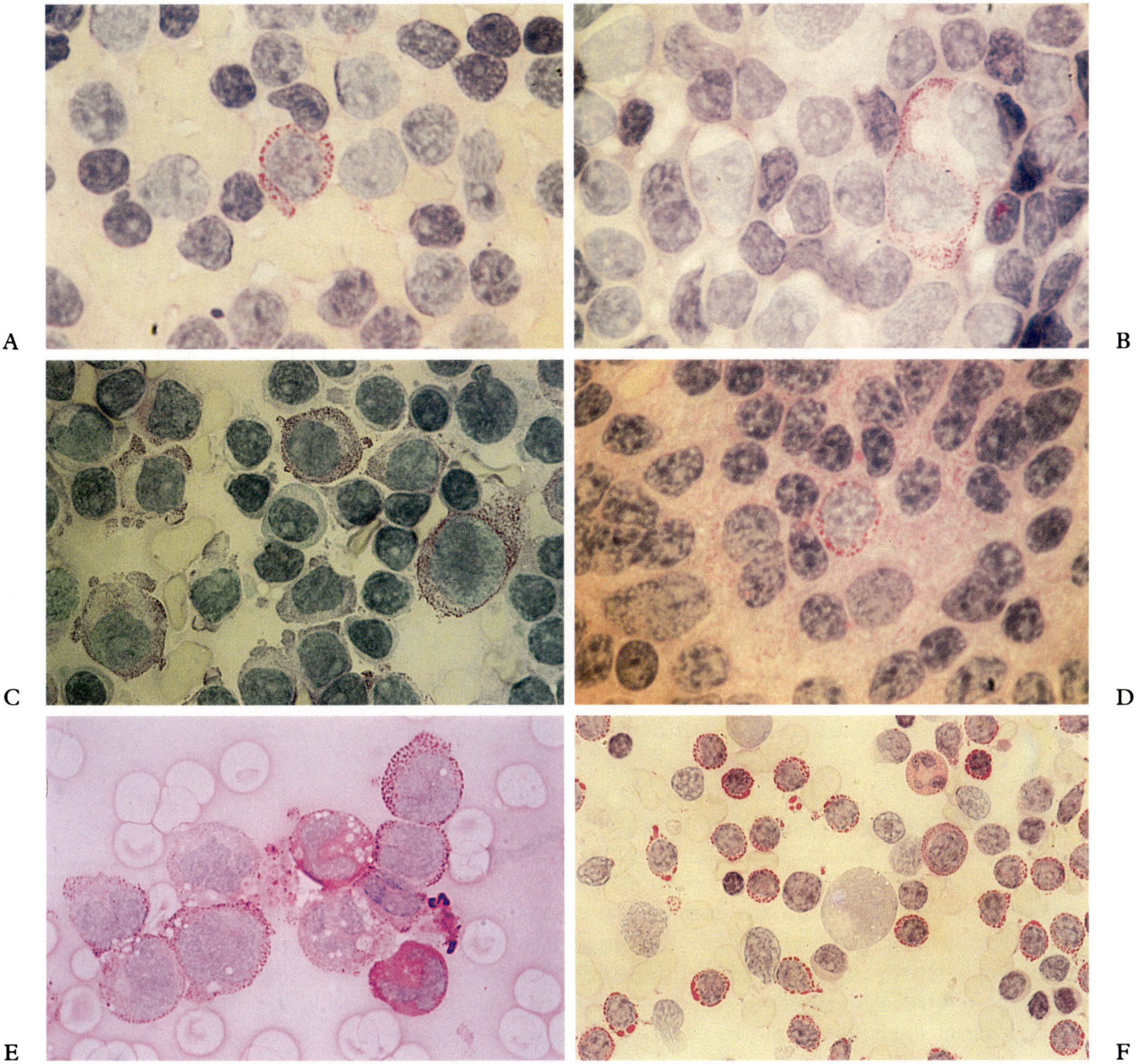

Plate 59

A. Lymph node imprint. Centrocytic–centroblastic lymphoma. PAS stain. A single centrocyte shows a ring of PAS positive granules. All the remainder are negative.

B. Lymph node imprint. Centrocytic–centroblastic lymphoma with chiefly centroblasts. PAS stain. Only a single centroblast shows granular PAS positivity; all other lymphoma cells are negative.

C. Lymph node imprint. Lymphoplasmacytoid lymphoma. PAS stain. Six of the larger and less mature tumour cells show coarse granular positivity, in contrast to the usual finding in secretory myeloma cells.

D. Lymph node imprint. Lymphoplasmacytoid lymphoma, of more mature cytology than in C. PAS stain. Only one of the tumour cells in this example shows granular positivity, while another cell shows a weakly PAS-positive Russell body; the remaining lymphoma cells are negative.

E. Leukaemic phase of immunoblastic lymphoma; buffy coat. PAS stain. A group of neoplastic immunoblasts show strong granular positivity. A neutrophil, a stab and an eosinophil show normal reaction.

F. Lymph node imprint. Sézary's syndrome. PAS stain. The Sézary cells show heavy and coarse granular cytoplasmic PAS positivity. A central immunoblast is negative. An eosinophil and a few residual lymphocytes show expected reactions.

intracytoplasmic lipids: the same finding, however, is, in our experience, infrequent in non-African Burkitt's lymphoma.

Finally, in Burkitt's lymphoma, semiquantitative studies of Wright and McAlpine (1966) indicated that the RNA content of tumour cells is particularly high, a finding which is presumably in relation to an extremely high proliferative rate, as confirmed by autoradiographic studies (Quaglino *et al.*, 1980), and is responsible for an intense cytoplasmic pyroninophilia.

In a third major group of B-cell lymphoid malignancies, the physiological counterpart of malignant cells appear to be represented by recirculating B-cells. B-prolymphocytic leukaemia, B-CLL and probably most cases of hairy cell leukaemia may be included in this group. B-prolymphocytes, both from the cytoimmunological and from the cytochemical point of view, seem to be similar to cells of the lymphocyte corona of the lymphoid follicles (Gobbi *et al.*, 1983). B-CLL cells display an immunological phenotype identical to that of the rare T1-positive lymphocytes, capable of rosetting with mouse erythrocytes, normally detectable on the edges of the germinal centres. The functional role of this B-cell subset, as well as its relationship with other B cell types, is still a matter for debate (Caligaris-Cappio *et al.*, 1986). As far as hairy cells are concerned, it has been suggested that they represent an activated state of the same cell type observed in B-CLL, since B-CLL cells may be induced by phorbol esters to express cytological features of hairy cells (Caligaris-Cappio *et al.*, 1985). The cytochemical features of chronic leukaemic B-cell disorders have been extensively dealt with in Chapter 14.

The neoplastic cells of the group of 'pre-B cell' malignancies display cytological characteristics corresponding to very early stages of normal B-cell development, which take place within the bone marrow: immunoglobulin gene rearrangements are demonstrable either in the absence of Ig synthesis, or together with the expression of cytoplasmic free (mostly μ) heavy chains (Vogler *et al.*, 1978; Korsmeyer *et al.*, 1981). Clinical and cytological features of acute lymphoblastic leukaemia (see Chapter 14) are observed in virtually all these cases. A slightly more advanced stage of differentiation is recognizable in the cell phenotype expressed by the rare cases of SIgM-positive non-Burkitt-like B-ALL.

T cell malignancies

At present, the basic criteria for the identification of neoplasms related to the T-cell differentiating line consist in the search for rearrangements of the T-cell receptor gene (O'Connor *et al.*, 1985) and in the demonstration of T-cell specific surface antigenic structures. The former method is not absolutely specific, since T-cell receptor gene rearrangements may be observed also in B-cell malignancies, but it appears to be the only safe way to assess the monoclonal nature of T-cell proliferations.

T-cell surface antigen expression, currently detected by monoclonal antibodies, characterizes several different malignant T-cell phenotypes, which in most instances are closely related to different stages of normal T cell differentiation.

Thus, blast cells from pre-T- and T-ALL express an immunological phenotype identical to that respectively of early and intermediate (cortical) thymocytes, while a late (medullary) thymocyte seems to be the physiological counterpart of cells from most cases of T-prolymphocytic leukaemia (Crockard, 1984). T-CLL cells present features related to those of normal effector T cells, more often with a 'suppressor' (Fcγ-receptor positive, $CD8^+$), and less frequently with a 'helper' (Fcμ-receptor positive, $CD4^+$) phenotype. All the leukaemic T-cell disorders have been dealt with in Chapter 14.

A mature 'helper' phenotype is usually expressed also by the cells of cutaneous T-cell lymphomas, of which mycosis fungoides represents the most typical example. Sézary's syndrome is currently regarded as the leukaemic variant of mycosis fungoides (Lennert, 1978), although some differences have been described between the immunological phenotype of lymph node cells in the two conditions (Willemze *et al.*, 1985).

In Sézary's syndrome, malignant cells have an atypical morphological appearance, best appreciated by electron microscopy (Lutzner and Jordan, 1968). Lutzner *et al.* (1973) in a further study described two different types of abnormal cells: a large-cell type with a serpentine and cerebriform nucleus, near-tetraploid DNA content and near-tetraploid chromosome counts, and a small-cell variant with indented nucleus, diploid DNA values and pseudodiploid or hyperdiploid chromosome counts.

The helper activity of neoplastic cells was confirmed by *in vitro* experiments, such as those carried out by Broder *et al.* (1976), which showed that Lutzner cells are capable of stimulating the production of immunoglobulins on the part of normal B lymphocytes.

Sézary's syndrome cells are rich in hydrolytic enzyme activities: acid phosphatase, α-naphthyl acetate esterase, β-glucuronidase and β-glucosaminidase reactions are strongly positive (Costello *et al.*, 1980; Machin *et al.*, 1980; Crockard *et al.*, 1982; Knowles *et al.*, 1982), although intriguingly the DAP IV reaction appears to be negative. A variable degree of PAS positivity has been usualy reported in Sézary's cells (Clendenning *et al.*, 1964), but in some cases the PAS reaction may be exceptionally intense in a high percentage of cells (Plate 59F).

Adult T-cell lymphoma/leukaemia (ATLL) was first described in Japan, but has now been identified in several racial groups including West Indian blacks and European and American Caucasians (Catovsky *et al.*, 1982). The malignant cells of this disorder are of variable morphological appearance, although irregular nuclear outlines appear to be a distinctive feature. Immunologically, ATLL cells show a mature post-thymic T helper-inducer phenotype (E-rosette$^+$, $T3^+$, $T4^+$, $T8^-$, TdT^-), although it has been suggested that these cells may represent the malignant counterpart of a subset of activated T helper-inducer cells (Tricot *et al.*, 1983). Although the phenotype is 'helper' the leukaemic cells may actively suppress lectin-induced proliferation in culture, and are $CD7^-$, so that they may in fact represent $T4^+$ suppressor cells (Miedema *et al*, 1984).

The cytochemical reactivity of ATLL cells in the different racial groups is fairly consistent (Usui *et al.*, 1979; Crockard *et al.*, 1983; Tricot *et al.*,

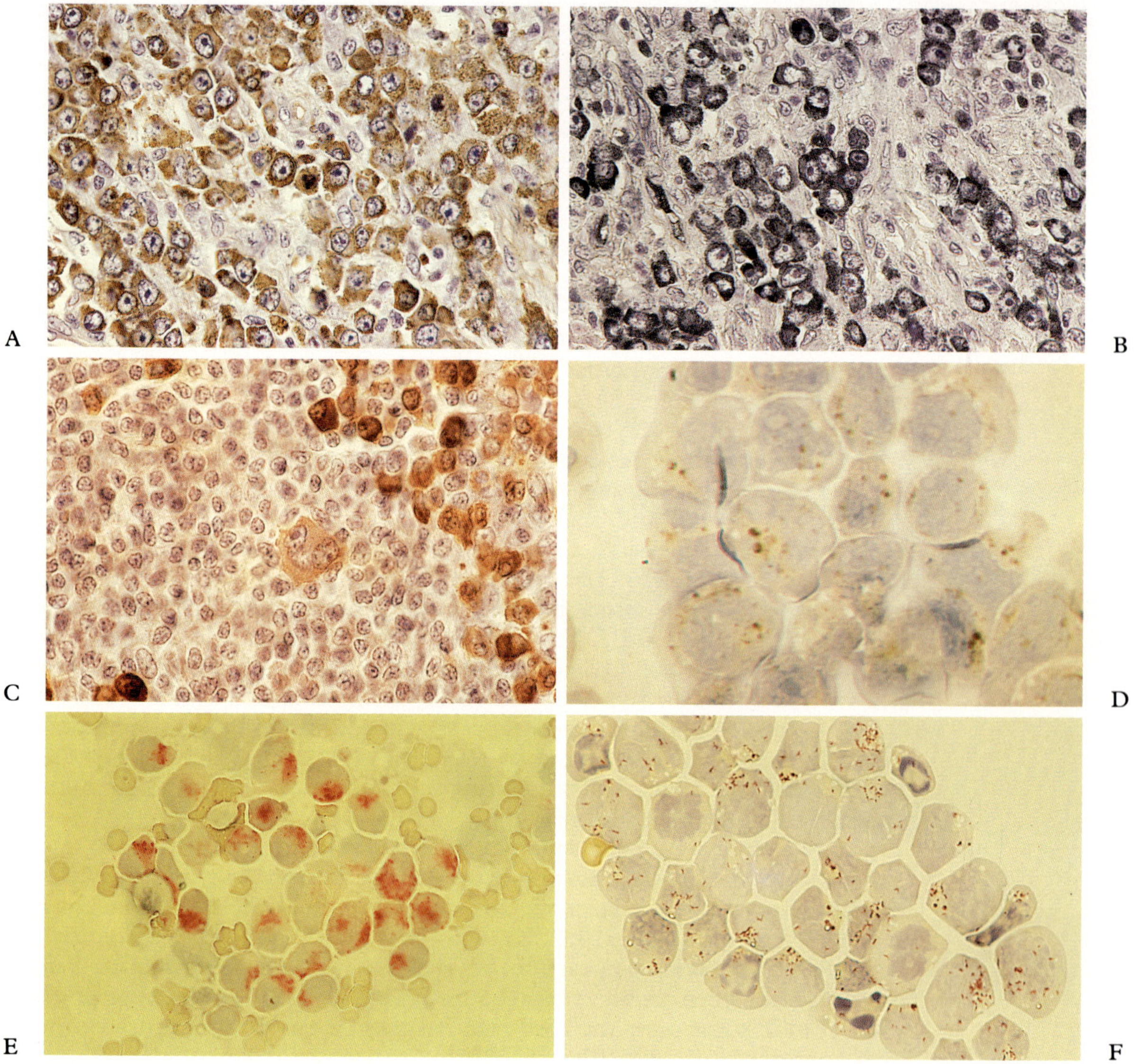

Plate 60

A. Immunoblastic lymphoma, lymph node section. Immunoperoxidase (PAP) reaction with anti-κ and the DAB method. Most cells show strong κ-chains expression.

B. The same specimen. PAP reaction with anti-κ, using 4CN as receptor. Equally striking positivity is shown.

C. Lymph node section. Hodgkin's disease. PAP reaction with anti-κ and DAB as receptor. Some plasmacytoid cells express κ chains and a central R-S cell is also moderately strongly positive.

D. Lymph node imprint. Centroblastic lymphoma. Acid phosphatase. Most centroblasts show a few coarse granules of positivity.

E. Lymph node imprint. Immunoblastic lymphoma. Acid phosphatase (with pararosaniline as receptor). Most immunoblasts show strong localised reaction, less sharply granular than with Fast Garnet as receptor.

F. Lymph node imprint. Burkitt lymphoma. Acid phosphatase. Many lymphoma cells show a few coarse granules of positivity.

1983). Positive reactions are observed for most hydrolases, although the intensity of the various reactions is less intense than that seen in other mature T cell leukaemias. Usui *et al.* (1982) have reported that tartrate-resistant acid phosphatase reactivity is frequent in Japanese patients, while this was not the case with West Indian subjects (Crockard *et al.*, 1983).

The histological and cytological features of lymph node T-cell neoplasias are extremely variable, thus making it very difficult to establish a simple reproducible classification of these forms (Lennert and Feller, 1985). The malignant cells probably represent the neoplastic counterpart of T-cells normally observed in the lymph node paracortex: the ability of these cells of producing lymphokines is presumably responsible in some histological varieties for the presence within the tumour tissue of a high percentage of non malignant cells, such as epithelioid cells or histiocytes.

T-zone lymphoma, T-immunoblastic lymphoma, lymphoepithelioid cell lymphoma and lymphohistiocytic T-cell lymphoma may be included into this group of neoplasms (Lennert and Feller, 1985): all of them represent quite rare forms. In T-zone lymphoma, according to Lennert (1978), neoplastic cells show strong acid phosphatase positivity in the absence of acid α-naphthyl acetate esterase activity. The PAS reaction is extremely variable in the different cases studied. No established cytochemical features are known at present for the other histological types of lymph node T cell malignancies.

Histiocytic malignancies

Histiocytic medullary reticulosis (malignant histiocytosis) and true histiocytic lymphoma are uncommon diseases, most of the lymphomas previously labelled histiocytic in earlier classifications of non-Hodgkin's lymphoma being now recognised as of B-cell origin. When they do arise, neoplastic states of histiocytic origin generally reflect in an enhanced form the functional and cytochemical characteristics of normal histiocytes, showing exaggerated phagocytic activity, for example, and strong positivity for acid phosphatase and other acid hydrolases, and also for acetate and butyrate esterases. Sudanophilia, PAS positivity and free iron staining, probably largely arising from phagocytosis, are also often increased. All these features are shown in Plate 61.

HODGKIN'S DISEASE

Histopathological evaluation of lymph nodes in Hodgkin's disease reveals a complex and variable association of cell types, many of which display normal morphological features, in contrast to a few, which have a frankly atypical appearance.

In fact, the only true neoplastic 'candidates', by definition or exclusion, are those cells which are designated eponymously as Hodgkin's cells (HC) and Reed-Sternberg cells (RSC) (Fiore-Donati *et al.*, 1985). HC are regarded as the precursors or more simply as the mononuclear variants of

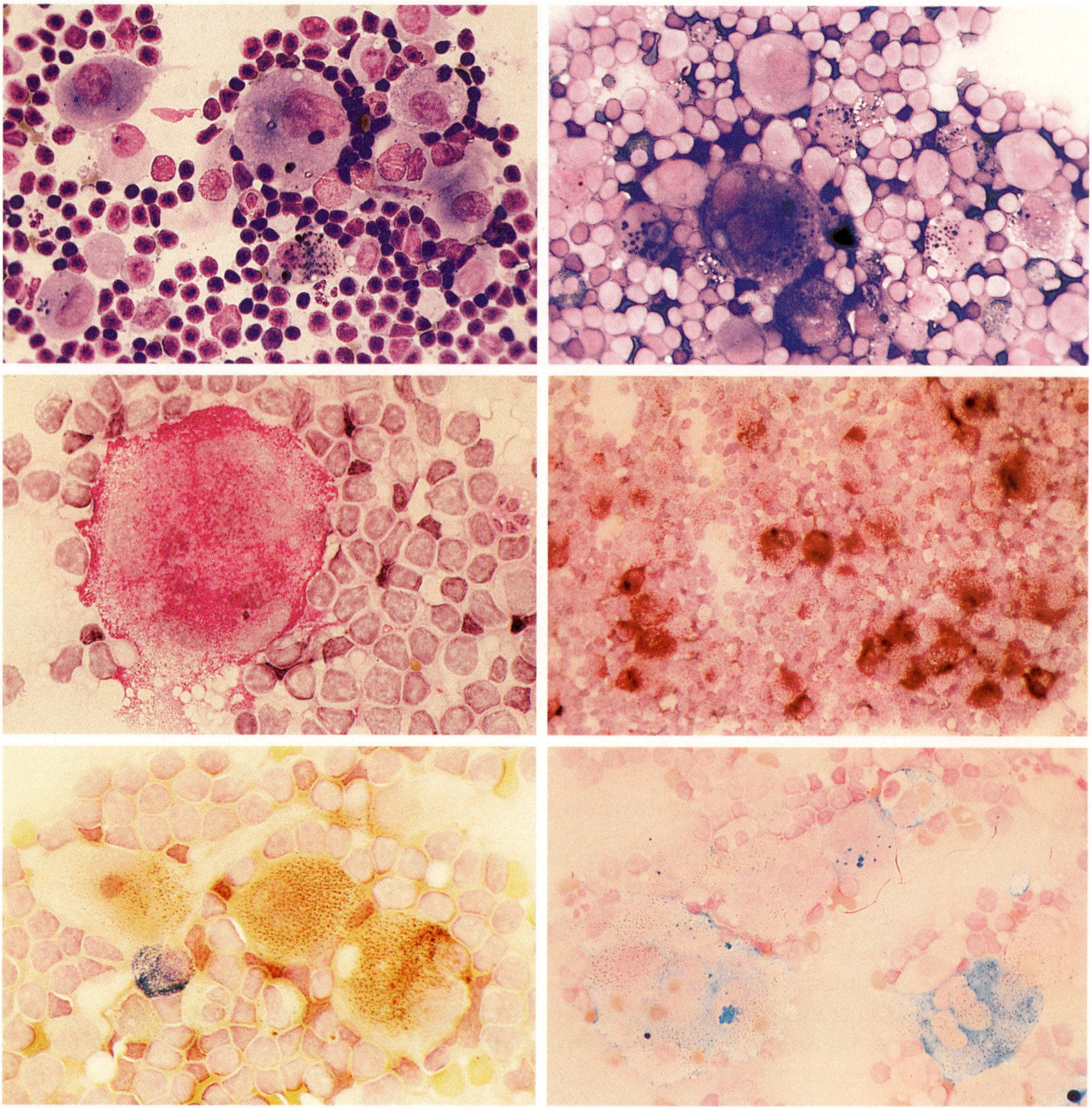

A B C D E F

Plate 61

A. Lymph node imprint. Histiocytic lymphoma. Leishman stain. The malignant cells show irregular size and shape and generally large amounts of cytoplasm and two contain ingested foreign material.

B. The same preparation. SB stain. Occasional granules of SB positivity, probably resulting from ingestion of sudanophilic cells, are found in several histiocytes.

C. The same preparation. PAS stain. A single giant histiocyte shows diffuse and finely granular PAS positivity.

D. The same preparation. Acid phosphatase. Numerous strongly positive histiocytes are seen.

E. The same preparation. Dual esterase. There is one CE-positive neutrophil present; the large histiocytic cells show moderately strong BE positivity.

F. The same preparation. Prussian blue stain. The malignant histiocytes contain phagocytosed cellular material including variable amounts of stainable free iron.

RSC, since they may share some of the morphological characteristics of the latter. HC may also be considered the actively proliferating cells of Hodgkin's disease, because they are usually found in higher number and more frequently in mitotic activity than RSC. When labelled with tritiated thymidine, HC also show a greater degree of incorporation of labelled DNA than RSC (Quaglino *et al.*, 1978), sometimes similar to that of immunoblasts. The problem is that HC are often not easily recognizable, due to the difficulty of distinguishing them with certainty on light microscopy from reactive immunoblasts or from other cell types (reticulum cells, epithelioid cells, etc.).

The matter is further complicated by the uncertainty which still surrounds the identity of HC and RSC. In fact, despite the many studies based on light microscopic, electron microscopic, immunologic, functional as well as tissue culture findings, the vexing problem of the nature and origin of HC and RSC still remains unsolved.

The conflicting results obtained by several authors on the mononuclear-phagocytic versus B or T lymphocytic derivation of RSC (Dorreen *et al.*, 1984) have been further complicated by the suggestion that these cells and their morphological variants might have different origins according to the different histological subtypes (Poppema, 1980; Stein *et al.*, 1985) or more than one derivation even in the same case (Hansmann and Kaiserling, 1981). At present it can only be said that although HC and RSC seem to display some features of cells derived from the mononuclear-phagocytic system (Strauchen, 1984), they are also capable of expressing lineage specific T or B cell markers including cytoplasmic immunoglobulin (Plate 60C). RSC have been found to express HLA-DR antigen (Kadin, 1982; Poppema *et al.*, 1982), the so called Tac antigen (CD25), associated with the binding of interleukin 2 (Fiore-Donati *et al.*, 1985), and an antigen detectable with Ki1 (CD30) monoclonal antibody, which is also expressed by immunologically activated lymphocytes (Stein *et al.*, 1985). In addition, HC and RSC share with the granulocyte lineage several antigens, such as those detected by monoclonal antibodies anti-Leu M1, TU5, TU6, TU9 and 3C4, and are capable of binding peanut agglutinin and concanavalin A (Stein *et al.*, 1982; Hsu and Jaffe, 1984; Strauchen, 1984; Pinkus *et al.*, 1985). According to Stein *et al.* (1985), T cell associated (T1, T3, T4, T11) and B cell associated (B1, B4, To15) membrane antigens are also frequently expressed by RSC of the nodular sclerosis, mixed cellularity and lymphocyte-depleted types of Hodgkin's disease, while J chains, suggestive of an active Ig synthesis (Mestecky *et al.*, 1977), are detectable in the cytoplasm of RSC in the nodular type of lymphocyte predominance.

In view of the controversial results obtained by immunocytological methods, it is of interest to recall the attempts made in the past to clarify the origin and nature of HC and RSC with cytochemical methods.

One of the first studies was that of Dorfman (1961) who, using cryostatic methods, demonstrated in lymph nodes and spleens from seven cases of Hodgkin's disease that binucleated and multinucleated RSC were consistently negative for acid phosphatase, alkaline phosphatase and non-specific esterases. It was postulated on the basis of these findings that RSC had suffered a loss of enzyme activity and were therefore different from normal histiocytes, or else that they originated from cells other than histiocytes.

A completely different cytochemical profile for RSC was obtained by Braunstein *et al.* (1962), who tested a battery of 21 staining reactions for enzymes on freeze-dried preparations, as well as on acetone-fixed, paraffin-embedded sections and formalin-fixed, frozen sections from 24 lymph nodes involved in Hodgkin's disease. The authors reported the presence of non-specific esterase activity and of positive reactions for acid phosphatase and aminopeptidase in RSC. The reactivity of oxidative enzymes was not characteristic but resembled that of histiocytes rather than that of germinal centre cells.

In another cytochemical study, carried out on RSC present in the peripheral blood of a patient with Hodgkin's disease in the terminal stage (Sinks and Clein, 1966), the cells were reported to be Sudan black, peroxidase, acid phosphatase and PAS negative (with the exception of a very small proportion of cells showing a diffuse type of positivity): moderate succinic and lactic dehydrogenase activity was present. However, when RSC were cultured in vitro, acid phosphatase activity became evident after 6–10 hours culture: the reaction gradually declined and was negative within 48 hours.

In another case, reported by Chan and McBride (1972), RSC were completely negative for Sudan black B and PAS reactions.

Nanba *et al.* (1975) investigated frozen sections of three cases of Hodgkin's disease. Acid phosphatase and non-specific esterase activity was only very weakly positive in RSC: alkaline phosphatase and β-glucuronidase were negative.

Yam and Li (1976) studied the cytochemistry of splenic lesions in Hodgkin's disease using imprints and frozen sections obtained from 45 spleens. They found that RSC showed insignificant quantities or no activity at all of the granulocytic marker enzymes and lysosomal enzymes. However, these cells were strongly pyroninophilic and were interpreted as stimulated lymphocytes. Similar findings were reported by Anagnostou *et al.* (1977), who investigated touch imprints and frozen sections from 21 patients with nodular sclerosing Hodgkin's disease, and by Poppema *et al.* (1979), who studied lymph node imprints in 3 patients with lymphocytic predominance.

In a study carried out by Stuart *et al.* (1974) on cell suspensions from four lymph nodes and one involved spleen, RSC were found uniformly negative for peroxidase, alkaline and acid phosphatase reactions. In a minority of RSC an appreciable non-specific esterase activity, producing a diffuse stain, was observed. The same finding was noted by Kaplan and Gartner (1977) on cultured RSC.

Hayhoe *et al.* (1978) reported the results of cytochemical, ultrastructural and immunological studies of RSC from the peripheral blood of a patient with Hodgkin's disease. RSC were negative for peroxidase, Sudan black B, alkaline phosphatase and also to non-specific esterase and naphthol AS-D chloroacetate esterase. After PAS staining only 3% of RSC had strong granular positivity: on the contrary the majority of these cells showed moderately strong granular positivity for acid phosphatase.

In a study carried out by Quaglino *et al.* (1978) on touch imprints from 12 cases of Hodgkin's disease, the authors reported in 9 cases a proportion of HC and RSC containing either a single block of glycogen or a scattering of moderately coarse PAS positive granules. The same cells presented occasional dispersed sudanophilia and showed evidence of acid phosphatase and non-specific esterase activity, especially prominent in the paranuclear area.

In 1980, Schaadt *et al.* described the morphological, cytochemical, immunological and karyotypic characteristics of two continuous cell lines (L428 and L439) obtained from the pleural effusions of two patients with nodular sclerosing Hodgkin's disease. In both cell lines, the cells displayed granular reactions for acid phosphatase and acid α-naphthyl acetate esterase, usually concentrated in the Golgi region. On the contrary, alkaline phosphatase, chloroacetate esterase and peroxidase were all negative in both established cell lines; however, peroxidase staining was found to be temporarily positive in some large cells shortly after the establishment of the L428 cell line. Chilosi *et al.* (1981) could demonstrate acid phosphatase and non-specific esterase activity in RSC, both on cryostat sections and paraffin-embedded lymph node samples from seven cases of Hodgkin's disease. The same cells were found unreactive with alkaline phosphatase, ATPase, myeloperoxidase and chloroacetate esterase. The authors also noticed that, unlike macrophages encountered in most reactive conditions, and similar to epithelioid cells, a number of macrophages in Hodgkin's disease displayed a tartrate-resistant acid phosphatase activity, while the enzyme activity of RSC was sensitive to inhibition.

More recently, Carbone *et al.* (1983) carried out a cytochemical study on lymph node imprints from 53 cases of histologically proven Hodgkin's disease, before treatment. The following cytochemical reactions were employed: periodic acid-Schiff, acid phosphatase, β-glucuronidase, alkaline phosphatase, naphthol AS acetate esterase and naphthol AS-D chloroacetate esterase. Histologically, Hodgkin's disease was classified according to the four main histological types: the concept of the 'cellular phase' of nodular sclerosis was taken into account, whereas those cases with a high concentration of focally aggregated epithelioid cells were subgrouped separately. The results obtained by the authors demonstrated in the majority of cases (40 out of 51) investigated that a proportion of RSC was positive for the acid phosphatase reaction. In a small number of cases PAS, naphthol AS acetate esterase and β-glucuronidase reactions were positive in a minority of RSC, whereas alkaline phosphatase and naphthol AS-D chloroacetate esterase reactions were completely negative. The intensity of naphthol AS acetate esterase positivity was less than that observed in histiocytes. Also the quality and quantity of the reaction product of acid phosphatase in RSC differed from that of neoplastic histiocytes of malignant histiocytosis and neoplastic lymphocytes of T cell lymphomas. However, the pattern of staining was similar in the diagnostic RSC and in their morphological variants, and was apparently unrelated to the histological subtypes.

The results of all the cytochemical studies reported suggest that HC and RSC display highly variable cytochemical features, which differ from case

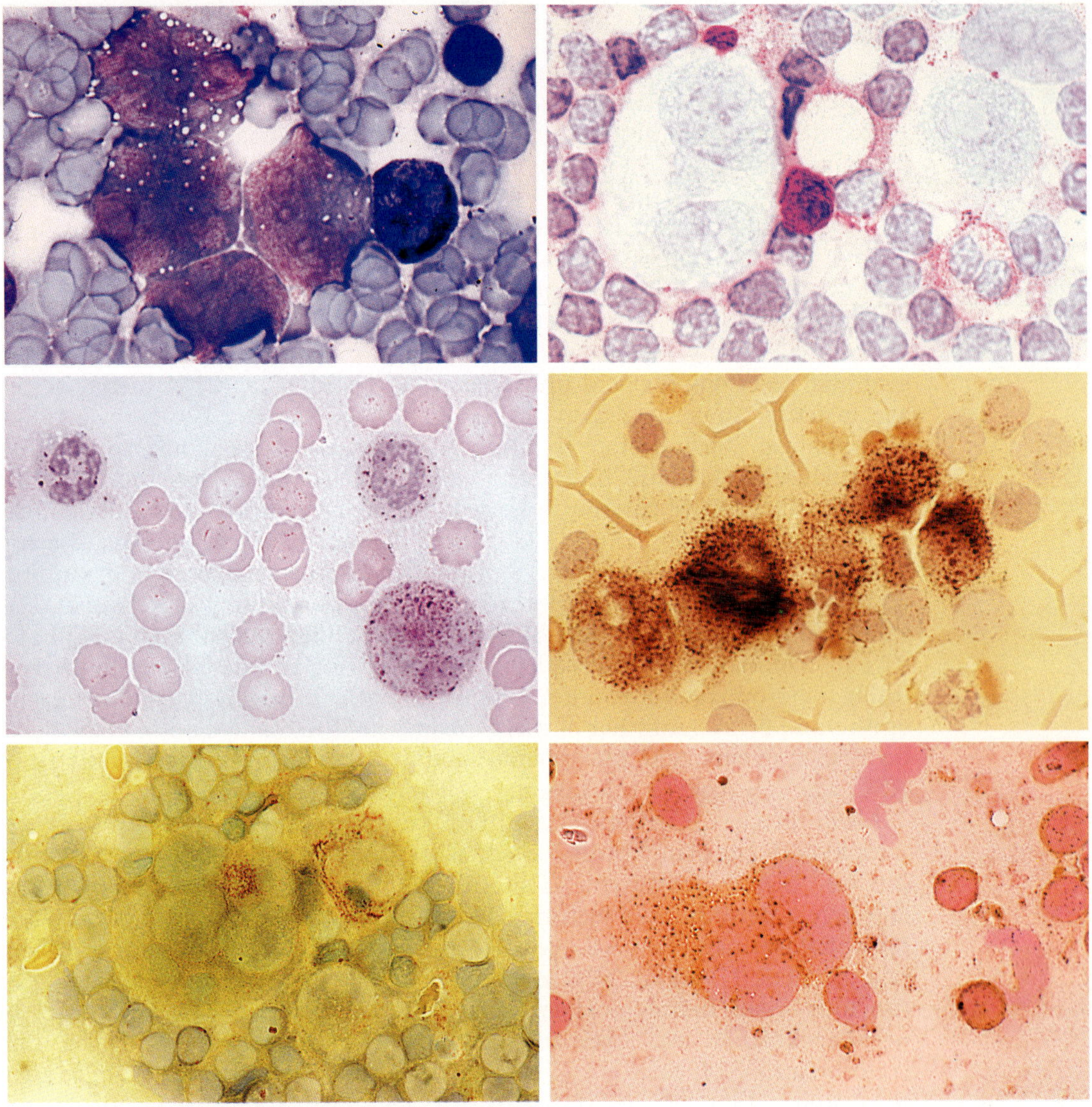

Plate 62

A. R-S cells in peripheral blood, SB stain. There is a possible neutrophil present but the R-S cells are negative.

B. R-S cells in a node imprint. PAS reaction. These cells are generally negative, as here, but may show occasional positivity, sometimes in the form of blocks.

C. R-S cells in peripheral blood. Acid phosphatase. Moderately coarse granular reaction is shown in the R-S cell and weaker reactions in two neutrophils.

D. R-S cells in a node imprint. Acid phosphatase. Very strong positivity is shown in these nodal R-S cells.

E. The same preparation. Acetate esterase. Granular positivity in two of three R-S cells.

F. R-S cells in another node imprint. Dual esterase. Several lymphocytes, some showing the T-cell form of BE positivity, and an R-S cell with weak, widely scattered, granular BE positivity.

to case and even between cells of the same population, and that they fail to show a cytochemical profile typical for either the monocyte-macrophage cell line or for normal lymphocytes. An illustrative set of findings from our own case material is shown in Plate 62. It is possible that the cytochemical characteristics of HC and RSC vary according to a combination of different factors which may interact erratically with each other (histological subtype, stage of the disease, aggressiveness of the same, the different morphological aspects of the neoplastic cell population, the particular functional and metabolic aspects of neoplastic cells, etc.).

A possible influence on the cytochemical behaviour of HC and RSC may be exerted by the different cellular composition of the various histological subtypes.

Payne *et al.* (1980) studied *in vitro* the RSC/lymphocyte interaction, and reported that lymphocytes adhering to RSC were mostly acetate esterase negative. However, this finding is in contrast with the cytochemical pattern observed in lymph node imprints, where Quaglino *et al.* (1978) reported an increase in non-specific esterase activity and glycogen content in the lymphocytes surrounding HC and RSC.

A study by Crocker *et al.* (1982), using α-naphthyl acetate esterase, showed in cases of Hodgkin's disease the marked variability in content of acetate esterase positive cells (other than T cells). In specimens of Hodgkin's disease, acetate esterase positive cells constituted 1.8–16% of all cells in the lymphocyte predominant group, appearing to cluster in two ranges, from 1.8 to 5.2% and from 12.5 to 16% respectively. Nodular sclerosis specimens contained a rather larger number of acetate esterase positive cells, from 13.6 to 22.1%, while enzyme positive cells constituted 6.5–14.6% of mixed cellularity specimens. The lymphocyte-depleted variety contained the greatest number of esterase positive cells, ranging from 39.8 to 47.7% of all cells. Two specimens of presumed true histiocytic lymphoma contained 95.2 and 97.5% acetate esterase positive cells respectively, the few negative cells appearing to be small lymphocytes.

REFERENCES

Abo T, Balch C M 1981 A differentiation antigen of human NK and K cells identified by a monoclonal antibody (HNK-1). Journal of Immunology 127: 1024

Abo T, Cooper M D, Balch C M 1982 Characterization of HNK-1$^+$ (Leu 7) human lymphocytes. I. Two distinct phenotypes of human NK cells with different cytotoxic capability. Journal of Immunology 129: 1752

Anagnostou D, Parker J W, Taylor C R, Tindle B H, Lukes R J 1977 Lacunar cells of nodular sclerosing Hodgkin's disease. An ultrastructural and immunohistologic study. Cancer 39: 1032

Armitage R J, Linch D C, Worman C P, Cawley J C 1982 The morphology and cytochemistry of human T-cell subpopulations defined by monoclonal antibodies and Fc receptors. British Journal of Haematology 51: 605

Basso G, Cocito M G, Semenzato G, Pezzutto A, Zanesco L 1980 Cytochemical study of thymocytes and T lymphocytes. British Journal of Haematology 44: 577.

Bataille R, Durie B, Sany J, Salmon S E 1980 Myeloma bone marrow acid phosphatase staining: a correlative study of 38 patients. Blood 55: 802

Beckstead J H 1983 The evaluation of human lymph nodes, using plastic sections and enzyme histochemistry. American Journal of Clinical Pathology 80: 131

HAEMATOLOGICAL CYTOCHEMISTRY

Beckstead J H, Warnke R, Bainton D F 1982 Histochemistry of Hodgkin's disease. Cancer Treatment Reports 66: 609

Bergui L, Campana D, Tesio L, Chilosi M, Pizzolo G, Janossy G, Caligaris-Cappio F 1985 Phenotypic heterogeneity of plasma cell disorders revealed by monoclonal antibodies. In: Quaglino D, Hayhoe F G J (eds). The cytobiology of leukaemias and lymphomas. Raven Press, New York, p 137

Bernard J, Dufer J 1983 Cytochemical analysis of human peripheral blood lymphocyte subsets defined by monoclonal antibodies. Scandinavian Journal of Immunology 17: 89

Bhan A K, Nadler L M, Stashenko P, McCluskey R T, Schlossman S F 1981 Stages of B cell differentiation in human lymphoid tissue. Journal of Experimental Medicine 154: 737

Bofill M, Janossy G, Janossa M, Burford G D, Seymour G J, Wernet P, Kelemen E 1985 Human B cell development. II. Subpopulations in the human fetus. Journal of Immunology 134: 1531

Braunstein H, Freiman D G, Thomas W, Gall E A, 1962 A histochemical study of the enzymatic activity of lymph nodes. III. Granulomatous and primary neoplastic conditions of lymphoid tissue. Cancer 15: 139

Breard J, Reinherz EL, Kung PC, Goldstein G, Schlossman S F 1980 A monoclonal antibody reactive with human peripheral blood monocytes. Journal of Immunology 124: 1943

Broder S, Edelson R L, Lutzner M A, Nelson D L, MacDermott R P, Durm M E, Goldman C K, Meade B D, Waldmann T A 1976 The Sézary syndrome. A malignant proliferation of helper T cells. Journal of Clinical Investigation 58: 1297

Caligaris-Cappio F, Gobbi M, Janossy G 1982 Immunological characterization of lymphoid follicles in relationship with B cell differentiation. In: Il centro germinativo del follicolo linfatico. Il Pensiero Scientifico Editore, Rome, p 55

Caligaris-Cappio F, Pizzolo G, Chilosi M, Bergui L, Semenzato G, Tesio L, Morittu L, Malavasi F, Gobbi M, Schwarting R, Campana D, Janossy G 1985 Phorbol ester induces abnormal chronic lymphocytic leukemia cells to express features of hairy cell leukemia. Blood 66: 1035

Caligaris-Cappio F, Bergui L, Tesio L 1986 New trends in the biology of B-chronic lymphocytic leukemia. Haematologica 71: 91

Carbone A, Micheau C, Caillaud J M, Carlu C 1981 A cytochemical and immunohistochemical approach to malignant histiocytosis. Cancer 47: 2862

Carbone A, Caillaud J M, Carlu C, Micheau C 1983 Cytochemistry of Reed-Sternberg cells in lymph node imprints. American Journal of Clinical Pathology 79: 553

Cassuto J P, Hammou J C, Pastorelli E, Dujardin P, Masseyeff R 1977 Plasma cell acid phosphatase: a discriminative test for benign and malignant monoclonal gammopathies. Biomedicine 27: 197

Catovsky D, Linch D C, Beverley P C L 1982 T cell disorders in haematological disease. Clinics in Haematology 11: 661

Chan B W B, McBride J A 1972 Hodgkin's disease and leukemia. Canadian Medical Association Journal 106 :558

Chilosi M, Pizzolo G, Menestrina F, Iannucci A M,Bonetti F, Fiore-Donati L 1981 Enzyme histochemistry on normal and pathological paraffin-embedded tissues. American Journal of Clinical Pathology 76: 729

Chilosi M, Pizzolo G, Menestrina F, Iannucci A M, Bonetti F, Fiore-Donati L 1982 Dipeptidyl (amino) peptidase (DAP IV) histochemistry on normal and pathologic lymphoid tissues. American Journal of Clinical Pathology 77: 714

Clendenning W E, Brecher G, van Scott E J 1964 Mycosis fungoides. Relationship to malignant cutaneous reticulosis and the Sézary syndrome. Archives of Dermatology 89: 785

Costello C, Catovsky D, O'Brien M, Morilla R, Varadi S 1980 Chronic T-cell leukaemias. 1. Morphology, cytochemistry and ultrastructure. Leukemia Research 4: 463

Crockard A D 1984 Cytochemistry of lymphoid cells: a review of findings in the normal and leukemic state. Histochemical Journal 16: 1027

Crockard A D, Catovsky D 1983 Cytochemistry of normal lymphocyte subsets defined by monoclonal antibodies and immunocolloidal gold. Scandinavian Journal of Haematology 30: 433

Crockard A D, Chalmers D, Matutes E, Catovsky D 1982 Cytochemistry of acid hydrolases in chronic B and T cell leukaemias. American Journal of Clinical Pathology 78: 437

Crockard A D, O'Brien M, Robinson D, Tavares de Castro J, Matutes E, Catovsky D 1983 Morphological and cytochemical features of adult T-cell lymphoma-leukaemia. Haematology and Blood Transfusion 28: 131

Crocker J, Jones E L, Curran R C 1982 A quantitative study of α-naphthyl acetate esterase-positive cells in Hodgkin's disease. Journal of Clinical Pathology 35: 1301

Curran R C, Jones E L 1977 Immunoglobulin-containing cells in human tonsils as demonstrated by immunohistochemistry. Clinical and Experimental Immunology 28: 103

De Martino C, Zamboni L 1967 Silver methenamine stain for electron microscopy. Journal of Ultrastructural Research 19: 273
De Pasquale A, Paterlini P, Di Francesco A, Vecchio L, Quaglino D 1986 Follicular lymphomas with prominent splenic involvement: report of two cases. Tumori 72: 109
De Waele M, Demey J, Moeremans M, Smet L, Broodiaerts L, Van Camp B 1983 Cytochemical profile of immunoregulatory T lymphocyte subsets defined by monoclonal antibodies. Journal of Histochemistry and Cytochemistry 31: 471
Diebold J, Reynes M, Kalifat R, Tricot G 1974 Les inclusions intranucléaires de la maladie de Waldenström. Nouvelle Presse Medicale 3: 1067
Dimitriu-Bona A, Burmester G R, Waters S J, Winchester R J 1983 Human mononuclear phagocyte differentiation antigens. I. Patterns of antigenic expression on the surface of human monocytes and macrophages defined by monoclonal antibodies. Journal of Immunology 130: 145
Dorfman R F 1961 Enzyme histochemistry of the cells in Hodgkin's disease and allied disorders. Nature 190: 925
Dorreen M S, Habeshaw J A, Stansfeld A G, Wrigley P F M, Lister T A 1984 Characteristics of Sternberg-Reed and related cells in Hodgkin's disease: an immunohistological study. British Journal of Cancer 49: 465
Dufer J, Trentesaux C, Desplaces A 1984 Differential effect of the serine protease inhibitor phenyl methyl sulfonyl fluoride on cytochemically detectable esterases in human leucocytes and platelets. Scandinavian Journal of Haematology 32: 25
Feller A C, Heijnen C J, Ballieux R E, Parwaresch M R 1982 Enzymehistochemical staining of Tμ lymphocytes for glycyl-proline-4-methoxy-beta-naphthylamide-peptidase (DAP IV). British Journal of Haematology 51: 227
Fiore-Donati L, Menestrina F, Chilosi M 1985 Morphofunctional aspects of Hodgkin's disease. In: Quaglino D, Hayhoe F G J (eds) The cytobiology of leukaemias and lymphomas. Raven Press, New York, p 1
Geniram A, Fiorini G, Spotti D, Frezzini G, Guarino P, Venturino M, Marini V 1984 Discrasie plasmacellulari benigne e maligne: metodi di indagine. Haematologica 69: 163
Gérard-Marchant R, Hamlin I, Lennert K, Rilke F, Stansfeld A G, Van Unnik J A M 1974 Classification of non Hodgkin's lymphomas. Lancet ii: 406
Gobbi M, Caligaris-Cappio F, Janossy G 1983 Normal equivalent cells of B cell malignancies: analysis with monoclonal antibodies. British Journal of Haematology 54: 393
Halper J P, Tolidjian B, Knowles D M 1982 Acid alpha-naphthyl acetate esterase (ANAE) in human B cells: correlation of expression with stages of B cell differentiation. Cellular Immunology 72: 367
Hansmann M L, Kaiserling E 1981 Electron-microscopic aspects of Hodgkin's disease. Journal of Cancer Research and Clinical Oncology 101: 135
Hayhoe F G J, Neuman Z 1976 Cytology of myeloma cells. Journal of Clinical Pathology 29: 916
Hayhoe F G J, Burns G F, Cawley J C, Stewart J W 1978 Cytochemical, ultrastructural and immunological studies of circulating Reed-Sternberg cells. British Journal of Haematology 38: 485
Hoffman J J M L, Breed W P M, Swaak-Lammers N 1985 Plasma cell acid phosphatase score in multiple myeloma and related disorders. British Journal of Haematology 59: 67
Hsu S M, Jaffe E S 1984 Leu M1 and peanut agglutinin stain the neoplastic cells of Hodgkin's disease. American Journal of Clinical Pathology 82: 29
Hsu S M, Hsu P L, McMillan P N, Fanger H 1982 Russell bodies, A light and electron microscopic immunoperoxidase study. American Journal of Clinical Pathology 77: 26
Hsu S, Cossman J, Jaffe E S 1983a Lymphocyte subsets in normal human lymphoid tissues. American Journal of Clinical Pathology 80: 21
Hsu S, Cossman J, Jaffe E S 1983b A comparison of ABC, unlabeled antibody and conjugated immunohistochemical methods with monoclonal and polyclonal antibodies. An examination of germinal center of tonsils. American Journal of Clinical Pathology 80: 429
Janossy G, Tidman N, Selby W S, Thomas J A, Granger S, Kung P C, Goldstein G 1980 Human T lymphocytes of inducer and suppressor type occupy different microenvironments. Nature 287: 81
Janossy G, Duke O, Poulter L W, Panayi G, Bofill M, Goldstein G 1981 Rheumatoid arthritis: a disease of T-lymphocyte/macrophage immunoregulation. Lancet ii: 839
Johansen P, Krogh Jensen M 1980 Enzymecytochemistry and immunohistochemistry in monoclonal gammopathy and reactive plasmacytosis. Acta Pathologica et Microbiologica Scandinavica (Section A) 88: 377
Kadin M E 1982 Possible origin of the Reed-Sternberg cell from an interdigitating reticulum cell. Cancer Treatment Reports 66: 601
Kaiserling E, Lennert K 1974 Die interdigitierende Reticulumzelle in menschlichen Lymphknoten. Eine spezifische Zelle der thymusabhängigen Region. Virchows Archiv B 16: 51

Kaplan H S, Gartner S 1977 'Sternberg-Reed' giant cells of Hodgkin's disease: cultivation in vitro, heterotransplantation and characterization as neoplastic macrophages. International Journal of Cancer 19: 511
Khaukin T 1984 More on the Russell body (Letter to the editor). American Journal of Clinical Pathology 81: 143
Knowles D M, Halper J P 1982 Human T-cell malignancies: correlative clinical, histopathologic, immunologic and cytochemical analysis of 23 cases. American Journal of Pathology 106: 187
Korsmeyer S J, Hieter P A, Ravetch J V, Poplack D G, Waldmann T A, Leder P 1981 Developmental hierarchy of immunoglobulin gene rearrangements in human leukemic pre-B-cells. Proceedings of the National Academy of Sciences USA 78: 7096
Korsmeyer S J, Arnold A, Bakhshi A, Ravetch J V, Siebenlist U, Hieter P A, Sharrow S O, Le Bien T, Kersey J H, Poplack D G, Leder P, Waldmann T A 1983 Immunoglobulin gene rearrangement and cell surface antigen expression in acute lymphocytic leukemias of T cell and B cell precursor origins. Journal of Clinical Investigation 71: 301
Lanier L L, Phillips J H, Warner N L, Babcock G F 1983 Subpopulations of human natural killer cells defined by expression of Leu 7 (HNK-1) and Leu 11 (NK-15) antigens. Journal of Immunology 131: 1789
Leder L D 1967 Der Blutmonocyt (Experimentelle Medizin, Pathologie und Klinik, Bd. 23), Springer-Verlag, Berlin
Leder L D 1978 Alkaline phosphatase and tartrate-resistant acid phosphatase activity in cells of prolymphocytic leukaemia. Klinische Wochenschrift 56: 313
Lennert K 1978 Malignant lymphomas other than Hodgkin's disease. Histology, cytology, ultrastructure, immunology. Springer-Verlag, Berlin
Lennert K, Feller A C 1985 Morphology and immunohistology of T-cell lymphomas. In: Quaglino D, Hayhoe F G J (eds). The cytobiology of leukaemias and lymphomas. Raven Press, New York, p 81
Lennert K, Stein H, Kaiserling E 1975 Cytological and functional criteria for the classification of malignant lymphomata. British Journal of Cancer 31 (suppl. II): 29
Lukes R J, Collins R D 1974 A functional approach to the classification of malignant lymphoma. In: Recent results in cancer research 46: 18. Springer-Verlag, Berlin
Lukes R J, Collins R D 1975 New approaches to the classification of the lymphomata. British Journal of Cancer 31 (suppl. II): 1
Lutzner M A, Jordan H W 1968 The ultrastructure of an abnormal cell in Sézary's syndrome. Blood 31: 719
Lutzner M A, Emerit I, Durepaire R, Flandrin G, Grupper C, Prunieras M 1973 Cytogenetic, cytophotometric, and ultrastructural study of large cerebriform cells of the Sézary syndrome and description of a small-cell variant. Journal of the National Cancer Institute 50: 1145
Machin G A, Halper J P, Knowles D M 1980 Cytochemically demonstrable β-glucuronidase activity in normal and neoplastic human lymphoid cells. Blood 56: 1111
Matamoros N, Horwitz D A, Newton C, Asherson G L, Webster A D B 1979 Histochemical studies of 5′-nucleotidase and alpha-naphthyl (non-specific) esterase in lymphocytes from patients with primary immunoglobulin deficiencies. Clinical and Experimental Immunology 36: 102
Mestecky J, Winchester R J, Hoffman T, Kunkel H G 1977 Parallel synthesis of immunoglobulin and J chain in pokeweed mitogen stimulated normal cells and in lymphoblastoid cell lines. Journal of Experimental Medicine 145: 760
Miedema F, Terpstra F G, Smit J W, Daenen S, Gerrits W, Hegde U, Matutes E, Catovsky D, Greaves M F, Melief C J M 1984 Functional properties of neoplastic T cells in adult T cell lymphoma/leukemia patients from the Caribbean. Blood 63: 477
Miliani A, Anichini P, Pellegrini G, Firorelli C, Cozzolino F 1979 Studi citochimici sulle plasmacellule. Haematologica 64 (suppl.): 35
Müller-Hermelink H K 1974 Characterization of the B-cell and T-cell regions of human lymphatic tissue through enzyme histochemical demonstration of ATPase and 5-nucleotidase activities. Virchows Archiv B 16: 371
Müller-Hermelink H K, Gaudecker B, Drenckhahn D, Jaworsky K, Feldman C 1981 Fibroblastic and dendritic reticulum cells of lymphoid tissue. Journal of Cancer Research and Clinical Oncology 101: 149
Naiem M, Gerdes J, Abdulaziz Z, Stein H, Mason D Y 1983 Production of a monoclonal antibody reactive with human dendritic reticulum cells and its use in the immunohistological analysis of human lymphoid tissue. Journal of Clinical Pathology 36: 167
Nanba K, Itagaki T, Iijima S 1975 Enzyme histochemical investigations of human malignant lymphomas. Beitrage zur Pathologie 154: 223
Nanba K, Jaffe E S, Braylan R C, Soban E J, Berard C W 1977 Alkaline phosphatase-positive

malignant lymphoma. A subtype of B-cell lymphomas. American Journal of Clinical Pathology 68: 535
O'Connor N T J, Wainscoat J S, Weatherall D J, Gatter K C, Feller A C, Isaacson P, Jones D, Lennert K, Pallesen G, Ramsey A, Stein H, Wright D H, Mason D Y 1985 Rearrangement of the T cell-receptor chain gene in the diagnosis of lymphoproliferative disorders. Lancet i: 1295
O'Conor G T 1961 Malignant lymphoma in African children. II. A pathological entity. Cancer 14: 270
Palutke M, Eisenberg L, Mirchandani I, Tabaczka P, Husain M 1982 Malignant lymphoma of small cleaved lymphocytes of the follicular mantle zone. Blood 59: 317
Palva T, Dammert K, Palva A 1973 On enzyme histochemistry of sarcoidosis. Acta Pathologica et Microbiologica Scandinavica (Section A) 81: 189
Papadimitriou C S, Müller-Hermelink U M, Lennert K 1979 Histologic and immunohistochemical findings in the differential diagnosis of chronic lymphocytic leukemia of B-cell type and lymphoplasmacytic/lymphoplasmacytoid lymphoma. Virchows Archiv A 384: 149
Payne S V, Newell D G, Jones D B, Wright D H 1980 The Reed-Sternberg cell/lymphocyte interaction. Ultrastructure and characteristics of binding. American Journal of Pathology 100: 7
Phillips J H, Babcock G F 1983 NKP-15: a monoclonal antibody reactive against purified human natural killer cells and granulocytes. Immunology Letters 6: 143
Pinkus G S, Thomas P, Said J W 1985 Leu M1. A marker for Reed-Sternberg cells in Hodgkin's disease. American Journal of Pathology 119: 244
Pizzolo G, Chilosi M 1984 Double immunostaining of lymph node sections by monoclonal antibodies using phycoerythrin labeling and haptenated reagents. American Journal of Clinical Pathology 82: 44
Poppema S 1980 The diversity of immunohistological staining pattern of Sternberg-Reed cells. Journal of Histochemistry and Cytochemistry 28: 788
Poppema S, Kaiserling E, Lennert K 1979 Hodgkin's disease with lymphocytic predominance, nodular type (nodular paragranuloma) and progressively transformed germinal centers. A cytohistological study. Histopathology 3: 205
Poppema S, Bhan A K, Reinherz A I, Posner M R, Schlossman S F 1982 In situ immunologic characterization of cellular constituents in lymph nodes and spleens involved by Hodgkin's disease. Blood 59: 226
Quaglino D, De Pasquale A, Artusi T, Piccinini L, Zanni G, Mauri C 1978 A cytochemical and autoradiographic study of lymph node populations in malignant lymphomas with particular reference to Hodgkin's disease. Haematologica 63: 241
Quaglino D, De Pasquale A, Piccinini L, Tonelli M, Mugnaini M 1980 Autoradiographic studies on lymph node populations from non-Hodgkin's B lymphomas. Acta Haematologica 64: 148
Reinherz E L, Kung P C, Goldstein G, Levey R H, Schlossman S F 1980 Discrete stages of human intrathymic differentiation: analysis of normal lymphocytes and leukemic lymphoblasts of T-cell disease. Proceedings of the National Academy of Sciences USA 77: 1588
Robb-Smith A H T, Taylor C R 1981 Lymph node biopsy. Miller Heyden, London
Rowe M, DeGast C G, Platt-Mills T A E, Asherson G L, Webster A D B, Johnson S M 1979 5′-nucleotidase of B and T lymphocytes isolated from human peripheral blood. Clinical and Experimental Immunology 36: 97
Schaadt M, Diehl V, Stein H, Fonatsch C, Kirchner H H 1980 Two neoplastic cell lines with unique features derived from Hodgkin's disease. International Journal of Cancer 26: 723
Schroff R W, Gottlieb M S, Prince H E, Chai L L, Fahey J L 1983 Immunological studies of homosexual men with immunodeficiency and Kaposi's sarcoma. Clinical Immunology and Immunopathology 27: 300
Seymour G J, Dockrell H M, Greenspan J S 1978 Enzyme differentiation of lymphocyte subpopulations in sections of human lymph nodes, tonsils and periodontal disease. Clinical and Experimental Immunology 32: 169
Sinks L F, Clein G P 1966 The cytogenetics and cell metabolism of circulating Reed-Sternberg cells. British Journal of Haematology 12: 447
Stein H, Kaiserling E, Lennert K, Parwaresch M R 1973 Makroglobulinbildende chronische lymphatische Leukämie ohne Makroglobulinämie. Klinische Wochenschrift 51: 389
Stein H, Gerdes J, Schwab V, Lemke H, Mason D Y, Ziegler A, Schienle W, Diehl V 1982 Identification of Hodgkin and Sternberg-Reed cells as a unique cell type derived from a newly detected small-cell population. International Journal of Cancer 30: 445
Stein H, Mason D Y, Gerdes J, O'Connor N, Wainscoat J, Pallesen G, Gatter K, Falini B, Delsol G, Lemke H, Schwarting R, Lennert K 1985 The expression of the Hodgkin's disease associated antigen Ki-1 in reactive and neoplastic lymphoid tissue: evidence that Reed-Sternberg cells and histiocytic malignancies are derived from activated lymphoid cells. Blood 66: 848

Strauchen J A 1984 Lectin receptors as markers of lymphoid cells. II. Reed-Sternberg cells share lectin-binding properties of monocyte macrophages. American Journal of Pathology 116: 370

Stuart A E, Williams A R N, Habeshaw J A 1974 Rosetting and other reactions of the Reed-Sternberg cells and other malignant reticulum cells. Lancet ii: 802

Sundstrom C, Nillsson K, Ranki A, Hayry P 1978 Presence of alpha-naphthylacetate esterase activity in human haemopoietic cell lines and in fresh biopsy specimens of lymphoma and myeloma. Scandinavian Journal of Haematology 21: 47

Tortarolo M, Cantore N, Grande M, Buonanno G, Dammacco F 1981 Plasma cell acid phosphatase as an adjunct in the differential diagnosis of monoclonal immunoglobulinemias. Acta Haematologica 65: 103

Tricot G J K, Van-Orshoven A, Den Ottolander G J, de Wolf-Peeters C, Meyer C J L M, Verwilghen R L, Jansen J 1983 Adult T-cell leukaemia: a report on two white cases. Leukemia Research 7: 31

Usui T, Kita K, Kimura K, Uchaea N, Sawada H, Takatsuki K, Uchino H 1979 Acid phosphatase and acid α-naphthyl acetate esterase activities in various lymphoid cells. Japanese Journal of Clinical Oncology 9: 459

Usui T, Konishi H, Sawada H, Uchino H 1982 Existence of tartrate-resistant acid phosphatase activity in differentiated lymphoid leukaemic cells. American Journal of Hematology 12: 47

Van der Valk P, te Velde J, Jansen J, Ruiter D J, Spaander P J, Cornelissen C J, Meyer C J L M 1981 Malignant lymphoma of true histiocytic origin: histiocytic sarcoma. Virchows Archiv A 391: 249

Velardi A, Grossi C E, Cooper M D 1985 A large subpopulation of lymphocytes with T helper phenotype (Leu 3/$T4^+$) exhibits the property of binding to NK cell targets and granular lymphocyte morphology. Journal of Immunology 134: 58

Veldman J E, Kaiserling E 1980 Interdigitating cells. In: Parr I, Daems W T (eds) The reticuloendothelial system. Plenum Press, New York, p 381

Vogler L B, Crist W B, Bockman D E, Pearl E R, Lawton A R, Cooper M D 1978 Pre B cell leukemia: a new phenotype of childhood lymphoblastic leukemia. New England Journal of Medicine 298: 872

Weisenburger D D, Nathwani B N, Diamond L W, Winberg CD, Rappaport H 1981 Malignant lymphoma, intermediate lymphocytic type: a clinicopathologic study of 42 cases. Cancer 48: 1415

Weisenburger D D, Kim H, Rappaport H 1982 Mantle-zone lymphoma: a follicular variant of intermediate lymphocytic lymphoma. Cancer 49: 1429

Willemze R, Scheffer E, Meijer C J L M 1985 Immunohistochemical studies using monoclonal antibodies on lymph nodes from patients with mycosis fungoides and Sézary syndrome. American Journal of Pathology 120: 46

Williams D, Williams W J, Williams J E 1969 Enzyme histochemistry of epithelioid cells in sarcoidosis and sarcoid-like granulomas. Journal of Pathology 97: 705

Wright D H, McAlpine J C 1966 Ribonucleic acid content of Burkitt tumour cells. Journal of Clinical Pathology 19: 257

Yam L T, Li C Y 1976 Histogenesis of splenic lesions in Hodgkin's disease. American Journal of Clinical Pathology 76: 976

Index

INDEX

INDEX